INTERATOMIC POTENTIALS AND SIMULATION OF LATTICE DEFECTS

INTERATOMIC POTENTIALS AND SIMULATION OF LATTICE DEFECTS

Edited by

Pierre C. Gehlen
Metal Science Group, Battelle Memorial Institute
Columbus Laboratories

Joe R. Beeler, Jr.
School of Engineering, North Carolina State University
Raleigh, North Carolina

Robert I. Jaffee
Department of Physics and Metallurgy,
Battelle Memorial Institute, Columbus Laboratories

**BATTELLE INSTITUTE
MATERIALS SCIENCE COLLOQUIA**

*Seattle, Washington, U.S.A.,
and Harrison Hot Springs,
B. C., Canada
June 14–19, 1971*

Robert I. Jaffee, Chairman

℗ PLENUM PRESS • New York – London • 1972

Library of Congress Catalog Card Number 72-77229
ISBN 0-306-30599-2

© 1972 Plenum Press, New York
A Division of Plenum Publishing Corporation
227 West 17th Street, New York, N. Y. 10011

United Kingdom edition published by Plenum Press, London
A Division of Plenum Publishing Company, Ltd.
Davis House (4th Floor), 8 Scrubs Lane, Harlesden, London,
NW10 6SE, England

All rights reserved

No part of this publication may be reproduced in any form
without written permission from the publisher

Printed in the United States of America

To **Dr. GEORGE VINEYARD**

who led the modern school of lattice simulation described in these proceedings

PARTICIPANTS

N. W. ASHCROFT *Laboratory of Atomic and Solid State Physics, Cornell University, Ithaca, New York, U.S.A.*

Z. S. BASINSKI *Division of Pure Physics, National Research Council of Canada, Ottawa, Canada*

J. R. BEELER, JR. *School of Engineering, North Carolina State University, Raleigh, North Carolina, U.S.A.*

R. BULLOUGH *Theoretical Physics Division, Atomic Energy Research Establishment, Harwell, Didcot, Berkshire, England*

R. CHANG *Science Center, North American Rockwell Corporation, Thousand Oaks, California, U.S.A.*

E. W. COLLINGS *Metal Science Group, Battelle Memorial Institute, Columbus, Ohio, U.S.A.*

J. M. COWLEY *Department of Physics, Arizona State University, Tempe Arizona, U.S.A.*

R. E. DAHL, JR. *Irradiation Analysis Section, WADCO Corporation, Richland, Washington, U.S.A.*

F. W. de WETTE *Department of Physics, The University of Texas at Austin, Austin, Texas, U.S.A.*

PARTICIPANTS

D. G. DORAN *Irradiation Analysis Section, WADCO Corporation, Richland, Washington, U.S.A.*

J. E. DORN *Department of Mineral Technology, Hearst Mining Building, University of California, Berkeley, California, U.S.A.*

M. S. DUESBERY *Division of Pure Physics, National Research Council of Canada, Ottawa, Canada*

G. EHRLICH *Coordinated Science Laboratory, University of Illinois, Urbana, Illinois, U.S.A.*

J. E. ENDERBY *Department of Physics, University of Leicester, Leicester, England*

A. ENGLERT *Université Libre de Bruxelles, Bruxelles, Belgium*

P. C. GEHLEN *Metal Science Group, Battelle Memorial Institute, Columbus, Ohio, U.S.A.*

L. A. GIRIFALCO *School of Metallurgy and Materials Science, University of Pennsylvania, Philadelphia, Pennsylvania, U.S.A.*

G. T. HAHN *Metal Science Group, Battelle Memorial Institute, Columbus, Ohio, U.S.A.*

W. A. HARRISON *W. W. Hansen Laboratories of Physics, Stanford University, Stanford, California, U.S.A.*

J. P. HIRTH *Department of Metallurgical Engineering, The Ohio State University, Columbus, Ohio, U.S.A.*

P. S. HO *Department of Materials Science and Engineering, Cornell University, Ithaca, New York, U.S.A.*

R. G. HOAGLAND *Metal Science Group, Battelle Memorial Institute, Columbus, Ohio, U.S.A.*

H. B. HUNTINGTON *Department of Physics and Astronomy, Rensselaer Polytechnic Institute, Troy, New York, U.S.A.*

D. P. JACKSON *Chalk River Nuclear Laboratories, Atomic Energy of Canada Limited, Chalk River, Ontario, Canada*

R. I. JAFFEE *Department of Physics and Metallurgy, Battelle Memorial Institute, Columbus, Ohio, U.S.A.*

R. A. JOHNSON *University of Virginia, Charlottesville, Virginia, U.S.A.*

M. F. KANNINEN *Advanced Solid Mechanics Division, Battelle Memorial Institute, Columbus, Ohio, U.S.A.*

R. KAPLOW *Department of Metallurgy, Massachusetts Institute of Technology, Cambridge, Massachusetts, U.S.A.*

G. L. KULCINSKI *Battelle Memorial Institute, Pacific Northwest Laboratories, Richland, Washington, U.S.A.*

A. A. MARADUDIN *Department of Physics, University of California, Irvine, California, U.S.A.*

N. H. MARCH *Department of Physics, University of Sheffield, Sheffield, England*

J. R. PARSONS *Chalk River Nuclear Laboratories, Atomic Energy of Canada Limited, Chalk River, Ontario, Canada*

L. B. PEDERSEN *Department of Structural Properties of Materials, The Technical University of Denmark, Lyngby, Denmark*

R. C. PERRIN *Theoretical Physics Division, Atomic Energy Research Establishment, Harwell, Didcot, Berkshire, England*

A. RAHMAN *Solid State Science Division, Argonne National Laboratory, Argonne, Illinois, U.S.A.*

R. RITTER *Institute Battelle, Carouge/Geneve, Switzerland*

M. T. ROBINSON *Solid State Division, Oak Ridge National Laboratory, Oak Ridge, Tennessee, U.S.A.*

A. SEEGER *Institut für Physik, Max-Planck-Institut für Metallforschung, Stuttgart, Germany*

F. SEITZ *The Rockefeller University, New York, New York, U.S.A.*

P. G. SHEWMON *Materials Science Division, Argonne National Laboratory, Argonne, Illinois, U.S.A.*

R. TAYLOR *Division of Physics, National Research Council of Canada, Ottawa, Canada*

I. M. TORRENS *Department de Metallurgie, Centre d'etudes Nucleaires de Saclay, Paris, France*

M. P. TOSI *Physics Department, University of Messina, Messina, Italy*

PARTICIPANTS

W. R. TYSON *Department of Physics, Trent University, Peterborough, Ontario, Canada*

G. H. VINEYARD *Brookhaven National Laboratory, Upton, L.I., New York, U.S.A.*

V. VITEK *Department of Metallurgy, University of Oxford, Oxford, England*

J. H. WEINER *Division of Engineering, Brown University, Providence, Rhode Island, U.S.A.*

M. J. WEINS *Department of Materials Engineering, University of Illinois at Chicago Circle, Chicago, Illinois, U.S.A.*

S. WILKINS *School of Physics, University of Melbourne, Parkville, Victoria, Australia*

W. D. WILSON *Sandia Laboratories, Sandia Corporation, Albuquerque, New Mexico, U.S.A.*

PAUL WYNBLATT *Metallurgy Department, Ford Motor Company, Dearborn, Michigan, U.S.A.*

H. H. YOSHIKAWA *Irradiation Analysis Section, WADCO Corporation, Richland, Washington, U.S.A.*

PREFACE

This book is the proceedings of the Sixth Battelle Colloquium on the Science of Materials. The Colloquium was devoted to a new field of materials science in which computers are used to conduct the experiments. Although the computer methods used have reached a high degree of sophistication, the underlying principles are relatively straightforward and well understood. The interatomic force laws — a vital input into these computations — however are less well understood. *Interatomic Potentials and Simulation of Lattice Defects* primarily discusses the validity of a variety of force laws — either from a theoretical point of view or through comparisons of experimental results and those obtained with computer simulation.

The format used in previous Battelle Institute Colloquia is followed. The opening session was aimed at providing an overall view of the field of interatomic forces and defect calculations by major contributors. It was led by Dr. G. H. Vineyard, one of the pioneers in this field. The second day was devoted to research papers on theoretical and experimental aspects of interatomic forces. The remaining days were devoted to research papers on computer simulation of the four types of defects: point defects, line defects, surface defects, and volume defects.

Each of the three sessions on computer simulation of defects was followed by an Agenda Discussion in which the participants had an opportunity to assess the issues involved somewhat more broadly than was possible after the individual research papers. Finally, two Concluding

Agenda Discussions were held — one dealing with computer techniques, and the other with critical issues to be resolved in the future.

The opening day in Seattle, Dr. Frederick Seitz gave an after-dinner address on implications of the SST cancellation, a topic of science policy of particular interest to the Seattle community, but of much broader implications.

As has become the custom in the Battelle Colloquia, one of the distinguished participants was honored at the concluding banquet. This was Dr. George Vineyard, who was introduced by his associate in his early work on computer simulation, Professor H. B. Huntington. Dr. Vineyard's remarks relating the circumstances of his involvement in the work are covered in the autobiographical sketch immediately following.

We are grateful for the support of the Colloquium to Dr. Sherwood L. Fawcett, President of Battelle Memorial Institute, and to Dr. Frederick J. Milford, Director of Research in the Physical Sciences of Battelle Institute, under whose program this Colloquium is a continuing activity. We wish to thank Dr. Tommy Ambrose, Director of the Battelle Seattle Research Center, and Dr. R. S. Paul, Director of the Battelle Pacific Northwest Laboratories, who were Battelle hosts for the Colloquium. We are grateful to the supporting staff from the Battelle Seattle Research Center, including Lou Bonnefond, in charge of local arrangements, Penny Raines, recreation and transportation, Kay Killingstad and Susie Armstrong, secretarial, and R. Wilton, general factotum. We also wish to thank Mrs. Robert Jaffee, who was in charge of the Ladies Program. The enthusiastic and efficient support by these and many other people made possible the success of the Colloquium.

The Organizing Committee is grateful to N. W. Ashcroft, R. Bullough, R. Chang, J. E. Enderby, and G. H. Vineyard for their advice and suggestions in the early stages of planning the Colloquium.

Finally we wish to thank the Office of Naval Research and particularly Dr. W. G. Rauch for their early recognition of the significance of this area of research.

This book is the first dealing with atomic simulation of lattice defects. We trust that it will not be the last, and that the field will grow as all new, viable fields of research grow, and will be reported through further such conferences as ours.

The Organizing Committee
R. I. Jaffee, Chairman
J. R. Beeler, Jr.
D. G. Doran
P. C. Gehlen
G. T. Hahn
J. P. Hirth

AUTOBIOGRAPHICAL REMARKS OF G. H. VINEYARD

Computer simulation of phenomena in crystal lattices has been of close concern to me for a long time, and it is gratifying to see such broad interest in the subject. I should like to take advantage of the occasion to reminisce on the early days of this work.

In the summer of 1957 at the Gordon Conference on Chemistry and Physics of Metals, I gave a talk on radiation damage in metals in which I explained current analytical theories of the damage cascade. The Snyder-Neufeld model, the Kinchin-Pease model, and the Harrison-Seitz model were of particular interest. A principal goal of the theories was the calculation of the number of lattice defects produced. After the talk there was lively discussion of how to verify the numerous assumptions and approximations that had to be made. Somewhere the idea came up that a computer might be applied to follow in more detail what actually goes on in radiation damage cascades. We got into quite an argument, some maintaining that it wasn't possible to do this on a computer, others that it wasn't necessary. John Fisher insisted that the job could be done well enough by hand, and was then goaded into promising to demonstrate. He went off to his room to work. Next morning he asked for a little more time, promising to send me the results soon after he got home. After about two weeks, not having heard from him, I called and he admitted that he had given up. This stimulated me to think further about how to get a high-speed computer into the game in place of John Fisher, and I drew up a memorandum proposing an atomic model for copper and a scheme for solving the classical equations of motion for the atoms of the model. Knowing nothing about computers, I had no idea whether the computation was truly feasible and took the memo to Milton Rose, then Chairman of Brookhaven's Department of Applied Mathematics. To my surprise and delight, he was highly encouraging, and remarked "It's a great problem. This was just what computers were designed for." Rose assigned an experienced mathematician-programmer, Martin Milgram, to the job of converting these general ideas into a program that would actually run on a machine. At about the same time, two young colleagues of mine in the solid-state group at Brookhaven became interested and joined the endeavor. One was John Gibson, who had come to Brookhaven about two years earlier from Iowa State. The other was Alan Goland who had come to us from Northwestern. Milgram wrote the basic program and supplied some very important ideas. He knew at once that the sidewise difference scheme, which, in my innocence I had proposed, was poorly convergent, and substituted a central difference scheme which looks similar but is far between. Then he devised an ingenious method for keeping track of neighbors of each atom so that, in computing the forces on one atom, it was necessary for the computer to look at only a few tens of other atoms which were actual or potential neighbors, rather than all of the N-1 remaining atoms, most of which had no influence at all. This stratagem, which is not simple for a problem in which arbitrarily large displacements are permitted, reduces the running time and the memory requirements substantially. Instead of being proportional to the number of pairs, $\frac{N^2-N}{2}$, the demands now become proportional to N itself. The four of us worked for a while, Gibson concentrating on devising the most plausible interatomic potential, and Goland on the intricacies of integrating the equations of motion with allowable error in an acceptable time. Gibson and Goland both learned programming and contributed to the refinement of the complex codes. About that time I went off for a month's stay in California, on another project. When I came back, John and Alan met me with the stimulating news "Things are going fine. We've finally got it running." And indeed they had. Primitive as the first results were, it was electrifying to us to commence computing orbits in a classical problem of 500 bodies in close interaction, and to puzzle over the results. It was like watching a chess game played by grand masters. Many of the moves were mystifying at first, but upon close study became plausible, then obvious. We concluded that the computer was smarter than we were.

From that time on we worked busily. We had to go in to the Mathematics Center at New York University where they had an IBM 704. It involved vacuum tubes and was, by today's standards, a primitive machine. Because most of our runs used large blocs of time, we had to go in late at night and run through the graveyard shift. At first the output was all numerical, and sorting and plotting to make sense out of these "telephone books" full of data became increasingly burdensome. We acquired a cheerful and efficient assistant, Barbara Garnier, who kept our books, reduced the data to more useful forms, and made the whole effort more pleasant.

The work progressed slowly, and looking back it is surprising how long it was before we were ready to make some statements. Having begun in September 1957, it was June of 1959 at a conference in Gatlinburg on radiation effects in semiconductors that we first got up publicly and said something about what we had been doing. This included explaining why we were reporting on copper at a conference on semiconductors. It was not until the following January that we presented the first formal papers at an APS meeting, and in 1960 the first paper appeared in the *Physical Review*.

Fairly early we conceived the idea of making moving pictures of the results, for a more dramatic display of what was happening. After buying the hardware to supplement that provided with the old 704, and with the help of Robert Walton of the BNL Photography Division, and more trips to the graveyard shift in New York, we made movies. The result was never entered at Cannes, but nevertheless generated so much demand that we arranged for a film supplier, B. and O. Film Associates, New York, to sell prints at a nominal cost, which they do to this day. B. and O. are mainly concerned with the world of television and its commercials, and they can't understand why this amateurish film of strangely moving specks, which they have entitled "Dots", continues to sell.

After the initial work on copper, we wished to try other substances with different types of lattices. As a representative of the bcc family, we chose iron. Anne Englert, who came to spend a semester at Brookhaven from E.R.A., Brussels, joined the work and was given the task of generating a suitable interatomic potential for pairs of iron atoms. David Erginsoy came from Turkey, by way of Vienna, and soon was handling a large share of the project, as I became sidetracked in administrative work. A Japanese physicist, Akinao Shimizu, spent a year with us, and helped with the second paper on iron, which was directed to knock-ons of higher energy. Midway in the work, the 704 was exchanged for the more advanced 7090, and another very able mathematician, Arlene Larsen, rewrote the program for the new machine. She also produced a comprehensive write-up of the various programs (there was by this time a considerable constellation of separate programs) for our archives and for the use of outside groups.

You will now appreciate how many able people have cooperated in this effort. We have also benefited from the superb supporting facilities at Brookhaven and N.Y.U., and from the unrestrained backing of the Atomic

Energy Commission. In particular, Donald Stevens, Assistant Director of Research for Metallurgy and Materials of the AEC, was remarkably patient, and, in our belief, farsighted in his sponsorship of this costly and slowly evolving work.

Two of the people most closely involved, John Gibson and David Erginsoy, are no longer living. Gibson died tragically in 1960, at the age of 34, just as the work to which he had contributed so much, was bearing real fruit. Erginsoy died suddenly in 1967 at the age of 43. Both were brilliant men in the prime of life, and much of the honor of this occasion is owed to them.

Reviewing the applications of computers to problems of crystal lattices, it is impressive to see how many extremely able people are now involved at so many different laboratories, and to see the wide variety of work that has been done. It is the highest tribute to a field of work when able people are willing to devote themselves to it.

This conference has been comprehensive and timely. It has emphasized the crucial problems, the difficulties and the pitfalls, and has illuminated many opportunities for further work. The flourishing of the field is a natural accompaniment of the explosive developments of computers in the last dozen years. I believe that the field is a long way from full maturity, let alone senility, and predict that the next decade will see equal or greater development.

It has been a privilege to be associated with so many stimulating colleagues. On their behalf, as well as my own, I am deeply grateful for this recognition.

Harrison Hot Springs, B.C., Canada
June 19, 1971

CONTENTS

Participants vii
Preface xi
Autobiographical Remarks by George Vineyard xiii

Part One INTRODUCTORY LECTURES

1. *Computer Experiments With Lattice Models* G. H. VINEYARD 3
   DISCUSSION 24
2. *Potential Functions and the Simulation of Defects in Lattice Dynamical Defect Problems* A. A. MARADUDIN 27
   DISCUSSION 67
3. *The Theory of Interatomic Potentials in Solids* W. A. HARRISON 69
   DISCUSSION 88
4. *Ion-Ion Interactions in Metals: Their Nature and Physical Manifestations*
   N. W. ASHCROFT 91
   DISCUSSION 110
5. *Kanzaki Forces and Electron Theory of Displaced Charges in Relaxed Defect Lattices* N. H. MARCH AND J. S. ROUSSEAU 111
   DISCUSSION 138

Part Two INTERATOMIC POTENTIALS

1. *Screening Functions in Simple Metals* M. P. TOSI 141
   DISCUSSION 151
2. *The Direct Construction of the Lattice Green Function* R. BULLOUGH AND V. K. TEWARY 155
   DISCUSSION 175
3. *Computer Simulation of Quantum Phenomena* J. H. WEINER AND A. ASKAR 177
   DISCUSSION 187

xviii   CONTENTS

4. *On the Validity of Two-Body Potentials in Metals*   L. A. GIRIFALCO AND
   T. M. Di VINCENZO 189
   DISCUSSION 213
5. *Experimental Techniques Used to Obtain Potentials*   J. E. ENDERBY AND
   W. S. HOWELLS   217
   DISCUSSION 229
6. *Molecular Dynamics Studies of Liquids*   A. RAHMAN 233
   DISCUSSION 246
7. *Interatomic Potentials; Aspects Which are Visible in Experimental Radial
   Pair Distributions*   ROY KAPLOW 249
8. *Derivation of Long-Range Interaction Energies From Diffuse Scattering
   in Diffraction Patterns*   J. M. COWLEY AND S. WILKINS 265
   DISCUSSION 277
9. *The Study of Interatomic Potentials by Planar Channeling Experiments*
   MARK T. ROBINSON 281
   DISCUSSION 297

Part Three POINT DEFECTS

1. *Defect Calculations for FCC and BCC Metals*   R. A. JOHNSON AND
   W. D. WILSON 301
   DISCUSSION 315
2. *Pseudopotential Calculation of Point Defect Properties in Simple Metals*
   P. S. HO   321
   DISCUSSION 335
3. *Impurity Atom Effects in Metallic Crystals*   J. R. BEELER, JR. 339
   DISCUSSION 371
4. *Rare Gases in Metals*   W. D. WILSON AND R. A. JOHNSON 375
   DISCUSSION 386
5. *On Pseudopotential Calculation of Point Defects in Metals*   R. CHANG . . . 391
   DISCUSSION 400
6. *Computer Simulation of the Short-Term Annealing of Displacement Cascades*
   D. G. DORAN AND R. A. BURNETT 403
   DISCUSSION 420
7. *Computer Simulation of Atomic Displacement Cascades in Solids*
   I. M. TORRENS AND M. T. ROBINSON 423
   DISCUSSION 437
8. *Temperature Dependence of the Vacancy Formation Energy in Krypton by
   Molecular Dynamics*   R.M.J. COTTERILL AND L. B. PEDERSEN 439
   DISCUSSION 448

   Agenda Discussion: Point Defects   Chairman, L. A. Girifalco 451

Part Four DISLOCATIONS AND STACKING FAULTS

1. *Influence of Dislocations on Electron Microscope Crystal Lattice Images*
   J. R. PARSONS 463
   DISCUSSION 472

2. *On the Motion of the $\frac{a}{2}<111>$ Screw Dislocation in Models of α-Iron*
   P. C. GEHLEN 475
   DISCUSSION 488
3. *On the Factors Controlling the Structure of the Dislocation Cores in BCC Crystals* V. VITEK, L. LEJCEK AND D. K. BOWEN 493
   DISCUSSION 507
4. *Extended Defects in Copper and Their Interactions with Point Defects*
   R. C. PERRIN, A. ENGLERT, AND R. BULLOUGH 509
   DISCUSSION 522
5. *Partial Dislocation Interactions in a Face-Centered Cubic Sodium Lattice*
   Z. S. BASINSKI, M. S. DUESBERY AND R. TAYLOR 525
   DISCUSSION 535
6. *The Motion of Screw Dislocations in a Model B.C.C. Sodium Lattice*
   Z. S. BASINSKI, M. S. DUESBERY AND R. TAYLOR 537
   DISCUSSION 551
7. *Atomistic Calculations of Peierls-Nabarro Stress in a Planar Square Lattice*
   W. R. TYSON 553

Agenda Discussion: Dislocations and Stacking Faults Chairman, J. P. Hirth . 561

Part Five SURFACES AND INTERFACES

1. *Experimental Studies of Atomic Behavior at Crystal Surfaces* G. EHRLICH . 573
   DISCUSSION 616
2. *Simulating Surfaces by the Summation of Pairwise Interatomic Potentials*
   D. P. JACKSON 621
   DISCUSSION 632
3. *Interaction Energy and Configuration of Ledges on (001) Copper Surfaces*
   P. WYNBLATT 633
   DISCUSSION 649
4. *Computer Calculations of Dynamical Surface Properties of Crystals*
   F. W. de WETTE 653
   DISCUSSION 669
5. *A Computer Simulation Study of Grain Boundaries in FCC Gamma-Iron and Their Interactions With Point Defects* R. E. DAHL, JR., J. R. BEELER, JR. AND R. D. BOURQUIN 673
   DISCUSSION 692
6. *Computer Simulation of the Structure of High Angle Grain Boundaries*
   M. J. WEINS 695
   DISCUSSION 711
7. *A Study of Crack Propagation in Alpha-Iron* M. F. KANNINEN AND P. C. GEHLEN 713
   DISCUSSION 722

Agenda Discussion: Surfaces and Interfaces Chairman, P. G. Shewmon . . . 725

Part Six COMPUTER TECHNIQUES

Agenda Discussion: Computer Techniques 735
Chairman, J. R. Beeler

Part Seven CRITICAL ISSUES

Concluding Agenda Discussion: Critical Issues 755
Chairman, A. Seeger

Index 767

Part One

INTRODUCTORY LECTURES

COMPUTER EXPERIMENTS WITH LATTICE MODELS

G. H. Vineyard

Brookhaven National Laboratory
Upton, Long Island

1 INTRODUCTION

High-speed computers have made possible an interesting class of investigations. Starting with an atomic model of a crystal lattice or liquid, the behavior of the model is directly computed to elucidate physical phenomena. Complex models with many degrees of freedom, displaying phenomena resistant to analytical treatment and to definitive study by experimental means, have been employed, and the results have been surprising and enlightening. The picturesque term "computer experiments" is appropriate for such work. In a sense this is a third estate of physics.

The bulk of the work has been done in the past 10 or 12 years. In this talk I shall briefly review, by selected sampling, computer experiments on crystal lattices, emphasizing work of interest to solid state and materials scientists. Recent reviews covering some of this territory very thoroughly have been written by Beeler.[1,13]

The typical model is a set of 100 to 1,000 (or more) mass points representing atoms interacting with realistic forces. In some work the requirements of rapid computation have led to the use of idealized forces, even hard spheres, but the increasing tendency is to employ forces that are as close as possible to the real forces, insofar as they are known. Illustrative calculations, particularly for questions of statistical mechanics, have been done in lattices of one or two dimensions (less often in more than three dimensions!). The adequacy of the force laws is usually the most serious question, and is an area in which much work remains to be done.

The laws of physics employed depend on the phenomenon of interest. Almost all the work has been done with classical mechanics*. For static properties one needs only the classical conditions of static equilibrium. For dynamic phenomena one needs the classical laws of motion. Properties of systems at thermal equilibrium can be computed by averaging over phase space, e.g., by Monte Carlo methods, without solving the equation of motion.

2 RADIATION DAMAGE

If one atom of the set is suddenly given momentum, a radiation damage cascade is simulated. Early work of this kind was done with a model of copper.[2] Primary knock-ons up to 400 eV in energy at a variety of directions initiated cascades that were followed fully dynamically until agitation became minimal and final defect configurations remained. Boundary forces, including damping, were applied to the fundamental set of atoms to simulate the response of surrounding material and to allow the kinetic energy to drain away. Focusing sequences along close-packed lines and also along <100> were observed. Interstitials and vacancies were produced. Another set of calculations was performed for a model of α-iron.[3,4] Fig. 1 shows the trajectories in one plane determined in a typical calculation. Effects rather similar to those in copper were seen, and the threshold energy for producing a stable defect was calculated for various directions of the primary knock-on. The results are shown in Fig. 2. The minimum threshold of about 17 eV is consistent with experimental results for electron bombardment at low temperatures[5], although the threshold is not determined very precisely by the experiments. A small number of

---

*The extent of quantum corrections is always of interest. Quantum effects have been introduced in a few calculations, but a fully dynamical quantum treatment of a system with many degrees of freedom still poses formidable difficulties to machine computation. Essentially, the numerical complexity of a time dependent classical problem is proportional to the number of degrees of freedom, while that of the corresponding quantum problem rises exponentially with the number of degrees of freedom.

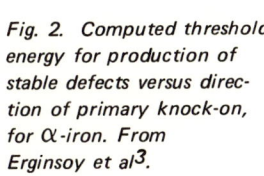

Fig. 1. Computed trajectories in a radiation damage cascade in iron. From Erginsoy et al[3].

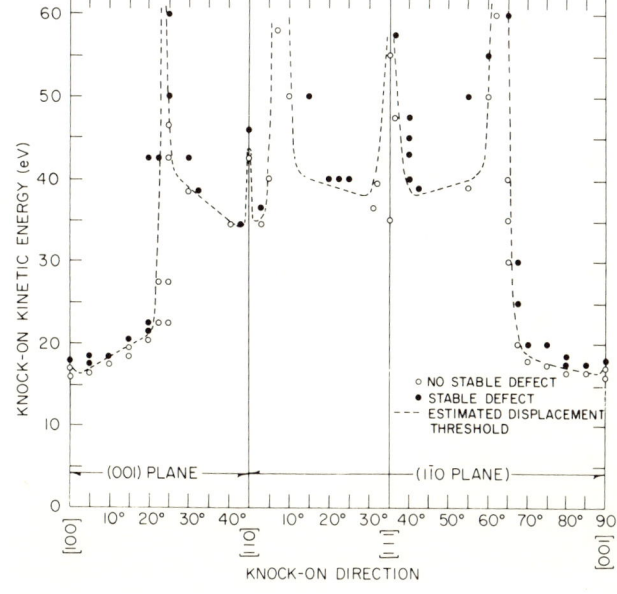

Fig. 2. Computed threshold energy for production of stable defects versus direction of primary knock-on, for α-iron. From Erginsoy et al[3].

calculations on ordered Cu_3Au, to test the disordering effects of radiation, were also performed[6]. Despite the large difference in the two masses, pronounced disordering tendencies in <110> rows, where the two kinds of atoms alternate, were seen.

Fully dynamical calculations of this type have been carried out for three ionic substances, PbI_2, NaCl, and KCl, by Chadderton and Torrens[7,9]. Short-range repulsive potentials were used (Born-Mayer for PbI_2, and power law for the others), supplemented by Coulomb interactions. With PbI_2, only individual planes of atoms were treated; in the alkali halides the calculations were fully three-dimensional. An example of a calculated cascade in KCl is shown in Fig. 3. In general, the results in the three ionic lattices followed a pattern similar to that observed in copper and α-iron, although PbI_2 is hexagonal, and KCl is pseudosimple cubic. In particular, low-energy damage was found to consist of vacancies and interstitials. Near threshold energy, the nature of the displacement process was strongly dependent on directional effects imposed by the lattice. In the alkali halides, focused collision sequences were observed in <100>, <110>, and <111> directions. The long-range Coulomb part of the potential was not found to be of primary importance during the energetic stages of a cascade, but it became important during relaxation when the crystal was settling down and selecting defect sites.

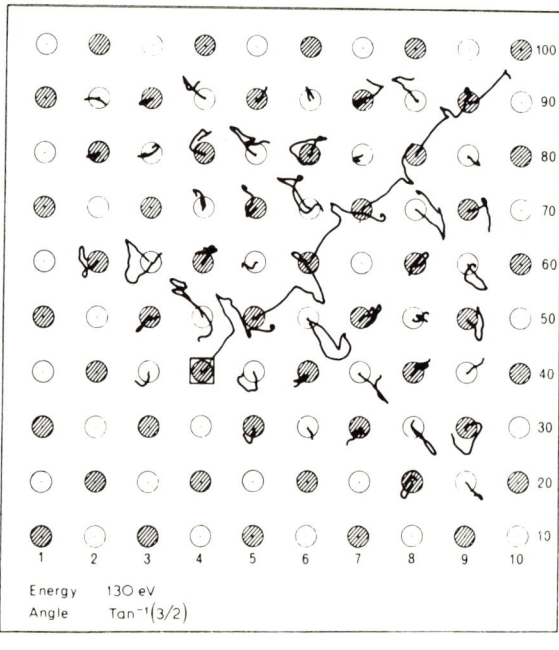

Fig. 3. Computed trajectories of a damage cascade in model of KCl. From Torrens and Chadderton[8].

These calculations, which may be described as fully dynamic since all degrees of freedom of the crystallite are allowed to vary at once in accordance with the requirements of the interaction forces and Newtonian dynamics, are rather severely limited by the size of the crystallite that can be accommodated. The limitation comes from the speed and memory capacity of computers. As a result, primary knock-on energies that will permit the cascade to be fully contained are limited to a few hundred or a thousand eV. Since real knock-ons produced by reactor irradiations and some other irradiations of interest extend up to energies of many tens of thousands of eV, other methods are needed. To this end "semidynamical", or what might be called combination approximations*, have been devised. Beeler and co-workers have pioneered and exploited this approach[10,11] (its earliest use was apparently by Yoshida[12]). Its basis is the assumption that the important interactions in a cascade are two-body collisions between a moving atom and a stationary lattice atom, that the moving atoms to be taken into account in the cascade are those that have received kinetic energy above some threshold, and that an interstitial will be produced by a moving atom whose energy drops below another threshold. The location of the split interstitial that is formed was assumed by Beeler to be a randomly selected neighboring site of the cell in which the moving atom finds itself when it first drops below threshold. A subsequent recombination of the interstitial with an existing vacancy is assumed to occur when the two are produced within a recombination spacing determined from separate fully dynamical calculations. The calculation starts with a single knock-on endowed with momentum in arbitrary direction. The atom is followed to its first encounter with a lattice atom, at which the collision is treated by two-body, classical considerations. For this, an interatomic pair potential is assumed. The products of this collision are then followed to their next collisions, and so on. A complex bookkeeping scheme is used to keep track of currently moving atoms in the cascade. Allowance is made for the damage already produced by faster members of the cascade when slower members come along, so that interference between different branches of the cascade can occur. After all atoms have fallen below critical energies, the cascade is considered to be terminated and the damage is tallied. Fig. 4 shows the trajectories of a cascade produced by a 5-keV knock-on in a model of α-iron. All knock-on trajectories have been projected onto the (001) plane. Alternate members of the cascade are depicted by alternate kinds of lines, the short heavy track being that of the initiating or primary knock-on. Fig. 5 is a perspective view of the region affected by the cascade of Fig. 4. Here an atom is considered to have been affected if it receives kinetic energy in a

---

*Referred to by Beeler as the "Branching binary collision model".

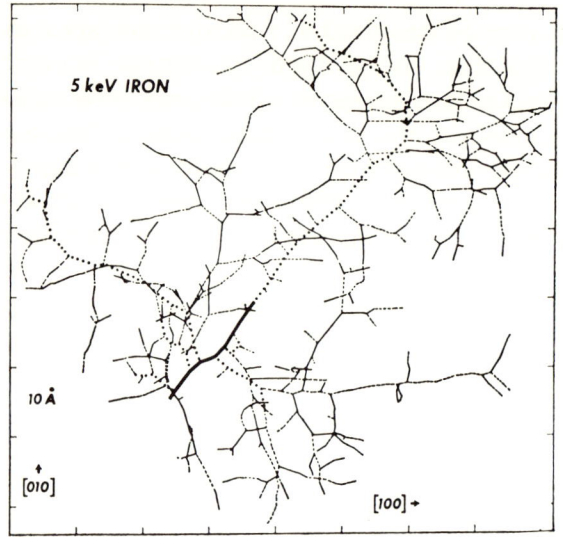

Fig. 4. Cascade produced by a 5-keV knock-on in α-iron, computed in semidynamical approximation by Beeler[11].

binary collision. Not all affected atoms will be displaced. The number displaced in this cascade is about 55.

Such calculations have given a reasonably reliable picture of the nascent damage in the kilovolt energy range for models of copper, α-iron, and tungsten. The number, spatial distribution, and tendency for clustering of the point defects (vacancies and interstitials) are determined.

Calculation of the subsequent annealing process by which vacancies and interstitials would be expected to recombine or to aggregate into larger clusters as the temperature is raised, have also been made by

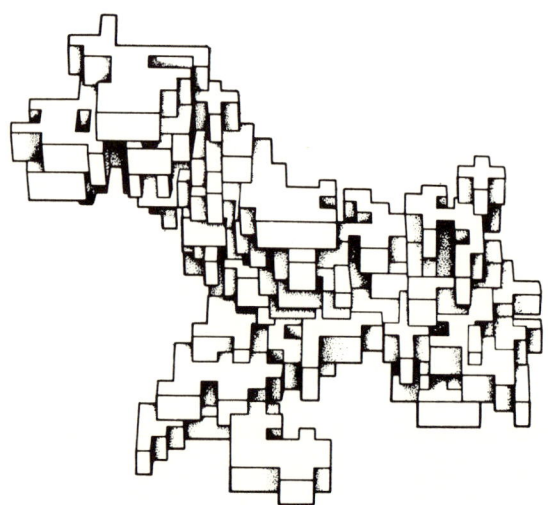

Fig. 5. Perspective view of the region affected by the damage cascade shown in Fig. 4.

Beeler.[13] Further work on the annealing of damage in iron has recently been published by Doran.[13a] In these studies, further assumptions must be made concerning activation energies of motion of the various defects and binding energies for their aggregation. These assumptions are partly supported by auxiliary calculation and other data, but they are rather speculative. The actual annealing process is then easily computed as a set of biased random walks of the mobile defects on the lattice to the sinks that are present. The many assumptions that have gone into the work must be remembered, but the overall picture of radiation damage provided by these calculations is useful and is probably close to reality.

3 ION PENETRATION

If a foreign atom is injected into the set of atoms, ion penetration phenomena are simulated. Robinson and Oen[14] initiated computer studies of this kind. For determining the course of an energetic incident ion, it is not necessary to take account of the full dynamics of the many-body system: as in the studies of Beeler it is usually sufficient to consider the lattice atoms to be at rest on lattice sites and to calculate the interaction of the incident ion with the lattice atoms as a succession of two-body collisions. To determine properly the stopping of the ion, it is necessary to allow for the inelastic part of the two-body interactions, which can be done reasonably well and is the more important the higher the energy of the ion. Fig. 6 shows the rather complex trajectories of copper ions

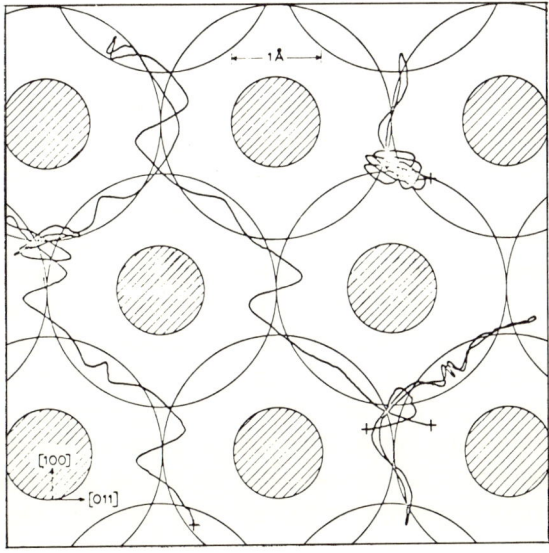

Fig. 6. Projections onto (011) plane of calculated orbits of copper ions moving in copper. Tendency for channeling in low index directions is evident. From Robinson and Oen[14].

slowing down in a copper crystal. The initial energy of each ion was 1 keV and the ions penetrate approximately 300 Å into the crystal. The tendency of the ions to move in "channels" parallel to low index atomic rows is evident. This channeling produces anomalously long ranges for ions properly directed into crystalline substances — ranges anomalously long compared with those of ions at arbitrary directions of incidence or ions penetrating the same substance in an amorphous state.

If the surface is realistically treated, and an energetic ion is introduced, sputtering phenomena may be simulated.

4 SHOCK WAVES

Another interesting perturbation of the system is brought about by driving the atoms in one face of the crystallite into the crystallite with a fixed velocity. Such a process sets up a plane shock wave propagating into the crystal in the direction of driving, and the amplitude of the shock wave is determined by the velocity of driving. Tsai and Beckett[15] have made an interesting series of calculations of this kind in linear chains, and in simple cubic and body-centered and face-centered cubic lattices with various potentials, including Morse potentials. The driving was along [100], and because of the high symmetry an essentially one-dimensional shock pattern was obtained in all cases. Anderson[16] and Manvi[17] have performed similar calculations in one-dimensional models consisting of mass points regularly spaced in a semi-infinite line. In Manvi's work, the points interact with a Morse potential, so their interactions are nonlinear, and in addition viscous damping is applied to each near neighbor pair to simulate dissipative processes at work in a real substance. Fig. 7 shows the velocities of the various atoms in a fully developed shock. The shock front is located momentarily at the right edge of the figure, and is moving to the right with a characteristic shock velocity. Behind the front, rapid oscillations in excitation are seen, settling down after about 50 atomic spacings to a uniform velocity, the well-known material velocity behind the shock. The oscillatory behavior in the intermediate region adds to the effective thickness of the shock front, and is a feature not found in the continuum models usually employed in shock wave theories.

Many further shock wave studies in fully dynamic lattice models can be imagined. As yet, no calculations for propagation in directions of low symmetry have been reported. All real shocks of high amplitude produce severe plastic distortion, with approximately hydrodynamic compression behind the front. The atomic processes by which a uniaxial compression is converted (by breakdown of the shear strength of the solid) into hydrodynamic compression are not known. One model envisions a wall of

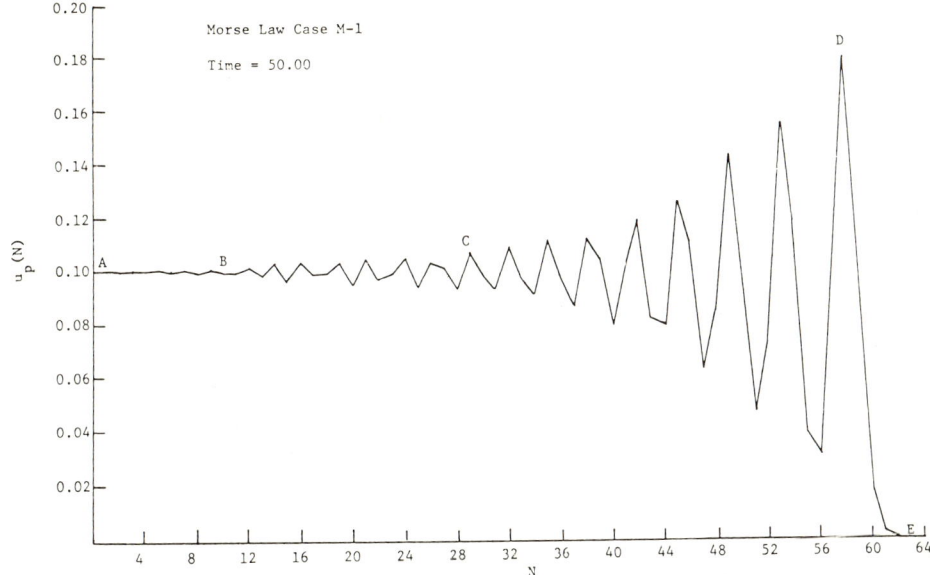

Fig. 7. Profile of computed shock wave moving in a one-dimensional atomic model. Velocities of atoms are plotted against position of atom. From Manvi[17].

dislocations moving with the shock front and accomplishing a more-or-less continuous relaxation of the shear stresses that are generated in a transient way by the shock front.[18] Whether this mechanism or something quite different prevails is an interesting and open question that may be settled some day by more sophisticated computer studies with atomic models.

A further kind of study with atomic models is the process of crack propagation. Gehlen and Kanninen have reported work in this direction.[19] The formation of voids in conditions of high-speed fracture can also, in principle, be investigated[20].

5 THERMAL MOTIONS

In most of the work described so far, the lattice was initially quiescent; thermal agitation was absent or, as in the studies of annealing of radiation damage, was introduced only through an artifice. Thermal agitation, which can also be introduced into the fully dynamic model in the beginning, brings about interesting changes that are susceptible to further analysis by computer. A large class of studies have been performed on simplified models (hard spheres or discs, hard spheres with attractive square wells, etc.). Realism was sacrificed deliberately to gain computing speed and allow attention to be focused on statistical mechanical phenomena, such as the attainment of equilibrium or the occurrence of phase changes, that are insensitive to the details of the interaction. Equations of state were determined by Monte Carlo methods by Wood and Parker[21] and by dynamical methods by Alder and Wainwright[22]. Alder and his collaborators have published a long series of important and interesting dynamical studies of dense gases, liquids, and solids. Computerized models (of liquids) with more realistic forces have been investigated extensively by Rahman[23] and Paskin[24]. This work is not discussed further since this paper emphasizes crystal lattices.

An especially interesting investigation of computerized models of real solids in which thermal agitation was introduced has been published by Dickey and Paskin.[25] This work dealt with fully dynamic models of rare-gas solids, particularly neon and krypton. Lennard-Jones interactions were assumed, and a set of 865 atoms arranged on a face-centered cubic lattice was used. Periodic boundary conditions employed minimize edge effects and allow the volume to be an independent parameter. Thermal agitation was introduced by giving the atoms a random initial distribution of velocities, with finite mean square. The equations of motion for the system were then solved numerically for a time sufficient to allow a distribution of displacements and velocities corresponding to thermal equilibrium to set in. At high temperatures this occurred quickly; at low temperatures, owing to the long lifetime of the normal modes, true equilibrium was slow to occur, and averages over a number of independent starts had to be taken. From the mean square kinetic energy of the atoms and the law of equipartition, the temperature, T, of each particular distribution was ascertained. Thus,

$$kT = \frac{1}{3N} \Sigma \, m v_i^2,$$

where v_i is the velocity of the ith degree of freedom, N is the number of atoms, and k is Boltzmann's constant.

The pressure, P, was determined from the virial expression

$$P = \rho k T - \frac{\rho}{6N} \sum_{i,j>i} r_{ij} \frac{\delta \varphi(r_{ij})}{\delta \, r_{ij}} . \tag{1}$$

Here ρ is the number density φ is the interatomic pair potential, and r_{ij} is the separation of the pair ij.

Having established that the system was in equilibrium, various thermodynamic properties were calculated. Fig. 8 shows the dependence on temperature at constant volume that was found for the pressure, energy, and mean-square amplitude of atomic vibration in the model of krypton. Anharmonicity produces curvature in these plots. Its effects are most pronounced in the pressure. Such data at two different volumes yield

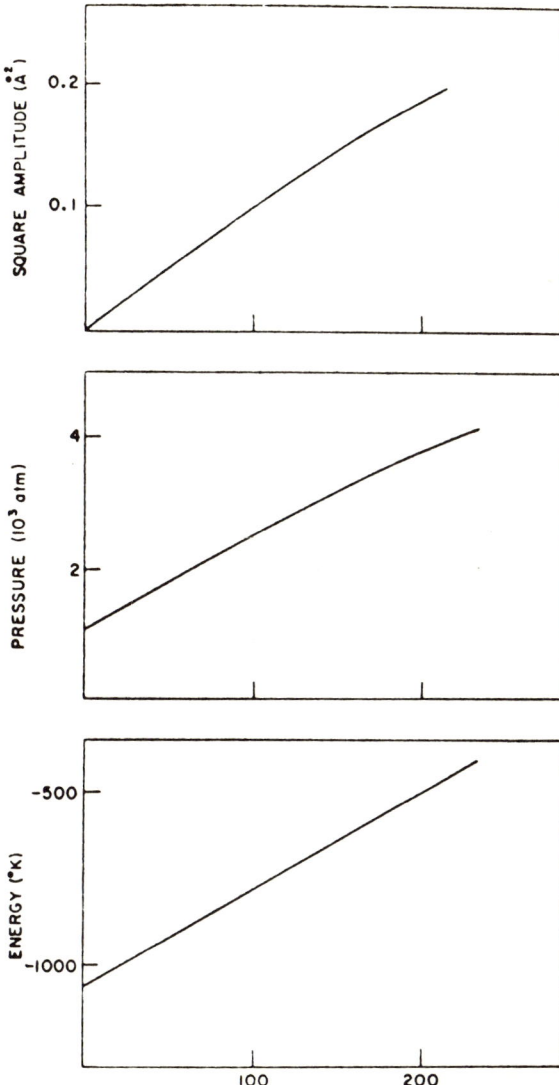

Fig. 8. Pressure, energy, and mean square amplitude of atomic vibration versus temperature for krypton by Dickey and Paskin[25].

the compressibility and thermal expansion, which agree well with actual experimental values, and the Grüneisen constant γ. Average phonon properties, including the effects of anharmonicity, are also readily found. It is easily shown that the normalized phonon frequency spectrum, $g(\omega)$, is simply related to the velocity auto correlation function $G(t)$.

The latter is defined by the relation

$$G(t) = <v_i(t)\, v_i(\mathrm{o})> / <v_i^2(\mathrm{o})> \quad , \tag{2}$$

and it can be shown that

$$g(\omega) = \frac{2}{\pi} \int_0^\infty G(t) \cos \omega t \, dt \quad . \tag{3}$$

Computer studies of the model and Eq. 2 readily allow $G(t)$ to be computed. The spectrum $g(\omega)$ is then found by performing the Fourier transformation (Eq. 3). This process is valid regardless of the temperature and anharmonicity. Fig. 9 shows the frequency spectrum found by Dickey and Paskin in this manner for krypton at low temperature, compared with the spectrum computed by Grindley and Howard[25a] by harmonic approximation and the more usual methods of lattice dynamics applied to a similar model. The agreement is seen to be quite good. Fig. 10 shows Dickey and Paskin's results for the phonon frequency spectrum at temperatures of 2.5 K, 91 K, and 186 K. The spectrum is seen to broaden and shift as the temperature rises. Dependence of the spectrum on volume was also calculated. The lifetime of a normal mode can be calculated directly by introducing a sinusoidally distributed perturbation on the displacements of the atoms at a particular time. The perturbation must be large enough

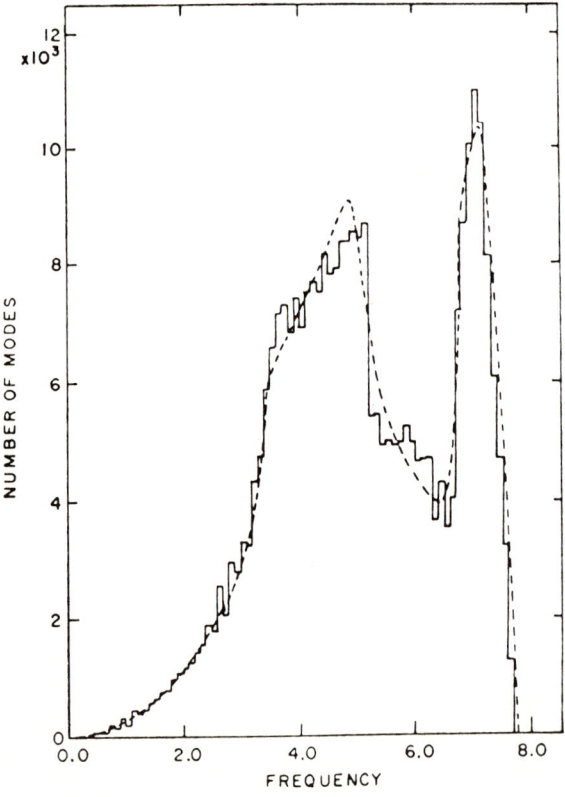

Fig. 9. Frequency spectra of lattice vibrations computed by Dickey and Paskin[25] for krypton at low temperatures, using fully dynamical lattice model (dashed curve) compared with Grindley and Howard[25a] results (solid curve).

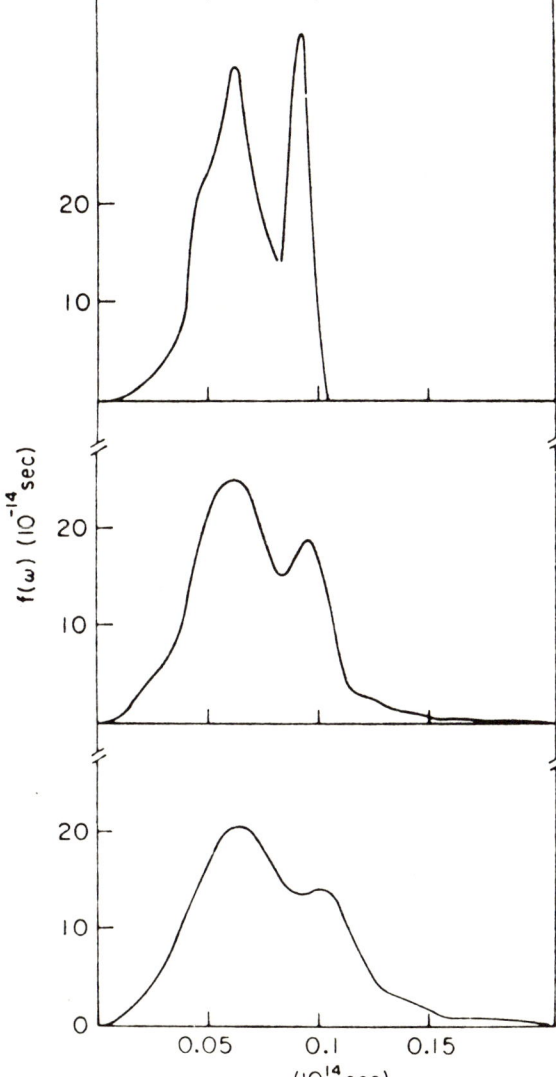

Fig. 10. Frequency spectra of lattice vibrations computed by Dickey and Paskin[25] for three different temperatures: 2.5 K (top curve), 91 K (middle curve), and 186 K (bottom curve).

to be visible, but so small that the temperature is not appreciably disturbed. The perturbation then decays, at a characteristic rate that is readily observable in the computer. The wavenumber of the phonon is the wavenumber of the perturbation.

It should be emphasized that most properties for which the harmonic approximation is valid do not need the elaborate machinery of this method for their computation. However, anharmonic effects are not accessible to the usual methods with high accuracy, and the dynamic lattice model appears to offer advantages for the direct computation of these effects.

Another possible application of dynamic models is to transport phenomena, such as heat conduction, and still another is to vibrations of disordered lattices. A combination of these two concerns is found in a study by Payton, Rich, and Visscher of thermal conductivity in a disordered, harmonic lattice.[26] For simplicity, only one and two-dimensional models were studied. Fig. 11 shows the two models. Two different masses, with a given concentration of light atoms, were assigned at random to the lattice points. A hot reservoir was simulated at one end by applying impulses to the atoms at that end, impulses distributed in a Maxwellian fashion with distribution parameter T_1 (i.e., temperature). A similar perturbation was introduced at the cold end with the smaller parameter T_2. The equations of motion for the system were solved to give its evolution in time. Upon equilibration, a gradient of local temperature was found along the chain, and a steady heat current from hot end to cold end existed. Fig. 12 shows the time-averaged temperature distribution along a chain in the equilibrium state. The quotient of heat current and temperature gradient gives the thermal conductivity. Fig. 13 shows the thermal conductivities determined in this manner for the chain with various concentrations of light atoms, for different mass ratios and for a case when

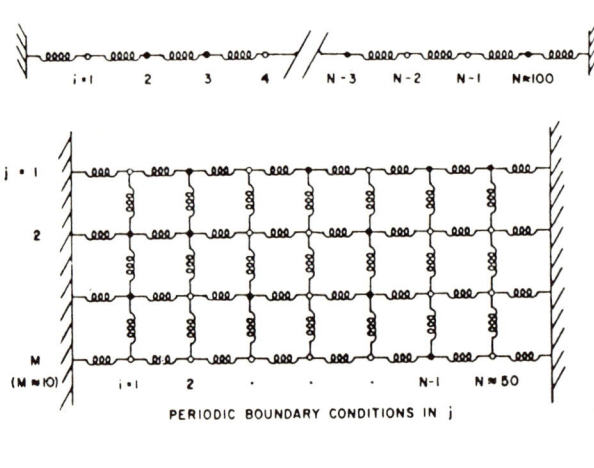

Fig. 11. One and two-dimensional models for computing thermal conductivity of lattices containing a random mixture of atoms of two different masses. From Payton et al.[26].

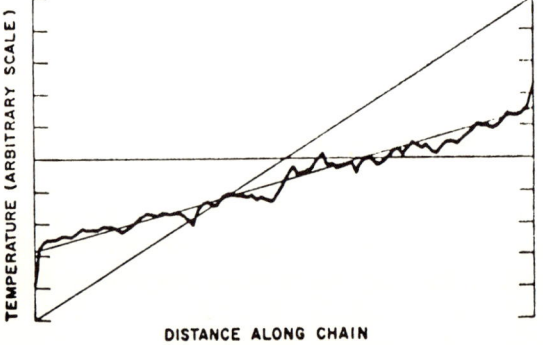

Fig. 12. Time-averaged distribution of temperature along chain found in one of the calculations of Payton et al.[26].

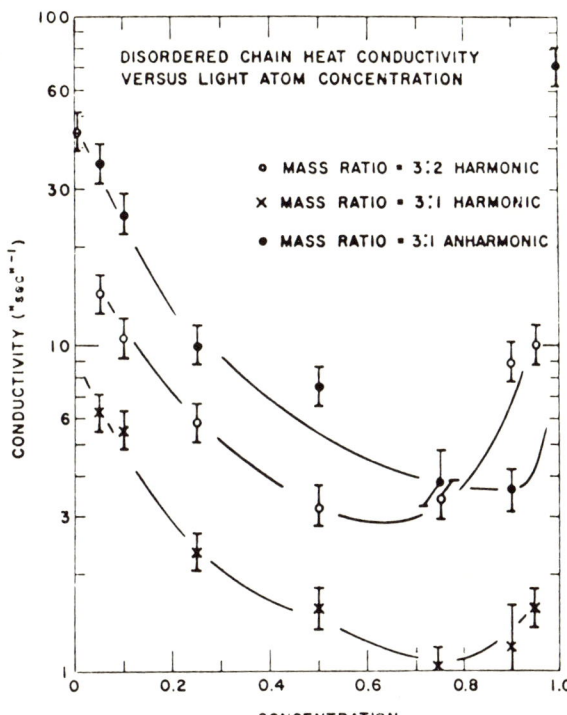

Fig. 13. Thermal conductivity versus concentration of light atoms computed by Payton et al.[26] for disordered one-dimensional lattice.

anharmonic forces were introduced. An interesting effect is the increase in conductivity that occurs when anharmonic forces are added to a disordered chain. They explain this anomaly on the basis that the disordered chain contains many localized or quasilocalized modes which cannot participate in the conductive process. Introduction of anharmonicity allows these modes to communicate with extended modes and thus to participate in the heat conduction.

6 LATTICE DEFECTS

Computerized models of lattices have been of great use in another large class of problems — the determination of the nature and properties of lattice defects. The various defects a lattice can accommodate are metastable equilibrium states of the lattice. Given a lattice model, these states can be determined in a number of ways. With a fully dynamic model, a configuration that approximates a defect configuration, when released from rest, will oscillate about the defect configuration. If the model is supplied with damping, it will eventually come to rest at the configuration itself. Various kinds of artificial damping can be supplied to hasten the convergence to the final configuration and minimize computer time. Alternatively, extremum-seeking programs which do not involve dynamics may be employed.

Without damping, the special character of the lattice vibrations caused by the defect (including local modes) can be determined in the fully dynamic models, although not much use has been made yet of this possibility. Static properties of interstitials, vacancies, and various clusters of interstitials and vacancies have been investigated in a variety of simple lattices. Since the dynamic aspects of these problems have not been of primary interest, the extensive accumulated literature of the last dozen years is not reviewed here. For quantitative purposes, the details of the interaction potential at separations near that of the perfect lattice are crucial. For most of these calculations the most searching question has not been how to compute the equilibrium configuration, or how to allow for the influence of the further reaches of the lattice, but what interactions to assume. This question is still far from finding a satisfactory resolution.

It should be remarked that the thermally induced migration of defects can be studied in essentially two ways: (1) A fully dynamic lattice model containing defects can be given thermal agitation, and the time development can be followed on the computer. Actual jumps of the defects can be observed and the frequency of jumping can be measured. On the other hand, the energy barrier to a jump can be computed by moving the jumping atom, or atoms, to a saddle point, with the aid of artificial constraints, and the calculated energy can then be inserted in a rate-theory formula to estimate the jump frequency. The entropy of activation can also be calculated. (2) The second method (use of rate theory) is of more general use and has been more widely applied. The first method (direct calculation), making much larger demands on computer time, has been seldom used, although it has been applied in interesting ways by Weiner and Adler[27] to test the predictions of rate theory and decide if its assumptions are sound. The accepted forms of rate theory were well supported.

Extended defects pose a more-difficult problem for dynamical models because of the limited size of the set of atoms that can be explicitly treated. Surfaces, stacking faults, and dislocations have their static and dynamic properties, all of which are of interest. A certain amount of work has been done with dislocations.[28] Highly schematic one-dimensional models and two and three-dimensional models have been studied. In three dimensions, the dimension parallel to an edge dislocation can be made infinite through use of periodic boundary conditions. The slow fall off of the strain field in the perpendicular directions, however, poses acute problems with regard to choice of the other boundary conditions.

7 INTERNAL DEGREES OF FREEDOM. MAGNETISM

All of the foregoing investigations have dealt with positions of atoms as the primary variables. The atoms were treated as mass points devoid of internal coordinates. Internal degrees of freedom, occurring in physical systems, provide a larger class of problems. The most prominent examples are magnetic systems. Others might be imagined: for instance, electronic excitation of the atoms (motion of excitons), nuclear excitations, a lattice of molecules, etc. A lattice populated with atoms possessing spin and magnetic moments forms a cooperative system even when the atomic positions are fixed, i.e., when no regard is paid to the translational degrees of freedom. While the equilibrium properties of such systems have long been under study, the dynamical properties have been investigated less carefully. Dynamic computer models offer an attractive means of studying these systems. Such work has been performed recently by Watson et al.[29] at Brookhaven and (for the high-temperature limit) by Windsor[30] at Harwell. The Brookhaven work concerned the dynamics of classical spins located on fixed lattices and endowed with Heisenberg interactions. Each spin behaves like a gyroscope precessing about an effective magnetic field that is the sum of the fields provided by its neighbors, and at a rate proportional to the magnitude of that field. Thus, if $\mathbf{S}_n$ is the spin on site n, the Hamiltonian in the absence of an external field is $-\frac{1}{2}J \sum_{[m,n]} \mathbf{S}_m \cdot \mathbf{S}_n$, leading to the equation of motion

$$\frac{d\mathbf{S}_n}{dt} = -J \left[\sum_m \mathbf{S}_m \right] \times \mathbf{S}_n, \quad n = 1, 2, ...N. \quad (4)$$

Here J is a coupling constant, the summation in the expression for the Hamiltonian is over all pairs of near neighbors, and in Eq. 4 the summation is over the near neighbors to site n.

The magnetization of the entire crystal is

$$\mathbf{M} = c \sum_{n=1}^{N} \mathbf{S}_n , \quad (5)$$

where c is a constant. The magnitude of each spin $|\mathbf{S}_n|$ is a constant, and the directions of the spins precess according to Eq. 4.

Starting with any set of spins at time 0, the computer solves Eq. 4 to determine the spins at subsequent times. A subtle difficulty is the determination of an appropriate initial configuration of spins to represent a state at arbitrary temperature T. This difficulty was met by first performing a Monte Carlo process, in which the temperature enters as an arbitrary parameter, to arrive at a "randomized" state characteristic of the temperature. This state is then used as the initial state for a dynamical calculation. A by-product of the Monte Carlo process is a determination of the magnetization and energy at that temperature. From the dynamical calculation, time-displaced correlations of various pairs of spins are determined. The time-displaced correlations show considerable structure. At low temperatures, the results are close to the prediction of analytical theories based on the spin-wave approximation. Fig. 14 shows one of many such correlations that were computed. This is for a simple cubic lattice, and shows the time-displaced self-correlation of the transverse component of spin on one site, $\langle \mathbf{S}_\perp(o) \cdot \mathbf{S}_\perp(\tau) \rangle$, where $\mathbf{S}_\perp = (S^{(x)}, S^{(y)})$. Magnetization is along the z-axis. The variable τ is the time expressed in Larmor periods

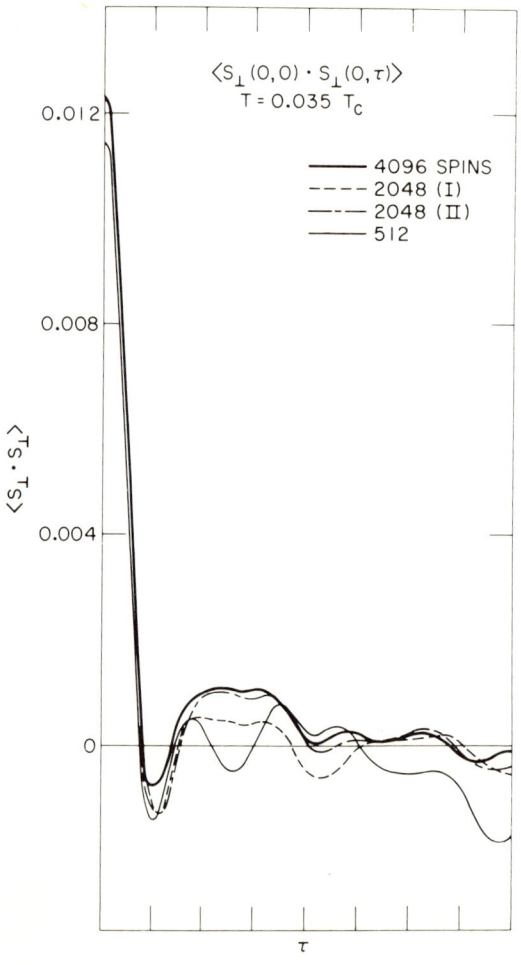

Fig. 14. Time-displaced self-correlation of spins, computed for a Heisenberg model of a ferromagnet on a simple cubic lattice. From Watson, et al.[29].

appropriate to the sample's magnetization. Results for four different calculations involving crystallites of three different sizes are shown. The rapid decline of correlations in a Larmor period and the subsequent oscillations are characteristic. The differences among the four curves show the influence of finiteness of the sample and finiteness of running time toward a steady state.

Time-displaced spin correlations, besides being of fundamental significance in models of magnetism, determine the inelastic magnetic scattering of neutrons from magnetic specimens. The momentum and energy distribution of the scattered neutrons is given by the Fourier transform of the time-displaced spin-correlation functions.

A second interesting result of the calculations is a depiction of the Green's function for spins, i.e., the influence of a single misaligned spin that is suddenly released in an otherwise ordered array of spins. Fig. 15 shows the results of such a calculation. Here a (100) planar section of a 16 x 16 x 16 spin array is shown at six successive times. At time 0, all spins are ferromagnetically aligned perpendicular to the plane, except one spin in the middle which is set at 90 degrees. Successive snapshots show the propagation of the disorder into the array.

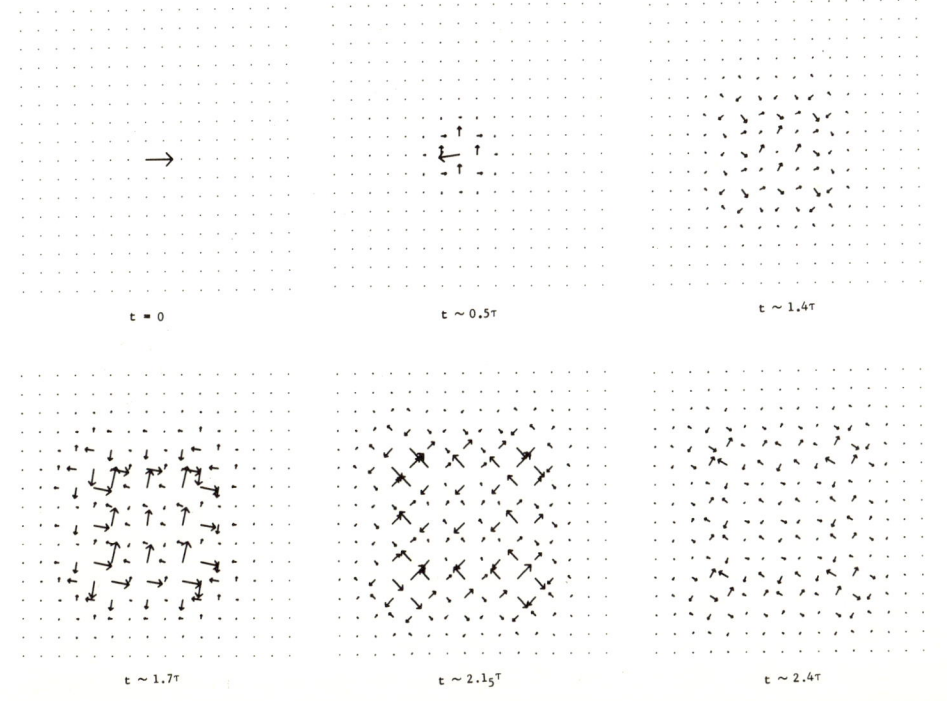

Fig. 15. A single spin is released at time 0 at 90 degrees to magnetic axis in perfectly magnetized Heisenberg model. Propagation of spin disturbances in (001) plane is shown. Time at lower right is 2.4 Larmor periods. From Watson, et al.[29].

Ramifications of this work have concerned two-dimensional simple square lattices and, most recently, a one-dimensional lattice of spins.[31] The more intricate problem in which lattice vibrations interact with spinning atoms (the magnon-phonon coupling situation) has not been attempted, but might be of interest.

8 COMMENTS AND CONCLUSIONS

In the relatively short time that computer experiments with lattice models have been underway, a remarkable range of work has been done. Looking over these results one could conclude that the cream has already been skimmed, or on the contrary, that a golden age is dawning. Reflection finds one more impressed with what remains to be done than with what has been done, interesting though the latter may be. In many potential areas of computer experimentation, the pioneering work has hardly been done, let alone the careful follow-ups needed for maturation of the field.

Worthwhile computer experiments require time and care. The easy understandability of the results tends to conceal the painstaking hours that went into conceiving and formulating the problem, selecting the parameters of a model, programming for computation, sifting and analyzing the flood of output from the computer, rechecking the approximations and stratagems for accuracy, and out of it all synthesizing physical information. Larger and faster computers are always a need. In the past 10 years the increase in available computer power has been enormous; more power is needed and will come. On the other hand, frontal assaults on many problems would place on computers demands that exceed the wildest imagination, and cleverness can be expected to play a crucial role in the future as it has in the past. Thus, a straightforward sampling of phase space for a many-dimensional system is inconceivably difficult. If each degree of freedom of the system is divided into a 10-point mesh, for example, N degrees of freedom requires sampling at 10^N points — a discouraging task if N is 100, let alone 1000! Nevertheless there is a stratagem, the Monte Carlo method, that provides surprisingly good averages over such a voluminous phase space. It works by recognizing that most of the points are of little importance, and that a random sampling biased by the "importance" of the points can be conducted in a reasonable time to yield good accuracy. A similarly successful stratagem is yet to be found for the fully dynamical treatment of quantum mechanical many-body systems.

Above all, better interatomic potentials are needed. The various computer experiments that are done with lattices require an enormous range of interatomic energies from the fractional-eV interaction at normal lattice spacings to the hundreds of kilovolt interactions called into play in radiation-damage cascades and ion bombardments. All of the interactions employed are approximate. Some uncertainties in the potentials are unimportant (or relatively unimportant) compared with the energies available. This is particularly so for the energetic interactions of radiation-damage events. Other uncertainties in potentials are quite comparable with the energies that determine the phenomenon. This is true for point defects near equilibrium. Formation energies, binding energies, and activation energies for motion of point defects are not determined with adequate precision, and alternative configurations and migration mechanisms often cannot be ruled out.

In such cases it may be better to invert the problem and to regard parameters in the interaction potential as the unknowns, designing the calculations and the confrontations with real experiments in such a way as to determine the parameters. The best means of doing this are not obvious, but quite possibly a significant part of the computer experimentation in years ahead will work in this mode and will produce still deeper insights into the many phenomena of crystal lattices.

ACKNOWLEDGMENT

This work was performed under the auspices of the U. S. Atomic Energy Commission.

REFERENCES

1. Beeler, J. R., Jr.: *Advances in Materials Research*, Vol. 4, ed. H. Herman, pp. 295-476 (John Wiley & Sons, Inc., 1970).
2. Vineyard, G. H., J. B. Gibson, A. N. Goland, and M. Milgram: Bull. Am. Phys. Soc. **5**: 26 (1960); Gibson, J. B., A. N. Goland, M. Milgram, and G. H. Vineyard: Phys. Rev. **120**: 1229 (1960).
3. Erginsoy, C., G. H. Vineyard, and A. Englert: Phys. Rev. **A 133**: 595 (1964).
4. Erginsoy, C., G. H. Vineyard, and A. Shimizu: Phys. Rev. **139A**: 118-25 (1965).
5. Lucasson, A., Lucasson, P. G., and R. M. Walker: *International Conference on the Properties of Reactor Materials and the Effects of Radiation Damage* (Butterworths Scientific Publications, Ltd., London, 1962).
6. Vineyard, G. H.: J. Phys. Soc. Japan **18**, Suppl. III: 144 (1963).
7. Chadderton, L. T. and I. McC. Torrens: Proc. Roy. Soc. **A 294**: 93 (1966).
8. McC. Torrens, I. and L. T. Chadderton: Phys. Rev. **159**: 671 (1967).
9. Chadderton, L. T. and I. McC. Torrens: *Fission Damage in Crystals* (Methuen & Co., Ltd., London, 1969).
10. Beeler, J. R., Jr. and D. G. Besco: J. Appl. Phys. **34**: 2873 (1963).
11. Beeler, J. R. Jr.: Phys. Rev. **150**: 470 (1966).
12. Yoshida, M.: J. Phys. Soc. Japan **16**: 44 (1961).

13. Beeler, J. R., Jr.: *Radiation Damage in Reactor Materials,* pp. 3-31, International Atomic Energy Agency, Vienna, 1969.
13a. Doran, D. G.: Radiation Effects **2**: 249 (1970).
14. Robinson, M. T. and O. S. Oen: Phys. Rev. **132**: 2385 (1963).
15. Tsai, D. H. and C. W. Beckett: J. Geophys. Res. **71**: 2601 (1966).
16. Anderson, G. D.: Ph.D. Thesis, Washington State University, 1964.
17. Manvi, R.N.R.: Ph.D. Thesis, Dept. of Mechanical Engineering, Washington State University, 1968.
18. Smith, C. S.: Trans. AIME **212**: 574 (1958).
19. Gehlen, P. C. and M. F. Kanninen: *Inelastic Behavior of Solids,* M. F. Kanninen, W. F. Adler, A. R. Rosenfield, and R. I. Jaffe (Eds.) (Mc-Graw-Hill Book Co., New York, 1970).
20. McClintock, F. C., private communication.
21. Wood, W. W. and F. R. Parker: J. Chem. Phys. **27**: 720 (1957).
22. Alder, B. J. and T. E. Wainright: J. Chem. Phys. **31**: 459 (1959); ibid **33**: 1439 (1960). See also the review by Beeler, ref. 1.
23. Rahman, A.: Phys. Rev. **136**: A405 (1964).
24. Paskin, A. and A. Rahman: Phys. Rev. Lett. **16**: 300 (1966); Paskin, A.: Advan. Phys. **16**: 223 (1967).
25. Dickey, J. M. and A. Paskin: Phys. Rev. **188**: 1407 (1969).
25a. Grindley, J. and Howard, R., in *Lattice Dynamics,* ed. R. F. Wallis (Pergamon Press, Inc., New York 1965), p. 129.
26. Payton, D. N., M. Rich, and W. M. Visscher: Phys. Rev. **160**: 706 (1967).
27. Weiner, J. H. and W. F. Adler: Phys. Rev. **144**: 511 (1966).
28. Weiner, J. H.: Phys. Rev. **136**: A863 (1964).
    Hartl, W. F. and J. H. Weiner: Phys. Rev. **152**: 634 (1966).
    Englert, A. and H. Tompa: J. Phys. Chem. Solids **21**: 306 (1961).
    Englert, A. and H. Tompa: J. Phys. Chem. Solids **24**: 1145 (1963).
    Doyama, M. and R.M.J. Cotterill: Phys. Letters **13**: 110 (1964).
    Doyama, M. and R.M.J. Cotterill: Phys. Letters **14**: 79 (1965).
    Doyama, M. and R.M.J. Cotterill: Phys. Rev. **150**: 448 (1966).
    Cotterill, R.M.J. and M. Doyama: Phys. Rev. **145**: 465 (1966).
29. Watson, R. E., M. Blume, and G. H. Vineyard: Phys. Rev. **181**: 811 (1969); Watson, R. E., M. Blume, and G. H. Vineyard: Phys. Rev. **B2**: 684 (1970).
30. Windsor, C. G.: Proc. Phys. Soc. (London) **91**: 353 (1967).
31. Watson, R. E. and M. Blume, to be published.

DISCUSSION on the Paper Presented by G. H. Vineyard

ASHCROFT: I hope I'm correct in assuming that the discussion of computer generation of spin-correlation function referred to three-dimensional spin arrays. What information is there on two-dimensional systems? In particular is it possible to test the theorems of Mermin and others on the *absence* of ferromagnetism (or magnetic ordering) for two-dimensional spin arrays described by a Heisenberg Hamiltonian (possibly in the thermodynamic limit)?

VINEYARD: The slides shown were indeed for spins in three-dimensional arrays.[1] We have also worked with two-dimensional models[2] and one-dimensional models (to be published). In two dimensions we find no ferromagnetic ordering at any temperature above absolute zero, in accordance with the theorem of Mermin[3] and Wagner[4], but also find evidence to confirm the conjecture of Stanley and Kaplan[5] that a state of "long-ranged short-range order" exists below a certain critical temperature.

SEEGER: There is now some experimental evidence that collision sequences may proceed in such a way that the dynamic interstitial defect configurations (e.g., dynamic crowdions) propagate along lattice rows bypassing the vacant lattice sites which they leave behind. I wonder what the evidence from the simulation studies was, and whether specific computer experiments could be performed that have a bearing on that point?

VINEYARD: Indeed the computer simulations of radiation damage events have showed this effect many times. A knock-on cascade energizes focused collision chains along several rows of atoms, and they are by no means confined to rows radiating from the site of the original knock-on. Figures 1 and 3 of my paper in this colloquium illustrate the effect clearly. Considerable quantitative information on the effect could be collected from the calculations already reported, and further specific calculations could be made if necessary.

REFERENCES

1. Watson, R. E., M. Blume, and G. H. Vineyard: Phys. Rev. **181**, 811 (1969).
2. Ibid., Phys. Rev. **B2**, 684 (1970).
3. Mermin, N. D.: J. Math. Phys., **8**, 1061 (1967).
4. Mermin, N. D. and H. Wagner: Phys. Rev. Let., **17**, 1133 (1966).
5. Stanley, H. E. and T. A. Kaplan: Phys. Rev. Let., **17**, 913 (1966).

POTENTIAL FUNCTIONS AND THE SIMULATION OF DEFECTS IN LATTICE DYNAMICAL DEFECT PROBLEMS

A. A. Maradudin

University of California
Irvine, California

ABSTRACT

The use of interatomic potential functions in the study of lattice dynamical defect problems has some differences from their use in other types of defect problems. In general, only the second derivatives of the potential function evaluated at the equilibrium separation between atoms in a crystal (the atomic force constants) are required. The latter can certainly be determined once an interatomic potential function is known, but can sometimes be determined indirectly without recourse to an assumed analytic expression for the potential function. For many substitutional impurities the potential of interaction between the atom or ion and the atoms of the host crystal is not known *ab initio*, and the atomic force constants have to be determined from experiment. In such cases the restrictions imposed on the force constants by the invariance and transformation properties of the force on an atom under infinitesimal, rigid-body translations and rotations of the crystal, as well as by the point symmetry of the defect site, can reduce considerably the number of independent, nonzero force constants that have to be determined. These conditions are described and their importance for a correct description of the dynamical properties of lattice defects is illustrated by applications drawn from the study of isolated point defects and of extended defects such as crystal surfaces. Finally, the results of attempts at first-principles calculations of the interactions between substitutional defects and the host crystal, in ionic crystals and in metals, are summarized and commented upon.

1 INTRODUCTION

The use of potential functions in lattice dynamical problems differs from their use in other types of problems in crystal physics in that it is not the entire interatomic potential that is required, but only limited information about it; namely, the values of its second (and in some cases, higher) derivative at a discrete set of distances — the equilibrium interatomic separations in a crystal lattice. These second derivatives are called the atomic force constants of the crystal. Together with the atomic masses, the atomic force constants determine the frequencies of the normal mode vibrations and the displacements of the atoms in each normal mode.

With some qualifications, the same is true in using potential functions in lattice dynamical defect problems. Here, it is not only the second derivatives of the interatomic potential, or potentials, describing the interactions of the atoms of the host crystal that are required. The second derivatives of the potential describing the interaction of a substitutional impurity, a vacancy, an interstitial atom, or an even more complicated defect, with the host crystal are also required.

The atomic force constants of the perfect host crystal are usually regarded as known in a lattice dynamical defect problem. The best values of these constants are usually obtained by fitting the phonon-dispersion curves for the crystal obtained by neutron spectroscopy along symmetry directions in the first Brillouin zone by the predictions of a phenomenological model of the crystal such as the deformation dipole model[1] or the shell model[2]. Consequently, in what follows, the emphasis is on the interaction between the defect and the host crystal, and little is said about the dynamical properties of the perfect host crystal.

At the present time *ab initio* calculations of the cohesive energy of perfect crystals are difficult to carry out with quantitative accuracy. It is only recently that *ab initio* calculations of phonon-dispersion curves of crystals (mostly metallic) have been carried out with any degree of success. Thus, although first-principles calculations of the interaction potential between substitutional or interstitial impurities and the ions of the host crystal have been carried out for a (very) few ionic crystals very recently, it is not surprising that in the overwhelming majority of lattice dynamical defect problems studied to date, no attempt has been made to determine, from first principles, the interaction of the defect with the host crystal. Rather, in the ideal situation the atomic force constants describing this interaction have been treated as parameters of the theory to be determined from a fit of the theory to one set of experimental data, and the results used to explain or interpret a different set of experimental data. In practice, what has happened, more often than not, has been that the first part of the preceding program has been carried out, it has then been pointed out that the theory is capable of explaining the experimental data, provided that the force constants describing the impurity-host crystal interaction have the experimentally obtained values, and the matter ends with some comment concerning the reasonableness of these values. This situation is not likely to change significantly until it becomes easier to calculate defect-host crystal interactions accurately.

The approach to lattice dynamical defect problems in which the perturbed force constants are treated as parameters of the theory has been widely used in the past (and will continue to be widely used for some time to come). Consequently, this paper begins with a discussion of the properties of atomic force constants in perturbed crystals. The results of such a discussion are important because they comprise the restrictions imposed on the number of independent force constants, and their nature, by general invariance conditions and the point symmetry of the crystal about the defect site. They assure that the theory will not contain more parameters than absolutely necessary, and that it satisfies all of the invariance conditions that it must. Examples of the use of this approach to the study of the defect-host crystal interaction are described, as well as recent attempts to calculate this interaction from first principles in some simple cases.

2 ATOMIC FORCE CONSTANTS IN A PERTURBED CRYSTAL

If we consider an isolated point defect in an otherwise perfect crystal and expand the potential energy of the crystal in powers of arbitrary displacements of the atoms from their equilibrium positions in the perfect crystal, the resulting expansion in general has the form

$$\Phi = \hat{\Phi}_0 + \sum_{\ell\kappa\alpha} \hat{\Phi}_\alpha(\ell\kappa)\, \xi_\alpha(\ell\kappa) + \frac{1}{2} \sum_{\ell\kappa\alpha} \sum_{\ell'\kappa'\beta} \hat{\Phi}_{\alpha\beta}(\ell\kappa;\ell'\kappa')\, \xi_\alpha(\ell\kappa)$$

$$\times\, \xi_\beta(\ell'\kappa') + \frac{1}{6} \sum_{\ell\kappa\alpha} \sum_{\ell'\kappa'\beta} \sum_{\ell''\kappa''\gamma} \hat{\Phi}_{\alpha\beta\gamma}(\ell\kappa;\ell'\kappa';\ell''\kappa'')\, \xi_\alpha(\ell\kappa)\, \xi_\beta(\ell'\kappa')$$

$$\times\, \xi_\gamma(\ell''\kappa'') + \ldots \tag{1}$$

In this expression $\xi_\alpha(\ell\kappa)$ is the α Cartesian component of the displacement of the $\kappa\underline{\text{th}}$ atom in the $\ell\underline{\text{th}}$ primitive unit cell from its equilibrium position in the perfect crystal, and the expansion coefficients $\hat{\Phi}_0$, $\{\hat{\Phi}_\alpha(\ell\kappa)\},\ldots$ are mixed partial derivatives of the potential energy with respect to the displacements $\{\xi_\alpha(\ell\kappa)\}$:

$$\hat{\Phi}_\alpha(\ell\kappa) = \left(\frac{\partial \Phi}{\partial \xi_\alpha(\ell\kappa)}\right)_0 \tag{2a}$$

$$\hat{\Phi}_{\alpha\beta}(\ell\kappa;\ell'\kappa') = \left(\frac{\partial^2 \Phi}{\partial \xi_\alpha(\ell\kappa)\, \partial \xi_\beta(\ell'\kappa')}\right)_0, \text{ etc.}, \tag{2b}$$

where the subscript o denotes that the derivatives are evaluated in the configuration in which all of the atoms are situated at their equilibrium positions. The caret on each expansion coefficient reminds us that these derivatives are calculated in the presence of the point defect. The terms linear in the displacements $\{\xi_\alpha(\ell\kappa)\}$ are not identically zero in Equation (1) because, in the presence of the defect, the atoms surrounding the defect in general are longer in static equilibrium when situated at the lattice sites of the perfect crystal: the defect exerts a force on them in this configuration which can be relieved only by the atoms moving to new equilibrium positions.

If the corresponding coefficients for the crystal without the point defect are denoted by a superscript o, i.e., by $\Phi_0^{(o)}$, $\{\Phi_\alpha^{(o)}(\ell\kappa)\}$, $\{\Phi_{\alpha\beta}^{(o)}(\ell\kappa;\ell'\kappa')\},\ldots$, we have that

$$\Phi_\alpha^{(o)}(\ell\kappa) = 0, \tag{3}$$

because no net force can be exerted on an atom of a perfect crystal when it and all the remaining atoms are at their equilibrium positions. We can also write the relations

$$\hat{\Phi}_{\alpha\beta}(\ell\kappa;\ell'\kappa') = \Phi_{\alpha\beta}^{(o)}(\ell\kappa;\ell'\kappa') + \Delta\Phi_{\alpha\beta}(\ell\kappa;\ell'\kappa'), \text{ and} \tag{4a}$$

$$\hat{\Phi}_{\alpha\beta\gamma}(\ell\kappa;\ell'\kappa';\ell''\kappa'') = \Phi^{(0)}_{\alpha\beta\gamma}(\ell\kappa;\ell'\kappa';\ell''\kappa'') + \Delta\Phi_{\alpha\beta}(\ell\kappa;\ell'\kappa';\ell''\kappa'') \,. \quad (4b)$$

Let us now assume that the displacement $\xi_\alpha(\ell\kappa)$ is compounded of two contributions,

$$\xi_\alpha(\ell\kappa) = d_\alpha(\ell\kappa) + u_\alpha(\ell\kappa) \,, \quad (5)$$

of which the first gives the static displacements of the atoms of the host crystal in accommodating to the presence of the point defect, and the second describes the dynamic displacements of the atoms about their new equilibrium positions. If Eq. 5 is substituted into Eq. 1, the static displacements $\{d_\alpha(\ell\kappa)\}$ are determined by the condition that the resulting expansion of the potential energy in powers of the dynamic displacements $\{u_\alpha(\ell\kappa)\}$ contain no term linear in the $\{u_\alpha(\ell\kappa)\}$:

$$\Phi = \Phi_0 + \frac{1}{2} \sum_{\ell\kappa\alpha} \sum_{\ell'\kappa'\beta} \Phi_{\alpha\beta}(\ell\kappa;\ell'\kappa') u_\alpha(\ell\kappa) u_\beta(\ell'\kappa') + \ldots, \quad (6)$$

where

$$\Phi_0 = \hat{\Phi}_0 + \sum_{\ell\kappa\alpha} \hat{\Phi}_\alpha(\ell\kappa) d_\alpha(\ell\kappa) + \frac{1}{2} \sum_{\ell\kappa\alpha} \sum_{\ell'\kappa'\beta} \hat{\Phi}_{\alpha\beta}(\ell\kappa;\ell'\kappa') d_\alpha(\ell\kappa) \times$$

$$d_\beta(\ell'\kappa') + \ldots, \text{ and} \quad (7a)$$

$$\Phi_{\alpha\beta}(\ell\kappa;\ell'\kappa') = \hat{\Phi}_{\alpha\beta}(\ell\kappa;\ell'\kappa') + \sum_{\ell''\kappa''\gamma} \hat{\Phi}_{\alpha\beta\gamma}(\ell\kappa;\ell'\kappa';\ell''\kappa'')$$

$$\times d_\gamma(\ell''\kappa'') + \ldots \,. \quad (7b)$$

The static displacements $\{d_\alpha(\ell\kappa)\}$ are obtained from the solution of the equation

$$0 = \hat{\Phi}_\alpha(\ell\kappa) + \sum_{\ell'\kappa'\beta} \hat{\Phi}_{\alpha\beta}(\ell\kappa;\ell'\kappa') d_\beta(\ell'\kappa') +$$

$$\frac{1}{2} \sum_{\ell'\kappa'\beta} \sum_{\ell''\kappa''\gamma} \hat{\Phi}_{\alpha\beta\gamma}(\ell\kappa;\ell'\kappa';\ell''\kappa'') d_\beta(\ell'\kappa') d_\gamma(\ell''\kappa'') + \ldots \quad (8)$$

If we introduce a matrix $U_{\alpha\beta}(\ell\kappa;\ell'\kappa')$ by

$$\sum_{\ell''\kappa''\gamma} \hat{\Phi}_{\alpha\gamma}(\ell\kappa;\ell''\kappa'') U_{\gamma\beta}(\ell''\kappa'';\ell'\kappa') = \delta_{\ell\ell'}\delta_{\kappa\kappa'}\delta_{\alpha\beta} \,, \quad (9)$$

the solution of Eq. 8 can be written as

$$d_\alpha(\ell\kappa) = - \sum_{\ell'\kappa'\beta} U_{\alpha\beta}(\ell\kappa;\ell'\kappa') \hat{\Phi}_\beta(\ell'\kappa') - \frac{1}{2} \sum_{\ell'\kappa'\beta} \sum_{\ell^2\kappa^2\gamma} \sum_{\ell^3\kappa^3\delta} \sum_{\ell^4\kappa^4\epsilon}$$

$$\sum_{\ell^5\kappa^5\zeta} U_{\alpha\beta}(\ell\kappa;\ell'\kappa') \hat{\Phi}_{\beta\gamma\delta}(\ell'\kappa';\ell^2\kappa^2;\ell^3\kappa^3) U_{\gamma\epsilon}(\ell^2\kappa^2;\ell^4\kappa^4)$$

$$\times \hat{\Phi}_\epsilon(\ell^4\kappa^4) U_{\delta\zeta}(\ell^3\kappa^3;\ell^5\kappa^5) \hat{\Phi}_\zeta(\ell^5\kappa^5) + \ldots \qquad (10)$$

In practice, it is extremely cumbersome to use all terms past the first in this solution.

The method of obtaining the displacements $\{d_\alpha(\ell\kappa)\}$ describing the static relaxation of the atomic positions about a point defect is called the method of "lattice statics". It was devised by Kanzaki[3], with subsequent developments due to Hardy and his collaborators[4] and others[5].

The matrix $U_{\alpha\beta}(\ell\kappa;\ell'\kappa')$ is familiar from the dynamical theory of imperfect crystals as the (negative of the) vibrational Green's function for the perturbed crystal, evaluated at zero frequency. Methods for its evaluation have been described in the literature.[6]

The preceding results show that when the equations of motion of a crystal perturbed by a point defect are written in the form

$$M_{\ell\kappa}\ddot{u}_\alpha(\ell\kappa) = - \sum_{\ell'\kappa'\beta} \Phi_{\alpha\beta}(\ell\kappa;\ell'\kappa')u_\beta(\ell'\kappa') , \qquad (11)$$

where $M_{\ell\kappa}$ is the mass of the atom $(\ell\kappa)$, the perturbed atomic force constants

$$\Phi_{\alpha\beta}(\ell\kappa;\ell'\kappa') = \Phi^{(o)}_{\alpha\beta}(\ell\kappa;\ell'\kappa') + \Delta\Phi_{\alpha\beta}(\ell\kappa;\ell'\kappa') +$$

$$\sum_{\ell''\kappa''\gamma} \left[\Phi^{(o)}_{\alpha\beta\gamma}(\ell\kappa;\ell'\kappa';\ell''\kappa'') + \Delta\Phi_{\alpha\beta\gamma}(\ell\kappa;\ell'\kappa';\ell''\kappa'') \right] d_\gamma(\ell''\kappa'') + \ldots \qquad (12)$$

differ from the unperturbed force constants not only by the second derivatives of the difference between the potential of interaction of the atom(s) comprising the point defect with the host crystal and the corresponding potential in the perfect crystal $\Delta\Phi_{\alpha\beta}(\ell\kappa;\ell'\kappa')$, but also include a contribution associated with the distortion of the host crystal about the defect as well.

In practice, the last contribution is neglected with the heuristic justification that, as long as the distortion of the host crystal by the point defect is not so severe that its structure is reconstructed in the vicinity of the defect, the effects of this contribution on the perturbed force constants can be absorbed into the effects coming from the second contribution, $\Delta\Phi_{\alpha\beta}(\ell\kappa;\ell'\kappa')$, which are unknown anyway, and are to be obtained from some kind of experimental data. This approach is convenient because it eliminates the need for knowing the (perturbed) cubic anharmonic force constants of the crystal, and because the determination of the static displacements of the atoms surrounding the defect is not a trivial calculation. However, the use of this approximation results in the neglect of an important physical effect: in the presence of lattice distortion about the point defect, the perturbation of the force constants of the perfect crystal $\left\{\Phi_{\alpha\beta}^{(o)}(\ell\kappa;\ell'\kappa')\right\}$ by the presence of the defect is not confined to the immediate vicinity of the defect, even if the $\left\{\Delta\Phi_{\alpha\beta}(\ell\kappa;\ell'\kappa')\right\}$ are nonzero only for the interaction of the defect with its nearest neighbors. Even the force constants for the interaction between a pair of atoms at a great distance from the defect will be perturbed by the introduction of the point defect due to the lattice distortion to which it gives rise. The perturbation of individual force constants far from the defect may be quite small, but this is a long-range effect, and the integrated consequences can be significant. Although methods for calculating the effects of extended defects on the vibrational properties of crystals have been described in the literature[6], the author knows of no calculation of a vibrational property of a point defect in a crystal considering the distortion of the crystal about it. Such a calculation would be difficult, but worthwhile.

Whatever approximations are made in defining the perturbed force constants $\left\{\Phi_{\alpha\beta}(\ell\kappa;\ell'\kappa')\right\}$, the results must possess certain general properties. Because they are the coefficients in a quadratic form in the atomic displacements they must be symmetric in the indices $(\ell\kappa\alpha)$ and $(\ell'\kappa'\beta)$:

$$\Phi_{\alpha\beta}(\ell\kappa;\ell'\kappa') = \Phi_{\beta\alpha}(\ell'\kappa';\ell\kappa) \ . \tag{13}$$

When a crystal is translated rigidly, no net force is exerted on any atom since the relative positions of the atoms comprising the crystal are not altered. This fact has the consequence for the atomic force constants that

$$\sum_{\ell\kappa} \Phi_{\alpha\beta}(\ell\kappa;\ell'\kappa') = \sum_{\ell'\kappa'} \Phi_{\alpha\beta}(\ell\kappa;\ell'\kappa') = 0 \ . \tag{14}$$

When a crystal is subjected to a rigid-body rotation, the force on an atom rotates with the rotation of the crystal. As a result, the atomic force constants must obey the conditions

$$\sum_{\ell'\kappa'} \Phi_{\alpha\beta}(\ell\kappa;\ell'\kappa') x_\gamma(\ell'\kappa') = \sum_{\ell'\kappa'} \Phi_{\alpha\gamma}(\ell\kappa;\ell'\kappa') x_\beta(\ell'\kappa'), \qquad (15)$$

where $x(\ell\kappa)$ is the vector to the equilibrium position of the atom $(\ell\kappa)$ from some conveniently chosen origin.

It should be emphasized that the derivations of the three conditions expressed by Eq. 13, 14, and 15 do not depend on the assumption that the atoms with whose interactions we are concerned occupy the lattice points of a crystal. They apply as well to any collection of a finite number of atoms, e.g., a molecule, whose total potential energy is a function of their instantaneous positions.

The conditions for Eq. 13, 14, and 15 must be supplemented by the conditions that are consequences of the particular symmetry and structure of the crystal about the defect site. These conditions take the form of the transformation law for the atomic force constants of a crystal containing an isolated defect when the crystal is subjected to a symmetry operation, i.e., to a rotation, reflection, translation, or a combination of these operations, which sends the crystal into itself. It is conceptually simpler to consider a point defect consisting of a single atom either present, as for a substitutional impurity or an interstitial defect, or absent, as for a vacancy. The result obtained is valid for more complicated point defects as well. It is clear that, for a crystal to be sent into itself by some operation performed on it, the operation must be such that it leaves the impurity site unmoved. The symmetry operations that leave invariant a crystal containing a defect are therefore those which belong to the point group G of the impurity site (one of the thirty-two crystallographic point groups). If we denote the symmetry operations comprising G by $\{S\}$, and the 3 x 3 real, orthogonal matrix representatives of these operations by $\{\mathbf{S}\}$, and choose the origin of a Cartesian coordinate system at the impurity site, the effect of a symmetry operation on a lattice position vector $x(\ell\kappa)$ can be written as

$$\mathbf{S} x(\ell\kappa) = x(LK). \qquad (16)$$

Since $\mathbf{S}$ describes a symmetry operation of the crystal it must send the lattice site $(\ell\kappa)$ into an equivalent site. The convention used here is to label with capital letters the lattice site into which a given site, labeled by the corresponding lower-case letters, is sent.

It can then be shown that the following transformation law is obtained for the force constants of a crystal containing a point defect[7]:

$$\Phi_{\alpha\beta}(LK;L'K') = \sum_{\mu\nu} S_{\alpha\mu}S_{\beta\nu}\Phi_{\mu\nu}(\ell\kappa;\ell'\kappa') \ . \tag{17}$$

These relations can be used to find the independent, nonzero force constants required by the symmetry and structure of a perturbed crystal. The general conditions stated by Eq. 13, 14, and 15 must then be imposed on the results to obtain the minimal set of perturbed atomic force constants for use in a lattice dynamical defect calculation.

It should be emphasized that the satisfaction of the conditions for these and Eq. 17 is not merely the satisfaction of some academic requirements on the atomic force constants whose neglect would have no serious consequences for the calculation in which the force constants are used. For example, if Eq. 14 is violated, the lower limit of the acoustic spectrum of the crystal would be shifted away from zero frequency — an unphysical result. As a less-trivial example, let us consider a semi-infinite crystal bounded by a free surface. Inasmuch as elasticity theory is the long-wavelength limit of a lattice theory of acoustic modes in a crystal, we expect that in the long-wavelength limit the results of a lattice theory must agree with the corresponding results obtained from elasticity theory. Ludwig and Lengeler[8] showed that, for the results of a lattice dynamical calculation of the dispersion relation for Rayleigh surface waves to agree with the results of a calculation based on elasticity theory, it is necessary for the rotational invariance condition (Eq. 15) to be satisfied for all the atoms of the crystal, including the atoms in the surface layers. Failure to ensure that Eq. 15 is obeyed has led to authors obtaining unphysical dispersion relation for surface waves.[9]

In the special (and unlikely) case that the atoms of the crystal are known to interact via a two-body, central potential $\varphi_{KK'}(r)$, where r is the separation between two atoms of types κ and κ', and the interaction between an impurity atom and one of the host atoms is known to be described by a potential $\varphi_{OK}(r)$, then all of the force constants entering Eq. 12 are directly calculable and Eq. 13, 14, 15, and 17 are automatically satisfied. The theory then has no adjustable parameters, and the study of the vibrational properties of the impurity atom poses only computational problems. This case is not considered further here.

The following section describes how experimental data of different types can be inverted to provide information about the interaction between an impurity atom and the host crystal.

IMPURITY ATOM-HOST CRYSTAL INTERACTIONS OBTAINED FROM EXCEPTIONAL VIBRATION MODES

When a hydride ion (U-center) is present substitutionally in the anion sublattice of an alkali-halide or alkaline earth-fluoride crystal, it gives rise to an exceptional vibrational mode whose frequency is approximately 2.5 times the maximum frequency of the unperturbed host crystal. The displacements of the ions vibrating in this mode decay faster than exponentially with increasing distance from the impurity site. In fact, calculations by Jaswal show that, for U-centers in NaCl and KCl, the vibrational amplitudes of the nearest neighbors to the impurity ion are only 1 to 2 percent of the amplitude of the impurity ion itself.[10] For these reasons this impurity-induced vibrational mode is called a localized mode. Because the site symmetry at the impurity site is O_h when the host crystal is an alkali-halide crystal, and T_d when it is an alkaline earth fluoride, the U-center localized mode is triply degenerate.

The localized mode has a first-order dipole moment (i.e., a dipole moment linear in the displacement of the U-center from its equilibrium position) and is consequently infrared active. It has been studied extensively in a wide variety of host crystals.[11] When the host crystal is an alkaline earth fluoride, a sharp peak is observed in the infrared absorption spectrum of the perturbed crystal at the frequency of the localized mode, a second peak is observed at essentially twice the localized mode frequency, and a doublet is observed at essentially three times the localized mode frequency.[12] These three peaks, overtones of the fundamental absorption peak, have their origin in the anharmonicity of the interaction between the U-center and the ions of the host crystal. (In the harmonic approximation electric dipole transitions are allowed only between consecutive harmonic oscillator energy levels.)

These experimental results, together with the qualitative description of the properties of the localized mode given above, make it possible to infer something about the interaction of the substitutional H^- ion with the ions of the host crystal. Because the vibration amplitudes of the atoms vibrating in the localized mode decay very quickly with increasing distance from the impurity ion, it is quite a good approximation to regard the localized mode as one in which only the H^- ion is vibrating in an anharmonic potential well due to all of the remaining ions at rest at their equilibrium positions. If we take account of the T_d site symmetry at the F^- sites in the alkaline earth fluorides, the Hamiltonian describing the motion of the H^- ion must have the form

$$H = \frac{p^2}{2M} + \frac{M\Omega^2}{2}(x^2+y^2+z^2) + Lxyz + M_1(x^4+y^4+z^4) +$$
$$+ M_2(x^2y^2+y^2z^2+z^2x^2) , \qquad (18)$$

through quartic anharmonic terms. Here M is the mass of the U-center, x,y,z are the Cartesian components of its displacement from its equilibrium position, and $p = (p_x,p_y,p_z)$ is its momentum.

The Hamiltonian 18 can be viewed in a different light, which removes the approximate character imparted to it by the manner in which it has been presented. If we work in the three-dimensional subspace of the three degenerate localized modes, and regard x, y, z in Eq. 18 not as particle displacement amplitudes but as the normal coordinates of the localized mode, then Eq. 18 is an exact Hamiltonian for the localized mode in the absence of any interactions with the wavelike band modes. The latter give rise to a finite lifetime for the localized mode[12,13], and to sidebands to the fundamental localized mode absorption peak[14]. However, if we are not concerned with these (interesting) effects, we can regard Eq. 18 as exact in the above sense. However, the interpretation of the coefficients L, M_1, and M_2 is different from what it is if we regard x, y, z as particle coordinates. In the latter case these coefficients are actually derivatives of the interaction potential between the U-center and the host crystal, but they are transforms of these derivatives with respect to mass-weighted unit eigenvectors if x, y, z are regarded as normal coordinates. Therefore, because of the direct connection between L, M_1, M_2 and potential derivatives when we regard x, y, z as particle coordinates, it is this identification that is used in interpreting Eq. 18.

The anharmonic terms can be regarded as a perturbation on the rest of the Hamiltonian 18, whose eigenvalues and wave functions are given by

$$E_{n_1 n_2 n_3} = \hbar\Omega(n_1+n_2+n_3 + \frac{3}{2}) \quad n_1, n_2, n_3 = 0, 1, 2, 3, \ldots \quad (19a)$$

$$|n_1 n_2 n_3\rangle = |n_1\rangle |n_2\rangle |n_3\rangle , \quad (19b)$$

where

$$|n_j\rangle = \left(\frac{\alpha}{\pi^{1/2} 2^{n_j} n_j!}\right)^{1/2} H_{n_j}(\alpha x_j) e^{-1/2 \alpha^2 x_j^2}, \quad \alpha = (M\Omega/\hbar)^{1/2} \quad (19c)$$

and $H_n(x)$ is the nth Hermite polynomial. The nth excited state of this three-dimensional, isotropic harmonic oscillator, where $n=n_1+n_2+n_3$, is $\frac{1}{2}(n+1)(n+2)$ – fold degenerate. The anharmonic terms in the potential energy lower the symmetry of the Hamiltonian from $U(3)$ to T_d, and split some of these degeneracies in the manner shown for the levels with $n=0,1,2,3$ in Fig. 1. The irreducible representations labeling the various energy levels are given in the notation of Koster et al.[15] The allowed electric dipole transitions between the ground state of the oscillator and the lower excited states (which must have Γ_5 symmetry) are indicated in Fig. 1, and have all been observed.

A perturbation theoretic calculation of the lower energy levels of the anharmonic oscillator to first order in the quartic anharmonic force

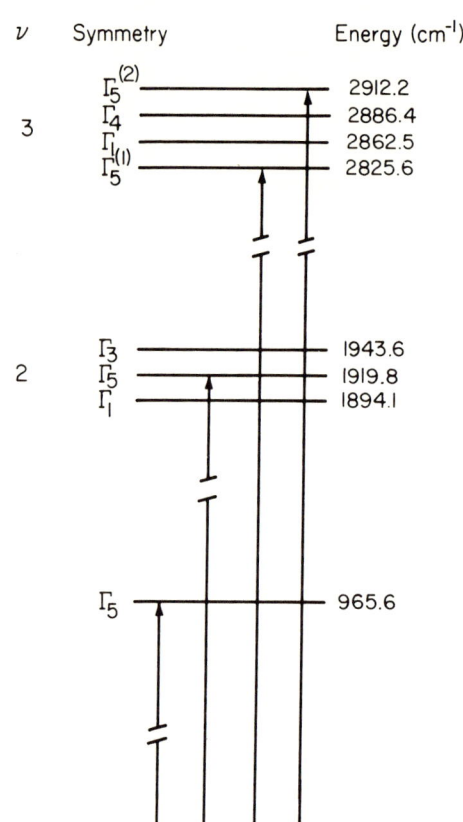

Fig. 1. The lower vibrational energy levels of H^- in CaF_2 at $20°K$. Observed first-order electric dipole transitions are indicated by vertical lines. The energies of other levels have been calculated. (R. J. Elliott et al.[12])

constants and to second order in the cubic anharmonic force constants yields the results that

$$\mathcal{E}_0^{\Gamma_1} = \frac{3}{2}\hbar\Omega + (9M_1+3M_2)\left(\frac{\hbar}{2M\Omega}\right)^2 - L^2 \frac{\hbar^2}{24M^3\Omega^4} \tag{20a}$$

$$\mathcal{E}_1^{\Gamma_5} = \frac{5}{2}\hbar\Omega + (21M_1+7M_2)\left(\frac{\hbar}{2M\Omega}\right)^2 - 5L^2 \frac{\hbar^2}{24M^3\Omega^4} \tag{20b}$$

$$\mathcal{E}_2^{\Gamma_1} = \frac{7}{2}\hbar\Omega + (45M_1+15M_2)\left(\frac{\hbar}{2M\Omega}\right)^2 - 21L^2 \frac{\hbar^2}{24M^3\Omega^4} \tag{20c}$$

$$\mathcal{E}_2^{\Gamma_3} = \frac{7}{2}\hbar\Omega + (45M_1+9M_2)\left(\frac{\hbar}{2M\Omega}\right)^2 - 3L^2 \frac{\hbar^2}{24M^3\Omega^4} \tag{20d}$$

$$\mathcal{E}_2^{\Gamma_5} = \frac{7}{2}\hbar\Omega + (33M_1+15M_2)\left(\frac{\hbar}{2M\Omega}\right)^2 - 13L^2\frac{\hbar^2}{24M^3\Omega^4} \tag{20e}$$

$$\mathcal{E}_3^{\Gamma_1} = \frac{9}{2}\hbar\Omega + (45M_1+27M_2)\left(\frac{\hbar}{2M\Omega}\right)^2 - 25L^2\frac{\hbar^2}{24M^3\Omega^4} \tag{20f}$$

$$\mathcal{E}_3^{\Gamma_4} = \frac{9}{2}\hbar\Omega + (57M_1+21M_2)\left(\frac{\hbar}{2M\Omega}\right)^2 - 15L^2\frac{\hbar^2}{24M^3\Omega^4} \tag{20g}$$

$$\mathcal{E}_3^{\Gamma_5(1)} = \frac{9}{2}\hbar\Omega + (69M_1+20M_2)\left(\frac{\hbar}{2M\Omega}\right)^2 - 20L^2\frac{\hbar^2}{24M^3\Omega^4}$$
$$-\frac{1}{2}\left\{\left[(-24M_1+10M_2)\left(\frac{\hbar}{2M\Omega}\right)^2 - 14L^2\frac{\hbar^2}{24M^3\Omega^4}\right]^2 + \right.$$
$$\left. + 24\left[2M_2\left(\frac{\hbar}{2M\Omega}\right)^2 - 6L^2\frac{\hbar^2}{24M^3\Omega^4}\right]^2\right\}^{1/2} \tag{20h}$$

$$\mathcal{E}_3^{\Gamma_5(2)} = \frac{9}{2}\hbar\Omega + (69M_1+20M_2)\left(\frac{\hbar}{2M\Omega}\right)^2 - 20L^2\frac{\hbar^2}{24M^3\Omega^4} +$$
$$+\frac{1}{2}\left\{\left[(-24M_1+10M_2)\left(\frac{\hbar}{2M\Omega}\right)^2 - 14L^2\frac{\hbar^2}{24M^3\Omega^4}\right]^2 + \right.$$
$$\left. + 24\left[2M_2\left(\frac{\hbar}{2M\Omega}\right)^2 - 6L^2\frac{\hbar^2}{24M^3\Omega^4}\right]^2\right\}^{1/2} \tag{20i}$$

In writing these results we have labeled the energies by a subscript giving the principal quantum number, n, of the level and a superscript denoting the irreducible representation of the point group T_d to which a given sublevel belongs in the presence of anharmonicity.

From a fit of the theoretical expressions for the energy differences of the lowest four infrared-active excited states of Γ_5 symmetry from the ground state of the anharmonic oscillator to the frequencies of the four absorption peaks in the infrared absorption spectrum of CaF_2, BaF_2, and SrF_2, Elliott et al.[12] were able to determine the values of the four constants Ω, L, M_1, M_2 entering the Hamiltonian 18. The results for CaF_2 (at 20 K) are

$$\Omega = 981.1 \text{ cm}^{-1}$$
$$L = -7.87 \times 10^{12} \text{ erg/cm}^3$$
$$M_1 = -2.32 \times 10^{19} \text{ erg/cm}^4$$
$$M_2 = -1.01 \times 10^{19} \text{ erg/cm}^4 \,. \tag{21}$$

In fact, only the magnitude of L can be inferred from such a fitting procedure. The negative sign indicated for this coefficient in Eq. 21 comes from a calculation on the basis of a shell-model description of the H^- ion[16], described below.

With the values of the coefficients now known, the energies of the remaining levels can be calculated from Eq. 20; the results are included in Fig. 1.

If the interaction between the U-center and the $\kappa^{\underline{th}}$ kind of ion of the host crystal can be represented by a function $\varphi_\kappa(r)$, where r is the separation between the two ions, the coefficients Ω, L, M_1, and M_2 can be expressed in terms of derivatives of this function according to

$$M\Omega^2 = \frac{1}{3} \sum_{\ell\kappa}{}' (\nabla^2 \varphi_\kappa(r))_{r = |x(\ell\kappa)|} \tag{22a}$$

$$L = -\sum_{\ell\kappa}{}' \left\{ \frac{xyz}{r^3} \left[\varphi_\kappa'''(r) - \frac{3}{r} \varphi_\kappa''(r) + \frac{3}{r^2} \varphi_\kappa'(r) \right] \right\}_{r = x(\ell\kappa)} \tag{22b}$$

$$M_1 = \frac{1}{24} \sum_{\ell\kappa}{}' \left\{ \frac{x^4 + y^4 + z^4}{3r^4} \left[\varphi_\kappa''''(r) - \frac{6}{r} \varphi_\kappa'''(r) + \frac{15}{r^2} \varphi_\kappa''(r) - \frac{15}{r^3} \varphi_\kappa'(r) \right] \right.$$

$$\left. + \frac{1}{r} \left[2\varphi_\kappa'''(r) - \frac{3}{r} \varphi_\kappa''(r) + \frac{3}{r^2} \varphi_\kappa'(r) \right] \right\}_{r = x(\ell\kappa)} \tag{22c}$$

$$M_2 = \frac{1}{4} \sum_{\ell\kappa}{}' \left\{ \frac{x^2 y^2 + y^2 z^2 + z^2 x^2}{3r^4} \left[\varphi_\kappa''''(r) - \frac{6}{r} \varphi_\kappa'''(r) + \frac{15}{r^2} \varphi_\kappa''(r) \right. \right.$$

$$\left. \left. - \frac{15}{r^3} \varphi_\kappa'(r) \right] + \frac{1}{3r} \left[2\varphi_\kappa'''(r) - \frac{3}{r} \varphi_\kappa''(r) + \frac{3}{r^2} \varphi_\kappa'(r) \right] \right\}_{r = x(\ell\kappa)}. \tag{22d}$$

In obtaining these results we have assumed that the equilibrium site of the U-center is the origin of coordinates, so that the prime on each lattice sum denotes that the term with $x(\ell\kappa) = 0$ is to be omitted.

If, now, an analytic expression is assumed for $\varphi_\kappa(r)$, the parameters in this expression can be determined by equating the right sides of Eq. 22 to the experimentally obtained values of Ω, L, M_1, and M_2.

It is worth remarking that in carrying out this fitting procedure it is not necessary to assume that the charges on the Ca^{++} and F^- ions of the host crystal have the nominal values of $2e$ and $-e$, respectively, where e is the magnitude of the electronic charge. In fact, it may be desirable not to do so. It has recently been pointed out that Kellermann's rigid-ion model of alkali-halide crystals, which leads to phonon-dispersion curves for the optical branches in poor agreement with experimental curves when the ions are assigned their nominal charges, is capable of yielding phonon-dispersion curves in good agreement with experiment if the ions are assigned effective charges that differ from their nominal values, and that are obtained from the frequencies of the long wavelength transverse and longitudinal optical modes.[17] The values of the effective charges obtained in this way are found to be quite close to the Szigeti effective charges[18] of the ions. While the result does not justify this way of bringing the rigid-ion model into agreement with experiment, it is gratifying that the values of the effective charges obtained in this way are not grossly unreasonable. Although the dynamics of crystals possessing the calcium fluoride structure have not been studied yet on the basis of such a modified rigid-ion model, it is reasonable to assume that when this is done the model will yield a satisfactory description of the phonon frequencies for such crystals as well.

If this is the case, then in determining the interaction of the U-center with the ions of the host crystal, for reasons of charge neutrality, it will be necessary to assume that the effective charge of the U-center is the same as that of the F^- ion it replaces. With the Coulomb interaction known from experiment, the only adjustable parameters in the expression for $\varphi_\kappa(r)$ would be those in the expressions for the short-range repulsive interactions. The assumption of a Born-Mayer potential $\lambda_+ e^{-r/\rho_+}$ to describe the repulsive interaction between the U-center and its six next nearest neighbor Ca^{++} ions, and a second such potential $\lambda_- e^{-r/\rho_-}$ to describe the short-range interaction between the U-center and its six next nearest neighbor F^-, ions, would yield a potential with four adjustable parameters — the number of pieces of experimental data available for the fitting procedure. This fitting would be simplified by the fact that the Coulomb part of $\varphi_\kappa(r)$ does not contribute to Ω, as is evident from Eq. 22a.

With an analytic expression for $\varphi_\kappa(r)$ it would then be possible to calculate the lifetime of the localized mode[13], of the one-phonon sidebands to the localized mode peak in the infrared absorption spectrum of a crystal containing U-centers[14], and of the temperature dependence of the integrated absorption under the localized mode peak[19] without any additional adjustable parameters entering the theory.

The rigid-ion model of a U-center just described is a special case of a more-general model[16] which takes into account some of the internal structure of the ion. In this model the U-center is represented by an ion core bearing a charge z_C and possessing a mass M surrounded by a massless shell representing the electrons surrounding the core and bearing a charge z_S. The net charge of the ion is the sum of the core and shell charges, $z_T = z_C + z_S$. The core and the shell are coupled to each other and to the ions of the host crystal, which are represented by point charges situated at their equilibrium positions, in a manner specified in detail below.

The displacement of the core from its equilibrium position will be denoted by **u**. The displacement of the shell relative to that of the core is **w**. For the harmonic terms in the potential energy expression the combination of cubic symmetry at the defect site and Laplace's equation has the consequence that the Coulomb interactions between the ions of the host crystal and the core and shell of the U-center do not contribute to the harmonic force constants. The short-range repulsive interaction is assumed to act only between the shell of the U-center and the shells of its four nearest neighbor (Ca^{++}) ions. In addition, the shell is assumed to be coupled isotropically to its core with a spring whose spring constant is k. Consequently, the harmonic contribution to the potential energy is given by

$$\Phi_2 = \frac{1}{2} \sum_{\alpha\beta} \Phi_{\alpha}^R (u_\alpha + w_\alpha)(u_\beta + w_\beta) + \frac{1}{2} k \sum_\alpha w_\alpha^2, \tag{23}$$

where u_α and w_α are the α-Cartesian components of the vectors **u** and **w**, respectively. In Eq. 23 the force constant $\Phi_{\alpha\beta}^R$ is given by

$$\Phi_{\alpha\beta}^R = \delta_{\alpha\beta} \frac{4}{3} \left[\varphi''(r_0) + \frac{2}{r_0} \varphi'(r_0) \right] \tag{24}$$

where $\varphi(r)$ is the short-range repulsive potential between two shells whose centers are separated by a distance r, and r_0 is the nearest neighbor distance.

To obtain the anharmonic terms in the potential energy we assume that the Coulomb forces act only between rigid ions and that the short-range repulsive interaction acts only between the shell of the U-center and the shells of its four nearest neighbors. With these assumptions we obtain for the cubic and quartic anharmonic contributions to the potential energy

$$\Phi_3 + \Phi_4 = \frac{1}{6} \sum_{\alpha\beta\gamma} \Phi_{\alpha\beta\gamma}^C u_\alpha u_\beta u_\gamma + \frac{1}{6} \sum_{\alpha\beta\gamma} \Phi_{\alpha\beta\gamma}^R (u_\alpha + w_\alpha)(u_\beta + w_\beta)$$

$$\times (u_\gamma + w_\gamma) + \frac{1}{24} \sum_{\alpha\beta\gamma\delta} \Phi_{\alpha\beta\gamma\delta}^C u_\alpha u_\beta u_\gamma u_\delta + \frac{1}{24} \sum_{\alpha\beta\gamma\delta} \Phi_{\alpha\beta\gamma\delta}^R (u_\alpha + w_\alpha)$$

$$\times (u_\beta + w_\beta)(u_\gamma + w_\gamma)(u_\delta + w_\delta), \tag{25}$$

where for an arbitrary crystal structure

POTENTIAL FUNCTIONS

$$\Phi^C_{\alpha\beta\gamma} = \underset{\ell\kappa}{\Sigma'} z_T z_\kappa \, 15\left\{\frac{x_\alpha x_\beta x_\gamma}{r^7} - \frac{3}{r^5}(x_\alpha \delta_{\beta\gamma} + x_\beta \delta_{\gamma\alpha} + x_\gamma \delta_{\alpha\beta})\right\}_{r=x(\ell\kappa)} \quad (26a)$$

$$\Phi^C_{\alpha\beta\gamma\delta} = \underset{\ell\kappa}{\Sigma'} z_T z_\kappa \left\{ 105\,\frac{x_\alpha x_\beta x_\gamma x_\delta}{r^9} \right.$$

$$-\,15\,\frac{x_\alpha x_\beta \delta_{\gamma\delta} + x_\alpha x_\gamma \delta_{\beta\delta} + x_\alpha x_\delta \delta_{\beta\gamma} + x_\beta x_\gamma \delta_{\alpha\delta} + x_\beta x_\delta \delta_{\gamma\alpha} + x_\gamma x_\delta \delta_{\alpha\beta}}{r^7} +$$

$$\left. +\,3\,\frac{\delta_{\alpha\delta}\delta_{\beta\gamma} + \delta_{\beta\delta}\delta_{\alpha\gamma} + \delta_{\gamma\delta}\delta_{\alpha\beta}}{r^5}\right\}_{r = x(\ell\kappa)} \quad (26b)$$

$$\Phi^R_{\alpha\beta\gamma} = -\underset{\ell\kappa}{\overset{n.n.}{\Sigma'}}\left\{\frac{x_\alpha x_\beta x_\gamma}{r^3}\left[\varphi'''(r) - \frac{3}{r}\varphi''(r) + \frac{3}{r^2}\varphi'(r)\right] + \right.$$

$$\left.\frac{x_\alpha \delta_{\beta\gamma} + x_\beta \delta_{\gamma\alpha} + x_\gamma \delta_{\alpha\beta}}{r^2}\left[\varphi''(r) - \frac{1}{r}\varphi'(r)\right]\right\}_{r = x(\ell\kappa)} \quad (26c)$$

$$\Phi^R_{\alpha\beta\gamma\delta} = -\underset{\ell\kappa}{\overset{n.n.}{\Sigma'}}\left\{\frac{x_\alpha x_\beta x_\gamma x_\delta}{r^4}\left[\varphi''''(r) - \frac{6}{r}\varphi'''(r) + \frac{15}{r^2}\varphi''(r) - \frac{15}{r^3}\varphi'(r)\right]\right.$$

$$+\,\frac{x_\alpha x_\beta \delta_{\gamma\delta} + x_\alpha x_\gamma \delta_{\beta\delta} + x_\alpha x_\delta \delta_{\beta\gamma} + x_\beta x_\gamma \delta_{\alpha\delta} + x_\beta x_\delta \delta_{\alpha\gamma} + x_\gamma x_\delta \delta_{\alpha\beta}}{r^3}$$

$$\times\left[\varphi'''(r) - \frac{3}{r}\varphi''(r) + \frac{3}{r^2}\varphi'(r)\right] + \frac{\delta_{\alpha\delta}\delta_{\beta\gamma} + \delta_{\beta\delta}\delta_{\alpha\gamma} + \delta_{\gamma\delta}\delta_{\alpha\beta}}{r^2}$$

$$\left.\times\left[\varphi''(r) - \frac{1}{r}\varphi'(r)\right]\right\}_{r = x(\ell\kappa)}, \quad (26d)$$

where z_κ is the net charge on the ion of type κ. Together with $z_T = z_F$ - it can be determined in the manner described above in the discussion of the rigid-ion model of the U-center. For the CaF$_2$ structure, the only independent coefficients are

$$\Phi^C_{xyz} = \underset{\ell\kappa}{\Sigma'} z_T z_\kappa \, 15\left(\frac{xyz}{r^7}\right)_{r = x(\ell\kappa)} \quad (27a)$$

$$\Phi^C_{xxxx} = \underset{\ell\kappa}{\Sigma'} z_T z_\kappa \left\{35\,\frac{x^4 + y^4 + z^4}{r^9} - \frac{21}{r^5}\right\}_{r = x(\ell\kappa)} \quad (27b)$$

$$\Phi^C_{xxyy} = \Phi^C_{xyxy} = \underset{\ell\kappa}{\Sigma'} z_T z_\kappa \left\{35\,\frac{x^2y^2 + y^2z^2 + z^2x^2}{r^9} - \frac{7}{r^5}\right\}_{r = x(\ell\kappa)} \quad (27c)$$

$$\Phi^R_{xyz} = -\frac{4}{3\sqrt{3}}\left[\varphi'''(r_0) - \varphi''(r_0) + \frac{3}{r_0^2}\varphi'(r_0)\right] \quad (27d)$$

$$\Phi^R_{xxxx} = \frac{4}{9}\left[\varphi''''(r_0) + \frac{12}{r_0}\varphi'''(r_0) - \frac{12}{r_0^2}\varphi''(r_0) + \frac{12}{r_0^3}\varphi'(r_0)\right] \quad (27e)$$

$$\Phi^R_{xxyy} = \Phi^R_{xyxy} = \frac{4}{9}\left[\varphi''''(r_0) + \frac{6}{r_0^2}\varphi''(r_0) - \frac{6}{r_0^3}\varphi'(r_0)\right]. \quad (27f)$$

The equations of motion of the ion core are

$$M\ddot{u}_\alpha = -\frac{\partial \Phi}{\partial u_\alpha}, \quad (28)$$

while the equations of motion of the shell are

$$0 = \frac{\partial \Phi}{\partial w_\alpha}, \quad (29)$$

and express the assumption that the electrons follow the nuclear motion adiabatically.

Eq. 29 can be solved for w_α in terms of the core displacements $\{u_\alpha\}$. When the result is substituted into Eq. 28 equations of motion for the ion cores are obtained which are of the form obtained from Eq. 18. Thus the coefficients Ω, L, M_1, and M_2 appearing in the latter expression can be expressed in terms of the constants k, z_s, λ and ρ which define the present model of the U-center, where λ and ρ are the parameters appearing in the short-range repulsive potential between shells, $\varphi(r) = \lambda \exp(-r/\rho)$. In fact, only k, λ, and ρ appear in the expressions for Ω, L, M_1, and M_2. Thus these three constants have to be chosen in such a way as to yield the best fit to these four experimental quantities. To obtain z_s, and to assess the goodness of the values so obtained for k, z_s, λ, and ρ, additional information about the defect can be used, namely its electronic properties. In the presence of an external electric field $\mathbf{E}$ whose frequency is so high that the ions cannot respond to it because of their heavier masses, only the electrons are driven by the field, and the left side of Eq. 29 is replaced by $z_s E_\alpha$. When the resulting equation is solved for w_α as a function of u_α and E_α, and the solution is substituted into the expression for the dipole moment of the crystal,

$$M_\alpha = z_T u_\alpha + z_s w_\alpha, \quad (30)$$

we obtain

$$M_\alpha = \sum_\mu M_{\alpha\mu} u_\mu + \frac{1}{2}\sum_{\mu\nu} M_{\alpha\mu\nu} u_\mu u_\nu + \ldots$$

$$+ \sum_\beta \left\{ P^{(0)}_{\alpha\beta} + \sum_\mu P_{\alpha\beta\mu} u_\mu + \ldots \right\} E_\beta + \ldots \quad (31)$$

The coefficients $M_{\alpha\mu}$ and $M_{\alpha\mu\nu}$ are the first- and second-order dipole moment coefficients for the U-center. The former, which is the transverse effective charge of the U-center, governs the strength of infrared absorption by the localized mode. It differs from z_T in general, and in principle is obtainable from the oscillator strength of the localized mode peak in the infrared-absorption spectrum of the crystal. The coefficient $P_{\alpha\beta}^{(0)}$ is the electronic polarizability of the U-center in the crystal. It is also possible to obtain the electronic polarizability of the free H^- ion from this model. It is given by $\alpha_{H^-} = z_S^2/k$. The coefficient $P_{\alpha\beta\mu}$ governs the intensity of Raman scattering by the U-center localized mode.

The advantage of this model of the U-center over the rigid-ion model discussed earlier is that, by taking into account the deformability and polarizability of the H^- ion and by allowing the electron shell to move with respect to the ion core, it permits the calculation of the physically important coefficients $M_{\alpha\mu\nu}$, $P_{\alpha\beta}^{(0)}$, and $P_{\alpha\beta\mu}$, which are identically zero for the rigid-ion model, and also allows for the departure of the transverse effective charge of the U-center from its rigid-ion value z_T.

A somewhat crude application of this model has been used to obtain the values of λ, ρ, k, z_S for a U-center in CaF_2, on the assumption that $z_{Ca^{++}} = 2e$, $z_{F^-} = -e = z_T$, and the use of Ω, L, and $P_{\alpha\beta}^{(0)}$ as input parameters.[16] The results obtained were

$$\lambda = 9.355 \times 10^{-14} \text{ erg}$$

$$\rho = 0.355 \text{ A}$$

$$k = 15.871 \times 10^4 \text{ erg/cm}^2$$

$$z_S = -1.43 \, e. \tag{32}$$

A more-careful fitting procedure would change the values of the parameters, perhaps by large amounts. Nevertheless, availability of these values made possible the quantitative study of dynamical properties of the U-center of experimental interest.[16,20]

More elaborate models of the U-center in alkali-halide crystals have been presented by Fieschi et al.[21] and by others[6], in which the ions of the host crystal are not treated as rigid ions but, with the U-center, are represented by shell models. However, in none of these treatments was an effort made to extract an analytic expression for the potential energy of interaction between the U-center and the ions of the host crystal.

When other types of defects are introduced into an alkali-halide crystal, a different kind of exceptional mode can be introduced into the spectrum of normal modes of the crystal. These modes, called resonance modes, are low-frequency modes, whose frequencies fall within the range of frequencies allowed the normal modes of the unperturbed crystal.

Because the density of vibrational states of the host crystal is nonzero at such frequencies, the resonance mode can decay into the continuum of band modes of the crystal, even in the harmonic approximation, and consequently acquires a width or finite lifetime. Thus, despite its name, a resonance mode is not a true normal mode of the perturbed crystal, in contrast with a localized mode, which is. However, if the frequency of the resonance mode is very low, the density of vibrational states at that frequency, which determines the lifetime of this mode, is very small, the mode has a long lifetime, and can be regarded as a well-defined mode of the crystal to a very good approximation. The kind of impurity ion which gives rise to resonance modes is one whose mass is much heavier than that of the ion it replaces (in the case of a substitutional impurity), and/or is very weakly coupled to the ions of the host crystal.

The experimental study of resonance modes can also provide useful information about the interaction of the impurity ion and the ions of the host crystal. Sievers and his coworkers[22] studied Cl^- impurities in NaI, Li^+ impurities in KBr, and Cu^+ impurities in NaCl. They studied the shifts in the frequencies of resonance modes, and the splittings of their degeneracies by the application of a strong, external, d-c electric field to the host crystal (the second-order Stark effect). For the NaI:NaCl, the observed shifts were found to be proportional to the square of the applied electric field. This result implies that the Cl^- impurities are situated at sites possessing inversion symmetry in the NaCl lattice. [The shifts would be linear in the electric field strength if the impurity site lacked inversion symmetry.] From the dependence of the results on the crystallographic axis along which the electric field was applied, and on the polarization of the incident infrared radiation, an impurity site possessing cubic symmetry was indicated. This immediately suggests that the Cl^- impurities substitutionally occupy I^- sites of the host crystal.

The Hamiltonian describing the motion of the Cl^- impurity in the resonance mode in the presence of the external electric field was taken to be

$$H = \frac{p^2}{2M} + \frac{M\Omega^2}{2}(x^2 + y^2 + z^2) + M_1(x^4 + y^4 + z^4) + M_2(x^2y^2 + y^2z^2 + z^2x^2) - eE(x\zeta_1 + y\zeta_2 + z\zeta_3), \qquad (33)$$

where e is the charge on the impurity (strictly speaking it is the Szigeti effective charge[18] of the impurity), E is the magnitude of the local electric field acting on the impurity ion (it is related to the strength of the applied electric field E_0 by $E = 2.4\, E_0$, where the factor 2.4 is a local field correction), and the $\{\zeta_i\}$ are the direction cosines of the electric field. The term $Lxyz$ present in the Hamiltonian 18 is absent here due to the presence of inversion symmetry at the impurity site. The representation of the resonance mode by an impurity ion vibrating in an anharmonic potential well (because all the other ions are at their equilibrium positions) is better the lower its frequency when the impurity becomes more and more decoupled from the host crystal.

The observed resonance mode peak in the infrared-absorption spectrum of NaI:NaCl is due to the electric dipole transition from the ground state of the Hamiltonian 33, of Γ_1^+ symmetry, to the triply degenerate (in the absence of the electric field) first excited state, of Γ_4^- symmetry. If we denote the shift in the energy of this transition due to the electric field by $\Delta\mathcal{E}$, then it is a general result that $\Delta\mathcal{E}$ has the form[23]

$$\Delta\mathcal{E}/E^2 = A_1 + A_2 (2\zeta_3^2 - \zeta_1^2 - \zeta_2^2)(2\eta_3^2 - \eta_1^2 - \eta_2^2) + 3A_3 (\zeta_1^2 - \zeta_2^2)$$

$$\times (\eta_1^2 - \eta_2^2) + 2A_5 (\zeta_1\zeta_2\eta_1\eta_2 + \zeta_2\zeta_3\eta_2\eta_3 + \zeta_3\zeta_1\eta_3\eta_1) , \qquad (34)$$

where the $\{\eta_i\}$ are the direction cosines of the polarization of the incident radiation. A second-order perturbation calculation relates the coefficients A_1, A_3, A_5 to the anharmonic force constants M_1 and M_2 (in units of $\hbar e^2/8M^3\Omega^5$)[22]:

$$A_1 = 16M_1 + \frac{16}{3} M_2, \ A_3 = 8M_1 - \frac{4}{3} M_2, \ A_5 = -12M_1 + 8M_2 . \qquad (35)$$

Consequently, from measurements of the electric field induced shifts in the frequency of the resonance mode for different experimental geometries, together with Eq. 34 and 35, the values of the quartic anharmonic coefficients M_1 and M_2 in the potential energy of interaction of the impurity with the host crystal can be obtained. The results obtained in this way for NaI:NaCl are[22]

$$\Omega = 9.61 \times 10^{11} \text{ sec}^{-1}, M_1 = 8.54 \times 10^{16} \text{ erg/cm}^4, M_2 = -0.16 \times 10^{16} \text{ erg/cm}^4. \qquad (36)$$

It is necessary to obtain these coefficients indirectly because it has not yet proved to be possible to observe the third harmonics of the resonance mode directly, and the second harmonics are forbidden by symmetry (for the same reason that $L = 0$ in Eq. 33).

As a test of the internal consistency of this model of a resonance mode we can compare the theoretical and experimental results for the isotope shift of the fundamental absorption frequency, i.e., the shift in the frequency of the transition from the ground state of the Hamiltonian 33 to the triply degenerate first excited state when ^{35}Cl$^-$ is replaced by ^{37}Cl$^-$

as the impurity giving rise to the resonance mode. From the expressions given by Eq. 20a and 20b for the energies of the ground and first excited states of the Hamiltonian 18, with $L = 0$, we find that this transition energy is given by

$$\Delta\mathcal{E}(M) = \hbar\Omega + (3M_1 + M_2)\frac{\hbar^2}{M^2\Omega^2} . \quad (37)$$

If we define a force constant Φ by

$$\Phi = M\Omega^2 , \quad (38)$$

the ratio of transition energies for two isotopes of masses M' and M'' is found to be

$$\frac{\Delta\mathcal{E}(M')}{\Delta\mathcal{E}(M'')} = \left(\frac{M''}{M'}\right)^{1/2} \frac{1 + (3M_1 + M_2)(\hbar/M'^{1/2}\Phi^{3/2})}{1 + (3M_1 + M_2)(\hbar/M''^{1/2}\Phi^{3/2})} . \quad (39)$$

The ratio $\Delta\mathcal{E}(M')/\Delta\mathcal{E}(M'') = (M''/M')^{1/2}$ is that expected for a particle moving in a harmonic potential, and Eq. 39 expresses the way in which the anharmonic terms in the potential energy modify this relation. For the system NaI:NaCl, the value of $\Delta\mathcal{E}(35)/\Delta\mathcal{E}(37)$ obtained from Eq. 39 and 36 is found to be 1.030, in excellent agreement with the measured value of 1.028 ± 0.008[24].

It is difficult to give much more than a qualitative significance to the results for the coefficients entering Eq. 33 obtained in this way, because, unlike the situation for a localized mode, the atoms of the host crystal do not provide a static potential in which the impurity ion vibrates: they are all vibrating themselves. On the other hand, if x, y, z are interpreted as normal coordinates rather than as particle displacements, Eq. 33 becomes more exact as Ω becomes lower and the resonance mode becomes better defined. However, in this case the coefficients M_1 and M_2 lose their simple interpretation as fourth derivatives of an interatomic potential. For this reason, expressions for the coefficients Ω, M_1, and M_2 in terms of derivatives of the Coulomb and short-range repulsive potentials have not been given here for an impurity at a site of O_h symmetry in a crystal of the rocksalt structure.

Nevertheless, the simple model of the resonance mode described by the Hamiltonian 33 is capable of interpreting experimental data for certain impurity-host crystal systems, and can describe qualitative features of the potential in which the impurity ion moves.

The situation is completely different for the resonance modes due to $^6\text{Li}^+$ and $^7\text{Li}^+$ in KBr. Here, it is found that the value of the anharmonic force constant $3M_1 + M_2$ required to explain the (anomalously large) value of $\Delta\mathscr{E}(6)/\Delta\mathscr{E}(7)$ is larger by at least a factor of 50 than that required to explain the electric-field induced frequency shifts on the basis of Eq. 34 and 35. To explain the large isotope effect and small second-order Stark effect for this system Clayman et al.[22] abandoned the Hamiltonian 33 and replaced it with a harmonic Hamiltonian to which a central barrier in the form of a Gaussian is added. In a one-dimensional model their Hamiltonian is given by

$$H = \frac{p^2}{2M} + \frac{M\Omega^2}{2}x^2 + Ae^{-bx^2}, \qquad (40)$$

which possesses anharmonic terms of all orders. There is some theoretical justification for a Hamiltonian of this form in the works of Quigley and Das[25] and Wilson et al.[26] on the determination of the minimum-energy configuration of Li^+ in KBr. These authors have shown that the potential energy minima for the impurity ion are displaced in the [111] directions from the center of the host ion (K^+) cavity. The potential barriers in the [110] directions between adjacent equilibrium sites are small (smaller than the vibrational zero point energy of the Li^+ impurity). In order that the dynamic dipole moment of the defect be small, as is suggested by the very small Stark splitting of the triply degenerate resonance mode, the central potential barrier must also be lower than the vibrational zero-point energy of the impurity. Otherwise, the impurity will spend a considerable portion of the time away from the center of the host-ion cavity, and will consequently possess a large dipole moment. The last term of Eq. 40 represents this low central barrier in a one-dimensional model of the defect.

A first-order perturbation calculation of the energy levels of the Hamiltonian 40 yields

$$E_0 = \frac{1}{2}\hbar\Omega + A[M\Omega/(b\hbar + M\Omega)]^{1/2} \text{ (ground state)} \qquad (41a)$$

$$E_1 = \frac{3}{2}\hbar\Omega + A[M\Omega/(b\hbar + M\Omega)]^{\frac{3}{2}} \text{ (first excited state)}. \qquad (41b)$$

Values of A, B, and Ω can be found which reproduce the measured isotope shift, and the resulting anharmonic potential is depicted in Fig. 2 for the particular choice of these parameters given by $A = 11.7$ cm^{-1}, $b = 1.19 \times 10^{17}$ cm^{-2}, and $\Omega(^7Li^+) = 20.5$ cm^{-1}. It is seen that the height of the perturbing central barrier is considerably less than the ground state energy of the perturbed Hamiltonian, so that the impurity can still be regarded as an on-center ion.

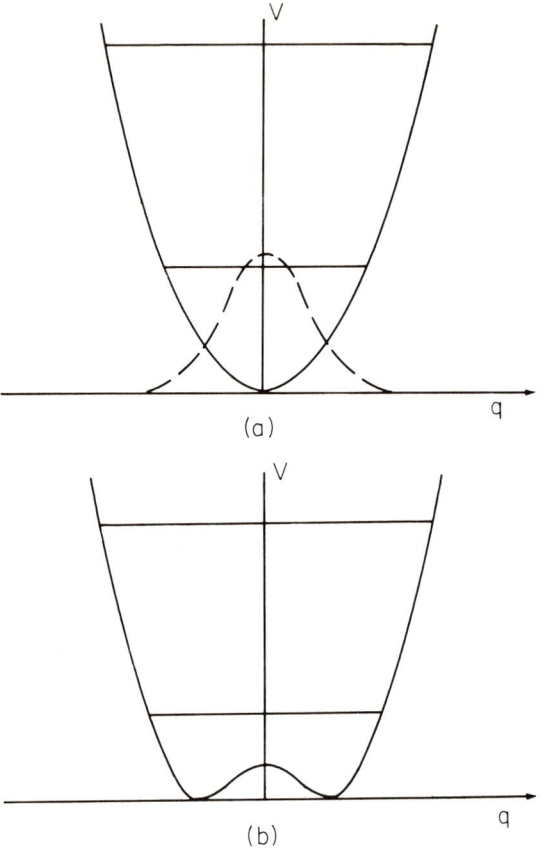

Fig. 2. One-dimensional central barrier model. (a) Harmonic well with two lowest states. The dashed curve shows the perturbing potential potential used. (b) Anharmonic well with two lowest states. Parameters are appropriate for KBr:Li$^+$ (B. P. Clayman et al.[22]).

When the electric field perturbation $H_e = -eEx$ is added to the Hamiltonian 40 it is found that the electric-field induced shift in the transition energy $E_1 - E_0$ is comparable to the observed shift. Thus a static potential well which has displaced energy minima, but barriers which are small compared to the impurity zero-point energy, causes a large isotope shift, but a small Stark effect.

Thus the results of this section show that the study of various vibrational properties of an impurity in a crystal can yield qualitative and even quantitative information about the potential function describing the interaction of the impurity with the host crystal.

4 FIRST-PRINCIPLES CALCULATIONS OF IMPURITY ATOM-HOST CRYSTAL INTERACTIONS IN IONIC CRYSTALS

Attempts to obtain the interaction potential between an impurity ion and the ions of the host crystal by fitting the derivatives of assumed analytic expressions to experimental data concerning the localized or resonance modes induced by the impurity (while useful for the help the results give in solving other kinds of lattice dynamical defect problems) have no fundamental justification, and in fact represent merely analytic interpolations among several pieces of experimental data. The only satisfactory way in principle of obtaining correct interatomic potentials describing the impurity-host crystal interactions is the difficult one of obtaining them from first principles, i.e., from the solution of Schrödinger's equation for the interacting system of electrons and nuclei in a crystal.

There exist only two examples of first principles calculations of the potential function describing the interaction of an impurity ion or atom and its surroundings in an ionic host crystal. These are the calculation of the U-center force constant (the coefficient Ω in Eq. 18) in KCl, KBr, and KI by Wood et al.[27], and the calculation of the interaction potential between an interstitial hydrogen atom and its nearest neighbor F$^-$ ions in alkaline-earth fluorides by Hartman et al.[28]

Both of these calculations are based on the adiabatic approximation, in which the potential energy for nuclear motion is the sum of the potential energy of the direct interaction between ion cores and the ground-state energy of the electronic subsystem, which depends parametrically on the nuclear coordinates. This should be a good approximation for the systems being considered.

The starting point for the calculations of Wood et al. is the following expression for the electronic part of the crystal Hamiltonian (in atomic units):

$$H = H_U + H_{int} + H_{cr} , \qquad (42)$$

with

$$H_U = \sum_{i=1}^{2} \left\{ -\frac{1}{2} \nabla_i^2 - \frac{1}{|\mathbf{r}_i - \mathbf{u}|} \right\} + \frac{1}{r_{12}} \qquad (43)$$

$$H_{int} = \sum_{i=1}^{2} \left\{ -\sum_{\ell\kappa} \frac{Z_\kappa}{|\mathbf{r}_i - \mathbf{R}(\ell\kappa)|} + \sum_{j=3}^{M} \frac{1}{|\mathbf{r}_i - \mathbf{r}_j|} \right\}. \qquad (44)$$

In these expressions $\mathbf{r}_i$ is the position of the ith electron, r_{12}^{-1} is the interaction of the two electrons of the U-center, $\mathbf{u}$ is the displacement of the proton, Z_κ is the charge on the κth kind of nucleus, and $\mathbf{R}(\ell\kappa)$ is the position of the ion $(\ell\kappa)$. The total number of electrons in the crystal is denoted by M. H_{cr} is the Hamiltonian for the rest of the crystal, including nuclear interactions. However, as it does not enter the calculation of the U-center force constant, it is not discussed further.

In calculating the U-center force constant, Ω, Woods et al. made the same assumption as underlies Eq. 18, viz., that the U-center moves in a static potential due to all the other ions at rest in their equilibrium positions. In the harmonic approximation, the Coulomb potential due to the ion cores makes no contribution to the constant Ω, as we have already noted, so that it is only the ground-state energy of the electronic Hamiltonian given by the sum of Eq. 43 and 44 that has to be calculated.

This was done in two steps. First, with the U-center at its equilibrium position ($\mathbf{u} = 0$), the ground-state energy of the electronic system was calculated in the Hartree-Fock approximation and minimized with respect to purely radial displacements of its six nearest neighbors, thus preserving the full cubic symmetry at the U-center site. With the nearest-neighbor ions fixed in their new positions, the same type of calculation was carried out for displacements of the H^- ion in the [100] direction.

The wave function used in the Hartree-Fock calculations had the form

$$\psi(1,\ldots,M) = \left\{\frac{M!}{2!\,(M-2)!}\right\}^{1/2} A\psi_U(1,2)\psi_C(3,\ldots,M) \qquad (45)$$

where $\psi_U(1,2)$ is an antisymmetrized two-electron wave function for the H^- ion and ψ_C is an antisymmetrized $(M-2)$ electron wave function for the rest of the crystal. A is an antisymmetrizer for the entire function. It exchanges electrons between, but not within ψ_U and ψ_C. It is also assumed that the condition of "strong orthogonality" holds, i.e., that

$$\int \psi_U(1,2)\psi_C(3,\ldots,k-1,1,k+1,\ldots,M)d\tau_1 = 0. \qquad (46)$$

The explicit expression used for ψ_C is

$$\psi_C = [(M-2)!]^{-1/2} A_c a_{1,1}(3) a_{1,2}(4)\ldots a_{\ell\kappa,j-1}(i-1) a_{\ell\kappa,j}(i)\ldots \qquad (47)$$

where $a_{\ell\kappa,j}(i)$ is shorthand for $a_j[\mathbf{r}_i - \mathbf{R}(\ell\kappa)]$ times a one-electron-spin function. Ideally, $a_j[\mathbf{r}_i - \mathbf{R}(\ell\kappa)]$ would be a Wannier function constructed from the jth band, describing the ith electron localized on the ion $(\ell\kappa)$. Since these Wannier functions were not available, the $\{a_j[\mathbf{r}_i - \mathbf{R}(\ell\kappa)]\}$ were approximated by free-ion Hartree-Fock orbitals.

The choice made for $\psi_U(1,2)$ was

$$\psi_U(1,2) = N\{\psi(1,2) + \eta\psi'(1,2)\}\theta(s_1,s_2), \qquad (48)$$

where $\theta(s_1,s_2)$ is a normalized singlet spin function, and

$$\psi(1,2) = N_U \{\psi_a(1)\psi_b(2) + \psi_a(2)\psi_b(1)\} \tag{49a}$$

$$\psi'(1,2) = N'_U \{\psi'_a(1)\psi'_b(2) + \psi'_a(2)\psi'_b(1)\} . \tag{49b}$$

N_U and N'_U are spatial normalization factors, while ψ_a, ψ_b are s-state spatial orbitals which are orthogonalized to the free-ion Hartree-Fock orbitals on the nearest-neighbor ions. For example, we have that

$$\psi_a = N_a \left[\varphi_a - \sum_{\ell\kappa,j} c_{\ell\kappa,j}\, a_j\,(\mathbf{r} - \mathbf{R}(\ell\kappa)) \right] , \tag{50}$$

with $c_{\ell\kappa,j} = (a_{\ell\kappa,j}|\varphi_a)$, $\varphi_a = (\beta_a^3/\pi)^{1/2} \exp(-\beta_a r)$, and $N_a = (1 - \sum_{\ell\kappa,j} c_{\ell\kappa,j}^2)^{-1/2}$.

The function ψ'_b is a p-function of the form of Eq. 50, but with

$$\varphi'_b = N'_b\, r\, e^{-\beta'_b r} \cos\gamma , \tag{51}$$

while ψ'_a has the same form as ψ_a. γ is the angle between the principal symmetry axis of the p-function and the vector $\mathbf{r}$.

The admixture of a function of p-like symmetry into the U-center wave function given by the second term on the right side of Eq. 48 is intended to represent the deformation of the electronic-charge cloud of the H⁻ ion as it is displaced from its equilibrium position. η is a variational parameter, as are the constants β_a and β_b in ψ and β'_a and β'_b in ψ'.

As remarked above, the determination of the U-center force constant proceeded in two steps. First, the expectation value of the electronic Hamiltonian $H_U + H_{int}$, with $\mathbf{u} = 0$, was evaluated with respect to the wave function (Eq. 45, 47, and 48) with $\eta = 0$. This evaluation was carried out for several values of the radial displacement of the six nearest neighbors of the U-center, and the values of β_a and β_b were determined by minimizing the expectation value of the Hamiltonian for each value of the radial displacement. The value of the displacement that makes the electronic ground-state energy an absolute minimum yields the new equilibrium positions of the nearest neighbors of the U-center. The results of this part of the calculation, in Fig. 3, show that the nearest neighbors are displaced inward toward the equilibrium position of the U-center. With the nearest neighbors fixed in their new positions, and the values of β_a and β_b determined in this way, the expectation value of $H_U + H_{int}$ was evaluated for displacements of the proton in the [100] direction equal to 2, 4, and 6 percent of the new nearest neighbor distance. For each value of $\mathbf{u}$ the expectation value was minimized with respect to variations of β'_a, β'_b, and η. In practice, to simplify the calculations, β'_a was chosen to be equal to

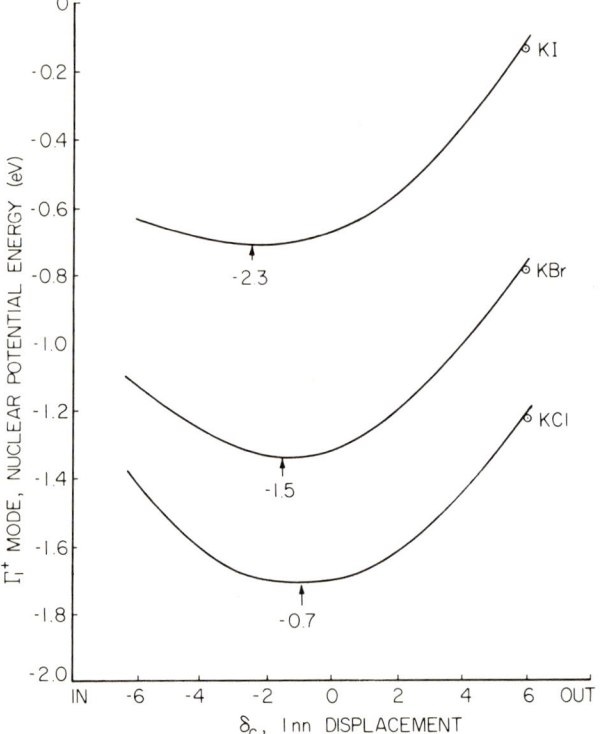

Fig. 3. The nuclear potential-energy curve for the six nearest-neighbor ions to a U-center in KCl in a totally symmetric mode of displacement. δ_c is given as a percent of the nearest-neighbor distance in the perfect crystal. (R. F. Wood and R. L. Gilbert[27]).

β_a and β_b' was chosen to be the value that minimized the electronic ground-state energy of KCl for an H⁻ displacement of 4 percent of the new nearest-neighbor separation. The values of β_b' for KBr and KI were determined from the KCl value by a scaling procedure. Thus the only remaining variational parameter was η. For comparison purposes, the second calculation was carried out with $\eta = 0$ to assess the importance of allowing the H⁻ ion to deform during displacement.

The effects of second-nearest-neighbor interactions were taken into account in an approximate fashion in these calculations. A Born-Mayer potential was used to represent these interactions, with parameters for the perfect crystal taken from the work of Tosi and Fumi[29]. The effects of the deformation of the H⁻ ion during its displacement were approximated by reducing the second-nearest-neighbor interaction in the same ratio as the nearest-neighbor interaction was found to be reduced by allowing for deformation.

Fig. 4 shows the potential-energy curves for the U-center in KCl calculated in various approximations, together with the experimental result. The several curves are shifted in such a way that they coincide when $\delta = 0$, where δ is the magnitude of the displacement **u** expressed as a

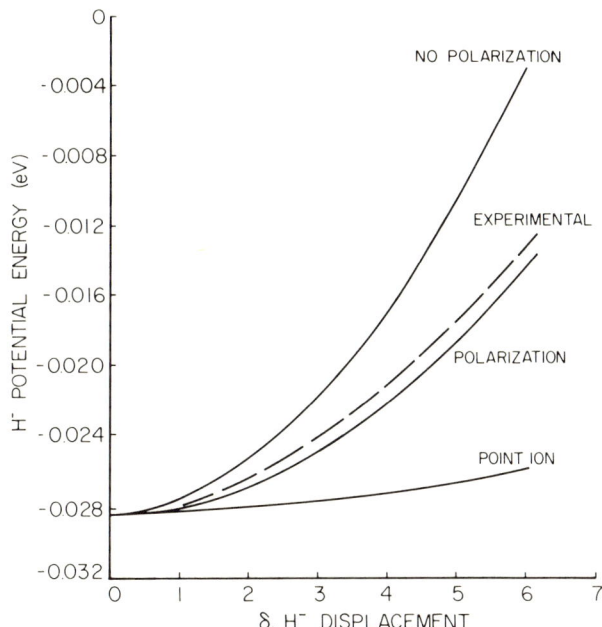

Fig. 4. Potential-energy curves for the displacement of an H⁻ ion in KCl in various approximations. The curves are adjusted to have the same values at zero displacement. Here δ is given as a percent of the nearest-neighbor distance in the crystal distorted by the presence of the U-center. (R. F. Wood and R. L. Gilbert[27].)

percent of the nearest-neighbor distance of the crystal distorted by the presence of the U-center at its equilibrium position. The frequency of the localized mode is simply related to the curvature of these curves at $\delta = 0$. From these results it is seen that treating the nearest neighbors as rigid ions, i.e., neglecting the electronic structure of all the ions, yields much too low a value for the U-center force constant, and hence much too low a value for the localized mode frequency. On the other hand, taking into account the electronic structure of the nearest neighbors, but neglecting the deformability of the H⁻ ion, yields a value of the localized mode frequency which is too high. Thus, it appears to be important to take into account both the electronic structure of the nearest neighbors and the deformability of the H⁻ ion in constructing phenomenological models of the U-center.

However, it should be kept in mind in comparing the theoretical and experimental results that the experimental curve shown in Fig. 4 is obtained from an expression for the potential energy of the U-center of the form

$$\Phi_2 = \frac{M\Omega^2_{\exp}}{2} (x^2 + y^2 + z^2) , \qquad (52)$$

which incorporates anharmonic contributions to the localized mode frequencies, whereas in fact it should be plotted on the basis of the expression

$$\Phi_2 = \frac{M\Omega^2}{2}(x^2 + y^2 + z^2),\qquad(53)$$

where Ω is the "bare" localized mode frequency, unrenormalized by anharmonic effects. Unfortunately, at the present time the difference between Ω_{exp} and Ω is not known for U-center localized modes in the alkali-halide crystals. In principle, the quartic anharmonic terms in the U-center potential energy could be obtained by fitting a fourth-degree polynomial to the numerical results. In U-center localized modes in CaF_2, for example, we have seen that Ω is greater than Ω_{exp}. If this is also true for the alkali-halide crystals, the experimental curve in Fig. 4 should be lowered, perhaps leading to better agreement between theory and experiment.

The calculations of Woods and his colleagues, representing a tremendous achievement in view of the complexity of the problem, are a significant contribution to the theory of U-center localized modes in alkali-halide crystals.

The calculations of Hartmann et al., in contrast, are less ambitious.[28] These authors have carried out a self-consistent field calculation of the interaction energy between a hydrogen atom and an F^- ion as a function of their separation, using the Hartree-Fock-Roothaan method[30]. The numerical results obtained were fitted by a least-squares procedure to an analytic expression of the form

$$\varphi(r) = -\frac{A}{r^4} + Be^{-r/\rho},\qquad(54)$$

where r is measured in bohrs (1 bohr = 0.52917A) and φ is given in hartrees (1 hartree = 27.211 eV). The first term in this expression is the monopole-induced dipole interaction, while the second is the repulsive interaction. The first term will not contribute to the localized mode frequency of an interstitial hydrogen atom in an alkaline earth-fluoride crystal because the net field due to all neighbors vanishes at a center of inversion, so that the hydrogen atom has no dipole moment in this configuration.

The values for the constants A, B, and ρ obtained by Hartmann et al. were

$$A = 1.931,\qquad B = 3.291,\qquad \rho = 9.624$$

with A and B in units of hartrees, and ρ in units of bohrs.

From the repulsive part of the potential (Eq. 54) the coefficients Ω, M_1, and M_2 in the Hamiltonian 18, which applies to the present problem as well as to substitutional impurities in alkali-halide crystals, were determined from Eq. 22. The frequency of the localized mode obtained in this way for CaF_2:H^o was found to be 602 cm^{-1}. When a correction was made for the small-amplitude motions of the nearest-neighbor F^- ions, this value was raised to 633 cm^{-1}, in excellent agreement with the experimental value of 630 cm^{-1} at room-temperature.[31]

Hartmann et al. have predicted localized mode frequencies of 498 and 422 cm^{-1} for interstitial hydrogen atoms in SrF_2 and BaF_2, respectively. (In these calculations the three crystals CaF_2, SrF_2, and BaF_2 are distinguished only by the different distances of the interstitial hydrogen atom from its nearest-neighbor F^- ions.) The predicted localized mode in SrF_2 has been seen[32], but at a frequency of 543 cm^{-1}, and no localized mode has been observed as yet in the system BaF_2:H^o.

These results suggest that, although the method used by Hartmann et al. may provide a good first approximation to the interaction between an impurity and an atom of the host crystal, an approach, such as that of Woods and his collaborators, which takes into account the fact that the impurity is situated in a crystalline environment (which gives rise to many-body interactions), is more likely to yield quantitatively accurate results. This is, perhaps, not surprising. Even in the comparatively simpler case of the inert-gas solids, interatomic potentials inferred from second virial coefficients in the gaseous phase are found not to describe vibrational properties of the crystalline phase with great quantitative accuracy.[33]

5 POINT DEFECTS IN METALS

In contrast with the difficult calculations required for the evaluation of the interaction potential between a substitutional H^- ion in an alkali-halide crystal or an interstitial neutral H atom in an alkaline-earth fluoride, the corresponding calculations for a point defect in a metal are formally much simpler, at least for nontransition metals. This is due to the circumstance that in such metals the valence electrons form a quasifree subsystem. Each ion occupies a comparatively small volume, approximately 10 percent of the volume per atom. As the ions do not overlap, the properties of isolated ions are preserved. In the major part of a unit cell, an individual electron interacts with the ion through a pure Coulomb potential. In the interior of the ion, however, the electron's wave function undergoes strong oscillations of the type appropriate to core electrons in an isolated ion. This leads to an effective repulsive interaction between the electron and the ion which largely cancels the strong Coulomb attraction between them. Consequently, it becomes possible to replace the true electron-ion interaction by a weak pseudopotential that can be treated as a perturbation on the free-electron gas in which the ion is embedded. This idea, which has been widely exploited in *ab initio* calculations of dynamical properties of perfect metals[34,35], and of static properties associated with point defects[36], can be extended to the study of the dynamical properties of defects in metal crystals. In this section is sketched a method of doing this within the Hartree approximation for describing the electron-electron interactions.

In addition, the assumed validity of the adiabatic approximation, justified for metals by the work of several authors[37], will not be discussed further here. In this approximation the potential energy describing nuclear motion is the sum of the potential energy of the direct interactions between the ions, and the ground-state energy of the electronic subsystem with the ions displaced arbitrarily from their equilibrium positions.

For simplicity, let us consider a monatomic metal possessing a Bravais lattice.* We will assume that a substitutional impurity atom is situated at the site which provides the origin of coordinates.

Since the ions do not overlap, they interact directly with each other as point charges interacting through the Coulomb potential. If Z is the charge on an ion of the host metal, and Z_O is the charge on the impurity ion**, the direct electrostatic energy of interaction between the ions is

$$E_{\text{electrostatic}} = \frac{1}{2} \sum_{\ell\ell'}{}' \frac{Z_\ell Z_{\ell'}}{|\mathbf{R}(\ell) - \mathbf{R}(\ell')|} \tag{55}$$

where

$$\mathbf{R}(\ell) = \mathbf{x}(\ell) + \mathbf{u}(\ell) \tag{56}$$

is the instantaneous position of the ℓth ion,

$$Z_\ell = Z_o \delta_{\ell o} + Z(1 - \delta_{\ell o}) \tag{57}$$

and the prime denotes that the terms with $\ell = \ell'$ are omitted.

Turning now to the electronic subsystem, the interaction potential of an electron with an ion of the host crystal is denoted by $U(\mathbf{r})$; the interaction of an electron with the impurity will be denoted by $U_o(\mathbf{r})$. For simplicity, we assume that both potentials are local. The total ionic potential seen by an electron is therefore

---

*The treatment which follows is clearly applicable to monatomic metals possessing nonprimitive lattices if the index ℓ is now regarded as labeling not only the primitive unit cells, but also the atoms in each unit cell.

**The sum of the nuclear charge plus the charge on the core electrons of the atom. In a more refined calculation it can be taken to be the renormalized ionic charge of Harrison[34], which contains the additional charge associated with the "orthogonalization hole" accompanying each ion.

$$V(\mathbf{r}) = \sum_{\ell} U_{\ell}(\mathbf{r} - \mathbf{R}(\ell)) \tag{58}$$

where

$$U_{\ell}(\mathbf{r}) = \delta_{\ell 0} U_0(\mathbf{r}) + (1 - \delta_{\ell 0}) U(\mathbf{r}). \tag{59}$$

If we Fourier analyze $U_{\ell}(\mathbf{r})$,

$$U_{\ell}(\mathbf{r}) = \sum_{\mathbf{k}} U_{\ell}(\mathbf{k}) e^{i\mathbf{k}\cdot\mathbf{r}}, \tag{60}$$

we can also Fourier analyze $V(\mathbf{r})$

$$V(\mathbf{r}) = \sum_{\mathbf{k}} V(\mathbf{k}) e^{i\mathbf{k}\cdot\mathbf{r}}, \tag{61}$$

where

$$V(\mathbf{k}) = \sum_{\ell} U_{\ell}(\mathbf{k}) e^{-i\mathbf{k}\cdot\mathbf{R}(\ell)}. \tag{62}$$

The system of electrons and ions is electrically neutral. The simplest way of incorporating this fact into calculations in an explicit fashion is to define the $\mathbf{k} = 0$ Fourier coefficient of every interaction potential appearing in them to be zero.

In the Hartree approximation the one-electron Hamiltonian is given by

$$H = -\frac{\hbar^2}{2m}\nabla^2 + V(\mathbf{r}) + V_H(\mathbf{r}) \tag{63}$$

where $V(\mathbf{r})$ is the potential due to the ions, and $V_H(\mathbf{r})$ is the Hartree potential

$$V_H(\mathbf{r}) = \int d^3r' u(\mathbf{r} - \mathbf{r}') n(\mathbf{r}') \tag{64}$$

where $u(\mathbf{r})$ is the Coulomb potential,

$$u(\mathbf{r}) = \frac{e^2}{r}, \tag{65}$$

and $n(\mathbf{r})$ is the electron number density,

$$n(\mathbf{r}) = \sum_{\mathbf{k}} 2\theta(\mu - E_{\mathbf{k}}) |\psi_{\mathbf{k}}(r)|^2. \tag{66}$$

In Eq. 66 μ is the chemical potential (Fermi energy) of the metal, and $E_{\mathbf{k}}$ and $\psi_{\mathbf{k}}$ are the one electron energies and wave functions of the Hamiltonian 63,

$$H\psi_k = E_k \psi_k. \tag{67}$$

We take as the unperturbed Hamiltonian

$$H_0 = -\frac{\hbar^2}{2m}\nabla^2 \tag{68}$$

so that the unperturbed wave functions and energies are

$$\psi_k^{(0)}(\mathbf{r}) = \Omega^{-\frac{1}{2}} e^{i\mathbf{k}\cdot\mathbf{r}} \equiv |\mathbf{k}\rangle \tag{69}$$

and

$$E_k^{(0)} = \frac{\hbar^2 k^2}{2m} \equiv \epsilon_k. \tag{70}$$

The ground-state energy in the Hartree approximation is

$$E_{\text{electron}} = \sum_k 2\theta(\mu - E_k)E_k - \tfrac{1}{2}\int d^3r \int d^3r' n(\mathbf{r})u(\mathbf{r}-\mathbf{r}')n(\mathbf{r}'). \tag{71}$$

We seek an expansion of this energy in powers of the ion potential $V(\mathbf{r})$. Although it suffices to terminate this expansion at the second-order terms, to obtain the harmonic-approximation for the vibrational Hamiltonian of the system, the expansion is carried to third-order terms for two reasons[35]. The first is that if it is reasonably assumed that the local pseudopotentials $U(\mathbf{r})$ and $U_0(\mathbf{r})$ are spherically symmetric, the truncation of the expansion of the electronic ground-state energy at second-order terms yields a vibrational Hamiltonian describing a metal in which the atoms interact pairwise with two-body, central forces. Except, perhaps, for the alkali metals, this is known to be a bad approximation for the interatomic forces in metals, as witnessed by the failure of the Cauchy relations for this class of solids. The inclusion of terms of third order in the pseudopotential yields harmonic force constants which arise from three-body forces, and hence are not central in character. The second reason for retaining third-order terms in the pseudopotential in the electronic ground-state energy is that there is an appreciable cancellation between the ionic contribution to the harmonic force constants and the electronic contribution arising from the part of the electronic ground-state energy which is of second order in the pseudopotential. Thus, although succeeding terms in the expansion of the ground-state energy decrease in proportion to $V(\mathbf{G})/E_F(\approx 0.1)$ where $\mathbf{G}$ is the smallest nonzero reciprocal lattice vector and E_F is the Fermi energy, the net contribution to the harmonic force constants from two-body interactions is, in a sense, anomalously small, and comparable with the contribution from the third-order terms in the pseudopotential. The higher-order terms in the pseudopotential which are neglected are conventionally small.

To obtain the desired expansion we expand the Hartree Hamiltonian formally in powers of V:

$$H = H^{(0)} + H^{(1)} + H^{(2)} + \ldots \tag{72}$$

where

$$H^{(0)} = -\frac{\hbar^2 \nabla^2}{2m} \tag{73a}$$

$$H^{(1)} = V(\mathbf{r}) + \int d^3r' \, u(\mathbf{r} - \mathbf{r}') n^{(1)}(\mathbf{r}') = V(\mathbf{r}) + V_H^{(1)}(\mathbf{r}) \tag{73b}$$

$$H^{(j)} = V_H^{(j)}(\mathbf{r}) = \int d^3r' \, u(\mathbf{r} - \mathbf{r}') n^{(j)}(\mathbf{r}') \quad j \geq 2. \tag{73c}$$

The single particle energies are similarly expanded:

$$E_k^{(0)} = \epsilon_k \tag{74a}$$

$$E_k^{(1)} = \langle k|H^{(1)}|k\rangle = 0 \tag{74b}$$

$$E_k^{(2)} = \sum_{k_1} \frac{\langle k|H^{(1)}|k_1\rangle \langle k_1|H^{(1)}|k\rangle}{\epsilon_k - \epsilon_{k_1}} \tag{74c}$$

$$E_k^{(3)} = \sum_{k_1} \frac{\langle k|H^{(1)}|k_1\rangle \langle k_1|H^{(2)}|k\rangle}{\epsilon_k - \epsilon_{k_1}} + \sum_{k_1} \frac{\langle k|H^{(2)}|k_1\rangle \langle k_1|H^{(1)}|k\rangle}{\epsilon_k - \epsilon_{k_1}}$$
$$+ \sum_{k_1 k_2} \frac{\langle k|H^{(1)}|k_1\rangle \langle k_1|H^{(1)}|k_2\rangle \langle k_2|H^{(1)}|k\rangle}{(\epsilon_k - \epsilon_{k_1})(\epsilon_k - \epsilon_{k_2})} \tag{74d}$$

The (normalized) wave function ψ_k can also be expanded in powers of $V(\mathbf{r})$

$$\psi_k = \psi_k^{(0)} + \psi_k^{(1)} + \psi_k^{(2)} + \ldots \tag{75}$$

where

$$\psi_k^{(1)} = \sum_{k_1} \frac{\langle k_1|H^{(1)}|k\rangle}{\epsilon_k - \epsilon_{k_1}} |k_1\rangle \tag{76a}$$

$$\psi_k^{(2)} = \sum_{k_1} \frac{\langle k_1|H^{(2)}|k\rangle}{\epsilon_k - \epsilon_{k_1}} |k_1\rangle + \sum_{k_1 k_2} \frac{\langle k_1|H^{(1)}|k_2\rangle\langle k_2|H^{(1)}|k\rangle}{(\epsilon_k - \epsilon_{k_1})(\epsilon_k - \epsilon_{k_2})} |k_1\rangle$$

$$- \sum_{k_1} \frac{\langle k_1|H^{(1)}|k\rangle \langle k|H^{(1)}|k\rangle}{(\epsilon_k - \epsilon_{k_1})^2} |k_1\rangle$$

$$- \frac{1}{2} \sum_{k_1} \frac{\langle k_1|H^{(1)}|k\rangle \langle k|H^{(1)}|k_1\rangle}{(\epsilon_k - \epsilon_{k_1})^2} |k\rangle + \ldots \tag{76b}$$

In expanding the electron number density and the ground-state energy we must also expand the electron occupation numbers in powers of $V(\mathbf{r})$:

$$\theta(\mu - E_k) = \theta(\mu^{(o)} - \epsilon_k) + \delta(\mu^{(o)} - \epsilon_k)[\mu^{(1)} - E_k^{(1)} + \mu^{(2)} - E_k^{(2)}$$
$$+ \mu^{(3)} - E_k^{(3)} + \ldots] + \frac{1}{2}\delta'(\mu^{(o)} - \epsilon_k)[\mu^{(1)} - E_k^{(1)} + \mu^{(2)}$$
$$- E_k^{(2)} + \ldots]^2 + \frac{1}{6}\delta''(\mu^{(o)} - \epsilon_k)[\mu^{(1)} - E_k^{(1)} + \ldots]^3 + \ldots \tag{77}$$

and impose the conditions

$$\sum_k 2\theta(\mu - E_k) = N = \sum_k 2\theta(\mu^{(o)} - \epsilon_k), \tag{78}$$

where N is the number of electrons in the normalization volume Ω.

The successive terms in the expansion of the electron number density are therefore given by

$$n^{(o)}(\mathbf{r}) = \sum_k 2\theta(\mu^{(o)} - \epsilon_k)\Omega^{-1} = n_o \tag{79a}$$

$$n^{(1)}(\mathbf{r}) = \sum_k f_k [\psi_k^{(o)*}\psi_k^{(1)} + \psi_k^{(1)*}\psi_k^{(o)}] \tag{79b}$$

$$n^{(2)}(\mathbf{r}) = \sum_k f_k [\psi_k^{(o)*}\psi_k^{(2)} + \psi_k^{(1)*}\psi_k^{(1)} + \psi_k^{(2)*}\psi_k^{(1)}]. \tag{79c}$$

In writing these expressions we have set $f_k = 2\theta(\mu^{(o)} - \epsilon_k)$.

It is convenient to Fourier analyze $n^{(j)}(\mathbf{r})$ according to

$$n^{(j)}(\mathbf{r}) = \sum_{\mathbf{q}} n^{(j)}(\mathbf{q})e^{i\mathbf{q}\cdot\mathbf{r}} \tag{80a}$$

$$n^{(j)}(\mathbf{q}) = \frac{1}{\Omega}\int d^3r\, e^{-i\mathbf{q}\cdot\mathbf{r}} n^{(j)}(\mathbf{r}) \,. \tag{80b}$$

Then from the preceding results we find that

$$n^{(1)}(\mathbf{q}) = \chi(\mathbf{q})\, V(\mathbf{q}) \tag{81}$$

where

$$\chi(\mathbf{q}) = \frac{\tilde{\chi}(\mathbf{q})}{1 - \dfrac{4\pi e^2}{q^2}\tilde{\chi}(\mathbf{q})} = \frac{\tilde{\chi}(\mathbf{q})}{\epsilon(\mathbf{q})} \tag{82}$$

$$\tilde{\chi}(\mathbf{q}) = \Omega^{-1} \sum_{\mathbf{k}} \frac{f_{\mathbf{k}} - f_{\mathbf{k}+\mathbf{q}}}{\epsilon_{\mathbf{k}} - \epsilon_{\mathbf{k}+\mathbf{q}}} \tag{83}$$

and

$$n^{(2)}(\mathbf{q}) = \frac{1}{2}\sum_{\mathbf{q}_1 \mathbf{q}_2} \chi(-\mathbf{q},\mathbf{q}_1,\mathbf{q}_2)\, V(\mathbf{q}_1)\, V(\mathbf{q}_2)\,, \tag{84}$$

where

$$\chi(\mathbf{q}_1,\mathbf{q}_2,\mathbf{q}_3) = \frac{2\delta_{\mathbf{q}_1+\mathbf{q}_2+\mathbf{q}_3,0}}{\epsilon(\mathbf{q}_1)\epsilon(\mathbf{q}_2)\epsilon(\mathbf{q}_3)}\frac{1}{\Omega}\Bigg\{\sum_{\mathbf{k}} \frac{f_{\mathbf{k}}}{(\epsilon_{\mathbf{k}} - \epsilon_{\mathbf{k}-\mathbf{q}_1})(\epsilon_{\mathbf{k}} - \epsilon_{\mathbf{k}+\mathbf{q}_2})}$$

$$+ \frac{f_{\mathbf{k}}}{(\epsilon_{\mathbf{k}} - \epsilon_{\mathbf{k}-\mathbf{q}_2})(\epsilon_{\mathbf{k}} - \epsilon_{\mathbf{k}+\mathbf{q}_3})} + \frac{f_{\mathbf{k}}}{(\epsilon_{\mathbf{k}} - \epsilon_{\mathbf{k}-\mathbf{q}_3})(\epsilon_{\mathbf{k}} - \epsilon_{\mathbf{k}+\mathbf{q}_1})} \Bigg\}. \tag{85}$$

The single-particle contribution to the electronic ground state energy is

$$E_{s.p.} = 2\sum_{\mathbf{k}} \theta(\mu - E_{\mathbf{k}})E_{\mathbf{k}} = E_{s.p.}^{(o)} + E_{s.p.}^{(2)} + E_{s.p.}^{(3)} + \ldots \tag{86}$$

The expression for $E_{s.p.}^{(2)}$ can be simplified to

$$E_{s.p.}^{(2)} = \Omega \sum_{\mathbf{q}} \tilde{\chi}(\mathbf{q}) \frac{V(-\mathbf{q})V(\mathbf{q})}{\epsilon(\mathbf{q})\epsilon(\mathbf{q})} \tag{87}$$

In a similar fashion we obtain for $E_{s.p.}^{(3)}$

$$E_{s.p.}^{(3)} = \Omega \sum_q \frac{4\pi e^2}{q^2} \tilde{\chi}(q) \frac{V(-q)}{\epsilon(q)} n^2(q)$$

$$+ \frac{\Omega}{6} \sum_{q_1,q_2,q_3} \chi(q_1,q_2,q_3) V(q_1) V(q_2) V(q_3) \quad . \tag{88}$$

The Hartree correction to the single particle contribution to the ground state energy of the electronic subsystem can be written in the form

$$E_H = -\frac{\Omega}{2} \sum_q \frac{4\pi e^2}{q^2} n(-q)n(q) = -\frac{\Omega}{2} \sum_q \frac{4\pi e^2}{q^2} \left[n^{(1)}(-q) n^{(1)}(q) \right.$$

$$\left. + 2n^{(2)}(-q) n^{(1)}(q) + \ldots \right] \tag{89}$$

where the n_0 term in $n(r)$ does not contribute because of charge neutrality.

Combining Eq. 86 and 89 we obtain finally for the ground-state energy of the electronic subsystem

$$E = \sum_k f_k \epsilon_k + \frac{\Omega}{2} \sum_q \chi(q) V(-q) V(q)$$

$$+ \frac{\Omega}{6} \sum_{q_1,q_2,q_3} \chi(q_1,q_2,q_3) V(q_1) V(q_2) V(q_3) + \ldots \tag{90}$$

Substituting Eq. 62 into Eq. 90 and expanding the result in powers of the atomic displacements, we obtain for the electronic ground-state energy

$$E = E_0 + \sum_{\ell\alpha} \Phi_\alpha^{(e)}(\ell) u_\alpha(\ell) + \frac{1}{2} \sum_{\ell\alpha} \sum_{\ell'\beta} \Phi_{\alpha\beta}^{(e)}(\ell\ell') u_\alpha(\ell) u_\beta(\ell') + \ldots \, , \tag{91}$$

where

$$\Phi_\alpha^{(e)}(\ell) = i \frac{\Omega}{2} \sum_q \sum_{\ell'(\neq \ell)} q_\alpha \chi(q) \left\{ U_\ell(-q) U_{\ell'}(q) e^{iq\cdot(x(\ell)-x(\ell'))} \right.$$

$$\left. - U_\ell(q) U_{\ell'}(-q) e^{-iq\cdot(x(\ell)-x(\ell'))} \right\} - i\frac{\Omega}{2} \sum_{qq_1q_2} \chi(q,q_1,q_2) \sum_{\ell_1\ell_2} q_\alpha U_\ell(q)$$

$$\times U_{\ell_1}(q_1) U_{\ell_2}(q_2) \times e^{-iq\cdot x(\ell) - iq_1\cdot x(\ell_1) - iq_2\cdot x(\ell_2)} + \ldots \tag{92}$$

$$\Phi^{(e)}_{\alpha\beta}(\ell\ell') = \Omega \sum_q \chi(q) U_\ell(-q) U_{\ell'}(q) e^{iq\cdot(x(\ell) - x(\ell'))} q_\alpha q_\beta$$

$$-\frac{\Omega}{3} \sum_{q_1 q_2 q_3} \chi(q_1, q_2, q_3) \sum_{\ell_1 \ell_2 \ell_3} U_{\ell_1}(q_1) U_{\ell_2}(q_2) U_{\ell_3}(q_3)$$

$$\times e^{-iq_1\cdot x(\ell_1) - iq_2\cdot x(\ell_2) - iq_3\cdot x(\ell_3)} \left[q_{1\alpha} \delta_{\ell_1 \ell} + q_{2\alpha} \delta_{\ell_2 \ell} + q_{3\alpha} \delta_{\ell_3 \ell} \right]$$

$$\times \left[q_{1\beta} \delta_{\ell_1 \ell'} + q_{2\beta} \delta_{\ell_2 \ell'} + q_{3\beta} \delta_{\ell_3 \ell'} \right] (\ell \neq \ell') \tag{93}$$

$$\Phi^{(e)}_{\alpha\beta}(\ell\ell) = - \sum_{\ell'(\neq \ell)} \Phi^{(e)}_{\alpha\beta}(\ell\ell') . \tag{94}$$

The contributions to the atomic force constants from the direct Coulomb interactions between ion cores are

$$\Phi^{(i)}_\alpha(\ell) = - \sum_{\ell'(\neq \ell)} Z_\ell Z_{\ell'} \frac{x_\alpha(\ell\ell')}{|x(\ell\ell')|^3} \tag{95}$$

$$\Phi^{(i)}_{\alpha\beta}(\ell\ell') = - Z_\ell Z_{\ell'} \left(\frac{3 x_\alpha(\ell\ell') x_\beta(\ell\ell')}{|x(\ell\ell')|^5} - \frac{\delta_{\alpha\beta}}{|x(\ell\ell')|^3} \right) (\ell \neq \ell') \tag{96a}$$

$$\Phi^{(i)}_{\alpha\beta}(\ell\ell) = - \sum_{\ell'(\neq \ell)} \Phi^{(i)}_{\alpha\beta}(\ell\ell') . \tag{96b}$$

There is a final contribution to the force constants from the interaction of the ion cores with the neutralizing background of negative charge, which is given by

$$\Phi^{(b)}_{\alpha\beta}(\ell\ell') = -\delta_{\ell\ell'} n_0 e Z_\ell \int d^3 r \left(\frac{3 x_\alpha x_\beta}{r^5} - \frac{\delta_{\alpha\beta}}{r^3} \right) . \tag{97}$$

These results formally solve the problem of obtaining the atomic force constants of a metal perturbed by the presence of a substitutional impurity.

In a perfect metal, whose lattice is parameter-free, $\Phi_\alpha(\ell)$ vanishes from symmetry considerations alone. In the presence of the substitutional impurity, however, $\Phi_\alpha(\ell)$ is no longer zero, and describes the forces on the atoms surrounding the point defect leading to a distortion of the crystal about the defect. Combined with the method of lattice statics these results make it possible to obtain both the static distortion of the crystal about the defect, and the perturbed normal mode frequencies. To my knowledge such a calculation has not been carried to completion by anyone at the present time. However, it is far from being an intractable program. The susceptibilities $\chi(q)$ and $\chi(q_1,q_2,q_3)$ can be evaluated analytically, in closed form, and bare ion pseudopotentials or model potentials which reproduce the phonon dispersion curves of the perfect host crystal exist. Nevertheless, the solution of the dynamical problem, by no means trivial, stands as a challenge to workers interested in the dynamical properties of metals perturbed by impurities.

In summary, the nature of the perturbed force constants in a crystal containing a point defect has been described, examples have been given of ways in which perturbed force constants can be inferred from experimental studies of localized and resonance modes induced by the defect, and descriptions have been given of ways in which the interaction between a point defect and the host crystal can be calculated from first principles in ionic crystals and in metals.

ACKNOWLEDGMENTS

I am grateful to A. Sievers for helpful discussions concerning the contents of Section 3. In the preparation of Section 5 I made use of unpublished work of L. J. Sham, which I acknowledge with thanks.

The research was sponsored by the Air Force Office of Scientific Research, Office of Aerospace Research, USAF, under Grant No. AFOSR 68-1448A. The United States Government is authorized to reproduce and distribute reprints for Governmental purposes notwithstanding any copyright notation hereon.

REFERENCES

1. J. R. Hardy, Phil. Mag. **7**, 315 (1962).
2. A.D.B. Woods, W. Cochran, and B. N. Brockhouse, Phys. Rev. **119**, 980 (1960).
3. H. Kanzaki, J. Phys. Chem. Solids **2**, 24 (1957).
4. J. R. Hardy, J. Phys. Chem. Solids **15**, 39 (1960). J. W. Flocken and J. R. Hardy, Phys. Rev. **B1**, 2447 (1970).
5. A. A. Maradudin, J. Phys. Chem. Solids **9**, 1 (1958). P. A. Flinn and A. A. Maradudin, Ann. of Phys. (N.Y.) **18**, 81 (1962).
6. J. B. Page, Jr., and D. Strauch, phys. stat. sol. **24**, 469 (1967). J. Oitmaa and A. A. Maradudin, Solid State Comm. **7**, 1371 (1969).
7. A. A. Maradudin, Repts. on Progr. in Phys. **28**, 331 (1965).
8. W. Ludwig and B. Lengeler, Solid State Comm. **2**, 83 (1964).
9. See, for example, S. Takeno, Progr. Theor. Phys. (Kyoto) **30**, 1 (1963).
10. S. S. Jaswal, Phys. Rev. **140**, A687 (1965).

11. For a survey see A. A. Maradudin, in *Solid State Physics* edited by F. Seitz and D. Turnbull (Academic Press, New York, 1966) vol. 18, p. 273; vol. 19, p. 1.
12. R. J. Elliott, W. Hayes, G. D. Jones, H. F. Macdonald, and C. T. Sennett, Proc. Roy. Soc. **A289**, 1 (1965).
13. E. Hanamura and T. Inui, J. Phys. Soc. Japan **18**, 690 (1963). A. A. Maradudin, Ann. Phys. (N.Y.) **30**, 371 (1964). M. A. Krivoglaz, Zhur. Eksper. i Teor. Fiz. **40**, 567 (1961) [English translation: Soviet Physics — JETP **13**, 397 (1961)]. M. A. Ivanov, M. A. Krivoglaz, D. N. Mirlin, and I. I. Reshina, Fiz. Tver. Tela **8**, 192 (1966). [English translation: Soviet Physics — Solid State **8**, 150 (1966)]. I. P. Ipatova and A. A. Klochikhin, Zhur. Eksper. i Teor. Fiz. **50**, 1603 (1966) [English translation: Soviet Physics — JETP **23**, 1068 (1966)].
14. T. Timusk and M. V. Klein, Phys. Rev. **141**, 664 (1966). Nguyen Xuan Xinh, Solid State Comm. **4**, 9 (1966); Phys. Rev. **163**, 896 (1967). T. Gethins, T. Timusk, and E. J. Woll, Jr., Phys. Rev. **157**, 744 (1967). J. B. Page, Jr., and G. B. Dick, Phys. Rev. **163**, 910 (1967).
15. G. F. Koster, J. O. Dimmock, R. G. Wheeler, and H. Statz, *Properties of the Thirty-Two Point Groups* (M.I.T. Press, Cambridge, Mass., 1963).
16. L. B. Humphreys, A. A. Maradudin, and R. F. Wallis, *Proceedings of the Tallinn Seminar on Point Defects in Crystals* (to appear).
17. K. V. Namjoshi, S. S. Mitra, and J. F. Vetelino, Solid State Comm. **9**, 185 (1971).
18. B. Szigeti, Trans. Faraday Soc. **45**, 155 (1949); Proc. Roy. Soc. **A204**, 51 (1950).
19. I. P. Ipatova, A. V. Subashiev, and A. A. Maradudin, Annals of Physics (N.Y.) **53**, 376 (1969). A. E. Hughes, Phys. Rev. **173**, 860 (1968).
20. L. B. Humphreys and A. A. Maradudin, Phys. Rev. (to appear).
21. R. Fieschi, G. F. Nardelli, and N. Terzi, Phys. Rev. **138**, A203 (1965).
22. B. P. Clayman, R. D. Kirby, and A. J. Sievers, Phys. Rev. **B3**, 1351 (1971).
23. W. Gebhardt, Phys. Rev. **159**, 726 (1967).
24. B. P. Clayman, I. G. Nolt, and A. J. Sievers, Solid State Comm. **7**, 7 (1969).
25. R. J. Quigley and T. P. Das, Phys. Rev. **164**, 1185 (1967); **177**, 1340 (1969).
26. W. D. Wilson, R. D. Hatcher, R. Smoluchowski, and G. J. Dienes, Phys. Rev. **184**, 844 (1969).
27. R. F. Wood and R. L. Gilbert, Phys. Rev. **162**, 746 (1967); R. F. Wood and U. Öpik, Phys. Rev. **162**, 736 (1967).
28. W. Hartmann, T. L. Gilbert, K. A. Kaiser, and A. C. Wahl, Phys. Rev. **B2**, 1140 (1970).
29. M. P. Tosi and F. G. Fumi, J. Phys. Chem. Solids **25**, 45 (1964).
30. A. C. Wahl, J. Chem. Phys. **41**, 2600 (1964). BISON, ANL Report No. 7271, 1968 (unpublished). T. L. Gilbert and A. C. Wahl, J. Chem. Phys. **47**, 3425 (1967).
31. R. E. Shamu, W. Hartmann, and E. L. Yasaitis, Phys. Rev. **170**, 822 (1968).
32. R. C. Newman, Adv. in Phys. **18**, 545 (1969).
33. See, for example, J. Grindlay and R. Howard, in *Lattice Dynamics*, edited by R. F. Wallis (Pergamon Press, New York, 1965), p. 129.
34. W. Harrison, *Pseudopotentials in the Theory of Metals* (W. A. Benjamin, New York, 1966).
35. Yu. M. Kagan and E. G. Brovman, in *Inelastic Scattering of Neutrons,* p. 3 (International Atomic Energy Agency, Vienna, 1968). E. G. Brovman, Yu. M. Kagan, and A. Kholas, in *Inelastic Scattering of Neutrons,* p. 145 (International Atomic Energy Agency, Vienna, 1968).
36. J. Callaway and A. J. Hughes, Phys. Rev. **156**, 860 (1967); **164**, 1043 (1967).
37. A. B. Migdal, Zhur. Eksper. i Teor. Fiz. **34**, 1438 (1958). [English translation: Soviet Physics—JETP **7**, 996 (1958)]. G. V. Chester, Adv. in Physics **10**, 357 (1961). E. G. Brovman and Yu. M. Kagan, Zhur. Eksper. i Teor. Fiz. **52**, 557 (1967). [English translation: Soviet Physics—JETP **25**, 365 (1967)].

DISCUSSION on the Paper by A. A. Maradudin

SEEGER: I should like to suggest as a possible application of the lattice dynamics to defect problems, the isotope effect of self-diffusion in lithium. The self-diffusion coefficients of Li^6 in Li^7 and of Li^7 in Li^6

differ much more than one would expect from considerations based on classical statistical mechanics. This difference is mainly in the pre-exponential factors, i.e., in the entropies of formation and migration of vacancies. More specifically, these data require a binding entropy between the light isotope and a vacancy. This does not come out of classical statistical mechanics. Since the Debye temperature of Li^6 is above the melting point, quantum statistics is applicable and should be capable of explaining the observations, together with the lattice vibration spectrum calculations.

THE THEORY OF INTERATOMIC POTENTIALS IN SOLIDS

W. A. Harrison

Applied Physics Department
University of Stanford
Stanford, California 94305

ABSTRACT

The use of interatomic potentials in studying lattice vibrations is reviewed, with the conclusion that long-range interactions are required in all systems, and the suggestion that empirical models can succeed only if they include the appropriate long-range form for the system in question. In simple metals, pseudopotentials provide a theory of the interactions, which to lowest order appear as two-body, central-force interactions. The long-range part corresponds to Friedel oscillations. While the reliability of the method in defect studies remains in question, current results are encouraging. The theory has been extended to noble and transition metals by direct addition of d-band and hybrid s-d contributions. This approach is promising, but the corresponding interatomic interactions have not been tested. In valence crystals, the low-order perturbation theory fails and, as a consequence, noncentral forces become essential to a representation of the interactions. The effects of higher-order terms are considered for ionic as well as valence crystals utilizing the concept of ionicity as formulated by Phillips. This suggests appropriate models for interatomic interactions in nonconductors and these are discussed.

1 INTRODUCTION

Interatomic interactions lie at the basis of all of the work discussed at this conference. In most cases, only a small fraction of the thought and effort in each program has gone into the interactions used. However, the significance and the reliability of the results depends in every case upon the appropriateness of the assumed interactions. Virtually all of this paper is devoted to that question.

Two aspects of this problem should be considered. First, some critique is given of the potentials which have been and are being used in the calculations on crystalline defects. Many of these have been justified principally by the fact that one did not know what better to do. Therefore, the reliability and range of applicability of these potentials are discussed as objectively as possible. Second, recognizing that one must proceed in any case if one is to make progress in the understanding of defects, those interactions that have the most convincing basis and some guidelines for the construction of empirical interactions for other systems are discussed. The similarities and the contrasts between the bonding forces in metallic, covalent, and ionic systems are included, but simple metals where the situation is clearest are emphasized.

2 HISTORICAL BACKGROUND

The study of lattice vibrations in crystals is very analogous to the study of lattice defects. In both, the theory must depend on a knowledge of the interatomic interactions. The study of lattice vibrations, however, is much older as an established field and a very complete experimental test has been possible. It is therefore very relevant to look at the history of that field. It is, in fact, a history of theories based on totally plausible, and frequently even rigorously correct, models of the interaction. In spite of the rigor, these models have frequently failed. We may avoid some of the pitfalls by reviewing that story. While this description may not qualify as history and detailed accuracy is not claimed, the spirit of the discussion is correct.

It might at first seem plausible to assume some form of radial interaction between atoms, such as a Lennard-Jones potential, with adjustable parameters fit for example to the elastic constants, and to proceed to calculate the vibration spectrum. This, in fact, never looked satisfactory since the Cauchy relations between the elastic constants, which follow from such a potential, are violated in almost all materials. This objection is not as serious as it initially looked since the Cauchy relations depend also on the atoms being in equilibrium under the influence of these forces alone. When additional volume-dependent terms are in the energy, this latter condition will not be satisfied and the Cauchy relations are not required. However, it seemed preferable in early times to abandon central forces and seek a more rigorous foundation.

This could be done by noting that the total energy of the system could, in principle, be written as some function of the coordinates of all of the atoms present. In the study of lattice vibrations we are concerned with small displacements from the equilibrium and we may systematically expand the energy in these displacements. The linear terms vanish since the system is in equilibrium with no displacements. The second-order terms are the harmonic terms that determine the vibrational frequencies, and the coefficients of these terms play the role of spring constants in the interaction between atoms. This provides a rigorous basis and it is natural to expect that these coupling constants describing the interaction between atoms drop off rapidly at large distances. Symmetry arguments could be used to drastically reduce the number of required constants. We could then use as many experimental parameters as we had to fit the constants for interactions between neighbors, and assume that the more distant interactions vanished. It was only at this last stage that the method failed.

Many of the early calculations were made on diamond, largely because of the dispute between Max Born and C. V. Raman over the nature of optical modes. (At this stage, consensus is that Born was correct in asserting the existence of optical frequency bands rather than discrete modes.) While this study was complicated by the subtlety of the symmetry conditions, a rather clear picture evolved. Using the elastic constants, one could fit parameters and estimate the Reststrahl frequency. This could be measured and did not agree with the estimate. Then one could add another interaction parameter and fit also the Reststrahl frequency and then estimate some other mode. This process proceeded until as many as seven interaction constants were fit to experiment. However, each time another constant was added, all of the earlier ones were changed and at any level of approximation there was little tendency for the interaction constants to drop to zero at large distance. The method was in principle correct but it turned out the interactions were so long ranged that it was not a useful starting point. The picture of a covalent crystal as a collection of atoms held together by short-range bonds has been misleading. Success could be had only by adding intrinsically long-range forces, such as those in the shell model, and patching up the error with short-range forces.

The experience in metallic crystals was similar though one expected long-ranged interactions to be screened. Once extensive data on the vibration spectra of metals was obtained with neutron diffraction, attempts were made to use a least-squares fit to obtain interaction constants between neighboring atoms. Again, the value of any one constant depended on how many interaction constants were being fit, and there was only a limited tendency for the interactions to decrease with distance. The long-range forces in this case are those associated with the now familiar Friedel oscillations. Though interactions that vary slowly in space are indeed screened, those which oscillate rapidly spacially are not, and can be of long range.

Two important lessons from this are relevant to the problem of defects. First, assumed short-range interactions between atoms are probably not appropriate to any physical system. Long-range forces are important in every system, but the forms they take may be very much different in different types of solids. Second, the fact that some model of the interactions gives a reasonable account of the vibration spectrum is no guarantee that it will properly describe the structure or properties of a defect. The difficulty in metals was not that one failed to fit the vibration spectrum fairly well with a limited number of force constants, but that the force constants obtained depended very much on how many were included. Thus, we might expect that different models could account for the vibration spectrum but could give very much different results on defect properties.

This does not mean that it is useless to carry out detailed calculations even with short-range potentials. Many defect properties seem to be very similar in metals, valence crystals, and insulators, and they presumably would be similar if a system with short-ranged interactions existed. Thus, a calculation based upon a simple model may tell us something about the structure of dislocation cores, but it very likely tells us nothing about the difference between dislocation cores in sodium, aluminum, and iron. If we adopt this approach we may still learn much about defects but we unquestionably drastically limit the scope of our theory. The other alternative is to seek a sufficient understanding of the theory of the system that we may incorporate the long-range as well as the short-range forces and learn about those aspects of the properties that distinguish different materials. Before returning to the assessment of the opportunities and limitations in that direction, another question is raised concerning the efficacy of computational efforts.

It seemed that a machine calculation on a model – or in fact the observation of a mechanical model of some crystal system – would be very much the same as doing an experiment. One could then hope for the new insights and concepts which often emerge from experiments. Many experiments, as many calculations, are designed to obtain some numerical quantity of interest. Very frequently, however, the observed behavior is not at all what was expected and in the course of sorting this out an entirely new effect has been discovered. The impression is that this most exciting aspect of experimental work has been almost totally absent from computer experiments and it is not understood why. Not really because the models that have been used are too limited! Calculations of the vibration spectrum in sodium included Kohn anomalies but they were noticed only after Walter Kohn, on the basis of formal arguments, argued that they must exist. Dislocations are easily observable in a soap-bubble raft but they seem to have been unnoticed until they were proposed to explain the properties of crystals. Discovery of an effect like channeling in radiation damage could have been discovered in a machine experiment, but, in fact, it was proposed before such an experiment was done. I am not certain that this is a property of machine experiments and if it is I do not know the solution. However, it may be worth worrying about along with other problems.

3 INTERACTIONS IN SIMPLE METALS

A reasonably full theoretical understanding exists only for the simple metals.[1] Use of the pseudopotential method has produced meaningful calculations of the entire vibration spectrum from first principles. It is important that these calculations include no experimental numbers except the fundamental constants. Only then is there a real test of the theory since it is possible to *fit* the vibration spectrum with a few adjustable parameters even if the nature of the interaction is totally misunderstood. The use of pseudopotentials in this connection rose directly from the experimental discovery that the energy bands were close to those expected for nearly free electrons. In the course of calculating the band structure and the Fermi surface of zinc it was noticed that in the context of a simplified band calculation the changes in the band structure with distortion could be immediately calculated.[2] It was then quickly apparent that the validity of the methods was not affected when entirely arbitrary configurations of the atoms were allowed.[3] In those more general circumstances, it was necessary to use perturbation theory to obtain eigenvalues rather than to diagonalize small matrices, but the perturbation theory was permitted because of the weak pseudopotential. Further, with the use of perturbation theory it became possible to sum the total energy of the system and to obtain that energy as a function of the positions of the ions. Thus we were led to an a priori theory of the interactions between atoms in a simple metal. It was a matter of choice as to whether these calculations were performed in wavenumber space using structure factors to describe the configuration of atoms or whether one used an equivalent effective interaction between ions in real space as suggested by Cohen.[4] Here, it is helpful to use both. A schematic derivation of the total energy in the metal provides the basis also for the understanding of interactions in more complex materials.

The interaction between the electrons and the ions in a metal may be represented by a weak pseudopotential, W. If W is weak it is sensible to calculate the energy by perturbation theory. Writing $\mathbf{k}$ as the wavenumber of the zero-order state, we obtain

$$E_k = \frac{\hbar^2 k^2}{2m} + \langle \mathbf{k}|W|\mathbf{k}\rangle + \sum_q \frac{|\langle \mathbf{k+q}|W|\mathbf{k}\rangle|^2}{\frac{\hbar^2}{2m}(k^2 - |\mathbf{k+q}|^2)} + \ldots \quad (1)$$

If a correct pseudopotential were used and the expansion were carried to all orders, this would be equivalent to a band calculation but we are interested in proceeding only to the second order. To that order the pseudopotential can be written as a sum of individual pseudopotentials arising from the individual ions, $W = \sum w(\mathbf{r}-\mathbf{r}_j)$. Thus each matrix element becomes a sum of matrix elements for the individual ions and may be factored, as in diffraction theory, into a structure factor

$$S(\mathbf{q}) = \frac{1}{N} \sum_j e^{-i\mathbf{q}\cdot\mathbf{r}_j} \quad (2)$$

(where the sum is over all N of the ion positions $\mathbf{r}_j$) and a form factor, $\langle \mathbf{k+q}|w|\mathbf{k}\rangle$. It is important to note that the structure factor does not depend upon the wavenumber of the state in question. It is unity when $q = 0$, and in a perfect crystal is nonzero only on a discrete lattice of wavenumbers, the wavenumber lattice, or reciprocal lattice. However, at this stage no assumption about the arrangement of the ions is needed.

We may now sum the energies of all of the occupied states, and, to the order computed, this is correctly obtained by summing over a Fermi sphere, $k < k_F$. We obtain

$$\sum_{k<k_F} E_k = \sum_{k<k_F} \left(\frac{\hbar^2 k^2}{2m} + \langle \mathbf{k}|w|\mathbf{k}\rangle\right) + \sum_q S^*(\mathbf{q})S(\mathbf{q}) \sum_{k<k_F} \frac{|\langle \mathbf{k+q}|w|\mathbf{k}\rangle|^2}{\frac{\hbar^2}{2m}(k^2 - |\mathbf{k+q}|^2)}. \quad (3)$$

The first two terms depend upon the volume of the system through k_F but are otherwise independent of the configuration of the ions. The final term includes the structure factors outside of the sum over $\mathbf{k}$, which then becomes a function only of the wavenumber q. It depends upon total volume but is otherwise independent of the configuration of the ions. Thus, for rearrangements of the ions at constant volume, all configuration dependence enters through the structure factors in the final term. That term, written as the energy per ion, called the *band-structure energy,* is written

$$E_{bs} = \sum_q S^*(\mathbf{q})\, S(\mathbf{q})\, F(q) \qquad . \qquad (4)$$

The only approximation upon the initial Schroedinger equation is the dropping of terms higher than second order in the expansion. Other terms that must be added to obtain the total energy are electron-electron interactions counted twice and the direct coulomb interaction between the ions. However, since all terms may be written in a form such as that above, they need not be considered explicitly in this discussion.

It is helpful first to discuss this energy in terms of the sum over wavenumbers $\mathbf{q}$, particularly since this will provide a good starting point for discussing valence crystals. Imagine for the moment a perfect metallic crystal and consider the band-structure energy of Eq. 4. The structure factors are nonzero only at the discrete lattice wavenumbers; each lattice wavenumber gives rise to a Bragg reflection plane — or in band structure language, a Brillouin zone face. This is illustrated in Fig. 1, showing the corresponding distortion of the Fermi surface due to each such plane. Other larger lattice wavenumbers also contribute to the energy but do not give a perceptible distortion of the Fermi surface. While the accuracy of the energy calculation is not guaranteed, experience has indicated that in simple metals the perturbation expansion works and we may meaningfully superimpose the effect of such lattice wavenumber.

Let us now transform the problem into real space, which is closer to our intuition and which may be more useful in defect calculations. The structure factors in Eq. 4 are written explicitly,

$$E_{bs} = \frac{1}{N^2} \sum_{\mathbf{q},i,j} e^{-i\mathbf{q}\cdot(\mathbf{r}_i - \mathbf{r}_j)} F(q) \qquad , \qquad (5)$$

and note that this can be written rigorously as

$$E_{bs} = \frac{1}{2N} \sum_{i,j} V(|\mathbf{r}_i - \mathbf{r}_j|) \qquad , \qquad (6)$$

where the effective interaction between ions is given by

$$V(\mathbf{r}) = \frac{2}{N} \sum_{\mathbf{q}} F(q) e^{-i\mathbf{q}\cdot\mathbf{r}} \qquad . \qquad (7)$$

Again, other terms must be included in the calculation, but it is true that, to second order in the pseudopotential, the energy can be written as a volume-dependent term plus a sum of central-force interactions between ions (which also depends upon the volume). Such an effective interaction between ions, calculated for aluminum, is shown in Fig. 2.

As noted earlier, the presence of volume dependence means that the ions need not be in equilibrium under the central-force interaction alone. Thus, these results do not imply the Cauchy relations for the elastic constants. Clearly the presence of volume dependence implies a rather complicated form of many-ion interaction, but, for rearrangements of the ions at constant volume, the two-body interaction suffices. This simplicity arises directly from restricting the calculation to second order; the interaction may be thought of as arising from the subsequent scattering of electrons from the two ions in question. The third order includes scattering from three atoms, thus explicit three-body interactions. The spherical symmetry of the interaction also arises from the use of second order. In higher orders states would be summed over the true Fermi surface, distorted as in Fig. 1, and asymmetry would occur. All of these complications are real, but it is hoped that the low-order calculation can give a meaningful theory.

A knowledge of this interaction can provide the basis for the theory of any structural property of a defect to be considered. It will in fact be helpful to draw some immediate conclusions. First, one of the simplest questions concerning a vacancy: what is the relaxation of the nearest neighbors to the vacancy? Clearly, in a perfect crystal the total force on any atom due to its symmetrically arranged neighbors will vanish. If one ion is removed, keeping all others fixed, the net force on the nearest neighbor will simply be the negative of the force between the two ions which were previously there. Fig. 2 shows that the nearest-neighbor distance lies beyond the minimum of the curve and that this interaction was therefore attractive; we conclude that the vacancy repels the nearest neighbors and they will move outward. The calculation of the magnitude of that displacement as well as the displacement of all other neighbors is complicated, but this primitive conclusion will certainly remain true for this effective interaction.

It is striking that this conclusion is contrary to almost all earlier models. It is not difficult to see why. Ordinarily, an interaction with a single minimum has been assumed and no volume-dependent terms are included. Then the interaction of one atom with all distant neighbors will be attractive and this can be counterbalanced only by a repulsive interaction between near neighbors; the volume will readjust itself to make this true. This leads immediately to the reverse conclusion for the neighbors to the vacancy.

How much confidence is there in this surprising result? Unfortunately very little, but at the same time it is realized that there is absolutely no

INTERATOMIC POTENTIALS

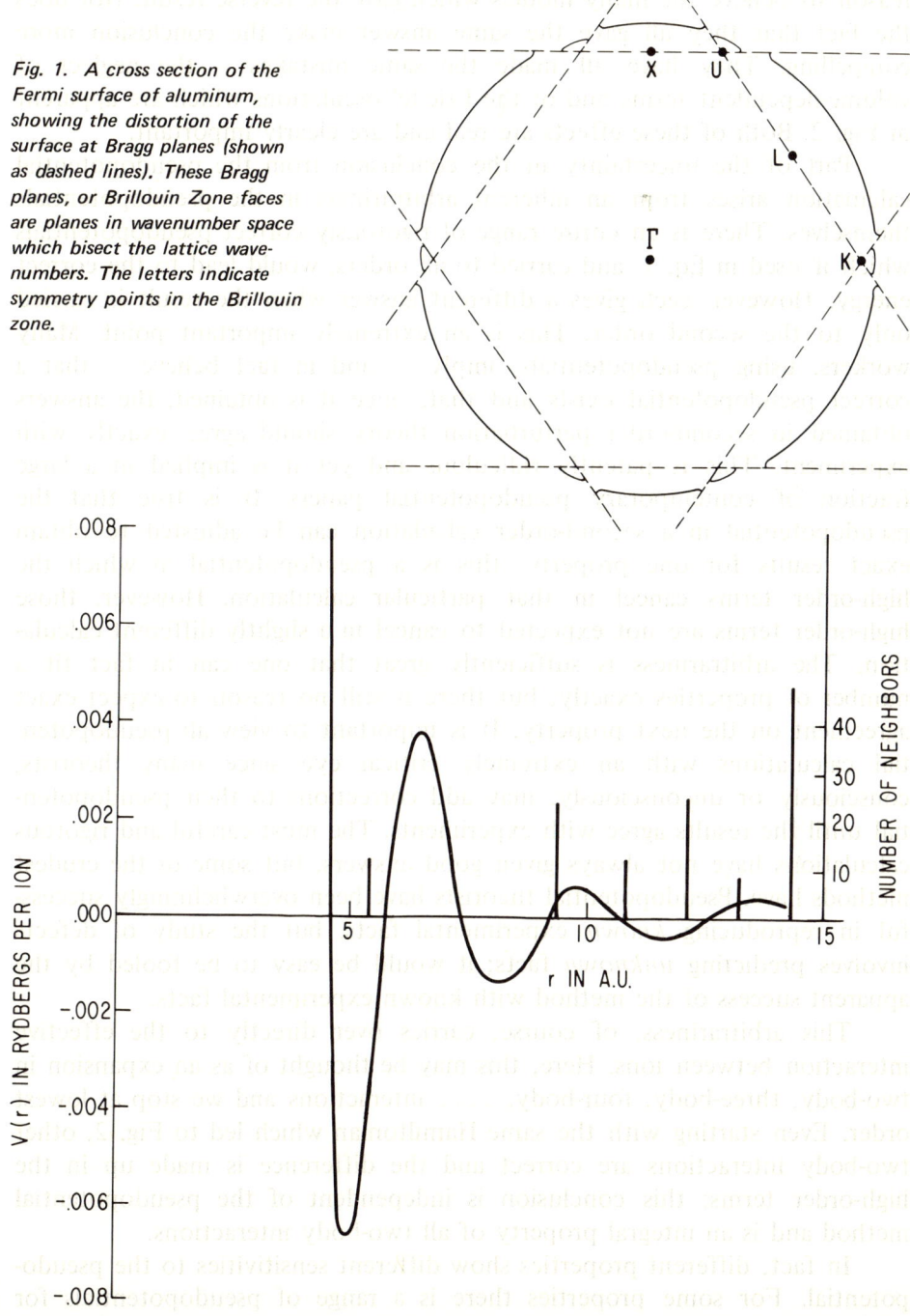

Fig. 1. A cross section of the Fermi surface of aluminum, showing the distortion of the surface at Bragg planes (shown as dashed lines). These Bragg planes, or Brillouin Zone faces are planes in wavenumber space which bisect the lattice wavenumbers. The letters indicate symmetry points in the Brillouin zone.

Fig. 2. A calculated effective interaction in metallic aluminum, showing Friedel oscillations at large distances. Also shown are the positions of the near neighbors in the aluminum structure. This, the earliest such calculation, included no exchange among valence electrons.

reason to believe the many models which gave the reverse result. Nor does the fact that they all gave the same answer make the conclusion more compelling. They have all made the same mistakes — the neglect of volume-dependent terms and of the Friedel oscillations which are apparent in Fig. 2. Both of these effects are real and are clearly important.

Part of the uncertainty in the conclusion from the pseudopotential calculation arises from an inherent arbitrariness in the pseudopotentials themselves. There is an entire range of rigorously correct pseudopotentials which if used in Eq. 1, and carried to all orders, would lead to the correct energy. However, each gives a different answer when the result is carried only to the second order. This is an extremely important point. Many workers, using pseudopotentials, imply — and in fact believe — that a correct pseudopotential exists and that, once it is obtained, the answers obtained in second-order perturbation theory should agree exactly with experiment. This is patently ridiculous and yet it is implied in a large fraction of contemporary pseudopotential papers. It is true that the pseudopotential in a second-order calculation can be adjusted to obtain exact results for one property; this is a pseudopotential in which the high-order terms cancel in that particular calculation. However, those high-order terms are not expected to cancel in a slightly different calculation. The arbitrariness is sufficiently great that one can in fact fit a number of properties exactly, but there is still no reason to expect exact agreement on the next property. It is important to view all pseudopotential calculations with an extremely critical eye since many theorists, consciously or unconsciously, may add corrections to their pseudopotential until the results agree with experiments. The most careful and rigorous calculations have not always given good answers, but some of the crudest methods have. Pseudopotential theorists have been overwhelmingly successful in reproducing *known* experimental facts, but the study of defects involves predicting *unknown* facts; it would be easy to be fooled by the apparent success of the method with known experimental facts.

This arbitrariness, of course, carries over directly to the effective interaction between ions. Here, this may be thought of as an expansion in two-body, three-body, four-body, ... interactions and we stop at lowest order. Even starting with the same Hamiltonian which led to Fig. 2, other two-body interactions are correct and the difference is made up in the high-order terms; this conclusion is independent of the pseudopotential method and is an integral property of all two-body interactions.

In fact, different properties show different sensitivities to the pseudopotential. For some properties there is a range of pseudopotentials for which the high-order terms beyond Eq. 1 would all be small and the pseudopotential method is reliable. A case in point is the vibration spectrum of the alkali metals. This is because the rigidity tends to be dominated by the electrostatic interaction between the ions, and the pseudopotentials provide only a small correction. At higher valence there is an increasing cancellation between electrostatic and band-structure terms, the sensitivity increases, and the pseudopotential method becomes less reliable. The resistivity of liquid metals, which can also be calculated by this method, is a case of extreme sensitivity in the alkalis as well as the polyvalent metals. This simply means that the expansion in the pseudopotential is not very reliable for the calculation of this property, although the expansion in the same pseudopotential may be reliable for the vibration spectrum.

A second important uncertainty comes from the electron-electron interactions. It is always necessary to include them in an approximate way – Fig. 2 was obtained using only the direct interactions between valence electrons – and it has not been possible to determine the error in the approaches used. This is a separate uncertainty, and even removing it completely would leave uncertainty due to the arbitrariness of the potential. However, there is evidence that a technique for including exchange and correlation among electrons (developed by Singwi et al.[5]) may at the same time increase the accuracy of the method and reduce the sensitivity to which particular pseudopotential is used. An interesting comparison may be made between the optimized pseudopotential which led to Fig. 2 and the optimized model potential due to Shaw[6]. (The model potential, formulated first by Heine and Akarenkov[7], is another rigorous method for obtaining interactions.) If electron-electron interactions in the model potential method are included as in the calculation of Fig. 2, the effective interaction does not show a minimum near the near-neighbor distance. This is not necessarily an incorrect result, only a different one. However, Shaw and Pynn found that the approximate addition of exchange effects both improved the accuracy of calculated vibration spectra and lowered the effective interaction at the near-neighbor distance.[8] When Shyu et al.[9], using Shaw's model potential, added interactions according to Singwi et al.,[10] a minimum similar to that of Fig. 2 appeared for lithium, sodium, potassium, and aluminum. (It is not known what these interactions would have done to the optimized pseudopotential calculation.) However, the minimum in all cases lay between the nearest and next-nearest neighbor distances. In general, calculations accorded well with experiment and on the particular question of near-neighbor minimum, they argued that it is supported by the change in sign of the tangential force constant between nearest and next-nearest neighbors derived from measured vibration spectra. The existence of such a minimum is not guaranteed a priori, but it fits comfortably with intuition. So also does the corresponding conclusion that the nearest neighbors relax toward the vacancy rather than away from it as suggested above.

Perhaps it is too early to be certain, but these recent findings fit together nicely and are very encouraging. They suggest that with an appropriate approximation for exchange and correlation, second-order perturbation theory in the pseudopotential is much more reliable than anticipated. It also supports the traditional optimization procedure for the selection of the pseudopotential.

From the point of view of defect studies there is an even more important conclusion — which again must be tentative at this stage. This suggests that there exists a rather well-defined, effective, two-body interaction between ions that describes defects as well as vibration spectra. It is legitimate to adjust a proposed form to fit experiment rather than relying entirely upon theory. However, it is essential that the form adjusted be an appropriate one. An interaction with the near minimum inside the near-neighbor distance can fit a vibration spectrum as well as one with a minimum outside or one with none at all, but the conclusion about defects will differ greatly.

The status of effective interactions between ions in the simple metals is encouraging. For this case, which is the best understood theoretically, there is reason to hope that available interactions will give reliable answers.

4 TRANSITION METAL INTERACTION

The procedure outlined for simple metals fails in the transition metals. It was based upon the fact that the electrons were only weakly disturbed by the ions present and the energy bands resembled those for free electrons. In transition metals the bands which arise from the atomic d-states bear essentially no relation to free-electron bands and a weak pseudopotential assumption is patently nonsense. However, the message that should have been learned is that if we start from a simple picture, in that case free electrons, and expand in a small correction we may hope to progress. For the transition metal, two starting points are required — free electrons for some states and atomic d-electrons for others. A look at the true calculated energy bands for a transition metal would suggest that this is not really a valid separation. On finding a state that seems very much like an atomic d-state, and following the band as wavenumber changes, it may slowly change to a free-electron-like state. It depends on the questions asked. A single sheet of a Fermi surface may contain portions originating from either starting point and the study must proceed carefully and completely. Similarly in the simple metals, if the interest is in the Fermi surface, the small cuts indicated in Fig. 1 become of overwhelming importance and the perturbation approach inappropriate. However, here total energies are discussed and it seems certain that the dual starting point is appropriate.

Starting with copper, where the bands with strong d-character are all occupied, it is most plausible to consider the addition of d-like states to an otherwise monovalent metal. The 3d-states in copper provide a shell of electronic charge which is considerably larger than that of the cores in the simple metals. Consequently, they provide a rather large additional potential well at each ion. Considering, for the moment, only the effect of this potential, it may be said that it will favor the contraction of the system which, by compressing the free-electron-like electrons, pushes them into the potential well. The corresponding, but smaller, effect of the ions in the simple metals was included in the term $\langle k|w|k \rangle$ which is responsible for the smaller cohesion of the simple metals. In the noble and transition metals, however, this tendency to contract is so large that the d-shells are brought into contact and a rather abrupt repulsion between neighboring d-shells is felt before the conduction-electron energy has reached its minimum value. Because this repulsion rises so abruptly with decreasing separation it contributes little to the total energy but dominates the atomic spacing and the elastic rigidity; infinitely hard shells would contribute nothing to the energy but would alone fix the spacing and give infinite rigidity. This explains the higher cohesive energy, compactness, and rigidity of copper in comparison to those of the polyvalent metals, and also provides a starting point for a theory of the electronic properties of copper[11]. The interatomic spacing can be fixed at the observed value and the states near the Fermi energy can be treated much as was done for the simple metals. However, complications arise from the inevitable distortion of the d-like states. The effect of their interaction with the conduction electrons, called hybridization, gives rise to additional terms entering with the ionic pseudopotentials. Physically, these may be thought of as arising from processes where the conduction electron is not directly scattered by the ion but rather scatters *through* the d-state; alternatively, it can be said that a d-electron jumps to the Fermi energy as another electron at the Fermi energy drops into the d-state, having the net effect of a scattering. The entire calculation can be carried through, expanding both in the pseudopotential and in the hybridization, leading again to an effective interaction between ions in exact analogy with the theory of simple metals. In lowest order, these hybridization terms also give a term depending only upon volume and not upon atomic arrangement which contributes significantly to the cohesion but is of little interest here. Finally, add an additional interaction between ions arising from the hard-core repulsion and this will dominate in many properties. This last part of the program is still incomplete[12], but something is to be learned by thinking it through.

Start with, perhaps largely by fiat, an isolated band of states with well-defined atomic origin. As the corresponding ions are brought together, the largest effect is the broadening of the degenerate states of the isolated ions into bands. The second effect is a slight shift in the center of gravity of the bands. In copper, since all states are occupied, the broadening has no effect on the total energy and it is only the second term that is important and causes the repulsion which fixes the spacing.

In true transition metals the d-states are partially occupied. This is most clearly understood in terms of the actual band structure where the bands are partially filled. It is also understandable by regarding the d-states as resonant states near the Fermi energy. (In this sense copper in the theory described above is to a small extent a transition metal.) For our purposes it is sufficient simply to regard the atomic d-state as only partially occupied. Thus, in the transition metal when the d-states broaden into bands, only the lower portions of those bands will be occupied and the broadening itself lowers the energy. This effect tends to be largest in the middle of the transition-metal series where half of the states are occupied. Thus, moving through the transition series to the left from copper, this additional attraction tends to more than counterbalance the reduced conduction-electron attraction as there are fewer and fewer electrons in the d-shell producing the potential well. Thus, properties such as the cohesive energy tend to peak in the middle of each transition metal series.[13]

It is not unreasonable to approximate these additional interactions between ions in the transition metal as central-force pairwise interactions though I am not aware of good calculations of them at present. It is not even unreasonable to regard them as independent of total volume and as of short range. However, a volume-dependent and long-range conduction electron contribution must be added. Thus all of the complication of simple metals is still with us along with the additional uncertainties arising from the atomic d-states.

5 VALENCE AND IONIC CRYSTALS

For materials such as diamond, germanium, and silicon, with the return to systems where the energy bands have a strong resemblance to free-electron bands, it would seem reasonable to return to a pseudopotential approach. It was in fact in connection with semiconductors that pseudopotentials were first used in solids.[14] We know, however, that the effect of the Bragg planes is not simply to distort a Fermi surface as shown in Fig. 1 but that the entire Fermi sphere disappears into these Bragg planes.

A very plausible and appealing description of this situation was given in the 30's.[15] The idea was to focus on a particular set of Bragg planes, which make up the "Jones zone" and which dominate the behavior; any effects of the others would be thought of as corrections. For the diamond structure, such a zone is made up of twelve (220) Bragg planes. These form a polyhedron, quite spherical in shape, and with precisely the volume to provide states for the four valence electrons per atom. In addition, the structure factor corresponding to these reflections is maximum. It seemed very reasonable to believe that the Fermi surface disappears at this zone face giving rise to the observed semiconducting behavior. Thus, the basis for understanding semiconductors would seem to be very much the same as that for simple metals. However, we are now in a position to calculate the individual ionic pseudopotential and it happens that matrix elements of the pseudopotential at the appropriate wavenumber for this Bragg reflection are very near zero. Thus with more information it would appear that this very appealing picture collapses.

However, Heine and Jones noted that when the leading term is accidentally small, the calculation must be carried to higher order.[16] This can be done in Eq. 1 by adding a second-order term to the pseudopotential wherever it appears. (Eq. 1 of course is correct as it stands and this simply includes selected higher-order terms with the lower-order sums.) Thus we obtain a second-order pseudopotential that has matrix elements such as

$$\langle k+q|W_{2nd}|k\rangle = \langle k+q|W|k\rangle + \sum_{k'} \frac{\langle k+q|W|k'\rangle\langle k'|W|k\rangle}{E_k - E_{k'}} \quad . \quad (8)$$

Heine and Jones find that the second-order terms in the case of the (220) reflections are indeed large and the essential features of the Jones-zone picture are reinstated.

The reinstatement occurs, however, with some fundamental changes concerning the problem of atomic interactions. Pair-wise interactions followed directly from the second-order calculation in which the structure factor appeared squared. The important terms in the total energy now involve four powers of the structure factor (from Eq. 8 squared) and, therefore, interactions depending upon the coordinates of as many as four atoms. To be sure, these fourth-order terms lead also to pair-wise interactions that may be thought of as four subsequent scattering events involving only two atoms. However, the three- and four-particle interactions become an integral part of the theory. Thus it is not surprising, as has been apparent in fitting vibration spectra of covalent crystals, that there are angular and directional contributions to the interactions which cannot be ignored.

It will be useful, before discussing interactions further, to extend the description in terms of the Jones-zone band gap to ionic crystals. This clarifies the relation between the different nonconductors and the relation to metals. It will also suggest models for the interatomic interactions which may be useful in the study of defects. Finally, a most promising a priori approach for these systems and its relation to the models indicated above are discussed.

Following Phillips[17] ionic crystals are approached gradually beginning with valence crystals. In both cases attention is focused on the single parameter, the band gap at the Jones zone. Then the band structure and virtually all of the properties become functions of the second-order matrix element given in Eq. 8. In simple metals, the first term dominates and is associated with metallic bonding. In covalent crystals, the second term dominates and is associated with covalent bonding. In any real system, both contribute and mixtures of the two types of bonding result.

One particular property that may be obtained in terms of this parameter is the dielectric response function of the system. Penn has, in fact, calculated the dielectric response as a function of frequency on this model, approximating the Jones zone by a sphere.[18] Phillips has chosen the static dielectric constant and Penn's formula to determine the band-gap parameter. Thus the input information for the discussion of properties comes entirely from the measured static dielectric constant. We may think then of the covalent material germanium and obtain its band gap from the dielectric constant.

Move now to the compound gallium arsenide, which has the same structure except that alternate atoms are gallium and arsenic (neighbors to germanium in the periodic table). The pseudopotentials are then different on the two atoms in the primitive cell, and these differences appear in the calculation of the matrix elements that enter Eq. 8. In evaluating a single-matrix element $\langle k+q|W|k \rangle$, where q is a lattice wavenumber, we let the pseudopotential form factor for the first atom in the primitive cell be $\langle k+q|w_1|k \rangle$ and that for the second atom be $\langle k+q|w_2|k \rangle$. If the second atom lies at a position δ with respect to the first, the two form factors enter with phases differing by a factor $e^{-i q \cdot \delta}$. Thus

$$\langle k+q|W|k \rangle = \frac{1}{2} \langle k+q|w_1|k \rangle + \frac{e^{-i q \cdot \delta}}{2} \langle k+q|w_2|k \rangle \quad , \quad (9)$$

still under the condition that q be a lattice wavenumber for the Bravais lattice.

Consider specifically the matrix elements giving rise to the Jones zone in the diamond (or zinc blende) structure; that is, a wavenumber of $q = (2\pi/a)[220]$. The separation of the two atoms in the primitive cell is $(a/4)[111]$ and, thus, the phase factor (unity in Eq. 9) leads simply to the average form factor. The first-order term in Eq. 8 depends only upon this average. The second-order term has an interesting contribution for $k' - k = (2\pi/a)[111]$ and $k+q-k' = (2\pi/a)[111]$, corresponding to the same total transfer $\mathbf{q} = (2\pi/a)[220]$. The phase factors may again be evaluated, and on writing $w = \frac{1}{2} (\langle\mathbf{k'}|w_1|\mathbf{k}\rangle + \langle\mathbf{k'}|w_2|\mathbf{k}\rangle)$ when the transfer $\mathbf{k'} - \mathbf{k}$ has a magnitude $2\pi\sqrt{3}/a$, and similarly writing $\delta w = \frac{1}{2} (\langle\mathbf{k'}|w_1|\mathbf{k}\rangle - \langle\mathbf{k'}|w_2|\mathbf{k}\rangle)$, it is found that the contribution of this term in Eq. 8 is given by $(1/2) (\overline{w}^2 + (\delta w)^2/(E_k - E_k'))$. A second contribution of exactly the same form arises from another $\mathbf{k'}$ corresponding to wavenumber transfers in the reverse order. The two contributions will have equal magnitude if $\mathbf{k}$ lies at the center of the Jones-zone face. These are presumably the dominant terms and the evaluation of Eq. 8 for gallium arsenide, for example, has given

$$\langle\mathbf{k}+\mathbf{q}|W_{2nd}|\mathbf{k}\rangle = \frac{\langle\mathbf{k}+\mathbf{q}|w_1 + w_2|\mathbf{k}\rangle}{2} + \frac{\overline{w}^2}{E_k - E_k'} + \frac{(\delta w)^2}{E_k - E_k'} . \quad (10)$$

There are three contributions to the gap at the zone face: the metallic term $\dfrac{\langle\mathbf{k}+\mathbf{q}|w_1 + w_2|\mathbf{k}\rangle}{2}$, the covalent term $\overline{w}^2/(E_k - E_k')$, and a new term $(\delta w)^2/(E_k - E_k')$, which is called the *ionic* contribution since it arises only when the two atoms in the primitive cell are different. It is striking that these three terms enter the band gap independently with no cross terms.

The metallic term is presumed to be negligible, and Phillips defines the ionicity of the system to be the ratio of the square of the ionic term to the square of the total. In gallium arsenide, $\overline{w}$ is approximately the same as the germanium value, and the ionic term may be evaluated from the dielectric function for gallium arsenide and that for germanium. Phillips has examined a wide range of properties demonstrating that there is a direct and simple correlation between properties which one intuitively associates with ionicity and with the ionicity determined in this way. The property of particular interest here is the interaction between atoms. One expects the forces arising from covalent character to be angular and those from the ionic character to be radial. In fact, the estimates based upon the vibration spectrum indicate a correlation of just this type.

This is all phenomenology and does not provide interatomic interactions to be used in a calculation on defects, though it can suggest a reasonable form if carried further. The distinction with metals, which has been emphasized by this approach, is the failure of two-body central force interactions in valence crystals. A further important distinction with simple metals can be made regarding the range of interactions and their behavior at large distances. In simple metals, the behavior was dominated by the Friedel oscillations, which depend upon the existence of a free Fermi surface and are, therefore, characteristic only of metals. In metals, the amplitude of the oscillations drops with distance as $1/r^3$; they cannot occur in valence crystals.

These oscillations disappear when a metal is gradually made into a semiconductor. This is done, as in Penn's treatment of the dielectric constant, by replacing the Jones zone by a sphere, which is in fact the free-electron sphere for the valence electrons. As a band gap is introduced on the sphere, an exponential decay of the Friedel oscillation appears. (At this stage, the individual ionic pseudopotentials are assumed to remain well localized.) The form of this damping is expected to be $e^{-\mu r}$ with μ related to the gap by $\hbar^2 \mu^2 / 2m = E_{\text{gap}}$. By the time the gap has reached 1 volt, the characteristic length for decay has dropped to 2 Å. Thus, the interactions are expected to be of quite short range if the pseudopotentials were well localized. However, the introduction of the gap has also made the dielectric constant finite at short wavelengths, rather than infinite as in a metal. The screening is not complete and a long-range coulomb or dipolar interaction can arise between different parts of the crystal.

At this stage the metallic description becomes quite artificial. If one really treated a displaced atom as a displaced charge and divided the field by a dielectric constant, the resulting fields would cause plasma-like oscillations (finite frequency at infinite wavelength) rather than sound-like oscillations (frequency inversely proportional to wavelength). This difficulty is avoided only with a much more complicated form for the dielectric response.[19]

6 MODELS AND THEORY FOR NONCONDUCTORS

An alternative to the use of a more-complete treatment of dielectric response is the use of a starting model that contains a part of the screening at the outset. Such an approach is the bond-charge model of Phillips.[20] In this model, as applied to silicon, a negative charge of two is placed midway between each pair of neighboring ions (each of charge plus four). Then the motion of one ion displaces no *net* charge, and acoustic waves that are obtained behave properly. Use of this model, with simple dielectric screening and additional short-range radial forces, can account for the observed vibration of silicon.[21] In this model the angular forces between atoms arise from charges localized midway between the ions. The full charge of two electrons does not enter these forces; the interaction is reduced by the static dielectric function (about twelve in silicon).

In contrast, the screening charge in metals is located quite isotropically around the ions, and the corresponding electrostatic interactions between atoms are radial. The origin of the difference comes from the crystal structure, not from differences in the pseudopotentials themselves.

If silicon atoms are forced into a close-packed structure, they form a metal, and metallic interatomic interactions are expected to obtain. In the open structure of valence crystals the symmetrically placed pseudopotentials are able to localize charge in the bond region between them.

As a crystal becomes ionic, the asymmetry of the pseudopotential shifts charge from the bond toward the atom of higher valence, and the angular forces become less strong in comparison to radial forces. With sufficiently high ionicity they are no longer sufficient to stabilize the open structure, and a more characteristically ionic structure, such as the NaCl structure, is formed. This correlation has been made quantitative by Phillips.[17]

These considerations suggest a simple model for interatomic interactions for the entire range of nonconducting solids. Long-range forces arise from coulomb interaction among bond charges and ionic charges. For any one compound, the numbers needed are the bond charge and the charge transfer between the two ion types; both numbers are presumably related to the ionicity, but at this stage they might best be taken as adjustable parameters. In addition a radial repulsive force is required to prevent collapse of the lattice. Martin's silicon calculation would suggest that a nearest-neighbor interaction would suffice.[21] This allows the fitting of vibration spectra but leaves an uncertainty in the formulation of defect problems. In the presence of a defect, some bond charges may be missing or modified and transfer charges may change. These uncertainties may well make the result of any defect calculation totally unreliable. Correct results may be obtained only if the correct guess is made concerning these new questions which do not arise in the studies of vibration spectra where the parameters of the model are fit.

This is a difficulty that is characteristic of a phenomenological approach. It can be fully resolved only by a more complete theory such as we seem to have for metals. The "orbital correction method", which the author is currently developing, may be the theory needed for nonconductors. This method grew out of the transition-metal pseudopotential theory described in this paper. In the course of that calculation it was necessary to obtain the energy of the d-like states. This was done by taking them in lowest order to be linear combinations of atomic d-states and to admix orthogonalized plane waves through hybridization. Similarly, starting states in valence crystals or ionic crystals might be taken to be appropriate linear combinations of atomic orbitals and corrected by the admixture of plane waves orthogonalized to these orbitals. This approach becomes a systematic expansion in the error in the starting linear combination of atomic orbitals. It can be carried to second order in the corrections to obtain the leading terms in the interatomic interactions. Some terms in the theory correspond to terms in the phenomenology described above, but there are also other terms and all parameters can be determined within the framework of the theory. It can be tested by studies of vibration spectra but, like the pseudopotential theory of metals, it will be equally applicable to systems with defects. Thus the outlook is good, but at present we must be satisfied with uncertain models for these systems.

ACKNOWLEDGMENTS

Supported in part by the National Science Foundation and in part by the Advance Research Projects Agency through the Center for Materials Research at Stanford University.

REFERENCES

1. Harrison, W. A., *Pseudopotentials in the Theory of Metals,* Benjamin, New York, 1966.
2. Harrison, W. A., Phys. Rev., **126**: 497 (1962).
3. Harrison, W. A., Phys. Rev., **129**: 2503 (1963).
4. Cohen, M. H., in *Metallic Solid Solutions,* J. Friedel and A. Guinier (eds.), p. XI-1, Benjamin, New York, 1962.
5. Singwi, K. S., A. Sjolander, M. P. Tosi, and R. H. Land, Phys. Rev., B, **1**: 1044 (1970).
6. Shaw, R. W., Jr., Phys. Rev., **174**: 769 (1968).
8. Shaw, R. W., Jr., and R. Pynn, J. Phys. C. (Solid State Phys.) **2**: 2071 (1969).
9. Shyu, Wei-Mei, John H. Wehling, Martin R. Cordes, and G. D. Gaspari, Phys. Rev. (in press).
10. Op cit.
11. Harrison, W. A., Phys. Rev., **181**: 1036 (1969).
12. Moriarty, J. M., unpublished.
13. See Kittel, C., *Introduction to Solid State Physics,* Third edition, p. 78, Wyley, New York, 1967.
14. Phillips, J. C., and L. Kleinman, Phys. Rev., **116**: 287 (1959).
15. See Mott, N. F., and H. Jones, *The Theory of the Properties of Metals and Alloys,* p. 159, Clarendon Press, Oxford, 1936, and Dover, New York, 1958.
16. Heine, V., and R. O. Jones, J. Phys. C. (Solid State Physics) [2], **2**: 719 (1969).
17. Phillips, J. C., Rev. Mod. Phys., **42**: 317 (1970).
18. Penn, D. R., Phys. Rev., **128**: 2093 (1962).
19. Pick, R. M., M. H. Cohen, and R. M. Martin, Phys. Rev., B **1**, 910 (1970).
20. Phillips, J. C., *Covalent Bonding in Crystals and Molecules,* University of Chicago Press, Chicago, 1969.
21. Martin, R. M., Phys. Rev., **186**: 871 (1969).

DISCUSSION on the Paper by W. A. Harrison

HO: I don't think that the effective interatomic potential of aluminum as shown is very reliable; in particular the sign of the slope of the potential at the first nearest neighbor is not correct. Phonon-dispersion-curve measurements give a negative slope at the nearest-neighbor position instead of a positive slope as given here. This may be due to the omission of the exchange and correlation effects in the screening function, which turn out to be important in determining the shape of the potential. This implies that the displacement of the first neighbor of a vacancy would be inward instead of outward. This is indeed what is found in my vacancy calculation of aluminum.

HARRISON: I agree. I attempted to make that point in my talk.

JOHNSON: From the point of view of those who wish to do calculations, a major difficulty arises from the sensitivity of the real-space pseudopotentials to the form factor and the dielectric function. This leads to problems of reproducibility, e.g., I cannot reproduce the Harrison aluminum potential. Using the input data given by Harrison and the Hartree dielectric function, I find no minimum at all in the vicinity of the first nearest neighbor.

As a second comment, I would like to point out that numerous parameters are not required to fit the dispersion curves of many transition metals. Since goodness of fit is somewhat subjective, I will show these later and let people decide for themselves.

HARRISON: There are many aspects to the question of reproducibility in pseudopotential methods. Some are fundamental, such as differences arising from different valid pseudopotentials or different approximations for the electron-electron interactions. The particular one you raise sounds to me (because you refer to the use of a Hartree dielectric function) like you began with a local model of the pseudopotential which I introduced in my book to treat phonon dispersion. My memory is that I did not get a near-neighbor minimum for that one either. The curve I showed in my talk came from a full nonlocal pseudopotential. I agree with your point however; the predictions from the method are meaningful only to the extent that they become somewhat reproducible. I am of the opinion now that there appears to be a reasonable consistency among the derived interactions from the best-founded pseudopotentials and the best treatments of the electron-electron interaction.

Concerning the transition metals, I did indicate that there are strong short-range forces arising from the unfilled d-shells. It is gratifying if these do dominate the elastic behavior sufficiently that one can get a good approximate account of the vibration spectrum with a few parameters. We should not forget, however, that there are certainly also long-range Friedel oscillations that can be important in other properties even if their effect on the vibration spectrum is small.

TORRENS: A number of published interatomic potentials which include electron-electron interactions in the dielectric screening appear to have long-range oscillations of substantially reduced amplitude compared to the form which you showed for aluminum. What experimental evidence do we have at the present time for long-range oscillations of any appreciable magnitude?

HARRISON: First, I am not aware that the inclusion of electron-electron interactions has reduced the calculated amplitude. It would have that effect if they were included completely enough to give a reduced Migdal discontinuity, but all treatments I have seen have left that discontinuity unchanged; thus the oscillation amplitudes should be unaffected. The most convincing experimental evidence is the observation of Kohn anomalies. Perhaps more direct, but less convincing, are the old work on the Knight shift in alloys and oscillatory force constants from fitting of the vibration spectra. It seems to me there is no doubt as to the existence of the long-range oscillations.

Note added in Proof

A more careful examination of Phillips' treatment of ionicity (Ref. 17) indicates that he did not use Eq. 10 which correctly applies to the Jones-zone band gap. He instead used the formula

$$\langle k+q|W|k\rangle^2 = 2\bar{w}^2 + 2\delta w^2$$

which would be appropriate on a (111) zone face but is inconsistent with the use of Penn's dielectric constant. This mistake appears to invalidate the relation between ionicity and the fundamental electronic structure which Phillips makes. On the other hand, viewed as a phenomenological prescription for defining ionicity in terms of dielectric constants, it remains a very effective way of correlating properties of different material.

ION-ION INTERACTIONS IN METALS: THEIR NATURE AND PHYSICAL MANIFESTATIONS

N. W. Ashcroft

Victoria University of Wellington
and
Cornell University (permanent address)

ABSTRACT

In calculating the total free energy of an arbitrary arrangement of charged ions immersed in a system of interacting electrons it is convenient to separate the structurally dependent terms into effective n-ion interactions. Normally the dominant contribution arises in simple metals from pairwise interactions in which the screening effects from the band electrons are incorporated within the framework of linear response. At finite temperatures, the energy of a crystal may be calculated as if the ions were static, provided the pair interaction is appropriately redefined. The dynamics of the crystal are thus transferred to the ion-ion potential, a simplification that may be useful for analyzing transformations between phases. Phase stability of metals and alloys is shown to be related almost entirely to the electron-response contribution to the total energy, expressed as indicated, in terms of pair potentials.

1 INTRODUCTION

Apart from a few oddities, most metals, and certainly the simple metals and their alloys, tend to crystallize in close-packed structures. These structures often result from the regular stacking of rigid spheres, and it is tempting to conclude at once that reasonably strong short-range forces act between ions (and perhaps less readily that weaker long-range forces differentiate the various breeds of close-packed structures). Of the rudimentary precepts associated with the field of crystalline defects, this notion of the atom as a rigid sphere remains particularly fundamental. It provides not only an easily visualizable construct for the perfect crystal, but also a model that has been used with remarkable ingenuity to produce crystalline abnormalities of apparently limitless intricacy.

Yet in metals, the bonds between ions are putatively long ranged, originating as they do with Coulomb interactions. To establish a complete picture of the ion-ion interaction (with, in particular, the hoped-for property of a "hard-core") requires, however, a systematic treatment of both ions *and* electrons. It is the primary purpose of this introductory contribution to reproduce the essential arguments and review the basic theory underlying the concept of the pair-interaction in simple metals. The emphasis is placed on the nature of the ion-ion interactions rather than on their applicability to the problem of defects.

It is helpful to proceed by analogy and first outline the development of the pair-potential concept familiar in the context of insulating solids. Then the ground-state energy of a simple metal (with fixed ions) is established. Only part of the latter depends specifically on structure and thus warrants interpretation in terms of a summation over pair potentials. The form, nature, and physical manifestations of these potentials is discussed. Their relation to problems connected with alloying and temperature effects are also given a cursory treatment.

2 THE PAIR-POTENTIAL CONCEPT

Although problems of detail remain, from the standpoint of the physical origin of cohesion in molecular solids it appears that the binding in a crystal of say, argon, is reasonably well understood. By virtue of their relative positions, an assembly of N such atoms* located at crystal sites

---

*The restriction to identical atoms is easily extended to more complex systems in what follows.

$R_1 \ldots R_N$ and confined to a volume V will have a potential energy* of interaction Φ which will be some function of the $R_1, \ldots R_N$:

$$\Phi = \Phi(R_1, R_2, \ldots, R_N) \tag{1}$$

This interaction is measured relative to the energy of the same atoms when they are well separated. On quite general grounds a detailed description of Φ involves a complex many-particle relationship in the positions R.

Almost by definition, the binding in molecular solids is weak, certainly when expressed (per atom) as a fraction of characteristic atomic or molecular energies. It follows that the electronic arrangement around each atom may be expected to suffer little (but a none-the-less critical) change in passing from the isolated to the condensed state. This being the case, an expansion:

$$\Phi(R_1, \ldots, R_N) = \sum_{i<j} \phi_{ij} + \sum_{i<j<k} \phi_{ijk} + \ldots, \tag{2}$$

is suggested in which the terms on the right in Eq. 2 represent interactions taken in pairs, triplets, etc., of atoms. For neutral atoms it is well known that the long-ranged parts of these interactions can be understood in terms of the resulting weakly attractive time averages of fluctuating and induced dipoles (van-der Waals or dispersion forces), whereas at short range the potentials** tend to be quite strongly repulsive, a consequence mainly of the exclusion principle. If the electron orbitals in the atoms are not easily polarizable, then, compared with the pair terms, the triplet and higher terms diminish rapidly in significance. Neglecting them entirely leads to the pair potential approximation:

$$\Phi(R_1, \ldots, R_N) = \sum_{i<j} \phi(R_i - R_j) = \tfrac{1}{2} \sum_{i \neq j} \phi(|R_i - R_j|), \tag{3}$$

a truncation of Eq. 2 often assumed to be sufficiently convergent for many applications of interest. Fig. 1 shows a familiar form for $\phi(r)$; it is radially symmetric and, for reasonable excursions in interparticle spacing, volume independent.*** As measured by a representative thermal energy, the rise in $\phi(r)$ at short range is normally quite rapid. The comparative weakness of the long-ranged part then leads quite naturally to the model of the atom (in the solid) as an impenetrable sphere weakly attracted to its neighbors. As a consequence, simple geometric considerations become

---

*Provided the temperature is sufficiently low and the atoms sufficiently massive the potential energy dominates the internal energy. However, as we will see the kinetic energy of the atoms can be an important term.

**Also known as Born-Mayer potentials. Note that the attempted spatial overlap of filled core-states forces the electrons to seek the nearest states available for them, a process which in a rare gas atom for example, is energetically expensive.

***It may be worth noting that certain of the 3-body (dipole-dipole-dipole) terms in Eq. 2 can be incorporated into effective 2-body terms, which as a result become volume and temperature dependent [see, for example, Casanova, et al., Molecular Physics 18, 589 (1970)].

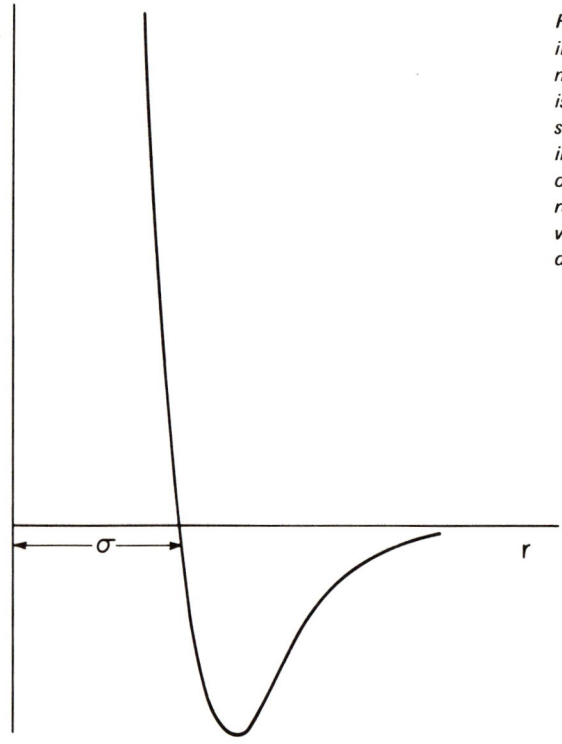

Fig. 1. Characteristic form of the interatomic potential $\phi(r)$ in a molecular solid. The dimension σ is a measure of the range of the strongly repulsive region originating in exchange interactions between outer electron orbitals. At longer range, the familiar dispersion (or van-der-Waals r^{-6}) attraction dominates.

important in deciding, for a given volume, which lattice energy, $\frac{1}{2}N\sum_{\mathbf{R}}\phi(\mathbf{R})$, is lowest.* Further, the notion of contiguity of neighbors implies elementary relationships between atomic sizes (essentially the σ of Fig. 1) and lattice constants. The presence of this strongly repulsive region is manifested quite clearly in diffraction studies not only of solids but also of liquids. Knowing the locations of the principal features in the static structure factor of a simple liquid, consistent estimates of σ can be inferred. Finally, although many of the common choices for $\phi(r)$ are quasiempirical, there is substantial support for their general shape from detailed quantum mechanical considerations. The advantage of simple parameterized forms for $\phi(r)$ (for example, the Lennard-Jones (12,6) form) is that the potential may often be determined by forcing agreement with a small set of equilibrium properties of the perfect solid. Thus armed we can embark on a treatment of dynamic properties of the crystal, and at least attempt to understand the connection between defects and pair potentials in both equilibrium and nonequilibrium situations.

---

*Here the **R** run over the sites of the chosen lattice. The comparison can be extended to liquids by considering $\int 4\pi r^2 \, dr \, g(r) \, \phi(r)$ where g(r) is the pair-distribution function.

3 STRUCTURAL ENERGIES IN SIMPLE METALS:

Turning to the metallic state, the central question raised by preceding arguments is to what extent can the energy and structure of metals be regarded as an expression of the existence of pair potentials similar in form to, say, Fig. 1. Unlike insulating solids, there are, of course, vast changes in the outer electron states of atoms brought together to form a metallic state. It is fundamental to the latter that electrons occupy states characterized by their propensity for distributing negative charge over an entire assembly of ions, the latter carrying localized net positive charges. In simple metals this negative charge has a substantial uniform components; but it also possesses components that mimic the structural arrangement of the ions. In the simplest picture we naturally expect energy Φ_S to be bound up in the nonuniform components of the electron distribution which will, accordingly, depend on the arrangement of the ions it is following. As we shall see, this is not difficult to demonstrate, but it is also expected (and straightforward to show) that there will be a large term Φ_V in the ground-state energy of the metal depending not on structure, but on the volume V of the system and having its origin in the uniform component of electron distribution. For N ions, we then have, in place of Eq. 1

$$\Phi = \Phi_V(V) + \Phi_S(V, \mathbf{R}_1, \ldots, \mathbf{R}_N), \qquad (4)$$

What characterizes a simple metal (Na, Mg, Al, etc.) is the fact that, in the independent-particle picture its valence band structure closely resembles a set of free-particle bands. Throughout much of the Brillouin zone the valence wave function of the occupied states may be taken as a single-plane wave. A reasonable starting point for the calculation of the ground-state energy of the metal is thus to assume a fixed distribution of charged ions in the presence of a neutralizing uniform distribution of electrons. An immediate estimate of the electrostatic energy involved results from surrounding each ion of charge Ze with a sphere of radius* r_o containing a total electron charge of $-Ze$ and assuming that the interaction between electrons and ion never deviates from Coulomb's law. Overlap of spheres is clearly implied by this approximation, but if the resulting multipole terms in the expansion of the energy are ignored, the potential

---

*The Wigner-Seitz radius r_o, defined by $\frac{4}{3}\pi r_o^3 a_o^3 = \frac{N}{V}$, is related to r_s, the familiar linear measure of electron density by $Z^{1/3} r_s = r_o$.

energy, in Ry per electron, is simply* $-1.8Z^{2/3}/r_s$ where r_s is the usual electron-spacing parameter. In terms of the same parameter, an interacting electron gas contributes an energy (again in Ry per electron)

$$u_{e.g} = \frac{2.21}{r_s^2} - \frac{0.916}{r_s} - u_c(r_s) .$$

the entries being respectively, the mean kinetic energy, the exchange energy, and the direct Coulomb correlation energy.** At this level of approximation we write the energy of the "metal" as

$$u = \frac{2.21}{r_s^2} - \frac{0.916 + 1.8Z^{2/3}}{r_s} - u_c(r_s) , \quad (5)$$

which produces an equilibrium r_s

$$r_s \doteq 4.42/(0.916 + 1.8Z^{2/3})$$

that decreases with increasing valence, but otherwise is obviously independent of the nature of the ions.

It is clear that although the model leading to Eq. 5 contains many of the physical requirements basic to a theory of cohesion in metals, it is also artificial in many respects, not the least of which is that there is no reference either to the arrangement of the ions or to their internal structure. Persisting with point ions and a uniform interacting electron gas, the potential energy can at least be improved (over the uniform-sphere estimate) by directly evaluating the energy of a given static array of point positive charges in a constant negative background. This Madelung energy*** is, for an *arbitrary* arrangement of ions at positions r_i:

$$U_M = \frac{N}{2V} \sum_{k \neq o} v_c^i(k) (S_i(k)-1) = \frac{N}{2V} \sum_{k \neq o} \frac{4\pi Z^2 e^2}{k^2} (S_i(k)-1) \quad (6)$$

---

*The energy per electron is $(-1/Z) \int_o^{r_o} \frac{1}{r} (4\pi r^2 dr \frac{NZe}{V})(1-(\frac{r}{r_o})^3)Ze$. Alternatively it is the energy $-3Z^{2/3}/r_s$ of a point charge Ze in the presence of a uniform sphere of negative charge, plus the energy, per electron, of the uniform sphere. The latter is half the average potential energy of interaction of any electron, i.e., half the average over the sphere of $\frac{3Ze^2}{2r_o} - \frac{Ze^2 r^2}{3r_o^3}$.

**Electrons with parallel spins tend to be kept spatially apart with the consequence that their potential energy is somewhat smaller than is calculated in footnote above. Electrons (with either spin) are also kept apart by the direct Coulomb repulsion reducing still further the Hartee energy. Of the various calculations of $u_c(r_s)$, the expression $u_c(r_s) = -(0.115 - 0.031 \ln r_s)$ due to Nozieres and Pines (*The Theory of Quantum Liquids*, W. A. Benjamin, Inc., N.Y., 1966) is representative over the range of r_s appropriate here.

***We specifically exclude the $k = o$ divergent contribution which is cancelled by the corresponding electron gas term.

where

$$v_c^i(k) = \int d\mathbf{r}\, e^{-i\mathbf{k}\cdot\mathbf{r}}\, \frac{Z^2 e^2}{r} = \frac{4\pi Z^2 e^2}{k^2}$$

and the structure factor

$$S_i(k) = \frac{1}{N} \langle \sum_{ij} e^{i\mathbf{k}\cdot(\mathbf{r}_i - \mathbf{r}_j)} \rangle .$$

In particular we have for fixed crystalline arrangement* $S(\mathbf{k}) = N\delta_{\mathbf{k},\mathbf{K}}$, where $\mathbf{K}$ belongs to the reciprocal lattice, and accordingly Eq. 6 will be expected to depend on structure. As it happens, however, for a fixed volume the differences in the Madelung energy (per electron) for the various structures turn out to be very small. It can be seen in Table I that (excluding the rather loose-packed simple cubic), the quantity (U_M/NZ) varies between the close-packed structures by less than a milli-Rydberg per electron** (for typical values of r_s, i.e. $2 \lesssim r_s \lesssim 6$). This energy difference is so slight that a model of point ions and a uniform compensating electron gas will produce an energy which for practical purposes is almost completely insensitive to structure. This conclusion remains valid if correction is made for the assumption that the electrons see the ions as point ions and the simple Coulomb interactions are replaced by a more realistic electron-ion interaction. We can preserve the notion of the electrons being essentially uniformly spread*** and at the same time substitute $-Ze/r$ by a potential accounting for the internal structure of the ion if the latter is taken**** to possess a weak pseudopotential V_{ps}. Evidently the potential energy calculated with the point-ion interaction is now corrected by an energy (per electron):

$$\Delta u = (ZN)^{-1} \int (n_e d\mathbf{r}) \sum_{\mathbf{R}} \Delta(\mathbf{r}-\mathbf{R}) = (-e)\frac{N}{V}\int d\mathbf{r}\, \Delta(\mathbf{r}) \qquad (7)$$

where the electron density is $n_e = NZ/V$, and

---

*The implied restriction to a Bravais lattice is straightforwardly extended to structures with a basis.

**The absolute value of U_M/ZN is remarkably close to the uniform sphere value and accordingly the equilibrium density will not change perceptibly over the value predicted by this model.

***There is a technical point here. It is the corresponding pseudo-wave function which now deposits the electron charge over the crystal. The pseudowave function has a slightly different normalization from the true wave function but the consequent effects of the change in the uniform component have been shown by Ballentine (Can. J. Phys. **46**, 2568 (1968)) to cancel out if the energy is subsequently calculated to second order in V_{ps}.

****At this point we are still concerned with the uniform component of the electron distribution and accordingly V_{ps} is a bare (i.e. unscreened) pseudopotential. In general it is also nonlocal and effects arising from these complications have been discussed by Shaw and Harrison, Phys. Rev. **163**, 604 (1967).

$$\Delta(\mathbf{r}) = V_{ps} - (-Ze/r).$$

This correction may be written

$$\Delta u = \beta/r_s^3 \quad (\text{Ry/electron})$$

where

$$\beta = \lim_{k \to 0} 6\left[\frac{1}{(a_0 k)^2} + \frac{V_{ps}(k)}{4\pi Zea_0^2}\right] \qquad (8)$$

and is a quantity* typically of order ~5. With the values of α listed in Table I, the energy per electron now becomes:

$$u = \frac{2.21}{r_s^2} - \frac{0.916 + \alpha Z^{2/3}}{r_s} - u_c(r_s) + \frac{\beta}{r_s^3}, \qquad (9)$$

and, as already noted the dependence on *structure* is virtually insignificant. However, with r_s in the range appropriate to simple metals (~2 < r_s < ~6) it is clear that Δu may be comparable with the kinetic energy term.[1] It therefore has a pronounced effect on the equilibrium density and immediately relates the corresponding r_s to the details of the internal structure of the ion.** Provided we are dealing with reasonably close packed structures the terms collected in Eq. 9 represent a volume-dependent energy and may thus be readily identified as contributors to the $\Phi_V(V)$ of Eq. 4.

The origin of the term Φ_S becomes clear when we consider the energy associated with the nonuniform components of the electron density. Departures in the valence bands from ideal free-electron bands*** give rise to additional terms in the energy referred to in a perfect solid as the band structure energy, U_{BS}. To calculate this term, suppose we have an arrangement of ions at positions $\mathbf{r}_i$: the Fourier transform of the ionic density function is $\rho_\mathbf{k}^{ion} = \sum_i e^{i\mathbf{k}\cdot\mathbf{r}_i}$. With each ion is associated an unscreened pseudopotential V_{ps} so that a Fourier component of the potential from the

---

*For the empty-core model of the pseudopotential (N. W. Ashcroft, J. Phys. C. **1**, 232 (1968)) the Fourier transform of V_{ps} is $V_{ps}(k) = (-4\pi Ze^2/k^2)\cos kR_c$ where R_c is a measure of the ion-core radius. This choice of pseudopotential leads to $\beta = 3(R_c/a_0)$.

**For example; in the alkali series from Eq. 9 the equilibrium r_s is given in terms of the ion-core radius[17] by

$$r_s \doteq 0.82 + 1.82 [R_c/a_0] [1 + 0(R_c^{-2})]$$

thus providing a simple explanation of the observed near linearity in r_s as a function of core size. In the special case of a point charge (for $Z = 1$ this would correspond to a metallic modification of hydrogen) the equilibrium value of r_s is 1.64.

***It is clear that the bands are taken as parabolic in Eq. 9.

Table I. Values of α used in Eq. 9 for crystals and liquids

Arrangement of Point Ions	Factor α multiplying $\dfrac{-Z^{2/3}}{r_s}$
Face centered Cubic[a]	1.79175
Hexagonal Close Packed[a,b]	1.79168
Body Centered Cubic[a]	1.79186
Simple Cubic[a]	1.76012
Liquid[c]	1.73

a. From C. A. Sholl, Proc. Phys. Soc. **92**, 434 (1967).
b. The Madelung energy depends on the c/a ratio. The value quoted here is the minimum for the hcp structure and falls close to the ideal c/a ratio.
c. Calculated with a hard-sphere structure factor for the liquid state (N. W. Ashcroft and D. C. Langreth, Phys. Rev. **155**, 682, 1967).

entire assembly of ions is $\rho_k^{ion} V_{ps}(k)$. The latter is responsible for the establishment of an induced electronic charge $\rho^{ind}(k)$, reflecting thereby the spatial arrangement of the ions. Through Poisson's equation and the dielectric function $\epsilon(k)$ for the electron system the induced charge is given by:

$$\rho_k^{ind} = \frac{k^2}{4\pi e}\left(\frac{1}{\epsilon_k} - 1\right) \epsilon \rho_k^{ion} V_{ps}(k) \qquad (10)$$

But this induced charge in the presence of the potential $\rho_k^{ion} V_{ps}(k)$ produces an average energy

$$U_{BS} = \frac{1}{2V} \sum_{k \neq 0} \rho_{-k}^{ind} \rho_k^{ion} V_{ps}(k) \qquad (11)$$

or, per electron:

$$u_{BS} = \left(\frac{1}{2NZV}\right) \sum_{k \neq 0} \frac{k^2}{4\pi} |V_{ps}(k)|^2 \rho^{ion}(k) \rho^{ion}(-k) \left(\frac{1}{\epsilon_k} - 1\right); \quad (12)$$

and, for a Bravais lattice this reduces to

$$u_{BS} = \frac{1}{3\pi^2} \sum_k \left(\frac{2k_F}{K}\right)^4 \left|\frac{K^2 V_{ps}(K)}{4\pi Ze}\right|^2 \chi(K)/\epsilon_K \qquad (13)$$

In Eq. 13 is written the dielectric function

$$\epsilon_k = 1 + 0.166 r_s \chi(k/2k_F) / (k/2k_F)^2$$

where the quantity χ (which is proportional to the polarizability) is dominated by the Lindhard function (viz; $\frac{1}{2} + \frac{1-x^2}{4x} \ln\left|\frac{1+x}{1-x}\right|$; $x = k/2k_F$)

but contains corrections for correlation and exchange.* Note that Eq. 13 may also be written

$$u_{BS} = \frac{1}{3\pi^2} \sum_K \left(\frac{2k_F}{K}\right)^4 \left|\frac{V_{ps}(K)}{V_c(K)}\right|^2 \chi(K)/\epsilon_K \quad \text{(Ry/electron)} \quad (14)$$

where $V_c(k)$ is the Fourier transform of the electron-ion potential *were* that ion a point charge. Since $\chi(x) \to \frac{1}{x^2}$ for large x, it follows that the sum in Eq. 13 is quite rapidly convergent** so that knowledge of $V_{ps}(K)$ for relatively few distinct values of K is sufficient for its computation. (This seemingly obvious point is an important factor favoring the formulation of the problem of the energy of a metal in reciprocal space.)

Although Eq. 13 is yet to be cast into an appropriate real space form, it is by far the major contribution to Φ_S in Eq. 4. That it is quite sensitive to valence follows from examining the relative magnitude of the first term in the summand. By way of example, in polyvalent metals $\left(\frac{2k_F}{K}\right) > 1$ whereas in the monovalent metals the reverse is generally true. When the pseudopotentials are known, u_{BS} and its volume derivatives are easily computed. It is then a straightforward matter to obtain theoretical estimates of the elastic constants, or by working at constant density, to test for configurations with the lowest energy.

The location (relative to $2k_F$) of the reciprocal lattice vectors **K**, their number and their weights ($\sim V_{ps}^2(K) \chi/\epsilon$) are clearly important, but although the contribution from u_{BS} can be appreciable (~ 0.1 Ry/electron) the differences *between* structures are generally very small (millirydbergs per electron or less). In the context of predicting the structures of metals the accuracy and validity of the arguments leading to Eq. 13 must be scrutinized with some care. The basic result is clearly a product of linear response theory and in fact can equally well be derived by summing the 1-electron band energies (diminished by half of the Hartree energy), these energies given to second order in V_{ps}. An obvious extension is consistently to pursue the calculation to third order in V_{ps} as has been done, for example, by Lloyd and Sholl[3], and Brovman and Kagan.[4] Not surprisingly the result now depends on the sign of $V_{ps}(K)$ and, if this quantity itself has a variable sign, considerable cancellation between terms with different K can result. It appears that the second-order term may

---

*As a result correct to second order in V_{ps}, (13) consistently includes many body effects incorporated through the dielectric function formulation (see, V. Heine and D. Weaire, Solid State Physics **24**, 249 (1970)).

**Although pseudopotentials are not known accurately for large k, they tend to fall off at least as quickly as k^{-2}.

suffice for many calculations of the ground state energy, but in derivative properties (elastic constants for example) contributions from the higher-order terms can be expected.[5]

4 EFFECTIVE ION-ION POTENTIALS

Bearing in mind the limitations just alluded to, Eq. 13 represents the most important term in the energy specifically related to the structure of a metal. From Eq. 12 the *total* band structure energy is:

$$U_{BS} = \frac{N}{2V} \sum_{k \neq 0} \frac{k^2}{4\pi} |V_{ps}(k)|^2 S_i(k) \left(\frac{1}{\epsilon_k} - 1\right)$$

whereas before $S_i(k)$ is the ionic structure factor, $N^{-1}\rho^{ion}(k)\rho^{ion}(-k)$. Now we can formally combine this U_{BS} with the total Madelung energy:

$$U_M = \frac{N}{2V} \sum_{k \neq 0} \frac{4\pi Z^2 e^2}{k^2} (S_i(k) - 1)$$

to get

$$\frac{N}{2V} \sum_{k \neq 0} \left[\frac{4\pi Z^2 e^2}{k^2} + \frac{4\pi Z^2 e^2}{k^2} \left| \frac{V_{ps}(k) k^2}{4\pi Z e} \right|^2 \left(\frac{1}{\epsilon_k} - 1\right) \right] (S_i(k) - 1) \quad (15)$$

plus a small term that depends only on volume and not on structure. Comparing Eq. 15 and 6, the structural energy can be interpreted as arising from an effective potential $\phi(r)$ whose Fourier transform is

$$\phi(k) = \frac{4\pi Z^2 e^2}{k^2} \left[1 + \left| \frac{V_{ps}(k) k^2}{4\pi Z e} \right|^2 \left(\frac{1}{\epsilon_k} - 1\right) \right]$$

If we choose to view the potential in real space then:

$$\phi(\mathbf{r}) = \frac{1}{V} \sum_{\mathbf{k}} e^{i\mathbf{k}\cdot\mathbf{r}} \left(\frac{4\pi Z^2 e^2}{k^2}\right) \left[1 + \left| \frac{V_{ps}(k) k^2}{4\pi Z e} \right|^2 \left(\frac{1}{\epsilon_k} - 1\right) \right] \quad (16)$$

and it is clear that in this combination we have taken the bare ion-ion potential (assumed Coulombic) and added to it contributions originating from the screening action of the interacting electron gas.

Some general comments follow. First, we reiterate that from the manner in which the ground-state energy has been derived, it cannot be reconstituted simply by summing over the $\phi(r)$. In fact, from Eq. 16 and 9 we have

$$U_{TOTAL} = \Phi_V(V) + \Phi_s(V)$$

where the structural part (which is also density dependent) is

$$\Phi_s(V) = 1/2 \sum_{R,R'} \Phi(R-R')$$

and the volume dependent part

$$\Phi_V(V) = NZ\left[\frac{2.21}{r_s^2} - \frac{0.916}{r_s} + \frac{\beta}{r_s^3} + u_s(r_s) + u_e(r_s)\right]$$

Here it is understood that all long-range components are compensated by the uniform component of the electron gas. The term $u_e(r_s)$ is a small structure-independent energy originating in linear response from the presence of the induced electron density associated specifically with a single ion.* Second; the behavior of the integrand in Eq. 16 is by no means as convergent as the summand in Eq. 13. An unfortunate consequence is that the details of $V_{ps}(k)$ at large k have an influence** on the resulting form of $\phi(r)$: usually $V_{ps}(k)$ is not well known at large k. Third; the pair potentials defined through Eq. 16 are volume dependent.

The characteristic form of $\phi(r)$ computed from Fermi surface determined pseudopotentials*** is shown for sodium in Fig. 2. For small separations of the ions, the potential is steeply rising, and there is a range where the potential is attractive. The location of the transition point between the two regions depends to some extent on V_{ps}. It is, however, considerably in excess of the usual ionic diameter. These features appear to be general in the simple metals. Furthermore, by structural rearrangement the nearest-neighbor positions frequently tend to fall near the minimum of the attractive region. This is not always true, since minimization of the total energy involves a consideration of the term $\Phi_V(V)$, and the requirement that the total energy (dominated by Φ_V) be a minimum may conflict with the location of the ions near the minima of the inter-ion potential. Note that for large separation of the ions the pair potential exhibits oscillatory features (Friedel oscillations) reflecting the weak singularity in the slope of the (zero temperature) ϵ_k, and this in turn is a manifestation of a sharp Fermi surface.

The nature of the potential or the bond is almost easier to classify by exclusion. It is clearly not simply heteropolar or ionic: if it were then as

---

*This term is proportional to $\sum_{k \neq 0} \frac{k^2}{4\pi}|V_{ps}(k)|^2\left(\frac{1}{\epsilon_k} - 1\right)$ and arises from replacing $S_i(k)$ in U_{BS} by $(S_i(k) - 1)$.

**As do the exchange and correlation corrections to ϵ_k. For large k, $\epsilon_k \approx 1$, and the quantity $\epsilon_k - 1$ (~χ) is sensitive to these terms.

***The $V_{ps}(k)$ for large k is extrapolated using the empty core potential.

Fig. 2. The form of the effective ion-ion interaction $\phi(r)$ in metallic sodium computed in the pseudopotential approximation (Ref. 11). The arrow denotes the nearest-neighbor location in the solid. In the range shown, the form is quite similar to Fig. 1, but for larger values of separation r, the potential $\phi(r)$ oscillates as can be seen in the inset in Fig. 3.

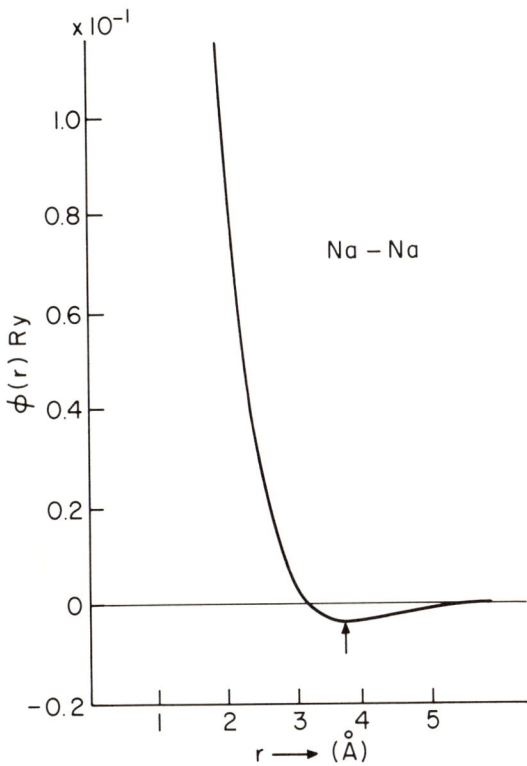

observed by Cohen[6] mass transport of the ions would also result in a net charge transport.* The requirements of charge neutrality and the effects of screening rule this out as a possibility. Nor is the bond simply covalent: if the $O(V_{ps}^2)$ theory is adequate then there is no directionality associated with the bonding (see Section 7). The possibility that dispersion forces are a serious contender has been disposed of already. In simple metals the bond represented by $\phi(r)$ has a character of its own; it is homopolar and specifically metallic.

Even at temperatures near the melting temperature, T_M, the short range part of a typical $\phi(r)$ remains steeply rising when expressed on an energy scale of, say, $k_B T_M$. Neutron or X-ray diffraction studies of simple insulating liquids (liquid argon is an example) and of liquid metals yield static structure factors $S_i(k)$ that are remarkably similar in form. We have already commented that for the insulating liquid $S_i(k)$ is explicable in terms of hard sphere theories[7] from which we may conclude that not only is the experimental evidence consistent with the short-range portion of the

---

*Cohen gives a similar argument to discount the possibility that the bond is covalent, viz; that if it were, then a group of ions covalently bonded would transport a charge differing from the total charge carried by the ions moving separately. Again it can be shown that the net charge transported is zero.

predicted ion-ion potentials, but that the ubiquitous hard-sphere approximation used also in the theory of liquid metal structure is physically reasonable.* (A simple measure of the hard-core dimension σ is given by the solution to

$$\frac{3}{2} k_B T + \phi_{min} \doteq \phi(\sigma) \qquad (17)$$

where ϕ_{min} is the minimum value of the effective ion-ion potential.**)

5 ALLOYS OF SIMPLE METALS

In many binary alloys of simple metals (as well as alloys of some not-so-simple constituents, e.g. the noble metals), structural changes are observed to occur at well defined concentrations which, as originally noted by Hume – Rothery[8], can be simply related to the average electron-to-atom ratio, being otherwise independent of the properties of the constituents. It has been suspected[9] that the stability of the alloys is connected with the singular properties of the dielectric function ϵ_k and recently Stroud and Ashcroft[10] have confirmed that this is indeed the case.

To calculate the ground-state energy of the alloy (composed, let us say, of fixed atoms of type A, charge Z_A, concentration c and atoms of type B, charge Z_B and concentration $(1-c)$, we retrace the steps leading from Eq. 5 to Eq. 13. The electron gas terms are now determined by the average electron density and the Madelung energy, per electron is:

$$\tfrac{1}{2} (NZ^*)^{-1} \sum_{R \neq R'} e^2 \frac{Z_R Z_{R'}}{|R - R'|} \qquad (18)$$

with $Z^* = cZ_A + (1-c)Z_B$ being the mean valence and the sum is taken over the sites of an assumed structure. On the further assumption that there is no short-range correlation (i.e., $\langle Z_R Z_{R'} \rangle = (Z^*)^2$) then the Madelung energy is simply that of a pure metal having the same structure but with an ion of valence Z^* on each site.

The band structure energy, on the other hand, is (by the arguments leading to Eq. 11):

$$\tfrac{1}{2} (NZ^* V)^{-1} \sum_{k \neq 0} \left(\rho_k^A V_{ps}^A (k) + \rho_k^B V_{ps}^B (k) \right) \rho_{-k}^{ind} , \qquad (19)$$

---

*Because of the presence of the Φ_v, however the ions are also moving relative to a large negative energy, a term not present in an insulating fluid.

**It is found that for most liquid metals slightly above their melting points that $(\pi/6) (N/V) \sigma^3 =$ constant, which simply expresses the fact that effective volume occupied by the ions (as determined by Eq. 17) is the same constant fraction of the atomic volume. (For a discussion of this point see D. Stroud and N. W. Ashcroft, Physical Review, to be published.)

where*:

$$\rho_{-k}^{ind} = \frac{k^2}{4\pi e}\left(\frac{1}{\epsilon_k} - 1\right)\left(\rho_{-k}^A eV_{ps}^A(-k) + \rho_{-k}^B eV_{ps}^B(-k)\right) \quad (20)$$

Combining Eq. 20 with Eq. 19 we arrive at an energy per electron:

$$u_{BS} = \frac{1}{3\pi^2}\sum_k \frac{(2k_F)^4}{K}\left|\frac{K^2\bar{V}_{ps}(K)}{4\pi Z^*e}\right|^2 (\chi(k)/\epsilon_k)|S_g(K)|^2 \quad (21)$$

where

$$\bar{V}_{ps}(k) = cV_{ps}^A(k) + (1-c)V_{ps}^B(k)$$

and the geometric structure factor $S_g(K) = n_b^{-1}\sum_b e^{i\mathbf{k}\cdot\mathbf{b}}$ is included to account for the possibility that the structure possesses an n_b atom basis.

The question of phase stability now rests principally on the relative magnitudes of Eq. 21 for different structures, and several systems have been analyzed from this standpoint.[10] On examining the outline of the method, it is again apparent that the problem is quite naturally tackled in reciprocal space. However within the limitations mentioned earlier, the behavior of effective ion-ion potentials may also be investigated in the context of alloys. Given ions of charge Z_A and Z_B the counterpart of Eq. 16 is now:

$$\phi_{AB}(r) = \frac{Z_A Z_B}{r}\left[1 - \frac{2}{\pi^2}\int dk \frac{\sin kr}{k}\left(\frac{k^2 V_{ps}^A(k)}{4\pi Z_A e}\right)\left(\frac{k^2 V_{ps}^B(k)}{4\pi Z_B e}\right)\left(\frac{1}{\epsilon_k} - 1\right)\right], (22)$$

where ϵ_k is calculated with a mean electron density appropriate to the entire assembly. From the form of Eq. 22 we expect the $\phi_{AB}(r)$ to display features similar to those shown in Fig. 2 (which refers to identical ions). In systems where the ϕ_{AB} have been calculated[11], the hard-core nature of the potential persists and the dimension characterizing the interaction between unlike atoms is nearly the average of the corresponding quantities for like pairs. The $\phi_{AB}(r)$, again being volume dependent, change with concentration**. This effect may be of some consequence in the theory of defects in alloys. Structural changes are anticipated to occur as the ions attempt (at fixed volume) to move away from relatively large mutual potentials produced in alloying.

---

*$\rho^{A,B}(k)$ are Fourier transforms of the density of ions of type A,B. The average indicated by $\langle\rangle$ is again taken assuming no short-range correlation.
**See Figures 3 through 6 of Ref. 10.

6 EFFECTS OF TEMPERATURE

To this point, the ions, if in a crystal, have been assumed to be fixed. In dealing with structural energies, the effects of temperature arise directly in the structure factor:

$$S_i(k) = N^{-1} \langle \rho_{\mathbf{k}}^{ion} \rho_{-\mathbf{k}}^{ion} \rangle = N^{-1} \sum_{i,j} \langle e^{i\mathbf{k}\cdot\mathbf{r}_i} e^{-i\mathbf{k}\cdot\mathbf{r}_j} \rangle$$

where the $\mathbf{r}_i$ are now the instantaneous position of the ions in a dynamic lattice and the average is taken over the various possible thermal excitations of the crystal. If the excursion of an ion from a given site, say $\mathbf{R}$, is $\mathbf{u}(\mathbf{R})$ then (assuming the electron response to the ionic motion is adiabatic)

$$S_i(k) = N^{-1} \sum_{\mathbf{R}\mathbf{R}'} e^{i\mathbf{k}\cdot(\mathbf{R}-\mathbf{R}')} \langle e^{i\mathbf{k}\cdot\mathbf{u}(\mathbf{R})} e^{-i\mathbf{k}\cdot\mathbf{u}(\mathbf{R}')} \rangle ,$$

and from Eq. 6 to Eq. 13 the combined Madelung and band structure energies are:

$$\frac{N}{2V} \sum_{\mathbf{k}} \left(N^{-1} \sum_{\mathbf{R}\mathbf{R}'} e^{i\mathbf{k}\cdot(\mathbf{R}-\mathbf{R}')} \langle e^{i\mathbf{k}\cdot\mathbf{u}(\mathbf{R})} e^{-i\mathbf{k}\cdot\mathbf{u}(\mathbf{R})} \rangle \cdot \frac{4\pi Z^2 e^2}{k^2} \left[1 + \left(\frac{V_{ps}(k)k^2}{4\pi Z e} \right)^2 \left(\frac{1}{\epsilon_k} - 1 \right) \right] \right) \quad (23)$$

Stroud and Ashcroft[12] show that if the $\mathbf{u}(\mathbf{R})$ are now expressed in phonon coordinates the resulting expression can be approximated by:

$$\frac{N}{2V} \sum_{\mathbf{k}} N^{-1} \sum_{\mathbf{R}\mathbf{R}'} e^{i\mathbf{k}\cdot(\mathbf{R}-\mathbf{R}')} e^{-k^2 \rho_{\mathbf{R}-\mathbf{R}'}^2} \cdot \left[\frac{4\pi Z^2 e^2}{k^2} + \frac{4\pi Z^2 e^2}{k^2} \left(\frac{V_{ps}(k)k^2}{4\pi Z e} \right)^2 \left(\frac{1}{\epsilon_k} - 1 \right) \right] ,$$

where, for ions of mass M,

$$\rho_R^2 = (2NM)^{-1} \sum_q \omega_q^{-1} (2n_q + 1)(1 - \cos \mathbf{q}\cdot\mathbf{R}) \quad (24)$$

In Eq. 22, ω_q and n_q are frequency and occupation number of a phonon* of wave-vector q. Now Eq. 23 can be recast in the form:

$$(U_M + U_{BS})_T = \tfrac{1}{2} \sum_{\mathbf{R}\neq\mathbf{R}'} \phi^T(\mathbf{R}-\mathbf{R}')$$

where

$$\phi^T(\mathbf{R}-\mathbf{R}') = \frac{1}{V} \sum_{\mathbf{k}} e^{i\mathbf{k}\cdot(\mathbf{R}-\mathbf{R}')} \left[\frac{4\pi Z^2 e^2}{k^2} e^{-k^2 \rho_{\mathbf{R}-\mathbf{R}'}^2} + \frac{4\pi Z^2 e^2}{k^2} e^{-k^2 \rho_{\mathbf{R}-\mathbf{R}'}^2} \left(\frac{V_{ps}(k)k^2}{4\pi Z e} \right)^2 \left(\frac{1}{\epsilon_k} - 1 \right) \right] , \quad (25)$$

---

*For simplicity it is assumed here that ω_q is independent of polarization.

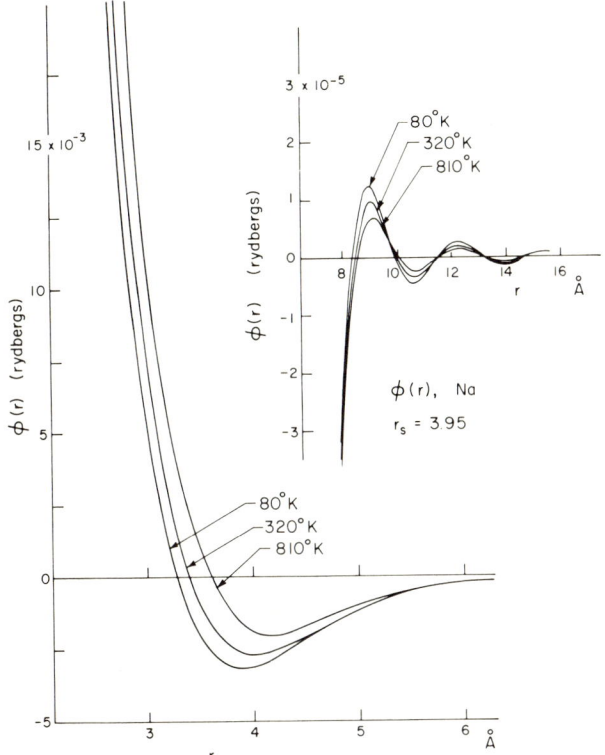

Fig. 3. Effective temperature-dependent ion-ion potentials in metallic sodium incorporating lattice dynamical effects but interpreted as acting between ions rigidly attached to the sites of a perfect lattice (Ref. 12). There is a substantial change in scale for the long-range oscillatory part of $\phi^T(r)$.

which formally has the appearance of an effective ion-ion interaction acting between ions in their static positions. The significant aspect of Eq. 25 is the appearance of a temperature dependence in the ion-ion potential which can be traced ultimately to the phonon occupation factor. Its effects are shown in Fig. 3 for the ion-ion interaction in sodium. Eq. 25 represents a crucial ingredient in the theory of melting. In constructing the Gibbs free energy we must add (as well as the electron gas terms) the phonon kinetic energy and entropy. The appropriate terms can also be calculated for the liquid, and it follows that it is possible to combine the information and predict melting curves for metals. In the present context, however, a more interesting question (bearing somewhat on aspects of the phonon-dislocation interaction) is to what extent dynamic effects in defected crystals can be incorporated into temperature-dependent ion-ion potentials.*

---

*From the example given in Fig. 3, as T increases the effective hard core seems to increase slightly but at the same time the interaction viewed as a whole appears to soften.

7 DISCUSSION

Within the limitations imposed by the use of linear-response theory, we have seen that the total structurally dependent energy of a metal can be constructed from a sum of pair-wise ion-ion potentials. Their form depends to some extent on the nature* of the pseudopotential V_{ps} but, where this has been taken from experiment (e.g., from fitted Fermi-surface data) and reasonably (but on occasion dubiously) extrapolated into the normally unknown region, the resulting $\phi(r)$ exhibit both the steeply rising and weakly attractive features normally associated with other (nonmetallic) solids. In the construction of the total energy it is apparent that the formulation proceeds straightforwardly in reciprocal space, and, indeed when this is done, the band-structure energy can be calculated with some precision since only a few of the lower Fourier components of V_{ps} are required. Formal resolution of this energy and the Madelung energy into a sum over pair terms can be given physical justification, but the conceptual gain of real-space functions has to be weighed against a possible loss in numerical accuracy — a point that has to be remembered in attempts to relate defect properties to fundamental ion-ion interactions. The nature of ϕ has been shown to be characteristically metallic. This conclusion is linked rather strongly to the foregoing analysis which casts out terms of higher order in V_{ps} than those of Eq. 13. Naturally, if the band gaps in the system are quite large, this result is expected to fail, and to lead to terms which even at third order in V_{ps} depend on angles between triplets of atoms. In the "bond" picture these effects are also encountered in covalent solids, and seen here is the kernel of an approach relating large band gaps to covalency[13].

The appearance of the "hard-core" in the effective ion-ion interactions in simple metals fits well with the notion of the packing of relatively rigid objects in the formation of metallic structures. Since it is partially related to the screening action of the electrons it may be expected, in a linear theory, that the ions will always carry this effective potential with them, and the $\phi(r)$ to be of (qualified) use in problems going beyond those of perfect or slightly disturbed lattices.

In the simple metals, the location of the strongly repulsive part of the ion-ion interaction is sufficiently far removed from the region in which core-core repulsion becomes important that the bare ion-ion interaction can simply be taken as a Coulomb law. This fortunate simplification is lost as soon as attention is turned to the transitional and noble metals. Compared with those for simple metals, the band structures of these metals are vastly more complex. To calculate their ground-state energies the single particle band energies are summed**, whether for complex bands

---

*The successful application of a given pseudopotential to the determination of, say, the phonon spectrum is no guarantee that the corresponding $\phi(r)$ can be viewed as necessarily correct.

**Corrected of course for one half of the Hartree energy.

this form can be expressed as a sum-over-pair potential in real space, has yet to be seen. If the d-bands are sufficiently narrow that they can be represented by Bloch tight-binding functions, then, as has already been suggested[14], it may be possible to incorporate (semi-empirically) the effects of core-core exchange and polarization. The energy contributions from the part of the bands readily identifiable as s-like can be treated by the methods already outlined. (Contributions from hybridization will cancel if the Fermi energy is sufficiently large that the mixed states are symmetrically disposed about the centroid of the d-bands.) To $\phi(r)$ in Eq. 16 should be added the potential arising from core-core repulsion and the longer-ranged time-averaged fluctuating dipole term, both corrected now for the screening action of the s-electrons. Over a limited range of ionic separation we may expect the leading terms* in an expansion of this energy to be given by $Cr^{-m} - Dr^{-n}$ the constants depending on the size and polarizability of the d-shell in question. In some instances[14] it may be possible to determine these parameters from the equilibrium properties of the metal (e.g., equilibrium density, compressibility, and other elastic constants) so that the ion-ion potentials can then be used in calculations in lattice dynamics and crystal defects.

ACKNOWLEDGMENTS

It is a pleasure to thank Dr. D. Stroud for many helpful and stimulating discussions. Part of this article was prepared at Victoria University and I wish to express my gratitude to the members of the Physics Department for their kind hospitality. This work was supported in part by the Advanced Research Projects Agency through the Materials Science Center at Cornell University (Report No. 1644).

REFERENCES

1. N. W. Ashcroft and D. C. Langreth, Phys. Rev., **155**, 682 (1967).
2. M. L. Cohen and V. Heine, Solid State Physics **24**, 37 (1970).
3. P. Lloyd and C. A. Sholl, J. Phys. C. (Proc. Phys. Soc.) **1**, 1620 (1968).
4. E. G. Brovman and Y. Kagan, J.E.T.P., **25**, 365 (1967).
5. E. G. Brovman and G. Solt, Solid St. Comm., **8**, 903 (1970).
6. M. H. Cohen, *Metallic Solid Solutions,* (J. Friedel and A. Guinier, Editors), W. A. Benjamin, Inc. (1963).
7. N. W. Ashcroft and J. Lekner, Phys. Rev. **145**, 83 (1966); N. W. Ashcraft, Physica, **35**, 148 (1967).
8. W. Hume-Rothery; *The Metallic State* (London, Oxford University Press), 1931.
9. A. Blandin, "Phase Stability in Metals and Alloys" Eds. R. S. Rudman, J. Stringer and R. I. Jaffee (McGraw Hill, New York) 1966; V. Heine, ibid, 1966; and "The Physics of Metals", Vol 1, Ed. J. M. Ziman (Cambridge University Press, London) 1969.

---

*This, of course, begs the question of the importance of the three-body (and higher) terms in a situation where the cores are in "contact".

10. D. Stroud and N. W. Ashcroft, J. Phys. F. (Proc. Phys. Soc.) 1971.
11. N. W. Ashcroft and D. C. Langreth, Phys. Rev. **159**, 500 (1967).
12. D. Stroud and N. W. Ashcroft, to be published.
13. J. C. Phillips, Phys. Rev. **166**, 832 (1966).
14. N. W. Ashcroft, Proceedings of the International Conference on Dislocation Theory, National Bureau of Standards, 1970.

DISCUSSION on the Paper by N. W. Ashcroft

TORRENS: Can you comment on the sensitivity of the real-space pair potential to small variations in the dielectric screening function. Supposing that a reliable pair interaction has been obtained for a perfect crystal, it seems that the use of this potential in a defect situation where the electron distribution is altered requires that V_s should be relatively insensitive to ϵ_k.

ASHCROFT: I think there are two aspects to consider here. First the $\phi(r)$'s are quite sensitive to uncertainties in exchange and correlation corrections, to ϵ_k particularly in the $2k_F \leq k \leq$ few k_F region. To give an example (admittedly rather extreme) it is possible in some instances to eliminate the principal minimum in ϕ by ignoring exchange and correlation corrections (i.e., using a simple Hartree dielectric function). Second, as has already been mentioned, the density dependence of ϵ_k shows itself in $\phi(r)$. The magnitude of this can be gauged from the curves given in Ref. 11. The effect can be substantial. I think that you may be hinting that near a defect the local background electron density may change from its average value and produce concomitant changes in $\phi(r)$. From the way the calculation has been formulated I think it is clear that this is not the case. It is worth keeping in mind the efficacy of the long-range Coulomb interaction and its role in maintaining overall charge neutrality.

KANZAKI FORCES AND ELECTRON THEORY OF DISPLACED CHARGE IN RELAXED DEFECT LATTICES

N. H. March and J. S. Rousseau

Department of Physics
The University
Sheffield
England

ABSTRACT

Almost all work concerned with the electron theory of point defects in metals has dealt with the displaced charge under the assumption that lattice relaxation is neglected.

This paper develops a theory whereby the displaced charge around a defect can be calculated in the presence of such relaxation. The calculations require knowledge of:

(a) Appropriate response functions for the perfect crystal. These, inevitably, can be found only approximately, and a procedure is adopted whereby direct use is made of the intensities of X-ray scattering at Bragg reflections for the perfect crystal.

(b) A defect potential. This defect potential is defined as follows. We first strain the perfect lattice so that all the atomic positions, except for those at which the point defects reside, are as in the final equilibrium position of the defect lattice. Let the density in this strained but otherwise perfect lattice (Kanzaki lattice) be the sum of the perfectly periodic lattice density $\rho_0(r)$ and a perturbation $\rho_1(r)$. The defect potential $V_d(r)$ is then, by definition, that scattering potential required to convert the density $\rho_0(r) + \rho_1(r)$ in the Kanzaki lattice into the final state density $\rho_f(r) = \rho_0(r) + \rho_1(r) + \rho_d(r)$ with the defect introduced.

The central result of this paper then is a form of $\rho_d(r)$ which is a sum of two terms:

(a) The displaced charge due to the potential $V_d(r)$ introduced into the perfect lattice, but with the nuclei held fast.

(b) A term correcting this, which is shown to be rather closely related to $\rho_1(r)$ above, but involves also the defect potential $V_d(r)$. It is also shown how a first approximation to this term presented here can be systematically refined.

Some preliminary numerical results for $\rho_1(r)$ for a vacancy in copper are presented.

1 INTRODUCTION

Especially in metals, there has been a good deal of interest in the theory of charged defects; for example a vacancy in copper, or magnesium atoms in dilute concentrations in lithium. So far, however, no very careful account of the detailed relaxations of the lattice round the defect or impurity site has been incorporated in such electron-theory calculations, although attempts have been made to include some account of relaxations by modifying the Friedel sum rule, which is frequently used in defect calculations to simulate self-consistency.

The electron theory of defects in metals has proved to make interesting predictions about, for example, charged defect interactions, especially in polyvalent metals, which appear to agree in general terms with experiment (see, for example, the review by the present writers[1]).

In a substantial body of work, a markedly different philosophy is adopted. Here, relaxation effects are studied as the prime objective, usually on the basis of pair forces. Unfortunately, such an approach often fails to include a proper account of electron redistribution caused by the introduction of defects or impurities into the lattice. Although this is a serious limitation, the approach has practical merit in that it allows local relaxations around defects to be estimated, in a way which matches these on to the correct, long-range lattice displacements given by elasticity theory, at large distances from the defect. Indeed, quite a body of work gives estimates of the displaced positions of lattice atoms about a vacancy, or a divacancy, in a number of metals.

Such calculations, made with plausible pair forces, show that even when a vacancy is created in an open body-centered cubic structure like sodium metal, the atomic movements are really quite small. This fortunate circumstance had been anticipated by Kanzaki[2] (see also Matsubara[3]) in his method of lattice statics. Here, the idea is to simulate the effect of a vacancy or impurity in a lattice by regarding the originally perfect lattice as strained until it takes up the relaxed configuration appropriate to the equilibrium atomic positions in the defect lattice. Obviously, external forces must be applied to hold the other atoms in their displaced positions. Here these external forces are called Kanzaki forces and the strained, but otherwise perfect, lattice is called the Kanzaki lattice. Kanzaki assumed that all the displacements, u_ℓ say, from the perfect lattice positions r_ℓ, were so small that one need work only to first-order in u_ℓ.

Thus, on the one hand, is the electron theory, in which a basic quantity characterizing the defect is the localized charge it displaces (as a specific example to be referred to again later, see the calculations of Stott, et al.[4], on the charge displaced around vacancies in close-packed polyvalent metals) and, on the other, the method of lattice statics, with the strained Kanzaki lattice held in the relaxed final configuration by external forces.

Although the present paper is concerned with the electronic structure of defects, the idea of the Kanzaki lattice plays a central role in setting up the theory. Specifically, this allows development of a means for treating the difference between the charge displaced by a given defect potential in the unrelaxed lattice and that displaced by the same defect potential inserted in the Kanzaki lattice. Some preliminary results for a vacancy in copper metal are given to illustrate the approach.

It can be argued that, to set up the Kanzaki lattice, the relaxations must be known at the outset. This is true, and starting estimates such as those referred to above from pair potential studies must be used. However, once the charge displaced in the Kanzaki lattice by the defect potential is known, the electric field at the ionic positions in the Kanzaki lattice can be calculated; and, hence, from Feynman's theorem, the Kanzaki forces. If these are inconsistent with the original displacements u_ℓ, Kanzaki's method can then be used to generate a new set of displacements. The new displaced charge must then be found. Clearly, in principle, a self-consistent result can be obtained through iteration.

Thus, at least in principle, the theory given here can solve the problem of the electron distribution around a defect in a correctly relaxed lattice, but a number of practical issues remain to be resolved, which will, almost certainly, involve major electronic computation. Of these, a prime problem is that, to obtain the electronic structure of a defect center, a great deal of information is needed about the perfect lattice. That information is not yet in a suitably explicit form for any metal; but, as shown in Section 3 below, a good deal of it can be gotten from well established methods in the band theory of perfect crystals. The second difficulty resides in the fact that methods of solving the problem of scattering off a defect potential in the unrelaxed lattice are still troublesome to apply. However, when both the perfect crystal, and the defect potential, can be constructed from muffin-tin potentials, we have the rigorous one-electron theory of Beeby[5], while for three classes of non-muffin-tin defect potentials we have available the explicit approximations given by Stoddart, et al.[6] The third difficulty is that we do not know precisely what potential to take to describe the defect. But at least, we can give an operational definition of what that potential must be. (see Section 4).

The paper concludes with a brief discussion of the way defect energetics might be studied within the present framework.

A convenient starting point from which to explain the basic ingredients needed in the electron theory is to study how the electron density $\rho_0(r)$ in the perfect crystal, which is accessible to experiment via the intensities of X-ray scattering at the Bragg reflections, is changed when we strain the crystal to form the Kanzaki lattice.

2 ELECTRON DENSITY IN KANZAKI LATTICE

Starting with a perfect crystal, with atoms at lattice points l, the lattice is strained, introducing neither foreign atoms, nor vacancies.

The method used to generate the electron density in this Kanzaki lattice is that used by Jones and March[7] (referred to subsequently as JM) in their theory of lattice dynamics. The theory is immediately useful within the framework of the Kanzaki method of lattice statics, provided only that the displacements from the sites r_ℓ, say u_ℓ, are sufficiently small that we need work only to first-order in the displacements. This is a basic assumption of the approach, and it will have to be checked for consistency in any application. As stressed above, however, even with a vacancy in an open structure, it is found in practice that relaxations are often quite a small fraction of the lattice parameters, and the present approach should be widely applicable.

Let the change in electron density when the lattice is 'strained' be $\rho_1(r)$. Then, to first-order in the displacements, the theory of JM allows us to write

$$\rho_1(r) = \sum_\ell u_\ell \cdot R(r-r_\ell) \tag{1}$$

The vector **R** satisfies an integral equation given by JM, which involves knowledge of the exchange and correlation energy of the inhomogeneous gas of electrons in the crystal (see Appendix 1 for some relevant details). However, it is important to note that **R(r)** evidently also determines the gradient of the perfectly periodic lattice density $\rho_o(r)$ through

$$\nabla \rho_o(r) = \sum_\ell R(r-r_\ell) \quad . \tag{2}$$

Information about **R(r)** in Fourier transform is therefore available at the lattice vectors **K**. Here, in fact R_K is determined uniquely by the intensity of X-ray scattering at the Bragg reflections, as discussed in JM. In particular

$$R_K = i K \rho_K \quad . \tag{3}$$

where ρ_K is the Fourier component at **K** of the charge density $\rho_o(r)$.

Thus, it is clear that if we know, from say a pair potential study, a first approximation to the u_ℓ's and if, as discussed above, these relaxations turn out to be sufficiently small, we can generate the density $\rho_o(r) + \rho_1(r)$ at any point in the Kanzaki lattice from a knowledge of the quantity **R** characteristic of the perfect lattice.

As an example to illustrate this method, we consider in Appendix 3 some preliminary numerical results for **R** in copper metal, from which $\rho_1(r)$ for a vacancy in Cu metal can be estimated using the displacements u_ℓ calculated by Tewordt.[8]

However, it is clear that, though $\rho_1(r)$ is a basic quantity needed in the theory, a good deal more information is required in order to generalize the approaches of Beeby[5] or Stoddart, et al.[6] to deal with perturbations in the Kanzaki lattice. Next, the Green function or density matrix describing the Kanzaki lattice is discussed. The discussion of the density above is, of course, a special case of this more general treatment. The above discussion is exact to first-order in the displacements if the properties of the perfect lattice, as summarized in **R(r)**, are known.

3 GREEN FUNCTION AND/OR DENSITY MATRIX IN KANZAKI LATTICE

As shown below, we can set up the Green function of the Kanzaki lattice, from that of the perfect lattice, $G_o(r\ r'\ E)$ say, the latter being constructed such that the exact electron density $\rho_o(r\ r\ E_f)$, with E_f the Fermi energy, is correctly incorporated.

Below we shall work with the canonical density matrix $C_o(r\ r'\ \beta)$ for the perfect lattice, defined by

$$C_o(r\ r'\ \beta) = \sum_{\gamma k} \psi^*_{\gamma k}(r)\ \psi_{\gamma k}(r')\ e^{-\beta E_\gamma(k)} \quad , \qquad (4)$$

where the Bloch wave functions $\psi_{\gamma k}(r)$ and corresponding eigenvalues $E_\gamma(k)$ are generated by the periodic potential $V_p(r)$ which gives by definition the exact ground-state density in the perfect unstrained lattice. If we take the Laplace transform of C_o with respect to β, and call the transformed variable $-E$, then we regain the Green function $G_o(r\ r'\ E)$ immediately, so that whether we work with C or G is purely a matter of mathematical convenience.

Following JM we can write for the strained lattice, a perturbing potential $\Delta V(r)$ having the form

$$\Delta V(r) = \sum_\ell u_\ell \cdot P(r-r_\ell) \quad . \qquad (5)$$

From JM (see also Appendix 1), we know that

$$R(r) = \int P(r)\ F(r\ r')\ dr' \quad , \qquad (6)$$

where F is a one-body response function given by (cf Appendix 1)

$$\frac{\partial F}{\partial E} = 2Re\left[G_o(r\ r'E)\ \frac{\partial \rho_o(r\ r'\ E)}{\partial E} \right] \qquad (7)$$

and $\rho_o(r\ r'\ E)$ is the Dirac density matrix of the perfect lattice. Once the periodic potential V_p is known, ρ_o and G_o can be calculated purely from the Bloch functions $\psi_{\gamma k}(r)$ and the energies $E_{\gamma k}$.

It is clear from Eq. 6 that knowledge of the one-body response function $F(r\ r')$ plus $P(r)$ is equivalent to knowing $R(r)$.

To generate the density matrix in the Kanzaki lattice, we now need to solve the Bloch equation

$$HC = -\frac{\partial C}{\partial \beta} \quad , \qquad (8)$$

where

$$H = -\frac{1}{2}\nabla^2 + V_p(r) + \Delta V(r) = H_o + \Delta V \quad , \qquad (9)$$

subject to the usual delta-function boundary condition $C(r\ r_o\ o) = \delta(r-r_o)$ expressing the completeness of the eigenfunctions. But H_o generates the density matrix C_o and ΔV is a perturbation, from Eq. 5, provided the u_ℓ's are small. Thus, we can write, to first-order in ΔV (per Ref. 9)

$$C_{Kanzaki}(r\,r'\,\beta) \equiv C_K(r\,r'\,\beta)$$

$$= C_o(r\,r'\,\beta) - \int_o^\beta d\beta_1 \int dr''\, C_o(r\,r''\,\beta-\beta_1)\, \Delta V(r'')\, C_o(r''\,r'\,\beta_1)\,, \quad (10)$$

which determines the density matrix or Green function of the Kanzaki lattice in terms of C_o and the perturbation ΔV, which is in turn given by Eq. 5.

Eq. 10 is the basic result from which we must build the solution for the defect lattice.

4 SOLUTION FOR DEFECT LATTICE

We now introduce the defect, and to be specific, we shall assume we create a vacancy at the origin. Let us suppose that the electron density $\rho_k(r) \equiv \rho_o(r) + \rho_1(r)$ in the Kanzaki lattice changes to $\rho_f(r)$, the density in the final state containing the relaxed vacancy.

4.1 Operational Definition of Defect Potential

We tacitly assumed above that we could generate the exact ground-state density $\rho_o(r)$, as observed say in X-ray scattering, from a one-body periodic potential $V_p(r)$: i.e.

$$\rho_o(r) = \sum_{\substack{\gamma k \\ \text{(occupied)}}} \psi^*_{\gamma k}(r)\, \psi_{\gamma k}(r)\,. \quad (11)$$

This essentially follows from the considerations of Hohenberg and Kohn[10] and Kohn and Sham[11]. An operational procedure to construct $V_p(r)$ from a given $\rho_o(r)$ has been discussed rather fully by Stoddart et al.[12]

The same argument, in essence, now enables us to define operationally the defect potential. We wish to find a potential $V_d(r)$ which, when added to the one-electron Hamiltonian H in Eq. 9, yields the *exact* final state density $\rho_f(r)$.

We want to stress that $V_d(r)$ is a potential to be added to H, *not* to H_o. Thus, $V_d(r)$ is generated in a system in which the ionic configurations differ only by the removal of the ion at the origin. If we had defined the defect potential as a change from H_o, all the ions would have moved and $V_d(\mathbf{r})$ would be a much more complex object.

For the vacancy, especially in polyvalent metals, we must not expect that the effect of $V_d(\mathbf{r})$ can be treated as a perturbation. The Bloch equation (Eq. 8) with H replaced by

$$H_f = H_o + \Delta V + V_d(\mathbf{r}) \tag{12}$$

can again be written as an integral equation

$$C_f(\mathbf{r}\,\mathbf{r}'\,\beta) = C_K(\mathbf{r}\,\mathbf{r}'\,\beta)$$
$$-\int_0^\beta d\beta_1 \int d\mathbf{r}''\, C_K(\mathbf{r}\,\mathbf{r}''\,\beta-\beta_1)\, V_d(\mathbf{r}'')\, C_f(\mathbf{r}''\,\mathbf{r}'\,\beta_1) \tag{13}$$

where now the object C_f which is required appears also on the right side of Eq. 13. In principle, knowing C_K from Eq. 10, we can obtain C_f from Eq. 13, with an assumed defect potential. In practice, an iterative scheme[13] would have to be used and the procedure is certainly very lengthy and somewhat troublesome. We shall therefore consider below two approximate methods which allow us to solve (or in the second method to circumvent) the integral equation (13).

The idea has already been made plain. Physically, we anticipate that the diagonal difference $C_f(\mathbf{r}\,\mathbf{r}\,\beta) - C_K(\mathbf{r}\,\mathbf{r}\,\beta) \equiv \Delta C_{fK}$ can be usefully split into two parts, $\Delta C_{uo} = C_u(\mathbf{r}\,\mathbf{r}\,\beta) - C_o(\mathbf{r}\,\mathbf{r}\,\beta)$, where C_u is the Bloch density obtained by introducing the defect potential into the perfect lattice with Bloch density C_o, plus another term δC taking relaxation into account.

4.2 Correction to Displaced Charge as Calculated in Unrelaxed Lattice

A rough approximation to estimate δC can be given as follows. Write, with the assumption that $V_d(\mathbf{r})$ varies slowly in space (cf Eq. 4)

$$C_f(\mathbf{r}\,\mathbf{r}\,\beta) = C_K(\mathbf{r}\,\mathbf{r}\,\beta)\, e^{-\beta V_d(\mathbf{r})} \tag{14}$$

and

$$C_u(\mathbf{r}\,\mathbf{r}\,\beta) = C_o(\mathbf{r}\,\mathbf{r}\,\beta)\, e^{-\beta V_d(\mathbf{r})} \tag{15}$$

Thus, from the definitions given above,

$$\Delta C_{fk} = C_K \left[e^{-\beta V_d} - 1 \right] \quad (16)$$

and

$$\Delta C_{uo} = C_o \left[e^{-\beta V_d} - 1 \right] . \quad (17)$$

Therefore it follows that

$$\delta C = \left[e^{-\beta V_d} - 1 \right] \left[C_K - C_o \right] = \left[e^{-\beta V_d} - 1 \right] \Delta C_{Ko} . \quad (18)$$

Here then, we have a rough, but quite practicable, way of estimating the correction to the displaced charge in the unrelaxed lattice, due to relaxation, since ΔC_{Ko} is given explicitly by Eq. 10.

In terms of densities, we could alternatively write

$$\rho_f (r\ E) \doteq \rho_K (r\ E - V_d(r)) \quad (19)$$

and

$$\rho_u (r\ E) \doteq \rho_o (r\ E - V_d(r)) . \quad (20)$$

We stress that we only use these forms to estimate the correction to the displaced charge $\Delta \rho_{uo} = \rho_u(r\ E) - \rho_o(r\ E)$ due to relaxation. Then we have

$$\Delta \rho_{fK} - \Delta \rho_{uo} = \delta \rho = \rho_K(r\ E - V_d(r)) - \rho_K(r\ E) - \rho_o(r\ E - V_d(r))$$

$$+ \rho_o(r\ E) = \int dr'\ \Delta V(r') \left[F(r\ r'\ E - V_d(r)) - F(r\ r'\ E) \right] \quad (21)$$

which follows from the definition of the response function F. Here then is a first approximation to correct the displaced charge in the unrelaxed lattice*. Since $\Delta V(r')$ is given by Eq. 5, it is clear that $\delta \rho$ is 0 (u_ϱ), as required.

Further comments on Eq. 21 include: through the potential $\Delta V(r)$, determined by perfect lattice properties plus the u_ϱ's, there is an obvious link with lattice dynamical properties. The displaced charge $\Delta \rho_{uo}$ due to $V_d(r)$ inserted into the unrelaxed lattice is to be corrected by the appropriate form (Eq. 21): this reduces to zero, as it must, when $V_d(r) = 0$.

Around the defect, however, we need to know $F(r\ r'\ E)$ as a function of E, in order to evaluate Eq. 21. As can be seen from Eq. 7, this is a

---

*A more precise, but far more complicated, way of correcting the displaced charge is given in Appendix 1.

problem in band theory, which, for a given periodic potential $V_p(\mathbf{r})$, is soluble by existing methods, though it will be an extensive computational task. Information about the response function $F(\mathbf{r}\,\mathbf{r}'\,E)$ will lead to progress in the field of lattice defects; another bonus one could gain then would be to map out the anisotropy of the displaced charge due to a weak perturbation, such as that due to beryllium in lithium.

Assuming that, for a muffin-tin model, $\Delta\rho_{uo}$ is found from the theory of Beeby, or for more general potentials from the approximation of Stoddart, March, and Stott, it is clear that Eq. 21 gives us an approximate method for estimating the effect of relaxation. This then constitutes a principal result of this paper.

Actually, when $\Delta V(\mathbf{r})$ does not vary too fast in space which is true away from the lattice sites, we can further simplify Eq. 21. In particular, for $\rho_1(\mathbf{r}\,E)$ given by

$$\rho_1(\mathbf{r}\,E) = \int d\mathbf{r}'\,\Delta V(\mathbf{r}')\,F(\mathbf{r}\,\mathbf{r}'\,E) \quad , \tag{22}$$

we can then write

$$\rho_1(\mathbf{r}\,E) \doteq \Delta V(\mathbf{r}) \int d\mathbf{r}'\,F(\mathbf{r}\,\mathbf{r}'\,E) \quad . \tag{23}$$

The integral over $\mathbf{r}'$ in Eq. 23 can now be carried out, to yield

$$\rho_1(\mathbf{r}\,E) \doteq -\Delta V(\mathbf{r}) \frac{\partial \rho_0(\mathbf{r}\,E)}{\partial E} \equiv -\Delta V(\mathbf{r})\,\sigma_0(\mathbf{r}\,E) \tag{24}$$

where $\sigma_0(\mathbf{r}\,E)$ is the local density of states in the perfect lattice. This is just the quantity Beeby[5] calculated from the KKR method, with the result

$$\sigma_0(\mathbf{r}\,E) = \frac{-1}{\pi} \sum_{LL'} R_L(\mathbf{r})\,R_{L'}(\mathbf{r})\,\frac{1}{\tau}\,\mathrm{Im}\int_\tau \left[t^{-1} - \tau\,G(\mathbf{m})\right]^{-1}_{LL'} . \tag{25}$$

t is the t matrix for a single muffin-tin potential, while all the structure dependence is in $G(\mathbf{m})$ for which Beeby gives explicit expressions. The integration in Eq. 25 is over the Brillouin zone of volume τ, while $R_L(\mathbf{r})$ is, as usual, the radial wave function. All the information needed in Eq. 25 is accessible through a KKR band structure calculation.

By comparing Eq. 24, used at the Fermi energy $E = E_f$, with the semiempirical procedure based on Eq. 1, it will be interesting to see whether the approximation (Eq. 24) in the interstitial region will allow $\Delta V(\mathbf{r})$ to be found explicitly. If so, we have, of course, through Eq. 24 the energy dependence of $\rho_1(\mathbf{r}\,E)$ required in the evaluation of Eq. 21, which in the approximation equivalent to Eq. 24 becomes

$$\delta\rho = \Delta V(\mathbf{r}) \left[\sigma_0(\mathbf{r}\ E_f - V_d(\mathbf{r})) - \sigma_0(\mathbf{r}\ E_f) \right] \quad . \tag{26}$$

Unfortunately, although numerical calculations are in progress, results are not as yet available for $\delta\rho$.

Later, when we have more knowledge of the basic response functions, refinement may be carried out via the method of Appendix 2, should it prove necessary.

If we can make the approximation that $\Delta V(\mathbf{r})$ is slowly varying, then we find Eq. 24, and if we apply this at the Fermi level, we have evidently the approximate result

$$\rho_1(\mathbf{r}\ E) \simeq \rho_1(\mathbf{r}\ E_f) \frac{\sigma_0(\mathbf{r}\ E)}{\sigma_0(\mathbf{r}\ E_f)} \quad . \tag{27}$$

Thus, in this approximation, we simply scale the displaced charge below the Fermi level E_f in the Kanzaki lattice with the local density of states in the perfect crystal.

Using the first approximation for $\delta\rho$ that

$$\delta\rho(\mathbf{r}\ E) = \rho_1(\mathbf{r}\ E - V_d(\mathbf{r})) - \rho_1(\mathbf{r}\ E) \quad , \tag{28}$$

we see that this takes the explicit form

$$\delta\rho(\mathbf{r}\ E) = \rho_1(\mathbf{r}\ E) \left[\frac{\sigma_0(\mathbf{r}\ E - V_d)}{\sigma_0(\mathbf{r}\ E_f)} - \frac{\sigma_0(\mathbf{r}\ E)}{\sigma_0(\mathbf{r}\ E_f)} \right] \tag{29}$$

and, hence, at the Fermi level

$$\delta\rho(\mathbf{r}) = \rho_1(\mathbf{r}\ E_f) \left[\frac{\sigma_0(\mathbf{r}\ E_f - V_d(\mathbf{r}))}{\sigma_0(\mathbf{r}\ E_f)} - 1 \right] \quad . \tag{30}$$

Thus, given the local density of states $\sigma_0(\mathbf{r}\ E)$ in the perfect crystal, from a band theory calculation, the defect potential $V_d(\mathbf{r})$ and the displaced charge in the Kanzaki lattice at the Fermi level, we can calculate the correction to the charge displaced by V_d in the unperturbed lattice. Of course, we have only circumvented the many-electron nature of the problem by using Eq. 1 with $\mathbf{R}(\mathbf{r})$ found empirically from the Bragg reflection intensities. We stress that we can include both many-electron effects and many-body forces via the presence of $\rho_1(\mathbf{r})$ in Eq. 30. But, of course, we are still involved in the assumption that V_d is slowly varying, and we expect that refinements of the theory will lead to a form

$$\delta\rho(\mathbf{r}) = \rho_1(\mathbf{r}\ E_f)\ f[\mathbf{r}\ E_f\ V_d(\mathbf{r})] \quad . \tag{31}$$

122 N. H. MARCH AND J. S. ROUSSEAU

The square bracket in Eq. 30 is a first approximation to the modulating function f.

4.3 Local Density of States for Mg in Li

Dr. J. E. Inglesfield (private communication) has recently made preliminary pseudopotential calculations of the local density of states $\sigma(\mathbf{r}E)$ for magnesium in lithium, and this section is concluded by mentioning these results briefly.

The results are shown for two energies $\frac{1}{2}E_f$ and E_f in Fig. 1 and 2. The variable used is r/r_s, where r_s is the radius of the lithium atomic cell (3.265 au). The Fermi energy of lithium (0.173 au) has been used.

The dashed curves show the original local densities of states for the pure materials, at each energy, and the solid curves show the local density of states for a magnesium impurity in lithium.

As Inglesfield has emphasized, something like von Laue's theorem is being recovered as the Fermi energy is approached.

It is quite clear that the local density of states for lithium thus obtained can be used to calculate the function in the square brackets in Eq. 30. Calculations are currently in progress, in conjunction with Drs. Bullough and Perrin, to obtain $\sigma_0(\mathbf{r}\, E)$ from Beeby's method for lithium, as well as the displaced charge around an unrelaxed vacancy in this metal. Eq. 30 can then be used to correct for relaxation.

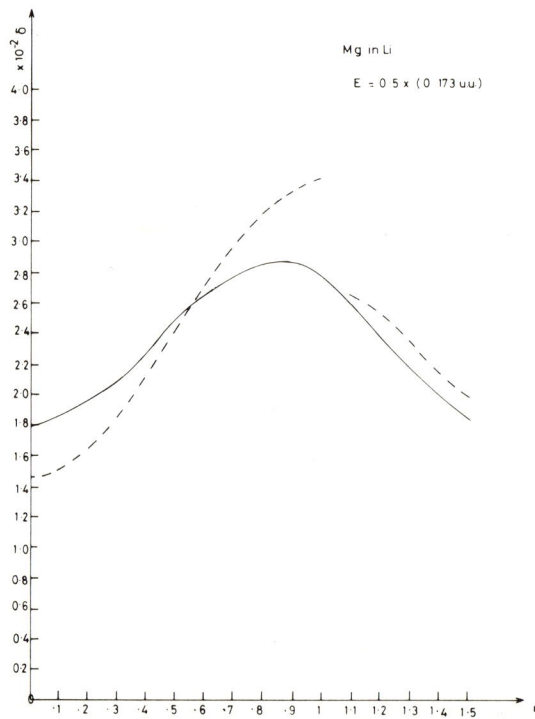

Fig. 1. Local density of states $\sigma(\mathbf{r}\, E)$ for Mg in Li (solid curve) for $E = \frac{1}{2} E_f$. Dashed curves show local densities of states in pure Mg and pure Li metal.

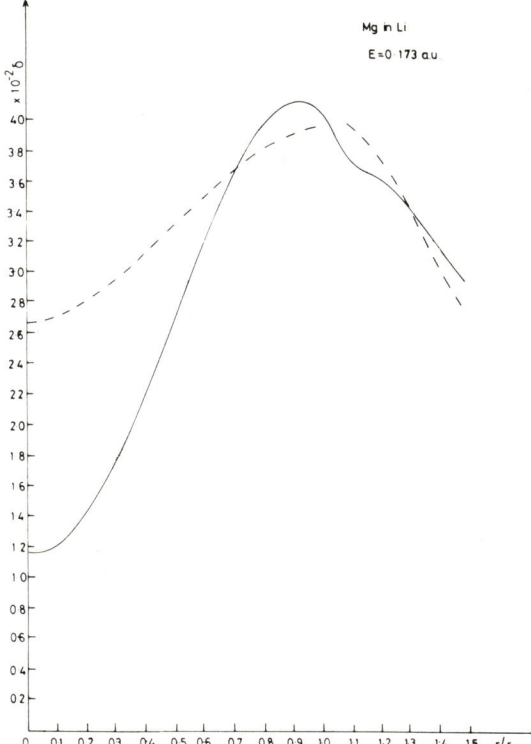

Fig. 2. Same as Fig. 1, but for $E = E_f$ (private communication from J. Inglesfield).

5 ELECTRIC FIELD IN KANZAKI LATTICE

Let us now consider how we can calculate the forces acting on the ions in the Kanzaki lattice from the electron theory developed above. Again, to be specific, we will consider the monovacancy, in which a single atom has been removed from the origin. Clearly, in this defect lattice, no forces act on the atoms, since the relaxed configuration is in equilibrium. This implies, from Feynman's theorem, that the total electric field $\mathbf{E(r)}$ in the relaxed defect lattice is identically zero when $\mathbf{r} = \mathbf{r}_\ell + \mathbf{u}_\ell$. Otherwise, there would be forces acting on the nuclei. This electric field is created by an electron density $\rho_0(\mathbf{r}) + \rho_1(\mathbf{r}) + \rho_d(\mathbf{r})$, plus the fields of the nuclei.

Now we put back the atom at the origin, in our vacancy example. The electron density in this Kanzaki lattice is $\rho_0(\mathbf{r}) + \rho_1(\mathbf{r})$. The electric field, $\mathbf{E}_{Kanzaki}(\mathbf{r}) \equiv \mathcal{E}(\mathbf{r})$, in this strained lattice is evidently determined by the electron density, $\rho_0(\mathbf{r}) + \rho_1(\mathbf{r})$, plus the nuclear configuration. It is then clear that the resultant electric field acting at the nuclei must be equivalent to that due to the difference between the electron densities in the Kanzaki lattice and in the relaxed defect lattice and to the difference between the nuclear configurations.

Immediately, for the monovacancy, we see that this field must be determined by the electrostatic potential due to the displaced electron charge around the vacancy, plus that due to the absence of the ion at the origin. This electron density is given by

$$\rho_f(\mathbf{r}) - \rho_K(\mathbf{r}) = \rho_d(\mathbf{r})$$

$$= \Delta\rho_{uo} + \left[\rho_1(\mathbf{r}\ E_f - V_d(\mathbf{r})) - \rho_1(\mathbf{r}\ E_f)\right] \quad , \tag{32}$$

and it is this density, together with the field of the "absent" ion, that determines the electric field at the ions in the Kanzaki lattice, and, hence, the Kanzaki forces. From our knowledge of potentials and displaced charges in free electron metals, it seems that, at large r, by Taylor expanding Eq. 26 in terms of V_d, $\delta\rho$ will be both smaller, involving the product of $\Delta V(\mathbf{r})$ and V_d, and shorter range than $\Delta\rho_{uo}$, though this will have to be verified by detailed numerical calculation. Thus, at large r, we shall tentatively assume $\Delta\rho_{uo}$ to dominate in Eq. 32.

5.1 Some Preliminary Results From Free Electron Model

However, we can readily estimate the field due to this term $\Delta\rho_{uo}$ and it is worth recording the result here by way of illustration. We have, from the Poisson equation

$$\nabla^2 \phi = -4\pi\Delta\rho_{uo} \quad , \tag{33}$$

where ϕ is the electrostatic potential. When $\Delta\rho_{uo}$ is spherically symmetric, we find

$$r^2 \frac{d\phi}{dr} = -\int_0^r 4\pi\Delta\rho_{uo}(r)\ r^2\ dr = -Q(r) \quad , \tag{34}$$

where $Q(r)$ is the total charge displaced inside a sphere of radius r. Thus, we find

$$\frac{d\phi}{dr} = -\frac{Q(r)}{r^2} \quad . \tag{35}$$

Hence, the electric field related to the displaced charge, $\mathcal{E}_{dc}$, in the Kanzaki lattice (neglecting the ρ_1 terms in Eq. 32) is given by

$$\mathcal{E}_{dc}(\mathbf{r}) = -\nabla\phi = \frac{\mathbf{r}}{r} \cdot \frac{Q(r)}{r^2} \quad . \tag{36}$$

Adding on the electric field of the ion, of resultant charge Z equal to the valency, we find

$$\mathcal{E}(\mathbf{r}) = \frac{\mathbf{r}}{r^3}[Q(r) - Z] \quad . \tag{37}$$

For free electrons, we know asymptotically that we have the Friedel oscillations represented by

$$\Delta\rho_{uo} \sim \frac{A\cos 2k_f r}{r^3} \tag{38}$$

where A is a constant and k_f is the Fermi wave number.* We can therefore write for the asymptotic form of $Q(r) - Z$

$$Q(r) - Z \sim -\int_r^\infty \Delta\rho_{uo}\, 4\pi r^2 dr$$

$$= -4\pi A \int_{2k_f r}^\infty \frac{\cos t}{t}\, dt$$

$$= 4\pi A\, \text{Ci}\,(2k_f r) \quad . \tag{39}$$

where Ci (x) is the usual cosine integral, as shown. Hence, we may write

$$\mathcal{E}(\mathbf{r}) \sim \frac{\mathbf{r}}{r^3}\, 4\pi A\, \text{Ci}\,(2k_f r) \quad . \tag{40}$$

An estimate of A is readily available from the work of Stott et al.[4] Some results for $Q(r)$ in copper are shown in Fig. 3. Hence, if we assume this electric field, evaluated at the atomic positions in the Kanzaki lattice, acts on the resultant ionic charge Ze at each site, we have a first estimate of the Kanzaki forces as $Ze\mathcal{E}(\mathbf{r})$.

---

*The introduction of a phase shift, necessary for strong scattering, is easily effected.

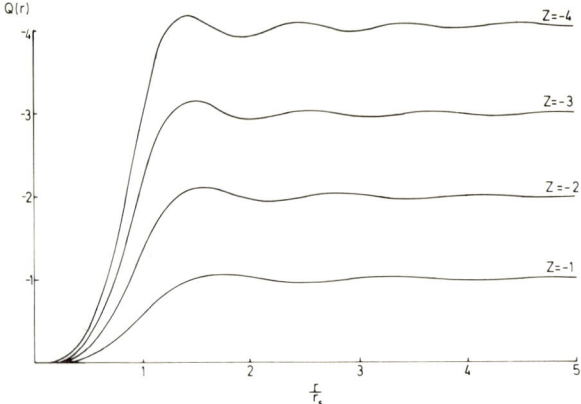

Fig. 3. Total displaced charge $Q(r)$ inside sphere of radius r in case of vacancies in Cu ($Z = -1$), Mg ($Z = -2$), Al ($Z = -3$) and Pb ($Z = -4$).
This is needed to calculate electric field in Kanzaki lattice. Results shown are from the work of Stott et al.[4]; see also March and Stoddart[19], p. 551).

We want to emphasize that this is only a very crude example. It will be necessary to estimate the contribution due to ρ_1 in Eq. 32 as well as to calculate $\Delta\rho_{uo}$ beyond a free-electron model. In both calculations, we need the perfect lattice solutions. Work is in progress to evaluate the displaced charge $\Delta\rho_{uo}$ in lithium for a monovacancy in the unrelaxed lattice, from the theory of Beeby[5], but results are not, as yet, available. However, Harris[14] has successfully applied Beeby's method to the impurity problem. Since the Beeby theory is based on the KKR method, the necessary response functions are contained in the KKR solutions for the perfect lattice. Thus we have here a basis for a rather precise evaluation of the displaced charge in the relaxed defect lattice.

Once the electric field is known, the Kanzaki forces can be found and Kanzaki's original method can be used to obtain a new set of displacements.

6 KANZAKI FORCES AND ENERGIES OF MONO- AND DIVACANCY

As a further illustration of the present approach, let us try to get the Kanzaki forces associated with a Hartree treatment of a vacancy in a metal. Then we can write for the total energy of the metal

$$U = -\frac{1}{2} \int d\mathbf{r} \left[\nabla_\mathbf{r}^2 \, \rho(\mathbf{r}\,\mathbf{r}') \right]_{\mathbf{r}' = \mathbf{r}} + \frac{1}{2} \int d\mathbf{r} \, d\mathbf{s} \, \frac{\rho(\mathbf{r})\,\rho(\mathbf{s})}{|\mathbf{r}\text{-}\mathbf{s}|}$$

$$+ \int d\mathbf{r}\,\rho(\mathbf{r})\,\Psi(\mathbf{r}) + \Phi \quad . \tag{41}$$

The first term in Eq. 41 is the kinetic energy, the second is the classical electron-electron interaction energy, the third is the interaction energy of the conduction electrons with the ion cores, and Φ represents the core-core interaction energies.

As the authors have argued, the energy can be written, when the metal is deformed, in terms of the components of the displacement field as

$$U = U_0 + \frac{1}{2} \sum_{\ell\ell'\alpha\beta} A_{\ell\ell'\alpha\beta} \zeta_{\ell\alpha} \zeta_{\ell'\beta} \qquad (42)$$

and if, at the same time, the lattice is subject to external forces, then the following term

$$K = - \sum_{\ell\alpha} F_{\ell\alpha} \zeta_{\ell\alpha} \qquad (43)$$

must be added to Eq. 42, where $F_{\ell\alpha}$ are the Kanzaki force components.

If we assume the existence of a pseudoatom, then we can write

$$-\sum_{\ell\alpha} F_{\ell\alpha} \zeta_{\ell\alpha} = E_{\text{pseudoatom}} + \int d\mathbf{r}\, \Delta(\mathbf{r})\, [V(\mathbf{r}) - \psi(\mathbf{r})]$$

$$- \int d\mathbf{r}\, \rho_v(\mathbf{r})\, \psi(\mathbf{r}) \qquad (44)$$

where

$$\Psi(\mathbf{r}) = \sum_{\ell} \psi(\mathbf{r}-\mathbf{r}_\ell) \quad . \qquad (45)$$

It will clearly be of interest to see whether the above approximation to the Kanzaki forces gives results similar to those of the previous section, based on direct calculation of the electric field at the ions.

The interest in the above calculation resides in its possible extension to estimate directly the effects of relaxation on the divacancy binding energy. Unfortunately, it is then necessary to know not only how the displaced charge around the two-centre problem is related to that around the one-center problem*, but also how to relate the positions of the relaxed atoms around a monovacancy to those when the second vacancy is brought up to the near-neighbor distance.

We want to conclude, because of the difficulty of this problem, by pointing out a possible approach related again to calculating the electric field $E(\mathbf{r})$ in the final equilibrium configuration.

We have, for a system in equilibrium under Coulomb forces, with kinetic energy T and potential energy U,

---
*Such a calculation for an impurity-vacancy complex in a metal was carried out by Alfred and March[15]. For a divacancy, it is practicable to find the displaced charge from the model of Seeger and Bross[16].

$$2T + U = 0 \quad . \tag{46}$$

But when forces are required to hold two defects at distance a, we can write

$$2T + U = -a\frac{dE}{da} \quad . \tag{47}$$

This equation can be integrated with respect to a, since $T + U = E$

$$a^2 E(a) = \int_a^\infty R\, U(R)\, dR \quad . \tag{48}$$

Thus, if the potential energy $U(R)$ can be obtained as a function of R, we could obtain the interaction energy as a function of distance. Certainly in a Hartree framework, the calculation of $U(R)$ is then a problem in electrostatics, knowing the $\mathbf{u}_\ell$'s and the density $\rho_o(\mathbf{r}) + \rho_1(\mathbf{r}) + \rho_d(\mathbf{r})$ in the final equilibrium state.

We want, however, to make the point that R in Eq. 48 is really to be restricted to lattice separations between the vacancies and is not therefore a continuous variable.

We conclude that, unless the use of pair potentials can be justified, which is unlikely around a divacancy in a polyvalent metal, it is going to be of considerable interest to map out the electric field $\mathbf{E}(\mathbf{r})$ (zero at $\mathbf{r} = \mathbf{r}_\ell + \mathbf{u}_\ell$) in the final state, and find the associated energy stored in the field. We suspect that Eq. 47 ought to be replaced eventually by difference equations in an exact formulation.

We turn finally to discuss how the energy in the Kanzaki lattice might be calculated within the present framework.

Let T_K and U_K be the kinetic and potential energies in the Kanzaki lattice for a given separation R between the two vacancies. Then we can write

$$2T_K + U_K = -\sum_\ell (\mathbf{r}_\ell + \mathbf{u}_\ell) \cdot \mathbf{F}_\ell$$

where the $\mathbf{F}_\ell$'s are the Kanzaki forces. Naturally, those forces, and also the displacements $\mathbf{u}_\ell$ depend on the separation R.

In one-body potential theory as used in this paper, where an attempt is made to incorporate some of the many-body effects into the potential, it seems best to relate the total energy to the potential energy since at least the classical part of this is calculable from the electron density plus the potential of the nuclear framework. Thus we can write the above equation in the form

$$E_K(R) + U_K(R) = - \sum_\ell (r_\ell + u_\ell) \cdot F_\ell .$$

Evidently, from the electron density and the electric field $\mathcal{E}(r)$, we can calculate $E_K(R)$ and $E_K(\infty)$, the difference giving us a part of the relaxation energy. Naturally $E_K(\infty)$ involves crucially the displacements around a single vacancy, while $E_K(R)$ involves those around a divacancy.

7 CONCLUSION

If X-ray scattering experiments could be refined to yield not only the electron density $\rho_0(r)$ in the perfect crystal, but the density around an impurity or an imperfection, band theory would allow the construction of the crystal potential $V_p(\mathbf{r})$ and the defect potential $V_d(\mathbf{r})$. Until such time as that becomes possible, we must use the best available methods, based on refinements of the Slater $\rho 1/3$ exchange to construct these potentials so as to incorporate at least some part of the electron-electron interactions.

Knowing the starting displacements $\mathbf{u}_\ell$, $\rho_1(\mathbf{r})$ can be estimated semi-empirically from Eq. 1 at the Fermi level. Furthermore its energy dependence is accessible through the approximate result (Eq. 24) in terms of the local density of states in the perfect crystal. The charge displaced in the Kanzaki lattice by the defect potential $V_d(\mathbf{r})$ can then be estimated from that displaced by V_d inserted in the unrelaxed lattice, corrected by Eq. 26. This displaced charge can then be used to find the Kanzaki forces and, hence, to calculate a new set of displacements.

Finally, it is emphasized that, within the present framework, calculations of defect energies should be attempted from the classical potential energy terms, plus, if necessary, estimates of the exchange energy from the electron density. The kinetic energy should not be calculated, however, from the density matrices discussed in this paper, for these have incorrect off-diagonal elements, although the diagonal elements agree with the correct many-body density matrix.

ACKNOWLEDGMENTS

We wish to thank particularly our colleagues Dr. W. Jones and Dr. J. C. Stoddart for numerous valuable discussions in the course of this work. We are grateful to Dr. J. Inglesfield for allowing us to use the results shown in Figs. 1 and 2. Dr. R. C. Brown kindly assisted us in the calculations shown in Fig. 6 and we wish to thank him for his generous help. Partial contractual support from the US Army through its European Office is acknowledged.

APPENDIX 1

**Relations Between Functions F, R and P
Characterizing the Perfect Lattice**

We want to summarize here some of the basic properties of the functions F, R and P which, in principle, can be found from the theory of the perfect crystal. Obtaining F, as stressed above, is a one-body problem, once the potential $V_p(\mathbf{r})$ is known. On the other hand, exact determination of R and P is not possible at present, as these quantities depend on the exchange and correlation energy of an inhomogeneous electron gas.

A1.1 Proof of Eq. 7 Between Response Function F and One-Body Green Function and Density Matrix

As remarked above, the Eq. 7 was given by Stoddart, et al.[6] An elementary proof of this will be sketched below.

Eq. 10 gives us the change in the Bloch density matrix $C_o(\mathbf{r}\,\mathbf{r}'\,\beta)$ due to a change in potential ΔV, a change we denote by $\Delta C(\mathbf{r}\,\beta)$ on the diagonal.

Then we have immediately from Eq. 10 that

$$\Delta C(\mathbf{r}\,\beta) = -\int d\mathbf{r}'\,\Delta V(\mathbf{r}') \int_0^\beta d\beta_1\, C_o(\mathbf{r}\,\mathbf{r}'\,\beta-\beta_1)\, C_o(\mathbf{r}'\,\mathbf{r}\,\beta_1) \quad . \tag{A1.1}$$

We now compare this result with the density change $\Delta\rho(\mathbf{r}\,E)$ caused by the same change in potential. This, from Eq. 21 with $V_d(r)$ set equal to zero is

$$\Delta\rho(\mathbf{r}\,E) = \int d\mathbf{r}'\,\Delta V(\mathbf{r}')\, F(\mathbf{r}\,\mathbf{r}'\,E) \quad . \tag{A1.2}$$

But we have the Laplace transform relation between C and ρ (provided we add a positive potential energy to bring all energies $\geqslant 0$)

$$\Delta C(\mathbf{r}\,\beta) = \beta \int_0^\infty \Delta\rho(\mathbf{r}\,E)\, e^{-\beta E}\, dE \quad . \tag{A1.3}$$

Integrating this equation by parts, we find

$$\Delta C(\mathbf{r}\,\beta) = \int_0^\infty e^{-\beta E}\, \frac{d}{dE}\, \Delta\rho(\mathbf{r}\,E)\, dE$$

$$= \int d\mathbf{r}'\,\Delta V(\mathbf{r}') \int_0^\infty e^{-\beta E}\, \frac{\partial F(\mathbf{r}\,\mathbf{r}'\,E)}{\partial E}\, dE \tag{A1.4}$$

where the last line follows by differentiating Eq. A1.2 with respect to E.

It is clear then from Eq. A1.1 and A1.4 that $\delta F/\delta E$ can be related to the integral over β_1 of the product of C_o's displayed explicitly in Eq. A1.1. Substituting the explicit form (Eq. 4) for C_o, the ρ_1 integration is readily accomplished. Similarly, the inverse Laplace transform required to obtain $\delta F/\delta E$ from Eq. A1.4 is easily completed. Hence, using the explicit definitions of C_o and ρ_o, Eq. 7 follows.

A1.2 Integral Equation for R

The determination of **R** from first principles presents more difficulty, though a semiempirical procedure can be devised to obtain a useful starting approximation (see Appendix A3).

The argument sketched below follows the discussion of JM and is given for completeness in that the potential $\Delta V(\mathbf{r})$ required to generate the Kanzaki lattice depends on **P** through Eq. 5. The essential point is that we can write the change in the one-body potential as we move the ions from $\mathbf{r}_\varrho$ to $\mathbf{r}_\varrho + \mathbf{u}_\varrho$ as

$$\Delta V = \Delta V_{\text{electrostatic}} + \int U(\mathbf{r}\ \mathbf{r}')\, \rho_1(\mathbf{r}')\, d\mathbf{r}' \quad . \tag{A1.5}$$

But ρ_1 is related to $\mathbf{R}(\mathbf{r})$ and we also have

$$\Delta V = \sum_\varrho \int \frac{\mathbf{u}_\varrho \cdot \mathbf{R}(\mathbf{r}'-\mathbf{r}_\varrho)d\mathbf{r}'}{|\mathbf{r}'-\mathbf{r}|} + \sum_\varrho \left[\frac{Ze}{|\mathbf{r}-\mathbf{r}_\varrho-\mathbf{u}_\varrho|} - \frac{Ze}{|\mathbf{r}-\mathbf{r}_\varrho|} \right]$$

$$+ \sum_\varrho \mathbf{u}_\varrho \int U(\mathbf{r}\ \mathbf{r}')\, R(\mathbf{r}'-\mathbf{r}_\varrho)\, d\mathbf{r}' \quad . \tag{A1.6}$$

This can now be expanded to $O(\mathbf{u}_\varrho)$ to yield Eq. 5 where,

$$P(\mathbf{r}) = \int \frac{\mathbf{R}(\mathbf{r}')\, d\mathbf{r}'}{|\mathbf{r}-\mathbf{r}'|} - \frac{Ze\mathbf{r}}{r^2} + \int U(\mathbf{r}\ \mathbf{r}')\, \mathbf{R}(\mathbf{r}')\, d\mathbf{r}' \quad . \tag{A1.7}$$

After some manipulation, the basic integral equation 6 of JM then follows.

APPENDIX 2

Approximate Form of Density Matrix for the Final State

To avoid the troublesome problem of solving Eq. 12 by numerical iteration, we shall now argue as follows. With the defect potential $V_d(r)$ introduced into the *perfect* lattice, let us suppose that by the methods of Stoddart, et al.[6] we have generated a density matrix C_u, where u stands for the unrelaxed defect lattice. Then we define

$$\Delta C_{uo} = C_u - C_o \qquad (A2.1)$$

and

$$\Delta C_{fK} = C_f - C_K \qquad (A2.2)$$

It seems reasonable to assume that a useful starting point for the calculation of ΔC_{fK} would be ΔC_{uo}, which at least, includes the full effect of the defect potential $V_d(r)$, though *not*, at first, the relaxations.

It is now a simple matter to show, from the integral form of the Bloch equation, that

$$\Delta C_{fK} = -\int_0^\beta d\beta_1 \int d\mathbf{r}'' \left\{ [\Delta C_{fK}(\mathbf{r}\,\mathbf{r}''\,\beta\text{-}\beta_1) + C_{fK}(\mathbf{r}\,\mathbf{r}''\,\beta\text{-}\beta_1)] \right.$$
$$\left. \times V_d(\mathbf{r}'')\, C_K(\mathbf{r}''\,\mathbf{r}'\,\beta_1) \right\} \qquad (A2.3)$$

and

$$\Delta C_{uo} = -\int_0^\beta d\beta_1 \int d\mathbf{r}'' \left[(\Delta C_{fK}(\mathbf{r}\,\mathbf{r}''\,\beta\text{-}\beta_1) + C_{uo}(\mathbf{r}\,\mathbf{r}''\,\beta\text{-}\beta_1)) \right.$$
$$\left. \times V_d(\mathbf{r}'')\, C_o(\mathbf{r}''\,\mathbf{r}'\,\beta_1) \right] \qquad (A2.4)$$

Subtracting these, we have for the error $\delta C = \Delta C_{fK} - \Delta C_{uo}$ the result

$$\delta C = -\int_0^\beta d\beta_1 \int d\mathbf{r}'' \, [\delta C + \Delta C_{uo} + C_K]\, V_d(\mathbf{r}'')\, C_K(\mathbf{r}''\,\mathbf{r}'\,\beta_1)$$
$$+ \int_0^\beta d\beta_1 \int d\mathbf{r}'' [\Delta C_{uo} + C_o]\, V_d(\mathbf{r}'')\, C_o(\mathbf{r}''\,\mathbf{r}'\,\beta_1) \qquad (A2.5)$$

Now we have also

$$C_K - C_O = \Delta C_{Ko} \quad , \tag{A2.6}$$

where ΔC_{Ko} is 0 (u_ϱ). Thus, to $0(u_\varrho)$ we find

$$\delta C = -\int_0^\beta d\beta_1 \int d\mathbf{r}'' \, [\delta C + \Delta C_{Ko}] \, V_d(\mathbf{r}'') \, C_o(\mathbf{r}'' \, \mathbf{r}' \, \beta_1)$$

$$-\int_0^\beta d\beta_1 \int d\mathbf{r}'' \, [\Delta C_{uo} + C_o] \, V_d(\mathbf{r}'') \, \Delta C_{Ko}(\mathbf{r}'' \, \mathbf{r}' \, \beta_1) \quad . \tag{A2.7}$$

If we further assume, as starting point, that $\delta C < \Delta C_{Ko}$ then we have as the first approximation to correct the charge "displaced" by $V_d(\mathbf{r})$ in the unrelaxed crystal

$$\delta C \simeq -2 \int_0^\beta d\beta_1 \int d\mathbf{r}'' \, \Delta C_{Ko} \, V_d(\mathbf{r}'') \, C_o(\mathbf{r}'' \, \mathbf{r}' \, \beta_1)$$

$$-\int_0^\beta d\beta_1 \int d\mathbf{r}'' \, \Delta C_{uo} \, V_d(\mathbf{r}'') \, \Delta C_{Ko}(\mathbf{r}'' \, \mathbf{r}' \, \beta_1) \quad . \tag{A2.8}$$

Here then, is an expression to correct the displaced charge for the effects of relaxation, in terms of the properties of the unrelaxed crystal containing the defect potential $V_d(\mathbf{r})$.

To get a self-consistent theory, it is clear that Eq. 20 must be used to refine the displaced charge in the unrelaxed crystal. If we work with the Green function, the β integration in Eq. 20 is immediately removed.

It may well be possible to calculate local changes around the vacancy by making the additional approximation in Eq. 20 that V_d varies slowly in space. Then for the diagonal element $\delta C(\mathbf{r} \, \mathbf{r} \, \beta)$ we find

$$\delta C(\mathbf{r} \, \mathbf{r} \, \beta) \approx -V_d(\mathbf{r}) \int_0^\infty d\beta_1 \int d\mathbf{r}'' \left[\Delta C_{Ko}(\mathbf{r} \, \mathbf{r}'' \, \beta_1) \right.$$

$$\left. \times \left\{ 2C_o(\mathbf{r}'' \, \mathbf{r} \, \beta - \beta_1) + \Delta C_{uo}(\mathbf{r}'' \, \mathbf{r} \, \beta - \beta_1) \right\} \right] \quad . \tag{A2.9}$$

The text gives a rather cruder form of Eq. A2.9. Clearly, however, Eq. A2.7 provides a perfectly proper starting point for refining the displaced charge calculated in the unrelaxed lattice.

APPENDIX 3

Preliminary Estimate of $\rho_1(\mathbf{r})$, the Charge "Displaced" in the Kanzaki Lattice, for a Vacancy in Copper

Since the evaluation of the displaced charge in the presence of relaxation is, via Eq. 2.1, intimately related to the charge displaced in Kanzaki lattice, namely $\rho_1(\mathbf{r})$, we have made some preliminary estimates of this quantity for a vacancy in copper.

Evidently, to calculate $\rho_1(\mathbf{r})$ from Eq. 1, we must have information on the displacements $\mathbf{u}_\ell$ and the response function $\mathbf{R}(\mathbf{r})$.

Tewordt[8] has dealt with the problem of the relaxations around a vacancy in copper by matching local relaxations around the vacancy, obtained using Born-Mayer pair potentials, on to the long-range relaxations which can be properly treated by replacing the discrete crystal by a continuum. This approximation becomes valid outside a sufficiently large region enclosing the defect or defect complex. For then the displacements $\mathbf{u}_\ell$ become small and vary slowly from atom to atom.

These displacements $\mathbf{u}_\ell$ as given by Tewordt[8] are used, even though they will eventually need refining to take account of the detailed nature of the displaced charge around the vacant site as discussed in this paper.

It only then remains to set up a suitable approximation to $R(\mathbf{r})$ in Eq. 1. We already have the Fourier components $\mathbf{R}_\mathbf{K}$ at the reciprocal lattice vectors $\mathbf{K}$ through Eq. 3. Batterman, Chipman and de Marco[17] have studied the X-ray scattering from copper and we therefore have approximate estimates available of the $\rho_\mathbf{K}$'s. Following JM, we have made the assumption that we can use Eq. 3 for all $\mathbf{k}$, smoothly interpolating between the $\rho_\mathbf{K}$'s at the reciprocal lattice vectors to obtain $\rho_\mathbf{k}$. Obviously this procedure is not unique, and eventually the integral equation of JM (see Appendix 1) must be solved to find $\mathbf{R}(\mathbf{r})$.

The above assumption, as emphasized by JM, is equivalent to the assumption of pair forces, and we can then write R as the gradient of a scalar density $\sigma(\mathbf{r})$, namely

$$\mathbf{R}(\mathbf{r}) = \nabla \sigma(\mathbf{r}) \quad , \tag{A3.1}$$

where, from Eq. 2 it then follows that

$$\rho_0(\mathbf{r}) = \sum_\ell \sigma(\mathbf{r}-\mathbf{r}_\ell) \quad . \tag{A3.2}$$

Provided $\sigma(\mathbf{r})$ has the correct Fourier components at the reciprocal lattice vectors, which is of course ensured by our procedure to within the accuracy of the experimental measurements, then Eq. A3.2 will reproduce the exact ground-state charge density of the crystal.

A3.1 Displaced Charge in Copper Metal

As an example, we shall compute the displaced charge in copper metal using Tewordt's displacements and making the preliminary approximation $R = \nabla \sigma$. We fit σ the 'pseudoatom' charge density to the X-ray scattering experiments of Batterman et al. (cf Jones et al.[18]) as shown in Fig. 4 and 5.

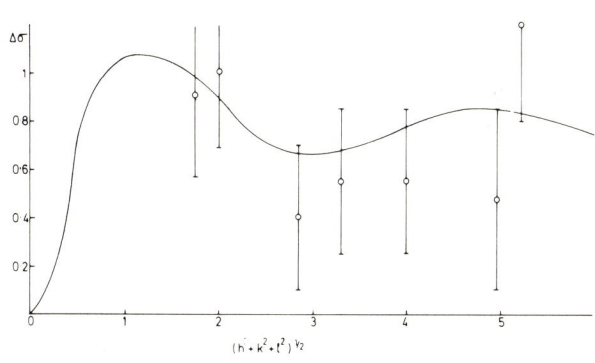

Fig. 4. $\Delta \sigma$ versus $(h^2 + k^2 + l^2)^{1/2}$ for Cu, used to fit X-ray results of Batterman et al. Form is given by $\Delta \sigma(k) = A_1 k\, e^{-a_1 k} + A_2 k^7 e^{-a_2 k}$
$A_1 = 3.24$, $a_1 = 1.11$,
$A_2 = 0.0083$, $a_2 = 1.35$.

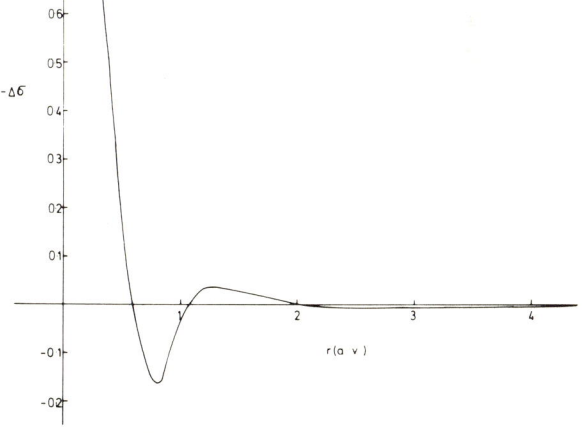

Fig. 5. $\Delta \sigma(r)$ corresponding to $\Delta \sigma(K)$ in Fig. 4.

Using the Hartree-Fock atom density as starting point, the correction $\Delta \sigma(k)$ required to fit the X-ray scattering at the Bragg reflections is shown explicitly in Fig. 4. This is not unique, we have simply drawn a smooth curve through the measurements, and, of course, eventually the integral equation of Jones et al. will have to be used to find $R(r)$. The Fourier transform $\Delta \sigma(r)$ is shown in Fig. 5.

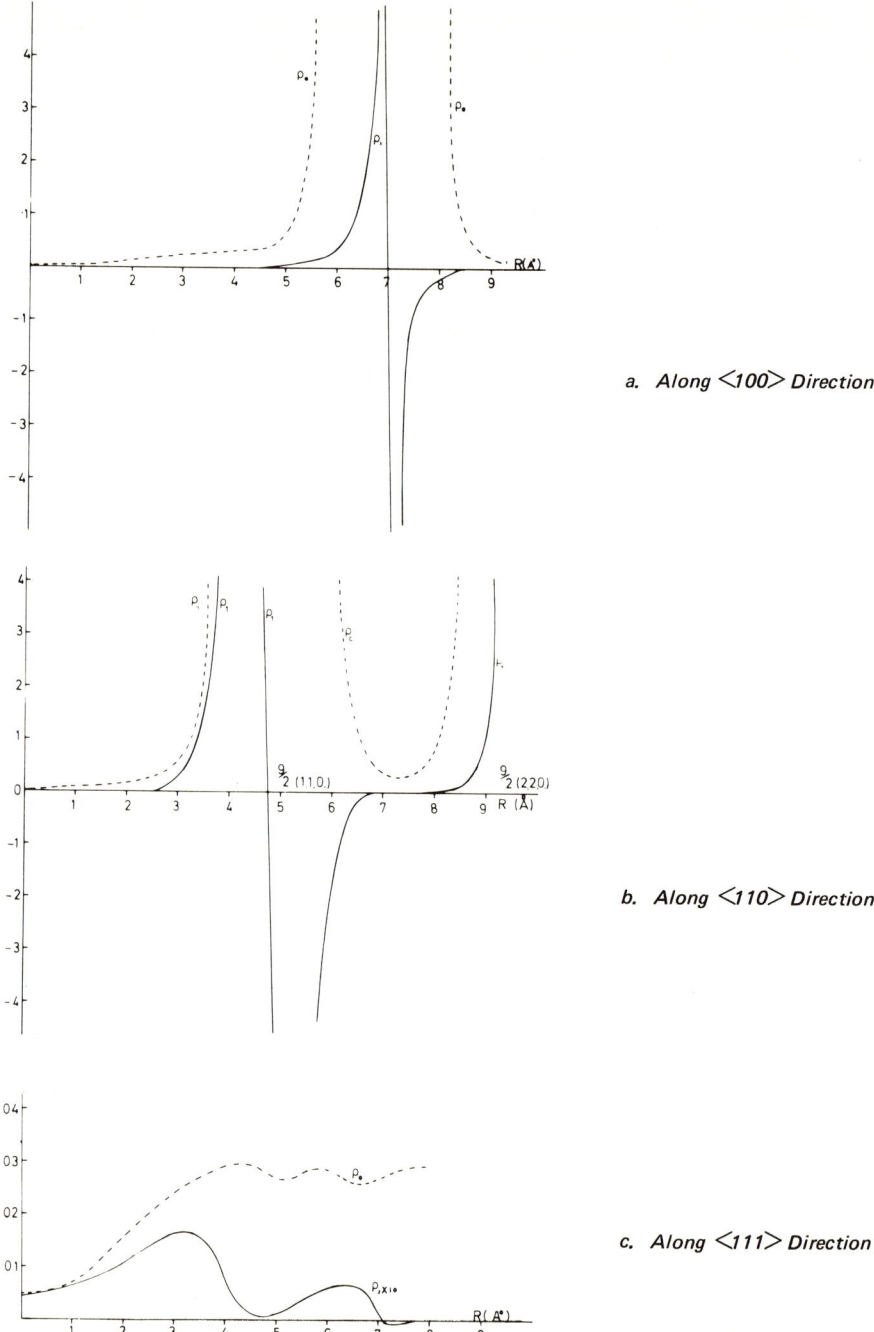

Fig. 6. Charge density ρ_0 in Cu, together with displaced charge ρ_1 in Kanzaki lattice when strained by a vacancy.

We then calculate

$$\rho_1(\mathbf{r}) = \sum_\ell \mathbf{u}_\ell \cdot \nabla \sigma(\mathbf{r}-\mathbf{r}_\ell)$$

for various directions in the crystal and the results are shown in Fig. 6. The very marked anisotropy evident there is a consequence of the fact that, as we go out along the <111> direction, we do not encounter an atom over the range shown. This then is a first estimate of the charge displaced in the Kanzaki lattice, strained to account for the displacements around a vacancy in copper.

APPENDIX 4

Relation Between Response Function Approach and Elasticity Theory in Long Wavelength Limit

The customary elasticity approach gives us the result that, in response to a localized body force, which in k space, in the long wavelength limit gives us the Fourier transform of the force **F** in the form

$$\mathbf{F}(\mathbf{k}) = i\mathbf{k}\lambda \quad , \tag{A4.1}$$

where λ measures the strength of the body-force, the displacements $\mathbf{u}_\ell(\mathbf{r})$ take the form

$$\mathbf{u}_\ell(\mathbf{r}) = \frac{c\mathbf{r}}{r^3} \tag{A4.2}$$

The multiplying constant in Eq. A4.2 is determined in Tewordt's method by matching to the "discrete" displacements in the immediate neighborhood of the defect.

It is evident in Kanzaki's method that F and u are related by the dynamical matrix $D_{\alpha\beta}(\mathbf{k})$, which in turn, as JM discuss, is related to the response function $R(\mathbf{r})$. It then becomes clear that elasticity theory gives us rather direct information on the response function $R(\mathbf{k})$ at small $\mathbf{k}$ and, in particular relates the small $\mathbf{k}$ behavior rather directly to the elastic constants. This will be discussed quantitatively elsewhere by Claesson, Jones and March (to be published) in connection with many-body forces in lattice dynamics.

We want only to make one other point relating to elasticity theory here. Corresponding to the displacements given by Eq. A4.2, it is well known that there is a volume change per defect given by

$$\Delta\Omega = 4\pi c\gamma \quad , \tag{A4.3}$$

where $\gamma = 3(1-\sigma)/(1+\sigma)$, σ being Poisson's ratio.

We anticipate that eventually it may be necessary to check that the defect potential adopted obeys some condition equivalent to a Friedel sum rule into which the volume change (Eq. A4.3) is incorporated, as we indicated in the Introduction. Our microscopic theory is not yet sufficiently well developed to give a precise understanding of this point, which is worth further study.

REFERENCES

1. March, N. H., and Rousseau, J. S., Crystal Lattice Defects, **2**: 1 (1971).
2. Kanzaki, H., J. Phys. Chem. Solids, **2**: 24 (1957).
3. Matsubara, T., J. Phys. Soc. Japan, **7**: 270 (1952).
4. Stott, M. J., Baranovsky, S., and March, N. H., Proc. Roy. Soc. **A316**: 210 (1970).
5. Beeby, J. L., Proc. Roy. Soc. **A302**: 113 (1967).
6. Stoddart, J. C., March, N. H., and Stott, M. H., Phys. Rev. **186**: 683 (1969).
7. Jones, W., and March, N. H., Proc. Roy. Soc. **A317**: 359 (1970).
8. Tewordt, L., Phys. Rev. **109**: 61 (1958).
9. March, N. H., Young, W. H., and Sampanthar, S., *The Many Body Problem in Quantum Mechanics* (Cambridge University Press) (1967).
10. Hohenberg, P. C., and Kohn, W., Phys. Rev. **136**: B864 (1964).
11. Kohn, W., and Sham, L. J., Phys. Rev. **140**: A1133 (1965).
12. Stoddart, J. C., Beattie, A. M., and March, N. H., Int. Journ. Quantum Chem. **4**: 35 (1971).
13. Hilton, D., March, N. H., and Curtis, A. R., Proc. Roy. Soc. **A300**: 391 (1967).
14. Harris, R., J. Phys. C. (Solid State Phys.), **3**, 172 (1970).
15. Alfred, L.C.R., and March, N. H., Phil. Mag. **2**: 985 (1957).
16. Seeger, A., and Bross, H., Z. Phys. **145**: 161 (1956).
17. Batterman, B., Chipman and de Marco, J. J., Phys. Rev. **122**, 68 (1961).
18. Jones, W., March, N. H., and Tucker, J. W., Proc. Roy Soc. **A284**: 289 (1965).
19. March, N. H., and Stoddart, J. C., Reports Prog. Phys. Vol **31**: 533 (1968).

DISCUSSION on the Paper by N. H. March and J. S. Rousseau

ASHCROFT: Can the last part of your calculation be performed entirely in reciprocal space and hence avoid the necessity of a Fourier transform of the data (which contains, as you mentioned, much scatter)?

MARCH: Certainly we can calculate the vector field **R**(r) in k space directly from the Bragg reflection intensities. For a rigid ion or pseudo-atom, with density $\sigma(k)$ in k space, the result is simply $\mathbf{R_k} = i\sigma(k)\mathbf{k}$ within this pair force model. Eventually, however, we want the electric field $\mathcal{E}(\mathbf{r})$ at the strained lattice positions $\mathbf{r} = \mathbf{r}_\ell + \mathbf{u}_\ell$, $\mathbf{r}_\ell$ being a direct lattice vector.

Part Two

INTERATOMIC POTENTIALS

SCREENING FUNCTIONS IN SIMPLE METALS

M. P. Tosi

Istituto di Fisica, Universita' di Messina
Gruppo Nazionale di Struttura della Materia
  del CNR, Messina

ABSTRACT

The essential ingredients entering the estimation of the effective interionic potential in simple metals are the form of the electron-ion potential and the treatment of the screening. The discussion is centered on the latter aspect of pseudopotential theory. Recent work aimed at developing a theory of the electron-electron interactions in the homogeneous electron liquid which is valid at metallic electron densities, and accounting for Umklapp effects in the screening, is reviewed. The consequences of these results with regard to the evaluation of the dynamical properties of solid and liquid metals, and of the cohesive properties of metals and atomic spectroscopic data, are discussed, stressing whenever possible their relevance to calculations of defect properties.

142 M. P. TOSI

1 INTRODUCTION

A microscopic theory of defects in metals is concerned with problems of formation and binding energies and with problems of lattice distortions in a sea of interacting electrons, and must make appeal to theories of metallic cohesion and of phonons in metals. The standard formulation of these problems[1] introduces an indirect ion-ion interaction potential via the conduction electrons, and hinges on the assumption that the electron-ion interaction is weak. In this case each ion in the metal can be regarded as an "external" perturbation which polarizes the electron sea, and the other ions can be regarded as "external" probes of the induced polarization. The electron-electron interactions are taken into account through the dielectric function describing the response of the sea to the perturbation.

In the determination of the dielectric function, for a first approximation we may study the screening properties of a homogeneous electron liquid, thereby neglecting the effect of the electron-ion interactions on the screening function. Obviously, it is necessary to develop a theory of the electron-electron interactions in the liquid which is valid for metallic electron densities, at which the mean kinetic energy and the mean potential energy per particle are comparable. Then, we would also like to include corrections for the fact that the electrons move in a periodic potential. Advances have been made recently on both these aspects of the problem, and it is the purpose of this paper to discuss them. Used to illustrate the results is the "empty-core" electron-ion potential proposed by Ashcroft[2], which combines extreme simplicity and a remarkable degree of success.

2 HOMOGENEOUS ELECTRON LIQUID

The dielectric function describes the response of the electron liquid to a weak external potential.[3] The distortion of the particle density, $\langle \rho(q,\omega) \rangle$, is related to the external potential V_{ext} through the density-density response function of the liquid, $\chi(q,\omega)$,

$$\langle \rho(q,\omega) \rangle = \chi(q,\omega) V_{ext}(q,\omega) \quad , \tag{1}$$

where q is the wave vector and ω, the angular frequency.

If we introduce the Hartree potential,

$$V_H(q,\omega) = V_{ext}(q,\omega) + \frac{4\pi e^2}{q^2} \langle \rho(q,\omega) \rangle \quad , \tag{2}$$

we find the dielectric constant,

$$\varepsilon^{-1}(q,\omega) \equiv V_H/V_{ext} = 1 + \frac{4\pi e^2}{q^2} \chi(q,\omega) \quad . \tag{3}$$

By standard linear response theory, the density-density response function, and therefore the dielectric function, is determined by the correlations between the density fluctuations in the liquid, according to

$$\chi(q,\omega) = \frac{i}{\hbar} \int d(r-r') \int_0^\infty d(t-t') e^{-iq\cdot(r-r')+i\omega(t-t')} <[\rho(r,t), \rho(r',t')]> . \tag{4}$$

where r and r' are spacial coordinates. From this relation the fluctuation-dissipation theorem follows,

$$F.T. \left\{ <\rho(r,t) \rho(r',t')> \right\} = -\frac{\hbar q^2}{2\pi e^2} Im\, \varepsilon^{-1}(q,\omega) \quad . \tag{5}$$

It is also convenient to introduce the screened response function, $\chi_{sc}(q,\omega)$, which relates the response to the Hartree potential,

$$<\rho(q,\omega)> = \chi_{sc}(q,\omega) V_H(q,\omega) \quad , \tag{6}$$

so that

$$\chi(q,\omega) = \chi_{sc}(q,\omega) \left[1 - \frac{4\pi e^2}{q^2} \chi_{sc}(q,\omega)\right]^{-1} \tag{7}$$

and

$$\varepsilon(q,\omega) = 1 - \frac{4\pi e^2}{q^2} \chi_{sc}(q,\omega) \quad . \tag{8}$$

We have already accounted in this way for the long-range Coulomb field, which is responsible for the plasmon resonance in the response. The screened response function contains residual effects of the interaction, appearing as short-range exchange and Coulomb correlations between the electrons.

Eq. 8 is the starting point of calculations of the dielectric function by perturbation techniques[4,5]. The screened response function is represented by diagrams (Fig. 1a) where an electron and a hole are excited, go through interactions with the many-body system, and are then de-excited. In the random phase approximation (RPA) one considers only the zero-eth order diagram (Fig. 1b), which represents the density-density response function $\chi_0(q,\omega)$ for the noninteracting electron gas[6]. In first order, one finds the first exchange diagram (Fig. 1c) as well as diagrams giving the first self-energy corrections for the electron and for the hole (Fig. 1d). Hubbard[4] neglected the self-energy diagrams and summed approximately the exchange diagrams to infinite order, finding the result

$$\chi_{sc}(q,\omega) = \chi_0(q,\omega)\left[1 + \frac{4\pi e^2}{q^2} G_H(q) \chi_0(q,\omega)\right]^{-1} \qquad (9)$$

where

$$G_H(q) = \frac{1}{2}q^2\left(q^2 + \xi k_F^2\right)^{-1} \qquad (10)$$

Originally Hubbard chose $\xi = 1$ but later suggested[7] (in order to account for screening of the interactions in the exchange diagrams) the value

$$\xi = 1 + k_{TF}^2/k_F^2 \qquad (11)$$

where k_{TF} is the Thomas-Fermi screening parameter and k_F the wave vector at the Fermi surface. Somewhat different values for this parameter have been suggested by other authors.[8,5]

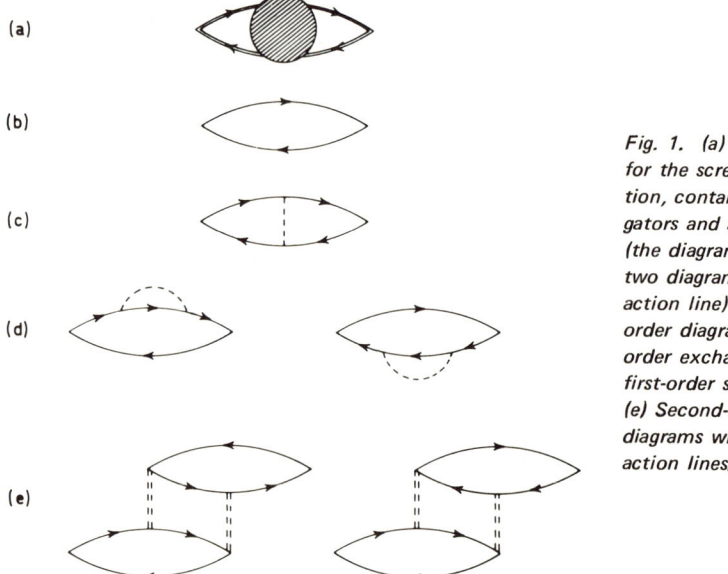

Fig. 1. (a) Schematic diagram for the screened response function, containing dressed propagators and a proper interaction (the diagram cannot be split into two diagrams by cutting an interaction line). (b) The zero-eth order diagram. (c) The first-order exchange diagram. (d) The first-order self-energy diagrams. (e) Second-order correlation diagrams with screened interaction lines.

A particularly critical test of a proposed dielectric function is the evaluation of the instantaneous short-range correlations between the electrons that it predicts through the fluctuation-dissipation theorem. In this respect the screened-exchange dielectric functions show great deficiencies[9,10], and one does not expect them to lead to reliable predictions on the rearrangement of the liquid around a perturbation, which is involved in the determination of the indirect ion-ion potential. As discussed by various authors[11-13], and as illustrated in Fig. 2 for metallic sodium, the various dielectric functions lead to wildly different predictions on the interionic potential. The ion-ion potential fails to have a main minimum if the screening is treated in the RPA, while even the gross features of the minimum found upon inclusion of exchange effects depend very greatly on how the electron-electron interactions in the exchange diagrams are screened. We should mention, however, that the results based on the value of ξ suggested by Animalu[8] ($\xi=1+\tfrac{1}{2}\,k_{TF}^2/k_F^2$; Curve 3) as well as those based on the suggestion of Geldart and Vosko[5] (fitted to the compressibility of the electron liquid; results not shown) are in fair agreement with the results based on more recent dielectric functions (Fig. 3).

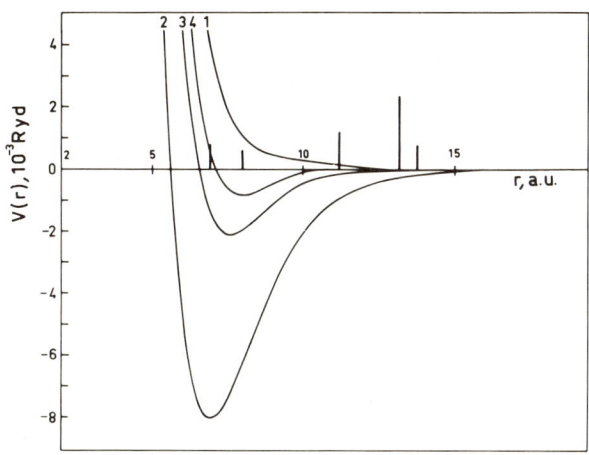

Fig. 2. The interionic potential $V(r)$ as a function of the interionic separation r in sodium metal, based on the Ashcroft electron-ion potential with $r_c = 1.694$ a.u. (from W. Shyu et al, Ref. 13). Curve 1, RPA screening; Curve 2, Hubbard-Sham screening with $\xi = 1$; Curve 3, Hubbard-Sham screening with $\xi = 1 + \tfrac{1}{2}\,k_{TF}^2/k_F^2$; Curve 4, Hubbard-Sham screening with $\xi = 1 + k_{TF}^2/k_F^2$. The vertical bars indicate the positions of the first few shells of neighbors.

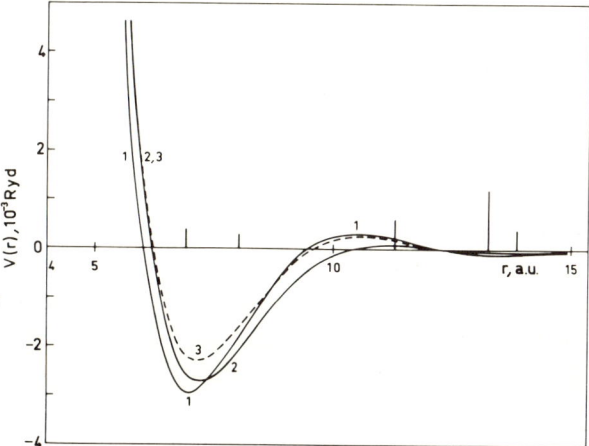

Fig. 3. The interionic potential $V(r)$ as a function of the interionic separation r in sodium metal, based on the Ashcroft electron-ion potential with $r_c = 1.694$ a.u. (from M. C. Abramo and M. P. Tosi, ref. 21). Curve 1, Geldart - Taylor screening; Curve 2, Singwi et al. screening; Curve 3, Toigo - Woodruff screening.

The recent work on the dielectric function of the electron liquid has focussed attention on the short-range Coulomb correlations, going beyond the screened exchange approximation as is necessary for the liquid at metallic densities. Geldart and Vosko[5] drew attention to the compressibility sum rule, which states that the long-wavelength limit of the static dielectric function,

$$\lim_{q \to 0} \mathcal{E}(q,0) = 1 + 4\pi n^2 e^2 \, K/q^2 \quad , \tag{12}$$

should yield a value of the compressibility K in agreement with the value obtained from the density dependence of the energy of the liquid. Satisfaction of this sum rule for the electron liquid has important consequences in the dynamical calculation of the elastic constants of metals. Geldart and Vosko pointed out that, in order to satisfy this sum rule to second order in the electron-electron interaction, it is necessary to include, besides the diagrams of Fig. 1c and 1d with screened interaction lines, the correlation diagram shown in Fig. 1e. Geldart and Taylor[14] have estimated these diagrams and discussed higher-order diagrams, suggesting the following approximation for the static screened response function:

$$\chi_{sc}(q,0) = \chi_0(q,0) \left[1 + \frac{g_1(q)}{1 + \lambda \chi_0(q,0)} \right] \quad . \tag{13}$$

Here, the function $g_1(q)$ is determined from the first-order diagrams, and the constant λ is adjusted so that the compressibility sum rule is satisfied. Since the frequency dependence of the dielectric function has not been estimated, one cannot evaluate the short-range correlations predicted by this type of approximation.

More complete estimates of the response, which also include the short-range Coulomb correlations, have been based on studies of the equation of motion for the density fluctuations, and involve approximate decouplings of higher-order correlation functions.* Singwi et al. have considered decoupling schemes which lead to a form of the screened response similar to Hubbard's,

$$\chi_{sc}(q,\omega) = \chi_0(q,\omega) \left[1 + \frac{4\pi e^2}{q^2} G(q,\omega) \chi_0(q,\omega) \right]^{-1} \quad , \tag{14}$$

where the function $G(q,\omega)$ is a functional of the two-body correlation function. The latter can be determined self-consistently from the proposed dielectric function by means of the fluctuation-dissipation theorem.[16,17] The numerical results (obtained neglecting the frequency dependence of the function G) satisfy approximately the compressibility sum rule and

---

*For the relation between decoupling approximations and perturbation theory in the classical plasma see Ref. 15.

yield a reasonable behavior for the short-range correlations.* An alternative decoupling scheme, which invokes satisfaction of the f-sum rule in decoupling, has been studied by Toigo and Woodruff.[19] Their approximation leads to Eq. 14 for the screened response, satisfies closely the compressibility sum rule, and yields[20] short-range correlations only slightly worse than those from the self-consistent approach of Singwi et al.

The interionic potential in sodium metal obtained with these three dielectric functions[21] is reported in Fig. 3. The gross features of the potential are now very well defined, and there remain only discrepancies of detail in the sharpness of the short-range rise and in the amplitude of the long-range oscillations. Of course, these features of detail are affected as well by the assumptions made on the electron-ion interaction.

3 LATTICE EFFECTS

The effects of lattice periodicity on the dielectric function in simple metals have been considered recently by several authors, with particular attention to calculations of lattice dynamics in the harmonic approximation.[22-25] Umklapp processes contribute to the screening, and the response function becomes a function of two wave vectors differing by a vector of the reciprocal lattice. As a consequence, the effective interaction between two ions depends not only on their separation but also on their positions relative to the other ions in the metal.

While the treatment of these effects in semiconductors, even in the simple RPA, requires a full knowledge of the energy-band structure and of the one-electron wave functions, in simple metals one can take advantage of the weakness of the electron-ion interaction to treat the effects of lattice periodicity by perturbation techniques. A full discussion in the long-wavelength limit has been given by Pethick[25], who includes up to two Umklapp processes. This order of approximation is required in order

---

*Very recently, Geldart and Taylor[18] have criticized the results of Singwi et al. on the counts that (i) they do not conserve the total number of particles, as expressed by the sum rule $n \int d^3 r [g(r) - 1] = -1$ where $g(r)$ is the pair correlation function, so that $g(r)$ must be inaccurate at large r; and (ii) they do not yield the correct value of the pair correlation function at small separations in the Hartree-Fock case ($\lim_{r \to 0} g_{HF}(r) = \frac{1}{2}$). With regard to the first criticism, we note that the sum rule expressing particle conservation is equivalent to $\lim_{q \to 0} S(q) = 0$, where $S(q)$ is the static structure factor. This sum rule is exactly satisfied by the approximation of Singwi et al. (see Eqs. 2 and 14 of Ref. 16; see also Eq. (A11) and related discussion). The argument of Geldart and Taylor involves the use of an analytic fit of the function $G(q)$ which, as clearly shown in Ref. 17, is approximately correct only for $q \lesssim 2k_F$ and to which Eq. 3 of Geldart and Taylor is obviously not applicable. With regard to the second criticism, we refer the reader to the discussion of the Hartree-Fock case in the approximation of Singwi et al. given in Ref. 16, section III (in particular, Eq. 23 which, combined with Eq. 15, implies $g_{HF}(0) = \frac{1}{2}$).

to satisfy, to second-order in the electron-ion interaction, the compressibility sum rule for the metal, which imposes agreement between the compressibility evaluated from the dispersion relations by the method of long waves and the compressibility evaluated from the density dependence of the cohesive energy. In a uniform compression the electron-phonon interaction disappears, and only the electron-perfect lattice interaction remains.

Calculations of the full phonon-dispersion curves which include periodicity effects have been carried out for aluminium by Johnson and Westin.[24] The results are quite sensitive to the electron-ion potential, but it appears that these effects appreciably affect the phonon frequencies at all wavelengths, contributing by 5 to 25 percent to the dynamical matrix. A characteristic effect is the appearance of new structure in the calculated dispersion curves[26], whose origin is similar to the Kohn effect, and which compares well with experimental observations.

4 LATTICE DYNAMICS AND COHESIVE ENERGY

Calculations of the phonon-dispersion curves in the simple metals have been one of the favorite testing grounds of pseudopotential theory. The frequencies are generally within 10 to 20 percent of those observed for the first-principle calculations, and somewhat closer for those containing adjustable parameters. The data reflect the behavior of the interionic potential in the region of the main minimum and in the long-range tail, and in principle, given a sufficiently accurate description of the screening function (inclusive of Umklapp effects), one may think of using these very accurate experimental data for the detailed determination of a parameterized electron-ion interaction. At present, it seems more fruitful to gloss over the more subtle effects entering the observed dispersion curves (such as nonlocal terms in the electron-ion interaction, Umklapp effects in the screening, and anharmonic terms) and to determine from a reasonably accurate dielectric function an effective ion-ion interaction that contains the least possible number of parameters.

Such a program, carried out for the alkali metals by Price et al.[27], leads for sodium, to the potential given by Curve 2 in Fig. 3. With only one disposable parameter, the effective core radius in the electron-ion potential of Ashcroft[2], one obtains very good agreement with the measured dispersion curves for the four alkali metals for which such data exist. Furthermore, the fitted values of the parameter for sodium and potassium are close to those derived from Fermi surface and liquid resistivity data, while the fitted values for lithium and rubidium lie between those obtained from the other physical properties. Of more direct interest from the point of view of the interionic potential is the closeness of the fit obtained in this way[13] for the interionic force constants for various orders of neighbors, which is reproduced in Table I for sodium. Similar results are obtained for the other alkali metals.

The usefulness of an interionic potential determined from a fit of the phonon-dispersion curves is, of course, rather limited. It may be expected to apply in situations that are rather close to those in which it has been determined. Such a situation is encountered in the description of the high-frequency and short-wavelength dynamical behavior of liquid metals, and in fact an application[28] to the study of inelastic neutron scattering from liquid sodium (based on theories well tested for liquid argon) leads to satisfactory results. However, the short-range rise of the potential is totally irrelevant to the phonon data. The behavior of the potential in this region is of importance for the study of diffusional motion, both in liquids and in solids. Rahman[29] has found that the interionic potential under discussion leads one to overestimate the diffusion coefficient in liquid sodium, as if the potential were too soft.

Because Umklapp effects in the screening have not been taken explicitly into account, the potential under discussion is also not applicable to a calculation of the cohesive properties of the metal[22,25,27]. As first shown by Ashcroft and Langreth[30], it is necessary to adjust the long-wavelength limit of the electron-ion potential, which essentially amounts to introducing a second parameter representing the binding energy of an electron at the bottom of the conduction band. By determining this parameter so that the cohesive energy of the metal has its minimum value at the correct density, one finds[31] good agreement with the experimental values of the cohesive energy and of the bulk modulus, as well as with those of the energy levels of the free atom, as illustrated in Table II for sodium. Similar results are obtained for the other alkali metals.

TABLE I. Interatomic force constants for sodium metal* (dyne/cm)

Shell	Force Constant	Shyu et al.	Exper.	Shell	Force Constant	Shyu et al.	Exper.
1	k_{xx}	1199	1178	4	k_{xx}	39	52
	k_{xy}	1350	1320		k_{yy}	2	-7
2	k_{xx}	453	472		k_{xy}	5	14
	k_{yy}	120	104		k_{yz}	14	3
3	k_{xx}	-66	-38	5	k_{xx}	13	17
	k_{zz}	-2	0		k_{xy}	13	33
	k_{xy}	-64	-65				

*From W. Shyu et al., Ref. 13.

TABLE II. Cohesive energy and bulk modulus of sodium metal, and energy levels of the free sodium atom*.

	E_c	B	E_{1s}	E_{IIs}	E_{Ip}	E_{Id}
Theory	0.460	0.065	0.377	0.144	0.223	0.111
Experiment	0.460	0.062	0.377	0.143	0.223	0.112

*From G. Cubiotti et al., ref. 31. Energies are in Rydbergs and B in 10^{12} dyne cm^{-2}.

5 CONCLUSIONS

The main points of the foregoing discussion may be summarized as follows:

(*i*) Recent improvements of the theory of the electron-electron interactions in the homogeneous electron liquid have bettered our knowledge of its dielectric function and greatly reduced the uncertainties in the ion-ion potential deriving from this aspect of pseudopotential theory.

(*ii*) The neglect of periodicity effects in the screening function implies the need of describing the cohesive properties and the dynamical properties by different electron-ion potentials. This point should be kept in mind in calculations of defect properties, such as vacancy formation energies and volumes, which involve consideration of both types of perfect-crystal properties.

(*iii*) A one-parameter potential is sufficient to fit the whole wealth of data contained in the experimental phonon dispersion curves. One additional parameter suffices to account for the cohesive properties at normal pressures and for the free-atom spectroscopic data.

(*iv*) The sharply rising part of the potential at short range, of interest in diffusional problems, appears to be imperfectly known.

REFERENCES

1. J. M. Ziman, Adv. Phys. **13**, 89 (1964).
2. N. W. Ashcroft, Phys. Lett. **23**, 48 (1966).
3. D. Pines and P. Nozières, *The Theory of Quantum Liquids* (W. A. Benjamin, New York and Amsterdam, 1966).
4. J. Hubbard, Proc. Roy. Soc. (London) **A243**, 336 (1957).
5. D.J.W. Geldart and S. H. Vosko, Canad. J. Phys. **44**, 2137 (1966).
6. J. Lindhard, Kgl. Danske Videnskab Selskab, Mat. Fys. Medd. **28**, no. 8 (1954).
7. L. J. Sham, Proc. Roy. Soc. (London) **A283**, 33 (1965).
8. A.O.E. Animalu, Phil. Mag. **11**, 379 (1965).
9. See, e.g. M. P. Tosi, Riv. N. Cimento **1**, 160 (1969).
10. L. Kleinman, Phys. Rev. **160**, 585 (1967) and **172**, 383 (1968) and P. R. Antoniewicz and L. Kleinman, Phys. Rev. **B2**, 2808 (1970).

11. V. Bortolani and A. Magnaterra, Phys. Lett. **28A**, 316 (1968).
12. W. Shyu and G. D. Gaspari, Phys. Lett. **30A**, 53 (1969).
13. W. Shyu, K. S. Singwi and M. P. Tosi, Phys. Rev. **B3**, 237 (1971).
14. D.J.W. Geldart and R. Taylor, Canad. J. Phys. **48**, 155 and 167 (1970).
15. T. O'Neil and N. Rostoker, Phys. Fluids **8**, 1109 (1965).
16. K. S. Singwi, M. P. Tosi, R. H. Land and A. Sjölander, Phys. Rev. **176**, 589 (1968).
17. K. S. Singwi, A. Sjölander, M. P. Tosi and R. H. Land, Phys. Rev. **B1**, 1044 (1970).
18. D.J.W. Geldart and R. Taylor, Solid State Commun. **9**, 7 (1971).
19. F. Toigo and T. O. Woodruff, Phys. Rev. **B2**, 395 (1970).
20. F. Toigo and T. O. Woodruff, to be published.
21. M. C. Abramo and M. P. Tosi, to be published.
22. D. C. Wallace, Phys. Rev. **182**, 778 (1969).
23. E. G. Brovman, Yu. Kagan and A. Holas, Zh. Experim. i Teor. Fiz. **57**, 1635 (1969) [Soviet Phys. JETP **30**, 883 (1970)].
24. R. Johnson, Kgl. Danske Videnskab Selskab, Mat. Fys. Medd. **37**, no. 9 (1970); R. Johnson and A. Westin, Report AE-365, AB Atomenergi (Stockholm, 1969).
25. C. J. Pethick, Phys. Rev. **B2**, 1789 (1970).
26. R. Johnson, Kgl. Danske Videnskab Selskab, Mat. Fys. Medd. **37**, no. 10 (1970).
27. D. L. Price, K. S. Singwi and M. P. Tosi, Phys. Rev. **B2**, 2983 (1970).
28. G. Cubiotti, K. N. Pathak, K. S. Singwi and M. P. Tosi, Lett. N. Cimento **4**, 799 (1970).
29. A. Rahman, private communication.
30. N. W. Ashcroft and D. C. Langreth, Phys. Rev. **155**, 682 (1967).
31. G. Cubiotti, E. Donato, E. S. Giuliano and M. P. Tosi, Lett. N. Cimento **1**, 743 (1971).

DISCUSSION on Paper by M. P. Tosi

TAYLOR: In referring to the work that Geldart and I[1] have done on the electron gas screening problem, you said that we had "estimated" some low order corrections to the RPA. I feel that, although this is a correct statement within the strict dictionary definition of the word, that is hardly a fair description of our work. We evaluated these low order terms in a consistent fashion; we summed the perturbation series by taking care to satisfy all of the known analytic properties of this series. In particular we pointed out that extremely large cancellations take place between the self-energy parts and the vertex corrections over the entire range of k. By examining the large k limit of the formal solution to the problem, we were able to show how these cancellations occur at each order in perturbation theory. We then constructed our sum to give the cancellations correctly.

You have pointed out that you get quite good agreement with the experimental phonon dispersion curves by adjusting a single parameter in the Ashcroft pseudopotential. I find it very disturbing that you get a good fit in lithium where the use of a pseudopotential only to second order in perturbation theory is certainly invalid. The $\ell = 1$ pseudo-wave function is badly misrepresented due to the rather strong potential that it sees. I would be exceedingly worried if I obtained a good fit to lithium with second order pseudopotential theory.

TOSI: In referring to *all* recent studies of the dielectric function as "estimates" of the effects of short-range Coulomb correlations, I am following the terminology adopted by Geldart and Taylor in reporting their work. For instance, section 4 of their second paper is entitled "Estimate of Lowest-Order Correlation Effects".

During my talk I illustrated the difficulties that one encounters with the Ashcroft potential for lithium. I showed that one finds rather different values for the core radius in this metal depending on the physical property that one fits.

CHANG: I do not understand Dr. Taylor's remarks that it is impossible for lithium to have a pseudopotential. In fact, "unique" pseudopotentials have been discussed recently by O'Keefe and Goddard.[2]

MARCH: In a simple classical fluid, the Ornstein-Zernike correlation function behaves like the negative of the pair potential $\phi(r)$ divided by the thermal energy $k_B T$, at large r. In an electron liquid the thermal energy is replaced by the zero-point energy of the plasma oscillations $1/2\ \hbar\omega_p$, where ω_p is the electron plasma frequency. Does the method of Singwi et al.[16,17] lead to this result?

TOSI: Yes. The self consistency condition imposed in our method is

$$S(q) = -\frac{\hbar q^2}{4\pi^2 e^2 n} \int_0^\infty d\omega\ \text{Im}\ \mathcal{E}^{-1}(q,\omega) \quad . \tag{1}$$

In the long-wavelength limit at a given frequency, our dielectric function reduces to the RPA dielectric function and the r.h.s. of the Eq. above reduces to the plasmon contribution:

$$\lim_{q \to 0} S(q) = \frac{3}{2}\frac{q^2}{q_{TF}^2}\frac{\hbar q_F^2}{2m}\left(\frac{\partial Re\mathcal{E}(0,\omega)}{\partial \omega}\right)^{-1}_{\omega=\omega_p} = \frac{1}{2}\hbar\omega_p/(4\pi n e^2/q^2) \quad .$$

This is equivalent to the sum rule you mentioned. In the above equation, n is the electronic density, m the mass, and $S(q)$ the structure factor. The other symbols are defined in the text.

CHANG: Although the one-parameter ion-electron interaction is very convenient to use, I do not see the validity of your conclusion regarding the choice of the core radius to fit the phonon-dispersion curve alone. Wouldn't it be more meaningful to choose a multiparameter ion-electron interaction and to fit these parameters to as many sets of experimental data as possible, such as phonon dispersion, diffusion, defect studies, including electronic properties?

TOSI: I agree, provided that one had (a) a fairly precise knowledge of the functional dependence of the electron-ion interaction on wave vector and energy, and (b) a rather sophisticated theoretical treatment of the various physical properties. As I showed, a one-parameter Ashcroft potential suffices to account, within present theoretical accuracy, for phonon dispersion and electronic properties in sodium and potassium, but not in lithium and rubidium.

ASHCROFT: You mentioned your work with Singwi et al.[3,4] led to predictions for $g(r)$ (pair distribution function) in the interacting electron gas. Do you distinguish between like and unlike spins and are both functions positive definite? Given $g(r)$ then the direct correction function should be easy to extract and hence would lead to an answer to Professor March's question.

TOSI: Our approach can be applied to any correlation function of one-particle properties, provided that one has sufficient physical insight in decoupling the appropriate three-point correlations. The simplest treatment of the spin-density correlations, that we have tried and reported in our paper, involves the heavy approximation of equal screening for interactions between parallel and between antiparallel spins, and does not lead to results as satisfactory as those obtained for the charge-density correlations. Specifically, the predicted long-wavelength magnetic properties are approximately correct; but the correlations between parallel spins are incorrect at short distances, as in all previous perturbation treatments.

The sum rule referred to by Professor March is a form of the particle-conservation sum rule, which may be written as

$$\lim_{q \to 0} S(q) = q^2/\omega_p ,$$

where ω_p is in units of the Fermi frequency and q in units of the Fermi wave vector. In the tabulation below I compare the values of the two sides of this equation, as read from our computer output at self-consistency for the smallest value of q that we considered (q = 0.05), for the theory developed in Ref. 16 and the refined theory developed in Ref. 4. The results are given for various electron density radii, r_s.

| | Value, X 10^{-2} | | |
| | Left Side According to | | |
r_s	Reference 3	Reference 4	Right Side
2	0.1878	0.1877	0.1878
4	0.1329	0.1328	0.1329
6	0.1085	0.1085	0.1085

REFERENCES

1. See Ref. 14.
2. P. M. O'Keefe and W. A. Goddard; III, Phys. Rev. **180**, 747.
3. K. S. Singwi, M. P. Tosi, R. H. Land and A. Sjölander, Phys. Rev. **176**, 589 (1968).
4. K. S. Singwi, A. J. Sjölander, M. P. Tosi, and R. H. Land, Phys. Rev. **B1**, 1044 (1970).

THE DIRECT CONSTRUCTION OF THE LATTICE GREEN FUNCTION

R. Bullough and V. K. Tewary

Theoretical Physics Division
Atomic Energy Research Establishment
Harwell, Berkshire, England

ABSTRACT

The harmonic distortion field in the neighborhood of a lattice defect can be calculated when the strength of the defect and the lattice response function (Green's function) are known. This paper discusses methods for the direct construction of the Green's function from the phonon-dispersion data. The procedure does not require the intermediate construction of a dynamical matrix based on a finite ranged interatomic potential and has the particular advantage that the long wavelength (continuum) response is explicitly contained in the total Green's function. The dispersion data is thus "fitted" with significantly fewer parameters than in the usual force constant method. Several applications of the Green function method are discussed.

1 INTRODUCTION

The static distortion field due to a point defect in a lattice can be easily calculated, within the harmonic approximation, when the strength of the defect and the lattice Green function are known. Two distinct procedures perform such a calculation: the so called Kanzaki method[1] and the direct Green function method. The precise relationship between these two methods is clarified by introducing the defect Green function. The Kanzaki method essentially involves using the source forces in the final atomic configuration with a perfect lattice Green function, whereas the direct Green function method involves the deliberate construction of a defect Green function and uses the initial source forces. The establishment of this simple relationship for defect calculations emphasizes the wide range of applications for which an accurately constructed lattice Green function can be used.

If the interaction potential between the atoms and, therefore, the force constants were known, the Green function could be calculated by inverting the dynamical matrix. In practice such interatomic potentials cannot be calculated easily from first principles, and one normally must resort to a phenomenological procedure in which first and second derivatives of the potential (force constants) are obtained by fitting to the phonon-dispersion data as provided by neutron-scattering experiments. Inevitably, to restrict the number of arbitrary parameters, this procedure necessitates a restriction on the range of the potential, and the force constants must be set to zero between atoms beyond a certain separation. Even for nonionic crystals this assumption is obviously unrealistic. In metals, for example, the free electrons act as carriers of interaction to very-long-distance neighbors and, in valence crystals, long-range quadrupole forces are known to exist. It is therefore not surprising that to get a very good fit to the phonon data it is necessary to include a large number of neighbors. In such calculations, the number of arbitrary parameters becomes so large that the fitted values of the force constants almost completely lose their physical significance. This paper discusses a procedure for the direct construction of the lattice Green function which does not require the intermediate calculation of the force constants or the dynamical matrix. The particular advantage of the method is that it imposes **no** prior restrictions on the range or form of the interatomic potential; in fact quite general interactions among all the atoms are automatically included. Furthermore, the Green functions thus obtained are likely to be more accurate than those calculated (by the usual method) from the force constants since the computational errors inevitably introduced in the intermediate process are avoided.

The method described here is called the semicontinuum Green function (SCGF) method since it is essentially a combination of the Born von

Karman model and a modified continuum lattice model. The spatial part of the phonon Green functions is decomposed into a sum of the continuum Green function and a lattice correction term; the latter, obtained from the Born von Karman model, is assumed to vanish beyond a certain distance. The correction terms are treated as parameters in the model with the constraint that the terms vanish for long waves when the total Green function is completely determined by the continuum part. Finally, in order that the correction term should not be unduly large or long ranged, the continuum Green function must be suitably modified to introduce the required dispersive effects but without affecting the long wavelength limit.

The SCGF method has been used to obtain the Green function for copper by fitting to the phonon frequencies in the high symmetry directions as observed by Sinha[2]; we find that correction terms only up to third neighbors in the total Green function are needed. In fact a total of only seven independent parameters are required to provide a good fit to the observed phonon frequencies compared with a fifteen-parameter model as used in Ref. 2.

The lattice Green function as obtained by the SCGF method has been used to discuss the static atomic configuration associated with a vacancy and a split interstitial in copper. To calculate the defect Green function in these cases we have used the model given by Bullough and Hardy[3]. For the interaction potential between the interstitials and the host atoms, the Born Mayer potential is used.

The SCGF method and its application to copper is given in section 2. The construction of the defect Green function is described in section 3 and its application to the cases of a vacancy and split interstitial is given in sections 4 and 5 respectively. Finally, a short discussion of the SCGF method and the uniqueness of the calculated Green function are given in section 6.

2 DIRECT CONSTRUCTION OF PHONON GREEN FUNCTION

We consider a monatomic Bravais lattice of N atoms subject to periodic boundary conditions and translation symmetry. We shall also assume that the usual adiabatic and harmonic approximations are valid. Let ℓ, ℓ' etc., label the lattice sites with position vectors $r(\ell)$, $r(\ell')$ etc., respectively, and α, β etc. denote the cartesian components x, y and z with respect to a chosen frame of reference. The dynamical matrix of the lattice is defined as

$$\mathbf{D} = \phi \tag{1}$$

where a system of units is assumed in which all the atomic masses are unity and ϕ is the force constants matrix. Both ϕ and $\mathbf{D}$ are square matrices of order $3N$ with their rows and columns labelled by the index doublets α,ℓ and β,ℓ'.

Translation symmetry ensures that $\mathbf{D}(\ell,\ell')$ depends on ℓ and ℓ' only through their difference and thus can be labelled by a single index $\ell-\ell'$. We can therefore define the Fourier transform of the dynamical matrix as follows

$$D_{\alpha\beta}(\mathbf{k}) = \sum_{\ell} D_{\alpha\beta}(\ell)\, e^{2\pi i \mathbf{k}\cdot\mathbf{r}(\ell)} \quad . \tag{2}$$

Similarly we can introduce the Fourier transform of the phonon Green function as

$$G_{\alpha\beta}(\mathbf{k},\omega^2) = \sum_{\ell} G_{\alpha\beta}(\ell,\omega^2)\, e^{2\pi i \mathbf{k}\cdot\mathbf{r}(\ell)} \quad , \tag{3}$$

where $\hbar\omega$ is the phonon energy. The Green function is given by[4]

$$G_{\alpha\beta}(\mathbf{k},\omega^2) = (\mathbf{D}(\mathbf{k}) - \omega^2 \mathbf{I})^{-1}_{\alpha\beta} \tag{4}$$

where $\mathbf{I}$ is the unit matrix.

In an analogous manner the Green function is defined for a continuum as

$$G^c_{\alpha\beta}(\mathbf{k},\omega^2) = (\Lambda(\mathbf{k}) - \omega^2 \mathbf{I})^{-1}_{\alpha\beta} \quad , \tag{5}$$

where $\Lambda(\mathbf{k})$, the Green-Christoffel matrix[5] is given by

$$\Lambda_{\alpha\delta}(\mathbf{k}) = \sum_{\beta,\gamma} c_{\alpha\beta\gamma\delta}\, k_\beta k_\gamma \tag{6}$$

and $c_{\alpha\beta\gamma\delta}$ are the elastic constants in suitable units. The static Green function, which is what we need for static defect calculations is simply the zero-frequency limit of the phonon Green function. For future reference we note that $\Lambda(\mathbf{k})$ is the low $|\mathbf{k}|$ limit of $\mathbf{D}(\mathbf{k})$ and, consequently, in the static case, $\mathbf{G}^c(\ell)$ is the asymptotic limit of $\mathbf{G}(\ell)$.

We shall now construct a parametric representation of the phonon Green function that can be related directly to the neutron-scattering data. For this purpose, the Fourier expansion method which has been described earlier by Aggarwal et al.[6] is used for the frequency spectrum, and Mahanty[7] and Tewary[8] for the Green function. In this method one expands the imaginary part of the Green function in terms of a set of orthonormal functions $f_n(\omega^2)$ as follows

$$\mathrm{Im}\, G_{\alpha\beta}(\ell,\omega^2 - i\,0^+) = \sum_n A_{n\alpha\beta}(\ell)\, f_n(\omega^2) \quad , \tag{7}$$

where ω^2 has been normalized so that

$$0 \leq \omega^2 \leq 1 \quad . \tag{8}$$

The Fourier coefficients A_n are given by[8]

$$A_{n\alpha\beta}(\ell) = (f_n(D))_{\ell,\alpha\beta} \quad , \tag{9}$$

where $f_n(D)$ represents a power series expansion of D within a circle of convergence provided that $f_n(\omega^2)$ are regular and have a power series expansion within the domain given in Eq. 8, and they obey the orthonormality condition

$$\int_0^1 f_n(\omega^2) f_m(\omega^2) \, d\omega^2 = \delta_{nm} \quad . \tag{10}$$

The real part of the Green function can be obtained by using the Kramers Kronig relations[7]. It may be remembered here that only the real part contributes to the static Green function since the imaginary part is zero at the ends of the band.

Taking the Fourier transform of both sides of Eq. 9 and suppressing the Cartesian indices we get

$$A_n(k) = f_n(D(k)) \quad . \tag{11}$$

The transformation that diagonalizes a matrix will also diagonalize all powers of that matrix. Since $f_n(D)$ represents a power series expansion of D, the transformation that diagonalizes $D(k)$ will also diagonalize $A_n(k)$. Thus we arrive at the result that the eigenvectors of $A_n(k)$ are the same as $D(k)$ and the eigenvalues of $A_n(k)$ are $f_n(\omega_j^2(k))$ where $\omega_j^2(k)$, the eigenvalues of $D(k)$, are normalized as in Eq. 8.

Thus we see that if $A_{n\alpha\beta}(\ell,\ell')$ were known, we can construct the matrix $A_n(k)$ from which the polarization vectors and the frequencies of the phonons can be calculated. In practice we can treat the $A_n(\ell)$ as parameters that can be determined so as to get the best fit between the observed phonon data and those calculated by Eq. 11. As the Green function is completely determined in terms of $A_n(\ell)$, Eq. 7 provides the required parametric representation of the Green function.

The transformation properties of $G(\ell,\ell')$ and therefore of $A_n(\ell,\ell')$ are exactly similar to those of $D(\ell,\ell')$. Further, if $f_n(\omega^2)$ are chosen in such a way that $f_n(0) = 0$ we have from Eq. 11

$$\sum_\ell A_n(\ell) = 0 \quad . \tag{12}$$

Eq. 12 is equivalent to the corresponding condition on the force constants $\sum_\ell D(\ell) = 0$ obtained from the condition of invariance of the potential energy against rigid-body translations. With this choice the matrices $A_n(k)$ and $D(k)$ will be exactly similar and the constants $A_n(\ell,\ell')$ can be used to interpret the lattice dynamical data in exactly the same way as the force constants.

The constants $A_n(\ell)$ and the matrix $A_n(k)$ in the present method play the same role as $D(\ell)$ and $D(k)$ in the usual method. In what follows, $A_n(\ell)$ and $A_n(k)$ are referred to as the response constants and the response matrix, respectively. To distinguish between the response constants for various values of n, $A_n(\ell)$ is the nth-order response constants and $A_n(k)$ is the nth-order response matrix. It may be mentioned here that once the first-order response constants have been obtained, all the higher ones follow directly with the help of recurrence relations appropriate to the function f_n[8]. It is, therefore, necessary to obtain only $A_1(\ell)$ by the fitting procedure (as the first-order response constants and the first-order response matrix play the most important part in the present method, for the sake of brevity, the words "first order", and the subscript 1 from $A_1(\ell)$ and $A_1(k)$ are henceforth omitted). The values of n will be explicitly given only for higher-order ($n \geqslant 2$) response constants and response matrices.

Since for large values of ℓ, the response constants can be calculated from the Green-Christoffel matrix we can write

$$A(\ell) = A^c(\ell) + A^d(\ell) \quad , \tag{13}$$

where

$$A^c(\ell) = \frac{1}{N} \sum_k f(\Lambda(k)) \, e^{2\pi i k \cdot r(\ell)} \tag{14}$$

and $A^d(\ell)$ are the correction terms that vanish for ℓ larger than a certain neighbor separation and which measure the deviation of the lattice response constants from the corresponding continuum values. Taking the Fourier transform of Eq. 13 the response matrix becomes

$$A(k) = f(\Lambda(k)) + A^d(k) \quad , \tag{15}$$

which is the sum of a nondispersive and a dispersive term.

For long wave length the lattice is nondispersive. We therefore impose the following condition

$$A^d(0(k^2)) = 0 \tag{16}$$

where $0(k^2)$ indicates that the terms in $A^d(k)$ have been expanded to include quadratic terms in $|k|$. This condition is equivalent to using Born's method of long waves which, for cubic crystals, gives three relations among the parameters $A^d(\ell)$.

Further, since a lattice is largely 'dispersive' for short waves, the nondispersive term in Eq. 15 is modified such that its contribution is reduced for large values of $|k|$, whereas its low $|k|$ behavior is unaffected. A convenient choice is an exponential multiplier, so the response matrix is finally written

$$A(k) = e^{-t^2 a^2 |k^2|} f(\Lambda(k)) + A^d(k) \quad , \qquad (17)$$

where t is a real and arbitrary constant.

The choice of the functions $f_n(\omega^2)$ is quite arbitrary apart from certain restrictions discussed earlier. For instance one can choose Legendre polynomials for $f_n(\omega^2)$ in which case Eq. 7 in the special case of frequency spectrum is equivalent to Montroll's moment trace method. The important criterion for deciding the choice of $f_n(\omega^2)$ is that of convergence in Eq. 7. From a practical point of view a convenient choice[6,8] is the following

$$f_n(\omega^2) = \sin n\pi \, \omega^2 \quad . \qquad (18)$$

With this choice, since $f_n(0) = 0$, we have

$$\sum_\ell A^d(\ell) = 0$$

and the form of $A^d(k)$ becomes identical to the dynamical matrix of the lattice.

To summarize, the present method consists of the following steps:
(i) Write down the general form of matrices $A^d(\ell)$, which is exactly similar to that of the force constant matrices, up to the required number of neighbors.
(ii) From the matrices $A^d(\ell)$ obtain the Fourier transform $A^d(k)$.
(iii) Construct the matrix $\sin \pi\Lambda(k)$ using the elastic constants of the crystal. The response matrix $A(k)$ is then given by Eq. 17 where t is treated as a parameter.
(iv) Diagonalize $A(k)$ and compare its eigenvalues with $\sin(\pi\omega_{oj}^2(k)/\omega_m^2)$ where ω_m is the normalization frequency and $\omega_{oj}(k)$ are the observed phonon frequencies. If measured values of the polarization vectors of the phonons are also available, compare these with the eigenvectors of $A(k)$. The values of the parameters $A^d(\ell)$ and t can then be determined by a least-square-fitting procedure.
(v) After the correction terms $A^d(\ell)$ and t have been obtained, the various elements of the Green function matrix and the frequency spectrum can be

obtained by using either of the following two methods:

2.1 Root Sampling Method

The usual root sampling technique can be used within the framework of the present method. The Green function is given by

$$G_{\alpha\beta}(\ell,\omega^2) = \frac{1}{N} \sum_{k,j} \frac{W_\alpha(k,j) W_\beta(k,j)}{\omega_j^2(k) - \omega^2} \exp[2\pi i k \cdot r(\ell)] \quad , \tag{19}$$

where $W(k,j)$ is the eigenvector corresponding to the eigenvalue $x_j(k)$ of $A(k)$ which is related to $\omega_j(k)$, the eigenvalue of $D(k)$ as follows:

$$\omega_j^2(k) = (\omega_m^2/\pi) \arcsin(x_j(k)) \tag{20}$$

2.2 Fourier Expansion Method

The Fourier expansion method is particularly suited to the SCGF model since the lowest Fourier coefficients are determined directly by the least-square-fitting procedure. The higher-order response matrices may be obtained from

$$[A_n(k)]_{\alpha\beta} = \sum_j W_\alpha(k,j) W_\beta(k,j) \sin n(\arcsin x_j(k)) \quad , \tag{21}$$

and the higher-order response constants follow by using the relation

$$[A_n(\ell)]_{\alpha\beta} = \frac{1}{N} \sum_k [A_n(k)]_{\alpha\beta} \exp[2\pi i k \cdot r(\ell)] \quad . \tag{22}$$

The imaginary part of the Green function can now be calculated by using Eq. 7, and the real part of the Green function can be obtained by using the Kramers-Kronig relations.

2.3 Application to Copper

We now apply the SCGF method to study the lattice dynamics of copper. A fairly extensive theoretical as well as experimental study of the lattice dynamics of copper has been carried out by Sinha.[2] He used a Born von Karman model to calculate the dispersion curves and the frequency spectrum. To obtain a good fit between the calculated and experimental dispersion curves in the high-symmetry directions, he finds that force constants between atoms up to sixth-neighbors separation have to be included. Eighteen independent parameters are thus required of which three can be obtained in terms of the elastic constants.

In the SCGF method we treat $A^d(\ell)$ as parameters. We find a reasonably good agreement with the experimental dispersion curves in the high-symmetry directions by including $A^d(\ell)$ only up to third neighbors. This gives us nine independent parameters of which three are determined by using Eq. 16. Since the constant t in Eq. 17 is also a parameter, we have, thus, a total of seven parameters which have been obtained by a least-square fitting of the calculated and experimental dispersion curves in the high-symmetry directions. We have also calculated[9] the temperature variation of the specific heat and the Debye Waller factor which agree quite well with the experimental results.

The general forms of the $A^d(\ell)$ matrices for the first three neighbors in a face-centered cubic lattice may be written

$$A^d(110) = -\begin{pmatrix} \mu_1 & \delta_1 & 0 \\ \delta_1 & \mu_1 & 0 \\ 0 & 0 & \lambda_1 \end{pmatrix}, \quad A^d(200) = -\begin{pmatrix} \mu_2 & 0 & 0 \\ 0 & \lambda_2 & 0 \\ 0 & 0 & \lambda_2 \end{pmatrix}$$

$$A^d(211) = -\begin{pmatrix} \mu_3 & \delta_3 & \delta_3 \\ \delta_3 & \lambda_3 & \nu_3 \\ \delta_3 & \nu_3 & \lambda_3 \end{pmatrix} \text{ and } A^d(0) = \mu_0 \begin{pmatrix} 1 & 0 & 0 \\ 0 & 1 & 0 \\ 0 & 0 & 1 \end{pmatrix},$$

where

$$\mu_0 = 8\mu_1 + 4\lambda_1 + 2\mu_2 + 4\lambda_2 + 8\mu_3 + 16\lambda_3.$$

The elements of the cubic Green Christoffel matrix (normalized) are[5]

and
$$\Lambda_{\alpha\alpha}(k) = \frac{2a^3}{\omega_m^2}\left[(c_{11}-c_{44})k_\alpha^2 + c_{44}k^2\right]$$

and
$$\Lambda_{\alpha\beta}(k) = \frac{2a^3}{\omega_m^2}(c_{12}+c_{44})k_\alpha k_\beta \quad (\alpha \neq \beta) \quad . \tag{23}$$

Using Eq. 16, we get the following relations

$$\mu_1 + \lambda_1 + 2\lambda_2 + 2\mu_3 + 10\lambda_3 = 0 \tag{24a}$$

$$\mu_1 - \lambda_1 + 2\mu_2 - 2\lambda_2 + 6\mu_3 - 6\lambda_3 = 0 \tag{24b}$$

$$\delta_1 + 8\delta_3 + 2\nu_3 = 0 \quad . \tag{24c}$$

The values of the parameters μ, ν etc., are chosen so as to get the best agreement between the calculated and the observed phonon frequencies in the $\langle 100 \rangle$, $\langle 111 \rangle$ and $\langle 110 \rangle$ directions together with consistency with Eq. 24. This was achieved by using a least-square-fitting procedure. The final values of the parameters are given in Table I.

We have calculated the static Green function on the SCGF model by using the root sampling technique in Eq. 19. These are given for the first few neighbors in Table II. We noticed at this stage the rather curious fact that the static Green function for all values of ℓ as calculated using the SCGF model is almost exactly the same as calculated on the basis of the Bullough and Hardy model[3] and also as calculated on Sinha's model. As the three models are entirely different, this observation suggests that the static Green function is not very sensitive to the details of the lattice model or to the nature and the range of the force constants. We shall return to this point in section 6.

Table I. Values of parameters for the SCGF method

Parameter	Value
μ_1	0.5795×10^{-1}
λ_1	0.2214×10^{-1}
δ_1	0.4647×10^{-1}
μ_2	-0.7579×10^{-2}
λ_2	-0.1677×10^{-4}
μ_3	-0.9504×10^{-2}
λ_3	-0.6099×10^{-2}
ν_3	-0.1464×10^{-3}
δ_3	-0.5768×10^{-2}
t^2	0.4332

3 ATOMIC DISPLACEMENTS IN A LATTICE DUE TO A POINT DEFECT

An otherwise perfect lattice containing a point defect is next considered. The defect, which may be, for instance, a substitutional defect like a vacancy or an interstitial atom, exerts a force on each atom in the lattice, and in general the force constants in the lattice will change. Let **G** and ϕ denote the static Green function and the force constant matrices, respectively, for the imperfect lattice. The corresponding matrices for the perfect lattice are distinguished with the superscript 'o'. If $\delta\phi$ denotes the change in the force constant matrix due to the defect, we can write

$$\phi = \phi^o - \delta\phi \tag{25}$$

and

$$\mathbf{G} = \phi^{-1} = (1 - \mathbf{G}^o \delta\phi)^{-1} \mathbf{G}^o \quad . \tag{26}$$

The last equation can be written as the familiar Dyson equation, i.e.

$$\mathbf{G} = \mathbf{G}^o + \mathbf{G}^o \delta\phi \mathbf{G} \quad . \tag{27}$$

Table II. Static Green functions G_i for a perfect copper lattice on Bullough and Hardy model in units of 10^{-4} dynes^{-1} cm. (Subscript i denotes the Cartesian components in Voigt notation).

Shell	ℓ	G_1	G_2	G_3	G_4	G_5	G_6
0	0,0,0	0.166154	0.166154	0.166154	0	0	0
1	1,1,0	0.052887	0.052887	0.041885	0	0	0.020455
2	2,0,0	0.023665	0.027906	0.027906	0	0	0
3	2,1,1	0.027684	0.024766	0.024766	0.004784	0.007474	0.007474
4	2,2,0	0.026334	0.026334	0.018348	0	0	0.012122
5	3,1,0	0.015699	0.017813	0.016306	0	0	0.003378
6	2,2,2	0.018212	0.018212	0.018212	0.006244	0.006244	0.006244
7	3,2,1	0.017240	0.016486	0.013741	0.003169	0.03559	0.006600
8	4,0,0	0.010430	0.012701	0.012701	0	0	0
9	3,3,0	0.015806	0.015806	0.010356	0	0	0.007738
9	4,1,1	0.010998	0.011860	0.011860	0.001009	0.001944	0.001944
10	4,2,0	0.011366	0.012292	0.010086	0	0	0.003874
11	3,3,2	0.013175	0.013175	0.011467	0.004229	0.004229	0.005626
12	4,2,2	0.011974	0.010645	0.010645	0.002916	0.003915	0.003915
13	4,3,1	0.011688	0.011493	0.008487	0.001934	0.001975	0.005303
13	5,1,0	0.007761	0.009374	0.009023	0	0	0.001241
14	5,2,1	0.008253	0.008860	0.007966	0.001020	0.001273	0.002518
15	4,4,0	0.010570	0.010570	0.006591	0	0	0.005526
16	4,3,3	0.010094	0.009223	0.009223	0.003587	0.004149	0.004149
16	5,3,0	0.008478	0.008896	0.006598	0	0	0.003656
17	6,0,0	0.005712	0.007282	0.007282	0	0	0
17	4,4,2	0.009616	0.009616	0.007427	0.002807	0.002807	0.004729
18	6,1,1	0.005878	0.006967	0.006967	0.000367	0.000872	0.000872
18	5,3,2	0.008680	0.008189	0.007094	0.002223	0.002519	0.003631
19	6,2,0	0.006025	0.007057	0.006304	0	0	0.001746
20	5,4,1	0.008361	0.008313	0.005578	0.001233	0.001225	0.004308
21	6,2,2	0.006281	0.006334	0.006334	0.001191	0.001771	0.001771
22	6,3,1	0.006352	0.006733	0.005444	0.000831	0.000894	0.002603
23	4,4,4	0.007620	0.007620	0.007620	0.003585	0.003585	0.003585
24	7,1,0	0.004397	0.005662	0.005532	0	0	0.000639
24	5,5,0	0.007531	0.007531	0.004421	0	0	0.004286
24	5,4,3	0.007670	0.007246	0.006350	0.002769	0.002988	0.003743
25	6,4,0	0.006427	0.006633	0.004460	0	0	0.003286
26	6,3,3	0.006557	0.005808	0.005808	0.001990	0.002578	0.002578
26	5,5,2	0.007140	0.007140	0.004953	0.001920	0.001920	0.003956
26	7,1,2	0.004592	0.005049	0.005409	0.000448	0.001290	0.000650
27	6,4,2	0.006481	0.006299	0.004825	0.001645	0.001731	0.003245

If $u_\alpha(\ell)$ and $F_\alpha(\ell)$ denote respectively the α-components of the atomic displacement from the position of equilibrium in the perfect lattice and the force on the atom at the lattice site ℓ, then from the definition of the Green function we can write

$$u_\alpha(\ell) = \sum_{\beta,\ell'} G_{\alpha\beta}(\ell,\ell') F_\beta(\ell') \quad,$$

or in the matrix notation

$$\mathbf{u} = \mathbf{G} \, \mathbf{F} \quad. \tag{28}$$

The above equation can also be derived by writing the potential energy of the lattice as a Taylor series in u(ℓ) and then minimizing it with respect to u(ℓ). Thus, we see that the atomic displacements due to a defect in a lattice can be calculated if the strength of the defect, i.e., F(ℓ) and the imperfect Green function are known. The latter can be calculated in terms of the perfect lattice Green function and the change in the force constants.

The formal equivalence of the present method and the Kanzaki method[1] is next established. Using Eq. 27 we can write Eq. 28

$$\mathbf{u} = \mathbf{G_o F^*} \tag{29}$$

where

$$\mathbf{F^*} = \mathbf{F} + \delta\phi \, \mathbf{GF} = \mathbf{F} + \delta\phi \, \mathbf{u} \ . \tag{30}$$

Since $\delta\phi$ is the second derivative of the change in the potential energy and, therefore, the first derivative of the force, Eq. 30 can be identified as the Taylor series expansion of $\mathbf{F^*}$ in terms of $\mathbf{u}$. Thus $\mathbf{F^*}$ can be interpreted as the force on the atoms in their displaced positions. This is exactly the force used in the Kanzaki method.

4 ATOMIC DISPLACEMENTS DUE TO A VACANCY IN COPPER

The main problem in the calculation of u(ℓ) is to calculate the imperfect Green function. For a localized defect, which is assumed to be the case, G can be calculated by using the matrix-partitioning technique as discussed by Maradudin[4]. To calculate $\delta\phi$ and F(ℓ) for a vacancy in copper, the model given by Bullough and Hardy[3] is used. In this model the force constants are assumed to be axially symmetric and are restricted to first and second neighbors of an atom. Taking the origin of coordinates at the vacant lattice site, the change in the force constant matrix is given by (summation convention is not used)

$$\delta\phi(\ell,\ell') = \phi(\ell,\ell') \, \delta_{\ell,0} \ . \tag{31}$$

The form of $\delta\phi(0,\ell)$ in the present model is given below

$$\delta\phi(0,110) = - \begin{pmatrix} \delta_1 & \delta_6 & 0 \\ \delta_6 & \delta_1 & 0 \\ 0 & 0 & \delta_3 \end{pmatrix} \tag{32a}$$

$$\delta\phi(0,200) = - \begin{pmatrix} \mu_1 & 0 & 0 \\ 0 & \mu_2 & 0 \\ 0 & 0 & \mu_2 \end{pmatrix} \tag{32b}$$

where the components of ℓ have been written in units of a, and $2a$ is the lattice constant. The form of $\delta\phi$ for other atoms in the same shell can be obtained by symmetry. The values of δ_ϱ and μ_ϱ are given below, in units of 10^4 dynes cm^{-1}

$$\delta_1 = 1.56220 \qquad \mu_1 = -0.05590$$
$$\delta_3 = -0.22475 \qquad \mu_2 = 0.00815$$
$$\delta_6 = 1.78695$$

For a calculation of the imperfect Green function we partition the matrices G, G^0, and $\delta\phi$ as follows

$$G = \begin{pmatrix} g & G_{12} \\ G_{21} & G_{22} \end{pmatrix} \tag{33a}$$

$$G^0 = \begin{pmatrix} g^0 & G^0_{12} \\ G^0_{21} & G^0_{22} \end{pmatrix} \tag{33b}$$

$$\delta\phi = \begin{pmatrix} \delta\phi & 0 \\ 0 & 0 \end{pmatrix} \tag{33c}$$

where the dimensions of the matrices g, g^0 and $\delta\phi$ are $3n \times 3n$, $G_{12} = \widetilde{G}_{21}$ and $G^0_{12} = \widetilde{G}^0_{21}$ are $3n \times (3N-3n)$ and G_{22} and G^0_{22} are $(3N-3n) \times (3N-3n)$, and $n-1$ is the number of atoms included in Eq. 31. We shall refer to the space of g, g^0 and $\delta\phi$ as the defect space. In the present case its dimensionality is 57, corresponding to the three Cartesian coordinates of the origin, its twelve nearest and six next-nearest neighbors.

Using the partitioned form of the matrices as in Eq. 33 we get the following equation for the Green function in defect space from Eq. 27

$$g = (1 - g^0\, \delta\phi)^{-1}\, g^0 \quad . \tag{34}$$

The preceding matrix equation can be considerably simplified with the help of group theoretical techniques as given by Maradudin[4]. In the present case only Γ_1 – irreducible representation contributes to the atomic displacements, the multiplicity of which is 2. We, therefore, have to solve only a 2 x 2 matrix equation. The results obtained are as expected, identical to those of Bullough and Hardy[3].

5 ATOMIC DISPLACEMENTS AROUND A SPLIT INTERSTITIAL IN COPPER

We shall now calculate the atomic displacements around a "dumbbell" interstitial in copper. In this configuration two copper atoms are symmetrically placed about a vacant lattice site in the <100> direction. Thus the defect is visualized as a vacancy and two interstitials. Let us call the interstitials I_1 and I_2. The vacancy is taken to be the origin and the coordinates of I_1 and I_2 are taken to be x,0,0 and $\bar{x}$,0,0, respectively, where x has to be determined on stability considerations. For the unperturbed case, a perfect lattice is assumed with two free atoms at the positions x,0,0 and $\bar{x}$,0,0. The defect configuration is obtained by removing the atom at the origin and switching on the interaction between I_1, I_2 and the host atoms.

We shall restrict the defect space to the vacancy, I_1, I_2, twelve atoms in the shell (110) and six atoms in the shell (200). The dimensionality of the defect space is 63. The force is also restricted to the defect space. This assumption about the range of force is quite justified in the present case since the interaction between the interstitials and the host atoms is usually assumed to be the Born Mayer type which falls off very rapidly.

To exploit the symmetry of the configuration we note that the lattice is invariant against the following operations

$$xyz, \bar{x}\bar{y}z, \bar{x}y\bar{z}, x\bar{y}\bar{z}, x\bar{z}y, xz\bar{y}, \bar{x}zy, \bar{x}\bar{z}\bar{y}$$

$$\bar{x}\bar{y}\bar{z}, xy\bar{z}, x\bar{y}z, \bar{x}yz, \bar{x}z\bar{y}, \bar{x}\bar{z}y, x\bar{z}\bar{y}, xzy \ .$$

These operators constitute what is known as the group X (e.g., Ref. 10). The relevant irreducible representation in the present case is X_1. By the usual character analysis we find that the multiplicity of X_1 in the defect space is 6. This means that the reduced matrices in Eq. 34 will be 6 x 6. To obtain these we shall need 6 transformation vectors which will be denoted by ψ^i where i stands for I,II... VI. These vectors which can be constructed by standard group theoretical techniques are given in Table III.

The perfect Green function matrix will have two additional matrix blocks corresponding to the two free-particle Green functions. They are given by

$$g^o(I_1,I_1) = g^o(I_2,I_2) = \begin{pmatrix} \Gamma^I & 0 & 0 \\ 0 & \Gamma^I & 0 \\ 0 & 0 & \Gamma^I \end{pmatrix} \qquad (35)$$

LATTICE GREEN FUNCTION

Table III. Components of the vectors ψ^i for dumb-bell configuration.

Atom	ℓ	α	$\psi_\alpha^I(\ell)$ × $1/\sqrt{2}$	$\psi_\alpha^{II}(\ell)$ × $1/\sqrt{8}$	$\psi_\alpha^{III}(\ell)$ × $1/\sqrt{8}$	$\psi_\alpha^{IV}(\ell)$ × $1/\sqrt{8}$	$\psi_\alpha^V(\ell)$ × $1/2$	$\psi_\alpha^{VI}(\ell)$ × $1/\sqrt{2}$
1	0	x	0	0	0	0	0	0
		y	0	0	0	0	0	0
		z	0	0	0	0	0	0
2	x00	x	1	0	0	0	0	0
		y	0	0	0	0	0	0
		z	0	0	0	0	0	0
3	$\bar{x}$00	x	-1	0	0	0	0	0
		y	0	0	0	0	0	0
		z	0	0	0	0	0	0
4	110	x	0	1	0	0	0	0
		y	0	0	1	0	0	0
		z	0	0	0	0	0	0
5	$1\bar{1}0$	x	0	1	0	0	0	0
		y	0	0	-1	0	0	0
		z	0	0	0	0	0	0
6	101	x	0	1	0	0	0	0
		y	0	0	0	0	0	0
		z	0	0	1	0	0	0
7	$10\bar{1}$	x	0	1	0	0	0	0
		y	0	0	0	0	0	0
		z	0	0	-1	0	0	0
8	$\bar{1}\bar{1}0$	x	0	-1	0	0	0	0
		y	0	0	-1	0	0	0
		z	0	0	0	0	0	0
9	$\bar{1}10$	x	0	-1	0	0	0	0
		y	0	0	1	0	0	0
		z	0	0	0	0	0	0
10	$\bar{1}0\bar{1}$	x	0	-1	0	0	0	0
		y	0	0	0	0	0	0
		z	0	0	-1	0	0	0
11	$\bar{1}01$	x	0	-1	0	0	0	0
		y	0	0	0	0	0	0
		z	0	0	1	0	0	0
12	011	x	0	0	0	0	0	0
		y	0	0	0	1	0	0
		z	0	0	0	1	0	0
13	$01\bar{1}$	x	0	0	0	0	0	0
		y	0	0	0	1	0	0
		z	0	0	0	-1	0	0
14	$0\bar{1}1$	x	0	0	0	0	0	0
		y	0	0	0	-1	0	0
		z	0	0	0	1	0	0
15	$0\bar{1}\bar{1}$	x	0	0	0	0	0	0
		y	0	0	0	-1	0	0
		z	0	0	0	-1	0	0
16	020	x	0	0	0	0	0	0
		y	0	0	0	0	1	0
		z	0	0	0	0	0	0
17	$0\bar{2}0$	x	0	0	0	0	0	0
		y	0	0	0	0	-1	0
		z	0	0	0	0	0	0
18	002	x	0	0	0	0	0	0
		y	0	0	0	0	0	0
		z	0	0	0	0	1	0
19	$00\bar{2}$	x	0	0	0	0	0	0
		y	0	0	0	0	0	0
		z	0	0	0	0	-1	0
20	200	x	0	0	0	0	0	1
		y	0	0	0	0	0	0
		z	0	0	0	0	0	0
21	$\bar{2}00$	x	0	0	0	0	0	-1
		y	0	0	0	0	0	0
		z	0	0	0	0	0	0

where Γ^I, the free-particle Green function which is of the form $1/m\omega^2$, is singular in the static limit. The singularity, however, cancels out in the imperfect Green function.

As the atoms I_1 and I_2 are free in the perfect case we have, for $\ell \neq I_1 \neq I_2$

$$g^o(I_1,I_2) = g^o(I_2 I_1) = g^o(I_{1,2},\ell) = g^o(\ell,I_{1,2}) = 0 \;,$$

and other elements of the type $g^o(\ell,\ell')$ are unaffected.

In the defect case the general forms of the force constant, force and displacement matrices, subject to X-symmetry are given below.

$$\phi(x00,\bar{x}00) = - \begin{pmatrix} \zeta_1 & 0 & 0 \\ 0 & \zeta_2 & 0 \\ 0 & 0 & \zeta_2 \end{pmatrix} \quad (36a)$$

$$\phi(x00,110) = - \begin{pmatrix} \eta_1 & \eta_6 & 0 \\ \eta_6 & \eta_2 & 0 \\ 0 & 0 & \eta_3 \end{pmatrix} \quad (36b)$$

$$\phi(x00,\bar{1}\bar{1}0) = - \begin{pmatrix} \nu_1 & \nu_6 & 0 \\ \nu_6 & \nu_2 & 0 \\ 0 & 0 & \nu_3 \end{pmatrix} \quad (36c)$$

$$\phi(x00,011) = - \begin{pmatrix} \rho_1 & \rho_4 & \rho_4 \\ \rho_4 & \rho_2 & \rho_3 \\ \rho_4 & \rho_3 & \rho_2 \end{pmatrix} \quad (36d)$$

$$\phi(x00,200) = - \begin{pmatrix} \chi_1 & 0 & 0 \\ 0 & \chi_2 & 0 \\ 0 & 0 & \chi_2 \end{pmatrix} \quad (36e)$$

$$\phi(x00,020) = - \begin{pmatrix} \kappa_1 & \kappa_6 & 0 \\ \kappa_6 & \kappa_2 & 0 \\ 0 & 0 & \kappa_3 \end{pmatrix} \quad (36f)$$

$$\phi(x00,\bar{2}00) = - \begin{pmatrix} \sigma_1 & 0 & 0 \\ 0 & \sigma_2 & 0 \\ 0 & 0 & \sigma_2 \end{pmatrix} \quad (36g)$$

$$F(x00) = \begin{pmatrix} f_I \\ 0 \\ 0 \end{pmatrix} \quad (37a)$$

$$F(110) = \begin{pmatrix} f_{II} \\ f_{III} \\ 0 \end{pmatrix} \quad (37b)$$

$$F(011) = \begin{pmatrix} 0 \\ f_{IV} \\ f_{IV} \end{pmatrix} \quad (37c)$$

$$F(200) = \begin{pmatrix} f_V \\ 0 \\ 0 \end{pmatrix} \quad (37d)$$

$$F(020) = \begin{pmatrix} 0 \\ f_{VI} \\ 0 \end{pmatrix} \quad (37e)$$

The $u(\ell)$ matrices are the same form as $F(\ell)$ with f_i replaced by ζ_i. The force constants for the vacancy have been given in Eq. 32. The force constants of the type $\phi(\ell,\ell)$ are obtained by imposing the condition

$$\sum_{\ell'} \phi(\ell,\ell') = 0 \quad . \quad (38)$$

For the interaction potential between the interstitials and the host atoms we have assumed the following form

$$V(r) = 0.053 \exp[-13.9 \, (r-a\sqrt{2})/a\sqrt{2}] \quad \text{eV.} \quad (39)$$

The force constants and the force matrices are obtained by taking the derivatives of the interaction potential at the unrelaxed positions of the atoms. For this purpose, the values of x are needed, i.e., the positions of the interstitials in the stable configuration. In the present model the stable configuration is obtained by minimizing the relaxation energy E_r, as defined below, with respect to x

$$E_r = \tfrac{1}{2} \sum_{\alpha,\ell} F_\alpha(\ell) \, u_\alpha(\ell) \quad . \quad (40)$$

We, therefore, calculate $\delta\phi, \mathbf{F}, \mathbf{u}$ and then E_r for various values of x and plot a curve between E_r and x. The minimum value of E_r is found to be 3.95 eV, and at x = 0.5255. This is taken to be the unrelaxed position for the interstitials. However, each interstitial is displaced by an amount 0.0436 along the <100> axis. Thus the final position of the interstitials I_1 and I_2 are at ± 0.5691, 0,0 and the equilibrium separation between the two, in units of a, is 1.14. This value is compared to that obtained by other authors on a semicontinuum model in Table IV, which also contains the present author's results for the relaxation energy as defined in Eq. 40 and the volume change associated with the "dumb-bell" which is given by

$$\Delta V = \frac{1}{3K} \sum_{\alpha,\ell} F^*_\alpha(\ell) \, r_\alpha(\ell) \quad ,$$

where K is the bulk modulus and $\mathbf{F}^*$ is the Kanzaki force.

Our results for the displacements of the atoms at 110, 011, 020 and 200 are given in Table V together with the values reported by Hoekstra

Table IV. Relaxation energy, volume change and the equilibrium separation between the interstitials in dumb-bell configuration in copper.

(Units: relaxation energy in eV, volume change in a^3 and the equilibrium separation in a)

	Relaxation Energy	Volume Change	Equilibrium Separation
Present work	3.95	1.57	1.14
Johnson and Brown[17]	–	1.10	1.2
Huntington[12]	4.43	–	1.16
Gibson et al.[18]	–	–	1.2
Hoekstra and Behrendt[11]	4.55	–	1.16

Table V. Near neighbor atomic displacements due to a split interstitial in <100> direction in copper.

(Units: a)

ℓ	Present Work	Huntington[12]	Hoekstra and Behrendt[11]
	u(ℓ)		
1,1,0	0.0282, 0.0945, 0	0.063, 0.15, 0	-0.0725, 0.1368, 0
0,1,1	0, -0.0154, -0.0154	–	0, -0.0468, -0.0468
2,0,0	0.0239, 0, 0	–	-0.0113, 0, 0
0,0,2	0, 0, -0.0280	0, 0, 0.0113	0, 0, 0.0097

and Behrendt[11] and Huntington[12] for the sake of comparison. However, a comparison with the results of Hoekstra and Behrendt is not very meaningful as they have used a different value for the interaction potential.

It is interesting to see that our result for the equilibrium separation between the two interstitials agrees reasonably well with those obtained by other authors on the basis of a semicontinuum model. This agreement probably can be attributed to the fact that, because of the assumed exponential nature of the interaction potential, the equilibrium separation between the interstitials depends mainly on their interaction with their near neighbors and is not very sensitive to the size of the core. The situation is different for the other quantities that depend upon the response of the lattice as a whole. This is reflected in the wide discrepancy between our results and those obtained by other authors for the atomic displacements, relaxation energy, and volume change.

6 DISCUSSION

The preceding sections have shown how the Green function can be used directly for static defect calculations. In this context the SCGF method for the direct construction of lattice Green function seems to be particularly useful. Perhaps the most important feature of this method is that it does not restrict the nature and the range of the force constants or the interatomic potential in the perfect lattice. As the force constants in most crystals are usually quite long range, the SCGF method in which all the force constants have been effectively included should provide a better physical representation than the usual force-constants models in which the long-range constants are somewhat artificially set to zero. The force constants can be calculated in the present model in terms of the eigenvalues and the eigenvectors of the response matrix $A(k)$ as follows

$$D_{\alpha\beta}(\ell) = \frac{\omega_m^2}{\pi N} \sum_{k,j} W_\alpha(k,j) W_\beta(k,j) \arcsin(x_j(k)) \exp 2\pi i k \cdot r(\ell) \quad . \quad (41)$$

Once the force constants have been obtained, it is possible to obtain some information about the crystal potential by using a technique such as that used by Cowley et al.[13] However, it is known that it is not possible to deduce a unique potential function even if all the force constants were known.[14] Actually the situation is worse because, in general, even the force constants are not known accurately enough. In practice one attempts to derive the force constants from the observed phonon frequencies. It has been shown by Leigh et al. that it is not possible to determine the force constants even if all the phonon frequencies were exactly known.[15] These authors have constructed certain transformations which can continuously change the force constants while keeping the phonon frequencies invariant.

To resolve this uncertainty it is essential to measure the polarization vectors of the phonons.

The transformations constructed by Leigh et al. can also be applied to the SCGF method. Thus it is not possible to determine the response constants uniquely by a measurement of phonon frequencies alone. Perhaps the uncertainty is somewhat reduced in the present case since the response constants for large values of ℓ are exactly determined in terms of the elastic constants.

Fortunately, in the static case the above uncertainty is very much reduced and the Green function is almost completely determined from the elastic constants. To see this we write the static Green function in Eq. 19 in the following form[4]

$$G_{\alpha\beta}(\ell) = \lim_{z \to 0} \int_o^{z_m} \frac{f_{\ell\alpha\beta}(\lambda) \, G_o(\lambda) \, d\lambda}{(\lambda-z)_p} \quad , \qquad (42)$$

where $f(\lambda)$ contains the eigenvectors and the exponential, $G_o(\lambda)$ is the spectrum of squared frequencies, $z = \omega^2$, and λ stands for the distinct phonon frequencies $\omega_j^2(k)$. We see that in the static case ($z=0$) the integrand in Eq. 42 has a pole at $\lambda=0$. The integral therefore is completely determined by the behavior of the integrand near the origin. In this region, $\omega_j^2(k)$ can be written as

$$\omega_j^2(k) = O(k^2) + O(k^4) + \ldots \quad . \qquad (43)$$

The dominant contribution to the integral comes from the terms $O(k^2)$ which can be written down exactly in terms of the elastic constants. The contribution of the terms $O(k^4)$, which are responsible for the dispersive effects, has been calculated by Siems[16] for an isotropic fcc lattice. His result for the static Green function, in our notation, is given below

$$G(\ell) \approx \frac{1}{r(\ell)} \left[1 + \left(3\text{-}5 \frac{\ell_1^4 + \ell_2^4 + \ell_3^4}{r^4(\ell)} \right) \frac{a^2}{64 \, r^2(\ell)} \right] \quad . \qquad (44)$$

We see from Eq. 44 that the contribution of the $1/r^3$ term arising from $O(k^4)$ terms in Eq. 43, is, even for nearest neighbors ($r = a\sqrt{2}$), less than 0.4 percent as compared to the $1/r$ term, which comes from $O(k^2)$ in Eq. 43 and which is completely determined in terms of the elastic constants. This situation is reflected by the fact that the static Green functions as calculated from the SCGF model, Bullough and Hardy model, and Sinha's model are almost the same, presumably because, in the low $|k|$ limit, all three models are exactly the same.

Thus, finally, we infer that as far as the static defect properties are concerned a simple lattice model like that of Bullough and Hardy may be quite adequate. However, even in this case it may be advisable to use the SCGF model since the Green function is explicitly fitted not only to the continuum phonon frequencies near $|\mathbf{k}| = 0$ but also to the continuum Green function. For phonon calculations the SCGF model seems to be definitely superior to the usual force-constant models since it is independent of the nature and range of the interatomic potentials.

REFERENCES

1. Kanzaki, H.: Phys. Chem. Solids **2**: 24 (1957).
2. Sinha, S. K.: P.R. **143**: 422 (1966).
3. Bullough, R. and Hardy, J. R.: Phil. Mag. **17**: 833 (1968).
4. Maradudin, A. A.: Rep. Prog. Phys. **28**: 331 (1965).
5. Federov, F. I.: *Theory of Elastic Waves in Crystals,* Plenum Press, New York (1968).
6. Aggarwal, K. G., Mahanty, J. and Tewary, V. K.: Proc. Phys. Soc. **86**: 1225 (1965).
7. Mahanty, J.: Proc. Phys. Soc. **88**: 1011 (1966).
8. Tewary, V. K.: Proc. Phys. Soc. **92**: 987 (1967).
9. Tewary, V. K. and Bullough, R.: J. Phys. F. (Metal Physics), **1**:554 (1971).
10. Callaway, J.: *Energy Band Theory,* Academic Press Inc., New York (1964).
11. Hoekstra, P. and Behrendt, D. R.: P.R. **128**: 560 (1962).
12. Huntington, H. B.: P.R. **91**: 1092 (1953).
13. Cowley, R. A., Woods, A.D.B. and Dolling, G.: P.R. **150**: 487 (1966).
14. Joshi, S. K. and Rajgopal, A. K.: Solid State Phys. **22**: 159 (1968).
15. Leigh, R. S., Szigeti, B. and Tewary, V. K.: Proc. Roy. Soc. **A320**: 505 (1971).
16. Siems, R.: Phys. Stat. Sol. **30**: 645 (1968).
17. Johnson, R. A. and Brown, E.: P.R. **127**: 446 (1962).
18. Gibson, J. B., Goland, A. N., Milgram, M. and Vineyard, G. H.: P.R. **120**: 1229 (1960).

DISCUSSION on paper by R. Bullough and V. K. Tewary

MARADUDIN: How did you determine the perturbation of the force constant matrix, $\delta\phi$, entering the Dyson equation for the defect Green's function for the split interstitial problem?

BULLOUGH: The detailed determination is given in the paper. The perfect Green function was constructed as for a perfect lattice with two free atoms at (xoo) and ($\bar{x}$oo); the defect Green function then follows from this by removing the atom at the origin and switching on the interaction between these two "free" atoms and the host atoms. It should be noticed that the singularities (in the static limit) associated with the free-particle Green functions disappear completely in the defect Green function. The interaction between the two interstitials and the host atoms is assumed to have the Born-Mayer form, and the force constants and force matrices are obtained from derivatives of this potential at the unrelaxed-atom positions.

ASHCROFT: On the question of the "continuum" versus "discrete" contribution to the lattice Green function, are the relative contributions significantly different for longitudinal and transverse disturbances (modes) in a metal? For shear distortion at constant volume one is not working against large electron gas terms, whereas in the longitudinal disturbances one is.

BULLOUGH: The dispersive contribution to the lattice Green function is equally significant in the longitudinal and shear modes.

COMPUTER SIMULATION OF QUANTUM PHENOMENA

J. H. Weiner and A. Askar

Brown University
Providence, Rhode Island

ABSTRACT

A new procedure, called the particle method, for the solution of the time-dependent Schrödinger equation is described. It uses the amplitude and phase of the complex wave function, and thus avoids some of the oscillatory problems encountered when the real and imaginary parts of this function are employed. The choice of dependent variables leads to the hydrodynamic analogy to quantum mechanics and the method is close in spirit to computation techniques for classical mechanics. Two examples of the method are presented. The first treats wave-packet dynamics on a two-dimensional quadratic barrier; the second involves proton tunneling in a hydrogen-bonded material.

1 INTRODUCTION

The computer simulation of lattice-defect phenomena has proved extremely valuable for obtaining insight into these complex processes. The technique has been employed both to obtain information about the static configuration of atoms in the defect neighborhood and also to study defect dynamics.

In almost all of this work, classical mechanics has been employed as the basis of the mathematical formulation. For many problems this is well justified. However, when we are concerned with light atoms and low temperatures, quantum effects may play a significant — and in some cases an over-riding — role. In dealing with defect dynamics it is then necessary to employ the time-dependent Schrödinger equation in place of the classical equations of motion. This step, of course, greatly complicates the whole procedure since we have now, for a system with N degrees of freedom, replaced a system of ordinary differential equations by a partial differential equation for the wave function $\psi(x_1,....,x_N,t)$.

In the study of defect motion, a significant dynamical problem is the crossing of a potential barrier in this N-dimensional configuration space. It is sometimes possible, as a simplifying step, to identify one or two directions in this space as particularly significant and to confine attention to the subspace spanned by these directions. Even with this simplification, the solution of the time-dependent Schrödinger equation presents difficulties when standard finite-difference procedures with a fixed mesh are employed. A principal source of difficulty is that the complex wave function has spatially oscillatory real and imaginary parts when it represents a moving wave-packet, with the wave-length in inverse ratio to the packet momentum. Therefore, the larger the momentum, the finer the space mesh which must be employed for accurate results. Also, the use of a fixed mesh for moving wave packets leads to continued and wasteful computation in regions in which $|\psi|^2$ has become negligible. Computations of moving wave-packet problems have been performed on the basis just described by McCullough and Wyatt,[1] and the computer time requirements are substantial (20 seconds per time step on the CDC 6600).

Recently, a new method for the numerical solution of the time-dependent Schrödinger equation has been developed which utilizes the amplitude, A, and phase, S, of ψ, rather than its real and imaginary parts. The oscillatory difficulties of the kind described above are thus by-passed. Furthermore, the equations satisfied by A and S lead to the hydrodynamic analogy to quantum mechanics (deBroglie[2] and Madelung[3]) and suggest a numerical procedure corresponding to the use of Lagrangean coordinates in fluid mechanics. Therefore, in moving wave-packet problems, the computational effort is automatically concentrated in the region where $|\psi|^2$ is greatest at any given time. The procedure, called the particle method, has been implemented in both one-spatial dimension (Weiner and Partom[4]) and two (Weiner and Askar[5]). It appears to require about 1/30 of the computer time required by the procedure of McCullough and Wyatt[1].

This paper presents a brief description of the particle method and the results of two examples illustrating its use. The basis of the method, the hydrodynamic analogy to quantum mechanics, is described in Section 2, together with a brief description of the numerical procedure it suggests. As the first example, in Section 3, the results of a computation involving wave-packet dynamics on a two-dimensional quadratic potential barrier is given; in Section 4, a second example treats hydrogen atom tunneling in the hydrogen bond model of Ibers[6]. Some concluding remarks are presented in Section 5.

2 HYDRODYNAMIC ANALOGY AND COMPUTATION PROCEDURE

Consider a particle of mass m subject to a potential $V(\mathbf{x})$. The time-dependent Schrödinger equation governing the evolution of the wave function $\psi(\mathbf{x}, t)$ is*

$$\frac{-\hbar^2}{2m} \nabla^2 \psi + V\psi = i\hbar \frac{\partial \psi}{\partial t} \tag{1}$$

To obtain the hydrodynamic analogy to this equation we write the complex wave function in terms of its amplitude and phase, that is

$$\psi(\mathbf{x}, t) = A(\mathbf{x}, t)e^{iS(\mathbf{x}, t)/\hbar} \tag{2}$$

If this form of ψ is substituted into Eq. 1 and real and imaginary parts equated, the following equations in terms of A and S are obtained:

$$\frac{\partial S}{\partial t} + \frac{1}{2m}(\nabla S)^2 = -V + \frac{\hbar^2}{2m}\frac{\nabla^2 A}{A} \tag{3}$$

$$\frac{\partial A}{\partial t} + \frac{1}{2} A \nabla^2 S + \nabla A \cdot \nabla S = 0 \tag{4}$$

By taking the gradient of Eq. 3 and multiplying Eq. 4 by A we may put these equations in the following form:

$$m\left(\frac{\partial \mathbf{v}}{\partial t} + \mathbf{v} \cdot \nabla \mathbf{v}\right) = -\nabla V - \nabla V_{qu} \tag{5}$$

$$\frac{\partial \rho}{\partial t} + \nabla \cdot (\rho \mathbf{v}) = 0 \tag{6}$$

---

*Here, and in what follows, **x** denotes a two-dimensional vector and the differential operators that appear are of the same dimension. However, the same derivation applies for a configuration space of arbitrary dimension.

Here we have introduced the following notation:

$$\mathbf{v} = \nabla S/m \tag{7}$$

$$\rho = A^2 \tag{8}$$

$$V_{qu} = \frac{-\hbar^2}{2m} \frac{\nabla^2 A}{A} \tag{9}$$

The hydrodynamic analogy follows from Eqs. 5 and 6. They describe the flow of a two-dimensional fluid of particles each acted upon by the prescribed potential $V(\mathbf{x})$ as well as by an additional "quantum" potential $V_{qu}(\mathbf{x},t)$ defined in Eq. 9. Eq. 5 is the equation of motion for this continuum of particles; the terms on the left side is recognized as the material derivative of the particle velocity in the form it assumes when Eulerian coordinates are used to designate the particle. Eq. 6 is the equation of continuity for this continuum.*

The computation procedure adopted for the solution of the time-dependent Schrödinger equation is based on the hydrodynamic analogy. Lagrangean, rather than Eulerian, coordinates are employed. A representative set of the "particles" of this continuum is selected and their trajectories are computed. At the end of each time interval in the numerical computation, the new value of the density $\rho(\mathbf{x},t)$ at each particle is computed by use of the equation of continuity (Eq. 6). This permits the determination of the new value of $V_{qu}(\mathbf{x},t)$ from Eq. 9 by numerical differentiation.** It is then possible to compute the change in velocity and position of each particle in the next time interval from the equations of motion (Eq. 5) and the process may be repeated. Further details regarding this procedure are given in Weiner and Askar.[5]

3 QUADRATIC POTENTIAL BARRIER

The particle method has been implemented thus far in one[4] and two[5] spatial dimensions. As an example of the procedure, some results of a test calculation[5] are reproduced here. The problem is that of the motion of a Gaussian wave packet on a two-dimensional potential surface with a saddle point (Fig. 1). The results of the particle-method calculation are shown in the computer-plotted (Calcomp system) Fig. 2. In Fig. 2(a) are seen the initial positions and velocities of the representative set of "particles" chosen to represent the wave-packet in the context of the hydrodynamic analogy. The subsequent portions of this figure show the

---

*It should be noted that in this analogy the particle mass m and the density ρ are unrelated.

**This step involves numerical differentiation based on the values of ρ at these irregularly positioned neighboring particles. This is a disadvantage in the use of Lagrangean coordinates which may outweigh the economy they introduce if highly distorted flows are encountered. In that case Eulerian coordinates would be preferable.

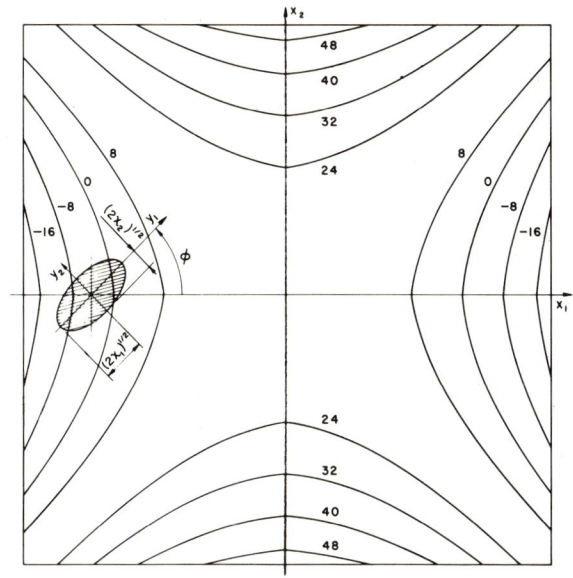

Fig. 1. Gaussian wave packet on a two-dimensional quadratic potential surface with a saddle point, $V(x_1, x_2) = E_b - x_1^2 + x_2^2$, $E_b = 16$. (Units in which $m = 1$, $\hbar = 1$ are employed.) Initial wave-packet conditions (superposed bars denote quantum means): $\bar{x}_1 = -4$, $\bar{x}_2 = 0$, $d\bar{x}_1/dt = 2(6)^{1/2}$, $d\bar{x}_2/dt = 0$; the initial wave-packet principal directions are at an angle $\phi = 45°$ with the x_1, x_2 axes, with principal values of the covariance matrix $\chi_{\alpha\beta}$ given by $\chi_1 = \frac{1}{2}$, $\chi_2 = \frac{1}{8}$.

evolution of this wave-packet. Some features of this process should be noted: (a) the wave-packet rotates in order to align itself with the direction of negative curvature of the saddle point, (b) the wave packet spreads greatly in the direction of the negative curvatures, and (c) a "stagnation region" in the flow develops which moves toward the saddle-point crest. In the last figure of the series, this region has almost reached the crest with substantially all of the particles beyond the crest going down the hill and others returning. The tunneling probability is therefore given, to good accuracy, by the fraction of the particles beyond the crest at this time.

The exact analytical solution is available for this problem.[7] A comparison[5] of the numerical results with this solution shows that the particle method can yield high accuracy with substantially less computer time required than in methods previously employed.

4 HYDROGEN BOND MODEL

As an example of the use of the particle method for a more realistic potential surface, one for which an analytical solution is not possible, we consider next the two-dimensional double-minimum model, Fig. 3, of a hydrogen bond (AH . . . A) introduced by Ibers[6] and used by Singh and Wood[8] in an extensive study of its spectral properties. The Hamiltonian of this model is

$$H = 1/2 \ (k_1\mu_1q_1^2 + k_3\mu_3q_3^2 + k_4\mu_3^2q_3^2 +$$
$$+ k_{133}\mu_1^{1/2}\mu_3q_1q_3^2 + p_1^2 + p_3^2) \quad . \tag{10}$$

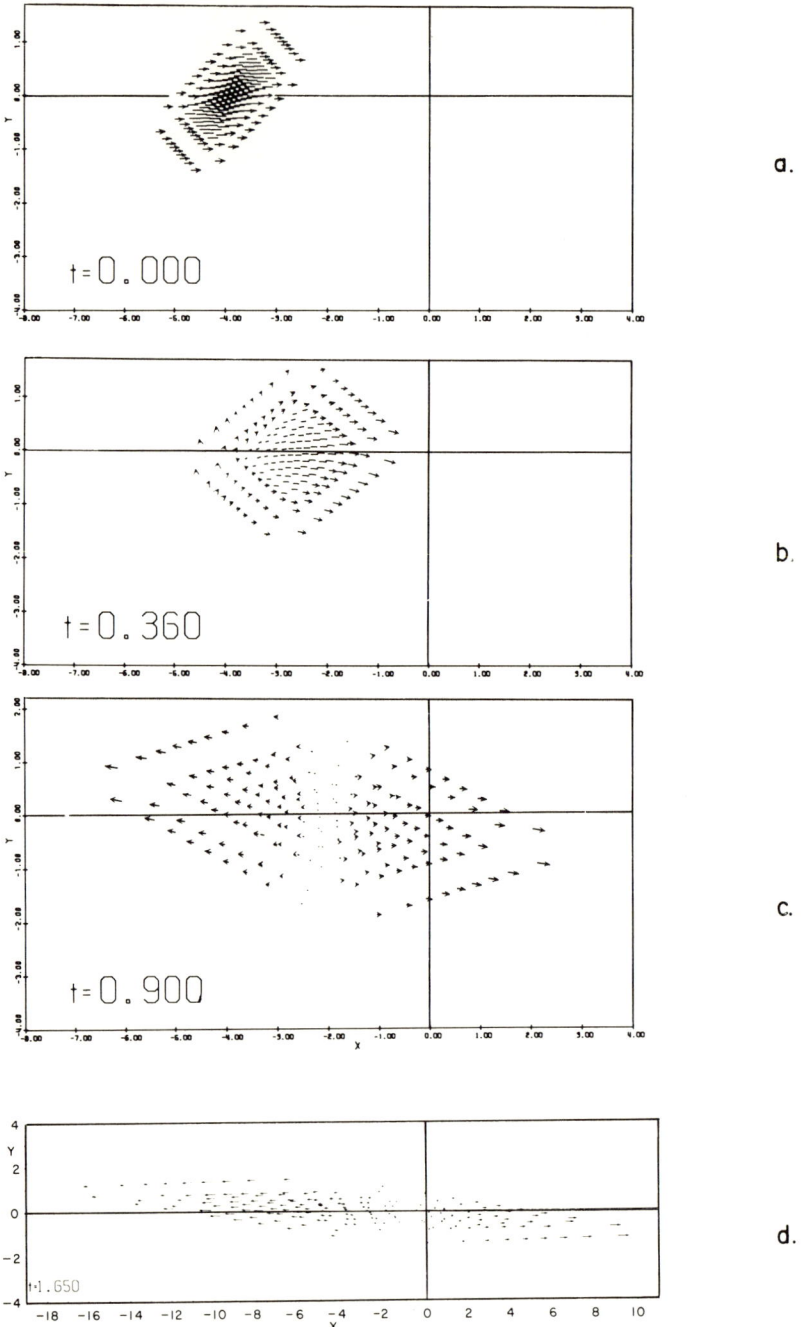

Fig. 2. Display of computer-plotted particle positions and velocities for Gaussian wave-packet problem described in Fig. 1. Intersecting horizontal and vertical lines pass through saddle point in each figure. Particle location is at origin of velocity vector. Note that Fig. 2d is shown to smaller scale than the preceding figures.

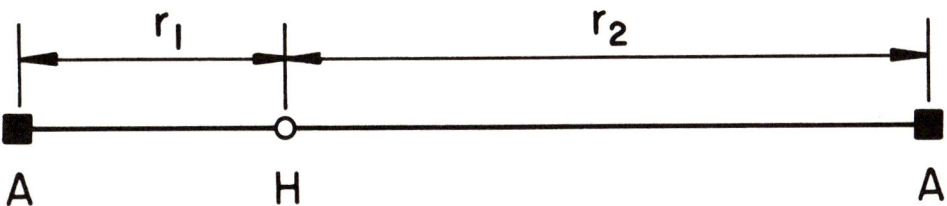

Fig. 3. Hydrogen-bonded system, AH ... A.

This model exhibits a double minimum in the antisymmetric stretching coordinate

$$q_3 = (\Delta r_1 - \Delta r_2)/(2\mu_3)^{1/2}$$

where $\mu_3 = 1/M_A + 2/M_H$, Δr_1, Δr_2 are the changes in distance of H from its two neighboring A's, and includes the effect of interaction with the symmetric stretching coordinate

$$q_1 = (\Delta r_1 + \Delta r_2)/(2\mu_1)^{1/2}$$

where $\mu_1 = 1/M_A$. The calculations are performed for the OH ... O system using the parameter values selected by Singh and Wood[8] as physically reasonable; these are as follows:

$$k_1 = 0.92 \text{ mdyn/Å}$$

$$k_3 = -1.56 \text{ mdyn/Å}$$

$$k_4 = 3.31 \text{ mdyn/Å}^3$$

$$k_{133} = -0.532 \text{ mdyn/Å}^2$$

The resulting potential surface is shown in Fig. 4 (with respect to actual, not mass-reduced coordinates). As initial condition, we choose a gaussian wave packet of the form

$$\psi(Q_1, Q_3, 0) = (2\pi/x_1 x_3)^{1/2} \exp[-Q_1^2/4x_1 - Q_3^2/4x_3 - i\bar{P}_1 Q_1 + \bar{P}_3 Q_3)] \quad (11)$$

Here Q_1, Q_3 are mass-reduced coordinates taken in the directions of principal curvature of the potential energy surface at the left minimum and are zero at that point;

$$x_j = \hbar/2\omega_j \quad , j = 1, 3$$

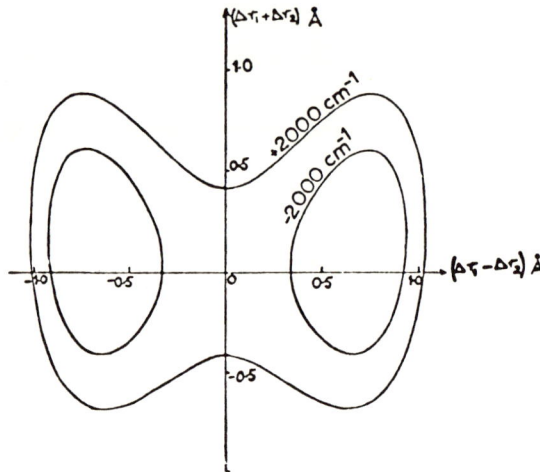

Fig. 4. Potential-energy surface (in actual, not mass-reduced coordinates) for hydrogen-bond model. Reproduced by permission from Singh and Woods[8].

where ω_j^2 are the eigenvalues of the potential-energy matrix expressed in mass-reduced coordinates at the potential minimum.

This choice of initial condition is suggested by an approach[9,4,7] to the theory of quantum rate processes in solids in which an ensemble of states of the form of Eq. 11 is employed, with the distribution of the wave-packet momentum components $\bar{P}_1$, $\bar{P}_3$ suitably chosen to duplicate the characteristics of thermal equilibrium. Based on the analytical solution for wave-packet dynamics on a quadratic potential surface with a saddle point, the assumption was made in this rate theory[7] that the tunneling probability depended only upon the component of initial wave-packet momentum perpendicular to the crest of the saddle point, that is, in the present notation, only upon $\bar{P}_3$. To examine this assumption for the hydrogen-bond model, three calculations were performed. In all three, the initial value of $\bar{P}_3$ was chosen to correspond to a classical kinetic energy of 0.9 E_b, where E_b = 0.02 au is the barrier height. The momentum component $\bar{P}_1$, on the other hand, was taken as 0, ± a, where a corresponds to a classical kinetic energy of 0.45 E_b for motion in the Q_1 direction. The tunneling probabilities obtained from the three computations are as follows:

$\bar{P}_1$	Tunneling Probability
+a	0.457
0	0.481
-a	0.494

It is seen that these preliminary calculations for a more realistic potential surface exhibit behavior similar to the analytical solution for quadratic potential surfaces, namely the tunneling probability is insensitive

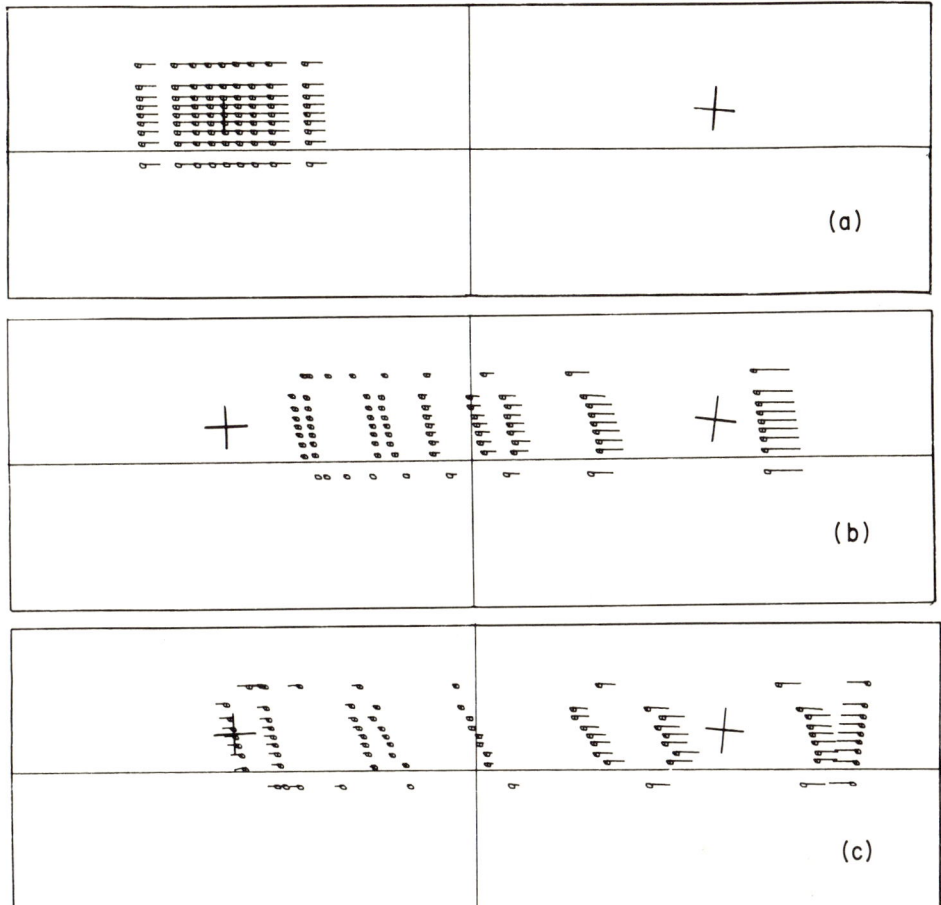

Fig. 5. Display of computer-plotted particle positions and velocities (in actual, not mass-reduced coordinates) for initially gaussian wave-packet moving on potential energy surface for hydrogen-bond model (Fig. 4). Intersecting horizontal and vertical lines pass through saddle point in each figure; crosses indicate location of potential minima. (a) shows the initial wave packet configuration and particle velocities (arrow heads omitted); (b) and (c) show the development at later times t = 172 and 258 (in au), respectively.

to the momentum components parallel to the critical surface through the saddle point.

The nature of the wave-packet evolution is shown in Fig. 5 for the case of $\bar{P}_1 = 0$. The character of the "flow" for this potential surface is quite similar to that for a quadratic surface (Fig. 2) although some differences may be noted; (i) the wave packet does not remain gaussian and some density concentrations occur (Fig. 5b), and (ii) the start of the return of some particles may be seen in Fig. 5c. The negligible effect of the value of $\bar{P}_1$ on the flow is shown in Fig. 6 in which the final stages in the flow for the three cases are compared.

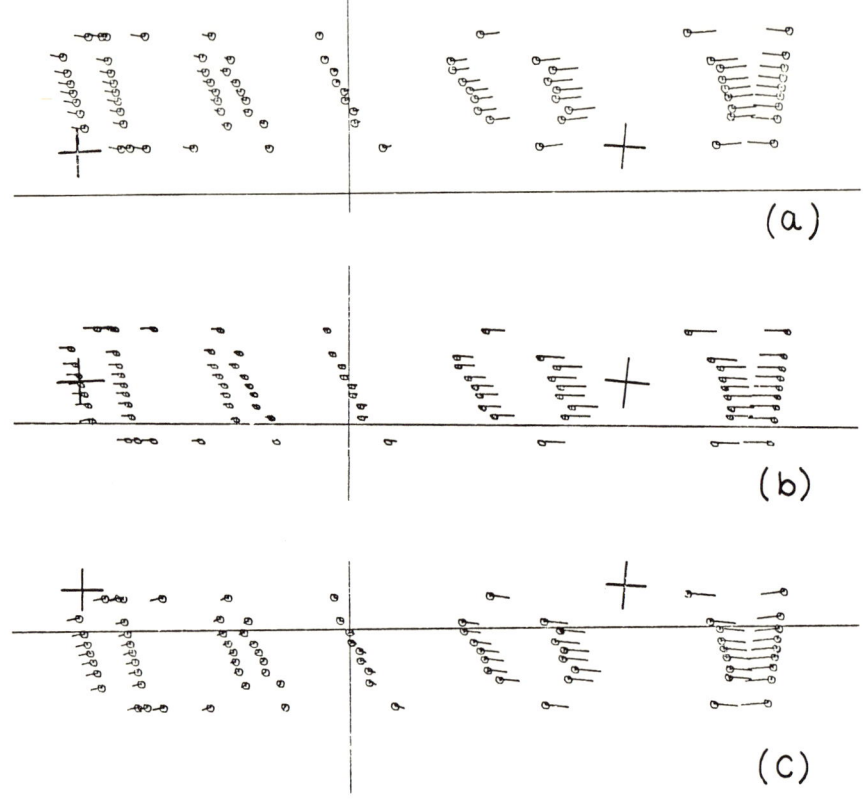

Fig. 6. Comparison of flow patterns at t = 258 (au) for the case of initial $\bar{P}_1$ positive (a), zero (b), and negative (c).

5 CONCLUSIONS

The method described for the use of quantum mechanics in computer simulation is very close in spirit to the classical approach. This should facilitate the utilization in the quantum regime of the experience gained in the extensive classical computations that have been performed. The similarity between the approaches also provides a clear indication of the increase in complexity and computer-time requirements introduced by the transition from classical to quantum mechanics; the trajectory computation of a single particle with N degrees of freedom is replaced by a similar problem for a set of "particles", with the number of such "particles" required being of the order of 10^N. This means that, at present, it is necessary to focus on one or two significant degrees of freedom to be treated on a quantum basis, with the remainder still treated classically.

REFERENCES

1. McCullough, Jr. E. A., and R. E. Wyatt: J. Chem. Phys., **54**: 3578 (1971).
2. de Broglie, L.: Compt. Rend., **183**: 447 (1926).
3. Madelung, E.: Z. Physik, **40**: 322 (1926).
4. Weiner, J. H., and Y. Partom: Phys. Rev., **187**: 1134 (1969).
5. Weiner, J. H., and A. Askar: J. Chem. Phys., **54**: 3534 (1971).
6. Ibers, J. A.: J. Chem. Phys., **41**: 25 (1964).
7. Weiner, J. H., and Y. Partom: Phys. Rev. B, **1**: 1533 (1970).
8. Singh, T. R., and J. L. Wood: J. Chem. Phys., **48**: 4567 (1968).
9. Weiner, J. H.: Phys. Rev., **169**: 570 (1968).

DISCUSSION on paper by J. H. Weiner and A. Askar

VINEYARD: Have you compared your results with calculations of averaged tunneling probability by use of Wigner distribution functions?

WEINER: No, we have not. We have preferred to consider averages over ensembles of minimum-uncertainty wave-packets since this ensemble can be described in terms of a probability distribution that is never negative, whereas this is not the case for the Wigner distribution.

VINEYARD: Have you looked at the effect of the initial shape of the wave-packet on the transmission probability?

WEINER: Since we are concerned primarily with ensembles of minimum-uncertainty wave-packets, we have only treated gaussian wave-packets, but we have considered the effect of the packet orientation.

WILSON: Does your calculation give the McCullough-Wyatt[1] result in the slow wave-packet limit?

WEINER: The McCullough-Wyatt method and the particle method are inherently different and do not coalesce in the slow wave-packet limit. Of course the results of the procedures should agree within the limits of numerical error.

WILSON: It seems that it should be possible to adjust the coefficients in your potential. Can you remove the tunneling by making the barriers artificially too high?

WEINER: Yes, it is easy to explore the effect of different potential shapes with this method and to observe the tunneling decrease by increasing barrier height and thickness.

SEEGER: Have you considered the possibility of applying your formalism to scattering problems, e.g., to nonseparable potentials?

WEINER: We have considered using the particle method for the solution of scattering problems. The nonseparability of the potential does not introduce any inherent difficulties, and the hydrogen-bond model is an example of such a calculation. However, if the "flow" gets too complex, it may be necessary to use Eulerian coordinates and give up the time-saving feature of Lagrangean coordinates.

SEEGER: In the diffusion of hydrogen in metal there are interesting and puzzling phenomena, e.g., the isotope effect for hydrogen, deuterium, and tritium in palladium. Have you considered any of these questions?

WEINER: The theory of rate processes including quantum effects provided the prime motivation for the development of this computation technique, and we plan to use it to investigate such questions as the anomalous isotope effect in hydrogen diffusion.

ASHCROFT: The classical limit ($\hbar \to 0$) is well known to be consistent with the hydrodynamic analogy. Under certain well-defined conditions an expansion in $\hbar$ leads to the quasiclassical (WKB) approach and I wonder if there are any WKB-type solutions to problems that you can check your computer results against?

WEINER: No, we haven't explored this, particularly since we had available the exact quantum solutions for gaussian wave-packets on quadratic surfaces to check the numerical procedure.

ASHCROFT: What (in terms of your V_{qu}) is a simple criterion for the important of quantum effects?

WEINER: The hydrodynamic analogy yields the following criterion: If, for a well-localized wave packet relative to significant system dimensions, $\nabla V_{qu} \ll \nabla V$ then the system will behave classically.

REFERENCE

1. See Ref. 1.

ON THE VALIDITY OF TWO-BODY POTENTIALS IN METALS

L. A. Girifalco and T. M. Di Vincenzo

School of Metallurgy and Materials Science
and
Laboratory for Research on the Structure of Matter
University of Pennsylvania
Philadelphia, Pennsylvania 19104

1 INTRODUCTION

The use of wave functions of the Bloch type for metals leads to a crystal energy in which the contribution of the valence electrons depends explicitly on the crystal volume.[1,2] Except for the contribution of the core-core interactions, the atomic positions do not occur explicitly in the crystal energy, although they enter as parameters defining the crystal potential. The reason for this is that the Bloch functions are extended throughout the crystal and are not preferentially associated with any particular atom. On this basis, therefore, the crystal energy of a metal cannot be expressed as a sum of interatomic potentials.

In spite of this, the convenience and conceptual simplicity of an interatomic potential representation of the crystal energy has led to its widespread use in both perfect- and defect-metal crystals. The methodology consists of choosing an empirical two-body potential with disposable parameters that are determined from experimental data. Fundamentally, the weaknesses of such an approach are obvious. However, the two-body potential model continues to generate useful results and enjoys a considerable calculational advantage over the quantum mechanical methods, particularly when applied to defect structures.

In view of the above, it is of considerable interest to investigate the extent to which the interatomic potential model in metals can be justified on sound quantum theoretic grounds. Some progress on this question has been made using pseudopotential theory in which it is shown that an effective atom-atom interaction can be defined.[3]

In this paper, the relation between the band and bond concepts in metals is studied by introducing the localized Wannier functions[4] into the theory of the crystal energy. It is shown that a purely volume dependent term always exists, but that, also, a two-body contribution depends explicitly on the crystal structure. Further, three- and four-body interactions also occur, although they are expected to contribute considerably less to the crystal energy than the volume-dependent and two-body terms.

2 WANNIER REPRESENTATION OF THE CRYSTAL ENERGY

While the basic formalism has been given previously[5], it is presented here in abbreviated form for the sake of completeness.

The energy for a monatomic one-electron metal is written as

$$E = E_{HF} + \frac{1}{2} \sum_{i,j}' \frac{e^2}{|R_i - R_j|} + E_c(V), \qquad (1)$$

where E_{HF} is the Hartree-Fock energy of the conduction electrons, the double sum is the interaction energy of all the ion cores at positions R_j, and $E_c(V)$ is the correlation energy which we assume depends only on the volume V. The HF energy is given by

$$E_{HF} = 2 \sum_k \langle k|\hat{H}_o|k\rangle + 2 \sum_{k,k'} \langle kk'|Q|kk'\rangle - \sum_{k,k'} \langle kk'|Q|k'k\rangle, \qquad (2)$$

where

$$Q \equiv e^2/|r_1 - r_2| \qquad (3)$$

and

$$\hat{H}_o \equiv -(\hbar^2/2m)\nabla^2 + V(r). \qquad (4)$$

$V(\mathbf{r})$ is the potential of an electron in the field of all ion cores, which can be written

$$V(\mathbf{r}) = \sum_j V_j(\mathbf{r}) ,\qquad(5)$$

$V_j(\mathbf{r})$ being the potential of an electron in the field of the jth ion core.

The second and third terms in Eq. 2 contain the Hartree and exchange matrix elements, respectively, and all sums are over occupied **k** states. The $|\mathbf{k}\rangle$ are the Bloch states that minimize E_{HF} and are determined by the HF equation

$$\hat{H}_o |\mathbf{k}\rangle + 2\sum_{\mathbf{k}'} \langle \mathbf{k}'|Q|\mathbf{k}'\rangle |\mathbf{k}\rangle - \sum_{\mathbf{k}'} \langle \mathbf{k}'|Q|\mathbf{k}\rangle |\mathbf{k}'\rangle = \epsilon_{\mathbf{k}} |\mathbf{k}\rangle ,\qquad(6)$$

where $\epsilon_{\mathbf{k}}$ is the energy required to remove an electron in state **k** from the crystal.

Wannier functions can be defined in terms of the Bloch states by*

$$|j\rangle \equiv \frac{1}{\sqrt{N}} \sum_{\mathbf{k}}^{BZ} e^{-i\mathbf{k}\cdot\mathbf{R}_j} |\mathbf{k}\rangle ,\qquad(7)$$

where the sum is over the Brillouin zone (BZ). Also

$$|\mathbf{k}\rangle = \frac{1}{\sqrt{N}} \sum_j e^{i\mathbf{k}\cdot\mathbf{R}_j} |j\rangle \qquad(8)$$

The Wannier function $|j\rangle$ is centered on the atom at $\mathbf{R}_j$ and falls off exponentially at large distances from this atom. Some lack of uniqueness is in the definition of Wannier functions in that they may be multiplied by a phase factor periodic in **k**. However, this does not affect the present problem, since the phase factors multiply to unity when Eq. 7 and 8 are substituted into Eq. 2. The result of this substitution is

$$E_{HF} = \sum_{i,j} F(j-i) \langle i|\hat{H}_o|j\rangle + \frac{1}{4} \sum_{i,j,n,m} B^{ij}_{nm} \langle ij|Q|nm\rangle ,\qquad(9)$$

where

$$B^{ij}_{nm} \equiv 2F(n-i) F(m-j) - F(m-i) F(n-j) ,\qquad(10)$$

$$F(\ell) \equiv \frac{2}{N} \sum_{\mathbf{k}} e^{i\mathbf{k}\cdot\mathbf{R}_\ell} \qquad(11a)$$

---
*See Ref 6 and 7 for discussions of Wannier Functions.

$$B^{ij}_{jj} = 2F(i\text{-}j) - F(j\text{-}i) = F(i\text{-}j) \;, \tag{11b}$$

the sum being over all occupied **k** states. Note that, from Eq. 11,

$$F(0) = 1 \;, \tag{12a}$$

$$F(\ell) = F(-\ell) \;, \tag{12b}$$

and that Eq. 10 gives the following symmetry relations:

$$B^{ij}_{nm} = B^{ji}_{mn} = B^{nm}_{ij} \;. \tag{13}$$

A trivial rearrangement of the terms in Eq. 9 shows how E_{HF} can explicitly be written as a sum of one-, two-, three-, and four-body atomic interactions. To do this, rewrite Eq. 9, using Eq. 4 and 5, as

$$E_{HF} = -\frac{\hbar^2}{2m} \sum_{i,j} F(j\text{-}i) \langle i|\nabla^2|j\rangle + \sum_{i,j,n} F(j\text{-}i) \langle i|V_n|j\rangle$$

$$+ \frac{1}{4} \sum_{i,j,n,m} B^{ij}_{nm} \langle ij|Q|nm\rangle \;. \tag{14}$$

Now collect terms of the type for which all four subscripts i,j,n,m are equal, for which three of the subscripts are equal, for which two of the subscripts are equal, and for which none are equal. This converts Eq. 14 to the form

$$E_{HF} = \sum_i U_i^{(1)} + \sum_{i,j}{}' U_{ij}^{(2)} + \sum_{i,j,n}{}' U_{ijn}^{(3)} + \sum_{i,j,n,m}{}' U_{ijnm}^{(4)} \;, \tag{15}$$

where

$$U_i^{(1)} = -\frac{\hbar^2}{2m} \langle i|\nabla^2|i\rangle + \langle i|V_i|i\rangle + \frac{1}{4}\langle ii|Q|ii\rangle \tag{16}$$

$$U_{ij}^{(2)} = -\frac{\hbar^2}{2m} F(j\text{-}i)\langle i|\nabla^2|j\rangle + \langle i|V_j|i\rangle + F(j\text{-}i)[\langle i|V_i|j\rangle + \langle i|V_j|j\rangle]$$

$$+ \frac{1}{4} B^{ii}_{jj}\langle ii|Q|jj\rangle + \frac{1}{4} B^{ij}_{ji}\langle ij|Qj|ji\rangle + \frac{1}{4} B^{ij}_{ij}\langle ij|Q|ij\rangle$$

$$+ B^{ii}_{ij}\langle ii|Q|ij\rangle \tag{17}$$

$$U_{ijn}^{(3)} = F(j\text{-}i)\langle i|V_n|j\rangle + \frac{1}{2}\left[B^{ii}_{jn}\langle ii|Q|jn\rangle + B^{ij}_{ni}\langle ij|Q|ni\rangle\right.$$

$$\left. + B^{ij}_{in}\langle ij|Q|in\rangle\right] \tag{18}$$

$$U^{(4)}_{ijnm} = \frac{1}{4} B^{ij}_{nm} \langle ij|Q|n\,m\rangle \tag{19}$$

The primes on the multiple sums in Eq. 15 indicate that none of the subscripts on the summands are equal.

These equations show that the Hartree-Fock energy can be decomposed into a set of interactions among atomic sites.

A pair-wise representation of the energy can be obtained if we assume that the Wannier functions are sufficiently localized to make the $U^{(3)}$ and $U^{(4)}$ terms negligible relative to $U^{(1)}$ and $U^{(2)}$. Then, since all the $U_i^{(1)}$ are equal, and the $U_{ij}^{(2)}$ depend only on the difference $(j-i)$, we can write Eq. 15 as

$$E_{HF} = N U_o^{(1)} + \frac{N}{2} \sum_j 2 U_{oj}^{(2)} , \tag{20}$$

where $U_o^{(1)}$ can be interpreted as the energy of a unit cell in the absence of interactions with other cells and $2U_{oj}^{(2)}$ can be interpreted as the "bond energy" between a central atom (labeled o) and the jth atom.

Several other results that display the connection between the Wannier (bond) and Bloch (band) representations can be written.[5] Thus, it was shown that the Wannier functions satisfy the following equation:

$$\hat{H}^F|\ell\rangle = \sum_i \mathcal{E}_{io}|\ell + j\rangle \tag{21}$$

where the $\mathcal{E}_{io}$ are the coefficients of the Fourier expansion of the energy

$$\mathcal{E}_{io} = \frac{1}{N} \sum_k^{BZ} \epsilon_k\, e^{i\mathbf{k}\cdot\mathbf{R}_i} , \tag{22}$$

$$\epsilon_k = \sum_j \mathcal{E}_{jo}\, e^{-i\mathbf{k}\cdot\mathbf{R}_j} , \tag{23}$$

and $\hat{H}^F$ is the Wannier representation of the Hartree-Fock operator

$$\hat{H}^F = \hat{H}_o + \sum_{i,j} F(j-i)\left(\langle i|Q|j\rangle - \frac{1}{2}\langle i|Q|\ell\rangle \frac{|j\rangle}{|l\rangle}\right) \tag{24}$$

Eq. 21 is just a generalization of Koster's equation[8] to a Hartree-Fock Hamiltonian, and is, in fact, valid for any one-electron Hamiltonian.

An alternative expression of the HF equation in the Wannier representation is

$$\sum_j F(j-i)\hat{H}_o|j\rangle + \frac{1}{2}\sum_{j,n,m} B^{in}_{jm}\langle n|Q|m|j\rangle = \sum_j \lambda_{ij}|j\rangle , \tag{25}$$

where

$$\lambda_{ij} \equiv \frac{2}{N} \sum_k^{occ.} \epsilon_k\, e^{i\mathbf{k}\cdot(\mathbf{R}_j - \mathbf{R}_j)} \tag{26}$$

The λ_{ij} and the $\mathcal{E}_{ij}$ are related by

$$\lambda_{ij} = \sum_n F(n-i)\, \mathcal{E}_{jn} \tag{27}$$

and λ_{oo} is the average energy of the electrons in the band given by

$$\bar{\epsilon} = \lambda_{oo} = \langle o|\hat{H}^F|o\rangle + \sum_{n \neq o} F(n)\, \langle o|\hat{H}^F|n\rangle \;, \tag{28}$$

which gives the average energy as an integral centered on one unit cell, plus a series of atomic interactions that decrease as the distance from the central atom decreases.

The width of the occupied portion of the band and the effective mass are related to the Wannier functions by[5]

$$\epsilon_{k_F} - \epsilon_o = -\sum_{n(\neq o)} (1 - e^{-i k_F \cdot R_n})\, \langle o|\hat{H}^F|n\rangle \tag{29}$$

$$\frac{1}{m^*_{\alpha\beta}} = -\frac{1}{\hbar^2} \sum_{n(\neq o)} R_n^\alpha R_n^\beta\, \langle o|\hat{H}^F|n\rangle \;, \tag{30}$$

where R_n^α and R_n^β are the vector components of $\mathbf{R}_n$.

If the Wannier formalism is applied to the HF energy of a full band, the result is

$$E_{\mathrm{HF}} = 2\sum_i \langle i|\hat{H}_o|i\rangle + \sum_{i,j} (2\langle ij|Q|ij\rangle - \langle ij|Q|ji\rangle) \tag{31}$$

Thus, for full bands, three- and four-body forces do not exist, at least in the Hartree-Fock approximation.

3 THE MATRIX ELEMENTS AND THE BAND ENERGY

In this section are presented the results of a model for the matrix elements of the kinetic energy and the electron-ion potential. Some work has been done on the electron-electron Coulomb matrix elements,[9] but is not presented here.

An estimate of the "unit cell" energy, $U_o^{(1)}$ can be made by adopting the model embodied in the following assumptions:

(1) The ion-core radius is small enough so that there is a negligible interaction among the electrons on different ion-cores.
(2) Outside an ion-core, which is assigned a definite radius r_I, the Wannier function can be constructed from plane-wave Bloch functions. Inside the ion-core, the Wannier function is identical to the atomic function of the valence electron in the free atom.

(3) Outside an ion-core, the electron-ion potential for that core is Coulombic. Inside the ion-core, the potential is identical to that in the free atom.
(4) The Brillouin zone and the Wigner-Seitz cell are spherical.

Because of Assumptions (2) and (3), the contribution of the ion-core region to the matrix elements is the same in the crystal as in the free atom. Since the free atoms are taken as the reference zero in defining the crystal energy, the ion-core contribution cancels out. Therefore, the one-body matrix elements can be computed by performing the integrations over the region outside the ion-core.

This model should be a fair representation of the alkali metals. It ignores the fuzziness of the ion-core boundary, and the required continuity of the Wannier function and the electron-ion potential in the core boundary region.

With these assumptions, the matrix elements in Eq. 16 can be evaluated.* The results are, for the kinetic energy and the electron-ion potential,

$$\langle o|\nabla^2|o\rangle = -\frac{3}{5}\left(\frac{9\pi}{2}\right)^{2/3}\left(1 - \frac{v_c}{v}\right)\frac{1}{r_s^2} \qquad (32)$$

$$\langle o|V_o|o\rangle = \frac{e^2 C(\gamma)}{3 \pi^2 \gamma^4 r_s} \qquad (33)$$

$$C(\gamma) \equiv \gamma k_B r_s \sin(2\gamma k_B r_s) - \sin^2(\gamma k_B r_s) + (\gamma k_B r_s)^2 \cos(2\gamma k_B r_s) - 2(\gamma k_B r_s)^2 \cos^2(\gamma k_B r_s) \quad . \qquad (34)$$

In these equations, r_s is the Wigner-Seitz radius, v_c/v is the ratio of the ion-core volume to the unit cell volume, k_B is the radius of the Brillouin zone and $\gamma = r_I/r_s$, where r_I is the ionic radius.

Assumptions (2) and (3) of the model cannot be applied so simply to the two-body matrix elements, since the model requires the exclusion of two separate spherical holes from the region of integration, and also requires the use of the atomic functions and potentials. It can be shown that relaxing these assumptions and performing the integrations over all space, using a hydrogenic potential and plane wave Wannier functions, introduces small errors in the matrix elements. However, when sums over the crystal are performed, it is not yet known what magnitude error is introduced in the crystal energy.

With these relaxed assumptions, the two-center kinetic energy matrix element is

---

*See Ref 9 for details of all calculations.

$$\langle o|\nabla^2|j\rangle = -\frac{2\,r_s^3}{3\,\pi R_j^5}\left[(3\,k_B^2\,R_j^2 - 6)\sin(k_B R_j)\right.$$

$$\left. - (k_B^3 R_j^3 - 6)\cos(k_B R_j)\right], \qquad (35)$$

and the electron-ion potential energy matrix elements are

$$\langle o|V_j|o\rangle = \frac{-e^2}{\pi R_j}\left[\frac{\sin^2(k_B R_j)}{2(k_B R_j)^3} + \frac{\sin(2k_B R_j)}{2(k_B R_j)^2} + \frac{5\cos(2k_B R_j)}{2\,k_B R_j}\right.$$

$$\left. - \frac{3\cos^2(k_B R_j)}{k_B R_j} + 2\,Si(2k_B R_j)\right] \qquad (36)$$

$$\langle o|V_j|j\rangle = \langle o|V_o|j\rangle , \qquad (37)$$

and

$$\langle o|V_o|j\rangle = -\frac{3e^2}{\pi R_j}\left[-\int_0^\infty \frac{W\sin(y + k_B R_j)}{(y + k_B R_j)}dy\right.$$

$$\left. + \int_0^{k_B R_j}\frac{W\sin(k_B R_j - y)}{(k_B R_j - y)}dy + \int_{k_B R_j}^\infty \frac{W\sin(y - k_B R_j)}{(y - k_B R_j)}dy\right], \qquad (38)$$

where

$$W \equiv \frac{1}{y^3}[\sin y - y\cos y] \qquad (39)$$

The integrals in Eq. 38 could not be reduced further and must be evaluated numerically.

To use the above results in crystal energy calculations, we also need $F(j)$ of Eq. 11. For a spherical Fermi surface, this is readily found to be

$$F(j) = \frac{3}{(k_F R_j)^3}\left[\sin(k_F R_j) - k_F R_j\cos(k_F R_j)\right], \qquad (40)$$

where k_F is the radius of the Fermi sphere.

Eq. 32 to 40 are sufficient to compute the energy of the Hamiltonian $[-\frac{\hbar^2}{2m}\nabla^2 + V(\mathbf{r})]$ within the limits of the model. Furthermore, they permit a separation of the energy into a unit cell volume-dependent term, and pairwise interactions. If a pairwise interaction energy between the zeroth and j^{th} sites is defined by

$$\frac{1}{2}g_{oj} = -\frac{\hbar^2}{2m}F(j)\langle o|\nabla^2|j\rangle + \langle o|V_j|o\rangle + 2F(j)\langle o|V_j|j\rangle , \qquad (41)$$

and a "unit cell" energy by

$$u = -\frac{\hbar^2}{2m} \langle o|V^2|o\rangle + \langle o|V_o|o\rangle , \qquad (42)$$

then the crystal energy can be written, exclusive of the electron-electron interactions, as

$$E' = Nu + \frac{N}{2}\sum_{j(\neq o)} g_{oj} + \sum_{j(\neq o)} \frac{e^2}{R_j} \qquad (43)$$

Half of the last term in Eq. 43 comes from the ion-ion interaction, and the other half from the coulombically divergent part of the electron-electron interaction. From Eq. 36, it is clear that $\langle o|V_j|o\rangle$ becomes coulombic and cancels the last term in Eq. 43 at large distances. In fact, numerical calculation shows that it is approximately coulombic even at short distances. The effective coulomb constant at the nearest-neighbor distance in sodium was found to be $-0.885\ e^2$, decreasing to e^2 at large R_j, as shown in Table IV.

Accuracy in computing $\langle o|V_j|o\rangle$ is particularly critical in crystal energy calculations because this term decays slowly. A small error has a serious effect on crystal sums, since it can cause an unreal, computational extension of the range over which sums must be carried before the $\langle o|V_j|o\rangle$ terms cancel the e^2/R_j terms in Eq. 43.

The other terms in Eq. 41 decay rapidly with distance, and while inaccuracies in their computation certainly introduce errors in crystal sums, they do not introduce divergence difficulties.

4 THE BAND ENERGY IN THE EFFECTIVE MASS APPROXIMATION

According to Eq. 23, we can write the average energy as

$$\bar{\epsilon} = \frac{2}{N}\sum_{k}^{k_F} \epsilon_k = \frac{2}{N}\sum_k \sum_j \mathcal{E}_{jo}\, e^{-k \cdot R_j} \qquad (44)$$

or

$$\bar{\epsilon} = \sum_j F(j)\, \mathcal{E}_{jo} . \qquad (45)$$

In simple quadratic form

$$\epsilon_k = \epsilon_o + \frac{\hbar^2 k^2}{2m^*}, \qquad (46)$$

where ϵ_o is the lowest energy in the band.

Eq. 46 is equivalent to taking the Bloch functions to be plane waves. This form can be used to evaluate $\mathcal{E}_{jo}$. Thus, substituting Eq. 46 into 22,

$$\mathcal{E}_{jo} = \epsilon_o \, \delta_{j,0} + \frac{1}{N} \sum_{k}^{k_B} \frac{\hbar^2 k^2}{2m^*} e^{i\mathbf{k}\cdot\mathbf{R}_j} \; . \tag{47}$$

Substituting this into Eq. 45 gives

$$\bar{\epsilon} - \epsilon_o = \frac{1}{N} \sum_j F(j) \sum_{k}^{k_B} \frac{\hbar^2 k^2}{2m^*} e^{i\mathbf{k}\cdot\mathbf{R}_j} \; . \tag{48}$$

In the plane-wave approximation, it is easy to show that

$$\frac{1}{N} \sum_{k}^{k_B} \frac{\hbar^2 k^2}{2m^*} e^{i\mathbf{k}\cdot\mathbf{R}_j} = -\frac{\hbar^2}{2m^*} \langle o|\nabla^2|j\rangle \tag{49}$$

where $\langle o|\nabla^2|j\rangle$ is given by Eq. 35. Therefore, we can write Eq. 48 as

$$\bar{\epsilon} - \epsilon_o = -\frac{\hbar^2}{2m^*} \sum_j F(j) \langle o|\nabla^2|j\rangle \tag{50}$$

Since the $\mathcal{E}_{j\ell}$ are just the Wannier matrix elements of the Hamiltonian, Eq. 45 can be written as

$$\bar{\epsilon} = \sum_j F(j) \langle o|\hat{H}_o|j\rangle \tag{51}$$

The Hamiltonian $\hat{H}_o$ neglects electron-electron interactions and is given by Eq. 4. Rewrite $\bar{\epsilon}$ as

$$\bar{\epsilon} = -\frac{\hbar^2}{2m^*} \sum_j F(j) \langle o|\nabla^2|j\rangle + \sum_j F(j) \langle o|V|j\rangle \; . \tag{52}$$

Now subtract Eq. 50 from 52 to get

$$\epsilon_o = \sum_j F(j) \langle o|V|j\rangle \tag{53}$$

Now if $V(\mathbf{r})$ is expressed as a sum of core potentials according to Eq. 5, then Eq. 53 becomes

$$\epsilon_o = \sum_{n,j} F(j) \langle o|V_n|j\rangle \tag{54}$$

The right side of this equation is just the contribution of the crystal potential to the total energy. Part of this, contained in $\langle o|V_j|o\rangle$ type terms cancels the coulombic part of the interelectron and interionic terms that were not explicitly included in the above treatment. Taking this into account, rewrite Eq. 54 as

$$\epsilon_o = \langle o|V_o|o\rangle + \sum_{j(\neq o)} 2 F(j) \langle o|V_j|j\rangle \; . \tag{55}$$

Therefore, an equation is obtained for the lowest band-state energy in terms of the electron-ion potential.

5 FEYNMAN'S THEOREM AND WANNIER REPRESENTATION OF INTERATOMIC FORCES

According to Feynman's Theorem,[10] the force on an atomic nucleus can be computed from the electrostatic field at the nucleus. Thus if $\mathbf{F}_j$ is the force on the jth nucleus of charge $q_j e$ then

$$\mathbf{F}_j = q_j e\, \mathbf{E}(\mathbf{R}_j) + e^2 \sum_{i(\neq j)} \mathbf{n}_{ij} \frac{q_i q_i}{|\mathbf{R}_i - \mathbf{R}_j|^2} \;, \tag{56}$$

where $\mathbf{E}(\mathbf{R}_j)$ is the electric field at $\mathbf{R}_j$ arising from the electronic charge density $-e\rho(\mathbf{r})$ so that

$$\nabla \cdot \mathbf{E} = 4\pi e\, \rho(\mathbf{r}) \tag{57}$$

and

$$\mathbf{E}(\mathbf{R}_j) = -e \int \mathbf{n}_j \frac{\rho(\mathbf{r})}{|\mathbf{R}_j - \mathbf{r}|^2}\, d\mathbf{r} \tag{58}$$

In Eq. 56, $\mathbf{n}_{ij}$ is a unit vector along the direction from nucleus j to nucleus i, and in Eq. 58, $\mathbf{n}_j$ is a unit vector from nucleus j to the position $\mathbf{r}$.

In Eq. 56, the first term on the right is the force on the jth nucleus arising from the ion-electron interaction, and the second term is the force from the ion-ion interaction.

In the one-electron approximation, as the electron density is a sum of electron densities for every electron in every occupied band, one may write

$$\rho(\mathbf{r}) = 2 \sum_n^{\text{core}} \sum_\mathbf{k}^{\mathbf{k}_B} \psi_{n\mathbf{k}}^* \psi_{n\mathbf{k}} + 2 \sum_\mathbf{k}^{\mathbf{k}_F} \psi_\mathbf{k}^* \psi_\mathbf{k} \;. \tag{59}$$

The electrons have been separated into two groups: those associated with the ion core and those in the outer valence band. $\psi_{n\mathbf{k}}$ is the wave function for an electron of state $\mathbf{k}$ in the nth band; the sum over n in Eq. 59 is taken over all core bands. $\psi_\mathbf{k}$ is the wave function for a k-electron in the valence band; for convenience, valence band index has been dropped.

Both the core and valence-band wave functions can be expressed in terms of Wannier functions. Thus,

$$\psi_{n\mathbf{k}} = \frac{1}{\sqrt{N}} \sum_j e^{i\mathbf{k} \cdot \mathbf{R}_j} |j\rangle_n \tag{60}$$

$$\psi_\mathbf{k} = \frac{1}{\sqrt{N}} \sum_j e^{i\mathbf{k} \cdot \mathbf{R}_j} |j\rangle \;, \tag{61}$$

where $|j\rangle_n$ are Wannier functions associated with the nth band, and $|j\rangle$ are Wannier functions for the valence band.

Using Eq. 60 and 61 in Eq. 59 gives

$$\rho(r) = \sum_{n}^{core} \sum_{i,j} S(i\text{-}j) |i\rangle_n \langle j|_n + \sum_{i,j} F(i\text{-}j) |i\rangle \langle j| \quad , \tag{62}$$

where

$$S(i\text{-}j) = \frac{2}{N} \sum_{k}^{k_B} e^{i\mathbf{k}\cdot(\mathbf{R}_i - \mathbf{R}_j)} \quad , \tag{63}$$

and $F(i\text{-}j)$ is given by Eq. 11.

Since the core bands are full (see Eq. 50 of Ref. 5)

$$\sum_{i} S(i - j) |i\rangle_n = 2|j\rangle_n \tag{64}$$

and Eq. 62 simplifies to

$$\rho(r) = 2 \sum_{n}^{core} \sum_{j} |j\rangle_n \langle j|_n + \sum_{i,j} F(i\text{-}j)|i\rangle \langle j| \quad . \tag{65}$$

Using this result, the electric field at nucleus m, as given by Eq. 58 is

$$E(\mathbf{R}_m) = -2e \sum_{n}^{core} \sum_{j} \int n_m \frac{|j\rangle_n \langle j|_n}{|\mathbf{R}_m - \mathbf{r}|^2} d\mathbf{r} - e \sum_{i,j} F(i\text{-}j) \int n_m \frac{|i\rangle \langle j|}{|\mathbf{R}_m - \mathbf{r}|^2} d\mathbf{r} \quad . \tag{66}$$

If the cores are assumed not to overlap, the first term on the right is readily simplified. For $j = m$, the integral in this first term represents the field at the center of a spherically symmetric charge distribution and is, thus, just the field of a point charge. Therefore, for small ion-cores, Eq. 66 reduces to

$$E(\mathbf{R}_m) = -e \sum_{j(\neq n)} n_{jm} \frac{q_j^c}{|\mathbf{R}_m - \mathbf{R}_j|^2} - e \sum_{i,j} F(i\text{-}j) \int n_m \frac{|i\rangle \langle j|}{|\mathbf{R}_m - \mathbf{r}|^2} d\mathbf{r} \tag{67}$$

where q_j^c is the number of electrons in the jth ion-core. Assuming a monatomic, monovalent metal, $q_j = q$ and $q_j^c = q - 1$; thus, substituting Eq. 67 into 56 for this case gives

$$F_m = -q e^2 \sum_{i,j} F(i - j) \int n_m \frac{|i\rangle \langle j|}{|\mathbf{R}_m - \mathbf{r}|^2} d\mathbf{r} + q e^2 \sum_{j(\neq m)} n_{jm} \frac{1}{|\mathbf{R}_m - \mathbf{R}_j|^2} \quad . \tag{68}$$

Eq. 68 can be rewritten in the form:

$$F_m = f_m^{(1)} + \sum_{j}' f_{mj}^{(2)} + \sum_{i \neq j}' f_{mji}^{(3)} \quad , \tag{69}$$

where the primes indicate the sums do not include terms for which i or j equal m, and

$$f_m^{(1)} = -q e^2 \int n_m \frac{|m\rangle \langle m|}{|R_m - r|^2} d r \quad . \tag{70}$$

$$f_{mj}^{(2)} = -2q e^2 F(m-j) \int n_m \frac{|m\rangle \langle j|}{|R_m - r|^2} d r - q e^2 \int n_m \frac{|j\rangle \langle j|}{|R_m - r|^2} d r$$

$$+ q e^2 \frac{n_j m}{|R_m - R_j|} \quad . \tag{71}$$

$$f_{mji}^{(3)} = -q e^2 F(i-j) \int n_m \frac{|i\rangle \langle j|}{|R_m - r|^2} d r \quad . \tag{72}$$

If the electron density associated with the Wannier function $|m\rangle$ is centrosysmetric, then $f_m^{(1)}$ is zero. The force on the mth nucleus is therefore expressed as sums of two- and three-body interactions with the other atomic sites in the crystal. This should be useful for the study of interatomic force constants.

6 BINDING ENERGY OF THE ALKALI METALS

The one-body matrix elements are readily computed using the model presented in Section 3. The values of r_s and r_I used in the calculations are shown in Table I, and the results for the alkali metals are shown in Table II.

Table I. Wigner-Seitz and Ionic Radii Used in Model Calculations of Matrix Elements (Atomic Units)

Metal	$r_s^{(a)}$	$r_I^{(b)}$
Li	3.252	1.13
Na	3.931	1.80
K	4.861	2.51
Rb	5.196	2.80
Cs	5.624	3.19

(a) From a compilation in Reference 12.
(b) From Pauling's univalent ionic radii.

Table II. Unit Cell Contributions to the Crystal Energy of the Alkali Metals (Rydbergs)

Metal	KE (1)[a]	PE (1)[b]	E (1)[c]
Li	0.318	-0.606	-0.288
Na	0.205	-0.446	-0.241
K	0.128	-0.334	-0.206
Rb	0.110	-0.303	-0.193
Cs	0.0907	-0.268	-0.177

[a] $KE\ (1) \equiv \dfrac{\hbar^2}{2m} \langle o|\nabla^2|o\rangle$

[b] $PE\ (1) \equiv \langle o|v_o|o\rangle$

[c] $E\ (1) \equiv KE\ (1) + PE\ (1)$.

The two-body terms in Eq. 41 were also computed at values of R_j determined by the bcc structure for the alkali metals, and summed out to the 38th neighbor shell. The results are given in Tables III to V. In these tables atomic units and the following abbreviations are used:

S = shell number (1 for 1st neighbor, 2 for 2nd, etc.)
J_S = number of atoms in Sth shell
$KE = F(j) \langle o|\nabla^2|j\rangle$
$KT = r_S^2 F(j) \langle o|\nabla^2|j\rangle$
$WE = J_S F(j) \langle o|\nabla^2|j\rangle$

$U(s) = r_S^2 \sum\limits_{\text{shells}}^{S} J_S F(j) \langle o|\nabla^2|j\rangle$

$X = R_j/r_S$

$F = \dfrac{R_j}{e^2} \langle o|V_j|o\rangle$

$VR = 2 F(j) \langle o|V_j|j\rangle$
$VI = 2 r_S F(j) \langle o|V_j|j\rangle$
$VIJ = 2 r_S J_S F(j) \langle o|V_j|j\rangle$

$VIS = 2 r_S \sum\limits_{\text{shells}}^{S} J_S F(j) \langle o|V_j|j\rangle$

The entries in these tables that contain r_S or r_S^2 as factors are general in the sense that they can be used to obtain the two-body terms for all the alkali metals by dividing by the appropriate r_S or r_S^2. The other entries, however (KE, WE, and VR) are specific to sodium.

KE is called the kinetic energy potential since it represents a pairwise interaction arising from the kinetic energy operator. $\langle o|V_j|o\rangle$ represents the interaction of the jth ion core with the Wannier electron density centered on the zeroth atom, and is called the unmixed electron-ion interaction.

Table III. The Kinetic Energy Potential and Its Crystal Sums for the BCC Structure

S	J_s	X	KE	KT	WE	U(S)
1	8	1.759	−0.00397	−0.0614	−0.0318	−0.491
2	6	2.031	−0.00358	−0.0552	−0.0215	−0.822
3	12	2.872	0.00085	0.0132	0.0103	0.663
4	24	3.368	−0.00071	−0.0109	−0.0170	−0.926
5	8	3.518	−0.00071	−0.0109	−0.0057	−1.013
6	6	4.062	0.00003	0.0005	0.0002	−1.010
7	24	4.426	0.00000	0.0000	0.0001	−1.009
8	24	4.541	−0.00008	−0.0012	−0.0019	−1.038
9	24	4.975	−0.00024	−0.0037	−0.0058	−1.127
10	32	5.277	−0.00013	−0.0020	−0.0041	−1.191
11	12	5.745	0.00000	0.0000	0.0000	−1.191
12	48	6.008	−0.00004	−0.0006	−0.0019	−1.220
13	30	6.093	−0.00006	−0.0009	−0.0017	−1.247
14	24	6.423	−0.00009	−0.0014	−0.0022	−1.281
15	24	6.659	−0.00006	−0.0010	−0.0015	−1.305
16	24	6.736	−0.00005	−0.0008	−0.0011	−1.322
17	8	7.036	−0.00000	−0.0000	−0.0000	−1.322
18	48	7.252	0.00000	0.0001	0.0002	−1.319
19	24	7.323	0.00000	0.0000	0.0000	−1.319
20	48	7.599	−0.00002	−0.0003	−0.0011	−1.336
21	72	7.800	−0.00003	−0.0005	−0.0024	−1.372
22	6	8.124	−0.00002	−0.0003	−0.0001	−1.374
23	24	8.312	−0.00001	−0.0001	−0.0001	−1.376
24	48	8.374	−0.00000	−0.0000	−0.0000	−1.377
25	36	8.617	0.00001	0.0001	0.0003	−1.371
26	56	8.794	0.00001	0.0001	0.0004	−1.365
27	24	8.853	0.00001	0.0001	0.0001	−1.363
28	24	9.083	−0.00000	−0.0001	0.0001	−1.365
29	72	9.252	−0.00001	−0.0001	−0.0007	−1.375
30	48	9.307	−0.00001	−0.0002	−0.0005	−1.383
31	24	9.526	−0.00001	−0.0001	−0.0002	−1.385
32	48	9.687	−0.00000	−0.0000	−0.0000	−1.385
33	24	9.950	0.00001	0.0001	0.0002	−1.382
34	72	10.104	0.00001	0.0002	0.0008	−1.370
35	30	10.155	0.00001	0.0002	0.0003	−1.365
36	72	10.356	0.00001	0.0001	0.0005	−1.358
37	72	10.504	0.00000	0.0000	0.0002	−1.356
38	32	10.553	0.00000	0.0000	0.0000	−1.354

Table IV. The Effective Coulomb Constant for the Unmixed Ion-Electron Interaction

S	X	F
1	1.759	−0.8854
2	2.031	−0.9002
3	2.872	−0.9313
4	3.368	−0.9408
5	3.518	−0.9436
6	4.062	−0.9515
7	4.462	−0.9552
8	4.541	−0.9563
9	4.975	−0.9603
10	5.277	−0.9626
11	5.745	−0.9655
12	6.008	−0.9670
13	6.093	−0.9675
14	6.423	−0.9693
15	6.659	−0.9704
16	6.736	−0.9708
17	7.036	−0.9719
18	7.252	−0.9727
19	7.323	−0.9730
20	7.599	−0.9740
21	7.800	−0.9747
22	8.124	−0.9757
23	8.312	−0.9762
24	8.374	−0.9764
25	8.617	−0.9771
26	8.794	−0.9775
27	8.853	−0.9777
28	9.083	−0.9783
29	9.252	−0.9787
30	9.307	−0.9788
31	9.526	−0.9793
32	9.687	−0.9796
33	9.950	−0.9801
34	10.104	−0.9805
35	10.155	−0.9806
36	10.356	−0.9809
37	10.504	−0.9812
38	10.553	−0.9813

Table V. The Mixed Ion-Electron Interaction and Its Crystal Sums for the BCC Structure

S	X	VR (Rydbergs)	VI	VIJ	VIS
1	1.759	-0.022548	-0.08864	-0.709	-0.709
2	2.031	0.001887	0.00742	0.045	-0.665
3	2.872	-0.004453	-0.01751	-0.210	-0.875
4	3.368	0.000522	0.00205	0.049	-0.825
5	3.518	0.001109	0.00436	0.035	-0.791
6	4.062	-0.000095	-0.00037	-0.002	-0.793
7	4.462	-0.000279	-0.00110	-0.026	-0.819
8	4.541	-0.000100	-0.00039	-0.009	-0.829
9	4.975	0.000482	0.00189	0.045	-0.783
10	5.277	0.000340	0.00134	0.043	-0.740
11	5.745	-0.000014	-0.00006	-0.007	-0.741
12	6.008	0.000045	0.00018	0.008	-0.732
13	6.093	0.000088	0.00035	0.010	-0.722
14	6.423	0.000205	0.00081	0.019	-0.703
15	6.659	0.000172	0.00068	0.016	-0.687
16	6.736	0.000133	0.00052	0.013	-0.674
17	7.036	0.000029	0.00011	0.001	-0.673
18	7.252	-0.000003	-0.00001	-0.001	-0.674
19	7.323	-0.000001	-0.00000	-0.000	-0.674
20	7.599	0.000044	0.00017	0.008	-0.665
21	7.800	0.000076	0.00030	0.021	-0.644
22	8.124	0.000063	0.00025	0.001	-0.642
23	8.312	0.000029	0.00012	0.003	-0.640
24	8.374	0.000018	0.00007	0.003	-0.636
25	8.617	-0.000013	-0.00005	-0.002	-0.638
26	8.794	-0.000014	-0.00005	-0.003	-0.641
27	8.853	-0.000010	-0.00004	-0.001	-0.642
28	9.083	0.000011	0.00004	0.001	-0.641
29	9.252	0.000024	0.00009	0.007	-0.634
30	9.307	0.000026	0.00010	0.005	-0.629
31	9.526	0.000022	0.00009	0.002	-0.627
32	9.687	0.000009	0.00004	0.002	-0.626
33	9.950	-0.000013	-0.00005	-0.001	-0.627
34	10.104	-0.000020	-0.00008	-0.006	-0.633
35	10.155	-0.000020	-0.00008	-0.002	-0.635
36	10.356	-0.000014	-0.00006	-0.004	-0.639
37	10.504	-0.000006	-0.00002	-0.001	-0.640
38	10.553	-0.000003	-0.00001	-0.000	-0.641

From its definition, F is seen to be an effective coulomb constant that varies with R_j; VR is an interaction involving Wannier functions from two sites and the core potential from one of these sites, and is called the mixed electron-ion interaction.

Table IV shows that the $\langle o|V_j|o\rangle$ term has not converged to the coulomb value, even at the 38th neighbor, although it must do so ultimately. This illustrates the sensitivity of this term to the errors in its calculation. These errors arise from the following sources: (a) use of a spherical Brillouin zone in constructing the Wannier functions; (b) use of a hydrogenic potential in the region of the ion core centered on j, rather than the atomic potential; (c) use of the plane-wave Wannier function in the region of the ion core centered on the zeroth atom, rather than the atomic wave function. Further work is required before the corrections to the simple model can be fully understood.

Plots of $-\dfrac{\hbar^2}{2m} F(j) \langle o|\nabla^2|j\rangle$ and $2F(j) \langle o|V_j|j\rangle$ are given in Fig. 1 and 2 as a function of position of the jth shell surrounding the zeroth atom. These quantities are the kinetic energy potential and the convergent electron-ion potential, respectively. Both of these potentials are oscillatory and decay to nearly zero at $x \approx 6$.

The individual terms in the two-body kinetic energy sum were computed by performing the integration to include the ion cores. However, in

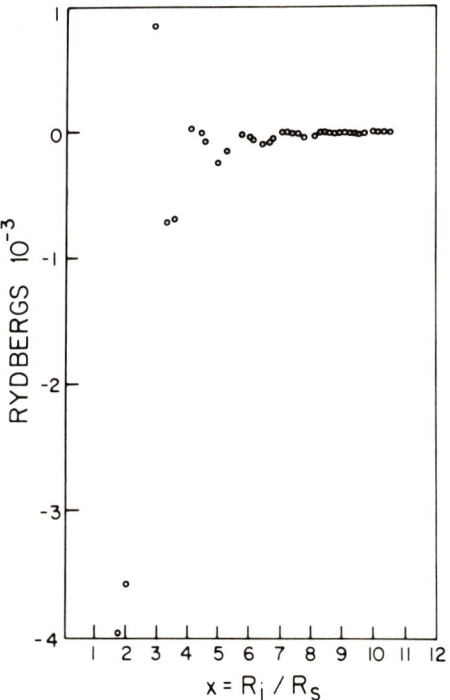

Fig. 1. The kinetic energy potential as a function of position in the bcc structure with $r_s = 3.93$.

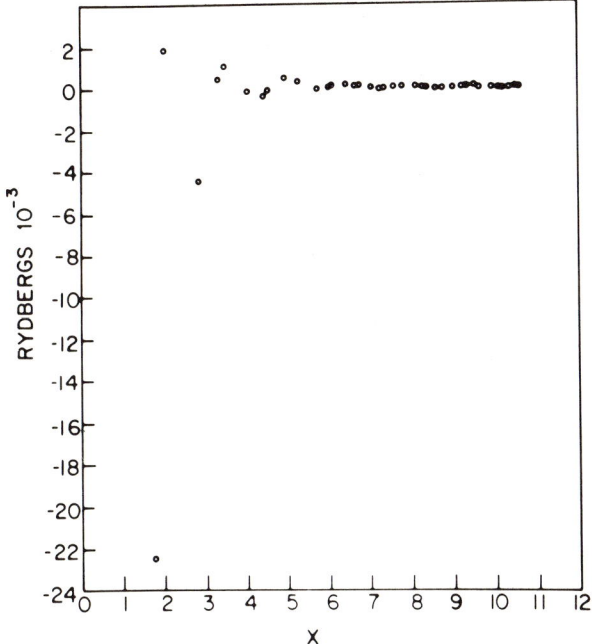

Fig. 2. The mixed electron-ion interaction as a function of position in the bcc structure with $r_s = 3.93$.

accord with the model, integration over the ion cores should be excluded. It can be shown that the kinetic energy sum should be reduced by a factor of $(1 - v_c/v)$ to be consistent with the model. Therefore, the contribution of this sum to the crystal energy is

$$\text{KE (2)} \equiv \left(1 - \frac{v_c}{v}\right) \sum_{j(\neq o)} -\frac{\hbar^2}{2m} F(j) \langle o|\nabla^2|j\rangle \quad , \tag{73}$$

and the two-body contribution of the mixed electron-ion energy to the crystal energy is

$$\text{PE (2)} \equiv \sum_{j(\neq o)} 2 F(j) \langle o|V_j|j\rangle \tag{74}$$

In addition to the errors inherent in the model, the results of Table VI are in error by approximately ± 0.003 Rydbergs because the sums in Eq. 73 and 74 were carried out only to the 38th-neighbor shell.

To compare these results with experiment, the effect of the Coulomb type matrix elements $\langle o|V_j|o\rangle$ in the pairwise interaction energy of Eq. 43 was taken to exactly cancel the ion-ion term.

The crystal energy (Eq. 43) then becomes

$$E' = Nu + \frac{N}{2} \sum_{j(\neq o)} g'_{oj} \tag{75}$$

where u is given by Eq. 42 and

$$g'_{oj} = -\frac{\hbar^2}{m} F(j) \langle o|\nabla^2|j\rangle + 4 F(j) \langle o|V_j|j\rangle \quad . \tag{76}$$

Using the results of Table II and Table VI, one can compute E' from Eq. 75. The results, along with the ratio of the unit cell and two-body contributions are shown in Table VII. The experimental binding energies, taken from a compilation in Ref. 2, are labeled E^c_{exp} in the table. Considering the errors inherent in the simple model used, the agreement between the calculations is remarkable. The error shows a definite trend, varying from -0.09 Rydbergs for lithium to +0.04 Rydbergs for cesium.

When comparison is made to the cohesive energies by subtracting the ionization energies, the results appear less satisfactory, as shown in Table VIII. This reflects the fact that the cohesive energy is a small difference of large numbers. A study of the detailed sources of error and more refined models is underway.

Table VI. Two-Body Contributions to the Crystal Energy of the Alkali Metals (Rydbergs)

Metal	KE (2)[a]	PE (2)[b]	E (2)[c]
Li	-0.124	-0.194	-0.318
Na	-0.0802	-0.1603	-0.241
K	-0.0500	-0.130	-0.180
Rb	-0.0428	-0.121	-0.164
Cs	-0.035	-0.112	-0.147

(a) $KE\ (2) \equiv \left(1 - \dfrac{v_c}{v}\right) \sum_{j(\neq 0)} -\dfrac{\hbar^2}{2m} F(j) \langle o|\nabla^2|j\rangle.$

(b) $PE\ (2) \equiv \sum_{j(\neq 0)} 2F(j) \langle o|v_j|j\rangle.$

(c) $E\ (2) \equiv KE\ (2) + PE\ (2).$

Table VII. Crystal Energy of the Alkali Metals (Rydbergs)

Metal	E'[a]	E^c_{exp}	E(1)/E(2)
Li	-0.606	-0.515	0.91
Na	-0.482	-0.458	1.00
K	-0.386	-0.390	1.14
Rb	-0.357	-0.374	1.18
Cs	-0.324	-0.360	1.20

(a) $E' = E\ (1) + E\ (2).$

Table VIII. Cohesive Energy of Alkali Metals
(Rydbergs)

Metal	E_{coh} (calc)	E_{coh} (exp)
Li	-0.208	-0.117
Na	-0.111	-0.088
K	-0.064	-0.069
Rb	-0.049	-0.066
Cs	-0.030	-0.066

7 ENERGY OF THE LOWEST BAND STATE IN THE ALKALI METALS

To compute ϵ_0, again assume that the $\langle o|V_j|o\rangle$ terms cancel out because of the ion-ion interaction, and write Eq. 53 as

$$\epsilon_0 = \langle o|V_0|o\rangle + \sum_{j(\neq o)} 2F(j)\langle o|V_j|j\rangle \qquad (77)$$

This allows ϵ_0 to be computed from results already obtained. The values calculated from the model, along with those obtained by Bienenstock[11] by a Wigner-Seitz method are shown in Table IX. The differences between the two sets of calculations show a regular trend varying from -0.11 Rydbergs for lithium to +0.04 Rydbergs for cesium. This agreement is as good as can be expected considering the simplicity of the model.

Table IX. Energy of the Lowest Band State in the Alkali Metals (Rydbergs)

Metal	ϵ_0 (calc)	ϵ_0 (Bienenstock)
Li	-0.799	-0.687
Na	-0.607	-0.609
K	-0.463	-0.486
Rb	-0.424	-0.459
Cs	-0.380	-0.421

8 INTERATOMIC POTENTIALS IN THE ALKALI METALS

The effective pairwise interaction energy given by Eq. 76 is readily constructed from the results given previously. These are shown in Fig. 3, 4, and 5, as continuous functions of the distance between two atoms. g'_{oj}

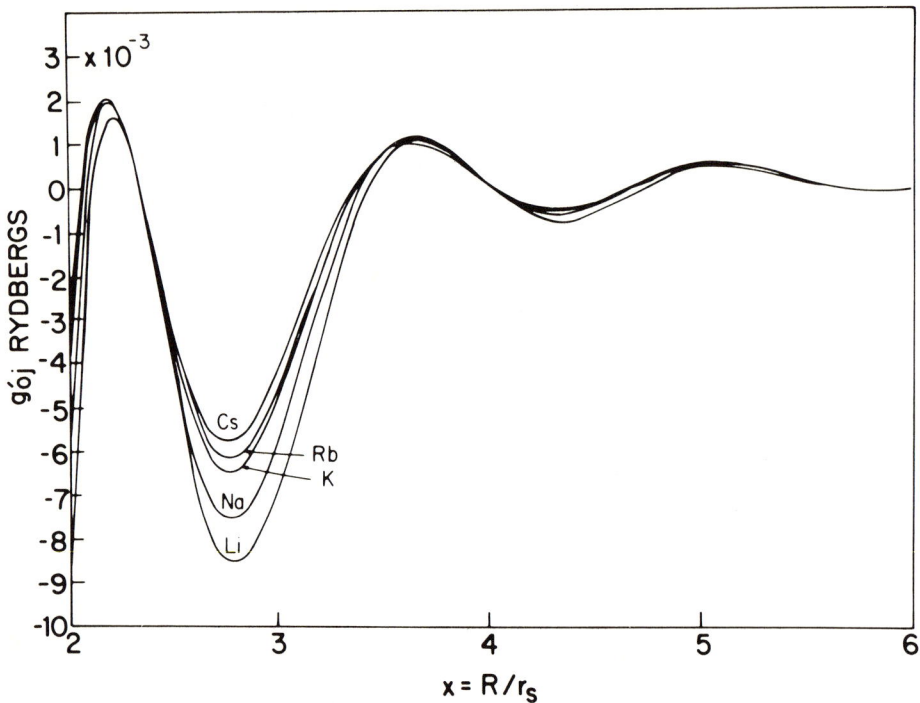

Fig. 3. Pairwise interaction functions in alkali metals.

Fig. 4. Pairwise interaction functions in alkali metals

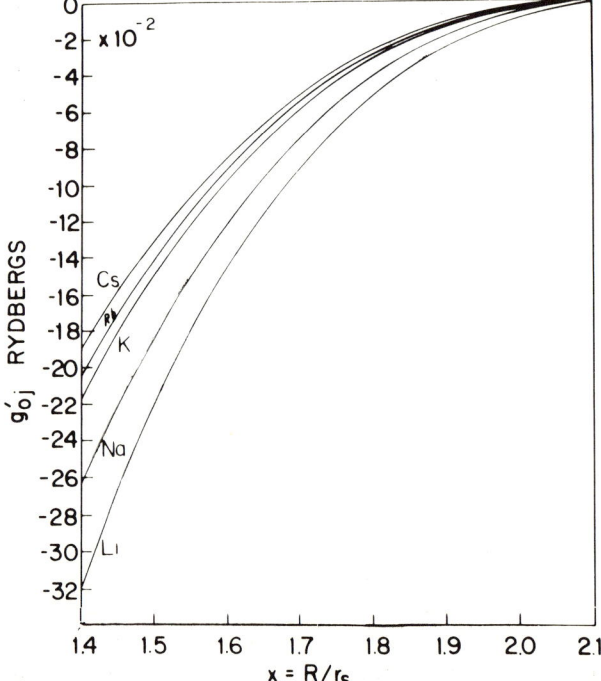

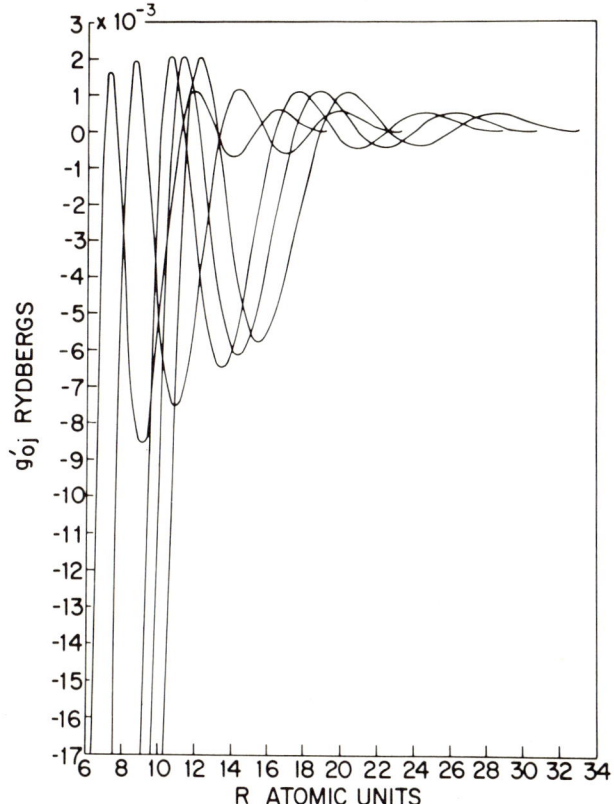

Fig. 5. Pairwise interaction functions in alkali metals

is, of course, defined only when the separation distance connects two lattice points. The extent to which these potentials change when the periodicity of the crystal is disturbed is not yet known.

In Fig. 3, g'_{oj} is plotted against the reduced distance scale $x = R/r_s$ where R is the distance between the two atoms. On this reduced scale, the pairwise interaction functions are rather similar for all the alkali metals, the differences among them being less pronounced at large R. The first minimum occurs at about the third-nearest-neighbor distance; the depth of the minimum decreases with increasing atomic volume. The variation of the interaction with distance for $x < 2$ is an order of magnitude higher than for $x > 2$. Accordingly, the results for $x < 2$ were plotted separately in Fig. 4.

Fig. 5 displays g'_{oj} as a function of R in atomic units. The effect of the atomic volume is clearly seen as we move from lithium to cesium. As the unit cell expands, the curve shifts to the right, and the minima become smaller.

The pairwise interaction decreases with decreasing distance at small x, and does not exhibit a minimum for any x less than that corresponding to third-nearest-neighbor separation. The stability of the crystal is maintained by the balance between the unit cell energy and the pairwise energy, and by the effect of crystal density on both these terms.

9 SUMMARY AND DISCUSSION

The localized wave functions in the Wannier representation permits the energy of a metal to be described in terms of a unit cell energy plus a sum of interactions among atoms. The basis for doing this has been developed and model calculations presented should apply reasonably well to the alkali metals. Matrix elements for the band Hamiltonian were computed for a model in which, outside the cores, the core potential is hydrogenic and the Wannier functions are constructed from plane waves. The total energy for this Hamiltonian was computed by performing a lattice sum on the pairwise interactions for the bcc structure and adding it to the volume-dependent contribution. This energy was found to account for most of the crystal energy of the alkali metals. This means that most of the crystal energy is in the terms used in the computation, and that the sum of the interelectron interactions, the ion-ion interactions, and $\langle o|V_j|o\rangle$ contribute only a small amount to the crystal energy.

It is of interest to note several points resulting from the above calculations. The first is that the Wannier representation throws half of the energy into a pairwise "bond energy" form. The second is that even the pairwise terms depend on the volume through k_B and k_F. Thus, a single pairwise potential can be used for calculations in which the volume is constant, but not for those in which the volume changes. However, the dependence of the pairwise interactions on k_B and k_F is explicit, so that they can be readily computed for various volumes.

From Eq. 35 and 40, we see that the kinetic energy potential (the first term in Eq. 41) is a universal function of $x_j \equiv R_j/r_s$, divided by r_s^2. This function is given as KT in Table III. Similarly, the third term in Eq. 41 (the mixed electron-ion potential) is a universal function of $x_j = R_j/r_s$ divided by r_s and is given as VIJ in Table V.

The ground-state energy of the band Hamiltonian is given as a lattice sum containing Wannier matrix elements of the crystal potential if the Bloch functions are plane waves. Calculation of the ground-state energy on this basis agrees well with the results of Wigner-Seitz calculations.

In comparing the method proposed here with the pseudopotential method of defining ion-ion interactions in metals, several points arise. In the pseudopotential method only the band-structure energy is distributed among ion-ion interactions, whereas in the proposed method, all the energy terms take part in the bond interactions. The pseudopotential method is based on second-order perturbation theory, and the results are sensitive to the method of constructing the pseudopotential. The Wannier method, on the other hand, does not depend on perturbation theory (at least for perfect crystals). The pseudopotential method has the advantage of not depending on the crystal periodicity in an essential way, and can readily be applied to imperfect crystals. To date, imperfect crystals have not been considered in the work performed. To handle imperfect crystals, it will be necessary to introduce perturbation theory into the Wannier representation. This can be done either by starting with the perturbation in the Bloch scheme using ordinary second-order perturbation theory, or by starting with the difference equations for a perturbed crystal in the Wannier representation. This problem is being studied.

ACKNOWLEDGMENT

This research was supported by the Advanced Research Projects Agency of the United States Department of Defense.

REFERENCES

1. Brooks, H., Nuovo Cimento, Suppl., 1958 ser. 10, Vol. 7 Ch. VII, pp 165-244.
2. Brooks, H., Trans. AIME, **227**, p 546 (1963).
3. Harrison, W. A., *Pseudopotentials in the Theory of Metals* (W. A. Benjamin, Inc., N.Y., 1966).
4. Wannier, G. H., Phys. Rev., **52**, p 191 (1937).
5. Girifalco, L. A., Phys. Rev., **179**, p 616 (1969).
6. Kohn, W., Phys. Rev., **115**, 809 (1959).
7. Blount, E. I., in *Solid State Physics,* edited by F. Seitz and D. Turnbull (Academic Press Inc., N.Y., 1962) pp 319-335.
8. Koster, G. F., Phys. Rev., **89**, p 67 (1959).
9. Di Vincenzo, T. M., *Wannier Representation of Energy in Metals,* Ph.D. Thesis, School of Metallurgy and Materials Science, University of Pennsylvania, 1970.
10. Feynman, R. P., Phys. Rev., **56**, 340 (1939).
11. Bienenstock, A., Ph.D. Thesis, Harvard (1962): referred to in Ref. 2.
12. Barrett, C. S., Acta. Cryst., **9**, 671 (1956).

DISCUSSIONS on paper by L. A. Girifalco and T. M. Di Vincenzo

HARRISON: It is difficult for me to see how your interaction potentials can be meaningful in defect problems since their construction depends so directly upon the translational symmetry of the lattice.

GIRIFALCO: You are correct in that the results so far are valid only for perfect crystals. However, the direction one must take for non-periodic systems is clear, particularly if the deviation from periodicity is not too large. One simply corrects the potential by perturbation theory in which the Bloch function matrix elements are converted to Wannier form. For large deviations from periodicity, one can start by using "modified Wannier functions" in which the transformation matrix (connecting them to the Hamiltonian eigenfunctions) contain the actual ion positions rather than lattice sites. This complicates the theory and makes it more difficult because of orthogonality failure.

HO: According to your two-body interatomic potential, there is very strong attraction at small separations; this does not appear to be physical. Is this due to the definition of the potential?

GIRIFALCO: The strong attraction I believe is real, and not an artifact of the calculation. Its physical origin is the fact that major components of the potential are the matrix elements representing the interaction of the Wannier electrons with ion cores. These must be attractive. It may be that, if we ultimately take into account the interelectron contributions, the potential will go positive at nearest-neighbor distances, but I do not think so.

DUESBERY: In previous papers we have heard that the use of a pseudo-potential formulation enables the definition of a two-body interaction which is valid for small deformations of the lattice. Could you explain why this is not so for your formulation?

GIRIFALCO: The Wannier functions depend on crystal periodicity for their definition, so the results we have presented are rigorously valid for perfect crystals only. It is rather straightforward, however, to deal with small deformations of the lattice by perturbation theory. This will give the potentials appropriate to the deformation.

HO: Have you checked your formulation by calculating the equilibrium lattice spacing for sodium based on the total lattice energy?

GIRIFALCO: No we have not. This is part of our program of study, but we wish to construct more accurate wave functions before proceeding too far with calculation of properties.

MARCH: While Hartree-Fock theory can give useful total energies in a metal, it has some unsatisfactory features because of the strong **k** dependence of the exchange potential. In the electron gas, this leads to a wrong prediction of zero density of states at the Fermi surface. We must screen the electron-electron interactions in the exchange terms to avoid this difficulty. Do such difficulties remain in the Wannier formalism?

GIRIFALCO: Yes they do; but I believe it will be worthwhile to reexamine the question of electron-electron interactions in the Wannier formalism. It is possible that the major part of this can then be put into the unit cell energy. In our model calculations, of course, we have neglected both exchange and correlation.

SEEGER: How well is the virial theorem satisfied in your numerical calculations?

GIRIFALCO: I do not know. Although the results for crystal energy and lowest energy of the conduction band are reasonable in our model, the wave functions used are certainly in error. This arises from the failure of the model wave functions to join smoothly across the region of the ion core radius.

WILSON: Is your method applicable only to alkali metals or is there a more general usage?

GIRIFALCO: The method is in principle applicable to any metal. There are, of course, the usual problems (d-electrons, transition metals, etc.), but there are no intrinsic limitations to treating crystals by the Wannier formalism. It is equally applicable to non-metals.

ASHCROFT: The crucial step in the reduction to two-body energies hinges on the range of the Wannier function. In a small-band-gap material one expects the range to be large (and to encompass many neighbors). Do you have a means of checking your assertion that n body terms ($n \geq 3$) are small?

A possible test for your method: take the potential inside the atom (or ion) radius equal to zero (leading to spherical Bessel function solutions). Can the free electron band energy then be reproduced (at $n = 2, 3, \ldots$ etc.)?

GIRIFALCO: We have not been able to compute the three- and four-body matrix elements, so we have no direct comparison of the relative magnitude of the pairwise and higher order terms. We are now developing a formalism in which the matrix elements need be evaluated only over unit cells, thereby greatly reducing the required computational effort. Perhaps this will enable us to make the direct comparison.

Thank you for your suggestion.

EXPERIMENTAL TECHNIQUES USED TO OBTAIN POTENTIALS

J. E. Enderby and W. S. Howells

University of Leicester
Leicester, England

1 INTRODUCTION

The electrons in a metal or alloy may be divided into two types: the **core** electrons which are tightly bound within the atom and the **conduction** electrons which are relatively loosely bound and which are responsible for most of the characteristic properties of the metallic state. The aim of the pseudopotential method is to transform the equation

$$[T+V(\mathbf{r})]\,\psi_\mathbf{k} = E_\mathbf{k}\,\psi_\mathbf{k} \qquad (1)$$

into

$$H\phi_\mathbf{k} = E_\mathbf{k}\,\phi_\mathbf{k}\,, \qquad (2)$$

where T is the kinetic energy operator, V the (Hartree) potential, and $\phi_\mathbf{k}$ is a **smooth** wave function that has none of the violent oscillations close to the nucleus which characterize $\psi_\mathbf{k}$, the set of conduction electron wave functions. It turns out that this transformation may be accomplished provided the new operator is written $H = T + W(r,k)$ where $W(r,k)$ is a pseudopotential.

The advantages of considering $W(r,k)$ rather than $V(r)$ have been discussed many times in the literature. The most significant is that W can be chosen to be much weaker than $V(r)$ and may be decomposed into spherically symmetric (momentum dependent) single ion-core potentials through

$$W(\mathbf{r}) = \sum_i{}' w(\mathbf{r}-\mathbf{r}_i) \qquad (3)$$

Moreover, each Fourier component of W can be screened independently. This forms the basis for the "dielectric constant" method of dealing with the problem of electron screening.

Pseudopotential theory enables the total energy per ion to be split up into a part (U_0) which depends only on the total volume of the conductor and a structure-dependent part which may be written as

$$U_s = U_E + U_{bs} \qquad (4)$$

where U_E is the Ewald electrostatic energy of point ions at sites $\mathbf{R}_j$ in a uniform negative background and U_{bs} is the band structure energy. It is then possible to express U_s as

$$\frac{1}{2N} \sum \phi(|\mathbf{R}_i - \mathbf{R}_j|) \qquad (5)$$

where ϕ, the **effective interionic potential**, takes the form

$$\phi(r) = \frac{Z^{*2} e^2}{r}\left[1 - \frac{2}{\pi}\int \frac{F(q)\sin qr}{q}\,dq\right] \qquad (6)$$

Here Z^* represents an effective valence and $F(q)$ is determined completely by $w_q(k)$, the Fourier transform of $w(r,k)$.[1]

Tabulated values of $w_q(k)$ obtained by different approximate schemes are available for most elements.[2,3] An effective interionic potential for Al, calculated by Harrison[1] is shown in Fig. 1.

Experimentally, in principle, two ways enable either $\phi(r)$ or $w_q(k)$ to be determined. The extent to which experiments yield useful data is the subject of this paper.

2 PAIR POTENTIALS DERIVED FROM LIQUID STRUCTURE

2.1 The Structure of Liquid Metals and Alloys

It is convenient to introduce the quantity $n_2(\mathbf{R}_k, \mathbf{R}_j)\, d\mathbf{R}_k\, d\mathbf{R}_j$ which is the probability that the volume elements $d\mathbf{R}_k\, d\mathbf{R}_j$ are both occupied.

In a liquid, $n_2(\mathbf{R}_k, \mathbf{R}_j)$ depends only on $r = |\mathbf{R}_k - \mathbf{R}_j|$. A **radial distribution function** may be defined through $g(r) = n_2(r)/n^2$ where n is the mean number density of atoms; since the occupancy of two distant volume elements is uncorrelated $g(r)$ tends to unity at large r. The connection between $a(q)$, the interference function, and $g(r)$ may be expressed in the form of a Fourier transform either as:

$$a(q) = 1 + \frac{4\pi n}{q} \int_0^\infty dr \left\{g(r)-1\right\} r\sin qr$$

or

$$g(r) = 1 + \frac{1}{2\pi^2 nr} \int_0^\infty dq \left\{a(q)-1\right\} q\sin qr \quad . \tag{7}$$

The quantity $\left\{g(r)-1\right\}$, often written as $h(r)$, is referred to as the **total correlation function**. In principle, measuring $a(q)$ by neutron or X-ray methods enables $g(r)$ to be evaluated. Difficulties in practice arise from the inherent uncertainties in $a(q)$ (see especially Ref. 4) and the existence of truncation errors coming from the finite upper limit (typically $\sim 12\text{A}^{-1}$) which restricts the evaluation of Eq. 7.

For small r, $g(r)$ tends to zero because of the finite size of atoms. It follows from Eq. 7 that $a(q)$ must satisfy the sum rule

$$-2\pi^2 n = \int dq \, q^2 \, (a(q)-1) \quad . \tag{8}$$

Reliable data are now available for a variety of metallic liquids. Fig. 2 illustrates the interference function for liquid lead (a typical quadrivalent liquid metal) at two temperatures, whilst the corresponding radial distribution functions are shown in Fig. 3.

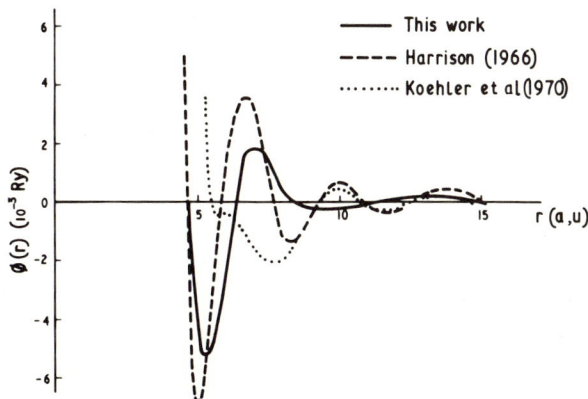

Fig. 1. Pair potentials calculated by Harrison[1] and Koehler et al[11] compared with that found in the present work.

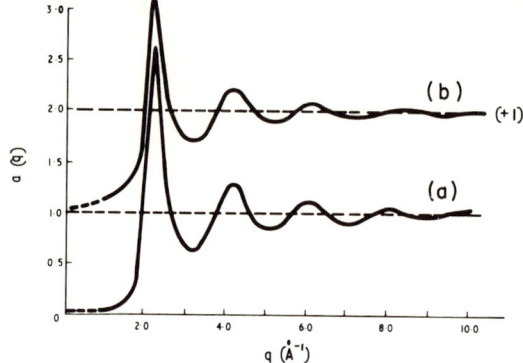

Fig. 2. Interference function for liquid lead
(a) 340 C (b) 600 C

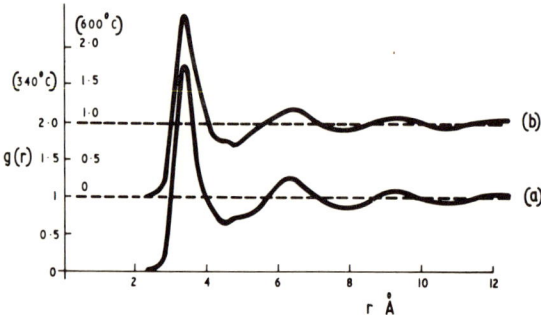

Fig. 3. Radial distribution function for lead
(a) 340 C (b) 600 C

In discussing liquid alloys the **partial radial distribution function** $g_{\alpha\beta}$ is first introduced to measure the average distribution of type β atom observed from an α atom at the origin. Here α and β are dummy suffices which may take the values 1, 2, ... j for a liquid containing j components. The differential scattering cross section for the coherent scattering of monochromatic radiation may be written as

$$\frac{d\sigma}{d\Omega} = \langle \Sigma \Sigma f_p f_p \exp i\mathbf{q}\cdot(\mathbf{r}_p - \mathbf{r}_q) \rangle$$

$$\equiv N \left\{ \sum_{\alpha=1}^{j} c_\alpha f_\alpha^2 + \sum_{\alpha=1}^{j} \sum_{\beta=1}^{j} c_\alpha c_\beta f_\alpha f_\beta (a_{\alpha\beta}-1) \right\}, \quad (9)$$

where c_α is the atomic concentration of the α atom type and $f_p, f_q, \mathbf{r}_p$ and $\mathbf{r}_q$ represent respectively the coherent scattering amplitudes and the position co-ordinate of the pth and qth atoms. $a_{\alpha\beta}(q)$ are known as the **partial interference functions** and are defined through

$$a_{\alpha\beta}(q) = 1 + \frac{4\pi n}{q} \int dr\, [g_{\alpha\beta}-1]\, r\sin qr = 1 + \frac{4\pi n}{q} \int dr\, h_{\alpha\beta}\, r\sin qr \quad . \quad (10)$$

The basic problem facing the experimentalist is to derive the fundamental quantities $a_{\alpha\beta}$ from diffraction data. Quite severe experimental difficulties arise in converting an observed intensity pattern in $a(q)$ even for a pure liquid; and, for multivalent systems, further difficulties are confronted.

For a liquid that contains two species denoted by 1 and 2, Eq. 9 reads

$$\frac{d\sigma}{d\Omega} = N\left\{c_1 f_1^2 + c_2 f_2^2 + I(q)\right\}, \qquad (11)$$

where $I(q) = c_1^2 f_1^2 (a_{11}-1) + c_2^2 f_2^2 (a_{22}-1) + 2c_1 c_2 f_1 f_2 (a_{12}-1)$.

Thus three partial interference functions are involved and, to determine them by experiment, three separate diffraction experiments on the same liquid must be carried out.

The first experiment designed to determine a_{11}, a_{22} and a_{12} was carried out by Enderby et al.[5] for a liquid alloy, Cu_6Sn_5 and involved neutron-diffraction experiments on alloys made with Cu^{63}, Cu^{65} and neutral copper.[5] An important conclusion from this work was that the main peak in a_{Cu-Sn} did not fall midway between that for a_{Cu-Cu} and a_{Sn-Sn}. A model based on hard spheres for this alloy is, therefore, inappropriate.

Fig. 4, 5, and 6 present new data, both in q and r space, for the alloy system Al-20 atomic percent nickel. In these experiments the techniques of neutron diffraction were again used, and the neutron-scattering amplitude for nickel was changed by alloying pure aluminum with nickel enriched to 99 percent ^{60}Ni, with nickel enriched to 99 percent ^{62}Ni, and with natural nickel. The scattering amplitude for thermal neutrons for these isotopic compositions are in the proportion 1:-0.86:0.29, and a full account of this work will appear elsewhere.[6]

2.2 Pair Potentials and the Theory of Liquids

Several approximate theories of the liquid state are in frequent use (see, for example, Ref. 7). Their common feature is that they attempt to relate the radial distribution function $g(r)$ to the interatomic pair potential $\phi(r)$. The Born-Green (BG) theory for a one-component system sets

$$\phi_{BG}(\mathbf{r}) = k_B T\left(-n \int E(|\mathbf{r}-\mathbf{r'}|) h(\mathbf{r'}) d\mathbf{r'} - \ln g(r)\right), \qquad (22)$$

where $E(t)$ is given by

$$E(t) = \int_\infty dr\, g(r) \frac{\phi'_{BG}(r)}{k_B T}, \qquad (13)$$

where k_B is Boltzmann's constant and $\phi'_{BG}(r)$ is the derivative of $\phi_{BG}(r)$ with respect to r. Another approach due to Percus and Yevick (PY) writes

$$\phi_{PY}(r) = k_B T \ln \left\{ 1 - [c(r)/g(r)] \right\}, \quad (14)$$

where $c(r)$, the direct correlation function is related to $a(q)$ through:

$$c(r) = (2\pi^2 nr)^{-1} \int_0^\infty dq \left(\frac{a(q) - 1}{a(q)} \right) q \sin qr \quad (15)$$

The basic Eq. 12 and 14 must, in general, be solved by numerical methods. It is unfortunately not yet possible to say which of these theories forms the most reliable means of relating $\phi(r)$ to $a(q)$. The difficulties involved have been discussed at length by Gehlen and Enderby[8] and need not be repeated here. All the theories seem to be capable of distinguishing between metals and insulators; some evidence indicates that the BG theory is to be preferred for metallic liquids whilst the PY theory is more suited to simple liquid insulators. Pair potentials for metallic lead calculated by the PY and the BG methods are shown in Fig. 7.

To discuss the alloy case, we introduce generalized direct correlation functions $c_{\alpha\beta}$ defined by equations of the form

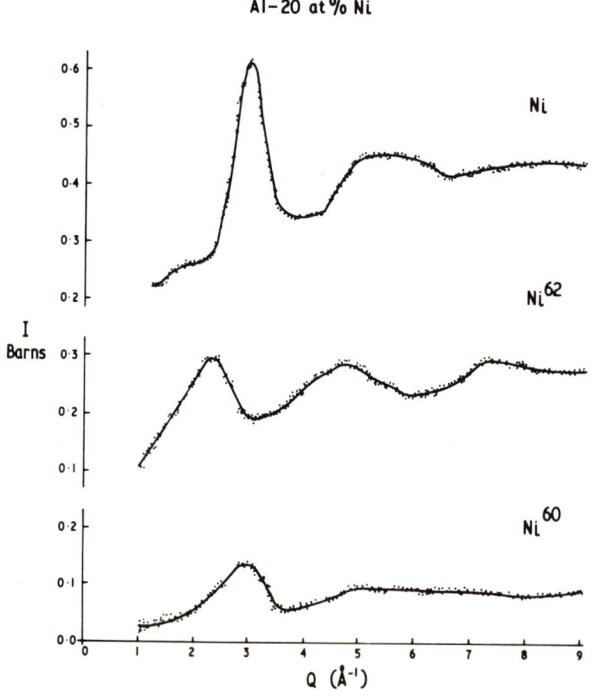

Fig. 4. Normalized intensity patterns for liquid Al - 20 atomic percent Ni for different isotopic compositions of nickel.

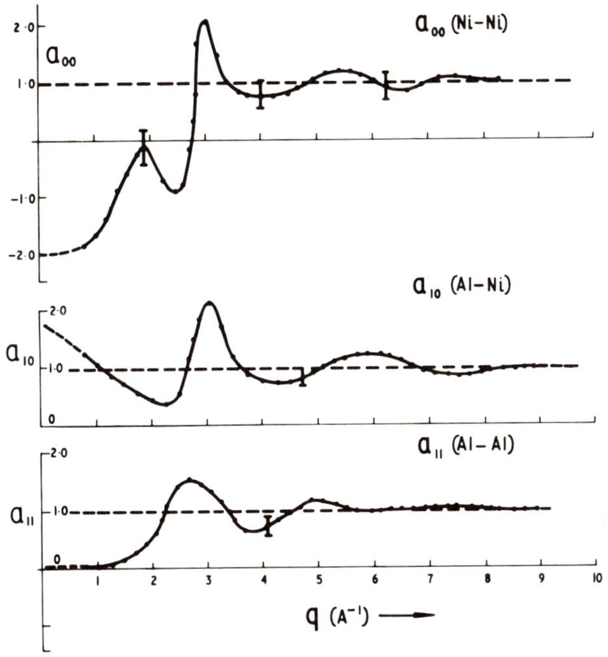

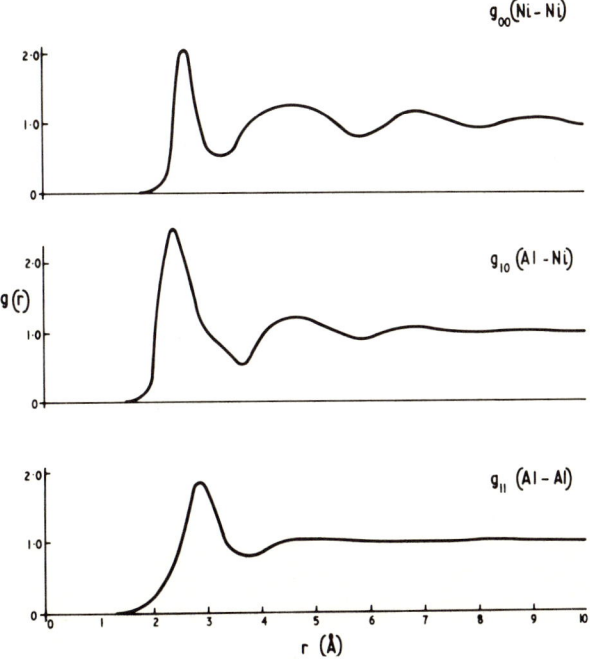

Fig. 5. Partial interference functions for the liquid alloy Al - 20 atomic percent Ni.

Fig. 6. Radial distribution functions for the liquid alloy Al - 20 atomic percent Ni.

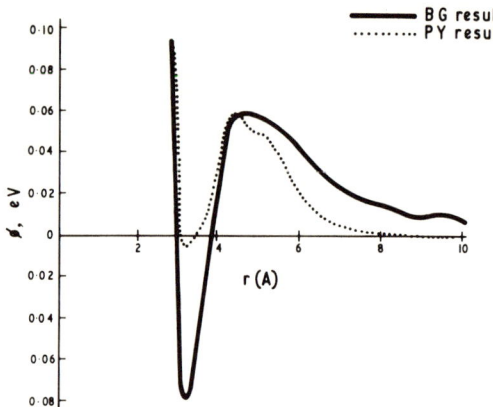

Fig. 7. Pair potential for liquid lead at 340 C calculated by Born-Green and the Percus-Yevick theories.

$$h_{\alpha\beta}(q) = c_{\alpha\beta}(q) + \sum_{j=1,2} n_j c_{\alpha j}(q) h_{j\beta}(q) , \qquad (16)$$

where n_j is number density of the jth atom type and, for instance,

$$c_{\alpha\beta}(q) = \frac{4\pi}{q} \int c_{\alpha\beta}(r) \, r \sin qr \, dr \qquad (17)$$

For a binary alloy, Eq. 16 represents four equations, but since $c_{\alpha\beta} = c_{\beta\alpha}$ only three have to be considered. The solutions for $h_{\alpha\beta}(q)$ are

$$h_{11}(q) = [c_{11}(q)(1 - n_2 c_{22}(q) + n_2 c_{12}^2(q))] \, P(q)^{-1}$$

$$h_{22}(q) = [c_{22}(q)(1 - n_1 c_{11}(q) + n_1 c_{12}^2(q))] P(q)^{-1}$$

$$h_{12}(q) = h_{21}(q) = c_{12}(q) \, P(q)^{-1}$$

with $P(q) = 1 - n_1 c_{11}(q) - n_2 c_{22}(q) + n_1 n_2 c_{11}(q) - n_1 n_2 c_{12}^2(q)$. (18)

The partial interference functions are related to $h_{\alpha\beta}(q)$ through

$$a_{\alpha\beta}(q) = 1 + n h_{\alpha\beta}(q) , \qquad (19)$$

and a generalization of the PY theory yields

$$\phi_{\alpha\beta} = k_B T \ln \left[1 - \frac{c_{\alpha\beta}(r)}{g_{\alpha\beta}(r)} \right] \qquad (20)$$

Thus effective interionic potentials for alloys can, in principle, be extracted from a knowledge of $a_{\alpha\beta}$.

The Born-Green theory for binary alloys has not yet been adapted for the present purposes and is not discussed further.

3.3 Results for Al-20 Atomic Percent Nickel

Preliminary results for the pair potentials derived with the aid of Eq. 20 from the neutron data given in Fig. 5 and 6 are shown in Fig. 8. It is important to emphasize that, whilst there may be some error in detail (and further experimental work is undoubtedly necessary), the data and the theory are probably sufficiently precise to yield the gross features of the effective interactions. As these are the only experimental potentials available for an alloy system at the present time, some general comments seem worth making. It should be emphasized that because of the random errors in the crucial region $q = 2k_F$ no statement can yet be made about the existence (or otherwise) of the Friedel oscillations.

Three points of significance are:

(a) The Al-Al interactions in the alloy are in moderate agreement with those calculated by Harrison for pure aluminum (Fig. 1). It is clearly of interest to see how ϕ_{Al-Al} depends on the concentration of nickel and such experiments are at present underway.

(b) The value of r for which ϕ_{Al-Ni} has its principal minimum, r^o_{Al-Ni}, is not $\frac{1}{2}(r^o_{Al-Al} + r^o_{Al-Ni})$. Thus calculations for alloys in which the cross

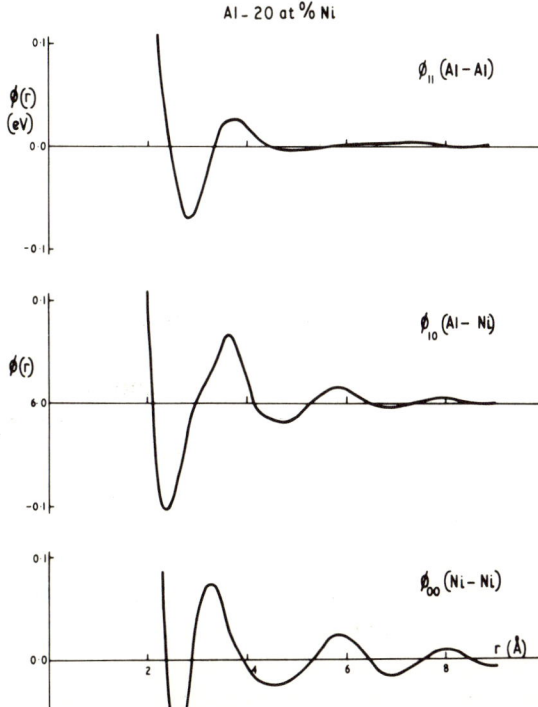

Fig. 8. Pair potentials for liquid Al - 20 atomic percent Ni calculated with the aid of the Percus-Yevick theory.

term in the total interatomic potential is assumed to be the average of those derived for the pure components are probably invalid.

(c) $\phi_{Ni\text{-}Ni}$ is different from $\phi_{Al\text{-}Al}$ in several respects: the short-range repulsive part is distinctly steeper whilst the long-range oscillations and the characteristic repulsive region which follows the first minimum are both more pronounced. We conclude that d-shell overlap is beginning to play a significant role.

3 LATTICE DYNAMICS

To measure the phonon dispersion relations of a material by inelastic neutron scattering, a monochromatic beam of neutrons is allowed to impinge on a single crystal, and the energy distribution of neutrons scattered through an angle θ is recorded. Coherent neutron scattering with the creation or destruction of a single phonon gives rise to sharp peaks in this distribution, from which the frequencies, ν, and wave vectors, $\mathbf{Q}$, of the phonons of mode j can be deduced from the equations

$$h\nu(\mathbf{Q},j) = E_o - E = h^2/2M_o\ (k_o^2 - k^2) \quad, \tag{21}$$

$$\mathbf{q} = \mathbf{k}_o - \mathbf{k} = 2\pi\tau \pm \mathbf{Q} \quad, \tag{22}$$

and

$$q^2 = k_o^2 + k^2 - 2k_o\ k\ \cos\theta \quad, \tag{23}$$

where $\mathbf{k}_o$ and E_o are, respectively, the incident neutron wave vector and energy, $\mathbf{k}$ and E the scattered neutron wave vector and energy, M_o the mass of the neutrons, and τ is any vector of the reciprocal lattice of the sample. It is customary to ensure that the plane of scattering contains the direction of the incident beam and is perpendicular to the axis of rotation of the sample. The angle θ between the incident beam and a symmetry direction of the crystal must also be known.

The two common experimental arrangements involve either time-of-flight techniques or the use of a triple-axis spectrometer.[9] An example of a dispersion curve obtained by the triple-axis technique is shown in Fig. 9.

Once the dispersion relations have been determined experimentally, it is then possible to construct models for the lattice dynamics which will predict ν for any $\mathbf{Q}$. A common approach for the metallic state is to use the Born-von Karman method, assuming that the forces are axially symmetric. The force constants are proportional to the first and second derivatives of the pair potential evaluated at the equilibrium lattice positions

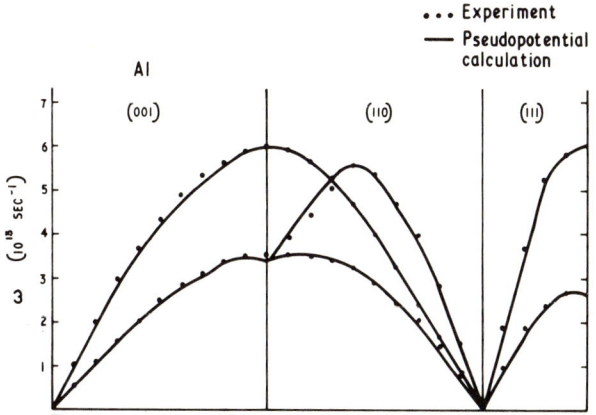

Fig. 9. Model potential fit of the phonon dispersion curves of aluminum at 80 K.

corresponding to a set of neighbors. This is not a unique procedure, however, and has been criticized by, inter alia, Leigh et al.[10]

More progress can be made if a functional form for the pseudopotential is assumed. The angular frequencies, ω, of vibration are found by solving the determinantal equation:

$$\det|D_{xy}(Q) - M\omega^2 \delta_{xy}| = 0 \quad . \tag{24}$$

Here $D_{xy}(Q)$ are the elements of the **dynamical matrix**, M the mass of the ion, and δ_{xy} the Dirac function. $D_{xy}(Q)$ contains both the electrostatic (Ewald) contribution and the band-structure contribution to the total energy. Prescriptions for the evaluation of $D_{xy}(Q)$ in terms of $w_q(k)$ are now part of the literature and are referred to by Ho in this book.

The technique of fitting a model pseudopotential to the experimental data by adjustment of a limited number of the pseudopotential parameters has been widely used, and the recent paper by Koehler et al.[11] deserves special mention. These workers obtained a rather close fit (Fig. 9) to the phonon data for aluminum of Stedman et al.[12] by a suitable choice of the two parameters that characterize the model potential of Harrison[1]. The effective interatomic potential derived from this potential with the aid of Eq. 1 is, unfortunately, quite different from that found in the present work (Fig. 1). It is important to emphasize, however, that quite different $w_q(k)$ can fit $\omega(Q)$ to within 10 percent, and it is far from clear at present whether a $w_q(k)$ adjusted to fit the phonons to (say) within 1 percent would necessarily be unique. As pointed out by Cohen and Heine, it is always desirable to fit $w_q(k)$ to as many physical properties as possible (including Fermi surface data).[13] For this reason we are unable to comment in detail on the apparent discrepancy between our potential and that due to Koehler et al.

4 NONCENTRAL FORCES

The significance of noncentral forces in metals is beginning to be noted, and some detailed work is reported by Johnson and Shaw at this conference. It is clearly of interest to see whether any gross deviation from simple pair forces can be detected by an examination of $a(q)$.

For hard spheres, the $g(r)$ has a sharp edge at the hard-sphere diameter. In contrast, $g(r)$ for a real liquid has a finite slope at the point where atoms overlap, and on taking a Fourier transform to obtain $a(q)$, one obtains a function that exhibits oscillations decreasing in amplitude with increasing q. The oscillations should die out more rapidly for real liquids than for the hard spheres. Moreover, it may be expected that these oscillations will move out of phase with respect to the hard-sphere oscillations as the damping increases. This effect has actually been demonstrated by Page et al.[14], who compared the interference function for liquid argon and liquid rubidium at temperatures and pressures where proper normalization of the data could be achieved. It was observed that the oscillations for rubidium were damped out more rapidly than those for argon, and there was a more pronounced phase shift. This is consistent with the rather soft interatomic potential that it is believed characterize the alkali metals.

If the experimental data for liquids like tellurium and selenium are considered (Hawker and Enderby, to be published), gross deviations are found from the behavior associated with hard spheres. For example, there is less damping in $a(q)$ for selenium than for hard spheres, and since the damping associated with hard spheres represents a minimum for simple liquids, it is concluded that neither tellurium nor selenium can be discussed in terms of central pair-wise interactions. This is entirely consistent with the electrical and other properties of these materials.

It is therefore instructive to examine carefully the damping of $a(q)$ about the asymptotic value of unity. For simple metals, a measure of the interpenetration of the atomic cores may be obtained. On the other hand, the absence of appreciable damping indicates that noncentral forces may be playing an important role. So far, data for Al-Ni appear to be consistent with purely central forces, but more data of higher accuracy for q values in excess of 4Å^{-1} are essential.

ACKNOWLEDGMENTS

We wish to acknowledge helpful conversations with Dr. P. C. Gehlen and the financial support of the Science Research Council.

REFERENCES

1. Harrison, W. A., *Pseudopotentials in the Theory of Metals* (New York: Benjamin, 1966).
2. Ashcroft, N. W., Phys. Letts., **23**, 529 (1966).
3. Shaw, R. W., J. Phys. C., **2**, 2335 (1969).
4. Enderby, J. E., *Advances in Structure Research by Diffraction Methods*, **4** (in press).
5. Enderby, J. E., North, D. M., and Egelstaff, P. A., Phil. Mag., **14**, 961 (1966).
6. Howells, W. S., Howe, A., Enderby, J. E., and Gehlen, P. C., to be published.
7. Enderby, J. E., and March, N. H., Adv. Phys., **14**, 453 (1965).
8. Gehlen, P. C., and Enderby, J. E., J. Chem. Phys., **51**, 547 (1969).
9. Egelstaff, P. A., *Thermal Neutron Scattering* (New York: Academic Press, 1965).
10. Leigh, R. S., Sizigeti, B., and Tewary, V. K., Proc. Roy. Soc., **A**, **320**, 505-526 (1971).
11. Koehler, T. R., Gillis, N. S., and Wallace, D. C., Phys. Rev., **B1**, 4521 (1970).
12. Stedman, R., Almquist, L., and Nilsson, G., Phys. Rev., **162**, 549 (1967).
13. Cohen, M. L., and Heine, V., Solid State Phys., **24**, 38 (1970).
14. Page, D. I., Egelstaff, P. A., Enderby, J. E., and Wingfield, B. F., Phys. Letts., **29A**, 296 (1969).

DISCUSSION on paper by J. E. Enderby and W. S. Howells

SEEGER: In my opinion, the fact that the experimental results do not show the Friedel behavior of the oscillations should be of considerable concern for those advocating the use of pseudopotentials in their present forms. So should be the fact that the nodes for the Ni-Ni, Ni-Al, and Al-Al potentials occur at somewhat different r-values. The prediction that, for large distance, the potential should behave as $\cos(2k_F r)/r^3$ with no additional phase factor is very fundamental to the whole pseudopotential approach using second-order perturbation theory. If the concept of pseudoatoms makes sense at all, the asymptotic law should be obeyed at least beyond the Wigner-Seitz radius.

MARCH: One would expect a Fermi surface which has more blurring when nickel is added to liquid aluminum than in pure aluminum. Is this perhaps the explanation for heavier damping of the Al-Al oscillatory interaction in the alloy than in pure aluminum?

ENDERBY: In reply to Dr. Seeger and Professor March, the data say nothing about the existence or otherwise of the Friedel oscillations in ϕ. This is a practical point — not one of principle — arising from the random errors in the data around the crucial region $q = 2k_F$.

TORRENS: I understand that the pair-correlation function $g(r)$ is reproduced with approximately the same accuracy by the Johnson, Hutchinson and March potential with large-amplitude, long-range oscillations and by a simple Lennard-Jones type potential. Is there a reliable method of distinguishing between these forms using experimentally measurable quantities?

ENDERBY: The differences between a "metallic" $\phi(r)$ and a Lennard-Jones $\phi(r)$ appear at the low q data in the structure factor. This gets smeared out over the whole r-space domain and, hence, the often made, but erroneous, statement that $g(r)$ is insensitive to $\phi(r)$. For this reason, all our calculations are carried out in q-space.

TORRENS: Both Born-Green and Percus-Yevick theories involve a superposition approximation for the triplet correlation function. Can you comment on the influence of this approximation on the transition from experiment to theory, particularly as regards the real-space pair potential?

ENDERBY: This is a difficult question to answer precisely. The Born-Green and Percus-Yevick equations do make some approximation for the triplet-distribution function and, therefore, the form of $\phi(r)$ we derive is not exact. On the other hand, various tests (e.g., the test for self-consistency) indicate that for many applications, the PY theory is satisfactory. It seems reasonable, therefore, to use ϕ_{PY} in a molecular-dynamics calculation and then improve on it by successive approximations.

VINEYARD: Are the results for Al-Ni sufficiently accurate to calculate the heat of solution and compare it usefully with the experimental value?

ENDERBY: I consider that it is worthwhile using these potentials to calculate thermodynamic properties like the heat of solution.

HO: What dielectric screening function did you use to combine with Harrison's pseudopotential of aluminum to produce such excellent agreement of the calculations with the measured phonon-dispersion curves?

ENDERBY: The screening function is described in the paper by Koehler, et al. (See Ref. 11 in text).

HO: Brown, et al.[1] recently calculated the displacements around a vacancy in sodium using liquid-metal interatomic potential. The results are quite different from several other calculations. Can you comment on that?

ENDERBY: Liquid sodium raises special problems from a neutron point of view and the original work by Johnson, Hutchinson and March in which ϕ for Na was calculated, involved the use of X-ray data.[2] These data are not now considered sufficiently reliable to enable ϕ to be obtained with very much confidence.

HO: Recently Shyu and Gaspari showed that the Friedel oscillations set in at a distance of a few r_c (r_c, the core-radius parameter in Ashcroft's empty-core model[3]). So the asymptotic behavior of $\phi(r)$ is well determined. But that does not agree with the $\phi(r)$ determined here for aluminum. Can this be a peculiar behavior of the liquid as it arises as a consequence of the experimental techniques used?

HARRISON: I am concerned about the meaning of these experimental potentials because of the great insensitivity of the radial distribution function to the details of the potential. I wonder whether the oscillations in your potentials are an artifact of the oscillations in the pair distribution function, with the scale set by the hard core repulsion, or whether they are the real oscillations in the potential with the scale set by the Fermi wavelength? One should be able to decide by a careful look at the phase and wavelength of the oscillations since the two lengths in the problem are not commensurate.

ENDERBY: Replying to Drs. Ho and Harrison, and as I point out in the paper, the experimental data are not yet accurate enough to yield the Friedel oscillations. They are, however, sufficiently accurate, to give the repulsive part of ϕ, to fix the position and depth of the first minimum, and to establish the existence (or otherwise) of a repulsive region in ϕ beyond the first minimum.

I cannot agree that the structure data are insensitive to changes in ϕ. The low-angle q-space data, particularly when expressed as the Fourier transform of the direct correlation function $c(q) = 1 - \frac{1}{a(q)}$ are very sensitive to the assumed form of ϕ.

ASHCROFT: What measure of confidence can you place on the ion-ion potentials (for example, how accurately in the Al-Ni system can multiple scattering corrections be eradicated?).

ENDERBY: The gross features of ϕ are, I think, established reasonably well. By "gross features" I mean those aspects of ϕ which I mention in my reply to Professor Harrison.

ASHCROFT: How accurately can the $c(r)$ and $g(r)$ be computed from the k-space data? In particular, do any truncation errors coincide with structure in the potentials?

ENDERBY: Truncation errors are not important for metals because of the damping in $a(q)$ at high q.

ASHCROFT: On Professor Seeger's point, we have seen already how sensitive the pair potentials are to the details of the dielectric function of the interacting electron gas. It is premature to conclude that the differences between Enderby's and Harrison's potentials signify the failure of the pseudopotential approach. Harrison's original calculation used simple Hartree-type screening as has been pointed out several times. It is also important to appreciate the fact that the two potentials shown are appropriate to different densities.

BULLOUGH: I think a direct indication of the noncentral nature of the interatomic potential could be obtained by fitting the data to the Green-function response constants (rather than to force constants, which as you have remarked are not unique). The Green function can then be inverted to give a dynamical matrix and the force constants will be exposed together with their range and possible noncentral nature.

ENDERBY: I agree that this will be worth trying.

REFERENCES

1. R. C. Brown, J. Worster, N. H. March, R. C. Perrin and R. Bullough, Phil. Mag. **23**, 555 (1971).
2. M. D. Johnson, P. Hutchinson and N. H. March, Proc. Roy. Soc. (London) **282A**, 283 (1964).
3. W. Shyu and G. D. Gaspari, Phys. Lett., **30A**, 53 (1969).

MOLECULAR DYNAMICS STUDIES OF LIQUIDS

A. Rahman

Argonne National Laboratory
Argonne, Illinois 60439

ABSTRACT

Over the past few years it has become apparent that certain properties of liquids can be calculated accurately with the interparticle potential as the only input requirement. The basic calculation is simply the solution of the classical equations of motion for a few hundred particles interacting with the given potential; the 'data' provided by such a computer 'experiment' can then be used to calculate a variety of liquid properties.

This presentation is concerned with three liquids: argon, sodium, and water; considerable work has been done on these liquids along the lines indicated above. For these three liquids the process of self-diffusion is dealt with in detail; a brief discussion of other properties is also given in appropriate places.

The results on self-diffusion show clearly that the notion of 'diffusion by jumps' or the 'quasicrystalline' behavior of particle motions in liquids has no validity in any of the three liquids; from the calculations on liquid sodium it is concluded that in liquid metals the constant of self-diffusion depends more sensitively on the details of the potential than some other properties of the liquid. In all three liquids the results show why the data on neutron inelastic scattering, when interpreted in terms of overly simple models of self-diffusion, will appear to support the erroneous notion of the quasicrystalline behavior of liquids.

1 INTRODUCTION

The dynamical correlations in systems of interacting particles are closely related to the interparticle potentials, and, under appropriate conditions, the relation can be investigated purely on the basis of classical mechanics. This presentation is not concerned with physical conditions in which quantum mechanics prevails (e.g., liquid and solid helium).

In recent years it has become possible to solve Newton's equations of motion for systems of several hundred interacting particles and to investigate the dynamical correlations from a knowledge, in complete microscopic detail, of the phase trajectory of the system.[1] This technique is usually referred to as Molecular Dynamics. This presentation is restricted to the process of self-diffusion in dense liquids even though, in appropriate places, mention is made of a variety of more complicated dynamical correlations, to indicate that liquid-like systems are being simulated with considerable fidelity.

In the following section, the nature of molecular dynamics calculations is explained, and the correlations on which this presentation is focussed are defined. Later sections show the quantitative information that has been gathered in recent years on three rather different liquids, namely: a 'Lennard-Jones' liquid which comes astonishingly close to the behavior of liquid rare gases like argon, a liquid sodium-like system, and a liquid water-like system.

Finally, the utility of these results in throwing light on the interparticle potentials on the one hand and theories of self-diffusion in liquids on the other is discussed.

2 MOLECULAR DYNAMICS

N particles, each of mass M, are enclosed in a cubic box of edge length L. Periodic boundary conditions are applied in the usual way by translations through distance L along the edges of the cube. This gives an infinite system, and the equation of motion of particle i of this infinite system is given to be Newton's equation:

$$M \frac{dv_i}{dt} = - \sum_{j \neq i} \frac{\partial \Phi(r_{ij})}{\partial r_i}, \quad (1)$$

where $r_{ij} = |r_i - r_j|$ and the pair potential $\Phi(r)$ is a given function of r and has a range R, i.e., $\Phi(r>R) = 0$; thus in the infinite system of particles the summation over j extends over only those, **finite** number of neighbors of i for which $r_{ij} \leq R$. As mentioned below, for long-range interaction potentials, another procedure has to be used.

Even though the system defined above is infinite, due to the periodic boundary conditions, only the equations of motion of N particles in a cube of edge length L need be solved and, as the particles move, the periodicity will be maintained indefinitely. The number of particles in any such cube will be N at all times anywhere in the infinite system; it is of course understood that the edges of this cube are parallel to the original directions in which periodicity was imposed.

As this is a mathematically well-defined problem, its solution is of particular interest to the theorist interested in the classical N-body problem. As has been shown over the past few years, much useful information can be obtained by integrating the above equations on a digital computer using a suitable integration step Δt. The initial conditions (i.e., the values at $t = 0$ of r_i, v_i for $i = 1, \ldots, N$) can be chosen in a variety of ways; it has been found that, if due care is taken to ensure that equilibrium has been reached, the results are quite independent of the particular starting conditions.

As time passes the system follows a trajectory in phase space and the points traversed in phase space can be thought of as a microcanonical ensemble for an N-particle system with number density $n_0 = N/L^3$ and mass density Mn_0. The quantity $N^{-1} \sum_{j=1}^{N} v_j^2(t)$, averaged over this ensemble, is denoted by $\langle v^2 \rangle$ and $M\langle v^2 \rangle/3k_B$ is called the temperature, T, of the system in degrees absolute.

A calculation along these lines gives, for a well-defined system, the positions and velocities of all the particles as a function of time, in the form of a table. Using this vast amount of tabulated data we can, at our ease, look for any desired information provided it can be expressed in terms of positions, velocities, and if necessary, higher time derivatives as well.

For such calculations those potentials $\Phi(r)$ that can be scaled in terms of an energy and a distance are particularly attractive because of the possibility of using the corresponding states relationship. But of course we are not in any way restricted in the choice of the potential. For long range potentials, one can still make such calculations by replacing the truncation procedure mentioned above by a complete Ewald type summation.

In dealing with systems of polyatomic molecules in which each molecule is considered to be a rigid body and in which the interaction depends on the mutual orientations of the molecules as well, the molecular dynamics technique remains, in principle, the same as described above. However, instead of Newton's equations for a system of point particles one has to solve a system of coupled Newton-Euler equations for a system of rigid bodies.

We shall consider in this presentation only a few of the large number of quantities that can be calculated through the technique of molecular dynamics.

2.1 Heat Capacity c_v

In a microcanonical ensemble in equilibrium at temperature T the fluctuation of the temperature around the value T is related to the heat capacity at constant volume. The relation, given by Lebowitz et al.[2] is

$$<(\Delta T)^2>/T^2 = 2/3N - 1/Nc_v \qquad (2)$$

where c_v is the heat capacity per particle in units of k_B. One test of equilibrium in the system is a nonsecular fluctuation in the temperature. For a system of rigid nonlinear polyatomic molecules one has $1/3N$ on the right side.

2.2 Pair Correlation $g(r)$

In a monatomic system with pair interactions, the basic correlation function for all thermodynamic properties is the pair-correlation function. If $n_o = N/V$ is the number density, in an isotropic system in equilibrium, then the average number of particles in a shell of radius r and thickness Δr, centered at a given particle, is given by

$$n_{r,r+\Delta r} = 4\pi r^2 \, \Delta r \, n_o \, g(r) \quad , \qquad (3)$$

where $g(r)$ is called the pair correlation or the radial-distribution function.

2.3 The Constant of Self-Diffusion and the Velocity Autocorrelation

The mean-square displacement of particles in the system, $<r^2>$, is defined as

$$<r^2> = <N^{-1} \sum_{i=1}^{N} [\mathbf{r}_i(t+\tau) - \mathbf{r}_i(\tau)]^2> \quad , \qquad (4)$$

where $\langle \ldots \rangle$ on the right indicates averaging over τ. A stringent test on the equilibrium of the system is a nonsecular dependence on τ of any time-displaced correlation, e.g., the one shown above.

The macroscopic law of diffusion states that, in a liquid, the probability for a particle to be displaced by $\mathbf{r}$ in time t, is normally distributed and the width of the distribution, or the mean-square displacement, $\langle r^2 \rangle$, is given by $6Dt$ where D is the constant of self-diffusion. This is a phenomenological law, well verified for macroscopic times t ($\gg 10^{-10}$ sec). However, it is obvious that for $t \to 0$, $\langle r^2 \rangle$ increases as t^2, being given by $\langle r^2 \rangle = (3k_BT/M)t^2$. As t increases, the motion is hindered due to the interactions between particles resulting in an eventual and permanent linear increase of $\langle r^2 \rangle$ with t. The mechanism for this hindrance is the central point in our lack of understanding of the process of self-diffusion, and it is clear why molecular dynamics is a useful method of getting precise, quantitative information to develop and verify our concepts and models for this aspect of fluid dynamics.

From the theory viewpoint, the essential difficulty lies in the fact that the particles in a liquid interact strongly with a fairly large number of neighbors; hence, a theory based on the concept of 'collisions' is hard to justify; however, it is not obvious why models based on solid-like pictures of diffusion (e.g., diffusion by jumps into holes) should be equally invalid.

Expressions for transport coefficients as integrals over time correlations are now well known. We have, e.g.,

$$D = 1/3 \int_0^\infty \langle \mathbf{v}(0) \cdot \mathbf{v}(t) \rangle dt \quad , \tag{5}$$

where

$$\langle \mathbf{v}(0) \cdot \mathbf{v}(t) \rangle = \langle N^{-1} \sum_{j=1}^{N} \mathbf{v}_j(\tau) \cdot \mathbf{v}_j(t+\tau) \rangle \quad , \tag{6}$$

is the velocity autocorrelation function. In fact

$$\langle r^2 \rangle = 2 \int_0^t (t-u) \langle \mathbf{v}(0) \cdot \mathbf{v}(u) \rangle du \quad , \tag{7}$$

or equivalently

$$\langle \mathbf{v}(0) \cdot \mathbf{v}(t) \rangle = 1/2 \frac{d^2}{dt^2} \langle r^2 \rangle \quad . \tag{8}$$

2.4 Complete Description of the Process of Self-Diffusion

The mechanism of self-diffusion cannot be described completely in terms of $\langle r^2 \rangle$ or its second derivative $\langle \mathbf{v}(0) \cdot \mathbf{v}(t) \rangle$. It is the Van Hove function $G_s(\mathbf{r},t)$ that gives the probability of finding a particle displaced by $\mathbf{r}$ in interval t. For a liquid, G_s is a function of $r = |\mathbf{r}|$. Only if G_s has a Gaussian shape, its second moment, namely $\langle r^2 \rangle$, is sufficient to determine the function completely, otherwise one needs to know the time behavior of all the higher moments of G_s. The shape of the function G_s can be exhibited conveniently[3] in terms of $\langle r^2 \rangle$, its second moment, together with the functions

$$\alpha_n(t) = \frac{\langle r^{2n} \rangle}{C_n \langle r^2 \rangle^n} - 1$$

$$C_n = 1 \times 3 \times 5 \ldots \times (2n+1)/3^n \quad . \tag{9}$$

In the special case of a Gaussian G_s, all $\alpha_n = 0$.

For very short times, the displacements are proportional to velocities, which, in equilibrium, are normally distributed. Hence, for such times G_s must be a Gaussian. On the other hand, for very large times the macroscopic diffusion equation again leads to a normal distribution of displacements. The virtue of molecular dynamics results is that they are able to give us useful information for intermediate times.

3 ARGON-LIKE LENNARD-JONES LIQUIDS

In this case the potential $\Phi(r)$ is $4\epsilon[(\sigma/r)^{12} - (\sigma/r)^6]$. The most convenient units are ϵ for energy, σ for length, and $(M\sigma^2/\epsilon)^{1/2}$ for time. In these units the number density is $\rho^* = n_0 \sigma^3$ and the temperature is $T^* = k_B T/\epsilon$. All statements below make use of dimensionless quantities defined in terms of the above units.

A molecular-dynamics calculation has recently been performed by the author for a Lennard-Jones system of 500 particles using a step of integration equal to .0075. The integration was continued for 3250 such steps (this representing an elapse of about 5×10^{-11} seconds, if M, ϵ, σ relevant for argon are considered). The temperature of the system was $T^* = .737$ (or ~ 89 K for argon) and the density $\rho^* = .811$ (or 1.37 g/cc for argon).

The fluctuations of the temperature gave a value of $c_v = 2.6$. The value quoted in the literature[4] for argon at 1.407 g/cc and 84.5 K is 2.32.

The pair correlation $g(r)$ for this system is shown in Fig. 1. While it is not the intention in this presentation to focus attention on this function, in passing it is mentioned that extensive work on the $g(r)$ of a Lennard-Jones fluid was performed by Verlet[5], and it is only recently that neutron-diffraction experiments have become accurate enough to make a comparison with the molecular dynamics calculation meaningful. The unpublished neutron diffraction work on liquid argon of Yarnell et al.[6] has given results remarkably close to the calculated values of Verlet and his collaborators (thus implying that the L-J potential is a very good approximation for the pair potential in heavy liquefied rare gases).

The main purpose of this presentation is to exhibit results on the process of self-diffusion. Fig. 2 shows the dependence of the mean-square displacement, $\langle r^2 \rangle$, on time, and also that of the velocity autocorrelation function.

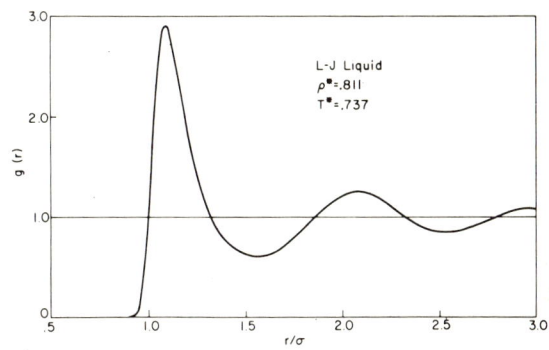

Fig. 1. Pair correlation of a Lennard-Jones liquid at $\rho^* = .811$ and $T^* = .737$. The data was collected for 3250 steps, each $.0075(M\sigma^2/\epsilon)^{1/2}$ long. The potential was truncated at $r/\sigma = 3$.

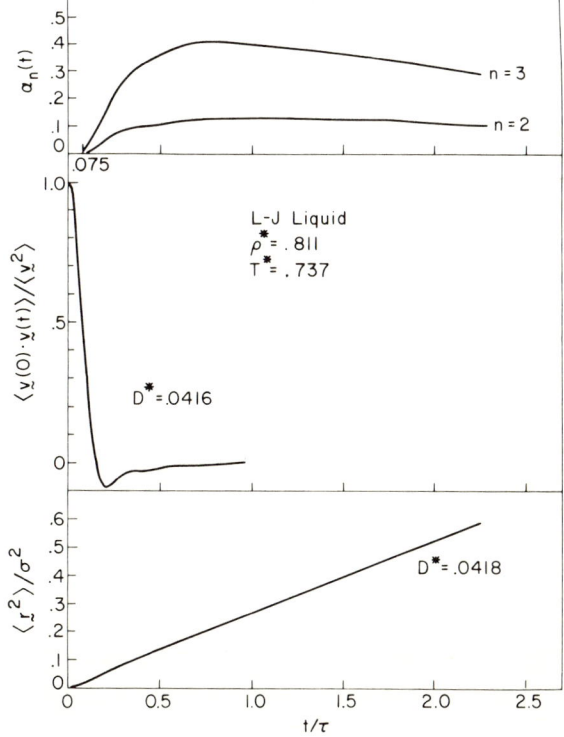

Fig. 2. Mean-square displacement $\langle r^2 \rangle$, velocity autocorrelation $\langle v(0) \cdot v(t) \rangle$ and non-Gaussian terms $\alpha_2(t)$ and $\alpha_3(t)$ for Lennard-Jones liquid at $\rho^* = .811$ and $T^* = .737$.

Apart from the fact that the diffusion constant obtained from this data is quite close to the values observed in the laboratory[4] for liquid argon near its triple point, the feature that is of great interest in this data is the complete absence of any quasicrystalline behavior in the time dependence of $\langle r^2 \rangle$, or to express the same fact in another way, $\langle v(0) \cdot v(t) \rangle$ is not very oscillatory in its time dependence. The quasicrystalline model for diffusion in liquids, due to Frenkel, presumes that in a liquid there is a characteristic oscillation time of the order of 1 (2 x 10^{-12} seconds for liquid argon) before diffusion occurs. As can be seen in Fig. 2 the curve for $\langle r^2 \rangle$ becomes linear at $t \approx 1$ which is presumed to be the oscillation time before diffusion; in other words, no such oscillations in fact exist.[7]

The non-Gaussian nature of $G_s(r,t)$ is shown in Fig. 2 in terms of the quantities $\alpha_n(t)$ defined in Section 2.4. The Gaussian behavior at short times is clearly seen. By extrapolating to the right one can say that at approximately $t = 4$ (or 10^{-11} seconds for liquid argon) the macroscopic diffusion equation becomes valid.

However for intermediate times the process of self-diffusion is not Gaussian at all, and this is clearly seen from the figure. The presence of non-Gaussian elements in the shape of $G_s(r,t)$ affects the manner in which neutrons are inelastically scattered by the moving particles. These have been dealt with, by Nijboer and Rahman[8] and by Levesque and Verlet[9]. Only very recently, such effects have actually been observed in neutron-scattering experiments[10].

4 DIFFUSION IN LIQUID SODIUM

In recent years, considerable attention has been given to the possibility of deriving an effective pair potential for the ion-ion interactions in metals. The purpose here is not to discuss the relative merits of the various models, but to report on the results obtained from molecular dynamics for one such effective pair interaction that was published by Shyu et al.[11] for sodium. This potential is shown in Fig. 3. The depth of the potential is 422.3 K; to the left of the minimum, the potential is zero at 3.301 A and to the right, are very weak oscillations arising out of Fermi surface effects in the motion of the conduction electrons. The potential is discussed again later in this section.

A molecular-dynamics calculation of 4000 steps in time, each step being .845 x 10^{-14} seconds, was performed on a 500-particle liquid sodium system at a density .927 g/cc at 397 K.

The fluctuations in temperature gave a value of $c_v = 3.4$, which is the value quoted in the literature but at the melting point 373 K.

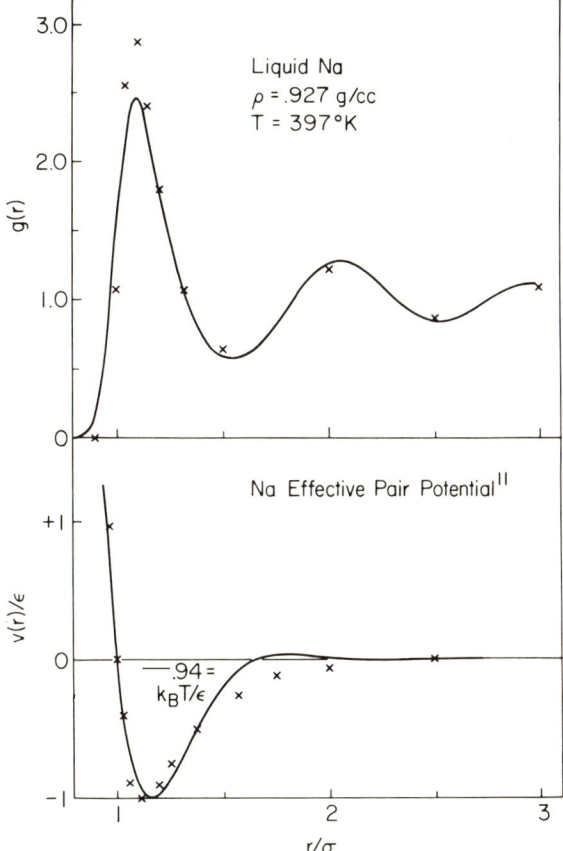

Fig. 3. Effective pair potential[11] in sodium and the pair correlation at 397 K and .927 g/cc. The depth of the potential is $\epsilon = k_B \times 422.3$ K. The short horizontal line is $k_B T/\epsilon$. The potential cuts the axis first at $\sigma = 3.301$ A. The data were collected for 4000 steps, each $.845 \times 10^{-14}$ sec. long. The potential was truncated at $r/\sigma = 3$.
For comparison a L-J potential and the g(r) from Fig. 1 is shown as x x x x.

A very detailed analysis of density fluctuations[12] gave a value of 2.7 × 10⁵ cm/sec for the velocity of sound, whereas in laboratory experiments[4] it is found to be 2.5 × 10⁵ cm/sec at 373 K.

The pair correlation arising out of the pair potential of Shyu et al.[11] is shown in Fig. 3. Because of the contribution of electronic motion to the overall thermodynamic properties, the knowledge of $g(r)$ alone is not enough to calculate the pressure. A detailed analysis of the electronic effects has been made by Price[13] which, together with the molecular dynamics data, gives a satisfactory account of the thermodynamic properties. He has shown, for example, that the pressure calculated using $g(r)$ alone is very high, but is brought down to a reasonable value when the electronic part is included in the relevant thermodynamic considerations.

The mean-square displacement and the velocity autocorrelation are shown in Fig. 4. Notice that the latter is more oscillatory than in the Lennard-Jones liquid (Fig. 2) but **not** enough to classify the process of self-diffusion as quasicrystalline. The relation between the oscillations in

$\langle v(0)\cdot v(t)\rangle$ and the 'hardness' of the pair potential has been investigated in great detail by Schiff[14].

The point emphasized here is the large value of D, the constant of self-diffusion. In the laboratory[4], the ions are found to diffuse (at 373 K) with a $D = 4.3 \times 10^{-5}$ cm^{2}sec^{-1} whereas the value we find is 5.5×10^{-5} cm^{2}sec^{-1}, almost 30 percent too large! This implies that, of all the properties investigated, it is D which seems to be most sensitive to whatever inaccuracies there are in the pair potential.

The non-Gaussian effects in diffusive motion are displayed in Fig. 2 (for an L-J liquid) in terms of the functions $\alpha_n(t)$; however these effects can be exhibited in another way by calculating $\langle \exp[i\mathbf{K}\cdot(\mathbf{r}(t+\tau) - \mathbf{r}(\tau))]\rangle$ which is usually referred to as $F_s(\mathbf{K},t)$, or, the intermediate scattering function for diffusive motions. The Fourier transform of F_s is called $S_s(\mathbf{K},\omega)$ and is related to the neutron inelastic scattering cross section. The function $F_s(\mathbf{K},t)$ for sodium at $\mathbf{K} = 11.35/3.301$ A, i.e., for a wavelength of about 1/2 the interparticle distance in the liquid, is shown in Fig. 5. If $G_s(r,t)$ has a Gaussian shape then F_s can be written as $\exp[-K^2\langle r^2\rangle/6]$; hence a comparison between $-\log F_s/K^2$ and $\langle r^2\rangle/6$ shows the degree to which non-Gaussian effects in G_s affect the shape of F_s. The top part of

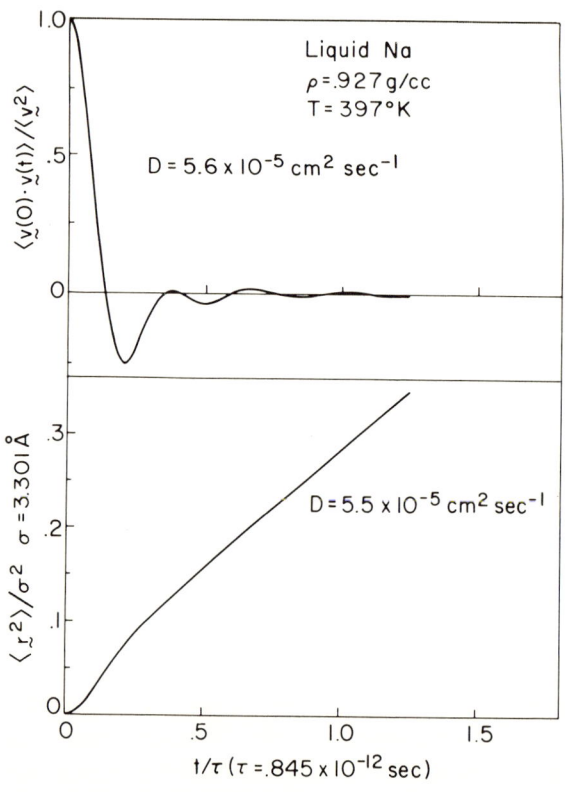

Fig. 4. Diffusion in liquid sodium. The observed value of D, at 373 K, is 4.3 x 10^{-5} cm^2/sec.

Fig. 5. Intermediate scattering function $F_s(K,t)$ for sodium at $K\sigma = 11.35$. Note the fallacious quasi-crystalline behavior obtained by considering $F_s(K,t)$ to be a Gaussian.

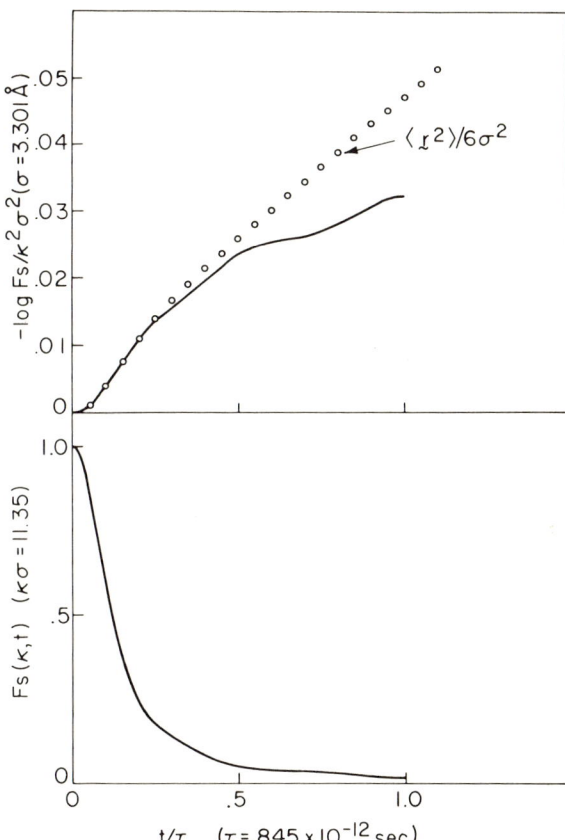

Fig. 5 shows that the 'width' obtained by forcing a Gaussian form for G_s will give rise to a mean-square displacement which, in a state of innocence, one can easily interpret as proof of the solid-like behavior of the motion of particles in a liquid!

5 MOTION OF MOLECULES IN WATER

Over the years, some attempts have been made to construct an effective pair interaction for molecules in water. The basic element is to assume that the electron distributions in the molecules can be looked upon as positive and negative point electric charges of a certain amount situated at certain specific points in the molecule.

Recently Ben-Naim and Stillinger have proposed a four-point-charge model for the water molecule, of which two positive charges are situated at the proton positions and two negative charges on the side of the oxygen atom away from the protons and placed in a way as to form a tetrahedral arrangement with the positive charges.[15] Since the model assumes a tetrahedral value for the angle HOH [$\cos^{-1}(-1/3)$], the positions of the four charges are made to form a perfect tetrahedron. The value chosen for these charges is 0.19 of the electron charge. Moreover since the molecule is isoelectronic with neon, they have assumed an appropriate L-J interaction between the positions of oxygens in the different molecules. Thus, there are very strong Coulomb forces between the molecules due to the four point charges distributed in each molecule. The neon-like L-J potential is quantitatively significant only upon close collisions. For further details the reader is referred to Ben-Naim and Stillinger[15].

The forces and torques arising out of this orientation- and distance-dependent interaction lead to correlated translatory and rotatory motion of the interacting molecules. A very detailed molecular dynamics study for this model of water has been made by Rahman and Stillinger[16]. Out of the large amount of data collected by these authors on the structural, dynamical, and dielectric properties of this model for water, brief mention is made of the oxygen-oxygen pair correlation function and the mean square displacement for the center of mass. (See Ref. 16.)

In Fig. 6 the $g(r_{O-O})$ is shown for a temperature 52.8 C and density 1 g/cc. The structure shown by this function is basically different from that in monatomic liquids. In Fig. 3 we see that the structure of a L-J liquid and of liquid sodium is qualitatively the same; both of these show the second maximum at a distance about twice that of the first-neighbor sharp maximum. However in Fig. 6 this occurs at 1.66 times the first-neighbor distance and shows the persistance of the second coordination shell of an ice-like, tetrahedral structure, even after melting. On the other hand, monatomic liquids do not show the parent crystalline structure so faithfully. The second-neighbor peak in a face-centered-cubic lattice for example occurs at a distance $2^{1/2}$ times the first-neighbor distance and this coordination shell does not appear in the liquid at all.

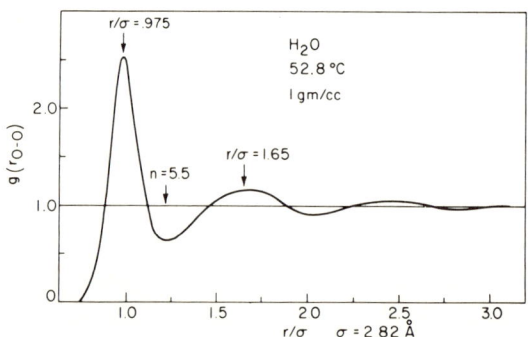

Fig. 6. Oxygen-oxygen pair correlation in water (from Rahman and Stillinger[16]). The average number of neighbors is 5.5 up to the distance shown. Tetrahedral coordination is quite apparent.

DYNAMICS OF LIQUIDS

The diffusion of the center of mass is shown in Fig. 7; note that even in a highly bonded liquid like water there is no indication of quasi-crystallinity in diffusive motion. For further details about the non--Gaussian aspects of proton motion and its effect on the scattering of neutrons from the system, the reader is referred to Rahman and Stillinger[16]. The only points emphasized here are firstly that the calculated value of D is close to the experimental value (4.1×10^{-5} cm^{2}sec^{-1}) and secondly that the diffusive process for the molecules is qualitatively the same as in the monatomic liquids treated above.

6 CONCLUSIONS

(i) In Fig. 3 are drawn together the potentials for the L-J case and for liquid sodium; the pair correlations are also shown together in this figure. The purpose of the drawing is to indicate that ordinary x-ray and neutron-diffraction experiments have to be extraordinarily accurate to give, after Fourier transformation of the data, a $g(r)$ that shows the subtle changes in the first peak of $g(r)$. Also, in the approximate integral equations that couple $g(r)$ and $V(r)$, one does not know the degree to which they are approximate. It is therefore evident that, at present, the diffraction data on the structure of monatomic liquids gives hardly any guidance regarding the pair potential.

(ii) In view of the success, using the potential of Shyu et al.[11] for sodium, in calculating several properties[12] (specific heat, sound velocity, compressibility) with fair accuracy **except** the constant of self-diffusion D, one should henceforth consider D as one of the critical quantities in judging the inaccuracies of an effective pair potential in liquid metals.

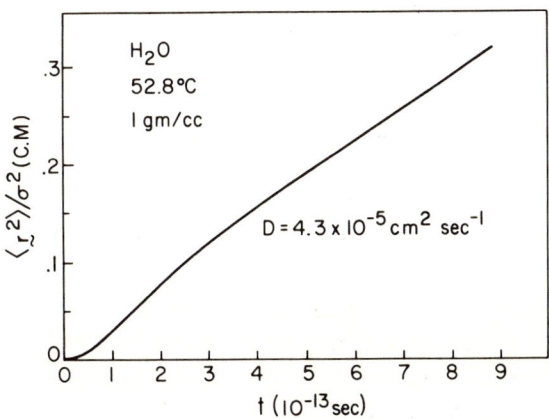

Fig. 7. Diffusion of the molecules in water. Quasicrystalline behavior is not indicated.

(iii) The quasicrystalline model for diffusion in liquids seems to be a completely chimeric notion even in highly bonded systems like water. Diffusion does not occur in 'long' jumps from momentarily stable particle positions; it is more akin to a process consisting of a series of displacements that are very small compared to particle distances in the liquid.

(iv) Neutron inelastic scattering experiments, if interpreted naively in terms of Gaussian models of $G_S(r,t)$, will lead to erroneous notions regarding the process of self-diffusion.

ACKNOWLEDGMENT

This work was performed under the auspices of the U. S. Atomic Energy Commission.

REFERENCES

1. A. Rahman, Rivista del Nuovo Cimento, Numero Speciale, Vol. I, 315 (1969).
2. J. L. Lebowitz, J. K. Percus and L. Verlet, Phys. Rev., **153**, 250 (1967).
3. A. Rahman, Phys. Rev., **136**, A405 (1964).
4. P. Egelstaff, *An Introduction to the Liquid State* (Academic Press, New York, 1967).
5. L. Verlet, Phys. Rev., **159**, 98 (1967), **165**, 201 (1968).
6. J. Yarnell and S. Koenig (private communication).
7. I. Z. Fisher, Ukrainskii Fiz. Zhur., **12**, 3 (1967), Section 8.
8. B.R.A. Nijboer and A. Rahman, Physica, **32**, 414 (1966).
9. D. Levesque and L. Verlet, Phys. Rev., **A2**, 2514 (1970).
10. K. Sköld, M. Rowe, P. Randolph and G. Ostrowski (private communication).
11. W. Shyu, K. S. Singwi and M. P. Tosi, Phys. Rev., **B3**, 237 (1970).
12. A. Rahman, Proceedings IUPAP Conference on Statistical Mechanics, Chicago, 1971 (under publication).
13. D. L. Price, Phys. Rev. (in press).
14. D. Schiff, Phys. Rev., **186**, 151 (1969).
15. A. Ben-Naim and F. H. Stillinger, "Aspects of the Statistical-Mechanical Theory of Water", in *Structure and Transport Processes in Water and Aqueous Solutions*, R. A. Horne, ed. (John Wiley, Interscience, New York, in press).
16. A. Rahman and F. H. Stillinger, Jour. Chem. Phys. (under publication).

DISCUSSION on Paper by A. Rahman

SEEGER: In the model you have been using for water, does the 4 C anomaly come out, i.e., does it account for polymerization?

RAHMAN: At this early stage it is not worthwhile to look for this subtle effect in the behavior of real water. The model needs to be extraordinarily successful in reproducing "simpler" properties of liquid water before one is justified in looking for this aspect.

SHEWMON: What is the mean-square displacement per oscillation of an atom in your sodium, or if you wish, the reciprocal of the characteristic frequency of the atoms in the liquid?

RAHMAN: I think that the basic trouble in answering your question arises from the use of concepts (and hence of terminology) which are completely invalid for self-diffusion in liquids. It is clear, since 1964, that the 'oscillation' you are thinking of is simply nonexistent. If the decay time of $\langle v(o) \cdot v(t) \rangle$ is called the oscillation time, then its value in liquid argon is about 10^{-12} sec and in this time a typical particle moves a distance of about 1.5 A which is only about half the near-neighbor distance.

TORRENS: Does the noncubic nature of your pair-correlation function $g(r)$ for water actually depend on the nature of the interaction between the oxygen and hydrogen ions, or is it simply a consequence of the nonspherical configuration of the H_2O molecule?

RAHMAN: It is the very strong angular dependence of the potential which leads to the structure. The Ben-Naim-Stillinger model was built in such a way as to favor the correct tetrahedral structure in the 'ordinary' Ice I structure. The point that appears from our calculations on the liquid is that the strong angular forces produce a structure that reminds you of the ice-type structure. Also it is not correct to use the phrase "noncubic" in this context.

VINEYARD: I would expect that at sufficiently high density the curve of $\langle r^2 \rangle$ versus time might behave more as Frenkel assumed. Have you any information on this?

RAHMAN: If I remember correctly, extensive work of the group at Orsay (Verlet and collaborators) shows no such behavior; in fact at high density $\langle v(o) \cdot v(t) \rangle$ shows not even the negative portion which occurs at normal densities. I believe that the picture of Frenkel is basically fallacious because it overlooks the fact that the so-called 'cage' is itself made up of particles with the same degree of agitation as the one in the cage.

SEEGER: How do the temperature dependences of the diffusion coefficients of liquids come out and compare with experiment?

RAHMAN: The work of Levesque and Verlet gives, at constant density, a linear rise of D with T. For the L-J liquid (argon) the comparison with experiments is quite good.

DE WETTE: In connection with Dr. Vineyard's question about the time behavior of the mean-square displacements at high densities, are there any results for hard spheres that indicate a flat part in $\langle r^2 \rangle$ as a function of time?

RAHMAN: The mean-square displacement in the hard-sphere fluid shows no such flat part. It shows, qualitatively, the same kind of marginal negative part in the velocity auto-correlation as in the data on Argon I have shown.

BULLOUGH: Are you able to get consistent melting temperatures?

RAHMAN: Systematic series of calculations need to be made as a function of temperature and density before one can pin down the melting point of a computer crystal. Moreover, considerable temperature fluctuations occur in a microcanonical ensemble of a few hundred particles, and this makes the investigation somewhat ambiguous.

ASHCROFT: As you know, Rowlinson and others have pointed out that the values for σ and ϵ normally used in the Lennard-Jones potential are "effective" density- and temperature-dependent parameters which, in fact, approximately include certain n-body terms (e.g., the Axilrod-Teller dipole-dipole interaction)[1]. I'm curious to know whether it is logically consistent to use these "effective" potentials in molecular-dynamics calculations.

One additional minor point; which ice structure are you referring to, there being at least nine that I have heard of?

RAHMAN: This point has been stressed by Verlet in much of his work. The success of the L-J potential with the popularly accepted values of σ and ϵ for argon means that the properties of this liquid can be well represented in terms of this simple two-body potential in the condensed state. It is well known that in the vapor this is not a good potential to use.

As for the ice structure, I did not mention during the presentation that I am talking of ice I which has a tetrahedral coordination.

REFERENCE

1. B. M. Axilrod and E. Teller, J. Chem. Phys. **11**, 299 (1943).

INTERATOMIC POTENTIALS; ASPECTS WHICH ARE VISIBLE IN EXPERIMENTAL RADIAL PAIR DISTRIBUTIONS

Roy Kaplow

Department of Metallurgy and Materials Science
Massachusetts Institute of Technology
Cambridge, Massachusetts 02139

ABSTRACT

The Fourier transform of the continuous X-ray intensity — in reduced form — $[(I - \langle f^2 \rangle)/\langle f \rangle^2]k$ yields an atomic radial pair-distribution function; $f(k)$ is the atomic X-ray scattering factor, $k = 4\pi \sin \theta/\lambda$ and 2θ, the diffraction angle. The analysis is applicable to crystalline as well as to amorphous solids. When examined in detail — and with proper regard for the effects of data termination — experiments yield information that relates closely to interatomic potentials: (i) the mean-force potential in the near-neighbor region; (ii) vibrational anharmonicity; (iii) coupled motions among near neighbors; (iv) the different scale of inter- and intramolecular vibrations; (v) deformation-caused displacements. Examples are presented in each instance, including previously unpublished data for cold worked nickel. The latter results indicate that a major effect of deformation is a displacement of some nearest neighbors to new equilibrium separations.

1 INTRODUCTION

The intensity variation, as of function of scattering vector, of the total coherently scattered X-radiation from an assemblage of atomic scatterers is related to the interscatterer distribution through a Fourier transformation.[1] If the average of the pair distribution, over the sample and time, is spherically symmetrical, that relationship may be expressed quite simply.[2,3]

$$F(k) = \int_0^\infty 4\pi r (\rho(r) - \rho_0) \sin kr \, dr \quad (1)$$

where:

$k = |\mathbf{k}| = 4\pi \sin\theta/\lambda$

λ = X-ray wavelength

θ = half the scattering angle

$\mathbf{k} = (\mathbf{S} - \mathbf{S}_0)$

$\mathbf{S}, \mathbf{S}_0$ = wave vectors of scattered and incident (plane wave) beams.

$|\mathbf{S}|, |\mathbf{S}_0| = 2\pi/\lambda$

$F = k[(I - \langle f^2 \rangle / \langle f \rangle^2]$

$I(k)$ = total coherently scattered intensity, in electron units

$\langle f^2 \rangle = \sum_{i=1}^{n} x_i f_i f_i^*$

n = number of different atomic species

x_i = atomic fraction, species i

$f_i(k)$ = atomic X-ray scattering form factor, i^{th} species

$\rho(r) = \sum \frac{x_i f_i f_j}{\langle f \rangle^2} \rho_{ij}(r)$

$4\pi r^2 \rho_{ij}(r)dr$ = the average number of j-type atom centers in a spherical shell between r and $r + dr$, centered on an average i-type atom.

The Fourier inversion of Eq. 1, over the total available range of data and with an optional weighting function ($W(K)$), yields an r-space function:

$$G'(r) = \frac{2}{\pi} \int^{\text{finite}} F(k)\, W(k)\, \sin kr\, dr \qquad (2)$$

$G'(r)$ generally contains a somewhat-distorted version of the true pair distribution in the sense that a δ-function peak in $4\pi r \rho_{ij}(r)$ is distorted into (i.e., convolved with)

$$Q_{ij}(r) = \int_{\text{finite}} x_i \frac{f_i f_j}{\langle f \rangle^2} W(k) \sin kr\, dk \qquad (3)$$

For this discussion, the data analysis complications thereby introduced are ignored. Note, however, that the problem for a multicomponent system is not particularly different from that for a pure material, providing that the separate (i-j) peaks in the total $G'(r)$ can be identified in the region of interest. Indeed, for most experimental situations (including light-element mixtures) the finiteness of the integral is a more significant factor than are variations in k of $f_i f_j/\langle f \rangle^2$. In either, the true shapes can be deduced (within experimental error) by trial-and-error fitting, even if not more directly.[2,3,4]

The true function is referred to as $G(r)$,

$$G(r) = 4\pi r\, (\rho(r) - \rho_0) \qquad (4)$$

For a pure material, $4\pi r^2 \rho(r) dr$ is the average number of atoms in a spherical shell around an average atom; most of the results discussed are for pure materials.

This paper reviews some of the aspects of those experimental pair-distribution results with which the author has been associated and which seem to be closely related to concepts of atomic interaction potentials. Indeed, if it is assumed that a breakdown into "interatomic" forces is a meaningful way to express the interactions among the electrons and nuclei, then the complete $G(r)$ is not only related to, but completely determined by, those potentials. The more illuminating aspects seem to occur in the near-neighbor region, however.

2 THE PAIR POTENTIAL, THE PAIR DISTRIBUTION, AND THE "MEAN FORCE" POTENTIAL

A further assumption regarding the forces is often made; they can be expressed in terms of pair potentials $V(r)$, that depend only on the interatomic separation. The relationship between such a pair potential and the pair distribution can be written in various approximate forms that are soluble for pure materials, but not necessarily accurate for the condensed, high-density state. As examples, the "hyper-netted chain" (HNC) and the

Percus-Yevick (PY) approximations are shown — both in Stell's formulation:[5]

(HNC) $\quad V(r) = \left\{ g(r) - 1 - \ln[g(r)] - c(r) \right\} k_B T$

(PY) $\quad V(r) = \ln\left\{ [g(r) - c(r)]/g(r) \right\} k_B T \qquad (5)$

where

$g(r) = \rho(r)/\rho_o$

k_B = Boltzmann's constant

T = absolute temperature

$c(r)$, the "direct-correlation function", is exactly related to $g(r)$ in terms of a three-dimensional convolution:

$$c(r) = [g(r) - 1] - \rho_o \int c(\mathbf{r}-\mathbf{R}) [g(\mathbf{R})-1] \, d\mathbf{R} \qquad (6)$$

The direct-correlation function may also be expressed directly in terms of the previously introduced reduced-intensity function.

$$c(r) = \frac{1}{2\pi^2 r \rho_o} \int_o^\infty \frac{F(k)}{1+F(k)/k} \sin kr \, dk \qquad (7)$$

This formulation shows more clearly that $c(r)$ and, hence, the derived $V(r)$ depends primarily on the small-k region of measurement. There, the intensities are small ($F(k) \approx -k$), and high experimental precision may be difficult to achieve for other reasons as well. Since $G(r)$ (for a high-density state) is relatively insensitive to the small-k region (compare Eq. 2, with infinite limits and $W(k) = 1$, to Eq. 7), one might guess that the relationship between $G(r)$ and $V(r)$ is not strongly unique within experimental imprecisions. Such an ambiguity has been demonstrated[6] (in the PY approximation), perhaps most vividly in the form reproduced in Fig. 1 and 2. The five "liquid-like" distribution functions in Fig. 1 (which most experimentalists would consider to be essentially identical) are derived from the five potentials in Fig. 2, for a density of 0.9 of ideal close packed. These five potentials were, in fact, designed to yield nearly identical first-peak-heights in $G(r)$. This was done by varying E for a given n in the four potentials of the Lennard-Jones type, $V(r) = E[(r_o/r)^n - 2(r_o/r)^6]$ (which $n = 11$, 12, and 13), and by varying the constant in the quadratic "soft-sphere" repulsive potential; it should not be construed therefore that other $V(r)$'s will also give the same result.

Different potentials (e.g., changing E for a given n) yield different shapes and peak heights for $G(r)$, but scarcely any variation in the

RADIAL PAIR DISTRIBUTIONS 253

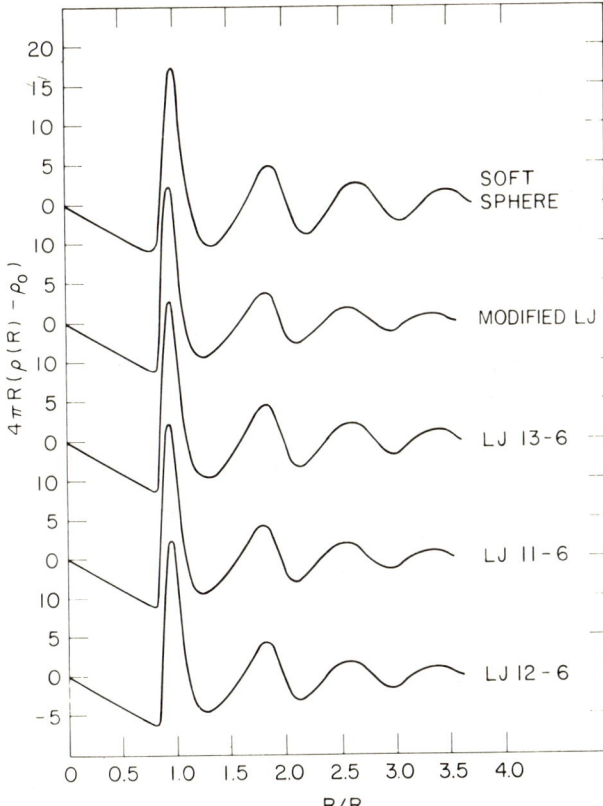

Fig. 1. Radial distribution functions derived from the five potentials shown in Fig. 2.[6]

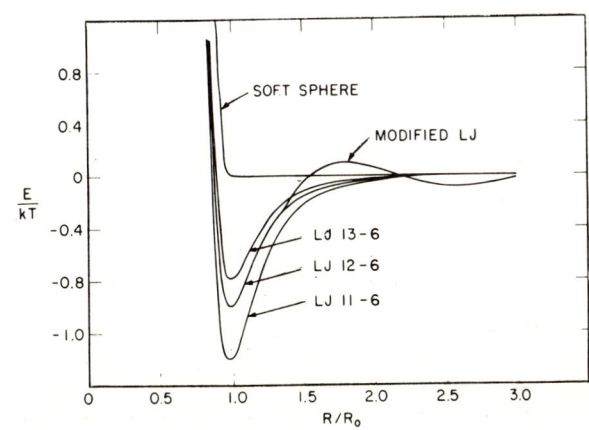

Fig. 2. Five potentials selected to yield equal nearest-neighbor peak heights.[6]

correlation positions, at a given density. Indeed, the higher the density, the less sensitive is the entire $G(r)$ to $V(r)$; at ~0.95 of close packed it seems necessary to maintain more than four significant figures in $I(k)$ to preserve the integrity of the relationship between V and G.[6]

It is important to note that the use of the necessarily approximate analytical expressions (Eq. 5) is not essential to consideration of the relationship between postulated interatomic forces and the structure (or $F(k)$). The feasibility of dynamic computer modelling, in which the newtonian equations of motion are solved numerically for many particles with assumed interaction potentials has been demonstrated.[7] Atomic spatial-distribution functions are thereby derived from inspection of the particle positions in the dynamically varying models. Such calculations also allow the study of materials that contain more than one component, and can also be set up (in principle) in terms of multibody potentials. It may be a feasible approach for theoretical investigation of directionally bonded atoms for which the pair-potential concept may be too restrictive. In any case, it would still appear that the more interesting cases to be examined are those which are significantly less than close packed, so that sensitivity to the potentials will be greater.

The foregoing paragraphs have discussed very briefly how one may attempt to derive (or to check) an interatomic potential model on the basis of diffraction data.* On the other hand, it is not entirely certain, even where such a breakdown may be feasible, that it provides more illumination for certain purposes than more gross information which is readily seen in the data. In particular, it is worth considering the distribution of nearest neighbors with respect to one another, just as they are in the sum of whatever interactions do occur in the material. That is, to examine $G(r)$ or $\rho(r)/\rho_o$ in the small-r region. Introducing another energy function, one may define a "mean force potential" in terms of the radial distribution (or vice versa):[8]

$$\rho(r)/\rho_o = \exp(-U(r)/kT) \quad . \tag{8}$$

A mean force, $\partial U/\partial r$ follows as a secondary definition; if U can be divided into pair-wise contributions, that force can also be expressed in terms of a pair potential

$$\frac{\partial U}{\partial r} = \frac{\partial V}{\partial r} + \int \frac{n_3(o,\mathbf{r},\mathbf{x})}{n_2(\mathbf{r})} \frac{\partial V(\mathbf{x}-\mathbf{r})}{\partial r} dx \tag{9}$$

where n_3 and n_2 are triplet and pair probabilities, respectively.

---

*More specifically, on the basis of diffraction data which is spherically averaged either naturally or artificially. In principle, however, one can certainly consider a three-dimensional $F(\mathbf{k})$ and its unaveraged relationship to the three-dimensional computer model of a dynamic calculation.

In the approximations of Eq. 5, the relationship between $V(r)$ and $U(r)$ can be expressed somewhat differently:

$$(PY) \quad V(r) = U(r) + kT \ln \left\{ g(r) - c(r) \right\}$$

$$(HNC) \quad V(r) = U(r) + kT \left\{ g(r) - 1 - c(r) \right\} . \quad (10)$$

These differences, in turn, can be expressed in terms of $F(k)$

$$(PY) \quad V(r) = U(r) + \frac{kT}{2\pi^2 r \rho_o} \ln \left\{ 1 + \int \left[\frac{F^2}{k+F} \right] \sin kr \, dk \right\}$$

$$(HNC) \quad V(r) = U(r) + \frac{kT}{2\pi^2 r \rho_o} \int \left[\frac{F^2}{k+F} \right] \sin kr \, dk . \quad (11)$$

It should not be inferred from Eq. 9, 10, and 11 that $U(r)$ and $V(r)$ are nearly identical — even in the near-neighbor region. Rather, the expressions are given to clarify the difference.

In a certain sense, the most important information relating to interatomic forces in $U(r)$ (or $\rho(r)$) is the structure itself — that is, the equilibrium three-dimensional arrangement. For theoretical work, that data has served mostly only as primary input for calculations regarding finer details such as the vibrational modes or electronic structures. However, there have been attempts and probably will be more sophisticated attempts to "predict" equilibrium structures from first principles.

3 EXPERIMENTAL DIFFRACTION EFFECTS OF ATOMIC DISPLACEMENTS

Beyond the structure, whatever information is in the pair-distribution function is related, in one way or another, to the atomic vibrations or to structural defects. If we had a random polycrystalline sample (so that $\rho(r)$ is spherically averaged), which had neither vibrations nor defects within the grains, and if we were to ignore the small intergrain effects, then the $\rho(r)$ would be simply a series of δ-functions. Those peaks would occur at each of the neighbor distances that occur in the structure (r_i), and each would have an area corresponding to the number of such pairs per average atom (c_i). For that idealized situation,

$$4\pi r \rho(r) = \sum_{i=1}^{\infty} \frac{c_i}{r_i} \delta (r-r_i) . \quad (12)$$

From such a situation, there is clearly little that can be said about the interactions, except for those aspects that led to the equilibrium structure itself. Leaving defects aside for the moment, note, therefore, that there will be no additional information in $\rho(r)$ that cannot be derived from a full knowledge of the vibrational modes, however that information is obtained.

From the point of view of X-ray scattering theory, it is possible to show that in a polycrystalline $F(k)$ the vibrations are manifested in two reasonably distinct ways:

(i) A damping of the Bragg crystalline reflections by the factor $\exp\left(-u_k^2 k^2\right)$. u_k^2 is the mean-square amplitude of vibration of an atom from its equilibrium site, along the crystallographic direction corresponding to the diffraction vector, **k**.

(ii) A diffuse scattering, the shape of which depends mainly on the fact that the interactions among the atoms, particularly nearest neighbors, do not allow them to vibrate independently of one another. If the atom "shells" in $G(r)$ were essential Gaussian-shaped, it would be possible to express this spherically-averaged "thermal diffuse scattering" (for a pure material) as

$$I_{TDS} = f^2 \left\{ \sum_{i=0}^{\infty} \frac{c_i}{kr_i} \left[\exp\left(-\frac{\sigma_i^2 k^2}{2}\right) - \exp\left(-\overline{u^2} k^2\right) \right] \sin kr_i \right\} \quad (13)$$

where $\overline{u^2}$ is 1/3 of the mean-square amplitude of vibration of an atom in three dimensions from its equilibrium site and σ_i^2 is the mean-square amplitude of relative displacement between ith neighbors along the line joining them. If, at a given r_i, the individual motions are effectively uncoupled, $\sigma_i^2 = 2\overline{u^2} \equiv \sigma_\infty^2$.

3.1 Gaussian Vibrations

In r-space, the given form of Eq. 13 is associated with a modification of Eq. 12

$$4\pi r \rho(r) = \sum_{i=1}^{\infty} \frac{c_i}{r_i} \frac{\exp\left(-\frac{(r-r_i)^2}{2\sigma_i^2}\right)}{\sqrt{2\pi\sigma_i^2}} \quad (14)$$

In fact, the Gaussian approximation is rather good; the deviations from that shape are not large even for nearest neighbors. This may be seen in Fig. 3, in which are reproduced data obtained by Stephen Strong on pure lead just below its melting point,[9] and a fit with Eq. 14.

3.2 Anharmonicity

The Gaussian assumption is not necessary, of course, for a direct analysis of the diffraction data, since the true peak shapes are visible in

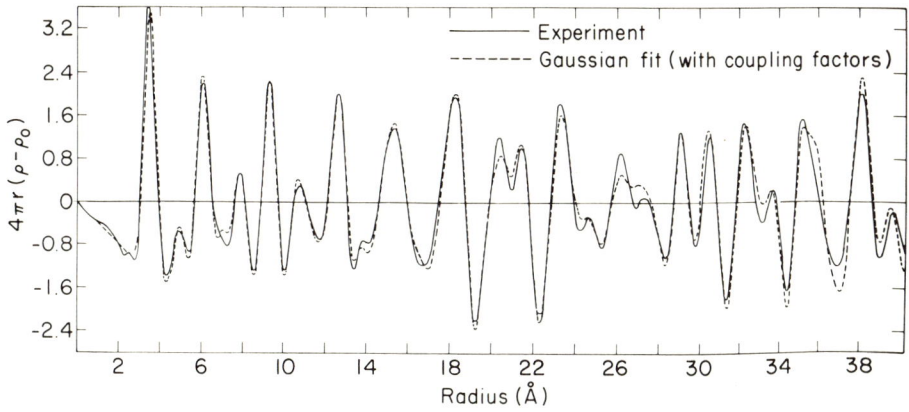

Fig. 3. Pair distribution, $4\pi r(\rho - \rho_o)$, in lead at 325 C.[9]

the derived $G(r)$. An example of anharmonicity may be seen in the nearest-neighbor peak ($r_i \approx 3.5$A, $c_i = 12$ atoms) of the same data.[9] Nearest neighbors were reasonably well isolated in the direct result; subtraction of slightly overlapping second and third neighbors (assumed Gaussian) leaves the result[9] shown in Fig. 4, a peak with an area of 12.1 atoms. The mean position of the peak agrees precisely with the fcc lattice parameter which best fits the large-r oscillation positions. The shift between the peak and the mean, 0.06A, is fairly small compared to the rms deviation, but approximately 60 percent of the total thermal expansion. It may be of interest to note, at the same time, that the most probable nearest-neighbor distance shifts another 0.10A smaller when lead melts. This latter effect, seen in Fig. 5, accompanies a reduction in density of about 3 percent.

The anharmonicity in the nearest-neighbor peak shape of solid lead is reflected in $U(r)$ which is shown in Fig. 6, along with its first and second derivatives. It might appear on first thought that any direction-dependent information in these functions will have been totally averaged out. That is not the case, as may be seen, for example, from the fact that in the fcc structure all nearest-neighbor radial vectors are in <110> directions. In the vicinity of r_1, therefore, all variations visible in $G(r)$ relate specifically to that direction (except insofar as they may be associated with large excursions from equilibrium positions, such as diffusive jumps).

3.3 Elastic Constants

The first derivative of $U(r)$ represents a force and the second derivative a force constant. In the sense of the foregoing paragraph $\partial^2 U/\partial r^2$

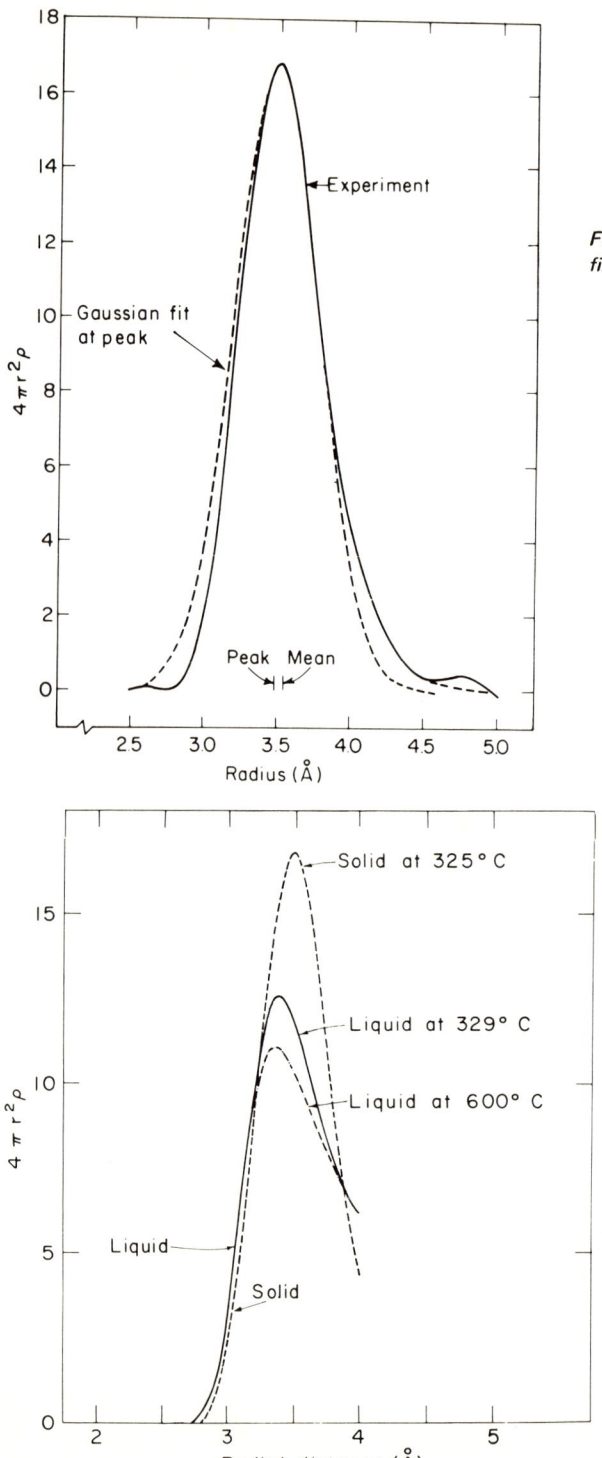

Fig. 4. Distribution of atoms in first shell, lead at 325 C.[9]

Fig. 5. Total radial density function for lead in vicinity of nearest-neighbor distance.[4]

Fig. 6. Relative density, mean potential, force curve, and "spring constant" in region of first shell.[9]

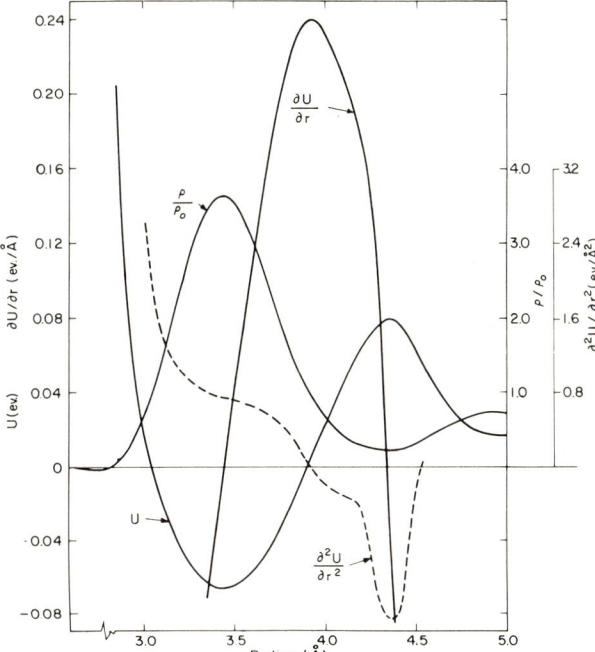

should be related to whatever elastic modulus is appropriate to the nearest-neighbor direction (<110> for fcc). Hooke's law, $F/A = E(\ell-\ell_0)/\ell_0$, differentiated and applied to atomic dimensions yields[10]

$$E = \frac{\partial^2 U(r)}{\partial r^2} \frac{r_1}{A}, \qquad (15)$$

where the energy of relative displacement has been identified with $U(r)$, r_1 is the nearest-neighbor distance and A is the area associated with one atom, perpendicular to the nearest neighbor direction. For the latter value, the area of the <110> cross section of the Wigner-Seitz zone seems appropriate, and is equal to $1.06\, r_1^2$ for both fcc and hcp structures. For crystalline lead[9] (at 325 C) and for both fcc and hcp cobalt (at their transformation temperature ~417 C[10]), <110> moduli obtained using Eq. 15 and the experimental values of $\partial^2 U(r)/\partial r^2$ at r_1 are in close agreement with values calculated from elastic constants obtained by other means.* It would therefore appear that the validity of interpreting $U(r)$ as a real

---

*For the hcp cobalt, the latter value was obtained using[11] $1/E = (1 - \ell_3^2)^2 S_{11} + \ell_3^4 S_{33} + \ell_3^2(1-\ell_3^2) \times (2S_{13} + S_{44})$, with ℓ_3 the projection of a nearest-neighbor unit vector on the c-axis, and $S_{11} = 4.72$, $S_{44} = 13.24$, $S_{13} = -0.67$, $S_{33} = 3.19$, $S_{12} = -2.31$ (all in 10^{13} cm^2/dyn)[12]. The average for the two nearest-neighbor directions $E_{<2\bar{1}\bar{1}0>}$ and $E_{<1\bar{1}01>}$, which differed by ~3 percent, was used.[10]

energy of relative displacement, and its maintenance of directional characteristics (in the sense referred to earlier) are both confirmed. It should also be mentioned that although there may be some fortuity in the closeness of those agreements (e.g., approximately 0.5 percent for cobalt), the experimental numbers are clearly sensitive to the detailed results. In particular, it may be noted that if the first peak were to be replaced with a Gaussian with $\sigma_1^2 = 2\overline{u^2}$, the derived E would be too small by a factor of 100 percent.

3.4 Coupling Coefficients

Indeed, for every material studied $\sigma_1^2 < 2\overline{u^2}$ and the ratio is roughly 0.5 for the simple metals at high temperature.[9,10,13,14,15,16,17] This effect is to be expected, and Walker and Keating[18] have estimated it quantitatively in terms of coupling coefficients

$$\gamma_i = \sigma_i^2/2\overline{u^2} \qquad (16)$$

using simplified models for the density of phonon states. In Fig. 7, their values and experimental results for lead at 325 C[9] are reproduced. Experimental results for fcc and hcp cobalt at 417 C[10] and for aluminum at 25, 325 and 655 C[13] are qualitatively similar. At lower temperatures, however, the γ's tend to be somewhat larger than they are at the melting point in lead, as one would intuitively expect. Fessler's aluminum results may be most indicative of the temperature dependence in spite of data-termination difficulties in that work.[19] For $\sigma_1^2/\sigma_\infty^2$ he finds 0.84, 0.72 and 0.55 at $T/T_M \cong .32, .64$ and 1.0, respectively.

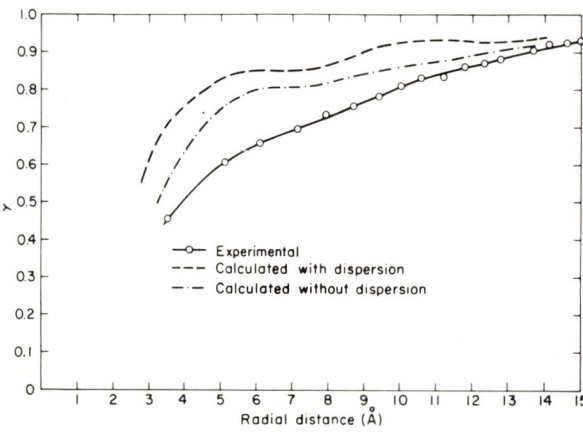

Fig. 7. Vibrational coupling factors in lead at 325 C. Calculations by Walker and Keating.[18,19]

In liquid metals it is a good deal less obvious from $G(r)$ what the vibrations are — as separated from diffusions. However, if the shape of the leading edge of the nearest-neighbor peak is used as a measure, then the experiments indicate that the vibrations in the liquid are roughly the same, or slightly smaller than in the solid at the same temperature (note the comparison in Fig. 5).[4,13] It may be noted, also, that the very asymmetric shape of the first liquid peak seems to be associated entirely with diffusions rather than with anharmonic vibrations. This guess is based on the fact that the simple metal liquids show essentially no expansion of the peak positions on increasing temperature, quite unlike amorphous or crystalline solids.

In principle, the variation of the γ's (about some smoothly increasing curve) should be related to anisotropy in the vibrations since each neighbor corresponds to a different direction. This sort of effect seems to be clearly beyond experimental uncertainty in molecular crystals, for which the inter- and intramolecular vibrations appear to separate naturally. In the stable form of crystalline selenium, the atoms are arranged in long chain-like molecules, which, in turn, crystallize in a 2-D hexagonal pattern.[20] Table I contains coupling coefficient results for that material, determined by Rowe by fitting the experimental $G(r)$ without consideration of molecular effects.[14] The intrachain relative motions are clearly more coupled — at short distances — than are the others. It is also of interest to note that the relatively small magnitude of near-neighbor distance deviations is retained in the amorphous form.[14]

Table I. Coupling coefficients for hexagonal selenium at room temperature $\sigma_\infty^2 = 0.055$ A.[14] Intra-molecular distances are starred (*).

Interatomic Separation, A	Coefficient $\sigma_i^2/\sigma_\infty^2$	
2.32	0.13	*
3.47	0.70	
3.68	0.55	*
4.36	0.65	
4.49	0.75	
4.93	0.85	
4.95	0.65	*
5.70	0.70	
6.06	0.90	
6.13	0.75	
6.59	0.95	

A similar situation is obtained in B_2O_3, which has a boron-centered oxygen triangle as its basic unit. The nearest-neighbor B-O and O-O bonds show much less relative vibration than do more distant or intermolecular pairs, and also show no measurable increase as the temperature is raised from ~295 to 673 K.[15,16] Thus, the evidence is that the intramolecular vibrational modes (or at least those that produce radial displacements) are not significantly excited beyond the zero point. This would be in qualitative agreement with interpretations of heat-capacity data in terms of several Einstein terms, $\theta_E > 1000$ K and a Debye contribution with a much smaller characteristic temperature.[21]

3.5 The Energy Function, U(r)

In at least two instances, $U(r)$ has been considered quite straightforwardly as a direct measure of relative energy.

In a comparison of the fcc and hcp phases of cobalt (both at 417 C), Lagneborg used the difference between the respective values of $U(r)$ at $r = r_1$ to estimate the difference in the internal energy and thereby the heat of transformation.[10] The value thus obtained is within 5 percent of the value measured by standard methods, 105 cal/g-atom.[22]

In quite a different application, refer to the solid-lead data again[9] as reproduced in Fig. 6. At the measurement temperature, just below the melting point, it may be noted that the potential barrier which keeps nearest neighbors in place is essentially equal to $3kT$; that energy "hump" is much larger at lower temperatures. A structural instability therefore seems evident from $G(r)$ alone. From geometrical considerations, it was suggested that the apparent instability was associated with the diffusive motion of an atom through the "hole" between four nearest neighbors in a (110) plane, to another nearest-neighbor site. It was further postulated that the restoring force against such motions would be inadequate when the amplitude associated with the maximum vibrational frequency (considered as a classical oscillator) became larger than the hole-size deficiency. This allowed an expression to be developed relating, at the melting point, the lattice parameter to the mean-square amplitude of vibrations (both parameters being temperature dependent).

$$a_0 (0.0946 - 0.6125\ \alpha T) = \overline{2u^2} \qquad (17)$$

where α is the (linear) expansion coefficient and a_0 the 0 K lattice parameter. With the introduction of a Debye characteristic temperature, it was possible to phrase Eq. 17 in the form of Lindemann's empirical rule (with no free parameters):

$$T_M = V^{2/3}\ \Theta^2\ W/1.71 * 10^4 \qquad , \qquad (18)$$

where V is the atomic volume, W is the atomic weight, and Θ is the Debye temperature.

3.6 Structural Defects

From Eq. 14 it may be surmised that structural defects will be visible in an experimental pair distribution if they are present in sufficient number. Deviations from the defect-free result can be discussed in terms of each of three "shell" parameters: r_i, c_i, and σ_i^2. In other words, the equilibrium mean position of a shell may change, the number of neighbors at or near a particular distance may change, and the mean-square deviation within a particular shell may change. For the latter, it may be convenient to speak of a σ_i^2 (total) = σ_i^2 (thermal) + σ_i^2 (defects), but with the recognition that the defect-free value for σ_i^2 (thermal) is not necessarily retained, particularly for close neighbors. More generally, it is entirely possible that the forms of the peaks are altered (becoming less, or more, Gaussian-shaped). It is also conceivable, indeed likely, that new equilibrium distances develop which may or may not overlap the equilibrium ones in $\rho(r)$.

From the experimental point of view, we expect that very precise reduced data is required for such a study, not the least-important region being at large scattering angles where the coherent scattering intensity is small.

In an attempt to study the experimental possibilities, Benquey has measured the pair distributions in abrasively filed and annealed nickel powders.[17] Although not entirely satisfactory in its precision, that work indicates quite clearly that measurable effects exist in the nearest-neighbor region – at least for such highly worked samples. In particular, the experimental comparison shows the following features in the region of the first two neighbors: (i) an increase of σ^2, for both shells of the order of .003 - .006 A^2; (ii) an increase in the density between the two peaks, which looks like a "tailing" of the first-neighbor distance, and (iii) a transfer of atoms from the region of the first shell to the region of the second-neighbor distance. Such information, assuming it can be verified and perhaps provided in greater detail through additional experiments, will clarify our understanding of the structures of defects in three-dimensions considerably.

Without considering three-dimensional arrangements, however, we have considered a direct relationship between the perturbed distribution function and the equilibrium $U(r)$. We postulate that energies that are input and stored during the deformation process are available, like the kinetic energy term kT, to excite atoms to higher levels of positional energy; the latter we take, as a first approximation, to be the equilibrium

$U^O(r)$ at the measurement (and working) temperature. This leads to an expression of the form:

$$\rho'(r) = A \exp[-U^O(r)/(kT+\delta)] \quad , \quad (19)$$

where A is a normalization constant which must preserve the total density, and δ is derived from the stored energy of cold-work.

This simple view surprisingly provides not only qualitative agreement with the features mentioned earlier, but also reasonable quantitative agreement, assuming a stored energy of the order of 0.0025 eV/atom.

ACKNOWLEDGMENTS

It is a pleasure to recall associations with Patrick Benquey, Raymond Fessler, Rune Lagneborg, Thomas Rowe, and Stephen Strong, who did so much of the work discussed herein. Professor B. L. Averbach has also been a close collaborator in much of the work since its inception.

I would like to acknowledge, also, the present grant support by the National Science Foundation, under Grant GK-1947.

REFERENCES

1. James, R. W.: *The Optical Principles of the Diffraction of X-rays*, G. Bell and Sons, London, 1954.
2. Warren, B. E.: *X-ray Diffraction*, Addison-Wesley Pub. Co., 1969.
3. Kaplow, Roy, S. L. Strong, and B. L. Averbach: "Local Order in Liquid Alloys", Chapter 5 in *Local Atomic Arrangements Studied by X-ray Diffraction*, J. B. Cohen and J. E. Hilliard (eds.), Gordon and Breach, New York, 1966.
4. Kaplow, Roy, S. L. Strong, and B. L. Averbach: Phys. Rev. **138**, No. 5A, Al336-Al345 (1965).
5. Stell, George: Physica **29**, 517 (1963).
6. Strong, S. L., and Roy Kaplow: J. Chem. Phys. **45**, No. 5, 1840-1842 (1966).
7. Rahman, A.: Phys. Rev. **136A**, 405 (1964).
8. Green, H. S.: *The Molecular Theory of Fluids*, North Holland, Amsterdam, 1952.
9. Kaplow, Roy, B. L. Averbach, and S. L. Strong: J. Phys. Chem. Solids **25**, 1195-1204 (1964).
10. Lagneborg, R., and R. Kaplow: Acta Met. **15**, 13-24 (1967).
11. Nye, J. F.: *Physical Properties of Crystals*, p. 1450, Oxford University Press, 1960.
12. McSkimin, H. J.: J. Appl. Phys. **26**, 406 (1955).
13. Fessler, R. R., Roy Kaplow and B. L. Averbach: Phys. Rev. **150**, No. 1, 34-43 (1966).
14. Kaplow, Roy, T. A. Rowe, and B. L. Averbach: Phys. Rev. **168**, 1068-1079 (1968).
15. Strong, S. L., and Roy Kaplow: Acta Cryst. **B24**, 1032-1036 (1968).
16. Strong, S. L., A. F. Wells, and Roy Kaplow: Acta Cryst. (in press).
17. Benquey, Patrick: *Radial Distribution Function in Annealed and Cold Worked Nickel*, M.S. Thesis (unpublished), Department of Metallurgy and Materials Science, M.I.T., 1968.
18. Walker, C. B., and D. T. Keating: Acta Cryst. **14**, 1170 (1961).
19. Fessler, Raymond R.: *Pair Correlations in Solid and Liquid Aluminum*, Sc.D. Thesis (unpublished), Department of Metallurgy, M.I.T., 1965.
20. Wyckoff, R.: *Crystal Structures*, Interscience Publishers, Inc., New York, 1963, 2nd ed., Vol. I.
21. Fajans, K., and S. W. Barber: J. Amer. Chem. Soc. **74**, 2761 (1952).
22. Hultgren, R., R. L. Orr, P. D. Anderson, and K. K. Kelley: *Selected Values of Thermodynamic Properties of Metals and Alloys*, p. 77, John Wiley, 1963.

DERIVATION OF LONG-RANGE INTERACTION ENERGIES FROM DIFFUSE SCATTERING IN DIFFRACTION PATTERNS

J. M. Cowley and S. Wilkins

Department of Physics,
Arizona State University
Tempe, Arizona 85281

ABSTRACT

Ordering theories for binary alloys valid above the critical temperature provide a relationship between the intensities of the diffuse scattering in X-ray, neutron or electron diffraction patterns and the Fourier transform of the pair-wise interaction energy function. Available diffraction data has been used to derive information on the form of the long-range oscillatory potentials attributable to conduction-electron energy terms. On this basis, the relationship of the diffuse scattering to the form of the Fermi surface is discussed. Consideration is given to the possibility of detecting and analyzing the effects of nonpair-wise interaction energies from diffraction data.

1 INTRODUCTION

The diffuse scattering intensity, $I_d(\mathbf{k})$, due to short-range ordering in a crystal is related to the atom-pair correlation parameters through the

Fourier transform relationship which for a binary A,B, alloy is

$$I(\mathbf{k}) = N(f_A - f_B)^2 m_A m_B \sum_i \alpha(\mathbf{r}_{oi}) \exp\{+i\mathbf{k}\cdot\mathbf{r}_{oi}\} \;, \quad (1)$$

where m_A and m_B are the fractions of A and B atoms, f_A and f_B are the atomic scattering factors for the radiation involved, and $\mathbf{r}_{oi} = \mathbf{r}_o - \mathbf{r}_i$ are the interatomic vectors measured from an origin at $\mathbf{r}_o$. These are the vectors between lattice points of the average lattice. Eq. 1 is valid only under the limiting conditions of the kinematical, single scattering approximation and in the absence of size-effect displacements and thermal motions of the atoms.

The atom-pair correlation parameters $\alpha(\mathbf{r}_i)$ may be defined in terms of the conditional probabilities such as $P_i^{A|B}|_j$, the probability that an A atom should be at site i given that a B atom is at site j.

$$P_i^{A|B}|_j = m_A\{1 - \alpha(\mathbf{r}_{ij})\} \;, \quad (2)$$

where $\mathbf{r}_{ij} = \mathbf{r}_i - \mathbf{r}_j$. Alternatively, they may be defined from the a priori probability for atom pairs such as the probability that there should be an A at i and B at j,

$$P_i^A{}_j^B = m_A m_B \{1 - \alpha(\mathbf{r}_{ij})\} \;. \quad (3)$$

Also we may write

$$\alpha_{ij} \equiv \alpha(\mathbf{r}_{ij}) = (4 m_A m_B)^{-1} \langle \sigma_i \sigma_j \rangle \;, \quad (4)$$

where σ_i is a site occupation parameter which is $2m_B$ for an A at i and $-2m_A$ for a B at i, and the bracket denotes an average as defined later.

The pair-correlation parameters are given by

$$\alpha(\mathbf{r}_i) = \frac{1}{\Omega_k} \int \alpha(\mathbf{k}) \exp\{-i\mathbf{k}\cdot\mathbf{r}_i\} d\mathbf{k} \;, \quad (5)$$

where the integral is over the unit cell of volume Ω_k in reciprocal space in which $\alpha(\mathbf{k})$ is periodic. This defines $\alpha(\mathbf{k})$ which is derived directly from $I(\mathbf{k})$, the intensity of scattering, corrected for the usual geometric, polarization, and scaling factors and for the contributions of Compton scattering, thermal diffuse scattering, size-effect scattering and scattering from defects.

The connection with atom pair-interaction energies is provided by theories of the statistical mechanics of short-range ordering. The problem posed in statistical mechanics is usually that of deriving the atom-correlation parameters given the interaction energies. In its simplest form involving nearest-neighbor energies only, this is the three-dimensional Ising model problem. For temperatures above the critical temperature for ordering, T_c, approximate solutions involving all pair-wise energy terms have been established as providing a solution to the ordering problem sufficiently accurate for most purposes. The results of these theoretical treatments are reviewed, then the problem is considered in reverse. The main interest is to investigate the extent to which the available experimental measurements of $\alpha(\mathbf{k})$, and so of $\alpha(\mathbf{r}_i)$ can be used to derive meaningful data on interactions between atoms, particularly for the long-range interactions such as those associated with conduction-electron energy terms.

Since our treatment is initially in terms of pair-wise correlations and pair-wise interactions only, it is inevitable that the results will be perturbed by any nonzero average contributions from nonpair-wise interactions. Progress towards the evaluation of such perturbations and an indication of the magnitude of many-atom interactions are discussed later.

2 ORDERING THEORIES

The pair-interaction Hamiltonian for the configuration energy is written

$$H = \frac{1}{4} \Sigma_{ij} V(\mathbf{r}_{ij}) \sigma_i \sigma_j \quad , \tag{6}$$

where

$$V(\mathbf{r}_{ij}) = \frac{1}{2} \left\{ V^{AA}(\mathbf{r}_{ij}) + V^{BB}(\mathbf{r}_{ij}) - 2V^{AB}(\mathbf{r}_{ij}) \right\} \tag{7}$$

and $V^{AB}(\mathbf{r}_{ij})$, for example, is the effective energy of interaction between an A atom on site i and a B atom on site j. These quantities are assumed to be symmetric with respect to interchange of sites. The $V(\mathbf{r}_{ij})$ functions of the vector separation of sites, may depend on temperature and composition, but are assumed to be independent of configuration. This assumption may give a good approximation for alloys when core overlap and size difference between atoms is small as in the case of Au-Ag and, to a lesser extent Cu-Au alloys.

The problem is then to find the thermodynamic averages $\langle \sigma_i \sigma_j \rangle$ which, in the grand canonical ensemble, are defined as

$$\langle \sigma_i \sigma_j \rangle = \sum_{\{\sigma_i\}} \sigma_i \sigma_j \rho \Big/ \sum_{\{\sigma_i\}} \rho \quad , \tag{8}$$

where $\rho = \exp\{-\beta H + \lambda \Sigma \sigma_i\}$, with $\beta = 1/kT$. The summation is over all possible sets $\{\sigma_i\}$, where each σ_i may take on the values $2m_B$ and $-2m_A$, independently, and λ is a chemical potential introduced to preserve the compositional constraint $\langle\sigma_i\rangle = 0$. The averages $\langle\sigma_i\sigma_j\rangle$ are related to the pair probabilities by the Eq. 2, 3 and 4.

From the high-temperature approximate solutions to the ordering problem, valid above the critical temperature, one may write the generalized mean-field expression[1].

$$\alpha(\mathbf{k}) = \frac{G_2(T)}{1+G_1(T) V(\mathbf{k})} \quad , \tag{9}$$

where G_1 is approximately proportional to T^{-1}, while $G_2(T)$ is the normalization factor chosen such that

$$\frac{1}{\Omega_k} \int \alpha(\mathbf{k}) \, d\mathbf{k} = 1 \quad , \tag{10}$$

where the integration is over the k-space unit cell.

From Eq. 5 we have

$$\alpha(\mathbf{k}) = \Sigma_i \alpha(\mathbf{r}_{oi}) \exp\{i\mathbf{k}\cdot\mathbf{r}_{oi}\} \tag{11}$$

and similarly we write

$$V(\mathbf{k}) = \Sigma_i V(\mathbf{r}_{oi}) \exp\{i\mathbf{k}\cdot\mathbf{r}_{oi}\} \quad . \tag{12}$$

Hence, Eq. 9 provides a direct connection between the interaction energies and the short-range order diffuse scattering in diffraction patterns. The minima of $V(\mathbf{k})$ will correspond to maxima of $\alpha(\mathbf{k})$ and, hence, the maxima of diffuse scattering intensity.

A form of Eq. 9, which is of more immediate use in practice, is given by including the critical temperature, defined as the temperature at which the denominator of Eq. 9 first goes to zero from above at the particular $\mathbf{k}=\mathbf{k}_m$ for which $V(\mathbf{k})$ is a global minimum in k-space. Then, since $V(\mathbf{k}_m)$ will be negative

$$-1 = G_1(T_c) V(\mathbf{k}_m)$$

and substituting in Eq. 9 gives

$$\alpha(\mathbf{k}) = \frac{G_2(T)}{1 - \frac{1}{X(T)} \left(\frac{T}{T_c}\right) \frac{V(\mathbf{k})}{V(\mathbf{k}_m)}} \quad , \tag{13}$$

where $X(T) = \left(\frac{T_c}{T}\right) [G_1(T_c)/G_1(T)]$ and this is close to unity for all the approximate ordering theories.

3 DETERMINATION OF PAIR-INTERACTIONS RATIOS

Eq. 13 provides a means whereby measurements of the diffuse scattering intensity, and so of $\alpha(\mathbf{k})$, can be used to derive values of $V(\mathbf{k})$ from which the pair interaction energies may be obtained. However the published results of diffuse-scattering studies take the form of lists of $\alpha(\mathbf{r}_{oi})$ values for the successive near-neighbor shells of symmetrically equivalent $\mathbf{r}_{oi}$ vectors. For the ith nearest-neighbor shell of equivalent vectors, the order parameter is α_i and the corresponding $V(\mathbf{r}_{oi})$ value is written V_i. Although it would seem more direct and ultimately desirable to use the measured $\alpha(\mathbf{k})$ in Eq. 13, we have chosen to use the real-space relationship given by Fourier transforms of Eq. 13 to derive V_i ratios from the published α_i. Because of the difficulties of attempting to place energy values on an absolute scale and the uncertainties concerning the reliability of the approximate theories, especially near T_c, the parameters fitted to the experimental data have been

$$X, V_2/V_1, V_3/V_1, \ldots \qquad (14)$$

From consideration of experimental factors it is clear that not all the α_i values will be determined with the same precision, so various sets of weighting factors have been applied to minimize the effects of possible errors on the resulting energy ratios.

Data of sufficient extent and accuracy to make a determination of the interaction ratios worthwhile are available only for one or two compositions in each of a small number of systems. Results considered here are from three systems that illustrate the range of results so far obtained.

The first accurate data was that of Moss[2] who published values for the first eleven α_i for Cu_3Au for temperatures of 405 and 450 C. This system appears to meet the requirements of the pair-interaction model reasonably well.

The values obtained for the V_i/V_1 have been reported elsewhere[1] but are reproduced here in Fig. 1 for purposes of comparison. The pair-interaction ratios are plotted against the interatomic distance $|\mathbf{r}_{oi}|$. The plot strongly suggests that the interaction energy is long ranged and oscillating with distance, following a curve similar to that drawn, appropriate to Friedel oscillations in the potential about an isolated impurity in a metal[3].

The second system considered here is that of gold plus 40 atomic percent palladium, recently studied at room temperature by Lin et al.[4] This alloy is fcc when disordered and shows an exceptionally small size effect, as is evident from the very small distortion of the short-range order diffuse peaks in the intensity contour map reproduced in Fig. 2. The diffuse maxima each appear to be split into two well-separated

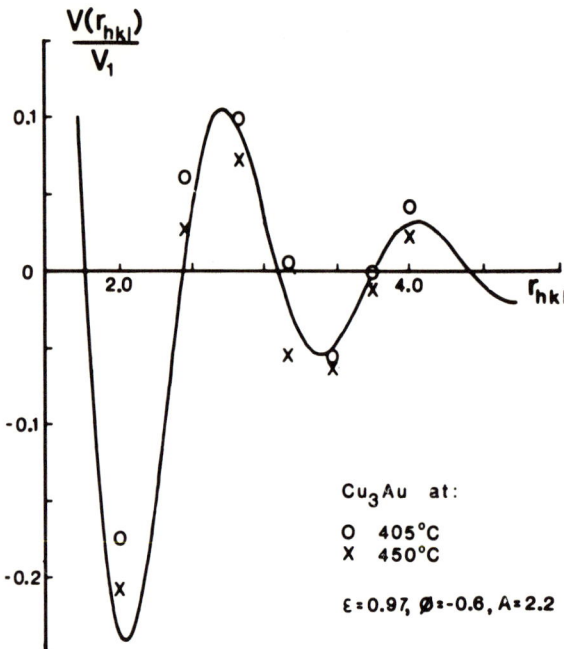

Fig. 1. Values for the pair-interaction ratios V_i/V_1 deduced from the α_i values obtained by Moss[2] for Cu_3Au at two temperatures, compared with the free-electron screening curve, Eq. 15, for the indicated parameters.

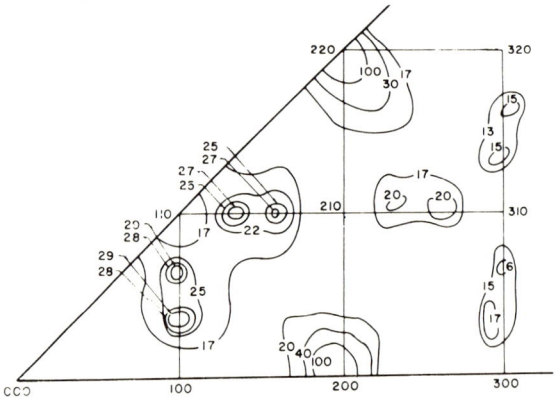

Fig. 2. Diffuse intensity in arbitrary units for the hk0 plane of reciprocal space deduced from the X-ray diffraction measurements by Lin et al.[4]

components such as those at positions close to (1, 3/8, 0) and (1, 5/8, 0). This suggests strong oscillations in $V(r)$ such that $V(k)$ will have well-defined minima at these points. From the first fourteen α_i values we have deduced values for the first eight interaction ratios. The results, plotted in Fig. 3, again show oscillations in the interaction energies, but the agreement with the form of the Friedel oscillations is not so apparent.

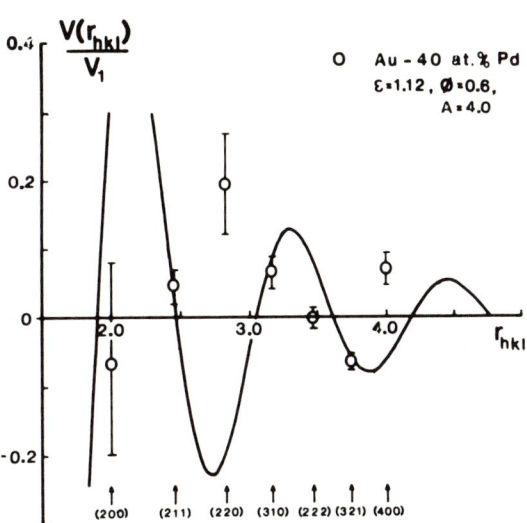

Fig. 3. Values for the pair-interaction ratios V_i/V_1 for Au-40 atomic percent Pd, deduced from the α_i values of Lin et al.[4] The error bars suggest the range of values obtained by varying the weighting factors on the α_i within reasonable limits. The free-electron screening curve is drawn for the electron/atom ratio appropriate for this alloy and suitably chosen ϕ and A. Indices for the various interatomic vectors are given below.

Finally we report the study of the results on the disordered α-phase of $LiFeO_2$, a nonmetallic phase undergoing an order-disorder transition at 720 C. Above this temperature the metal atoms are disordered on a fcc sublattice, and the oxygens remain on the ordered interpenetrating fcc cation lattice of the rock-salt structure. Brunel and de Bergevin[5] determined the first five α_i values from powder samples quenched from 742 and 875 C. Reconstruction of the diffuse scattering configuration from these α values shows a maximum at ½, ½, ½. Electron-diffraction observations on disordered $LiFeO_2$ by Allpress[6] have confirmed that maxima occur at these points, but suggest a rather complicated total configuration of diffuse scattering. Brunel and de Bergevin found a significant displacement of the oxygen ions toward the iron ions and also inferred that the oxygens are strongly polarized. Hence, it may be difficult to justify the use of a theoretical analysis based on the assumptions of pair-wise interactions and small size effect. Nevertheless, applying the same method as for the alloys gives the values for the first four pair-interaction ratios shown in Fig. 4. It appears that the interactions fall off smoothly within the limited range shown, and the decrease is faster than the r^{-1}, which would be given by an unscreened Coulomb interaction, and slower than r^{-3}, the dependence for full screening.

4 THE FORM OF THE PAIR-INTERACTIONS

The method used to deduce the pair-interaction ratios has been independent of any assumed model for their origin. Hence the values

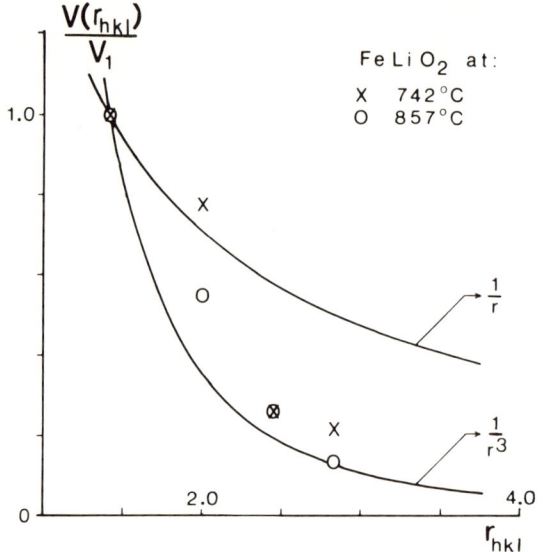

Fig. 4. Values for the apir-interaction ratios V_i/V_1 deduced from the α_i values obtained by Brunel and de Bergevin[5] for $FeLiO_2$ at two temperature, compared with r^{-1} and r^{-3} curves through the first point.

obtained offer an unprejudiced test of the various models for their origin.

The immediate comparison to be made is with the free-electron screening model[7,8] which gives an effective pair-interaction energy $V(r_i)$ between ions at large distances as

$$\frac{V(r_{hk\ell})}{V_1} = \frac{A \cos(2k_F a_o r_{hk\ell} + \phi)}{r_{hk\ell}^3} \qquad (15)$$

where k_F is the Fermi wave number, A is assumed to be a constant, and ϕ is a phase factor. The unit-cell dimension is a_o and $r_{hk\ell}$ is the interatomic distance in units of half the unit-cell axes for lattice sites with integral coordinates h, k, ℓ. Comparison with the more exact result for a small-impurity potential suggests that at metallic-electron densities this form should be good beyond the first nearest-neighbor distance[7]. The Fermi wave number is proportional to ϵ the cube root of the average number of conduction electrons per atom.

Comparison with results for Cu_3Au in Fig. 1 shows that the curve for $\epsilon = 0.97$ and $\phi = -0.6$ appears to fit the experimental results quite well. This is understandable because the Fermi surface for both copper and gold[9] is close to the free-electron spherical form for most directions.

For the Au-Pd alloy, however, there is only partial agreement with the curve drawn on the basis of two conduction electrons for each palladium atom and one for each gold. In this case one might suppose that by increasing the electron-atom ratio from 1.0 for gold to 1.4, the Fermi surface would be considerably more perturbed as it approaches the Brillouin zone boundary in the {100} direction and possibly extends to the next zone in the {111} direction.

Roth et al.[10] have shown how the effective interaction between ions may be modified for a nonspherical Fermi surface, at least for weak perturbing potentials. They showed that the interaction in the direction of a unit vector $\hat{r}$ oscillates with wave-number $(k_{F_1} - k_{F_2})\cdot\hat{r}$ where k_{F_1} and k_{F_2} are vectors to the points on the Fermi surface where the group velocities are antiparallel along $\hat{r}$ as suggested in Fig. 5. Also the strength of the interaction depends on the curvature of the Fermi surface at these two points. For flat portions of the surface, the interaction falls off in proportion to

$$\frac{1}{r}\cos\left[\left(k_{F_1} - k_{F_2}\right)\cdot r\right] . \qquad (16)$$

Hence the interpretation of the interaction is more complicated for nonspherical Fermi surfaces.

For Au-Pd we see that the large deviations from the simple screening interaction curve occur for points in directions of high symmetry such as {110} and {100}. In particular the greatest deviation is for the fourth nearest-neighbor interaction for which the interatomic vector is in the {110} direction and this is the direction in which near-planar regions of the gold Fermi surface are known to occur[9].

The equivalent consideration in reciprocal space has been used by Moss[11] and by Krivoglaz[12] who discussed the function

$$V(k) = \Sigma_K \, v_0 \, (k + K) \quad , \qquad (17)$$

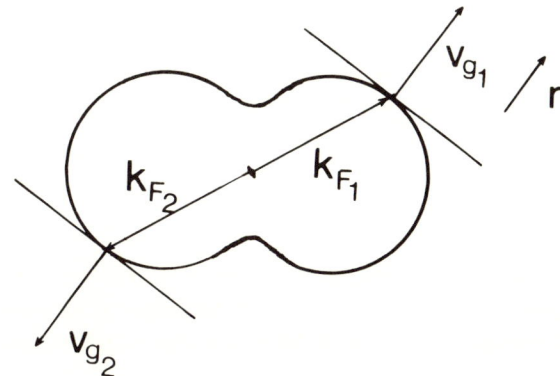

Fig. 5. Section of an imagined Fermi surface illustrating the form of the vector $k_{F_1} - k_{F_2}$ for regions having antiparallel group velocities.

where **K** is a reciprocal lattice vector and

$$v_o(\mathbf{k}) = \int V(\mathbf{r}) \exp\{-i\mathbf{k}\cdot\mathbf{r}\} d\mathbf{r} \quad .$$

For a spherical surface, there is a logarithmic singularity in the derivative of $v_o(\mathbf{k})$ at $2k_F$, but this produces a modification of $V(\mathbf{k})$ which is scarcely detectable. For flat areas of the Fermi surface however, there is a logarithmic singularity in $v_o(\mathbf{k})$ itself for $\mathbf{k} = \mathbf{k}_{F_1} - \mathbf{k}_{F_2}$. Such singularities give rise to the well-known Kohn anomalies in phonon spectra. The repetition of $v_o(\mathbf{k})$ around each reciprocal lattice point indicated in Eq. 17 has the effect of superimposing these singularities at particular points, and, at these points, pronounced minima in $V(\mathbf{k})$ may be produced giving in turn, by Eq. 13, maxima in $\alpha(\mathbf{k})$ and in the diffuse-scattering intensity. Hence, from observations on diffraction patterns, it is possible to experimentally caliper the Fermi surface in several directions.

For example if the split diffuse peaks in Fig. 2 are identified with large Kohn anomalies due to flat regions of the Fermi surface for Au-Pd, then, for the $\{110\}$ directions, the ratio of k_F to the free-electron value is 0.934. This value agrees within 1 or 2 percent with the current experimental and theoretical values for the Fermi surface of gold[9].

One aspect of the above discussion that may be queried is the use of the concept of a Fermi surface for a disordered alloy since all derivations and most applications of the Fermi surface concept depend on the assumption of a periodic lattice. However, as for phonons, periodicity of the average lattice seems to provide a sufficient basis for the application, and the results reported here, with others[11], suggest that it is a useful concept even when the perturbations from the average lattice are considerable.

For fully ordered structures, it has been proposed[13] that stability of a particular superlattice structure may be provided when the electron-energy term in the configurational energy is minimized by introducing a new Brillouin zone boundary, associated with a superlattice reflection, in such a position that it coincides with a flat portion of the Fermi surface for the basic lattice. The equivalent situation for the disordered lattice is provided by Fourier transforming of Eq. 6 to give

$$\langle H_{\text{config.}} \rangle = \frac{N m_A m_B}{(2\pi)^3} \int \alpha(\mathbf{k}) \, V(\mathbf{k}) \, d\mathbf{k} \quad , \tag{18}$$

where the integration is over one Brillouin zone of the average lattice. The configurational energy will be minimized when a deep minimum (greatest negative) value of $V(\mathbf{k})$ coincides with a maximum of $\alpha(\mathbf{k})$, in accord with the indications of Eq. 9 and 13. This is the condition that the flat areas of the Fermi surface, giving rise to the most important Kohn anomalies through the folding of $v_0(\mathbf{k})$ with the reciprocal lattice as in Eq. 17, should coincide with the diffuse Brillouin zone boundaries associated with the diffuse peaks of $\alpha(\mathbf{k})$.

5 MANY-ATOM INTERACTIONS

The restriction of our considerations to pair-wise interactions and two-site correlation parameters may well constitute a serious limitation. Certainly, when there are directed homopolar bonds between atoms, such a simple treatment cannot be adequate, and increasing evidence suggests that nonpair-wise interactions may be of importance even for alloys of noble metals[14].

As a first step toward understanding the nature and importance of many-atom interactions, the statistical mechanics of short-range ordered alloys has recently been reformulated to include three- and more-atom interactions and three- and more-atom correlation parameters[15]. The Hamiltonian (Eq. 6) is replaced by

$$H = \frac{1}{4} \Sigma_{ij} V(\mathbf{r}_{ij})\sigma_i\sigma_j + \frac{1}{6} \Sigma_{ijk} W(r_i, r_j, r_k)\sigma_i\sigma_j\sigma_k \ldots \quad , \qquad (19)$$

and a solution is sought in terms of the correlation parameters $\langle\sigma_i\sigma_j\sigma_k\rangle$ and so on, as well as $\langle\sigma_i\sigma_j\rangle$. Results of interest have been obtained showing the influence of three-atom interactions on two- and three-atom correlations.

The difficulty then arises of obtaining experimental measurements that offer confirmation of the theoretical predictions and, preferably, provide a means for deriving multiatom interactions from multiatom correlation parameters along the same lines as has been done for two-atom interactions. The authors have been studying the possible use of diffraction experiments for this purpose.

The simple kinematical scattering theory of Eq. 1 gives intensities as the modulus squared of amplitudes which depend linearly on atom-scattering factors and are, therefore, of second order and cannot involve correlations between more than two atoms. Correlations between higher numbers of atoms can only affect terms of higher order in atomic-scattering factors or site-occupation parameters and these can arise in two different ways. First, higher-order terms may be introduced by coherent multiple scattering as in "dynamical" diffraction processes. These should be most prominent and accessible in electron diffraction, and Cowley and Murray[16] have shown that the intensities should in fact be dependent on multiatom correlation parameters although preliminary estimates suggest that significant measurements may be difficult and would provide information on combinations of restricted types of parameters only. On the other hand, displacements of atoms introduce exponential phase factors into the intensity expression, and if the displacements depend on the occupancy of the lattice sites as in the size-effect displacements of atoms, terms of higher order in the site-occupation parameters can arise from the exponential terms. It was shown by Cowley that the intensities of both the sharp Bragg reflections and the diffuse scattering then depend on summations over multiatom correlation parameters.[17] But since there seems little hope of deriving values for individual correlation parameters from diffraction data it may be necessary to resort to the comparison with experiment of intensity distributions calculated using sets of correlation parameters derived from various models of multiatom interactions.

There seems to be some possibility for deriving multiatom correlation parameters from observations with the field-ion microscope but so far there is very little indication that this can provide the range or accuracy of data that would seem to be required to allow significant deductions to be made.

It has been suggested by Clapp[18] and by Cohen and Gragg[19] that the range of possible variation of multiatom correlation coefficient is limited by the specification of two-atom correlations. Hence the influence of multiatom interactions may be detected only by the measurement of relatively small changes in a large number of many-atom correlation parameters.

ACKNOWLEDGMENT

This work was supported in part by the National Science Foundation Area Development Grant in Solid State Science (No. GU3169).

REFERENCES

1. Wilkins, Stephen (1970) Phys. Rev. B, **2**, 3935.
2. Moss, S. C. (1964) J. Appl. Phys. **35**, 3547.
3. Kohn, W. and Vosko, S. H. (1960) Phys. Rev. **119**, 912.
4. Lin, Wen, Spruiell, J. E., and Williams, R. O. (1970). J. Appl. Cryst. **3**, 297.
5. Brunel, M. and de Bergevin, F. (1969) J. Phys. Chem. Solids **30**, 2011.
6. Allpress, J. G. (1971) to be published in J. Mat. Sci.
7. Blandin A. and Deplanté, J. L. (1963) in *Metallic Solid Solutions,* edited by J. Friedel and A. Guinier (Benjamin, New York).
8. Harrison, R. J. and Paskin, A. (1963) in *Metallic Solid Solutions,* edited by J. Friedel and A. Guinier (Benjamin, New York).
9. Roaf, D. J. (1962) Phil. Trans. Roy. Soc. (London), Ser. A **255**, 85 and Kupratakuln, S. (1970) J. Phys. C Supplement, **3**, S109.
10. Roth, L. M., Zeiger, H. J., and Kaplan, T. A. (1966) Phys. Rev. **149**, 519.

11. Moss, S. C. (1969) Phys. Rev. **22**, 1108.
12. Krivoglaz, M. A. and Hao, T'u in *Defects and Properties of the Crystal Lattice* (Izd. Naukova dumka, Kiev, 1968).
13. Sato, H. and Toth, R. S. (1961) Phys. Rev. **124**, 1833 and (1962) **127**, 469.
14. Clapp, P. C. (1969) Technical Report Number 210, Ledgemont Laboratory, Lexington, Mass.
15. Shirley, C. G. and Wilkins, Stephen, in press.
16. Cowley, J. M. and Murray, R. J. (1968) Acta Cryst. **A24**, 329.
17. Cowley, J. M. (1969) Acta Cryst. **A24**, 557.
18. Clapp, P. C. (1969) J. Phys. Chem. Solids **30**, 2589.
19. Cohen, J. B. and Gragg, J. "Battelle Colloquium on Critical Phenomena" (1970).

DISCUSSION on paper by J. M. Cowley and S. Wilkins

SEEGER: You showed that one could account for some of the long-range interaction effects in CuAu alloys in terms of the Friedel oscillations associated with what you termed the "flat parts" of the Fermi surface. Considering this problem from a purely theoretical standpoint I must say that I find this result somewhat surprising. After all, those "flat parts" are not so flat in copper or gold, and they'll still have considerable curvature. On the other hand, it has been shown that the amplitudes of the Friedel oscillations are proportional to the reciprocal square root of the Gaussian curvature.[1] They diverge at those points of the Fermi surface where the Gaussian curvature changes its sign, and it is in the corresponding directions in r-space that I feel the long-range interactions should occur.

COWLEY: Our basis for the interpretation of the fine structure of diffuse-scattering intensity distributions (not resolved in the X-ray measurements of Cu-Au alloys) is that maxima of $\alpha(k)$ will be produced at the position where the minima occur in $V(k)$ as a result of the electron-energy terms associated with the flatter parts of the Fermi surface. On this basis it is possible to predict with some accuracy where the diffuse-scattering maxima will be formed[2] and how the positions will vary with electron-atom ratio[3]. The number of systems to which these concepts have been applied is so far very limited, but there are no indications to date of cases for which this rough correlation fails. Undoubtedly it would be desirable to make a complete three-dimensional transformation to relate the variations of curvature of the Fermi surface in detail to the electron contributions to interatomic potentials for the various real-space lattice vectors or the corresponding diffraction intensities. At the moment, the limitations of the experimental methods and the uncertainties of some aspects of the theory (e.g., the importance of nonpair wise contributions to the energies) suggest that such an ambitious program might be premature.

ASHCROFT: (1) In view of the fact that LEED is very much a surface probe, do you expect the information on the V's (if it can be obtained) to be similar to that obtained from bulk probes? (e.g., because of free-energy considerations it is by no means clear to me that the relative "concentration" of the species at the surface is the same as in bulk).

(2) I think you may have implied in your talk that a spherical Fermi surface produces no Kohn anomalies. Since the electron response term in the dynamic matrix involves a wave-number dependent dielectric function with singular slope, the phonon dispersion should also exhibit this structure. This should be true even in the absence of the Umklapp type terms that you mention, and hence for systems with spherical or near spherical Fermi surfaces.

(3) Can you briefly describe how, in the X-ray case, one extracts $I(k)$ for an alloy? Is some simple assumption being made about the atomic scattering factors?

COWLEY: Low-energy electron diffraction should be, in principle, very sensitive to the potential distribution between atoms near the surface. If an adequate practical procedure for calculating LEED intensities were available, the method should reveal the modification of the potential distributions which undoubtedly occur near surfaces. Unfortunately, meaningful calculations of LEED intensities have been made only for over-simplified idealized models of surfaces since strong n-beam dynamical diffraction effects are involved including both forward and back scattering. Our experience with the much simpler case of high-energy-transmission electron diffraction suggests that "bulk" potential distributions can be investigated with high accuracy for a few cases of perfect single crystals[4], but diffuse scattering studies are much more complicated[5] and the possibility of deducing V_i/V_r ratios, as done from X-ray studies, is remote. However, the evidence is that transmission-electron-diffraction studies of diffuse scattering may still be used as a valuable qualitative indication of the presence and form of long-range oscillatory potentials. Diffuse scattering in LEED may perhaps be used in the same way although the degree of uncertainty involved is greater.

(2) As Dr. Ashcroft has indicated, the discontinuity associated with the spherical Fermi surface is one in the differential of the function of interest rather than in the function itself. For a discontinuity in the phonon-dispersion curve or, in our case, in the $V(k)$ and so the $\alpha(k)$ functions, it is necessary to have a vanishing curvature of the Fermi surface, and the "flat" regions are particularly effective. For the short-range order diffuse scattering, the correlation between the positions of the maxima in the intensities or $\alpha(k)$, and the "flat" regions of the Fermi surface is very good for all cases investigated so far.

(3) The observed diffuse-scattering intensities in the X-ray case are corrected for several instrumental effects such as the Lorentz and polarization factors. The contributions of Compton, thermal diffuse and size-effect scattering are subtracted. Then the remainder is divided by $(f_A - f_B)^2$, where f_A, f_B are the atomic scattering factors. Mostly, the data are collected in a range of scattering angles for which the f-values are known with sufficient accuracy to ensure that no serious errors are introduced from this source.

GEHLEN: From your talk, I conclude that you consider the size-effect modulation as a nuisance since you have to eliminate it before you can analyze the remainder of the diffuse scattering. On the contrary, a wealth of information is contained in the size-effect coefficients and I wonder if it would not be helpful to combine it with the information that you already have.

COWLEY: I would agree that the size-effect information available from diffraction patterns from alloys with short-range order could possibly be of considerable value. Those who have analyzed the experimental data for α coefficients have removed the size effects as a matter of convenience (I claim no responsibility for this), but it is just as easy to use the same symmetry arguments to deduce the size-effect terms and eliminate the short-range order contribution. This has been done to a limited extent, but, for this purpose, the diffraction data used would preferably come from larger scattering angles than have been used to date. From the first-order diffraction theory for size effects, the displacements of neighbor atoms around any given atom should be derived by relatively straightforward methods, given the order parameters. Higher-order approximations to the diffraction theory suggest complications arising from three- and many-atom correlation parameters and higher-order terms in displacements, but there should be ways to reduce these effects or they could be investigated separately. It is my impression that if much more extensive diffraction data were collected and they were analyzed appropriately, a great deal more valuable information could be deduced for any of the alloys studied so far.

GEHLEN: From a practical point of view, the quantities that are of importance are V^{AA}, V^{BB}, and V^{AB} rather than the quantity $1/2 (V^{AA} + V^{BB} - 2V^{AB})$. Is the information necessary for a separation contained in the diffuse scattering, and if so, how would your theory have to be modified?

COWLEY: The energy terms $V_{ij} = \frac{1}{2}(V_{ij}^{AA} + V_{ij}^{BB} - 2V_{ij}^{AB})$ are those determining the order-dependent component of the configuration energy when only pair-wise interactions, independent of configuration, are assumed as in most current treatment of the statistical mechanics of ordering. Hence, these energy terms are the ones that are derived from the order parameters within the theoretical framework now available. Dependence on the individual V_{ij}^{AA}, etc., may arise if a more complete statistical mechanics theory is used and if the details of size-effect diffraction contributions are considered but such a dependence is well beyond the present capabilities for analysis of the diffraction data.

REFERENCES

1. Mann, E. Phys. stat. sol. **13**, 293 (1966).
2. Moss, S. C. Phys. Rev. Lett. **22**, 1108 (1969).
3. Hashimoto, S. and S. Ogawa, J. Phys. Soc. Japan **29**, 710 (1970).
4. Cowley, J. M. Acta Cryst. **A25**, 129 (1969).
5. See Ref. 16.

THE STUDY OF INTERATOMIC POTENTIALS BY PLANAR CHANNELING EXPERIMENTS

Mark T. Robinson

Solid State Division
Oak Ridge National Laboratory
Oak Ridge, Tennessee 37830

ABSTRACT

A model is described which permits experimental data on the energy losses of ions transmitted through planar channels in thin monocrystalline targets to be used to study interatomic potentials in the separation region near 1 A. The method depends upon the empirical observation that the rate of energy loss of such ions is a linear function of their transverse oscillation frequency in traversing the channel. The techniques are illustrated by application to some recent data on the channeling of several ions in gold, silver, and silicon.

1 INTRODUCTION

When a beam of energetic ions is incident upon a crystalline target in a direction sufficiently close to one of its principal crystallographic axes or planes, a number of interesting effects are observed, collectively described as *channeling*.[1] Compared to particles incident in arbitrary directions or on structureless targets, channeled ions experience unusually small rates of energy loss and penetrate unusually deeply into the targets. These effects, first predicted in computer simulations[2], have been extensively studied in the past several years, primarily because of their connection with sputtering, ion implantation, and radiation damage. They come about because of correlations in the positions of successive scattering centers. The cumulative effect of these is to steer an approaching ion away from a row or plane of atoms and to direct it back into the adjacent parallel channel. Since it is constrained always to move at comparatively large distances from the atomic centers, a channeled ion will lose energy at a rate reduced from that of randomly directed particles, which make at least occasional close encounters with lattice atoms. Furthermore, in favorable circumstances, its trajectory may be highly regular, offering an opportunity to study the forces between the ion and the atoms of the target. It has recently become possible to do this using experimental data on the energy losses of ions transmitted through planar channels in thin monocrystalline targets. This article describes the experiments, the model on which their interpretation is based, and some of the results obtained. They provide access to the interaction potential between the ions and the lattice atoms in the region of separation near 1 A, which is otherwise difficult to examine.

2 EXPERIMENTAL PROCEDURE

The experimental procedure is outlined in Fig. 1. A beam of ions is extracted from an accelerator at an initial kinetic energy E_0, usually about 0.1 to 1 MeV/amu. After collimation to ~0.01 degree, the beam is incident upon one surface of a thin single-crystal target at a small angle, ψ_0, with a selected planar channel. This angle, commonly about 0.5 degree, is controlled by one axis of the two-axis goniometer in which the targets are mounted. The pathlength of the ions through the crystal can be varied by adjusting the second goniometer angle. The target crystals are typically 0.5 μm thick, and the experimental pathlength may range up to about twice this value. The ions transmitted through the crystal impinge upon an energy-sensitive detector, fitted with collimators so that only a small angular range of particle directions is sampled. The detector can be caused to scan the spectrum of energies, E, of the ions emerging from the crystal at a particular angle, ψ, from the channel plane.[3] Alternatively, the angular distribution of ions emerging from the target with a particular energy may be recorded.[4] Both sorts of experiment have been used to study interaction potentials.

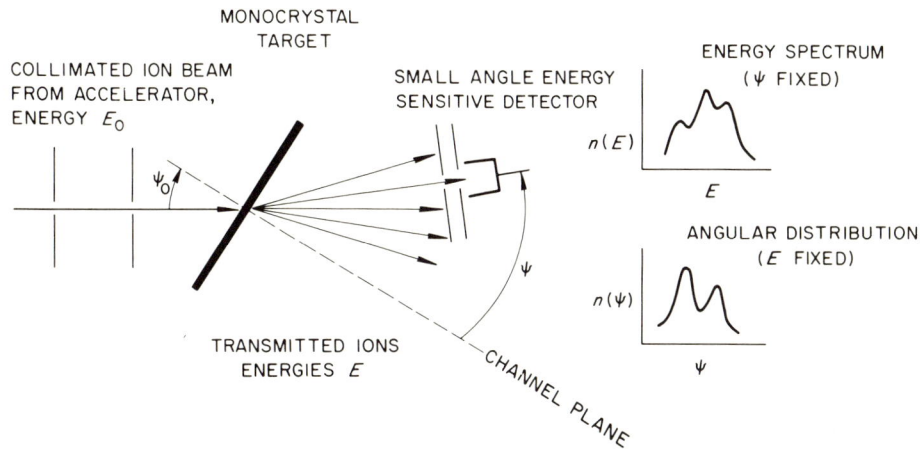

Fig. 1. Outline of the experimental arrangements for studying the energy losses of ions in thin single-crystal targets.

In either experiment, the observations consist of the energies E of particles which have traversed a pathlength z in the target. The ions lose energy in the target at a rate

$$-dE/dz \propto E^p \qquad (1)$$

where the parameter p may be assumed to be constant, at least over a limited energy range. It will decrease in value from about 1/2 at low energies through zero to -1 at high energies. If Eq. 1 is integrated, the result may be used to correct the rate of energy loss to the initial energy of the ions. This quantity, the *initial stopping power*, is[5]

$$(-dE/dz)_{E=E_o} = E_o^p(E_o^{1-p} - E^{1-p})/(1-p)z, \quad p \neq 1 \qquad (2)$$

A second integration of Eq. 1 allows the evaluation of the time that the ion spends in traversing the crystal, t, as

$$t = (2m)^{1/2}(1-p)z(E_o^{1/2-p} - E^{1/2-p})/(1-2p)(E_o^{1-p} - E^{1-p}), \quad p \neq \tfrac{1}{2}, 1,$$

$$= (2m)^{1/2}z[\ln(E_o/E)]/4(E_o^{1/2} - E^{1/2}), \qquad p = \tfrac{1}{2}, \qquad (3)$$

where m is the mass of the ion. Eq. 2 and 3 are needed in analyzing the experimental data. The required values of p can be taken from experiments on the stopping power of randomly directed ions.

3 EXPERIMENTAL RESULTS AND A QUALITATIVE DESCRIPTION OF THE MODEL

When the direction of the small-angle detector is collinear with the incident ion beam, that is, when $\psi = \psi_0$, the spectrum of emergent ion energies consists of a series of well-defined groups.[3,6,7,8,9] These spectra are illustrated by Fig. 2, which shows the energies observed[6] in a beam of 60-MeV ^{127}I ions transmitted through a {111} planar channel in a thin gold crystal.[6] The emergent energy of randomly directed ions in this case would be about 44 MeV. The origin of the structure in the spectrum may be understood by reference to Fig. 3. Here, the closely spaced atomic planes that border the channel are represented by fixed walls between which the channeled ions are subjected to the influence of an anharmonic potential. They execute transverse oscillations with wavelengths (or periods) that depend upon their amplitudes of oscillation. At the same time, the ions lose energy by excitation of the electrons of the target, the rate of energy loss increasing as the ion approaches the channel wall. Those ions that make an integral number of oscillations in passing through the channel will emerge from the crystal in the same direction that they entered and will be counted by the detector. Examples are trajectories A and C in Fig. 3, which execute one and two oscillations, respectively. Ions

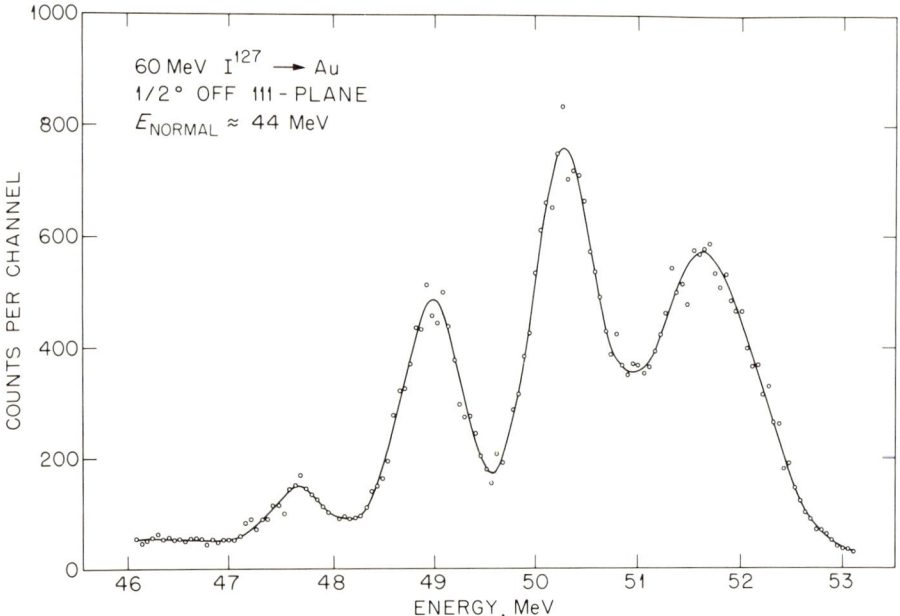

Fig. 2. The energy spectrum observed[6] in 60-MeV ^{127}I ions transmitted through a {111} planar channel in a 0.7-μm-thick gold crystal. The beam was incident 0.5 degree from the plane. The detector was collinear with the incident beam.

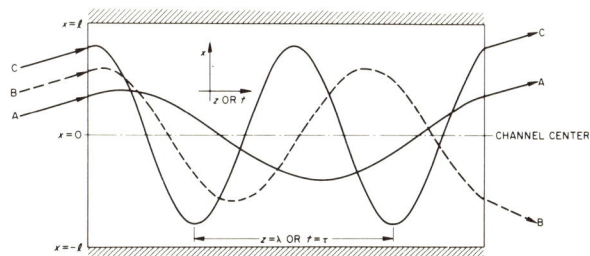

Fig. 3. Determination of the wavelengths (or periods) of oscillation of planar channeled ions by measurement of the energy spectra of transmitted ions in a collinear detector.

on intermediate trajectories, such as B in Fig. 3, emerge from the crystal at angles different from that at which they entered and will not enter the detector. Thus, the maxima in Fig. 2 correspond to ions that made integral numbers of oscillations in passing through the crystal. The energy dispersion in the transmitted beam comes about because the particles of large oscillation amplitude and small wavelength make closer encounters with lattice atoms than do ions of small amplitude and large wavelength, and, therefore, experience greater energy losses. As is illustrated in Fig. 4, when the pathlength through the crystal channel is increased, each spectral maximum shifts to a lower value of the energy-loss rate.[3] The use of the energy-loss rate (stopping power) as the abscissa in this figure avoids those energy changes that are due merely to the increase in thickness of the target. The peak shifts are easily understood in terms of the model: since the *number* of oscillations is fixed for each spectral group, the increased pathlength requires an increased wavelength and, consequently, a reduced oscillation amplitude and a reduced stopping power. There is a well-defined minimum energy-loss rate (maximum energy) in the spectrum which is independent of the target thickness. The corresponding spectral maximum is marked "A_o" in Fig. 4. When the stopping powers corresponding to the various spectral maxima are plotted against the target thickness, a series of curves is obtained, as illustrated by Fig. 5 which shows data for 60-MeV ^{127}I ions in two planar channels of a silver monocrystal.[10] The stopping powers have been corrected to the initial energy using Eq. 2 with $p = 1/2$. The figure also shows the values of the minimum stopping power at the channel center, s_o, deduced from the low energy loss (high energy) edge of spectra similar to those in Fig. 2 and 4. Each of the curves in Fig. 5 represents particles with a particular value, n, of the number of oscillations executed in traversing the crystal. The wavelength corresponding to each stopping power is simply the pathlength interval between two of the curves at the desired stopping-power value. Division of a pathlength by the wavelength yields the required value of n. It is convenient to present the experimental data as plots of the initial stopping power of the channeled ions as a function of their *transverse oscillation frequency*. The "frequency" ω is defined as

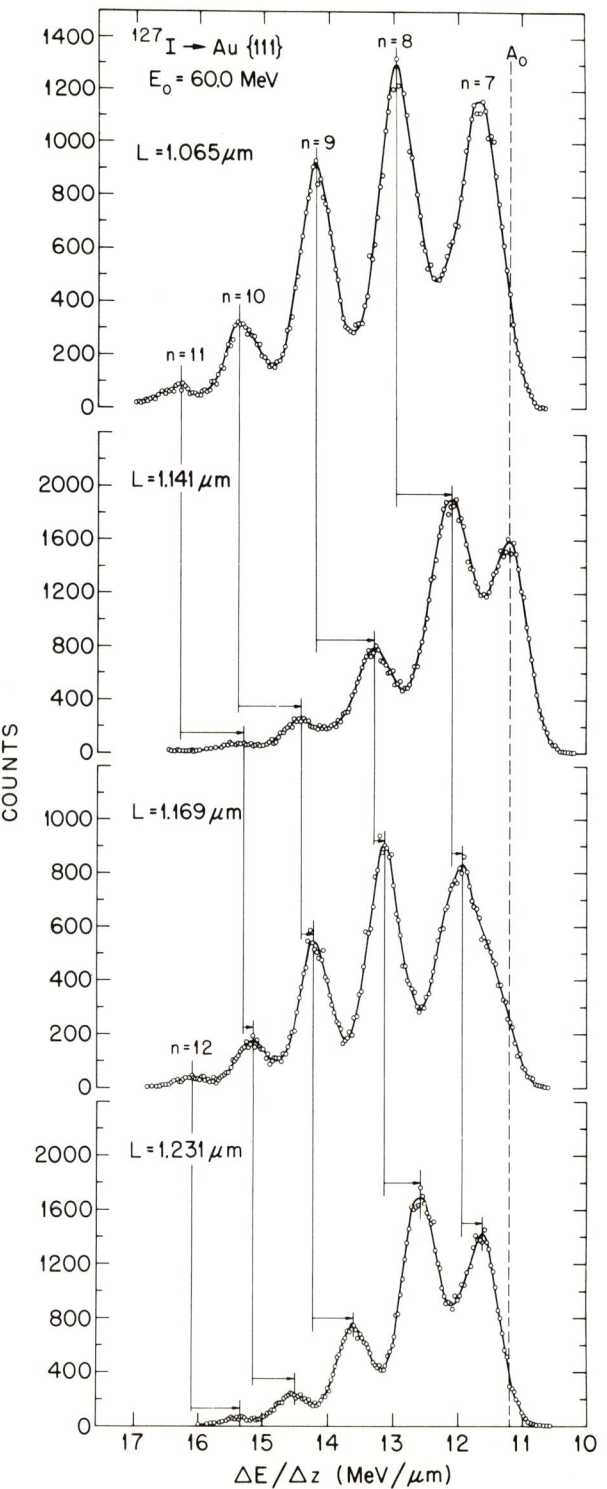

Fig. 4. Energy spectra observed[3] in ^{127}I ions transmitted through $\{111\}$ channels of a gold crystal as a function of the pathlength.

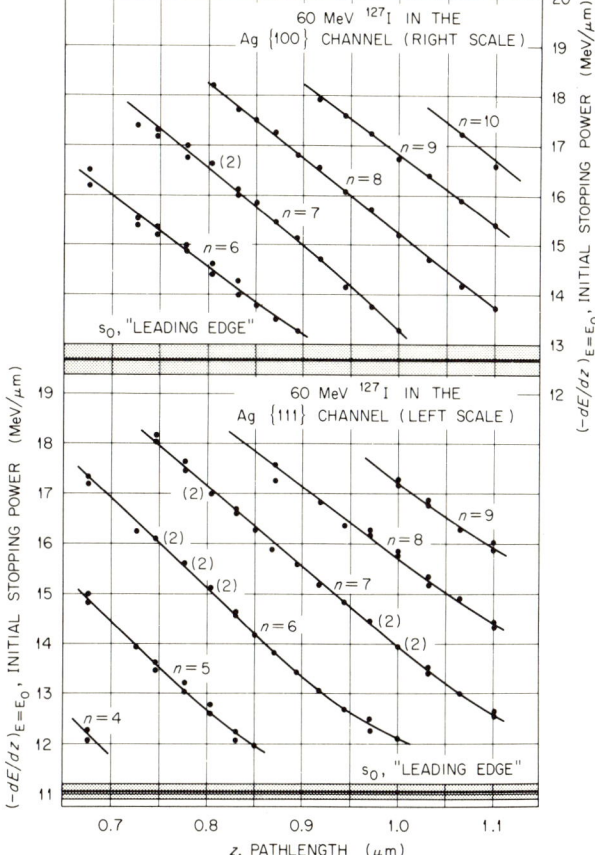

Fig. 5. The dependence of the energy loss rate spectra of 60-MeV ^{127}I ions on their pathlength through thin silver monocrystals.[10]

$$\omega = n(2m)^{1/2}/t , \qquad (4)$$

where the crystal transit time t is given by Eq. 3. (Note that because of the mass factor $(2m)^{1/2}$, ω is not a true frequency.) The dependence of the initial stopping power on the transverse oscillation frequency is typified by Fig. 6, showing some data[9] for 10-MeV ^{16}O ions in two channels of gold monocrystals. The linear dependence displayed in the figure appears to be quite general, having been found also for 60-, 21.6-, and 15-MeV ^{127}I ions in gold[3,5,8,9], for 3-MeV ^{4}He in gold[3,5], and for 21.6- and 60-MeV ^{127}I ions in silver[10], using the techniques described here.

When protons traverse the planar channels of thin crystals, only a single-channeled energy-loss group is observed. Here, however, study of the angular distribution of monoenergetic transmitted ions[4] makes possible the determination of wavelengths, using a technique explained by Fig. 7. There are always two sets of trajectories with the same amplitude which differ in entering the crystal on opposite sides of the channel center. If

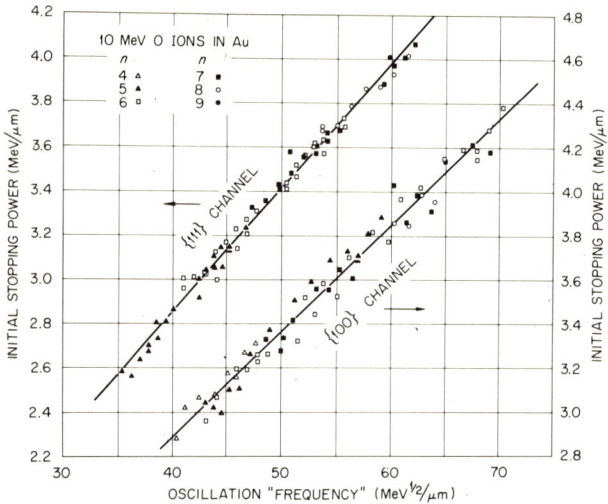

Fig. 6. The dependence of the stopping power of 10-MeV ^{16}O ions on their transverse oscillation frequencies in two planar channels of gold.[9]

the beam is incident on the channel at a slight angle ($\psi_0 \neq 0$), these two trajectories will emerge from the target at equal angles with the channel plane as long as the ions execute an odd number of quarter-oscillations in traversing the crystal. If the detector is set to respond to the correct emergent energy, the angular distribution will consist of two equally populated peaks, symmetrically disposed about the channel center. If an odd number of quarter-oscillations is not made, the populations in the two groups will be different and the channel center may not occur at the centroid of the distribution. Some angular distributions reported for 0.4-MeV protons in silicon $\{111\}$ channels are shown in Fig. 8. From such

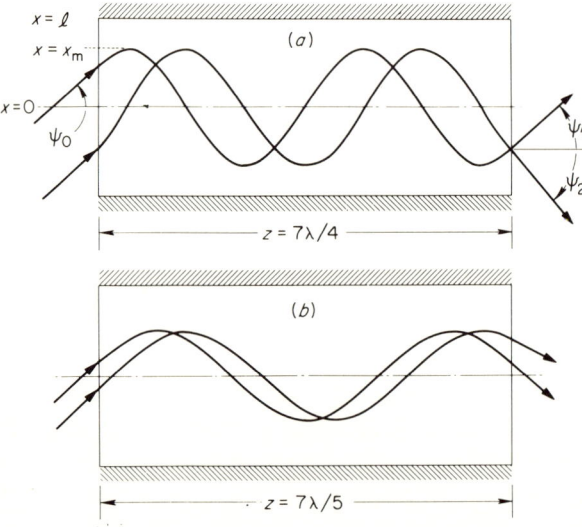

Fig. 7. Determination of the wavelengths (or periods) of oscillation of planar channeled ions by measurement of the angular distributions of monoenergetic transmitted ions.

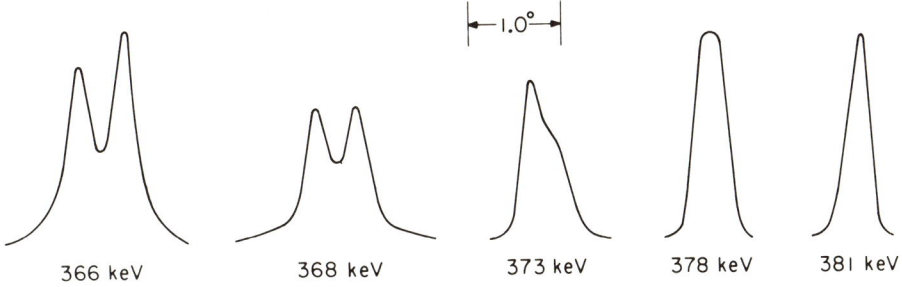

Fig. 8. Some angular distributions reported[4] for protons transmitted through the $\{111\}$ channels of thin silicon crystals. The initial energy of the ions was 400 keV.

measurements, plots similar to Fig. 5 may be constructed, except that the n values obtained are no longer integral but are of the form $n = k \pm 1/4$, k an integer. The dependence of the stopping power on the transverse channel oscillation frequency is obtained in the same way as previously. Fig. 9 illustrates the procedure with some data[4] on 0.4-MeV protons in silicon planar channels.

The model outlined here makes a number of predictions about the energy spectra observable with the detector in general positions, $\psi \neq \psi_0$. These are discussed in detail elsewhere.[3] The influence of such quantities as target mosaic spread and thickness variations, divergence in the incident beam, and the angular acceptance of the detector have also been considered.[3,4,5] As they are not important in the present application, they are

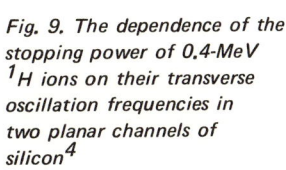

Fig. 9. The dependence of the stopping power of 0.4-MeV ^{1}H ions on their transverse oscillation frequencies in two planar channels of silicon[4]

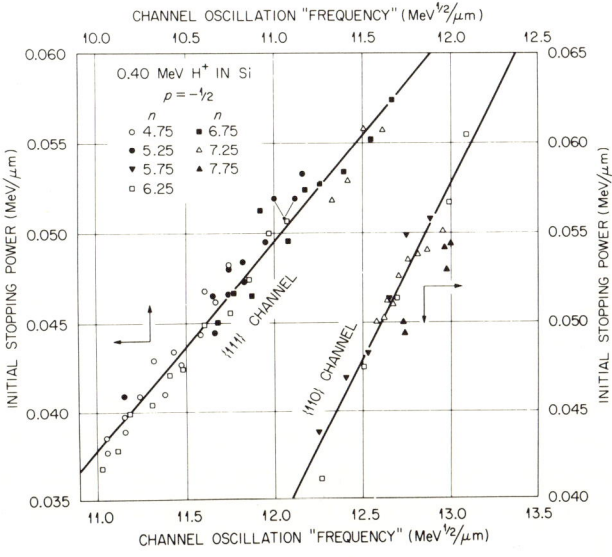

not considered further. It should be noted, however, that widespread application of the channeling technique is seriously restricted by the requirement for very thin, rather uniform, and fairly perfect target crystals.

4 QUANTITATIVE DESCRIPTION OF THE MODEL

The energetic channeled ions slow down by inelastic collisions with the atoms of the medium. It is generally an excellent approximation to separate these collisions into a quasielastic part describing the deflection of the projectile and an electron excitation part which contributes only to the energy loss. The quasielastic collisions may be described with satisfactory accuracy by classical mechanics. Because of the high velocities of channeled tons and the large impact parameters associated with the individual collisions, the conditions for application of the classical impulse approximation are met. The ions are deflected only slightly in individual collisions and are turned away from the atomic planes by the cumulative effect of many encounters. In this circumstance, as long as incidence parallel to principal crystallographic axes is avoided, the atomic nature and regular structure of the planes bordering the channel may be ignored and the ions represented as interacting with the planar continuum potential

$$V_1(\bar{x}) = 4\pi\kappa\rho\ell \int_{\bar{x}}^{\infty} rV(r)dr \qquad (5)$$

where $V(r)$ is the interaction energy between an ion and a lattice atom separated by a distance r, $\bar{x}$ is the length of the normal from the ion to the plane of lattice atoms, ℓ is the half-width of the planar channel, ρ is the atomic density of the target, and κ is a factor allowing for the possibility that the atomic density in a plane differs from $2\rho\ell$. The channeled ions move between two such planes, oscillating in the planar channel potential

$$V_2(x) = V_1(\ell+x) + V_1(\ell-x) , \qquad (6)$$

where the origin is midway between the two planes, a distance ℓ from each. Since the angle between the direction of motion of the ions and the bordering atomic planes is small, $\sin\psi = \tan\psi = \psi$ and $\cos\psi = 1$ are sufficiently accurate approximations. The velocity of the ion, $(2E/m)^{1/2}$, may then be identified with its longitudinal component. The inelastic energy losses may be considered to affect only this component of the motion, leaving the transverse motion to continue without damping. In this approximation,[5] the transverse oscillation "frequency" ω is defined by

$$\omega^{-1} = 2 \int_0^{x_m} [V_2(x_m) - V_2(x)]^{-1/2} dx , \qquad 0 < x < x_m , \tag{7}$$

where x_m is the oscillation amplitude. The model stopping power is assumed to be

$$S(x,E) = s_0 + s_1 [\sigma(x) - 1] , \tag{8}$$

where s_0 and s_1 depend on the ion energy. The spatial dependence of the stopping power is described by the function $\sigma(x)$, normalized so that $\sigma(0) = 1$. The value of s_0, the stopping power at the channel center, may be taken directly from the experiments, as illustrated in Fig. 5. The stopping power of the channeled ions is found by averaging Eq. 8 over the oscillatory motion, weighting each portion of the channel cross section by the time which the ions spends there.

Empirically, it has been found[3,4,5,6,8,9,10] that the data are well represented by the straight line

$$(-dE/dz)_{E=E_0} = \alpha + \beta\omega \tag{9}$$

as illustrated by Figs. 6 and 9. Because of the apparent generality of this observation, it has been adopted as an empirical first principle, the consequences of which are to be explored.[11] When Eq. 8 is averaged over the motion of the ion and the result is compared with Eq. 9, it is found that as long as the number of half-oscillations executed by the ion is integral (and approximately otherwise[4]), the intercept and slope of the straight line are given by

$$\alpha = s_0 - s_1 , \tag{10}$$

$$\beta = 2s_1 \int_0^{x_m} \sigma(x) [V_2(x_m) - V_2(x)]^{-1/2} dx . \tag{11}$$

From Eq. 11 it is easily found that, as long as β is constant,

$$\sigma(x) = \frac{d}{dx} \left\{ \frac{2}{V_2''(0)} [V_2(x) - V_2(0)] \right\}^{1/2} , \tag{12}$$

where the primes represent differentiation with respect to x. The function $\sigma(x)$ is a measure, not only of the stopping power of the ions, but also of the anharmonicity of the planar channel potential: it deviates from unity only insofar as $V_2(x)$ deviates from harmonicity. In the derivation of Eq.

12, it has been assumed only that $V_2(x)$ is even and that $V_2''(0)$ does not vanish, the first property following from the construction of the function.

The experimental data may be used to evaluate the *curvature parameter*,

$$y = 2\pi^2(s_0 - \alpha)^2/\beta^2 \ell \tag{13}$$

$$= V_2''(0)/\ell = 2V_1'(\ell)/\ell$$

$$y = -8\pi\rho\kappa[V(\ell) + \ell V'(\ell)] \quad , \tag{14}$$

where the empirical form is given on the first line and three equivalent theoretical forms are given on the following lines. Thus, the experiments determine the curvature of the planar channel potential at the channel center directly. Given data for a sufficient number of channels with distinct values of ℓ, the potential itself can be evaluated by integration of the last of Eq. 14. For more limited data, the parameters of an assumed potential function can be evaluated. In either case, but especially in the latter, it is desirable to have a further test of the suitability of the resulting potential. This may be found by evaluating the *random* stopping power. If Eq. 8 is averaged over the channel cross section using equal weighting for all intervals, thus corresponding to a particle crossing the planar channel at a large angle, the random stopping power obtained is

$$(-dE/dz)_{\text{random}} \equiv \hat{S} = \alpha + (\beta/\pi\ell)[V_2(\ell) - V_2(0)]^{1/2} \quad , \tag{15}$$

where Eq. 10 and 12 have been used. The potential derived from the channeling data can be used to evaluate $\hat{S}$ and the result can be compared with experimental observations. Application of these techniques to the available experiments is made in the following section.

5 APPLICATION OF THE MODEL

Experimental curvature data have been obtained[9] for ^{4}He, ^{16}O, and ^{127}I ions in the {111} and {100} channels of thin gold crystals. It is characteristic of the data that the curvature parameters are strongly dependent on the ion and its energy, but that the ratio of values for a particular case is independent of these variables. If it is assumed that the interaction potential between the ions and the lattice atoms is of the form

$$V(r) = (Z_1 Z_2 e^2/r)\, \phi(r/a) \quad , \tag{16}$$

where Z_1e and Z_2e are the (effective) nuclear charges of the projectiles and targets, respectively, and the screening function ϕ depends only on the screening length a, then the curvature of the planar channel potential at the channel center is

$$y = -(8\pi\rho\kappa Z_1 Z_2 e^2/a) \, \phi'(\ell/a) \quad . \tag{17}$$

It is clear from Eq. 17 that the experimental results for the various ions in gold can be accounted for if the screening length a is independent of the ion and represents a property of the gold target, while the effective charge on the ion Z_1 is a strong function of its nature and energy. A similar conclusion can be drawn from some recent data[10] for ^{127}I ions in silver crystals. The ions are regarded as test charges of energy-dependent magnitude which sample the electron distribution in the target crystal. It is thus possible to compare the results obtained with calculated electron distributions, or, more conveniently, with the self-consistent potentials that determine these.

Fig. 10, 11, and 12 show the result of applying this procedure to gold[11], silicon[11], and silver[10], respectively. For each, the empirical screening length was found to be in excellent agreement with that calculated for the isolated neutral atoms using relativistic Hartree-Fock-Slater

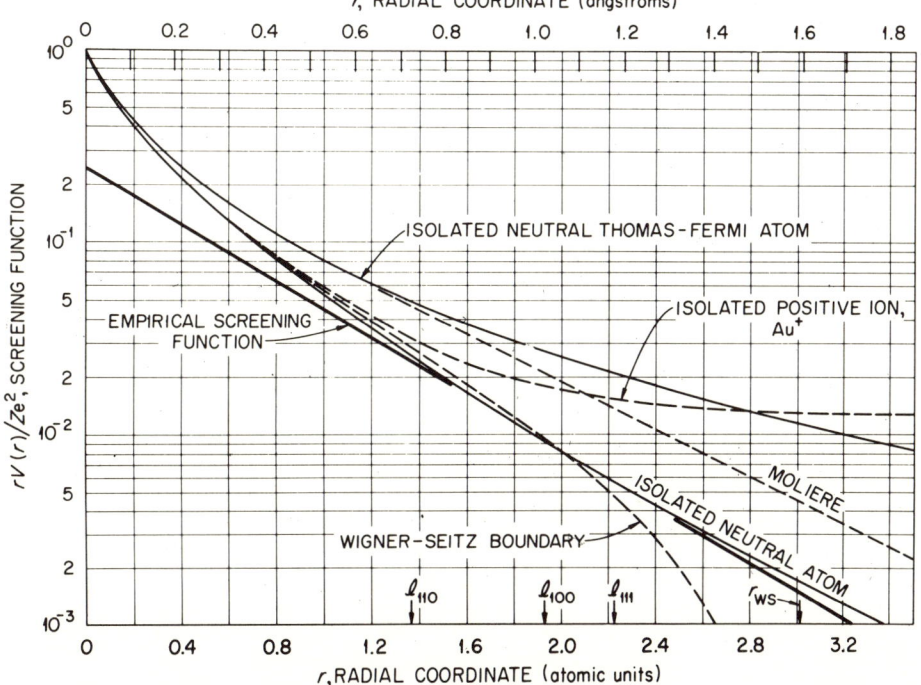

Fig. 10. Comparison of a screening function deduced from planar channeling experiments[9] in gold with Thomas-Fermi[13] and relativistic Hartree[12] calculations.

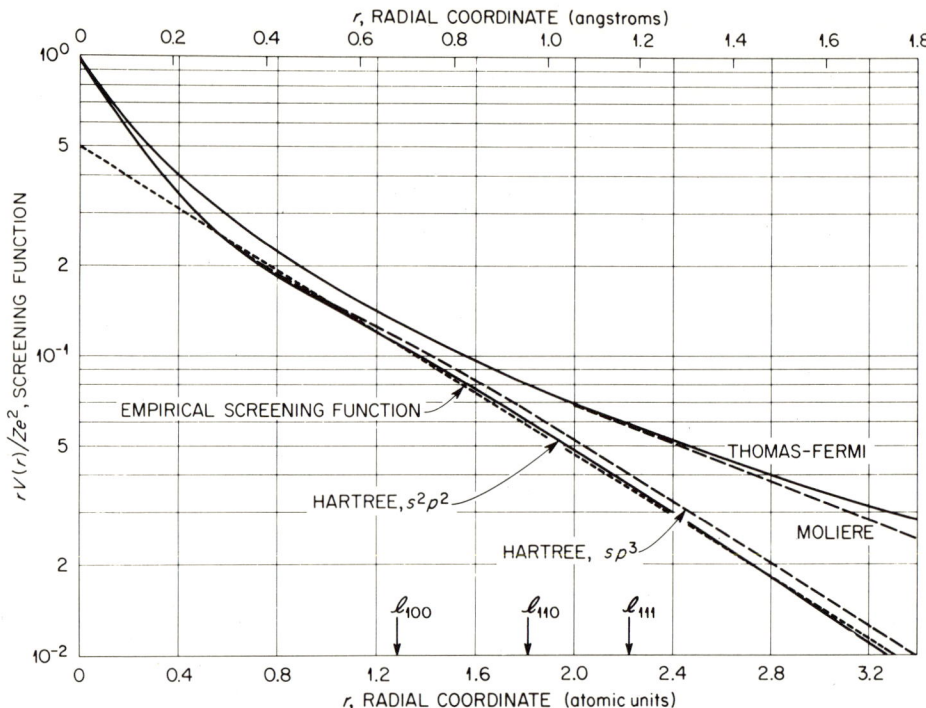

Fig. 11. *Comparison of a screening function deduced from planar channeling experiments*[4] *in silicon with Thomas-Fermi*[13] *and relativistic Hartree*[12] *calculations.*

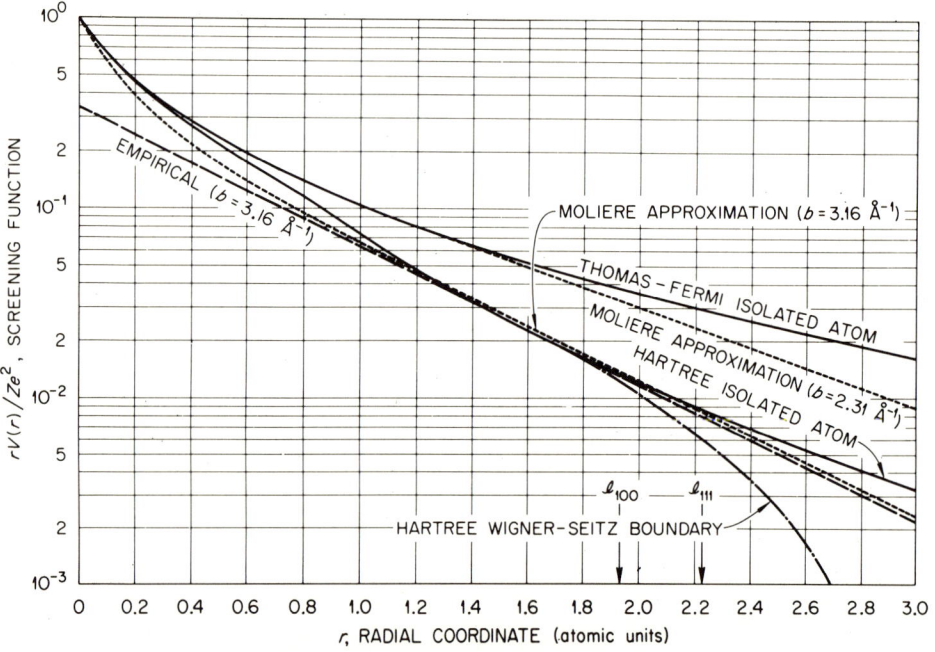

Fig. 12. *Comparison of a screening function deduced from planar channeling experiments*[10] *in silver with Thomas-Fermi*[13] *and relativistic Hartree*[12] *calculations.*

atomic wave functions.[12] In the three figures, the empirical screening function is drawn tangent to the relativistic Hartree function in the region near the centers of the channels used in the measurements. The general superiority of the Hartree function over the Thomas-Fermi screening function[13] and the widely used Molière approximation[14] to it will be evident. By extrapolating the empirical screening function to the origin, the effective ionic charge can be evaluated. These values are shown in Table I along with the screening lengths. For the proton and alpha particle, this procedure yields the expected charges. For ^{16}O ions, the ionic charge is in fair agreement with the equilibrium charge-state value 5.5, measured in beams of oxygen ions transmitted through gold crystals.[15] For the ^{127}I ions in both silver and gold, the charges calculated are roughly twice the values expected from experimental measurements of equilibrium charge states,[16] although they depend on energy in the correct manner. This implies that as the iodine ions approach the channel walls, the screening by their own electrons decreases appreciably. The interaction potential must rise more rapidly close to the channel walls than expected from the test charge viewpoint. If the relativistic Hartree potentials are used to calculate random stopping powers, using Eq. 15, the results[11] are in fair accord with observation for the lighter ions, when a simple model of the thermal vibrations of lattice atoms is included. For the iodine ions, the calculated stopping powers are too low, another reflection of the changing screening of these ions in close encounters.

6 CONCLUSION

The technique presented allows planar channeling experiments to be used to study interatomic potentials. The results of applications thus far

Table I. Screening Lengths and Ionic Charges Deduced for Various Ions in Silicon, Silver, and Gold

Target	Screening Length, Å	Ion	Energy, MeV	Ionic Charge
Si[4]	0.45	^{1}H	0.4	0.9
Ag[10]	0.31	^{127}I	21.6	33
			60	41
Au[9],[11]	0.31	^{4}He	3	2.0
		^{16}O	10	4.8
		^{127}I	15	26
			21.6	30
			60	40

are quite encouraging for the viewpoint on which the model was based. Nevertheless, this basis is still empirical and research is necessary to provide a theoretical understanding of the linear dependence of the stopping power of channeled ions on their transverse oscillation frequencies. Furthermore, there is evidence in recent experiments[10] that this relationship may not hold for large amplitude oscillations, thus limiting the use of Eq. 15. Experimental exploration of this point is needed.

ACKNOWLEDGMENTS

I am grateful to my colleagues B. R. Appleton, S. Datz, F. H. Eisen, C. D. Moak, and T. S. Noggle for many fruitful and stimulating discussions of energy-loss spectra, planar channeling, thin-crystal preparation, and other topics involved in this work.

This research was sponsored by the U.S. Atomic Energy Commission under contract with Union Carbide Corporation.

REFERENCES

1. For general reviews, see Datz, S., C. Erginsoy, G. Leibfried, and H. O. Lutz: Ann. Rev. Nuclear Sci. **17**: 129 (1967) and Mayer, J. W., L. Eriksson, and J. A. Davies: *Ion Implantation in Semiconductors,* Academic Press, New York, 1970.
2. Robinson, M. T., and O. S. Oen: Appl. Phys. Letters **2**: 30 (1963); Phys. Rev. **132**: 2385 (1963).
3. Datz, S., C. D. Moak, T. S. Noggle, B. R. Appleton, and H. O. Lutz: Phys. Rev. **179**: 315 (1969).
4. Eisen, F. H., and M. T. Robinson: Phys. Rev. B **4**: 1457 (1971).
5. Robinson, M. T.: Phys. Rev. **179**: 327 (1969).
6. Lutz, H. O., S. Datz, C. D. Moak, and T. S. Noggle: Phys. Rev. Letters **17**: 285 (1966).
7. Gibson, W. M., J. B. Rasmussen, P. Ambrosius-Olesen, and C. J. Andreen: Can. J. Phys. **46**: 551 (1968).
8. Datz, S., C. D. Moak, B. R. Appleton, M. T. Robinson, and O. S. Oen: *Atomic Collision Phenomena in Solids,* D. W. Palmer, M. W. Thompson, and P. D. Townsend (eds.), p. 374, North-Holland, Amsterdam, 1970.
9. Appleton, B. R., S. Datz, C. D. Moak, and M. T. Robinson: Phys. Rev. B **4**: 1452 (1971).
10. Appleton, B. R., S. Datz, C. D. Moak, and M. T. Robinson: unpublished data.
11. Robinson, M. T.: Phys. Rev. B **4**: 1461 (1971).
12. Carlson, T. A., C. C. Lu, T. C. Tucker, C. W. Nestor, and F. B. Malik: U.S.A.E.C. Report ORNL-4614, 1970.
13. Gombas, P.: *Handbuch der Physik,* S. Flügge (ed.), Vol. 36, p. 109, Springer-Verlag, Berlin, 1956.
14. Moliere, G.: Z. Naturforsch. **2a**: 133 (1947).
15. Martin, F. W., B. R. Appleton, L. B. Bridwell, M. D. Brown, S. Datz, and C. D. Moak: to be published.
16. Moak, C. D., H. O. Lutz, L. B. Bridwell, L. C. Northcliffe, and S. Datz: Phys. Rev. **176**: 427 (1968).

DISCUSSION on paper by Mark T. Robinson

TORRENS: Is there any evident theoretical reason for the direct linear relationship between the initial stopping power and oscillation frequency of the ion in the channel?

ROBINSON: No. It might be expected that the stopping power of an ion would depend (among other things) on the local electron density encountered along its path. It is easily shown, however, that this is not alone the source of the observed dependence [cf. Ref. 11]. In some sense, the observed proportionality between stopping power and transverse oscillation frequency maps out the transition probabilities for the excitations of the target electrons. The enumeration of the various excitations is very difficult, however, and requires considerable further work.

Part Three

POINT DEFECTS

DEFECT CALCULATIONS FOR FCC AND BCC METALS

R. A. Johnson

University of Virginia
Charlottesville, Virginia 22901

and

W. D. Wilson

Sandia Laboratories
Livermore, California 94550

ABSTRACT

Interatomic potentials employing non-central forces have been developed for fcc metals and used to determine the minimum energy configurations of interstitial defects in these materials. In the bcc metals, central potentials were developed and used to calculate vacancy migration energies and divacancy binding energies. Phonon dispersion curves were also calculated using these models.

1 INTRODUCTION

In an attempt to simulate defects in metals, extensive lattice-model calculations have been carried out on the basis of two-body interatomic potentials. Behind all such calculations is the principle that changes in the lattice energy can be written as a function of the positions of the atoms comprising the metal. Various schemes for determining the parameters of this energy function have been based on empirically fitting known physical properties of the metal, on theoretical considerations, or on combinations of the two (semiempirical methods).

When any empirical fitting is used to help determine the parameters of the energy function, there is no uniqueness to the result: in most cases, any number of quite different models can yield satisfactory agreement. In addition, the physical properties to which the fit is made apply primarily to the perfect lattice, although some defect parameters have been used. Thus the energy function is matched to several known points of configuration space, and then investigated in an entirely different region: the whole process may be thought of as very complicated curve fitting to one part of configuration space and extrapolation to another region. There is no a priori way of knowing whether this extrapolation is reasonable, i.e., whether the forces which describe the crystal at the fitted points are adequate to describe the defect state.

In the present work (both this paper and the paper **Rare Gases in Metals**), a combination or semiempirical approach has been used to develop interatomic potentials. Calculations have been carried out over a wide range of materials and defect configurations in an effort to obtain as complete an understanding of the applicability of those potentials as possible. Since this work essentially involves an extension of earlier calculations (e.g., Johnson[1], Wilson and Johnson[2], and Wilson and Bisson[3]), much of the background material and rationale for those studies is not repeated here. Accordingly, this paper is divided into two basic sections; the development of the interatomic potentials, and the results of calculations with these potentials. The calculational methods used to find the appropriate minima or saddle points in the energy function corresponding to defect configurations is also outlined.

2 INTERATOMIC POTENTIALS

Interatomic potentials for both fcc metals (Ni, Cu, Pd, Ag, and Au) and bcc metals (V, Fe, Mo, Ta, and W) are developed in this section. Although the details are different for the two metal structures, the outline for the development of the potentials is the same: The potentials are short ranged (first nearest neighbor for fcc, first and second nearest neighbor for bcc), are adjusted to reproduce exactly the elastic constants (which requires noncentral forces for fcc, but not for bcc), and are matched to theoretically derived core-core interactions (at approximately 0.5 to 0.6 of the nearest-neighbor distance). In both cases, the depth of the potential (as well as other parameters in the model) is related to the single-vacancy formation and migration energies, and some adjustments were made in this regard. Finally, the spectrum of vibrational modes predicted by the resulting potentials have been compared to experimental phonon dispersion curves.

2.1 Fcc Potentials

The primary character of the potentials is determined by the fit to the elastic constants. If a nearest-neighbor central interaction is used for the fcc lattice, two parameters, ϕ_1' and ϕ_1'', are available to fit the three elastic constants, where ϕ_1' and ϕ_1'' are first and second derivatives of the potential, evaluated at the first-neighbor distance (indicated by the subscript). Although reasonable agreement can be obtained for nickel, copper, and silver, the agreement for palladium and gold is quite unsatisfactory, and necessitated the development of a noncentral potential.

A noncentral interaction recently introduced by Wilson and Johnson for calculations in LiH is given by[2]:

$$\psi(\mathbf{r}_{ij}) = \phi(r_{ij}) f(\theta_{ij}, \varphi_{ij}) \quad , \tag{1}$$

where

$$f(\theta_{ij}, \varphi_{ij}) = 1 + K \left(\frac{x_{ij}^4 + y_{ij}^4 + z_{ij}^4}{r_{ij}^4} - \frac{3}{5} \right) \quad ,$$

$\mathbf{r}_{ij}$ is the vector distance between atoms i and j, $r_{ij} = |\mathbf{r}_{ij}|$, and x_{ij}, y_{ij}, and z_{ij} are the cartesian components of $\mathbf{r}_{ij}$. Here $\phi(r_{ij})$ is a radial pairwise interaction between atoms i and j, while the term $f(\theta_{ij}, \varphi_{ij})$, which is invariant under all the cubic symmetry operations[4] contains the noncentral parameter K.

The relationships between the elastic constants and the parameters of this potential are given by:

$$B = \frac{4}{3a} \left(\phi_1'' - \frac{2}{r_1} \phi_1' \right) \left(1 - \frac{K}{10} \right)$$

$$C = \frac{1}{a} \left(\phi_1'' + \frac{3}{r_1} \phi_1' \right) \left(1 - \frac{K}{10} \right) - \frac{4}{a^3} K \phi_1$$

$$C' = \frac{1}{2a} \left(\phi_1'' + \frac{7}{r_1} \phi_1' \right) \left(1 - \frac{K}{10} \right) + \frac{12}{a^3} K \phi_1 \quad , \tag{2}$$

where a is the lattice constant, r_1 is the first-neighbor distance, B is the bulk modulus, C is the C_{44} shear and C' is the $1/2(C_{11} - C_{12})$ shear. It should be noted that two additional parameters, K and ϕ_1, have been introduced by the use of this potential. In practice, a value for ϕ_1 was chosen, and then K, ϕ_1', and ϕ_1'' were determined from the elastic constants by using Eq. 2.

The radial term ϕ in Eq. 1 was written as a sixth order polynominal:

$$\phi = A_6 r^6 + A_5 r^5 + A_4 r^4 + A_3 r^3 + A_2 r^2 + A_1 r + A_0 \tag{3}$$

The seven coefficients $A_0, A_1 \ldots A_6$ were then determined by matching to $\phi_n, \phi_n', \phi_1, \phi_1', \phi_1'', \phi_c$, and ϕ_c'. Here ϕ_n and ϕ_n' are the value and slope of the potential at r_n, the inner cutoff, taken as $r_n = 0.6\, r_1$ for all the fcc metals. These values for the potential are determined from the modification of Wedepohl method developed by Wilson and Bisson, is discussed in greater detail in **Rare Gases and Metals**. The effective valence for the core-core interaction is a physical input required for obtaining the Wedepohl potentials. The outer cutoff value and slope, ϕ_c and ϕ_c', were both set to zero and r_c was taken as somewhat less than the lattice constant (second-neighbor distance). The vacancy calculations were not sensitive to the choice of r_n, but were quite sensitive to the choice of r_c.

Thus, four adjustable parameters enter into the development of these potentials: (i) the effective valence for the core-core interaction, (ii) the choice of r_n, (iii) the choice of ϕ_1, and (iv) the choice of r_c. Following Hume-Rothery[5], the valences for nickel and palladium were taken as zero, while copper, silver, and gold were taken as plus one valence. As mentioned above, r_n was taken as $0.6\, r_1$, and there were no problems associated with this parameter. No unique procedure was found to determine the final two parameters, ϕ_1 and r_c. The single vacancy formation energy, E_{1V}^F, and activation energy for migration, E_{1V}^M, were calculated using various choices of E_{1V}^U (the unrelaxed single vacancy formation energy) and r_c until what was considered to be reasonable agreement with experiment was obtained. The appropriate data for the fcc potentials are listed in Table I and the potentials are shown in Fig. 1.

Experimental phonon dispersion curves are available for both nickel[6] and copper[7]. As pointed out earlier[8], a near-neighbor interaction can give satisfactory agreement to the nickel and copper dispersion curves. In the present case, the elastic constants were fitted exactly; the predicted dispersion curves, shown in Fig. 2, are somewhat higher than the experimental values. However, the general pattern of the predicted curves is in reasonable agreement with experiment.

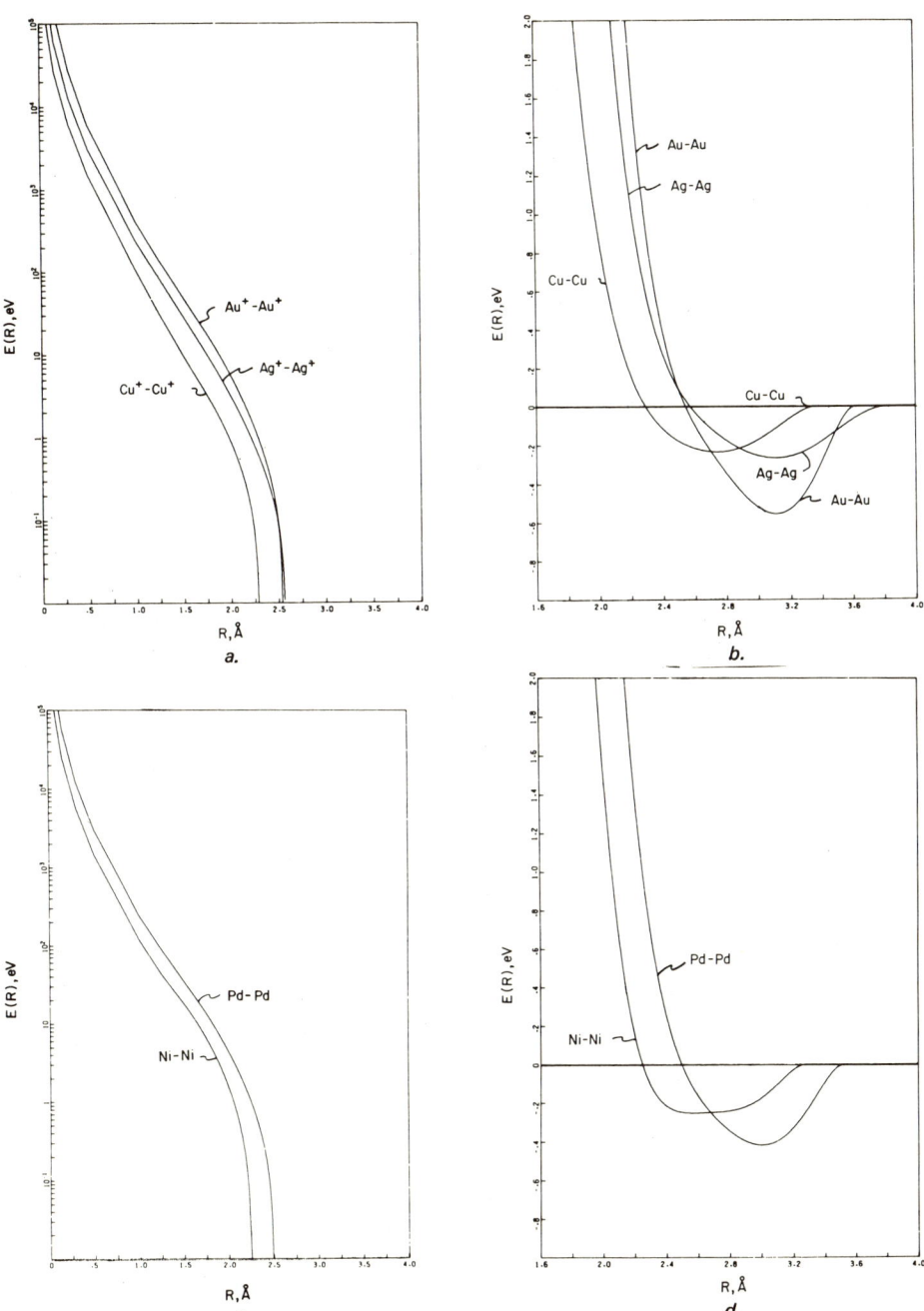

Fig. 1. Short range ($r \leqslant r_n$) interatomic potentials given in a and c determined by the modified Wedepohl method. Note the Born-Mayer behavior at intermediate separations and the 1/R behavior at small internuclear distances. Central parts of fcc potentials valid in the range $r_n \leqslant r \leqslant r_c$ in b and d are sixth-power polynomials fitted to the elastic constants and the short-range potentials given in a and c.

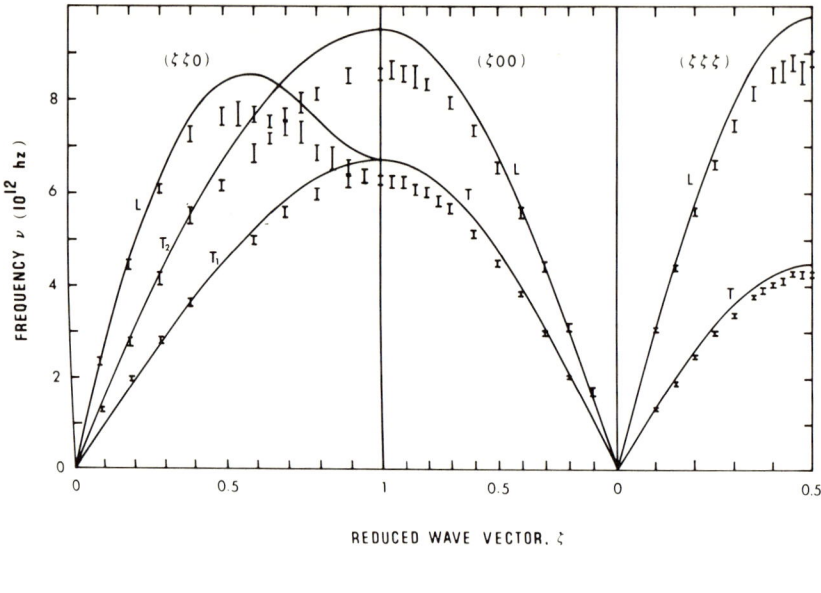

a. Nickel, Ref. 6

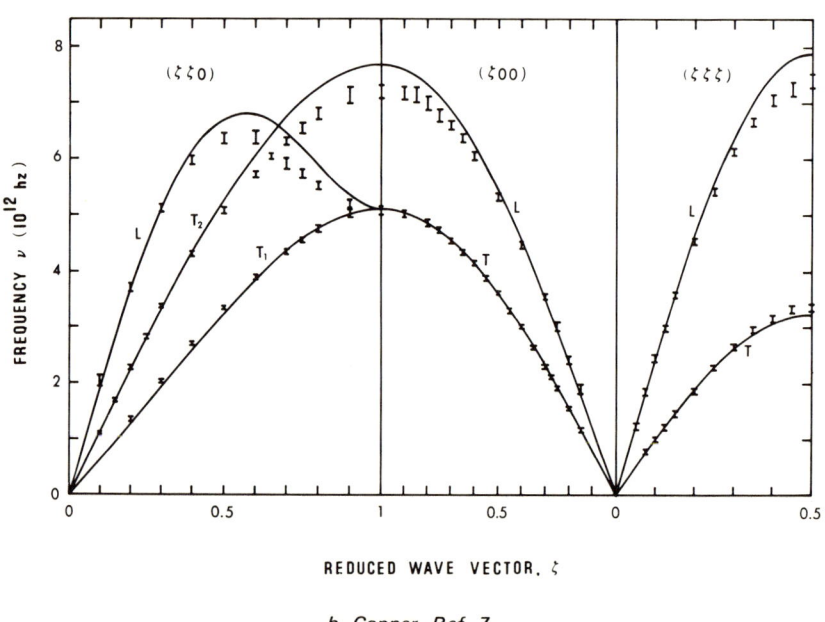

b. Copper, Ref. 7

Fig. 2. Phonon-dispersion curves for nickel and copper predicted by the noncentral force model.

Table I. Data appropriate to the fcc metals. The coefficients A_i are for the potential, $\phi = A_6 r^6 + A_5 r^5 + A_4 r^4 + A_3 r^3 + A_2 r^2 + A_1 r + A_0$, which is valid in the range $r_n \leq r \leq r_c$. ϕ is in eV and r in A; elastic constants are in 10^{12} dyne/cm^2; a, r_n and r_c are in A.

	Cu[a]	Ni[a]	Ag[a]	Pd[b]	Au[a]
C_{11}	1.7	2.465	1.24	2.341	1.86
C_{12}	1.225	1.473	0.935	1.761	1.57
C_{44}	0.758	1.247	0.461	0.712	0.42
a	3.6204	3.52	4.08	3.8718	4.07
r_n	1.7	1.495	1.732	1.64	1.728
r_c	3.366972	3.2736	3.8352	3.523338	3.6223
E_{1V}^U	1.2	1.4	1.39	2.15	3.0
E_{1V}^F	1.19	1.40	1.14	1.60	0.35
E_{1V}^M	0.81	1.22	0.68	1.34	2.51
K	0.274083	0.598732	−0.608165	−1.800036	−1.681869
A_0	444.08761	180.13441	1343.2286	990.14454	1578.6313
A_1	−818.49496	159.03629	−2551.4703	−1534.2540	−2643.6909
A_2	613.65431	−635.35226	2011.9952	888.60145	1755.7795
A_3	−236.39382	539.07973	−841.85754	−212.82476	−573.17697
A_4	48.165321	−207.75799	196.87046	5.4613105	89.710084
A_5	−4.6748698	38.661857	−24.372712	5.9048906	−4.6568404
A_6	0.14481872	−2.8269336	1.2471687	−0.7080931	−0.1498279

(a) Huntington, H. B.: *Solid State Physics, Volume 7*, F. Seitz and D. Turnbull (ed.), p. 274, Academic Press, Inc., New York, 1958.
(b) Rayne, J. A.: Phys. Rev., **118**: 1545 (1960).

2.2 Bcc Potentials

As with the fcc potentials, the primary character of the bcc potentials is determined by the fit to the elastic constants. However, since the first- and second-neighbor distances are similar in the bcc lattice, the potential was extended to include both first- and second-neighbor interactions. Whereas 80 percent of the total solid angle from a lattice site is subtended by the 12 nearest neighbors in the fcc lattice (the atoms being considered as touching hard spheres), only 54 percent is subtended by the 8 nearest neighbors in the bcc lattice. When a first- and second-neighbor model is used in the bcc case, close to 83 percent is subtended, i.e., using a first- and second-neighbor model in the bcc lattice is roughly equivalent to using a nearest-neighbor model in the fcc lattice.

Since a nearest-neighbor model was used for the fcc calculations, the two parameters required for a central potential were overdetermined by the three elastic constants. In the bcc case with first- and second-neighbor interactions, four parameters are required for a central interaction. Thus, if one parameter, say ϕ_1', is chosen, then the others are given by:

$$\frac{1}{r_2} \phi_2' = -\frac{1}{r_1} \phi_1' - \frac{r_2}{4}\left(B - C - \frac{2}{3}C'\right)$$

$$\phi_1'' = +\frac{1}{r_1} \phi_1' + \frac{3r_2}{4}\left(B + C - \frac{2}{3}C'\right)$$

$$\phi_2'' = -\frac{1}{r_1} \phi_1' + \frac{r_2}{4}\left(B - C + \frac{10}{3}C'\right) \quad , \tag{4}$$

where subscripts 1 and 2 indicate the value at first- and second-neighbor distances, respectively, and the elastic constant terms are as defined in the fcc case.

As was done in the development of the fcc potentials, the value and slope of the potential, ϕ_n and ϕ_n', determined by the Wedepohl method, are matched at an inner cutoff distance r_n. The effective valences for the core-core interactions were taken as zero for all the bcc metals. The outer cutoff value and slope, ϕ_c and ϕ_c', were both set to zero, and r_c was taken as midway between the second- and third-neighbor distances, i.e., $r_c = \frac{1}{2} r_2 (1 + \sqrt{2})$.

The bcc potentials were divided into two segments:

$$\phi = a_4 r^4 + a_3 r^3 + a_2 r^2 + a_1 r + a_0 \quad , \quad r_n \leq r \leq r_m$$
$$\phi = b_5 r^5 + b_4 r^4 + b_3 r^3 + b_2 r^2 + b_1 r + b_0, \quad r_m \leq r \leq r_c \tag{5}$$

where r_m is midway between first- and second-neighbor distances. The two segments of the potential are joined at r_m with continuous value, slope, and curvature. A value of $E_{1V}^U = -(4\phi_1 + 3\phi_2)$, the vacancy formation energy prior to relaxation, was held constant for the development of the potential for each metal. Since the vacancy properties of these metals are not well known, this choice was rather arbitrary. Then, for a given r_n, only one ϕ_1' fits all these conditions, i.e., for each value of r_n, a potential that reproduces the elastic constants is uniquely determined. The shape of the potentials varied quite rapidly with r_n, and it was a straightforward procedure to choose the potential with the minimum number of inflection points for use in the defect calculations. Fig. 3 shows the bcc potentials determined in this way using the data given in Table II. The Fe-Fe potential shown in Fig. 3 is for a value of r_n of 1.45 A. In Fig. 4, the Fe-Fe potential for r_n = 1.50 A is shown to illustrate the sensitivity of the shape of the potential to this inner radius.

Experimental phonon dispersion curves are available for all these bcc metals [9-13]. The dispersion curves predicted by the present potentials compare with the experimental data with varying degrees of success: Iron, shown in Fig. 5a (for r_n = 1.45 A), and tungsten yield satisfactory

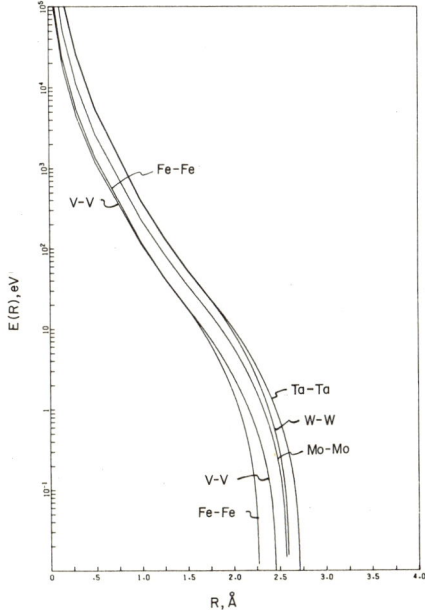

Fig. 3a. Short range $(r \leq r_n)$ interatomic potentials determined by the modified Wedepohl method. Note the Born-Mayer behavior at intermediate separations and the 1/R behavior at small internuclear distances.

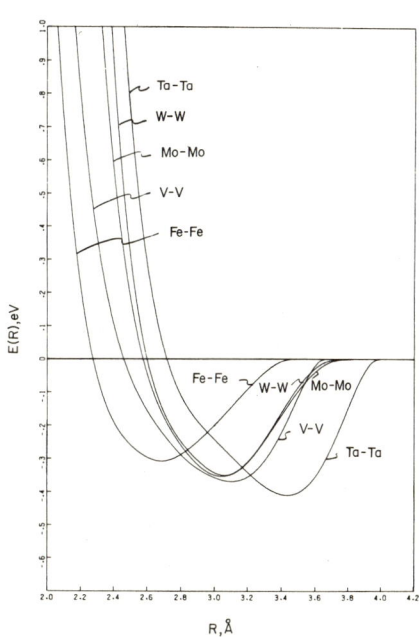

Fig. 3b. BCC potentials valid in the range $r_n \leq r \leq r_c$. These potentials have fourth- (fifth) power polynomial form for values of $r \leq (>) r_m$, the midpoint between the first- and second-neighbor distances.

Table II. Data appropriate to the bcc metals. The coefficients a_i are for potential $\phi = a_4 r^4 + a_3 r^3 + a_2 r^2 + a_1 r + a_0$, which is valid in the range $r_n \leq r \leq r_m$. The coefficients b_i are for the potential, $\phi = b_5 r^5 + b_4 r^4 + b_3 r^3 + b_2 r^2 + b_1 r + b_0$, which is valid in the region $r_m \leq r \leq r_c$. The value of r_m is the midpoint of the first and second nearest-neighbor distances in each metal. The units are the same as in Table I.

	$V^{(a)}$	$Fe^{(b)}$	$Mo^{(c)}$	$Ta^{(c)}$	$W^{(c)}$
c_{11}	2.324	2.431	4.5002	2.6632	5.3255
c_{12}	1.1936	1.381	1.7292	1.5816	2.0495
c_{44}	0.4595	1.219	1.2503	0.8736	1.6313
a	3.03	2.86645	3.14	3.3	3.16
E_{1V}^U	1.8	1.8	1.8	1.8	1.8
r_n	1.5	1.45	1.7	1.75	1.72
r_c	3.6575336	3.4601112	3.7903153	3.9834524	3.8144574
a_4	7.628262	13.04390	9.549910	8.289647	13.73968
a_3	-82.91610	-133.9749	-108.7556	-100.9184	-158.4247
a_2	338.5989	517.0894	465.5270	460.7722	685.8136
a_1	-616.3056	-889.1483	-888.2687	-935.6271	-1321.530
a_0	422.0486	574.6311	637.5701	713.0222	956.4773
b_5	-7.042406	-2.173018	2.960482	-4.810546	5.078514
b_4	109.4269	33.05986	-49.55566	80.28783	-85.06017
b_3	-678.1437	-102.8342	329.1443	-533.6104	566.3906
b_2	2096.422	618.2276	-1083.838	176.6369	-1873.738
b_1	-3234.785	-949.7110	1768.996	-291.4093	3079.346
b_0	1993.803	584.6021	-1145.079	191.7840	-2011.494

(a) Alers, G. A.: Phys. Rev., **119**: 1532 (1960).
(b) Huntington, H. B.: *Solid State Physics, Volume 7*, F. Seitz and D. Turnbull (ed.), p. 274, Academic Press, Inc., New York, 1958.
(c) Featherston, F. H., and J. R. Neighbours: Phys. Rev., **130**: 1324 (1963).

DEFECT CALCULATIONS 311

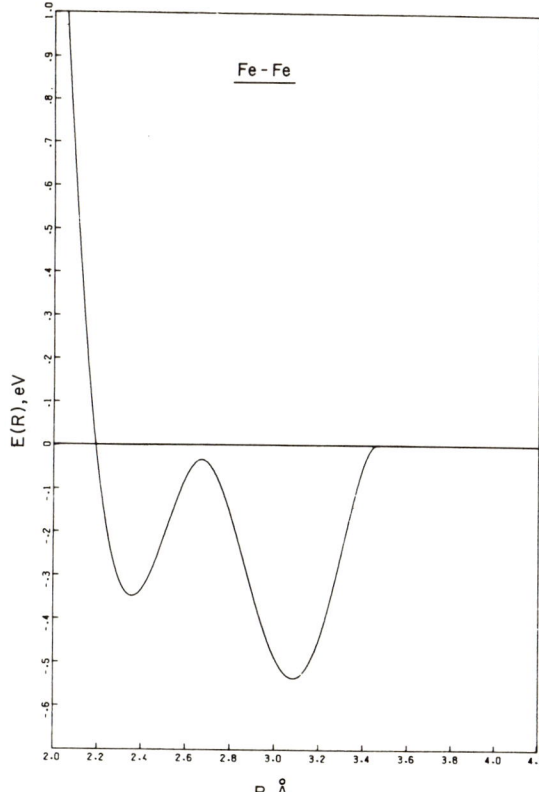

Fig. 4. Fe-Fe potential for r_n = 1.50 A. Note the dual minima in the potential unlike the r_n = 1.45 A potential given in Fig. 3.

agreement, and, as with nickel and copper, could be made to fit even better with the present type of model if the elastic constants were not matched exactly. The agreement for molybdenum is quite good with the exception of several branches near symmetry points where the experimental data indicate anomalous behavior. The agreement is poor for both vanadium and tantalum (shown in Fig. 5b).

3 DEFECT CALCULATIONS

As the method of calculation of the defects is presented in **Rare Gases and Metals**, the discussion here is restricted to the results.

As discussed earlier, the single-vacancy formation and migration energies were used as inputs to the fcc potentials so that these potentials reproduce reasonable values for these quantities. The various interstitial configurations predicted by these potentials in copper, nickel, silver, and palladium were then calculated; the results are given in Table III. In silver

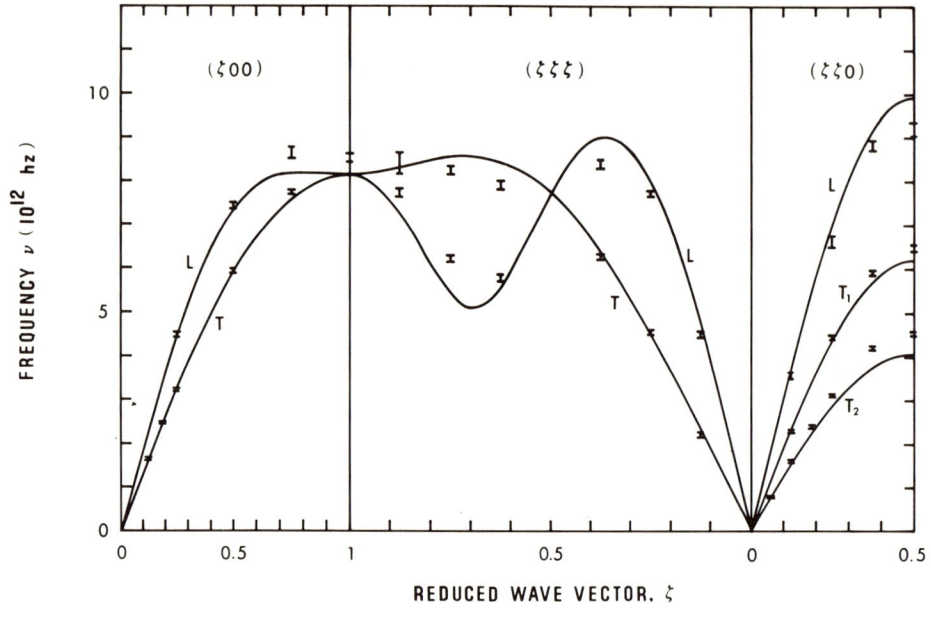

a. Iron, r_n = 1.45 A, Ref. 10

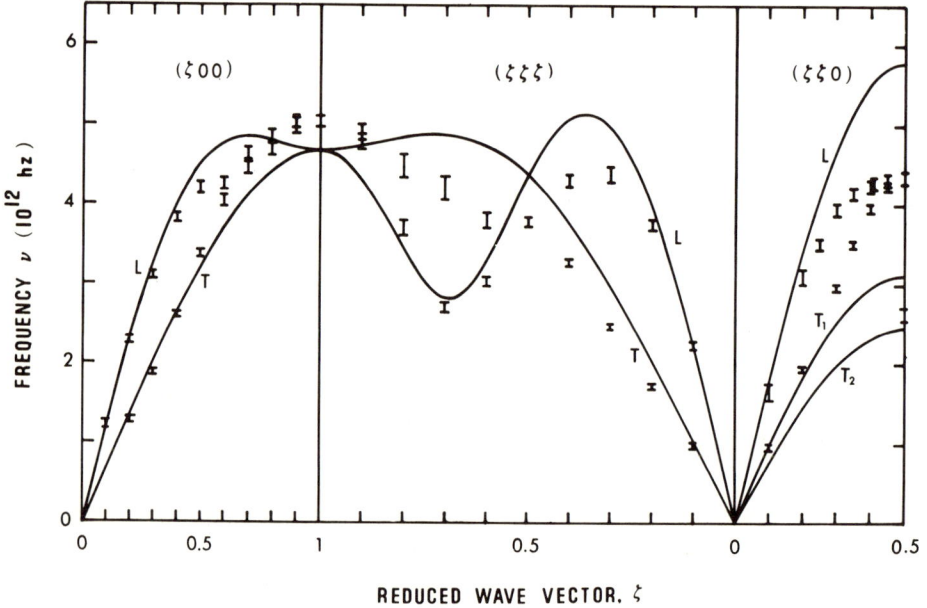

b. Tantalum, r_n = 1.75 A, Ref. 12

Fig. 5. Phonon-dispersion curves predicted by the bcc model.

Table III. Formation energies, in eV, of interstitial defect configurations in fcc metals using noncentral forces. Coordinates are given in units of the half-lattice constant.

Metal	Formation Energy					
	Split Interstitials			Metal Atom at		
	<100>	<110>	<111>	(1,0,0)	(½,½,0)	(½,½,½)
Cu	3.44	3.55	2.74	3.45	3.28	2.72
Ni	3.34	4.21	3.53	3.78	3.71	3.78
Ag	3.48	3.72	2.83	3.70	3.83	3.01
Pd	8.06	9.47	7.10	8.78	10.09	8.09

and palladium, the material most requiring the noncentral forces, the <111> split interstitial has the lowest energy configuration of the six interstitials investigated. In nickel, the <100> split interstitial lies lowest in energy and in copper, the <111> split and the (1/2, 1/2, 1/2) position have similar energies. The calculated energy difference between these two configurations (~0.02 eV) is too small to be physically meaningful. It is interesting to note that the inclusion of the noncentral forces does not in itself force a particular defect to be most stable. Indeed copper and palladium have essentially the same stable interstitial configuration but the most widely differing values of K, the noncentral force parameter. Nickel, which also has a rather small noncentral term, has an interstitial configuration different from that of copper.

An estimate of the activation energy for interstitial motion can be made if one makes certain assumptions about the saddle points. In nickel, the value of 0.44 eV (the difference in energy between the <100> split and the (1,0,0) configurations) is only an upper limit to the activation energy (previous calculations[8] have shown that the saddle point is not the (1,0,0) position). In copper, the ~0.02-eV difference between the <111> split and (½,½,½) interstitial positions is not inconsistent with the known rapid motion (~0.01 eV) of a copper interstitial. Silver has an interstitial activation energy of 0.18 but the palladium interstitial has an activation energy value of nearly 1 eV, a surprisingly high value. As shown in **Rare Gases and Metals**, the helium interstitial activation energy in palladium is also very high.

It was shown in previous calculations[8] that the relative stability of the three pairs (i) <100> split and octahedral, (ii) <110> split and crowdion, and (iii) <111> split and tetrahedral, is quite sensitive to the parameters in the potential function. It was pointed out, however, that the activation energy for migration of each of the three pairs is similar, independent of their relative stability. The same situation is found in the present case, where the difference in energy between the two members of pairs (i), (ii), and (iii) is 0.01, 0.27, and 0.02 eV for copper; 0.44, 0.50, and 0.25 eV for nickel; 0.22, 0.11, and 0.18 eV for silver; and 0.72, 0.62, and 0.99 eV for palladium.

Noticeably absent from these interstitial calculations is gold. The relaxation energy, even about a single vacancy in gold, is so great that it was found difficult to simultaneously fit the vacancy formation and migration energies in this material. The potentials for gold given in Fig. 3 are preliminary only — further work is being done to obtain a potential that can reasonably reproduce the defect energies. The potential shown gives too low a formation energy (~0.35 eV) and too high an activation energy (~2 eV) for single-vacancy motion, but it was included for comparison purposes.

In the bcc metals, the single-vacancy migration energy was not used as an input parameter. The calculated vacancy migration energies are given in Table IV. These calculations were performed by fixing the appropriate metal atom at many positions along a $<111>$ direction between two vacancies and allowing the entire region surrounding the defect to relax. Although firm experimental values are not available, the activation energies are noticeably smaller than one might expect. The iron value of 0.71 eV is consistent with a previous calculation[14] using a similar type potential of 0.68 eV, and a position near (0.4,0.4,0.4) was to be found to be the saddle point with a local minimum occurring at the symmetrical (0.5,0.5,0.5) position in both the present and the earlier calculations. The results of calculations by Wynblatt[15] using a Morse potential and a different relaxation scheme are also given in Table IV. Although the Morse potential is long ranged, the agreement is seen to be quite good. The Wynblatt activation energies tend to be higher than the present, which is consistent with his smaller relaxation region surrounding the defect.

The binding energies of several divacancy configurations in bcc metals are given in Table V. Again in agreement with the previous calculations for iron[14], a second-neighbor pair of vacancies was consistently found to be more stable than a first-neighbor pair.

Table IV. Calculated single vacancy migration energies in eV, for bcc metals.

Metal	Vacancy Migration Energy	
	This Work	Wynblatt
Fe	0.71	0.88
W	1.44	2.26
Mo	1.26	1.31
Ta	0.66	0.66
V	0.54	—

Table V. Calculated divacancy binding energies, in eV. A negative energy indicates binding. Coordinates are given in units of the half-lattice constant.

Metal	Divacancy Binding Energy for Vacancy Pair at (0,0,0) and		
	(1,1,1)	(2,0,2)	(2,0,0)
Fe	−.14	.04	−.28
W	−.05	.05	−.42
Mo	−.02	.06	−.44
Ta	−.33	.05	−.84
V	−.19	.08	−.58

ACKNOWLEDGMENTS

The authors are grateful to Dr. Walter Bauer of Sandia Laboratories, Livermore, for many helpful and fruitful discussions. We also wish to thank C. L. Bisson for his untiring help with the calculations.

This work was supported by the U.S. Atomic Energy Commission.

REFERENCES

1. Johnson, R. A.: Phys. Rev., **145**: 423 (1966).
2. Wilson, W. D., and R. A. Johnson: Phys. Rev. **B, 1**: 3510 (1970).
3. Wilson, W. D., and C. L. Bisson: Phys. Rev. **B** (in press).
4. Von der Lage, F. C., and H. A. Bethe: Phys. Rev., **71**: 612 (1947).
5. Hume-Rothery, W., and G. V. Raynor: *The Structure of Metals and Alloys,* p. 194, Institute of Metals, London, 1956.
6. Birgeneau, R. J., J. Cordes, G. Polling, and A.D.B. Woods: Phys. Rev., **136**: A1359 (1964).
7. Svensson, E. C., B. N. Brockhouse, and J. M. Rowe: Phys. Rev., **155**: 619 (1967).
8. Johnson, R. A.: Radiation Effects, **2**: 1 (1969).
9. Colella, R., and B. W. Batterman: Phys. Rev. **B, 1**: 3913 (1970).
10. Brockhouse, B. N., H. E. Abou-Helal, and E. D. Hallman: Solid State Communications, **5**: 211 (1967).
11. Woods, A.D.B., and S. H. Chen: Solid State Communications, **2**: 233 (1964).
12. Woods, A.D.B.: Phys. Rev., **136**: A781 (1964).
13. Chen, S. H., and B. N. Brockhouse: Solid State Communications, **2**: 73 (1964).
14. Johnson, R. A.: Phys. Rev., **134**: A1329 (1964).
15. Wynblatt, P. J.: J. Phys. Chem. Solids, **29**: 215 (1968).

DISCUSSION on paper by R. A. Johnson and W. D. Wilson

HUNTINGTON: Where you have the choice of fitting the potential to neutron data or elastic constants, wouldn't the former be more suitable for defect work?

JOHNSON: The elastic constants are known with considerably greater accuracy and over a wider range of materials than are known for neutron scattering data. The potentials derived from either type of data are similar and yield similar results when used in point-defect calculations.

HUNTINGTON: Have you run into particular problems with the hexagonal materials?

JOHNSON: Yes, especially for nonideal c/a ratios. So far we find the central potentials are quite unsatisfactory for these purposes.

BULLOUGH: I wonder if you can say why it is preferable to use noncentral interactions with first neighbors only in fcc instead of central interactions out to second neighbors. In bcc materials you are content to use first- and second-neighbor central interactions — why not in fcc?

I would like to comment that the noncentral potentials you have constructed are precisely analogous to the void-void potentials required to explain the existence of the void superlattice.

JOHNSON: When we try to use second-neighbor interactions in the fcc lattice, especially in materials like palladium and gold, we fine it very awkward to obtain a smooth fit. Likewise, when we try to use a first-neighbor noncentral potential in the bcc lattice, we obtain poor results. This is really what I would expect in the first place because, from a pragmatic point of view, the "close" neighbors in the fcc lattice are the nearest neighbors, while in the bcc lattice, they are the first and second neighbors. From another point of view, if you allow for thermal motion and look at what is equivalent to a radial distribution function in liquids, the first major peak, of course, comes from the 12 nearest neighbors in the fcc lattice, while the peaks from the 8 first and 6 second neighbors combine to form the first major peak in the fcc lattice.

In regard to your second point, yes, we are aware of this relation.

WILSON: To comment on Dr. Bullough's question about why we included a noncentral term instead of more distant neighbors: in LiH, where we know the central potential from molecular orbital theory it was found to be impossible to fit the elastic properties, binding energy and interionic separation with a central force alone. The inclusion of our simple potential form made this fit quite easy.

TAYLOR: While looking at your tantalum results it occurred to me that there were almost certainly at least one and perhaps two Kohn anomolies present in the phonon-dispersion curves. These, of course, are just modifications of the long-range oscillations that we have been hearing about. Hence you don't stand a chance of obtaining a fit to tantalum with your form of potential. Does it worry you that the long-range oscillations are not included in your potential?

JOHNSON: We certainly stand no chance of fitting Kohn anomalies. In general we find that we cannot fit the general form of the dispersion curves in materials with Kohn anomalies.

HARRISON: I am not certain but it appears to me that your angular potential violates the kind of rotational invariance condition which Maradudin pointed out yesterday. You could check this by seeing if a pure rotation (with respect to the crystal axes) gives an energy change proportional to the square of the angle of rotation. Such an interaction is physically unrealistic and would lead, for example, to the result that the energy stored in a bar bent in the middle would depend on the length of the free ends of the bar.

JOHNSON: No, I think it is elasticity theory which is unphysical in this regard. In your bending bar example, if it is an anisotropic material, the elastic constants clearly remain aligned with the bar, not with space, far from the area of bending, i.e., the bar is not invariant to the rotation you describe.

ASHCROFT: Your potential parameters appear to be fixed by comparison with derivative properties (e.g., elastic constants). Bearing in mind the delicacy of a total-energy calculation, can you none-the-less test your potentials at constant volume (say the equilibrium volume) to examine whether the correct structure is obtained?

CHANG: For a metal that exists in two forms (e.g., iron) do you have potentials for each allotropic form? In other words, does your potential for a given metal predict its structural stability?

JOHNSON: The change in energy in comparing different structures is extremely small, and we do not use this quantity in the development of our potentials. However, with the boundary conditions we use, our models are always stable with respect to structure.

ENGLERT: Could you extend your noncentral potential to more-distant neighbors, and how would this affect the stability of various lattice structures?

JOHNSON: We could extend our noncentral potentials to more-distant neighbors, but this would, of course, bring more parameters into the potential model. The structural stability is discussed after the questions by Ashcroft and Chang.

ENGLERT: Do you include a volume term in the bcc metals when you use a central potential?

JOHNSON: Yes, we include the volume-dependent term that arises in a natural way when we fit the experimental elastic constants.

ENGLERT: Could you give more details regarding your potential in hcp structures and the stability of these structures with respect to fcc structures?

JOHNSON: We are presently working on the development of potentials for calculations in hcp metals, but these potentials are not yet ready for use.

DORAN: Please comment on how your new nickel results compare, in general, with your previously published γ-Fe/Ni results.[1]

JOHNSON: The two nickel calculations compare quite well with the earlier work. In the present case we fit to vacancy data, and so the vacancy values are close but not identical. The interstitial results are quite similar.

SHEWMON: You chose your coefficients by fitting small displacement data, i.e., elastic coefficients and phonon spectra. What basis have you for the validity of your potential at the large displacements needed for vacancy energies or vacancy-motion energies?

JOHNSON: As I stressed at the beginning of the talk, the basis of the semiempirical approach is to fit known physical data and extrapolate to unknown properties. We do use some defect data in deriving these potentials, but we make no claim to a theoretical basis for the extrapolation to large displacements.

BASINSKI: Since in this work you are considering a very specific type of potential, in particular, by taking into account only very short-range interactions, how significant is the observed insensitivity to the values of the parameters used?

It appears at least possible that the properties not considered to be sensitive to the potential might become very sensitive if a different type of the potential was used.

JOHNSON: For the types of properties we are calculating, we have as yet to find a reason to include longer-range forces. In general we feel that these effects will be small.

TORRENS: Did you make any allowance for electronic contribution to the experimental elastic constants before adjusting the interatomic potential parameters?

JOHNSON: No, the full experimental elastic constants were used for all these potentials.

REFERENCE

1. See Ref. 1.

PSEUDOPOTENTIAL CALCULATION OF POINT DEFECT PROPERTIES IN SIMPLE METALS

P. S. Ho

Cornell University
Ithaca, New York

ABSTRACT

This paper shows a point-defect calculation formulated completely within the framework of the pseudopotential theory. From the functional form of the total energy of the perfect lattice, the explicit expressions for the formation energy and formation volume of a vacancy have been derived. It is shown how the volume-dependent part of the lattice energy can be taken into account in defect calculation. This energy is found to have significant effect on the vacancy properties. To calculate the relaxed configuration around the vacancy, the method of lattice statics was reformulated in the context of pseudopotential theory. The present method was applied to calculate the vacancy-formation energy and formation volume for alkali metals and aluminum, and the results are reported.

1 INTRODUCTION

There are three basic problems in the atomistic calculations of point-defect properties in simple metals. The first is how to construct a realistic interatomic potential that can represent the nature of metallic bond. The second is how to calculate the relaxed atomic configuration around the defect in the presence of the long-range and oscillatory interatomic forces. The last problem is how to take into account the effect of the volume-dependent part of the lattice energy in the calculation of the characteristic energy and volume of point defects (such energy can have significant influence on defect properties, as is shown). All these problems can be attributed to the many-body nature of the interaction between conduction electrons and the ions in metals.

Fig. 1 shows schematically the procedure[1] used in most of the previous calculations for determining defect properties. The crystal lattice is divided into three regions. In Region I, the atoms interact through a two-body empirical potential, and their relaxed displacements around the vacancy are calculated by an iteration method. The atoms in Region II are used to match the discrete atomic displacements in Region I to the elastic relaxation in Region III. In this matching procedure, the defect volume is also determined. This scheme would be difficult to apply for materials with large elastic anisotropy such as the alkali metals because of the convergence problem in obtaining elastic solutions.[2] The uniform pressure P is applied to the boundary of Region II to account partly for the effect of the volume-dependent energy and partly for the noncentral portion of the interatomic potential if a central interatomic potential were used in Regions I and II. For metals, this parameter P has an important effect on the vacancy formation energy, and because it is somewhat ad hoc in nature, it causes considerable uncertainty in the results.

This paper presents a method developed on the basis of the pseudopotential theory for calculating point-defect properties in simple metals. It is different from the majority of previous works in the interatomic potential used; in the method of calculating defect relaxed configuration; and in the treatment of the volume-dependent part of the lattice energy. The purpose of this paper is to show that a point-defect calculation can be developed completely within the framework of the pseudopotential theory. It shows explicitly how the volume-dependent part of the lattice energy can be taken into account in defect calculation. For the relaxed configuration around the defect, the lattice-statics method has been reformulated also in the framework of pseudopotential theory. The present method is applied to calculate the vacancy-formation energy and formation volume of the alkali metals and aluminum. The complete results for the alkali metals and part of the results for aluminum are presented.

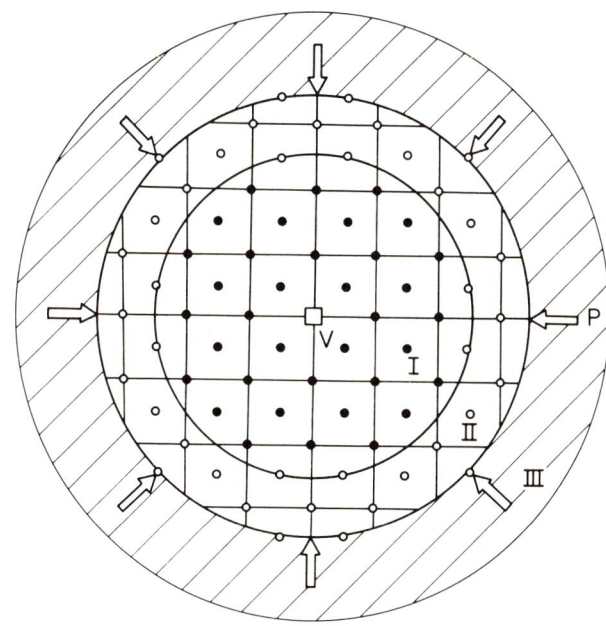

Fig. 1. Representation of a defect lattice used in previous calculations. This shows the simple case of vacancy formation.

Region I   Discrete Atoms
Region II   Semidiscrete Atoms
Region III   Elastic Continuum

2 FORMULATION OF THE DEFECT CALCULATION

2.1 Outline of the Defect Calculation

As shown in Fig. 2, there are three steps in the vacancy calculation. First, the vacancy is introduced by removing the ion from the center of a finite solid containing N atoms and placing it on one of the surface lattice sites. In this step, there is no relaxation of individual atoms from their lattice positions, but the electron density decreases by a factor of $1/(1+N)$ due to the addition of a lattice site. The structural energy required for changing the atomic arrangement in order to create the vacancy is E_S. The second step is to apply the lattice-statics method to calculate the relaxed configuration around the vacancy. The calculation is carried out under constant atomic volume condition. The energy gained in this relaxation process is E_R. The last step is to allow the defect-lattice complex to relax uniformly to an equilibrium configuration. The amount of relaxation δ, defined as the fractional change of the lattice parameter, is determined by minimizing the total energy of the defect lattice with respect to δ. This energy includes the energy change due to variations of the atomic volume,

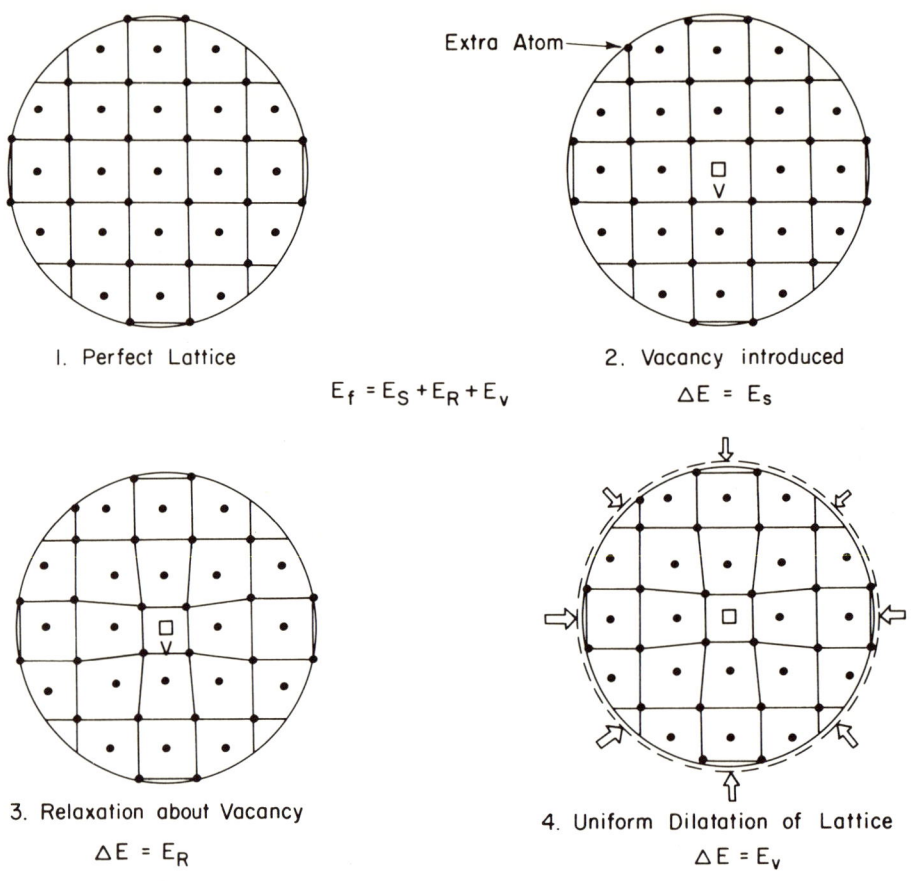

Fig. 2. Steps used in the present work for calculating vacancy formation energy and formation volume.

electron density, and the Fermi wave vector. The energy gained in this relaxation process is E_V. The formation energy of the vacancy is simply

$$E_f = E_S + E_R + E_V, \qquad (1)$$

and the formation volume Ω_f of the vacancy is

$$\Omega_f = \Omega_o (1 - 3N\delta) \qquad (2)$$

where Ω_o is the atomic volume.

2.2 Structural Energy of a Vacancy

In the pseudopotential theory, the total lattice energy per atom U_T can be written as the sum of the volume-dependent energy U_o, the electrostatic energy U_e, and the band-structure energy U_b,[3] i.e.

$$U_T = U_o + U_e + U_b \qquad (3)$$

The term U_o has the following form in atomic units:

$$U_o = Z \left[\frac{1.105}{r_s^2} - \frac{0.458}{r_s} - (0.0575 - 0.0155 \ln r_s) \right] + \sum_k \langle k|w|k \rangle \quad , \qquad (4)$$

where the terms in the bracket are respectively: the average kinetic energy of the electron, the exchange energy, and the correlation energy of the free electron gas*, and the last term is the Hartree energy. The charge density parameter r_s is defined in terms of the atomic volume Ω_o by

$$\frac{4\pi}{3} r_s^3 = \frac{\Omega_o}{Z} \quad ,$$

where Z is the valence number.

The expression of U_e can be written as[5]

$$U_e = \underset{q_o}{\Sigma'} S(q)S^*(q) F_e(q) + \sum_\ell \phi(r^\ell) - \frac{Z^{*2}\xi^{1/2}}{\pi^{1/2}} - \frac{\pi Z^{*2}}{2\xi\Omega_o} \qquad (5)$$

where

$$F_e(q) = \frac{2\pi Z^{*2}}{\Omega_o} \frac{e^{-q^2/4\xi}}{q^2}$$

and

$$\phi(r^\ell) = \frac{Z^{*2}}{(\pi r^\ell)^{1/2}} \int_{\xi^{1/2} r^\ell}^{\infty} e^{-x^2} dx \quad .$$

Z^* is the effective valence charge which is generally different from Z in the pseudopotential theory due to correction of the orthogonal hole.[6] ξ is a parameter to be chosen to optimize the convergence of the sums of the two series. $S(q)$ is the structure factor which is defined for a lattice containing N atom as

$$S(q) = \frac{1}{N} \sum_\ell^N e^{-i\mathbf{q}\cdot\mathbf{r}^\ell} \qquad (6)$$

For a perfect lattice, $S(q)$ vanishes due to the periodic condition, except when q equals a reciprocal lattice vector q_o, then $S(q_o) = 1$.

The band-structure energy U_b can be expressed as

$$U_b = \underset{q_o}{\Sigma'} S(q) S^*(q) F_b(q) \qquad (7)$$

---
*The form for correlation energy selected is suggested by Ref. 4.

where $F_b(q)$ is the energy-wave number characteristic. The detailed expression of $F_b(q)$ depends on the pseudopotential chosen (given in the next section).

The separation of the lattice energy into the volume-dependent part (U_o) and the structure-dependent part (U_e and U_b), is a very convenient starting point for point-defect calculations. Harrison was the first to point out this particular property of the pseudopotential and to apply it for some preliminary calculations of defect properties.[7] Here, this means that for the first two steps one has to consider only changes in the structure-dependent energy since the volume of the defect-lattice complex is kept constant. In the last step, when one calculates the amount of the uniform relaxation of the whole lattice, then the contribution from the volume-dependent energy must also be taken into account.

The vacancy is introduced through the structure factor of the following form for the defect lattice

$$S'(q) = \frac{1}{N} \sum_\ell^{N+1} e^{-i\mathbf{q}\cdot\mathbf{r}^\ell} - \frac{1}{N} \quad . \tag{8}$$

To calculate U_s, the structural energy of the vacancy, we have to determine only the changes in both U_e and U_b due to the change in the structure factor; the energy change due to atomic relaxations around the defect is not included. For simplicity, let us first consider U_b. The total band-structure energy for the defect-lattice complex is

$$NU_b' = N \sum_q{}' S'(q) S'^*(q) F_b(q) - \sum_{q_o} F_b(q) \quad . \tag{9}$$

The last term, which sums only over the reciprocal lattice vectors q_o, accounts for the decrease in U_b for that atom was removed from the origin and placed on a surface lattice site.

Substituting Eq. 8 into Eq. 9 obtains

$$NU_b' = N \sum_q{}' S_o'(q) S_o'^*(q) F_b(q) - \sum_q{}' [S_o'(q) + S_o'^*(q)] F_b(q)$$

$$+ \frac{1}{N} \sum_q{}' F_b(q) - \sum_{q_o}{}' F_b(q) \quad , \tag{10}$$

where

$$S_o'(q) = \frac{1}{N} \sum_\ell^{N+1} e^{-i\mathbf{q}\cdot\mathbf{r}^\ell} \quad .$$

The physical significance of the various terms in the above expression is not difficult to see. The last two terms consist of the band-structure energy required in creating the vacancy; hence, they are part of U_S. The first term gives the "band-structure" interactions between atoms in the defect lattice. The second term, a result of the cross product of the defect term $(1/N)$ and U_b of the rest of the lattice, represents the interaction between the vacancy and the other atoms. These two terms are part of the relaxation energy E_R, which must be minimized in order to obtain the relaxed configuration around the defect.

To complete the calculation for U_S, one must include the contribution from the electrostatic energy. The treatment of this term is identical to that of U_b except for the additional series sum in real space. The detailed derivation is not shown here but is given in Ref. 8.

The final expression of E_S turns out to be

$$E_S = \frac{\alpha Z^{*2}}{2r_S} + \frac{\Omega_o}{(2\pi)^3} \int F_b(q)\, d^3q - \sum_{q_o}{}' F_b(q) \quad , \tag{11}$$

where the first term is the negative electrostatic energy for an ion, and α is the Madelung's constant. For convenience, the third term in Eq. 10 has been converted to the integral term in E_S. In actual calculation, the first two terms in E_S are found to be opposite in sign but both are of order-of-magnitude larger than the last term. Since the electrostatic energy can be calculated exactly, E_S is very sensitive to the exact form of the pseudopotential and the screening dielectric function used.

2.3 Relaxed Configuration Around a Vacancy

To calculate the atomic configuration around the defect, we adopt the approach of the method of lattice-statics which was first applied to point-defect calculation by Kanzaki[9] and later reformulated by Flinn and Maradudin[10]. The difficulty of calculating the defect configuration is mainly due to the mutual coupling of all the atoms in the lattice. The lattice-statics method utilizes the idea that if the atomic displacements in the periodic lattice are small, their Fourier components are completely decoupled in the harmonic approximation and can be solved by standard techniques similar to that used in lattice dynamics. The advantage of using this method is that the displacements from the equilibrium positions are solved simultaneously for all the atoms around the defect. Without the need of iterating calculation, this method is efficient, and the result is guaranteed to yield a configuration of minimum relaxation energy. Furthermore, its application is not limited by the elastic anisotropy of the crystal lattice. However, as the solution is exact only in the harmonic approximation, its validity is open to questions for large relaxation displacements.

This paper only outlines the procedure of the calculation, a more detailed description can be found in Ref. 8. Let ξ^ℓ be the displacement vector from the equilibrium position r_o^ℓ. We first substitute r^ℓ by $r_o^\ell + \xi^\ell$ into the general expression of the relaxation energy E_R and expand E_R to the harmonic term ξ^ℓ. Then ξ^ℓ is converted by a Fourier transform in terms of the normal coordinates a_Q, i.e.

$$\xi^\ell = \sum_Q{}' (a_Q e^{-i\mathbf{Q}\cdot\mathbf{r}^\ell} + a_Q^* e^{-i\mathbf{Q}\cdot\mathbf{r}^\ell}) \quad (12)$$

This step also changes E_R into a function of a_Q to the second order. a_Q can be determined by minimizing E_R according to

$$\frac{\partial E_R}{\partial a_Q} = 0 \text{ for all } Q\text{'s.}$$

This condition yields a set of equilibrium equations for each a_Q, which can be written as

$$\sum_\beta V_{\alpha\beta}(Q) a_Q^\beta = G_\alpha(Q) \quad \alpha,\beta = 1,2,3 \quad , \quad (13)$$

where

$$G_\alpha(Q) = \sum_{q_o} [(q_o-Q)_\alpha F(q_o-Q) - (q_o+Q)_\alpha F(q_o+Q)] - 2Q_\alpha F(Q) + \left.\frac{\partial \phi}{\partial x_\alpha^\ell}\right|_{r^\ell} e^{-i\mathbf{Q}\cdot\mathbf{r}^\ell}$$

$$V_{\alpha\beta}(Q) = (N+1)\left\{\sum_{q_o} [(q_o-Q)_\alpha (q_o-Q)_\beta F(q_o-Q) + (q_o+Q)_\alpha (q_o+Q)_\beta F(q_o+Q) - 2q_{o\alpha} q_{o\beta} F(q_o)] + 2Q_\alpha Q_\beta F(Q)\right\}$$

$$+ (N+1) \sum_\ell \phi_{\alpha\beta}^{o\ell} e^{-i\mathbf{Q}\cdot\mathbf{r}^\ell}$$

and

$$\phi_{\alpha\beta}^{o\ell} = \left.\frac{\partial \phi(r^\ell)}{\partial x_\alpha^\ell \partial x_\beta^\ell}\right|_{r=r^\ell} ; \phi_{\alpha\beta}^{oo} = -\sum_\ell \phi_{\alpha\beta}^{o\ell} ; F(q) = F_e(q) + F_b(q) \quad .$$

After solving the a_Q's from Eq. 13, one can obtain the relaxation energy as

$$E_R = -\sum_{Q,\alpha} G_\alpha(Q) a_Q^\alpha \quad , \quad (14)$$

and calculate the atomic displacement ξ^ℓ according to Eq. 12.

2.4 Uniform Relaxation of the Defect Lattice

The last step is to calculate the uniform relaxation for the defect lattice. This is obtained by minimizing both the structure-dependent and volume-dependent energies of the defect lattice with respect to δ. If the uniform lattice relaxation is expressed as

$$r^\ell = r_o^\ell (1-\delta) \quad ,$$

then

$$q = q^o (1+\delta) \quad ,$$

but

$$k_F = k_F^o (1+\delta)\left(1 - \frac{1}{3N}\right) \quad .$$

The additional change in k_F comes from the fact that, when a vacancy is created, there is one extra lattice site but the same number of electrons. One can substitute the above expressions into the total energy of the defect lattice and expand it to the second order in δ. δ can then be determined by minimizing the total energy of the defect lattice. The result is

$$\delta = \frac{A}{B} \tag{15}$$

$$A = Z\left(\frac{2}{3}\frac{2.21}{r_s^2} - \frac{1}{3}\frac{0.458}{r_s}\right) + 3 \sum_k \langle k|w|k\rangle - \frac{\alpha Z^{*2}}{2r_s} - \frac{\Omega_o}{(2\pi)^3}\int \frac{\partial F_b}{\partial q} d^3q +$$

$$\sum'_{q_o} \frac{\partial F_b}{\partial q} q + \frac{1}{3}\sum_{q_o} \frac{\partial^2 F_b}{\partial(k_F/q)\partial q} k_F + \sum_{Q,\alpha}\left[G_\alpha(Q) - \frac{\partial G_\alpha(Q)}{\partial Q} Q^\alpha\right] a_Q^\alpha$$

$$B = Z\left(3 \frac{2.21}{r_s^2} - \frac{0.916}{r_s} - 0.0155\right) + 12 \sum_k \langle k|w|k\rangle - \frac{\alpha Z^{*2}}{r_s} + \sum'_{q_o} \frac{\partial^2 F_b}{\partial q^2} q^2$$

After δ is determined, we can calculate how much energy is gained during uniform relaxation of the lattice. This energy, E_V, is found to be

$$E_V = \left(\frac{1}{3} - N\delta\right)\left[Z\left(-\frac{2.21}{r_s^2} + \frac{.458}{r_s} + .0155\right) - 3 \sum_k \langle k|w|k\rangle\right]$$

$$+ N\delta\left[\sum'_{q_o} \frac{\partial F_b}{\partial q} q - \frac{\alpha Z^{*2}}{2r_s}\right] - \frac{1}{3}\sum'_{q_o} \frac{\partial F_b}{\partial(k_F/q)}\left(\frac{k_F}{q}\right) \tag{16}$$

One can see that E_V is equivalent to the $P\Delta V$ term in the formation energy of previous calculations. Here we are able to derive an explicit expression for this term since we know the exact functional form of the volume-dependent energy. We found that E_V compensates a large portion of E_S, therefore, it is an important term in the vacancy-formation energy and should be carefully calculated.

3 DETERMINATION OF THE PSEUDOPOTENTIALS

For the pseudopotential, we have taken the simplified Heine-Abarenkov form for the bare ion potential,[14] i.e.

$$V^{ion}(r) = -V_0 \text{ for } r < R_M$$
$$= -Z^*/r \text{ for } r > R_M . \tag{17}$$

We use the following form for the dielectric screening function

$$\epsilon(q) = 1 - \frac{8\pi}{\Omega_0 q^2} [1-G(q)]\chi(q) \tag{18}$$

where

$$\chi(q) = -\frac{m^*}{m} \left(\frac{3Z}{4E_F}\right) \left(\frac{1}{2} + \frac{4k_F^2 - q^2}{8k_F q} \ln\left|\frac{2k_F + q}{2k_F - q}\right|\right),$$

m^* is the band-structure effective mass. The factor $1-G(q)$ is the approximate correction for the exchange and correlation effects. For alkali metals, we used the Hubbard-Sham form for $G(q)$[11]:

$$G(q) = \frac{1}{2(1+\alpha\eta^2)} \qquad \eta \equiv \frac{q}{k_F} \tag{19}$$

For aluminum, we used the self-consistent correction function suggested by Shaw[12]:

$$G(q) = 1 - \frac{2\alpha}{\eta} D\left(\frac{\eta}{2\alpha}\right) \tag{20}$$

where $D(x)$ is the Darwin function.

For local pseudopotentials used, the energy-wave number characteristic can be written as

$$F_b(q) = [W(q)]^2 \chi(q)/\epsilon(q) , \tag{21}$$

where $W(q)$ is the Fourier transform of the bare ion potential.

The parameters V_O, R_M and α are determined according to the measured elastic constants.[13] In addition, we have imposed the equilibrium condition at Ω_O, i.e., $\partial U_T/\partial \Omega |_{\Omega_O} = 0$. For aluminum, this condition was used to determine the Hartree energy. For alkali metals, this step was not necessary and the Hartree energy can be determined directly from V_O and R_M[8]. The phonon dispersion curves calculated for the alkalis and aluminum according to the parameters determined, show good agreement with the neutron measurements. The results for aluminum are shown in Fig. 3. Table I compares the calculated atomic properties of aluminum with the experimental values.

4 RESULTS AND DISCUSSIONS

The results of the defect calculation for alkali metals are summarized in Table II. There is good agreement between the calculated and measured E_f and Ω_f. Our results clearly show the importance of the volume-dependent lattice energy in calculating E_f and Ω_f. For sodium, the

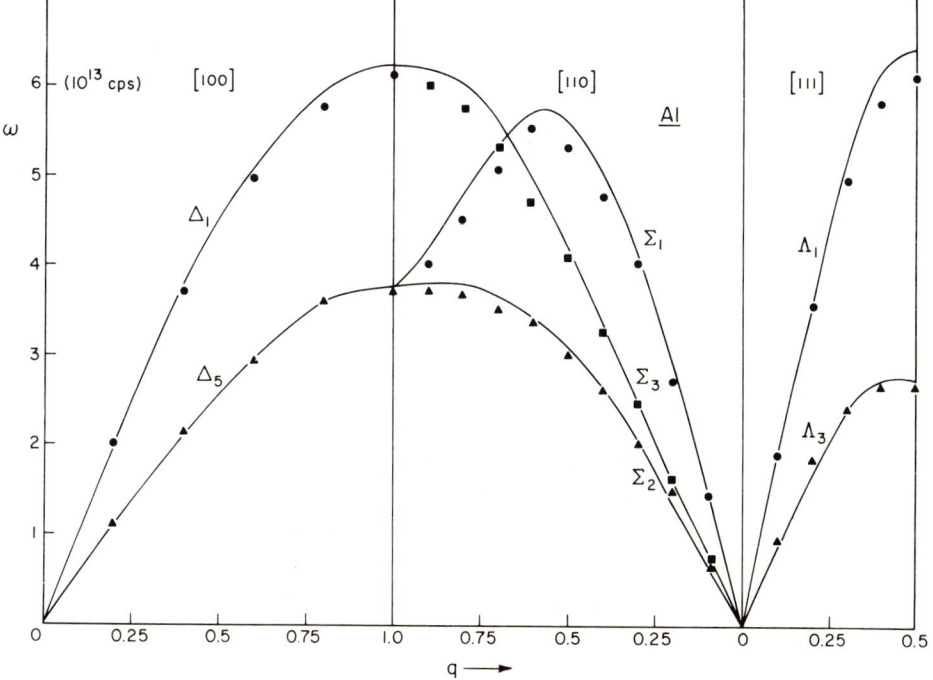

*Fig. 3. Comparison of the calculated and the measured phonon dispersion curves of aluminum. Data taken from Stedman, R. and Nilsson, G., Phys. Rev. **145**, 492 (1966). The units for q and b, b$\sqrt{3}$ along [100], [110], and [111], respectively.*

Table I.

Calculated Atomic Properties of Aluminum

Pseudopotential Parameters	Lattice Parameter
$V_o = 0.564$ Ryd	$a = 7.62\ a_o$
$R_M = 1.14\ a_o$	
$\alpha = 0.84$	Effective Charge
$m^*/m = 0.93$	$Z^* = 3.21$

	Calc.	Exp.
Lattice energy/electron	1.57	1.41
$C_{11}(10^{11}\ \text{dyne/cm}^2)$	11.02	11.43[a]
C_{44}	3.41	3.16
$\frac{1}{2}(C_{11}-C_{12})$	2.00	2.62

Interatomic Force Constants (Unit: dyne/cm)

	$K_t\ (\frac{1}{r}\frac{d\phi}{dr})$		$K_r\ (\frac{d^2\phi}{dr^2})$	
Atom Position	Calc.	Exp.[b]	Calc.	Exp.
(110)	−1228	−1616	22836	21232
(200)	−139	−515	2613	2494
(211)	57	−75	−1225	−621
(220)	−43	465	295	−411
(310)	23	95	141	565
(222)	−1	137	−344	15
(321)	−11	−77	147	−33
(400)	5	−63	121	−756

(a) Ho, P. S. and Ruoff, A. L., J. Appl. Phys. **40**, 3151 (1969).
(b) Gilat, G. and Nicklow, R. M., Phys. Rev. **143**, 487 (1966).

structural energy E_S required for creating the vacancy is 2.67 eV, a major part of which, 2.17 eV, is gained back due to the change in volume-dependent energy during uniform relaxation of the lattice. The volume-dependent energy is found to be equally important in determining Ω_f. We can conclude that in calculating characteristic energy for point defects, if there is an associated defect volume, one should extend the calculation to include the defect volume also. The defect volume is needed for determining the contributions from volume-dependent energies. This may be particularly important for calculating the formation energy of a divacancy or other types of vacancy cluster.

Table II. Results of Vacancy Calculation for Alkali Metals

	Li	Na	K	Rb	Cs
E_s(eV)	2.93	2.59	2.25	2.07	1.93
E_v	-2.56	-2.17	-1.84	-1.68	1.56
E_R	-0.047	-0.055	-0.06	-0.055	-0.05
E_f	0.032	0.36	0.35	0.33	0.32
Ω_f	0.53	0.54	0.53	0.52	0.52
E_f(exp.)	0.34 ± 0.04 [a]	0.42 ± 0.03 [b]			
	0.40 [c]	0.39	0.39		
Ω_f(exp.)	0.28 [d]	0.41			

(a) Feder, R., Phys. Rev. **2B**, 828 (1970).
(b) Feder, R. and Charbnau, H., Phys. Rev. **149**, 464 (1966).
(c) MacDonald, D.K.C., J. Chem. Phys. **21**, 177 (1953).
(d) Hultsch, R. A. and Barnes, R. G., Phys. Rev. **125**, 1832 (1962).

The vacancy relaxation in alkali metals was found to be quite large and highly anisotropic. The displacements are inward and as large as 4 percent of the radial distance for the nearest neighbors and outward, 3 percent for the second neighbors. However, the relaxation energy is only about 0.06 eV, which is about 15 percent of the formation energy. This indicates that, for alkali metals, even though the relaxation around the vacancy is large, it is not the determining factor for calculating the formation energy and volume of a vacancy.

The detailed vacancy configuration for sodium has been tabulated in Ref. 8.*

The authors are in the process of calculating E_f and Ω_f for aluminum. The atomic relaxation around the vacancy is considerably smaller than for the alkalis, only about inward 2 percent of the radial distance for the nearest neighbors and inward 1 percent for the second neighbors. The relaxation energy is about 0.10 eV. Fig. 4 compares the vacancy relaxed configurations of aluminum and sodium. Table III shows part of the atomic displacements near a vacancy in aluminum. Reliable values for E_f and Ω_f have not yet been obtained. The difficulty is partly due to the larger valence number of aluminum ($Z=3$) and partly due to the larger contribution of the band-structure energy in comparison to the alkali metals. The valence difference implies that the calculations for E_S and E_V have to be order-of-magnitude more accurate for aluminum than sodium

---

*There one can also find comparisons in vacancy configuration, formation energy, and formation volume between the present work and other theoretical calculations on sodium. In addition, the results based on the optimized model potentials determined by Shaw and Pynn[15] were presented.

Table III. Results of the Vacancy Relaxed Configuration for Aluminum
(Displacements are expressed in percent of lattice parameter.)

Atom Position (Unit = a/2)	ξ^ℓ		
(110)	−0.9	−0.9	0
(200)	−0.13	0	0
(211)	−0.27	−0.18	−0.18
(220)	−0.38	−0.38	0
(310)	0.02	−0.01	0
(222)	−0.04	−0.04	−0.04
(321)	−0.11	−0.06	−0.04
(400)	−0.01	0	0
E_R (eV)	0.10		

since both these energies are roughly proportional to Z^2. The large contribution from the band-structure energy requires more accurate pseudopotential for aluminum also. Hopefully, the results can be reported in the near future.

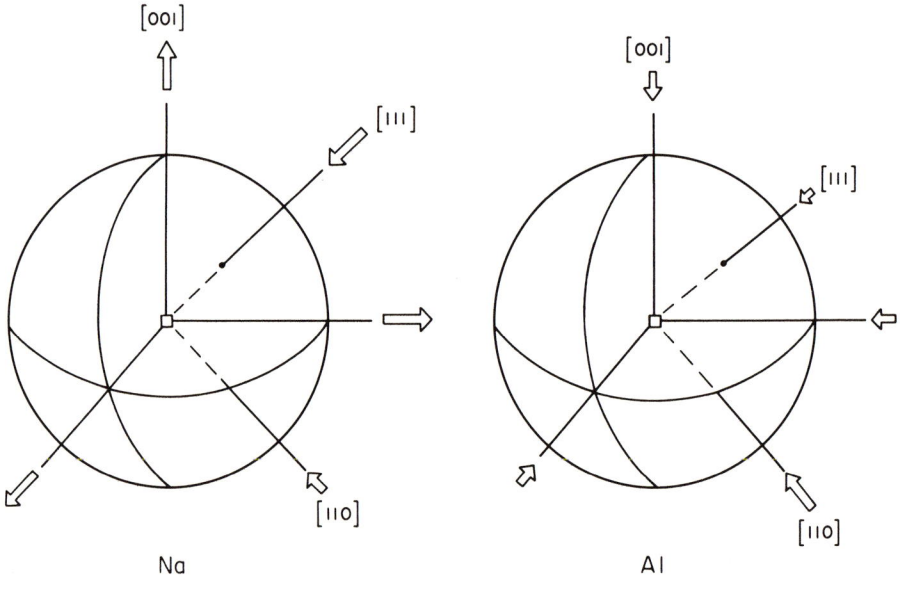

Fig. 4. Schematic comparison of the anisotropic relaxations around a vacancy in sodium and aluminum. The arrows indicate the direction and rough magnitude of the atomic displacement.

ACKNOWLEDGMENT

This work was supported by the United States Atomic Energy Commission.

REFERENCES

1. See, for example, Johnson, R. A. and Brown, E., Phys. Rev. **127**, 446 (1962) and Johnson, R. A., Phys. Rev. **145**, 423 (1966).
2. Lie, K. C. and Koehler, J. S., Adv. in Phys. **20**, 421 (1969).
3. Harrison, W. A., *Pseudopotentials in the Theory of Metals,* Chp. 2, W. A. Benjamin, Inc., New York, 1966.
4. Pines, D. and Nozieres, P., *The Theory of Quantum Liquids,* Vol. 1, Chp. 5, W. A. Benjamin, Inc., New York, 1966.
5. Fuchs, K., Proc. Roy. Soc. (London) **A151**, 585 (1935).
6. Shaw, R. W., J. Phys. C. (Solid St. Phys.) **2**, 2335 (1969).
7. Section 6.4 in Ref. 3.
8. Ho, P. S., Phys. Rev. **B3**, 4035 (1971).
9. Kanzaki, H., J. Phys. Chem. Solids **2**, 24 (1957).
10. Flinn, P. A. and Maradudin, A. A., Annals of Phys. **18**, 81 (1962).
11. Hubbard, J., Proc. Roy. Soc. (London) **A193**, 299 (1958); Sham, L. J., Proc. Roy. Soc. (London) **A283**, 33 (1965).
12. Shaw, R. W., J. Phys. C. (Solid St. Phys.) **3**, 1140 (1970).
13. Ho, P. S., Phys. Rev. **169**, 523 (1968).
14. Heine, V. and Weaire, D., *Pseudopotential Theory of Cohesion and Structure* in Solid State Phys. Vol. 24, Academic Press, New York, 1970.
15. Shaw, R. W. and R. Pynn, J. Phys. C. (Solid St. Phys.) **2**, 2071 (1969).

DISCUSSION on papers by P. S. Ho

TORRENS: How does your real-space pair interaction for aluminum compare with previously proposed potentials? In particular, how significant are the long-range oscillations?

HO: The interatomic potential obtained here for aluminum is quite different from previous potentials such as Harrison's[1] and Ashcroft's[2]. I suspect that the difference comes mainly from the different corrections used for correlation and exchange effects in the dielectric function. The long-range oscillations are about the same as those of Ashcroft's potential. I don't know how significant those oscillations are with regard to the displacement field around the defect since my calculation is done in k space. I would expect, however, that they are more important than in the sodium case since the aluminum interatomic potential is of longer range as shown by the force constants determined from the phonon-dispersion curves.

BEELER: What was the recipe used to represent reposition of the extracted atom to the crystal surface in the vacancy-formation calculations?

HO: This is accounted for in my calculation by subtracting the energy difference between a surface atom and an interior atom. For example, the treatment of the band-structure energy for that atom is shown in Eq. 9. The last term is exactly such an energy correction. A similar correction has to be made for the electrostatic energy. This kind of energy correction also affects the formation volume. That is to say, in calculating the defect volume, the change in the energy correction for the surface atom has also to be taken into account in the relaxation process.

VITEK: Is the division of the formation energy into three independent parts, $E_f = E_S + E_R + E_V$, justified? I think that these parts are in fact interdependent, e.g., E_R must change after the volume relaxation.

HO: Yes, what you said is right. In fact, E_R should be varied with all other energies in the relaxation process. It has a certain contribution in determining the formation volume.

SEEGER: With regard to your failure to obtain sensible results for vacancies in aluminum, I would like to make the following point. If you form a vacancy in a trivalent metal you introduce a perturbation that is essentially equivalent to removing three conduction electrons. If the resulting perturbation potential is localized within, say, the Wigner-Seitz cell, it is much too strong to be handled by second-order perturbation theory. If not, it is so much spread out that it becomes very unlikely that the concept of superposition of pseudoatoms is applicable. I feel that this is a basic difficulty which cannot be circumvented within the present approximate (second-order perturbation theory) framework of pseudopotential theory. By contrast, the situation in sodium and similar cases (strong relaxation of the neighboring atoms into the vacant sites of alkali metals) is such that one may obtain reasonable numerical results.

HO: That may well be the case. However, in the pseudopotential formulation, the cancellation of the electrostatic energy and the band-structure energy is a more natural way to look at the problem. I will not elaborate on this point since Taylor makes this point clear in his comment. I think that the relaxation energy and configuration of aluminum as reported here are reliable, since they are determined essentially by the measured phonon dispersion curves.

DEFECT PROPERTIES

TAYLOR: I think the problems in aluminum are due to the fact that you have first of all a direct ion-ion Coulomb repulsive force and, cancelling against that, the screening of this force by the conduction electrons. If you compare the phonon frequencies you would obtain from the direct ion-ion interaction with the experimental ones, you see that in sodium the cancellation is about 30 percent, in aluminum the cancellation is about 80 percent, and in lead, you really get hit on the head, where it is about 95 percent. Hence, you need to calculate the screening in aluminum to a much higher degree of accuracy than that in sodium to obtain the net forces in the two materials to the same precision.

DUESBERY: I would like to comment on the relative magnitudes that you calculate for the structural, relaxation, and volume energies. We made a similar calculation just for the case of sodium, and would agree that the structural and volume terms make the largest contributions to the formation energy. However, we found the relaxation energy to be about 1/3 of the structural energy. I wonder if our different assessment of the importance of the relaxation energy is due in part to the method of calculation. We used a real-space calculation, using a two-body effective ion-ion potential, and could therefore take account of changes in the force constants during relaxation. In the method of lattice statics, the force constants are assumed to remain unchanged, and this may lead to an underestimate of the relaxation around the defect.

HO: I think that the difference is probably not a result of the use of different methods but different energy separations in the calculation. I. Torrens and M. Gerl[3], using a nonlattice statics method, obtained good agreement with my results. It should be pointed out that the effective interatomic potential does not include all the energy terms in the structural energy, but the latter is completely accounted for in the lattice statics calculation. This may be one explanation. Another explanation is that we have somewhat different form factors for the pseudopotential.

HARRISON: What you have said seems to violate the fact that the activation energy for formation of a vacancy, for a system in equilibrium is the same at constant volume and at constant (vanishing) pressure. The difference goes to zero in a large system. There are certainly terms in the energy which do not go to zero with volume change, but they must cancel identically. If you obtained a difference in formation energy you must have left something out. Isn't that correct?

HO: No. I think that Professor Hirth's comment in the agenda discussion on point defects on different schemes of relaxation in forming a vacancy will answer your question.

338 P. S. HO

JOHNSON: Does your pseudopotential yield a negative divacancy binding energy? It seems it would, because the energy is positive at nearest neighbor distance.

HO: I have not calculated the divacancy binding energy. It may be negative as you suggested.

ASHCROFT: When the atom is "placed on the surface" (if it ever gets there) there will be a change in surface free energy. Where does it appear in your theory and how big is the term?

HO: When I remove an "atom" from the crystal interior to the surface, I really remove a pseudoatom, i.e., the ion plus its self-consistent screening charge. It is not a metallic ion in the conventional sense. If a charge redistribution is induced due to the additional surface atom, it will be different (probably less) in my calculation. I have not included this correlation in my results. Nevertheless, since the alkali results appear to be reasonable, I would think it is probably small for that case. On the other hand, this correlation may be considerably higher for aluminum due to its large valence charge and it may be partially for this reason that the results are not good.

ASHCROFT: Can you explain why an effective mass $m^*_m = 0.93$ is used in your aluminum calculation. What is its physical origin?

HO: It is the "band-structure" effective mass as defined in Eq. 18.

EHRLICH: The energy change in removing an atom from inside a lattice and placing it at a kink site is equal to the heat of vaporization. It would seem that your quantity E_S should be related to this experimentally determined quantity. Do you find such a correlation?

HO: Yes, it is related to the heat of vaporization but it also depends on how one divides the total lattice energy into structure and volume dependent energies. In fact, Professor Hirth showed in his comment in the Agenda Discussion that the E_S should be about twice the heat of vaporization, and the magnitude really depends on the relaxation process used. (For detail, see Hirth's comment in the Agenda Discussion.)

REFERENCES

1. See Ref. 3.
2. Ashcroft, N. W. and Langreth, D. C. Phys. Rev. **159**, 500 (1967).
3. Torrens, I. M. and Gerl, M. Phys. Rev. **187**, 912 (1969).

IMPURITY ATOM EFFECTS IN METALLIC CRYSTALS

J. R. Beeler, Jr.

The Ohio State University
Columbus, Ohio 43210
Present address:
North Carolina State University
Raleigh, North Carolina 27607

1 INTRODUCTION

This paper is concerned with compound defects in metals which consist of impurity atoms and crystal lattice defects. These compound defects are often referred to as complexes.[1] First, the paper deals with complexes of metalloid interstitial impurity atoms with monovacancies, divacancies, split-interstitials, and free surfaces. Initial results on how a metalloid interstitial impurity atom affects the production of a Frenkel pair are discussed briefly. Second, the paper takes up the migration of large and small substitutional impurity atoms for which the principal interest is the role of impurity atoms as traps for crystal lattice defects in displacement spike annealing.

The objective of the work described was to define the most stable configuration for each complex of interest. One bcc metal, one hcp metal and two fcc metals were treated. Johnson's α-iron potential[2] was used as the metal-metal interaction function for each of the four model metals which are termed "α-iron", "γ-iron", "nickel", and "cobalt". These model metals are defined by Table I. The differences among the four model metals arise from the assignment of different lattice constants and crystal lattice structures. The rationale for using a common potential function for several different model metals, in this manner, has been discussed by Johnson[3] for the cubic metals concerned here. The hcp model was an arbitrary construction. The interaction function for carbon in α-iron of Johnson et al.[4] was adopted to describe the metalloid-metal interaction in each case. For convenience we will refer to the metalloid atom as being "carbon".

Table I. The Half-Lattice Constant in Angstroms and the Crystal Structure for the Model Metals

Metal	hlc, A	a, A	Structure
α-Iron	1.43	2.86	bcc
γ-Iron	1.823	3.646	fcc
Nickel	1.76	3.52	fcc
Cobalt	—	2.50[(a)]	hcp

(a) Interatomic distance.

On this basis, a systematic study of vacancy-carbon and split-interstitial-carbon complexes in four model metals was performed. Free surface-carbon complexes of α-iron were also studied.

Initial results from a more restricted study of substitutional impurity vacancy complexes are also reported. In this work, the effects of substitutional impurity size on the complex configuration and its migration process were investigated.

2 CONFIGURATION AND BINDING ENERGIES

Configuration energies for all the point defects considered are listed in Table II. All configuration energies quoted were computed using the atomic interaction functions and host metal interatomic distances used in the present study.

A summary of the computed binding energies of the complexes formed by these defects with one to four carbon atoms is given in Table III.

Table II. Configuration Energies of Point Defects

Point Defect	Energy, eV			
	α-Iron	γ-Iron	Nickel	Cobalt
V_1	2.91	3.00	2.76	2.80
$V_2(1)$	5.67	5.74	5.07	5.19
$V_2(2)$	5.62	6.00	5.48	—
CAR(Oct)	−1.32	−1.94	−1.68	−1.69
CAR(Tet)	−0.46	−0.31	0.13	—
Interstit.	3.12	2.56	3.93	3.19

Table III. Binding Energies of Divacancies, Vacancy-Carbon Complexes, and Split Interstitial-Carbon Complexes

(Energies given in eV.)

Defect	α-Iron	γ-Iron	Nickel	Cobalt
$V_2(1)$	0.15	0.26	0.45	0.41
$V_2(2)$	0.20	0	0.04	—
V_1-C	0.42	0	0.31	0
V_2-C	0.62	—	0.86	—
V_2-2C	1.03	—	0	—
I-C	0.56	0.36	0.46	—
I-2C	0.76	0.70	0.92	—

3 CARBON COMPLEXES IN NICKEL

3.1 Atomic Interaction Functions Used

The metal-metal and metalloid-metal interaction functions used in this study are plotted in Fig. 1.

3.2 Atom Site Nomenclature and Geometry for Nickel

The discussion of the configuration states for the carbon-vacancy complexes in nickel and in γ-iron centers mainly on the different ways in which the first neighbors of the vacancy can interact with a carbon atom. The lattice site numbers for the vacant site and its neighbors are listed in Table IV for the computational cells of the two computer programs used

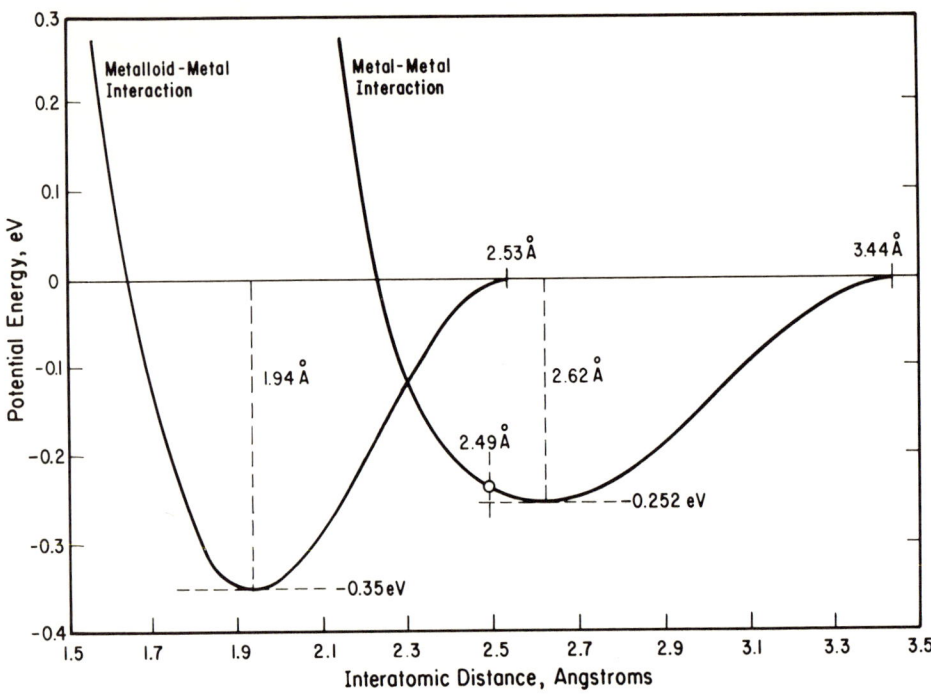

Fig. 1. Interaction functions used. The metal-metal interaction is Johnson's α-iron potential. The metalloid-metal interaction is his carbon-iron potential.

Table IV. Site Numbers for the First-Neighbor Sites of the Vacancy in the Carbon-Vacancy Complex, and One Particular Second-Neighbor Site

(The vacancy is located at the origin of the coordinate system and the unit of length is the half lattice constant.)

Atom Site Number		Site Coordinates		
DYNAMIC	DEFECT	x	y	z
1076	267	0	-1	-1
1220	327	-1	-1	0
1221	328	1	-1	0
1365	388	0	-1	1
1084	272	-1	0	-1
1085	273	1	0	-1
1373	393	-1	0	1
1374	394	1	0	1
1093	278	0	1	-1
1237	338	-1	1	0
1238	339	1	1	0
1382	399	0	1	1
1246	344	0	2	0
1229	333	0	0	0

in the study. The coordinates of the neighboring sites are in units of half-lattice constants (hlc) and the vacant site is taken to be the origin. The two programs used were the DYNAMIC Program and the DEFECT Program. The DYNAMIC Program is a dynamical-method program based on the ideas of Gibson et al.[5]. It can be used to simulate the dynamical response of an array of atoms to a perturbation in either the structure or the composition of a crystal. It can also be used to generate static mechanical equilibrium defect configuration states. The DEFECT Program is based on the variational method ideas of Johnson and Brown.[6] It generates static mechanical equilibrium defect states associated with a perturbation in either the structure or the composition of a crystal. The use of both the dynamical and variational methods in a given defect configuration problem was found to be highly advantageous. As usually programmed, each method induces a different latent symmetry of *initial* atom-relaxation response to a given perturbation of the crystal. In the study of complexes, different initial responses to the same perturbation often lead to different "final" states. In such instances, experience rapidly nurtures an appreciation for the usefulness of a dual-method approach.

Fig. 2 describes the geometrical arrangement of the sites listed in Table IV in terms of three square rings of four atoms each, lying in the $y = -1$, $y = 0$ and $y = 1$ planes. Most of the discussion on strain fields concerns planar atom relaxations in these ring planes and the bodily translation of these rings along the y-axis in response to changes in the carbon-atom position on the y-axis.

3.3 Monovacancy-Carbon Complexes

The lowest energy monovacancy-carbon complex configuration found in the computer experiments for nickel calculations is that shown in Fig. 3. The open square represents the vacancy and the filled circle the carbon atom. The strain field and characteristic energies for this configuration are summarized in Table V. The relaxation of the atoms in the $y = 0$ ring is a radial relaxation toward the vacancy and is restricted to the $y = 0$ plane.

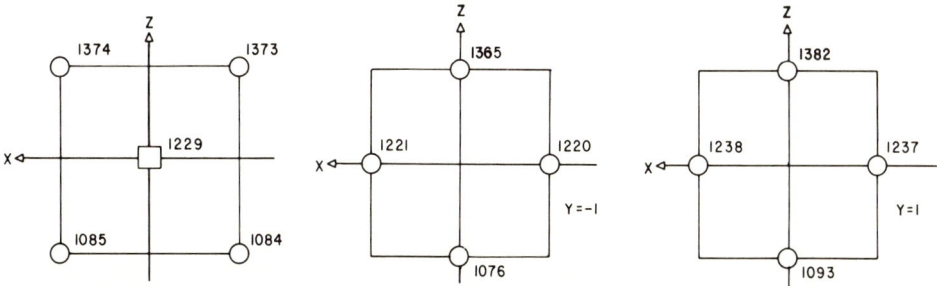

Fig. 2. First-neighbor sites of a vacancy at site 1229 in the DYNAMIC Program fcc computational cell. (The coordinates of these sites relative to site 1229 are given in Table IV.)

344 J. R. BEELER

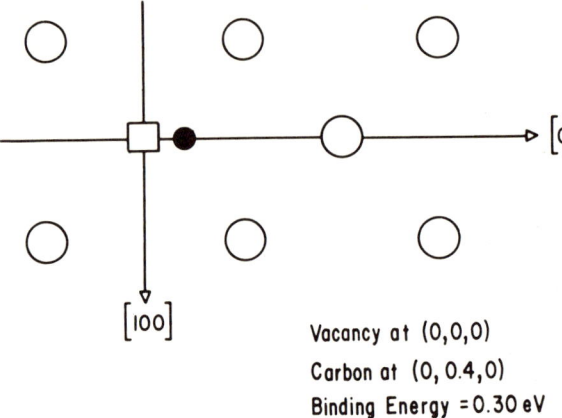

Fig. 3. The most stable vacancy-carbon atom complex configuration in "nickel". The open square designates the vacancy position. The filled circle designates the carbon atom position. Open circles denote normal atom positions in the fcc crystal lattice.

Vacancy at (0,0,0)
Carbon at (0, 0.4, 0)
Binding Energy = 0.30 eV

Table V. Carbon-Vacancy Complex; Lowest Energy State
(Data from DYNAMIC Run RAN 7B.[a])

| | Strain Field for First-Neighbor Atoms | | | | | |
| | Atom Coordinates | | | Strain Component | | |
Site	x	y	z	Δx	Δy	Δz
1373	−0.784	0.0	0.824	0.216	0.0	−0.176
1084	−0.784	0.0	−0.824	0.216	0.0	0.176
1374	0.784	0.0	0.824	−0.216	0.0	−0.176
1085	0.784	0.0	−0.824	−0.216	0.0	0.176
1237	−0.943	1.078	0.035	0.057	0.078	0.035
1382	0.038	1.078	0.962	0.038	0.078	−0.038
1093	−0.038	1.078	0.962	−0.038	0.078	0.038
1238	0.943	1.078	−0.035	−0.057	0.078	−0.035
Carbon	0.0	0.422	0.0	—	—	—

Characteristic Energies

Configuration Energy: 0.783 eV   Carbon Atom Energy: −2.350 eV

Carbon-Metal Atom Interactions:

Atoms	Distance, A	Interaction, eV
1084, 1085, 1373, 1374	2.134	−0.265
1237, 1238	2.023	−0.33
1093, 1382	2.050	−0.32

(a) In DYNAMIC Run RAN 7B the carbon atom relaxed to (0, 0, 0.422) with a vacancy at (0,0,0). The above strain field is a transposition of the original data to correspond to the strain field for a carbon atom which had relaxed to (0, 0.422, 0).

The y = 1 ring of atoms is displaced 0.078 hlc in the positive y-direction. At the same time, this ring is rotated counterclockwise (when viewed looking along the negative y-direction), and also distorted into an isosceles parallelogram. This transformation is illustrated in Fig. 4.

The carbon-metal interaction has a much shorter range than the metal-metal interaction, in this study, and is significantly stronger than the metal-metal interaction at the maximum bonding atom separation distance. The minimum in the carbon-metal interaction energy occurs at the separation distance r = 194A. The minimum in the metal-metal interaction energy occurs at r = 2.62A. The cutoff for the carbon-metal interaction occurs at 2.53A, just 0.04A beyond the first-neighbor distance (2.94A) for the nickel host metal. In this range of interaction the carbon-metal bond, being about 40 percent stronger than the metal-metal bond, clearly dominates the local configurational energy. This allows the carbon atom to induce the formation of a tough "molecule" consisting of a few metal atoms arranged about the carbon atom and tightly bound to it. This molecule can assume several strongly bound metastable configurations each of which can produce large localized strains. The differences in the range of the two interaction functions concerned, in the location of their minima and in the strength of these minima, combine to produce a tetragonal strain field about the monovacancy-carbon complex in nickel.

In addition to the lowest energy configuration, there are three important metastable monovacancy-carbon complexes in nickel. Fig. 5 describes the three carbon locations at y = 0.42, y = 0.49 and y = 0.79 relative to

Fig. 4. Rotation and distortion of the y = 1 ring of atoms in the carbon-vacancy complex strain field in nickel.

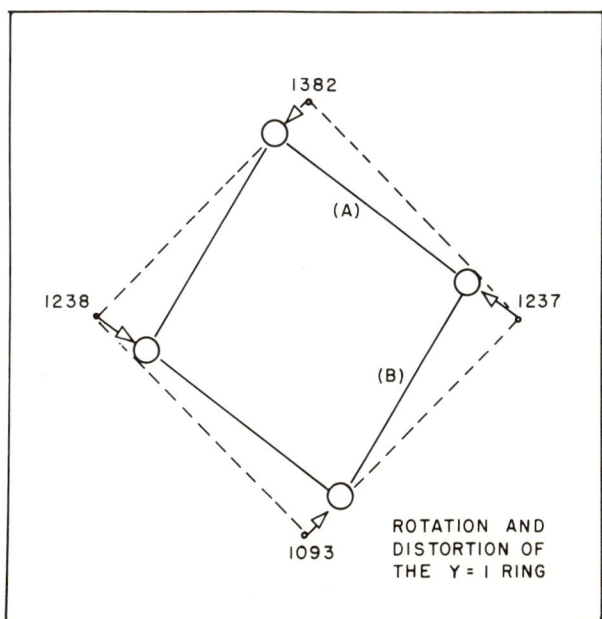

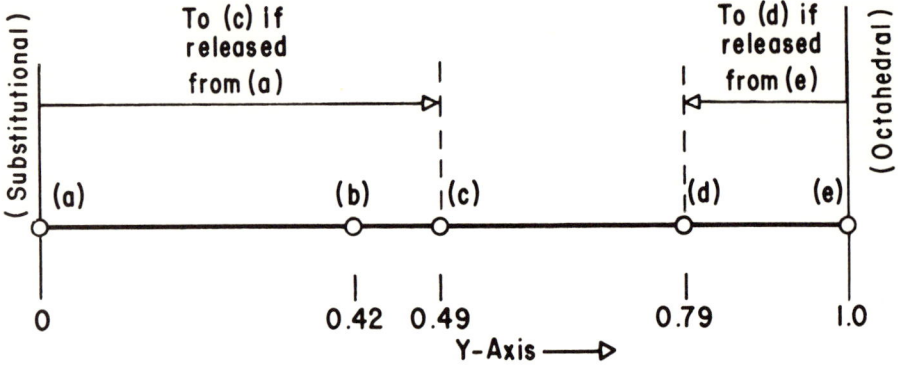

Fig. 5. Reference States for the carbon-vacancy complex in a fcc metal. (a) Substitutional carbon limiting case. Carbon held in vacancy. Carbon energy is -1.35 eV; (b) Lowest energy state. Carbon interacts with eight atoms. The carbon energy is -2.35 eV; (c) Metastable state y = 0.49. Carbon interacts only with the four atoms 1093, 1237, 1238, and 1382 in the y = 1 ring. The carbon-metal atom distance is 1.94 A and the carbon-metal atom pair interaction at this distance is -0.35 eV. Carbon energy is -1.40 eV. Carbon goes to this state if released from substitutional limiting; (d) Metastable state y = 0.79. Carbon interacts with the four atoms in the y = 1 ring, cited in (c) above, and with atom 1246 at (0,2,0). Carbon energy is -1.64 eV. Carbon goes to this state if released from octahedral limiting; (e) Octahedral limiting. Carbon held in an octahedral site adjacent to a vacancy. Carbon vacancy is -1.54 eV.

the vacancy, for these complexes. First, a second low-energy configuration associated with carbon at y = 0.42 exhibits a symmetrical strain field in which the shape of the y = 1 ring is preserved. This symmetrical configuration is described by Table VI and Fig. 6. As explained in Fig. 5, two higher-energy metastable states are associated with a carbon atom at y = 0.49 and at y = 0.79.

Table VI. Low-Energy State With Carbon at (0, 0.4, 0). Strain Field for First-Neighbor Atoms (From DEFECT Run R036.)

Site	Atom Coordinates			Strain Component		
	x	y	z	Δx	Δy	Δz
1084	-0.818	-0.008	-0.819	0.182	-0.008	0.181
1085	0.818	-0.008	-0.821	-0.182	-0.008	0.179
1373	-0.821	-0.008	0.818	0.179	-0.008	-0.182
1374	0.820	-0.008	0.820	-0.180	-0.008	-0.180
1093	0.0	1.057	-0.957	0.0	0.057	0.043
1237	-0.957	1.057	0.0	0.043	0.057	0.0
1238	0.957	1.057	0.0	-0.043	0.057	0.0
1382	0.0	1.057	0.957	0.0	0.057	-0.043
Carbon	0.0	0.404	0.0	—	—	—

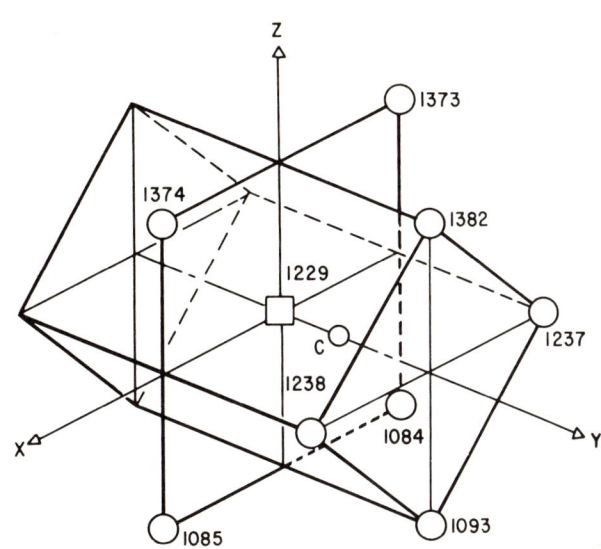

Fig. 6. Low Energy State y = 0.42. The carbon-metal atom distance for atoms 1084, 1085, 1373, and 1374 is 2.159A. At this distance the carbon-metal atom pair interaction is -0.241 eV. The carbon-metal atom distance for atoms 1093, 1237, 1238, and 1382 is 2.039A. At this distance the carbon-metal atom pair interaction is -0.323 eV. The carbon energy is -2.243 eV. DEFECT Program Run R036. A slightly lower energy state was obtained in DYNAMIC Program Run RAN7B using a randomized initial strain field.

The monovacancy-carbon complex configuration energy was found to be a double-valued function of the carbon atom position on the y-axis as illustrated in Fig. 7. Different branches evolve depending upon the initial conditions from which the atom-relaxation process developed. Branch 1 results when the initial state corresponds to "substitutional" carbon, and Branch 2 when it corresponds to "octahedral" carbon.

The "octahedral" carbon initial conditions would be appropriate, for example, to an irradiation experiment in which a metal atom situated adjacent to an octahedral carbon atom was displaced by the primary radiation. The "substitutional" carbon initial conditions would be appropriate when an energetic carbon atom, displaced during irradiation, passed through a vacancy near the end of its trajectory. Dynamical simulations of such effects in an irradiation computer experiment are described in Section 3.6.

3.4 Divacancy-Carbon Complexes

Usually, two vacancies in a metal exhibit a binding energy for certain separation-distance vectors with magnitude less than about three lattice constants.[7,8] The two divacancy configurations of interest in connection with divacancy-carbon complexes in cubic metals are the first-neighbor divacancy, $V_2(1)$, and the second-neighbor divacancy, $V_2(2)$. Two bound configurations were found for $V_2(1)$-C complexes in nickel. Their configurations are shown in Fig. 8. In Case (1), the carbon atom lies to the

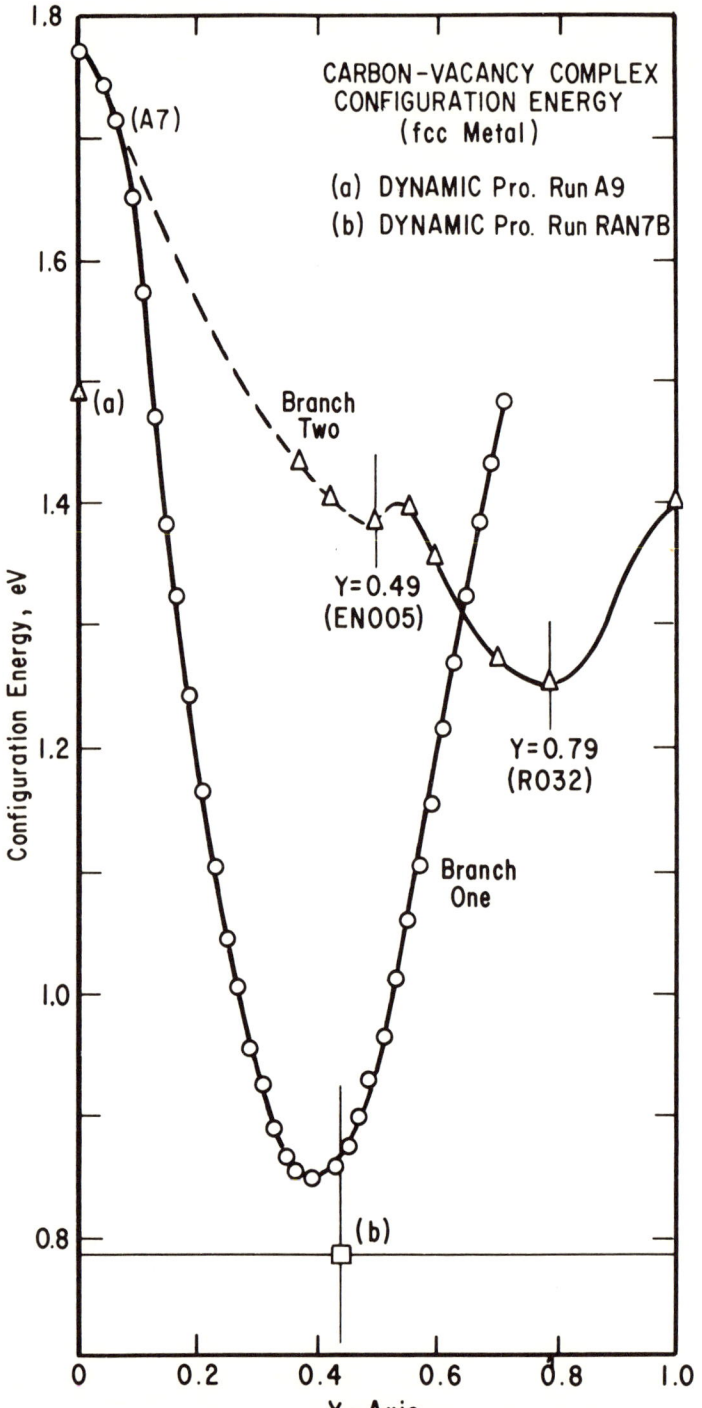

Fig. 7. Two-branched configuration energy. Carbon-vacancy complex in a fcc metal. First branch evolves from substitutional limiting, second branch from octahedral limiting.

IMPURITY ATOM EFFECTS   349

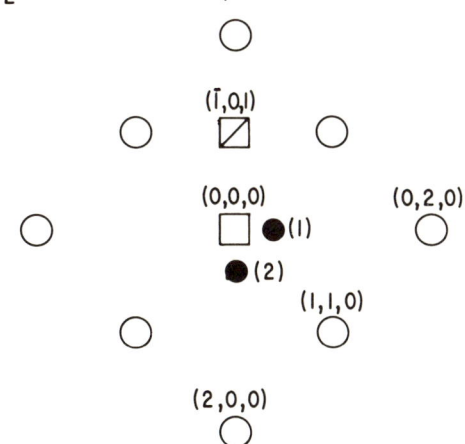

Fig. 8. First-neighbor divacancy-carbon atom complexes in the "nickel" model metal. $V_2(1)$ designates a first-neighbor separation distance divacancy configuration.

(1) Carbon at (0.02, 0.38, -0.02)   Binding Energy = 0.17 eV

(2) Carbon at (0.44, 0, 0.02)   Binding Energy = 0.72 eV

side of the divacancy axis; in the Case (2), the carbon atom is aligned with the divacancy axis. The strong binding energy of the second configuration indicates that interstitial impurity atoms can more strongly trap divacancies than monovacancies. No bound configurations were found for the $V_2(2)$ divacancy and a carbon atom.

3.5 Split-Interstitial-Carbon Complexes

Fig. 9 is a schematic of the strain field of a [010] split-interstitial in nickel. As first shown by Gibson et al.[5] and by Johnson[3,6], this self-interstitial configuration appears to be the lower-energy configuration for fcc metals. The atoms at each site shown in Fig. 9 are displaced at least 1 percent of the interatomic distance from their perfect-lattice position. Atoms marked with a cross are displaced at least 10 percent of an interatomic distance from their perfect-lattice positions. The two cross-hatched central atoms are centered about the origin at (0, -0.6, 0) and (0, 0.6, 0). The greater-than-10-percent-strain-field-site locations are used as a reference in designating the carbon atom position(s) in split-interstitial-carbon complexes. This method is useful because the strain-field shape shown in Fig. 9 is typical for fcc transition metals. Fig. 10 shows the complex configuration for one carbon atom adjacent to the interstitial and two carbon atoms adjacent to the interstitial. The binding energies are 0.46 eV and 0.92 eV, respectively. Configurations with three and four carbon atoms at the interstitial beltline have binding energies of 1.38 eV

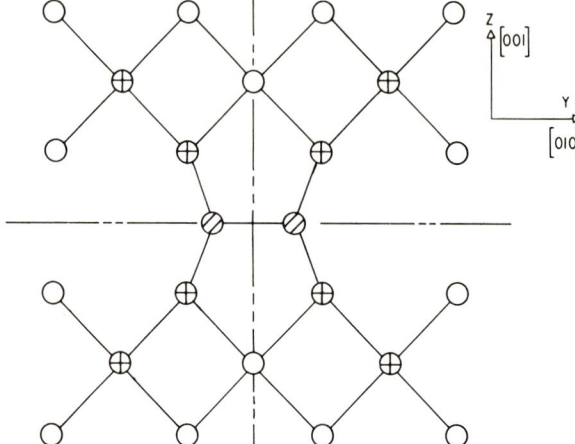

Fig. 9. Atom sites associated with strains of 1 percent or more about the [010] split-interstitial in "nickel". The hatched circles in the center represent the two central atoms of the split-interstitial. Sites denoted by ⊕ are those experiencing 10 percent strain or more. They represent the core of the interstitial strain field; plane x = 0.

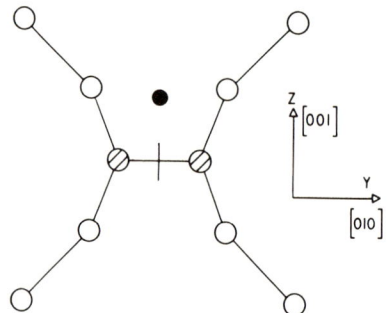

Fig. 10. A carbon atom at (0,0,0.91) above the center (0,0,0) of a [010] split-interstitial in "nickel" form the most stable complex with this interaction for one carbon atom. The atom sites shown are the core sites in the strain field of the interstitial.

and 1.84 eV, respectively. This suggests that the [010] split interstitial can collect up to four carbon atoms in a fcc metal.

Summarizing the case for carbon complexes with a crystal lattice point defect in "nickel", we find that the most strongly bound complex containing a single carbon atom is the $V_2(1)$-C complex shown in Fig. 8. The binding energy for this complex (0.86 eV) is nearly three times that for the monovacancy-carbon complex and nearly twice that for the split-interstitial-carbon complex.

3.6 Effects of Carbon on Frenkel Pair Production

The strong interactions between carbon and metal atoms in the carbon-vacancy complexes suggests that the displacement threshold energy should be affected by either an octahedral carbon atom or a carbon-vacancy complex. It also suggests that the Frenkel pair stability criteria (recombination-region shape and size) could be significantly affected by the presence of carbon.

Exploratory calculations were made using the DYNAMIC Program in the fully dynamic mode to simulate Frenkel pair production in the neighborhood of an octahedral carbon atom. These simulations show in a rewarding manner the usefulness of the detailed work done on the initial-state dependence of carbon-vacancy complex metastable states in nickel. In particular, they show that the lowest-energy complex configuration is not produced when a neighboring metal atom is displaced. Rather the high-energy metastable state associated with the carbon atom at y - 0.79 (State d in Fig. 5) is produced.

The Frenkel pair production computer experiment proceeded as follows: A carbon atom was placed at the (1,0,0) octahedral site, with respect to site 1230, as shown in Fig. 11a. The metal atom at site 1230, and all other metal atoms, were initially situated at their relaxed positions in the octahedral carbon defect configuration strain field. The metal atom at 1230 was then ejected in the negative x-direction with an energy of 28.2 eV. A stable Frenkel pair-carbon complex was formed. The parts of the complex were a [100] split-interstitial centered on site 1228 and a metastable carbon-vacancy complex of the (0.79,0,0)-type at site 1230. This run furnishes an example of a metastable carbon-vacancy configuration being formed directly in a fully dynamic damage event simulation

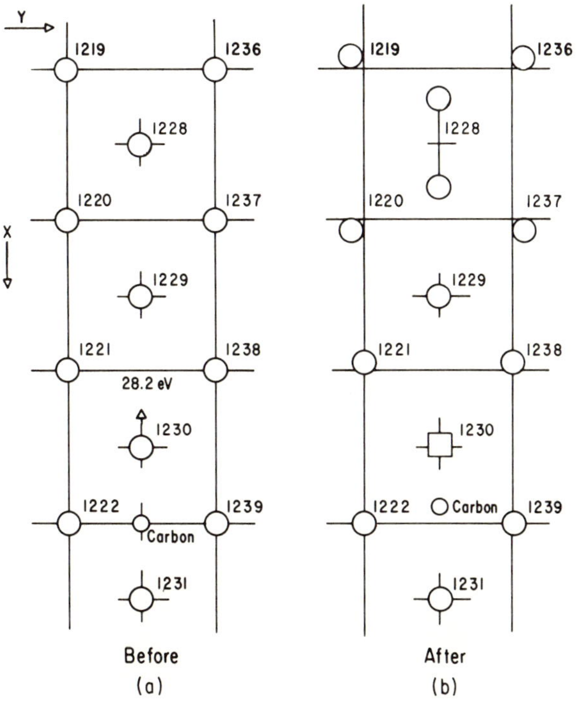

Fig. 11. (a) Initial state at the instant a dynamic event was initiated by shooting atom 1230 out along the negative x-axis with a kinetic energy of 28.2 eV. (b) The result of this dynamic event was a split-interstitial at site 1228 and a carbon-vacancy complex at site 1230. Note that the metastable y = 0.79 complex was formed in this dynamic event rather than the lowest energy complex state y = 0.42. DYNAMIC Program series RAN23 to RAN 30.

rather than the lowest-energy carbon-vacancy configuration. The etiology of this complex is explained in terms of the double-valued configurational energy curve for the carbon-vacancy complex. Briefly, from the standpoint of carbon-atom behavior, the natural beginning state in this damage event process is the octahedral carbon initial state. The carbon atom was restricted to "sampling" only those bonding opportunities afforded by the octahedral initial conditions during the lifetime of the damage production event. At no time during this event did it have the opportunity to establish bonding with atoms in the four-atom first-neighbor ring lying in the $x =$ constant plane containing the vacancy, i.e., the substitutional limiting environment. A prerequisite for the formation of the lowest-energy state, is either a "substitutional" carbon initial state or one quite similar to it.

The viewpoint of a variable complex structure, depending upon the initial state from which its formation proceeded and consideration of the dynamical prerequisite conditions for accessibility to other possible configuration states, should be helpful in devising mechanisms for impurity atom-lattice defect production and annealing behavior. That the metastable state formed did not transform into the lowest-energy state is clearly of interest considering the intense local energetic perturbation to the crystal lattice. This circumstance directly indicates that thermally induced transitions between the metastable states concerned are not possible.

4 CARBON COMPLEXES IN γ-IRON

4.1 Vacancy-Carbon Complexes

No bound configurations were found for either monovacancy or divacancy complexes with carbon in γ-iron. The interatomic distance in γ-iron is 2.58A. This exceeds the 2.53A cutoff distance for the carbon-metal atom interaction function by 0.05A. Hence, as expected, the vacancy and carbon atom strain fields do not interact in the γ-iron and carbon model.

4.2 Split-Interstitial-Carbon Complexes

The geometry of the split-interstitial-carbon complex configuration in γ-iron is similar to that in nickel. Hence, Fig. 10 also describes the γ-iron results. Again, a ring of up to four carbon atoms can be arranged about the interstitial beltline. The binding energy per carbon atom in this sequence of complexes is 0.36 eV.

5 CARBON COMPLEXES IN α-IRON

5.1 Vacancy-Carbon Complexes

Monovacancy-carbon complexes in α-iron were originally treated by Johnson.[9,10,11] His runs for α-iron were repeated in the present study to check technique and were then extended in scope.

Computer experiment runs on the monovacancy-carbon complex in α-iron did not give the complicated behavior observed in the runs on that complex in nickel. It appears that this is due largely to the fact that both first- and second-neighbor interactions between metal atoms exist when the α-iron interatomic distance is assumed. The most-stable monovacancy-carbon complex in α-iron had a binding energy of 0.42 eV. The configuration for this complex appears in Fig. 12. Present results were the same as those of Johnson. Other vacancy-carbon complexes are shown in Fig. 13.

The most stable $V_2(1)$-C complex observed has a binding energy of 0.55 eV. When a second carbon atom was added, a binding energy of 0.97 eV was observed. The fact that the binding energy for the two-carbon complex is less than twice the binding energy for a single-carbon atom indicates that the single-carbon configuration is the most probable case.

The configuration for the most stable $V_2(2)$-C complex is 0.62 eV. Again, attachment of a single carbon atom to the divacancy was preferred over multiple attachment. For attachments involving two carbon atoms the binding energy was 1.03 eV, less than twice the binding energy for single-carbon atom attachment. Putting aside consideration of carbon-carbon interactions, the results for the $V_2(2)$-2C complex indicate that as

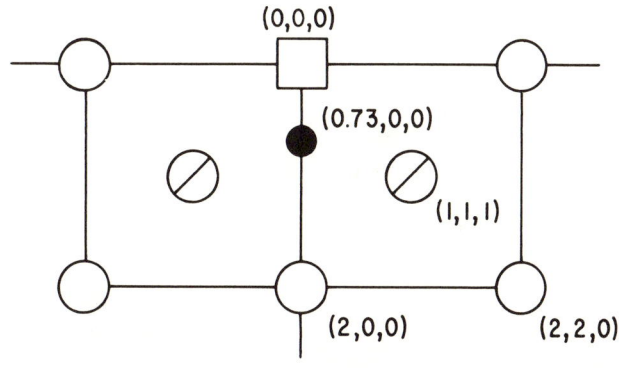

Fig. 12. The most stable vacancy-carbon atom configuration in α-iron.

354 J. R. BEELER

many as 10 carbon atoms could be attached to a $V_2(2)$ divacancy in α-iron.

5.2 Split-Interstitial-Carbon Complexes

Two tightly bound configurations for complexes for carbon with the [110] split-interstitial in α-iron were found. One of these, defined in Fig. 14a (initially found by Johnson[4]) gives an asymmetric strain field. In this instance the carbon atom is positioned at (0,0,1) above the center of the [110] split-interstitial. When a second carbon atom was added,

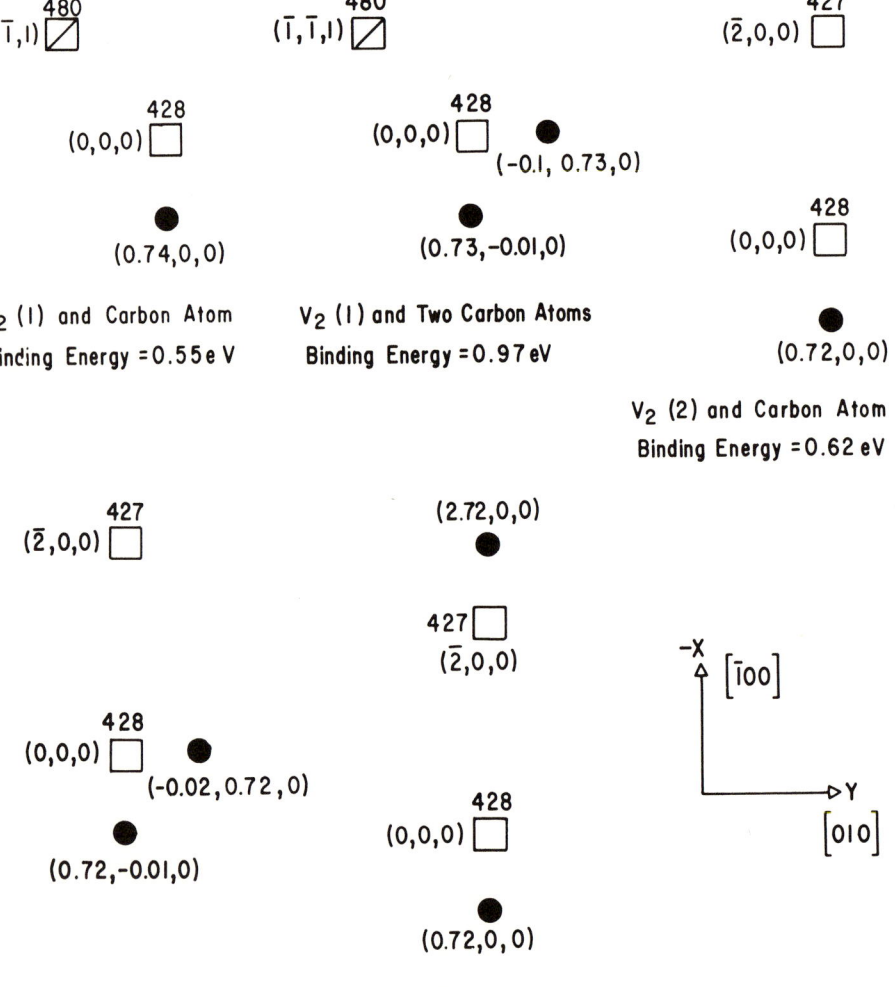

Fig. 13. Other vacancy-carbon configurations in α-iron.

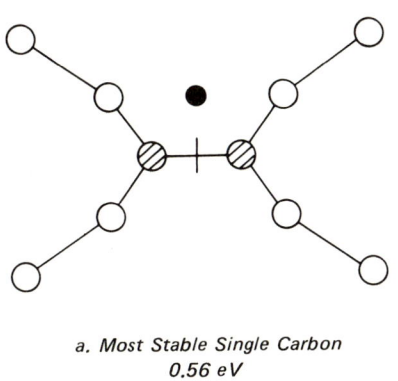

a. Most Stable Single Carbon
0.56 eV

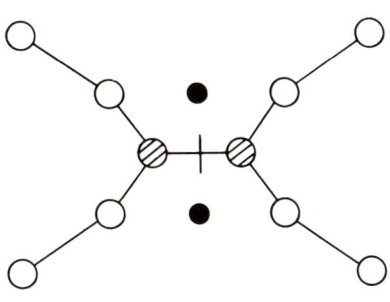

b. Most Stable 2-Carbon
0.76 eV

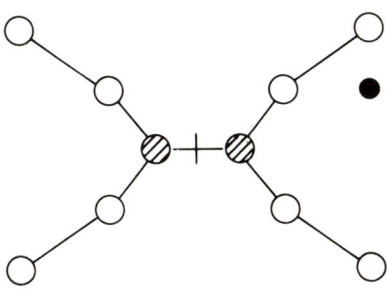

c. Second Most Stable Single Carbon
0.29 eV

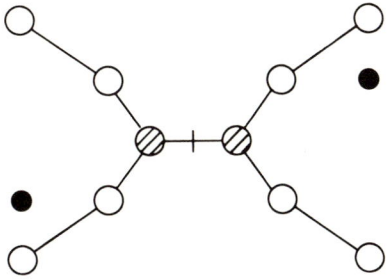

d. Second Most Stable 2-Carbon
0.57 eV

Fig. 14. One- and two-carbon complexes with [110] split-interstitial in α-iron.

symmetrically, at (0,0,-1), another bound complex was formed. The two-carbon complex is shown in Fig. 14b. The binding energy for the single carbon atom attachment is 0.56 eV and that for the two-carbon atom attachment is 0.76 eV. The binding energy for the attachment of two carbon atoms is markedly smaller than twice the binding energy for the attachment of a single carbon atom. Hence, in contrast to the results for nickel (fcc) in which up to four carbon atoms could be attached to a split-interstitial, only single carbon atom attachment is favored in α-iron (bcc).

Placement of carbon atoms in octahedral sites at the ends of the split-interstitial was also investigated. In Fig. 14c, a carbon atom occurs at the octahedral site (2,2,1) on the right end of the interstitial. The binding energy for this configuration is only 0.29 eV, compared with 0.56 eV for the (0,0,1) placement of the carbon atom. When a second carbon atom was added, symmetrically, as (-2,-2,-1) the binding energy was 0.57 eV, effectively twice the single carbon atom attachment binding energy. This configuration appears in Fig. 14d.

Carbon atoms placed at the side of the planar strain field of the [110] split-interstitial did not form bound configurations.

Single carbon atom attachment to a $V_2(2)$ divacancy occurs with a binding energy of 0.62 eV and that to a [110] split-interstitial occurs with a bind-energy of 0.56 eV. Hence, as for nickel, the attachment to the divacancy is favored over attachment to a split-interstitial.

5.3 Carbon at a (110) Free Surface

An earlier study[12] reported some simple aspects of single carbon atom behavior at a (110) free surface in α-iron. The major concern in this study was the most-stable position for a single carbon atom near a (110) free surface. The indication of this preliminary work was that a carbon atom seeks to position itself between the free surface atom plane and the immediately interior atom plane parallel with the free surface. The present study amplifies the results of the initial work for a single carbon atom and extends the investigation to a systematic examination of carbon atom clusters at (110) and (001) free surfaces. It turns out that the natural starting point for a discussion of carbon at a free surface in "α-iron" is a review of the salient features of the octahedral interstitial carbon defect in bulk α-iron. This is true because only short-range interactions are involved in the model and because the carbon-metal interaction dominates the metal-metal interaction. Consequently, adjustments of the position of the six metal atoms, nearest to the carbon atom, to the free surface perturbation almost completely determine the character of the stable configurations. Table VII contains the pertinent information required. In this table, the term *tetragonal distance* refers to the distance from the carbon atom to either of the two host metal atoms lying adjacent to it on the tetragonal axis of the octahedral carbon defect strain field. Fig. 15 illustrates this geometry, with the tetragonal axis lying along [100]. The term *ring distance* refers to the distance from the carbon atom to any one of the four host metal atoms lying in a (010) plane through the carbon atom position. In bulk iron, the ring distance is 1.94A, the separation distance for maximum bonding between a carbon and a metal atom. The tetragonal distance in bulk iron is 1.75A. The tetragonal strain is along the [100]

IMPURITY ATOM EFFECTS 357

Table VII. Characteristics of the Octahedral Interstitial Carbon Defect in α-Iron[a]

Energy of the carbon atom	−1.842 eV
Configuration energy of the interstitial configuration	−1.31 eV
Tetragonal distance	1.75 A
Ring distance	1.94 A
Tetragonal strain	0.22 hlc
Ring strain	0.06 hlc

(a) Johnson's[4] potential functions were used in computing these results.

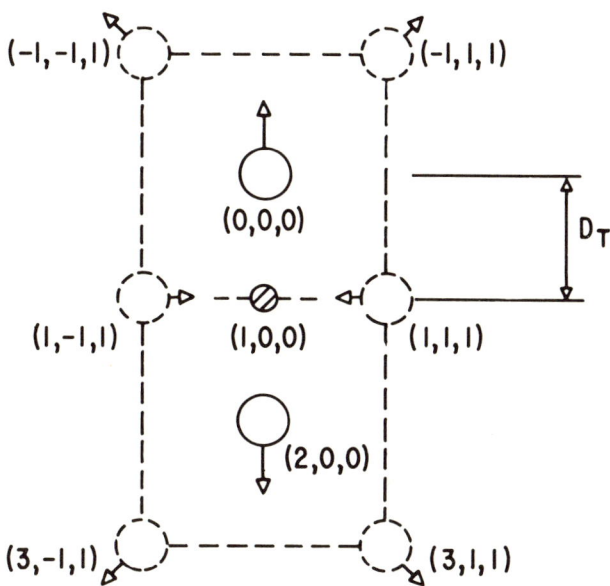

Fig. 15. Atom sites and characteristic distances of carbon in α-iron. The small hatched circle represents the carbon atom. The half-lattice constant (hlc) is the unit of length. D_T is the "tetragonality" distance and D_R is the ring distance.

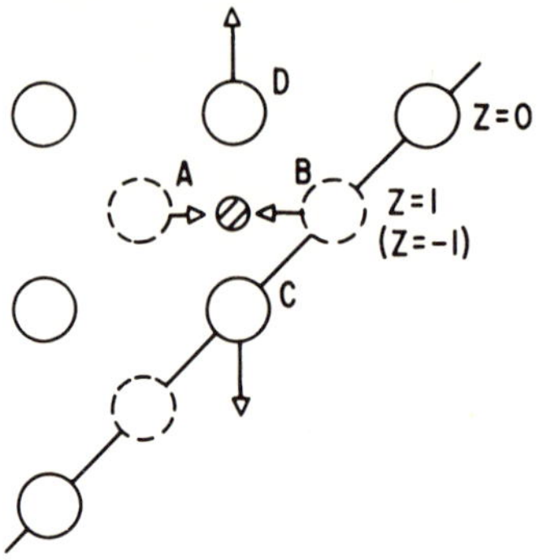

358   J. R. BEELER

Atom	Strain Vector		
	ΔX	ΔY	ΔZ
A, Z=1	0.039	0.050	-0.047
B, Z=1	-0.004	-0.049	-0.050
C	0.345	0.121	0.0
D	-0.176	0.0	0.0
Carbon	0.089	-0.005	0.0

BE = 0.26 eV

(a)

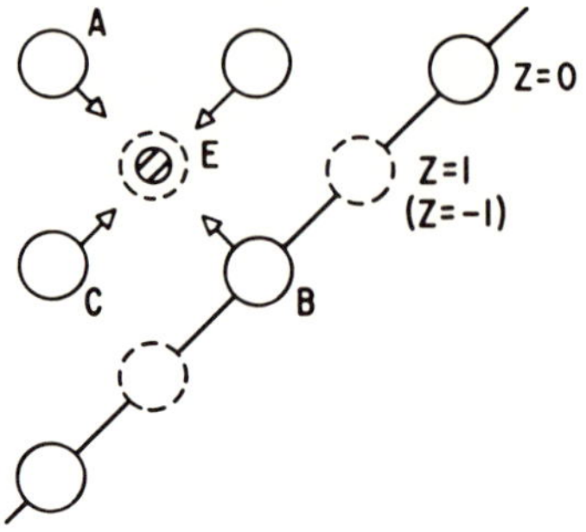

Atom	Strain Vector		
	ΔX	ΔY	ΔZ
A	0.035	0.035	0.0
B	-0.066	-0.066	0.0
C	-0.043	0.038	0.0
E, Z=1	0.009	0.009	0.229
Carbon	-0.019	-0.019	0.0

(b)

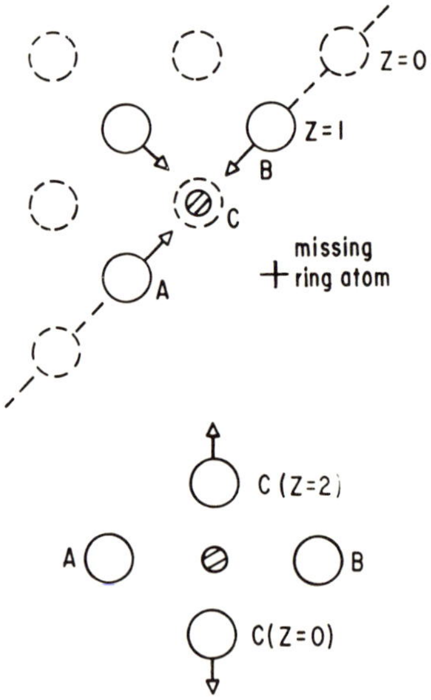

(c)

Fig. 16. Three carbon-atom positions at a (110) free surface in α-iron. (a) Most stable position near the surface. Strain is given as a fraction of a hlc. Run R275; (b) At surface with its tetragonality axis parallel to the surface. D_T is nearly the same as that for the bulk octahedral and hence the binding energy is small. Run R278; (c) "In" the surface. Shows the three "ring" atoms and shows the tetragonality atoms, both configurations as seen looking into the (110) free surface. Run R282.

and [$\bar{1}$00] directions emanating away from the carbon atom. The ring strain is along < 110 > directions directed radially toward the carbon atom. The eight atoms at (3,1,1) distances from the carbon atom experience a 0.059 hlc displacement, but they are not present on the side of the carbon atom nearest the free surface and do not concern us.

The guiding principal observed in the calculations was that the lowest-energy configuration tended to be that for which the ring distance was least disturbed while, at the same time, the tetragonal distance was lengthened toward the 1.94A separation distance for maximum carbon-metal atom bonding. In addition, the lowest-energy configuration also tended to be that in which the displacement vectors for the surface atoms were normal to the free surface. This guiding principal holds for both the (110) and (001) free surfaces.

Three different carbon atom positions at a (110) free surface are described in Fig. 16. The lowest energy configuration is the one shown in Fig. 16a. The dashed-circle atom positions A and B correspond to the four metal atom ring positions in the planes $z = 1$ and $z = -1$. All other metal atom positions and the carbon atom are located in the $z = 0$ plane. The strain vectors for the ring atom neighbors (A,B) and the tetragonal atom neighbors (C,D) of the carbon atom are listed in the insert table. These strain vectors are expressed as fractions of a hlc. Zero strain corresponds to either a perfect lattice position or an octahedral position.

Considering the free surface as one lattice defect and the interstitial carbon atom as a second lattice defect, the binding energy of the carbon atom to the free surface is 0.26 eV. This binding energy is less than that for attaching a carbon atom to a vacancy (0.42 eV), a divacancy (0.62 eV), or an interstitial (0.56 eV) in α-iron.

The tetragonality axis of the octahedral carbon interstitial strain field cuts across the free surface at a tetragonal atom neighbor position. This allows an adjustment of the tetragonal distance, D_T toward the separation distance for maximum carbon-metal atom bonding. In this instance, D_T increased from 1.75A in bulk material, to $D_T = 1.8$A at the free surface.

When the carbon atom is placed in the plane immediately interior to the free surface, the tetragonality axis is, perforce, oriented parallel with the free surface. This configuration is described in Fig. 16b. No significant lengthening of D_T occurs in this instance and the binding energy of the carbon atom to the free surface is negligible.

Location of the carbon atom in the free surface is not favored because (1) the tetragonality axis is parallel to the surface and (2) one of the ring atoms is lost. This unbound configuration is depicted in Fig. 16c.

Because the runs for a single carbon atom at a (110) free surface showed that the lowest energy configuration occurs for carbon atom insertion between the free surface plane and the immediately interior and parallel plane, only two-carbon atom configurations of this nature need be considered.

The two carbon pair configurations which were considered near a (110) free surface are shown in Fig. 17a and 17b. The configuration shown in Fig. 17a is the more stable of the two, with a binding energy of 0.88 eV. Since the binding energy of a single carbon atom to the free surface is 0.26 eV, one sees that the two carbon atoms in the lowest-energy pair configuration are bound together with an energy of 0.36 eV. Johnson's model for iron-carbon systems does not include a carbon-carbon interaction; hence, the carbon-carbon binding arising here comes solely from strain field interactions. Chemical binding between carbon atoms could serve to further increase the binding energy of the pair configuration. There is some evidence, however, that carbon atoms repel each other in iron[9]; hence, the question of the net effect of strain field binding, chemical binding and electrostatic repulsion between carbon atoms remains open. Johnson discusses carbon-carbon interaction effects in Ref. 11 (p 515) and concludes that these effects are probably not important. The carbon atom pair in Fig. 17b has a binding energy of 0.52 eV relative to the (110) free surface. Since the single carbon atom binding energy is 0.26 eV, the two carbon atoms in this configuration do not experience binding to one another via their strain fields. Strain field binding is maximized when carbon atoms share a tetragonal atom at the free surface, such as atom C in Fig. 17a. This leads to a displacement vector for the shared atom that is normal to the surface.

A detailed drawing of the strain field for the most-stable carbon atom pair is given in Fig. 17c. The atom displacements in this figure are plotted on the same scale as the interatomic distance. Stability is promoted by the presence of a shared tetragonal neighbor metal atom (856) at the free surface. Shared interior tetragonal neighbors also promotes stability as is demonstrated later. The strain field for the lowest-energy carbon atom pair at a (110) free surface is listed in Table VIII.

One of the most interesting computer experiment results for a free surface is that defect pair configurations that are metastable in bulk material either dissociate or exhibit a smaller binding energy near a free surface.[12] Moreover, the binding energy of the lowest-energy defect configuration in bulk material is often increased when this configuration is positioned near a free surface.[12] Such is the case with carbon atom pairs. Johnson[11] found that the binding energy of the first-neighbor carbon atom pair, shown in Fig. 17a, was 0.14 eV in bulk α-iron. At a (110) free surface, the present study finds that the binding energy for this pair configuration increases to 0.36 eV and at a (110) free surface, the binding energy for the second-neighbor pair vanishes. Johnson[11] found that the binding energy for the more weakly bound second-neighbor carbon atom

IMPURITY ATOM EFFECTS 361

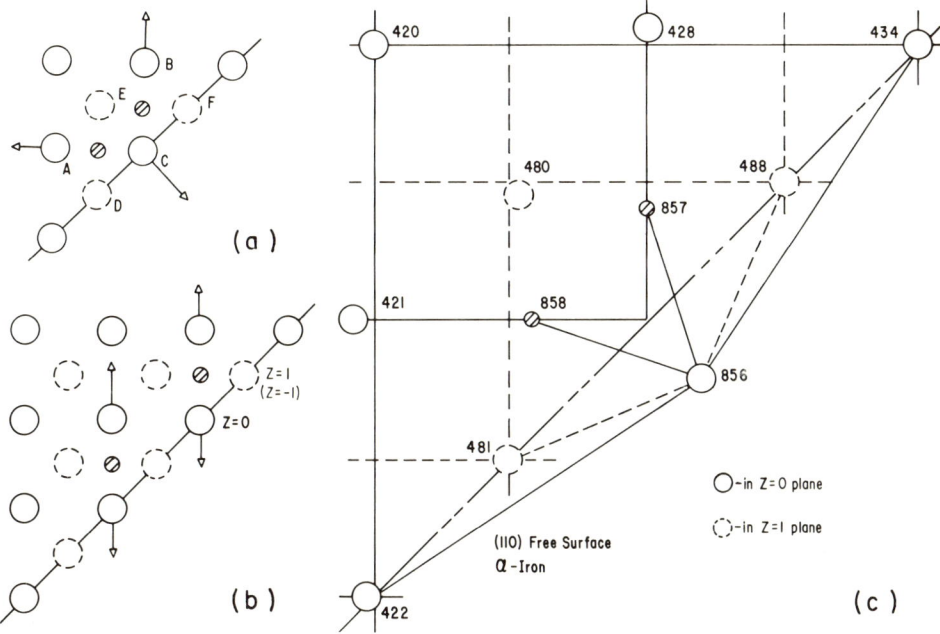

Fig. 17. Carbon atom pair configuration and strain field at (110) free surface, (a) Most stable configuration in α-iron. The combination of the two tetragonality strains opposing those of atoms A and B into that for atom C give the enhanced binding energy; Run R276; (b) Each atom with its its tetragonality axis parallel to the free surface. The carbon atoms in this configuration are not bound to each other, as they are in the configuration (a), since their tetragonality axis do not interact, Run R301; (c) Strainfield for (a). Scale of atom displacement from perfect lattice sites is the same as for the lattice constant. Atom numbers shown pertain to the atom number tale used in Run R276 for this configuration.

Table VIII. Strain Field for the Most-Stable Carbon Pair Configuration Near a (110) Free Surface in α-Iron. See Fig. 17.

Atom Number[a]	Perfect Lattice Coord.	Strain[b]			Atom Type
		Δx	Δy	Δz	
480	(−1, −1, 1)	0.058	0.058	−0.061	Iron
488	(−1, 1, 1)	−0.011	−0.034	−0.044	Iron
367	(−1, −1, −1)	0.058	0.058	0.060	Iron
375	(−1, 1, −1)	−0.011	−0.034	0.044	Iron
856	(0, 0, 0)	0.413	0.414	0.0	Iron
420	(−2, 0, 0)	−0.130	0.002	0.0	Iron
421	(0, −2, 0)	−0.002	−130	0.0	Iron
857	(−1, 0, 0)	0.174	0.017	0.0	Carbon
858	(0, −1, 0)	0.016	0.175	0.0	Carbon

(a) Atom number in the computational cell lattice.
(b) Strain in fraction of the half-lattice constant (hlc).

pair, shown in Fig. 17b, was 0.11 eV. The significant implication of present results is that estimation of stable defect pair binding energies from bulk specimen data are likely to be lower bounds for the binding energies of these defect pairs in the tension region both near an edge dislocation core and near the open disordered part of a grain boundary.

Investigation of carbon atom cluster configurations at a (110) free surface was concluded with calculations of four-atom clusters. Both lineal and platelet configurations were considered. The stability of the platelet configurations was superior to that of the lineal forms.

Runs were made for the lineal forms shown in Fig. 18. The configuration in Fig. 18a has two interior shared tetragonal atoms and one shared tetragonal atom in the free surface. The binding energy is 1.4 eV (0.35 eV/carbon atom). The configuration in Fig. 18b has one interior shared tetragonal atom and two shared tetragonal atoms in the free surface. The binding energy for this configuration is 1.73 eV (0.425 eV/carbon atom). The latter configuration exhibits a binding energy per carbon atom equal to that for a monovacancy-carbon complex (0.42 eV).

On the basis of the computed binding energies, one can assign binding-energy contributions associated with single (100) tetragonal bonds, shared interior $(110)_{int.}$ tetragonal bonds, and shared exterior $(110)_{ext.}$ tetragonal bonds. A 'pair' of single (100) tetragonal bonds is shown in Fig. 16a. The shared interior and exterior (110) tetragonal bonds are defined in Fig. 18. The results are:

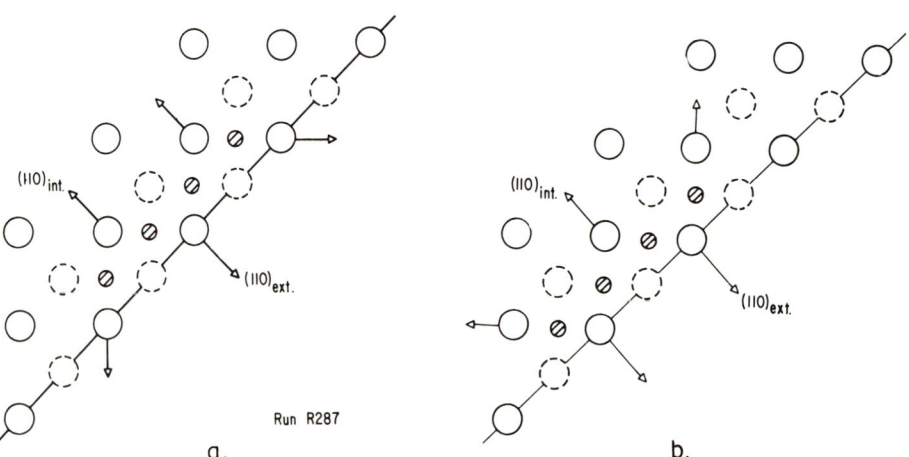

Fig. 18. Lineal array of four carbon atoms at a (110) free surface in α-iron. (a) compare with the lineal array in (b).; (b) Most stable lineal form for four carbon atoms since it has the greater number of $(110)_{ext.}$ tetragonal atoms in the free surface.

Bond Type	Energy, eV
(110)	0.13/bond
(110)$_{int.}$	0.27/bond
(110)$_{ext.}$	0.60/bond

Thus, for a given number of carbon atoms, the lineal form that exhibits the greatest number of (110)$_{ext.}$ bonds is the most stable.

Carbon atom platelets built up of first-neighbor pairs (Fig. 17a) exhibited a binding energy of 0.48 eV/carbon atom and those built up of second-neighbor pairs (Fig. 17b) exhibited a binding energy of 0.31 eV/carbon atom. Johnson[11] found an average binding energy of 0.31 eV/carbon atom for carbon atom clusters in bulk alpha-iron. One concludes from this comparison of binding energies for carbon platelets, in bulk and free surface regions, that free carbon atoms should tend to aggregate at internal free surfaces (void facets) and that both lineal and platelet forms should exist. This feature is significant in nuclear reactors where free carbon atoms are produced by irradiation induced attrition of precipitates. It is noteworthy that for four or more carbon atoms, the binding energy per carbon atom is nearly the same for both forms. Dissociation-energy calculations have not yet been made for carbon aggregates near a (110) free surface. (1.34 eV)

5.4 Carbon at a (001) Free Surface

It is easier to observe the correlation between an increase in the tetragonal distance and enhanced carbon atom cluster stability for a (001) free surface than for a (110) free surface. The data for the most significant instances are presented in Table IX. Simulations for a single carbon atom placed in the (001) free surface and then at interior positions show, again, that the carbon atom is most strongly bound when it lies between the free surface atom plane and the immediately interior parallel atom plane.

The immediate strain field for the most stable configuration for a single carbon atom is drawn in Fig. 19. The carbon atom in plane $z = -1$ drops downward slightly toward the free surface at $z = -2$, but the ring atoms surrounding it do not do so. Because of this, a "dimple" occurs in the plane holding the carbon atom, at the carbon atom position. This dimpling effect also occurs in the (110) free surface cases but it is irregular and more difficult to illustrate clearly than in the (001) free surface cases.

Strain field characteristics for the most stable carbon pair configuration are depicted in Fig. 20. Again there is a dimpling effect in the plane containing the carbon atoms.

Table IX. Binding Energy, Tetragonal Distance (D_T) and Ring Distance (D_R) for Carbon Atom Defects Near a (001) Free Surface in α-Iron. The Free Surface is the Plane z = −2. The Crystallite Lies Above This Plane.

Defect	Run	Binding Energy, eV	Carbon Atom Energy, eV	D_R, Å	D_T, Å	Comment
Octahedral Carbon (bulk)	D502	–	−1.84	1.94	1.75	Bulk Reference. See Fig. 15.
Carbon at (0,0,−1)	R272	0.39	−2.00	1.93	1.82	First interior position. Most-stable position. See Fig. 19.
Carbon at (0,0,1)	R283	0.06	−1.87	1.93	1.76	Third interior position. Tetragonal axis normal to free surface.
Carbon at (0,1,0)	R284	0.01	−1.84	1.93	1.75	Second interior position. Tetragonal axis parallel with free surface.
Carbon at (0,1,−2)	R300	−1.37	−1.31	1.93	1.82	On free surface.
Pair: (0,0,−1), (0,2,−1)	R277	0.85	−2.02	1.93	1.83	First interior position. Most stable pair. See Fig. 20.
Platelet: (2,0,−1), (2,2,−1)	R295	1.77	−2.02	1.93	1.83, 1.84	Most stable quartet. Two pairs of type in R277.
Platelet: (2,0,−1), (−2,0,−1)	R294	1.69	−2.01	1.93	1.83	Pairs do not interact. Less stable than quartet of R295.

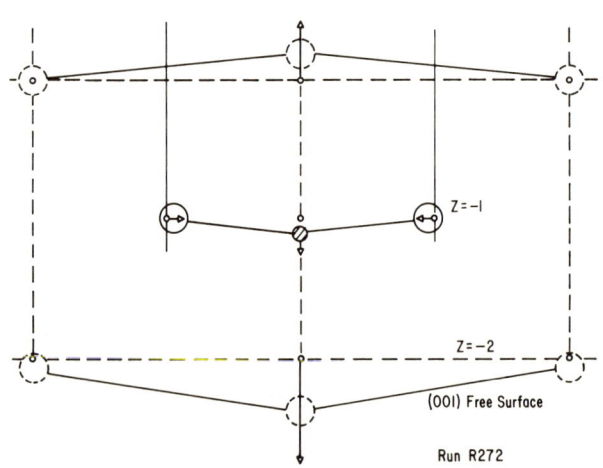

Fig. 19. The most stable carbon atom position near a (001) free surface in α-iron. The schematic character of the associated strain field.

When a platelet of four carbon atoms is considered, one finds that the tetragonal distance above the platelet and that below the platelet are no longer equal; the tetragonal distance between the platelet and the free surface is the larger distance. In addition, different classes of ring distances

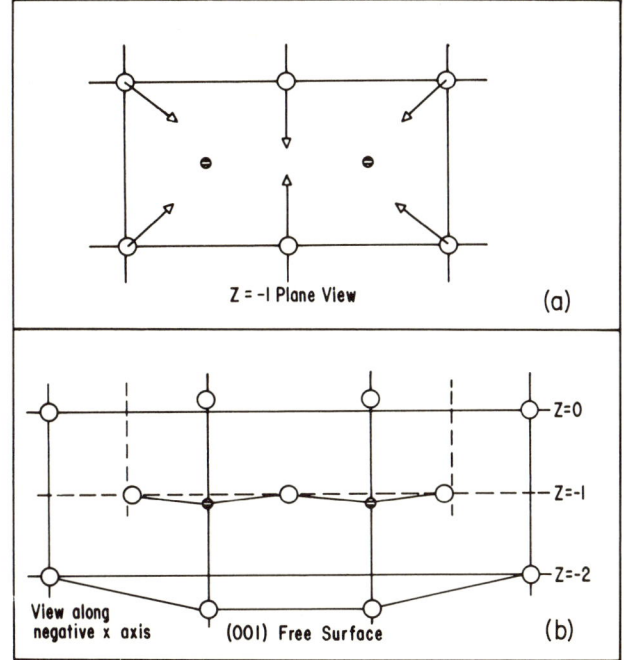

Fig. 20. Carbon atom pair at a (001) free surface in α-iron. Schematic character of the associated strain field. Note dimple structure in the plane containing the carbon atoms. Run R277.

also appear for the first time. The three ring distance classes are defined in Fig. 21.

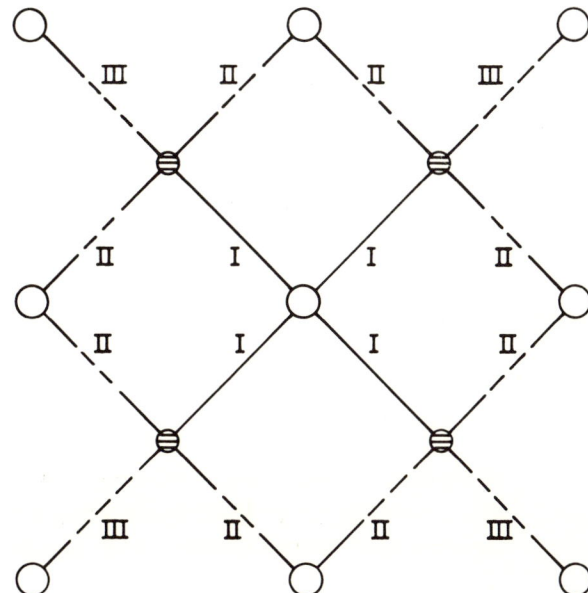

Fig. 21. The three classes or ring distances associated with a (001) platelet of four carbon atoms near a (001) free surface in α-iron.

A comparison of the elementary results for carbon atoms at (110) and (001) free surfaces in α-iron shows some important differences. The binding energy data tabulated below illustrates these differences.

Carbon Configuration	Binding Energy, eV	
	(110)FS	(001)FS
Single carbon	0.26	0.39
Carbon pair	0.88	0.85
Carbon quartet	1.91	1.77

Binding of a single carbon atom to a (001) free surface is 50 percent larger than that to a (110) free surface. On the other hand, the pair binding energy is somewhat less for the (001) free surface than that for the (110) free surface. Finally, the carbon-carbon interaction for a pair of a (110) free surface is 0.36 eV. This exceeds the carbon pair binding energy of 0.14 eV in bulk iron. The carbon-carbon binding energy at a (001) free surface is very weak, being only 0.07 eV. On this basis, one would expect that carbon atoms would place themselves at (001) free surfaces sufficiently far apart that no significant clustering activity would be observable since a carbon-carbon attachment either at a (110) free surface or in bulk iron leads to a greater decrease in the crystal energy. Carbon aggregates should form at (110) free surfaces since platelet binding at this free surface exceeds that in bulk iron.

6 CARBON IN COBALT

The properties of carbon in hcp model metal "cobalt" were studied using the DYNAMIC Program exclusively. The interatomic distance d_1 was set at 2.54A. In the ideal hcp lattice, second-neighbor distance is $d_2 = \sqrt{2d_1}$. This gives $d_2 = 3.53A$ for cobalt which exceeds the cutoff distance (3.44A) for the metal-metal atom interaction function.

Interstitial atom and defect migration saddle-point locations are conveniently described in terms of five interstitial points in the ideal hcp lattice. These points are as follows:

Point   Definition

A   Midpoint between two first-neighbor sites in adjacent basal planes. This point is immediately surrounded by six atoms which form the apexes and edge corners of a rectangular-base octahedron. (See Fig. 22a.)

Point	Definition
B	Midpoint between two second-neighbor sites in adjacent basal planes. This point is immediately surrounded by six atoms which form the apexes and edge corners of a square-base octahedron. It is analogous to the octahedral position in the bcc lattice. (See Fig. 22b.)
C	The center of a tetrahedron whose apexes are the three apex atoms of an equilateral triangle in a basal plane and the atom positioned above the center of this triangle in an adjacent basal plane.
D	Midpoint between two first-neighbor sites in a basal plane.
E	Center of the equilateral triangle formed by three mutually first-neighbor sites in a basal plane.

The configuration energies for the five possible carbon interstitial positions are given in Table X. By far, the most favorable position is the Point B octahedral position. Carbon migration took place by the conversion sequence B → A → B with a migration energy of 1.08 eV. Provided that carbon occupied the Point D position initially, carbon migration could also take place via the conversion sequence D → E → D with a migration energy of 0.33 eV. The activation energy required for a conversion of carbon from Point B to Point D is 1.53 eV. The converse process, Point D to Point B takes place with an activation energy of 0.502 eV. Thus at low temperatures, occupancy of Point D positions could be established, for example, by either irradiation or cold work and would give rise to a low-energy migration process with $E^m = 0.33$ eV. Upon a rise in

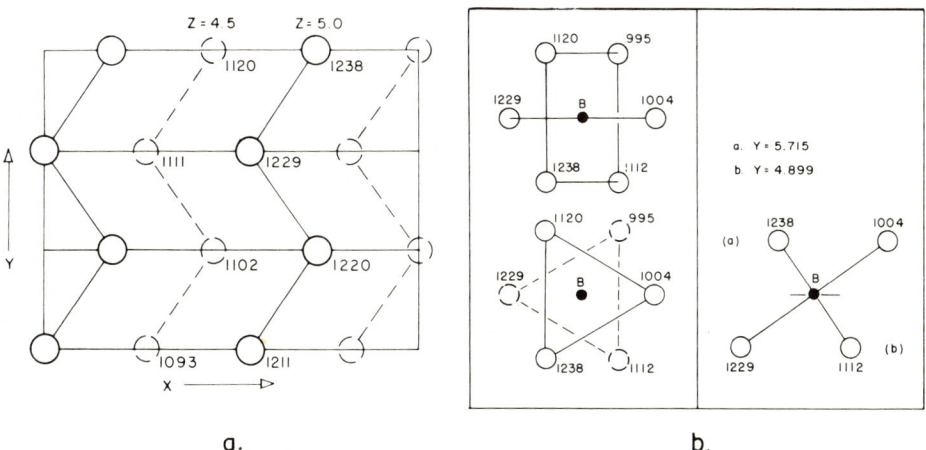

Fig. 22. Interstitial Positions in the ideal hcp structure. (a) The Type A (A-Point); (b) The Type B (B-Point).

Table X. *Configuration Energies of the Five Possible Interstitial Carbon Configurations in an Ideal HCP Structure*

Point	Configuration Energy, eV	Relative Magnitude	Run Number
A[a]	−0.609	3rd	EH093
B[a]	−1.691	5th	EH078H
C	0.109	1st	EH091
D	−0.663	4th	EH073
E	−0.369	2nd	H16R01

(a) See Fig. 22.

temperature, however, carbon atoms at Point D positions would be thermally converted to Point B positions and the effective migration energy would then increase sharply to 1.08 eV.

No bound vacancy-carbon complexes were found in cobalt. Table XI lists the configuration energies for the three closest carbon-vacancy separation distances. The configuration energy for a noninteracting carbon atom and a vacancy is 1.11 eV in the model metal cobalt. No energy listed in Table XI is less than 1.11 eV. One concludes that carbon atoms and vacancies would migrate independently in the ideal hcp structure treated here.

No runs were made of split-interstitial-carbon complexes. The split-interstitial strain field in the hcp structure is not symmetrical, as it is in cubic structures, and the investigation of split-interstitial-carbon complexes was set aside as a separate future study.

Table XI. *Configuration Energies for the Three Carbon-Vacancy Complexes in an Ideal HCP Structure With the Smallest Separation Distance Between Defects*

Carbon-Vacancy Distance, $\underline{a}$	Configuration Energy, eV	Run
0.567	1.301	EH081
1.224	1.123	EH083
1.347[a]	1.107	EH082

(a) At this separation distance the interaction energy is the same as that for infinitely separated defects, i.e., there is no interaction between the carbon atom and the vacancy.

7 SUBSTITUTIONAL IMPURITY ATOMS

The migration of substitutional impurity atoms was simulated in the model metal "α-iron". The primary intent was to examine impurity atom-size effects on vacancy migration and impurity-atom migration. The impurity atom-host atom interaction function was constructed so that the elastic constants of the impurity atom metal and the host atom metal were the same. Both oversized and undersized substitutional impurity atoms were then considered; the atomic radius deviation was 10 percent in each instance.

As expected, the substitutional impurity atom migration proceeded via the vacancy mechanism; hence a description of impurity atom migration, in this case, is also a description of substitutional impurity-vacancy complex formation, dissociation, and migration. Attention, therefore, is focussed on the transformation of the complex from one configuration to another. This characterization has the advantage that it simultaneously describes the effect of the impurity atom on monovacancy migration, the effect of the monovacancy on impurity atom migration and the behavior of the complex as an entity per se.

Fig. 23 displays the configuration energies of the impurity atom-vacancy complex for different separation distance vectors, in the form of an energy level diagram. All configuration energies are referred to pure α-iron. The horizontal dashed line indicates the energy level that corresponds to zero interaction between the impurity atom and the vacancy. All levels below the dashed line signify a bound complex state, and all levels above the dashed line signify an unbound complex state. One immediately notices that a configuration that is bound for a small impurity atom is unbound for a large impurity atom in these size-effect computer experiments.

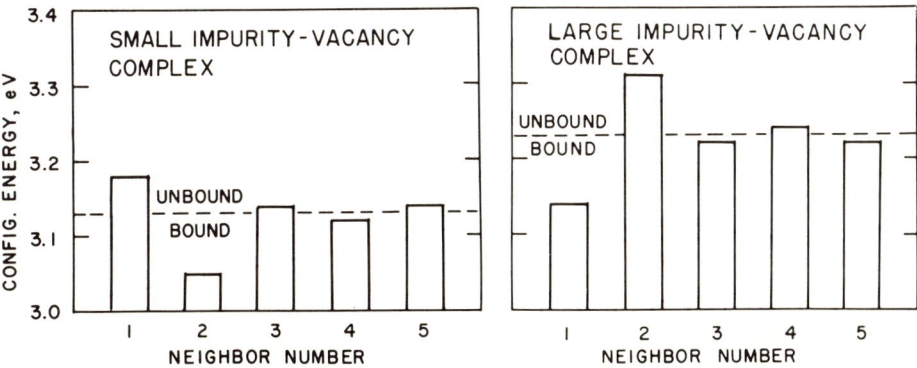

Fig. 23. Configuration energies for first-neighbor to fifth-neighbor complex configurations comprised of a substitutional impurity atom and a vacancy in α-iron. 10% misfit.

The activation energies required to initiate a transformation from one complex configuration to another are displayed in Fig. 24 for the small and large impurity atom complexes, respectively. In each figure, the encircled numbers designate the separation distance between defects in terms of the 1st, 2nd, 3rd, 4th and 5th-neighbor separation vectors in the lattice. The numbers at each arrowhead indicate the activation energy in eV, for a transformation into the configuration designated by the arrowhead, from that indicated by the opposing arrowhead at the opposite end of the connecting line between boxes.

For the small impurity atom, the most tightly bound complex is the second-neighbor complex. The binding energy is 0.08 eV. In the limit of a zero radius "small" substitutional atom (a vacancy) the second-neighbor configuration is also the most tightly bound complex (divacancy). Migration of the 10 percent misfit small impurity could proceed by alternative transformations between second- and first-neighbor configurations with an intermediate first-neighbor interchange between the impurity and the vacancy. According to Fig. 24, the migration energy for this process would be 0.71 eV.

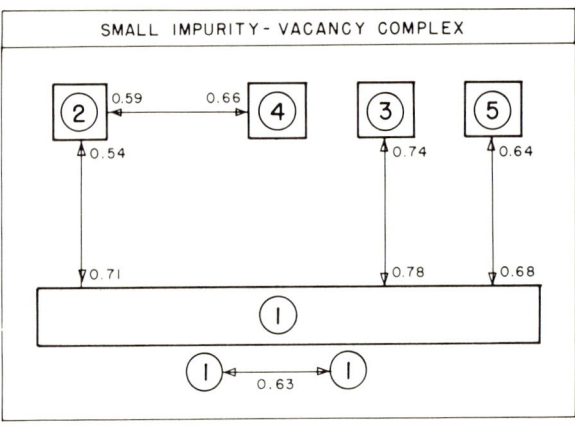

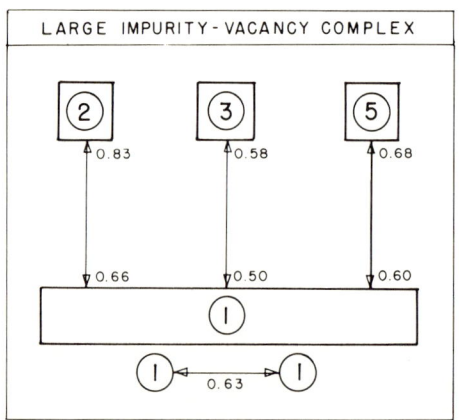

Fig. 24. Activation energy chart for the configuration change of substitutional impurity atom-vacancy complexes in α-iron. 10% misfit.

The first-neighbor configuration is the most tightly bound configuration for the large impurity atom. This complex could also migrate via alternative transformations between first- and second-neighbor configurations. According to Fig. 24, the migration energy for this process is 0.83 eV.

A monovacancy moves with an activation energy of 0.68 eV in the pure host metal.[2] Hence, relative to vacancy migration, both impurity complexes as an entity have a larger migration energy than the monovacancy. The alternative transformation from first- to second-neighbor configurations is not, however, the only way the complexes can migrate through the host metal. The necessary transformation for any net transport of the impurity atom component of the complex is the first-neighbor interchange transformation. In each case, the activation energy for this interchange transformation is less than the monovacancy migration energy. Hence, statistically speaking, a distribution of more complicated transformation sequences exists wherein each sequence has associated with it an activation energy smaller than the migration energy for the first- to second-neighbor transformation sequence (mechanism).

ACKNOWLEDGMENT

This work was supported by the U. S. Atomic Energy Commission under contract number AT-(40-1)-3912.

REFERENCES

1. A. C. Damask and G. J. Dienes: *Point Defects in Metals,* p. 21, Gordon and Breach, New York, 1963.
2. R. A. Johnson: Phys. Rev., **134**: A1329 (1964).
3. R. A. Johnson: Phys. Rev., **145**: 423 (1966).
4. R. A. Johnson, G. J. Dienes and A. C. Damask: Acta Met., **12**, 1215 (1964).
5. J. B. Gibson, A. N. Goland, M. Milgram, and G. H. Vineyard: Phys. Rev., **120**, 1229 (1960).
6. R. A. Johnson and E. Brown: Phys. Rev., **127**, 446 (1962).
7. R. A. Johnson: Phys. Rev., **152**, 629 (1966).
8. J. R. Beeler, Jr. and R. A. Johnson: Phys. Rev., **156**, 677 (1967).
9. R. A. Johnson and A. C. Damask: Acta Met., **12**, 443 (1964).
10. R. A. Johnson: Acta Met., **13**, 1259 (1965).
11. R. A. Johnson: Acta Met., **15**, 513 (1967).
12. J. R. Beeler, Jr.: *International Conference on Vacancies and Interstitials in Metals,* Jül-Conf-2(Vol. II), p. 598, Kernforschungsanlage Jülich, Germany, 1968.
13. J. R. Beeler, Jr.: Defect Interaction and Metallurgical Process Simulation in Metals and Alloys, USAF Report AFML-TR-70-260, November 1970.

DISCUSSION on the Paper by J. R. Beeler, Jr.

WILSON: How large was the void in your calculation? Perhaps there are electron-redistribution effects on the inner metallic surface.

BEELER: The void sizes treated ranged all the way from the smallest faceted void nucleus, a bipyramid, six-vacancy cluster, to the limiting case of a free surface of infinite extent. Electron-distribution effects certainly do occur in real metals, from elementary considerations. They were not accounted for in our simulations. All interactions and effects in the model we used are those intrinsically contained in Johnson's potentials for iron-iron and for carbon-iron interactions.

WILSON: We have found that a single vacancy can act as a nucleation site for as many as 30 helium atoms. Did you happen to try an extended calculation of this sort for carbon in α-iron in the region of a single vacancy or a divacancy?

BEELER: Yes. The strongest effect observed among the four metals considered was the collection of up to 10 carbon atoms at a $V_2(2)$ divacancy in α-iron. This feature was alluded to in the talk but was not emphasized there as strongly as it is in the written version. In addition, up to four carbon atoms were collected at the beltline of a $<100>$ split-interstitial in nickel. In the written version we extend ourselves somewhat and generalize these particular effects to the stable divacancy in bcc metals and the stable interstitial in fcc metals.

Perhaps the most significant of our results, from the standpoint of multiple collection of carbon atoms, is the collection of carbon into platelet clusters at a (110) void facet or free surface. Carbon atom binding at these particular surfaces was signifiantly stronger than carbon atom binding in bulk metal.

BULLOUGH: As a point of information, the variational procedure is not the method of lattice statics. Lattice statics enables the defect-displacement field to be obtained by performing a summation over points of the reduced Brillouin zone of the defect superlattice. It is thus a reciprocal lattice procedure and not a direct-square procedure.

HO: How does the formation of a carbon atom platelet on a void surface affect the migration of vacancies to that surface? This question may have some bearing on understanding the effect of carbon impurities on swelling in nuclear-reactor materials.

BEELER: Extensive carbon plating at a void surface would slow the rate at which vacancies were assimilated into the void because each approaching vacancy would be trapped for a time at the carbon atom platelet surface which is farthest from the void surface. Trapping would occur with roughly the same attachment geometry as occurs in the carbon-vacancy complex, i.e., with a $<100>$ orientation. But since the presence of a trapped vacancy raises the carbon platelet energy, the vacancy would eventually permeate into the void. Equally important would be the rejection of interstitials by a carbon platelet at a void surface. The strong tetragonal strain field of the platelet would repel mobile interstitials and thus diminish the rate at which interstitials initiated annihilation events at a void surface. This would work against void shrinkage by diffusion of interstitials and void shrinkage by radiation attrition.

HO: It would also be interesting to supplement the present calculation with a similar calculation for carbon behavior at dislocations. Again the problem would be important relative to either aiding or impeding void formation.

BEELER: Dahl has worked with tilt boundaries. To some degree the results might be carried over to dislocations, especially their results for low-angle boundaries. This is discussed by Dahl in his paper. For one thing, carbon in a dislocation will strongly trap any vacancy making its way via pipe diffusion at the dislocation core. Two opposite effects occur, thereafter, from the standpoint of void formation. On one hand, vacancy trapping impedes vacancy transport to voids via a dislocation road. On the other hand, trapping of vacancies in a dislocation core by carbon atoms, could produce a void nucleation site. The effect of this competition is determined by the irradiation conditions and irradiation history.

TORRENS: What prevents the carbon atoms just beneath the surface layer from passing through the surface?

BEELER: On the basis of the carbon-iron interaction used in our simulations (Johnson's potentials), the carbon-iron interaction is about 40 percent stronger than the iron-iron interaction. Consequently, the increment by which the crystal energy is lowered when the carbon is placed between atom layers at the surface, as described in the text, is about twice that which occurs when the carbon atom is placed in or on the surface. Hence, although all states are indeed possible, the Boltzmann factor for the between-layers position dominates those for the other positions at or on the surface. Moreover, the stability of the platelet in the between-layer form at the surface is greater than of any carbon cluster in the bulk, as well.

SHEWMON: You've discussed the ability of vacancies to stabilize carbon interstitial clusters. Have you considered the ability of carbon atoms to stabilize vacancy clusters?

BEELER: Actually our work has been reported as concerning the stabilization of voids by carbon atoms rather than the stabilization of carbon clusters by vacancies. We find that carbon atoms effect an "external" stabilization of voids by locating themselves between the two atom layers at the void surface. For carbon, on the basis of Johnson's interaction model, it is true that the surface also stabilizes the carbon atom platelet. They exert a mutual stabilizing influence upon each other. Until your question, I had not considered the fact that vacancies could stabilize a carbon platelet, but it does truly happen in our simulations. Yes, even a small vacancy cluster will contribute a stabilizing influence on a carbon platelet in bulk α-iron, within the framework of Johnson's carbon-iron interaction model.

ENGLERT: Could you treat the problem of the distribution of substitutional impurity atoms around various defects?

BEELER: Yes. A short account of 10 percent misfit impurity atoms bound to a vacancy is given in the last section of the written version of this paper. We are, in fact, doing the substitutional impurity problem along the same lines described here for a small interstitial impurity. Carbon was given first priority because of technological dictums.

KULCINSKI: Do you have any feeling on how your results might change if you were to use other interstitial impurity atoms such as nitrogen or oxygen?

BEELER: Yes. According to Johnson's ideas, the effect of nitrogen should be qualitatively similar to the effect of carbon.[1] The chemical aspects of oxygen effects appear to be too complicated for the simple type of treatment we have been using.

REFERENCE

1. R. A. Johnson, Acta Met. **12**, 1215 (1964).

RARE GASES IN METALS

W. D. Wilson

Sandia Laboratories
Livermore, California 94550

and

R. A. Johnson

University of Virginia
Charlottesville, Virginia 22901

ABSTRACT

Using models for fcc and bcc metals developed in an earlier paper, the properties of He interstitial and substitutional defects were studied. The He-metal potentials were obtained by a modification of the Wedepohl method. Interstitial activation energies for helium motion are found in general to be quite low, being of the order of 0.5 eV for fcc materials and 0.25 eV for the bcc cases with the exception of palladium ($\sim$ 1.7 eV). Substitutional detrapping is found to require an activation energy of from $\sim$2-5 eV, the higher energies found mostly for the bcc materials.

1 INTRODUCTION

The great upsurge of interest in the behavior of rare gases in solids has been motivated to a large extent by nuclar-reactor technology. Neutron irradiation of reactor materials creates a fission gas product buildup in the lattice and subsequently causes a swelling of the material. Proton irradiation has been used to study the voids in the absence of gas. Helium-ion implantation of metals, such as nickel, before proton irradiation has been found to significantly affect void formation.[1] It seems that the helium atoms can prevent the collapse of large vacancy clusters into loops, thus increasing the number of three-dimensional voids. Experimental helium-ion implanation results in copper, nickel, and palladium have been presented by Bauer.[2]* These results are an outgrowth of the 5.6–MeV low-dose implanations done earlier by Holt, et al.[3] and Thomas, et al.[4].

There have been very few atomistic calculations of rare gas behavior in metals. The early work of Rimmer and Cottrell[5] and, more recently, Anderman and Gehman[6] on rare gas configurations in Cu are notable exceptions. These calculations employed interatomic potentials for the gas-host atom interaction which were obtained by averaging the known gas-gas and Cu-Cu potentials. Wilson and Bisson[7] have calculated these necessary interactions by an approximate quantum mechanical method (originally due to Wedepohl[8]) which eliminates averaging or fitting procedures. The interactions were found to compare well with all known rare-gas diatomic potentials and were therefore used for the gas-metal atom interactions. The short-range Cu-Cu potential determined in this way was also found to be in excellent agreement with the well-known Gibson et al.[9] potential used in so many defect calculations.

In this work, the same method has been applied to the calculation of interatomic potentials between helium atoms and metal host atoms or ions in fcc and bcc materials. These potentials were then used to calculate various helium-atom configurations in Cu, Ni, Ag, Pd, Fe, Mo, V, W, and Ta. The lattice atom potentials for these materials determined in **Defect Calculations for Fcc and Bcc Metals** were employed.

The He-host atom potentials are presented and the method of calculation of the defects discussed. The results of calculations of He interstitial migration and also of the detrapping of helium atoms from substitutional sites are given.

---

*These irradiations were performed at liquid nitrogen temperatures at an energy of 300 keV in a chamber allowing for in situ mass spectrometric analysis of He outgassing as a function of temperature. A very dramatic fluence dependence is found — there is very little low temperature (below 0C) outgassing until doses of $\sim 10^{18}$ He atoms/cm^2 are reached. Most of the gas release occurs in a higher temperature region (500 C) and is less dose dependent than its low-temperature counterpart. Gas release is also detected at intermediate temperatures (50 C).

2 INTERATOMIC POTENTIALS AND METHODS OF DEFECT CALCULATION

In Fig. 1, the He-metal interatomic potentials determined by the modified Wedepohl method are given. Note that the magnitude of the energies at any particular distance increase as expected with increasing atomic number (with the exception of nickel, which crosses the palladium curve). Note also that at short range, $R \leqslant 1$ A, there is a deviation from pure exponential or Born-Mayer behavior – the curves resemble the $1/R$ dependence known to exist for diatomic rare-gas interactions. At large (2.5-A) internuclear separations, the curves exhibit an increasingly more negative slope due to the fact that a small binding or negative interaction energy is obtained at these larger separations which goes asymptotically to zero for even larger separations. Very elaborate molecular-orbital computations show this effect when correlation energy is included. In this approximate treatment, these negative energies are the result of the semiclassical nature of the calculation as discussed by Wedepohl[8]. In actual practice, the potentials are cut off at the zero of energy.

The potentials given were calculated in the Hartree-Fock-Slater (HFS) approximation to the exchange and include the so-called "tail correction"[7]. This approximation was found to give the best agreement with molecular-beam-scattering results and hence was used throughout this work. The slight roughness of the curves in Fig. 1 is a consequence of fitting straight lines (on the semilog plots) between the ~24 points calculated. This logarithmic interpolation is the same scheme used in the defect calculations and, hence, the figures were not artificially smoothed.

Two completely independent methods were developed for the helium-defect calculations. In Method A, N atoms surrounding the defect configuration are allowed to relax, symmetry being employed to reduce the $3N$ possible displacement parameters to a more tractable number. The number of atoms allowed to relax is decided upon by a series of calculations involving an increasing number of atoms. It turns out that ~200 atoms is sufficient to determine the energy to ~0.01 eV – the inclusion of relaxations beyond this region has virtually no effect on the energy. Some care was taken to choose the same number of relaxed atoms for various defect calculations (as far as is possible from symmetry) so that relative numbers, such as activation energies, will be better determined. These relative numbers are generally less sensitive to the number of atoms included than are individual formation energies.

Table I shows the number of atoms allowed to relax and the corresponding number of independent displacement parameters. Also listed are the coordinates of a typical first-nearest neighbor (*1nn*) to the defect and the definition of the *1nn* displacement parameter that is given in the results later. For convenience, the multiplicities of the "shells" of atoms surrounding the defect are given in the last column.

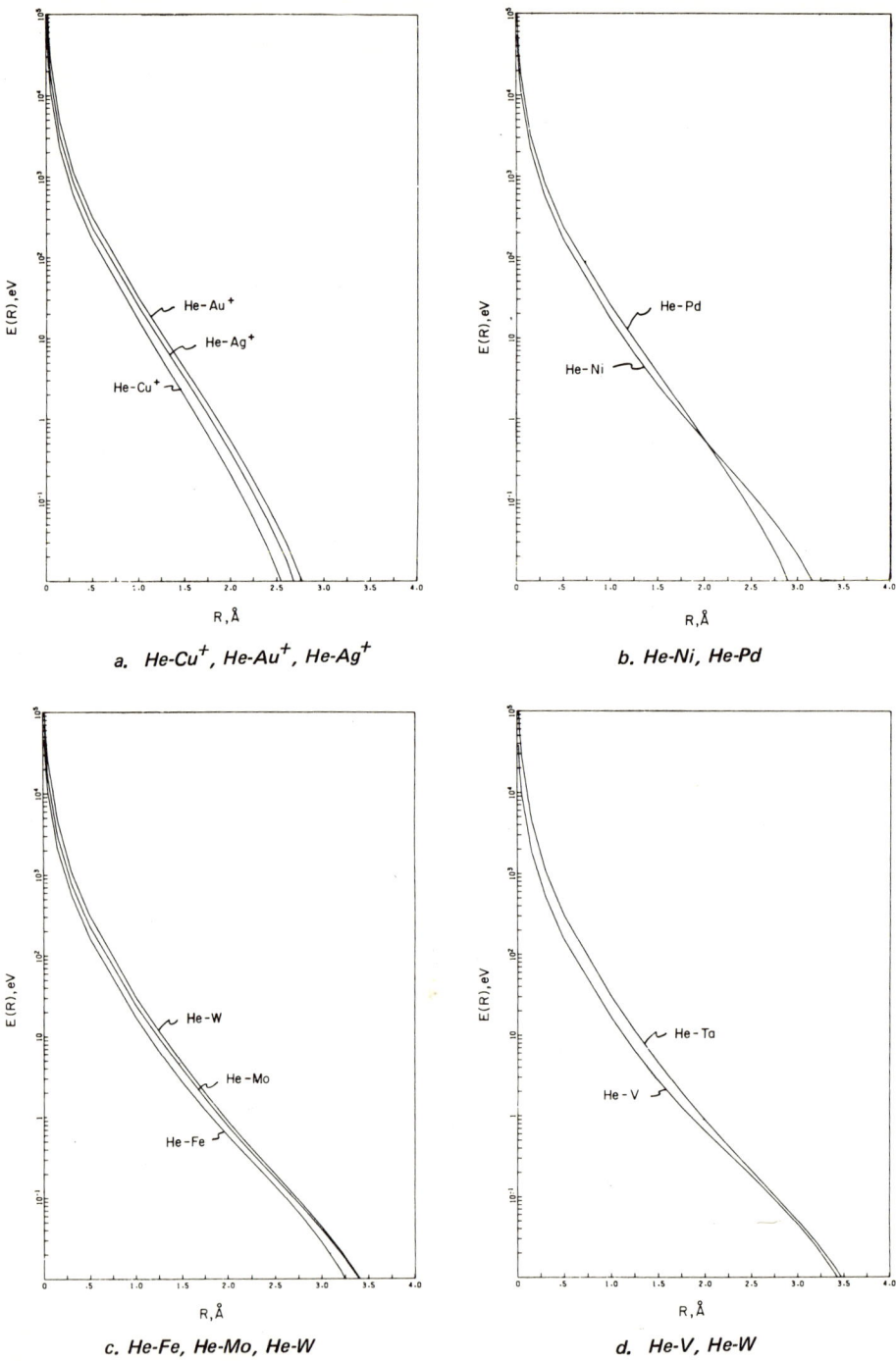

Fig. 1. Potentials determined by the modified Wedepohl method. The Hartree-Fock-Slater approximation to the exchange was used in all cases.

Table I. Number of atoms allowed to relax and the corresponding number of independent displacement parameters in the fcc case. The definition of the displacements of a typical first-nearest-neighbor atom are given in Column 4. The number of atoms in each movable shell of atoms surrounding the defect are given in the last column.

Defect Configuration	Number of Relaxed Atoms	Number of Displacement Parameters	Coordinates of a Typical 1nn Relaxed Atom	Multiplicities of Atoms in Shells About the Defect
He in a Metal Vacancy	224	19	$-1+p_1, -1+p_1, 0$	12, 6, 24, 12, 24, 8 48, 6, 24, 12, 24, 24
He at (1, 0, 0)	236	20	$-1+p_1, 0, 0$	6, 24, 8, 6, 24, 24, 24, 24, 24, 48, 24
He at (½, ½, 0)	232	97	$-½+p_1, -½+p_1, 0$	2, 2, 8, 4, 4, 4, 4, 8, 8, 4, 8, 4, 2, 8, 4, 4, 8, 4, 8, 4, 4, 8, 8, 8, 8, 8, 8, 8, 4, 4, 4, 4, 8, 8, 28, 4, 4, 8, 4, 8
He at (½, ½, ½)	228	35	$-½+p_1, -½+p_1, -½+p_1$	4, 12, 12, 12, 12, 24, 4, 12, 24, 12, 4, 12, 24, 24, 12, 12, 12

Having assigned the displacement parameters, the energy relative to the perfect lattice is next expressed as a function of these displacements. Then, each parameter is varied by some fraction, f, of the half-lattice constant, r_0, using a direct search. After going through the entire set of parameters for a given value of f, this fraction is then reduced and the entire procedure repeated. Usually, six iterations ending with $f = 0.001$ were sufficient to insure that the minimum energy is known to within ~0.01 eV. Several variations of starting parameters were chosen to gain confidence that the results are not dependent upon the initial configuration.

A second method (Method B) was also developed. Here a cubic lattice is generated but the $3N$ displacements are not reduced. There is, therefore, no difference in the number of degrees of freedom between a highly symmetrical defect such as a single vacancy and a defect involving, e.g., multiple helium atoms and vacancies. The forces on each atom are computed from the interatomic potentials; then the relaxed coordinates at some time $\Delta t = t - t_0$ later are computed from a set of equations of the form:

$$x(t) = x(t_0) + \frac{F}{m}\bigg|_{t_0} (\Delta t)^2 \ldots \quad (1)$$

The forces are again computed and the process repeated as many times as is necessary to obtain the zero-force condition. Generally, ~100 iterations were required, but even at that, Method B is computationally faster than Method A. When little or no symmetry exists, Method B is over 5 times faster than Method A.

Physically speaking, there is a great difference between the two methods. In Method A, each shell of atoms is moved to its approximate minimum and then held fixed while the next set is relaxed, etc. Method B allows all atoms to be moved simultaneously – a more physically appealing process. Adjustments of the Δt value enable the minimization to take place.

As the noncentral forces have not yet been included in Method B, a test case involving central forces alone was made to determine the effect of these minimization procedures on the energy. No difference whatever was found – both methods give the same results. Therefore, Method A was used for the fcc calculations (which require noncentral forces) and Method B for the bcc metals.

It should be mentioned that the technique of including enough atoms to fix the energy within some reasonable limit eliminates the need for an elastic continuum. The merits and the problems associated with such continuum methods are not discussed here. Note that a fixed boundary of atoms surrounds the movable region in both Methods so that essentially a volume dependent force is holding the lattice together.

3 RESULTS OF He DEFECT CALCULATIONS

In Fig. 2, the symmetrical interstitial positions for a helium atom are shown for fcc and bcc metallic lattices. Table II contains the results of the helium interstitial calculations in the fcc metals. The vacancy migration saddle-point configuration (a metal atom between two vacancies along a <110> direction) is highly sensitive to the cutoff, r_c, in the host-atom

Table II. Formation energies, in eV, for helium atoms in interstitial positions in fcc metals. Activation energies are given in the last column. The numbers in parentheses are the first-nearest-neighbor displacement parameters, p_1, in units of the half-lattice constant (see Table I).

Solid	He (1, 0, 0)	He (½, ½, 0)	He (½, ½, ½)	Activation Energy, eV
Ni	4.60 (-.121)	4.52 (-.246)	5.39 (-.092)	0.08
Cu	2.03 (-.069)	2.60 (-.215)	2.96 (-.074)	0.57
Pd	3.68 (-.071)	5.42 (-.182)	5.43 (-.062)	1.74
Ag	1.53 (-.052)	2.39 (-.200)	2.60 (-.058)	0.86

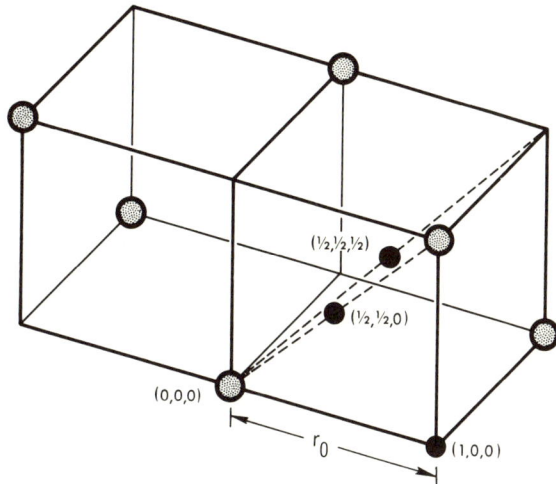

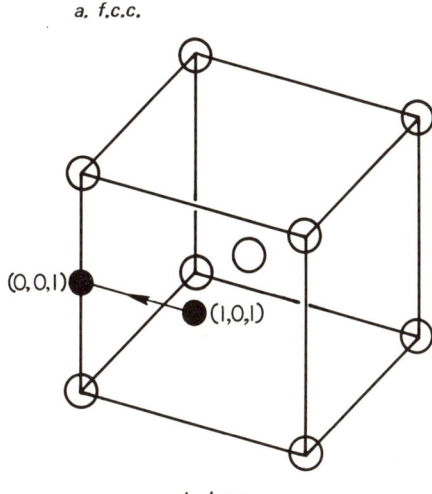

Fig. 2. Symmetrical interstitial positions for a helium atom in metal lattice. (a) in fcc; (b) in bcc. The (1,0,1) and (0,0,1) configurations are identical.

potential. The energy of the helium at the (½, ½, 0) position is also somewhat sensitive to this cutoff because it is also a close-packed configuration. The value of r_c chosen for the calculations given in Table II is the value that gives reasonable agreement with the experimental single-vacancy migration energy, E_{1V}^M, just as was done for the interstitial configurations in **Defect Calculations for Fcc and Bcc Metals**.

Several notable characteristics are contained in Table II. First, it is encouraging to note that the activation energy for helium interstitial migration in copper (0.57 eV) lies within the range predicted by the five central-force models of Wilson and Bisson (0.45 to 0.71 eV[7]), in spite of the fact that the model for copper used here has a different functional form, cutoff value, matching radius (to the Wedepohl potential at short range) and includes the noncentral terms. Nevertheless, the helium interstitial activation energy remains almost invariant. In silver, calculations show that helium interstitials will migrate with an activation energy of 0.86 eV, only a few tenths of an eV higher than in copper. Silver also behaves like copper in its insensitivity to variations in potential parameters. For example we varied the cutoff, r_c, from 0.85 to 0.95a (a is the lattice constant) and found less than 0.2-eV variation in helium interstitial activation energy.

Palladium stands out as the only metal in which a helium interstitial will require a high activation energy for motion (of the order of 1.5 eV). This anomaly naturally led to varying many of the parameters in the potential. First, we calculated the $Pd^+\text{-}Pd^+$ and $He\text{-}Pd^+$ potentials to determine the sensitivity of the results to the core valence. This resulted in an activation energy of 1.67 eV as compared to 1.74 eV given in Table II for the zero-valence case. The lattice constant was increased by 5 percent to see what effect this might have on helium motion. The formation energies in each configuration were reduced by a factor of 1.5, but the helium activation energy was determined to be 1.48 eV. Even varying the cutoff, r_c, did not have a significant effect on the helium migration energy in palladium. The situation in nickel is not quite so clear. Both the single-vacancy-migration energy and the helium interstitial activation energy are rather insensitive to a variation of parameters for cutoff values less than 0.91a. Up to this point, the vacancy migration energy is 1.6 eV according to the calculations, and helium motion has an activation energy of 0.9 eV. Increasing the cutoff to 0.93a will lower the value of E_{1V}^M to 1.20 eV, in better agreement with experiment, but also lowers the helium interstitial motion energy to less than 0.1 eV. The minimum energy configuration even changes from the (1, 0, 0) to (½, ½, 0) position of the helium, but this energy change is so small that no physical significance can be attached to it.

Table III contains the results of analogous calculations in bcc materials. A helium atom was fixed in each of eight different interstitial positions (and a ninth position which is redundant) to determine the minimum energy configuration and activation energy for helium mobility. Note that the energy of the helium atom in its (1,0,1) and (0,0,1) configurations are the same in all the solids studied (thus serving as an internal check on the calculations). The formation energies are all rather high, consistent with the low solubility of helium in these metals. If the saddle point for helium activation is assumed to be the tetrahedral (½, 0, 1) position, then the activation energy for helium migration is found to be given by the last row in Table III. These energies are much lower, in general, than those for the fcc materials owing to the more open structure of the bcc lattice. In tantalum, the saddle point configuration was lower in

Table III. *Formation energies, in eV, of helium atoms in interstitial positions in bcc materials. The origin is taken as a normal lattice atom.*

Coordinates of Helium Atom	Vanadium	Iron	Molybdenum	Tantalum	Tungsten
1, 0, 1	4.61	5.36	4.91	4.23	5.47
0.5, 0, 1	4.74	5.53	5.14	4.22	5.71
0, 0, 1	4.61	5.36	4.91	4.23	5.47
0.75, 0.25, 1	4.90	5.76	5.55	4.60	6.19
0.5, 0.5, 1	5.38	6.28	6.52	5.04	7.35
0.25, 0.75, 1	4.90	5.76	5.55	4.60	6.19
0.75, 0.25, 0.75	5.02	5.80	5.67	4.62	6.32
0.50, 0.50, 0.50	5.18	6.16	6.56	5.15	7.44
0.25, 0.75, 0.25	5.02	5.79	5.66	4.61	6.32
Activation Energy	0.13	0.17	0.23	~0.0	0.24

energy than the (1, 0, 1) position, but the magnitude (~0.01 eV) of this effect is too small to attribute any physical significance to it. It is likely, however, that the helium activation energy in tantalum is extremely small.

Next let us turn our attention to the question of substitutional motion of a helium atom – the next rare gas defect in order of complication. Fig. 3 shows the paths of motion of the helium atom during this substitutional detrapping process – a "popout" mode of diffusion – in both fcc and bcc lattices. Positions A and B are the nearest interstitial positions in which the helium atom can reside without falling back into the vacancy. The paths A_i are possible directions the helium may follow while diffusing toward position A; path B_1 seems a reasonable direction for diffusion to position B. Using a central potential for copper developed in a previous work*, the helium was placed at various positions along these paths and the saddle point determined to be the (1, ½, ½) position. The saddle-point energy was found to be 2.15 eV, and the difference in energy between the helium atom in a vacancy and an interstitial position was 1.84 eV. That is, a reasonable estimate of the popout energy can be obtained by calculating the relative energy of the helium in a vacancy and interstitial configuration (within ~0.3 eV in the case cited).

For the bcc materials the calculations were performed both by mapping the path (see Fig. 3b) and also by finding the relative energies discussed above. For the fcc metals the relative energy approximation was used to conserve computer time except for the central-force result given above. Table IV gives the results of these calculations. Note that without actually calculating the saddle-point configuration in bcc metals, the

*(Potential III of Wilson and Bisson[7]).

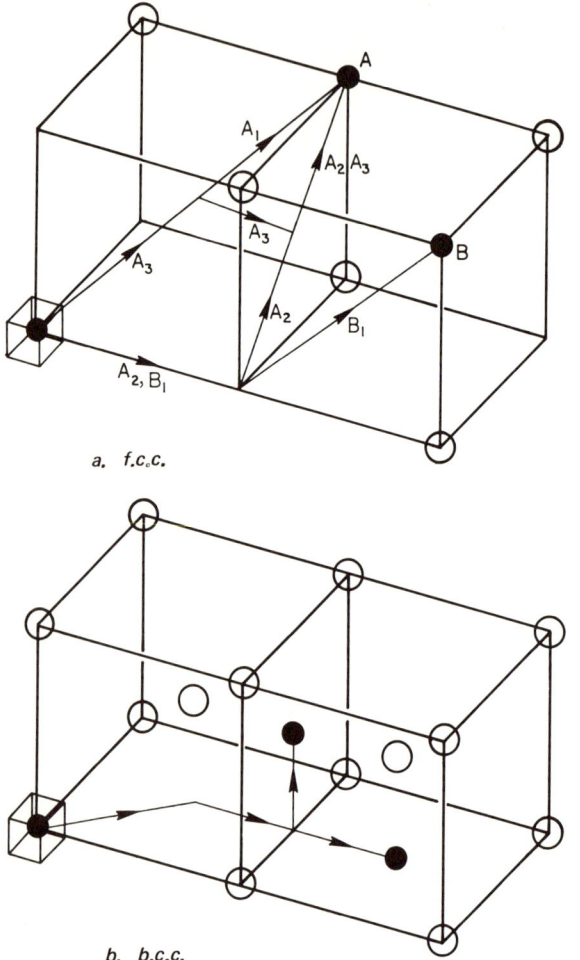

Fig. 3. Substitutional detrapping of an intert gas atom ("popout") in a metal lattice. (a) For fcc, the positions A and B are found to be stable positions of the He atom relative to recombination. Possible paths to each of these positions are indicated by subscripts. (b) For bcc metal lattice, the path of motion is indicated by the arrow.

substitutional detrapping energy would be determined to within 10 percent. Table IV also gives the energies of formation of a helium atom in a vacancy as a necessary constituent of the popout calculations. It is interesting to note that a helium atom goes into a silver vacancy without any change in strain energy whatever and into copper with only a small increase in energy.

To summarize, the activation energies for motion of a He atom in several fcc and bcc metals have been calculated. The potentials used for the host atoms, including noncentral forces in the fcc cases, enabled a fit to the elastic properties of these materials. The activation energies for interstitial motion were, in general, quite low except for palladium where several variations in the potential parameters involved were performed and the anomalously high activation energy remained constant within ~0.2 eV.

Table IV. *Substitutional detrapping energies in fcc and bcc metals. Values in parentheses are the relative energies of a helium atom in a vacancy and in an interstitial position.*

	Formation Energy, in eV, of a He Atom in a Vacancy	Formation Energy, in eV, of a He Interstitial	Substitutional Detrapping (Popout) Energy, in eV
Cu	0.15	2.03	(1.88)
Ni	1.36	4.52	(3.16)
Pd	0.52	3.68	(3.16)
Ag	0.0	1.53	(1.53)
Fe	1.61	5.36	3.98 (3.75)
Mo	1.04	4.91	4.19 (3.87)
W	1.05	5.47	4.75 (4.42)
V	1.65	4.61	3.20 (2.96)
Ta	0.93	4.23	3.44 (3.30)

For substitutional detrapping, the activation energy is consistently high, being somewhat higher in bcc materials than in the fcc materials.

Finally, it should be mentioned that calculations in gold are not yet completed because of the large relaxation energies of the defects. However, the He-Au$^+$ potential determined by the Wedepohl method has been included in Fig. 1a.

ACKNOWLEDGMENTS

It is a great pleasure to thank Dr. Walter Bauer of Sandia Laboratory for his many contributions and motivations during the course of this work. The authors are also grateful to C. L. Bisson, also of Sandia Laboratory, for invaluable programming assistance.

This work was supported by the U. S. Atomic Energy Commission under contract numbers AT(29-1)-789 and AT-(40-1)-3108.

REFERENCES

1. R. S. Nelson, D. J. Mazey and J. A. Hudson, J. Nucl. Mat. **37**, 1 (1970).
2. W. Bauer and W. D. Wilson, *He Migration in Metals,* Proceedings of the 1971 International Conference on Radiation Induced Voids in Metals (Albany, New York, June 9-11, 1971).
3. J. B. Holt, W. Bauer and G. J. Thomas, Radiation Effects, **7**, 269 (1971).
4. G. J. Thomas, W. Bauer and J. B. Holt, Radiation Effects (to be published, March 1971).
5. D. E. Rimmer and A. H. Cottrell, Phil. Mag. **2**, 1345 (1957).

6. A. Anderman and W. G. Gehman, Phys. Stat. Sol. **30**, 283 (1968).
7. W. D. Wilson and C. L. Bisson, Phys. Rev. (to be published, June 1971).
8. P. T. Wedepohl, Proc. Phys. Soc. **92**, 79 (1967).
9. J. B. Gibson, A. N. Goland, M. Milgram and G. H. Vineyard, Phys. Rev. **120**, 1229 (1960).

DISCUSSION on Paper by W. D. Wilson and R. A. Johnson

ROBINSON: The method which you ascribe to Wedepohl was developed around 40 years ago by Lenz and Jensen. It is described, for instance, in Ref. 1. The entire procedure, including $\rho^{5/3}$ kinetic energy and $\rho^{4/3}$ exchange contributions is described there for two-center systems.

WILSON: I suggest that questions about the originality of Wedepohl's work[2] be referred to Wedepohl.

ROBINSON: The first application of this technique using quantum-mechanical (rather than statistical) theory wave functions was by K. Günther[3] for Ar-Ar – how do your results compare to his?

WILSON: We restricted our comparisons for the most part to (i) molecular beam scattering results and (ii) molecular-orbital calculations because of the accuracy of these techniques. Abrahamson's work[4] is available (and in English) for all rare-gas pairs, so we included it because of the convenience of doing so without making any judgments about its correctness. Again, since I do not claim originality for the basic method, I refer questions of originality to the authors or perhaps to the historians who are usually the people who concern themselves with this.

ROBINSON: The high results obtained by Abrahamson have been attributed by Günther to a combination of a theoretical and a numerical error. Since Abrahamson has never publicly dealt with this question, why compare with his results?

WILSON: Perhaps Günther is right – the evidence seems to indicate he is. I compared with Abrahamson's results for the reason I have already stated above.

ROBINSON: Do you not worry about using the $\rho^{5/3}$ free-electron (Thomas-Fermi) kinetic energy contribution in your application? Could changes in this not be comparable to exchanges in importance?

WILSON: We did not vary the $\rho^{5/3}$ free-electron kinetic energy term as we did the $\rho^{4/3}$ exchange term. It is hard for me to be worried in the light of our excellent agreement with experiment for so many interactions.

ROBINSON: What was the source of your atomic wave functions?

WILSON: Hartree-Foch-Slater, i.e., the Herman-Skillman program.

DAHL: What mechanisms do you believe cause the reversal in the temperature-dependent release of helium shown in the ion-implantation results?

WILSON: The reversal takes place at higher temperatures and is believed to be due to plastic deformation caused by the buildup of gas pressure in the lattice. Scanning electron microscopy of the surface of the palladium samples by Thomas, et al.[5] show an actual raising of the grains after the high fluence reversal takes place. I refer you also to J. B. Holt et al.[6]

DAHL: What was the time in the isochronal annealing?

WILSON: It was a 20-minute isochronal release schedule.

DAHL: What is the nature and extent of the strain field caused by a cluster of helium atoms with and without the presence of a vacancy? Would the stress levels cause plastic deformation?

WILSON: I do not have the details here but the introduction of a vacancy to a helium-atom cluster should tend to lower the strains about the cluster. It is difficult to say on an atomistic basis if these stresses cause a plastic deformation.

DORAN: Rimmer and Cottrell, in an old paper, gave a prescription for estimating a gas-metal potential given the gas-gas and metal-metal potentials.[7] Does this method look reasonable in the light of your more sophisticated calculations?

WILSON: We did not reproduce the Rimmer-Cottrell potential for comparison. Our feeling is that extended molecular-orbital calculations and molecular-beam experiments give the most accurate potentials; so we restricted ourselves to comparisons with these. I agree that we should look at this, however, because we have done defect calculations which are similar to those of Rimmer and Cottrell.

HO: Your results on helium in nickel are very interesting. On one hand, the migration energy required is very small (0.08 eV) compared to other fcc metals, e.g., copper and silver (0.45 eV). On the other hand, the formation energy of an interstitial or a substitutional helium atom is very high (~4 eV) compared to other fcc metals (1 to 2 eV). It seems to be that this implies that helium is difficult to be put into nickel, but once it goes into nickel, it migrates rapidly. Can one correlate these results with the role of helium in void or bubble formation in nickel?

WILSON: You are quite correct in your statement about our nickel results. We are presently doing calculations of the effect of helium in preventing the collapse of three-dimensional vacancy clusters produced by neutron damage (the void problem). It is hoped that these results will shed some light on this important technological question.

JACKSON: E. Kornelsen[8] (NRC, Ottawa) has recently done some interesting experiments on He release from metal crystals. Are you familiar with this work and, if so, would comment on it as compared to your calculations?

WILSON: I have recently become aware of the measurements of Kornelsen on tungsten. He feels that he sees a He interstitial motion which is very rapid at room temperatures and also He atom detrapping modes which have activation energies as high as 8 eV. Our calculations in W indicate a 0.24 eV activation energy for interstitial motion and a 4.75 eV substitutional detrapping energy; not in disagreement with these measurements.

JACKSON: Predamaging the crystal with heavy ions before implanting the He produces a much larger population of defects. Therefore the He release profile consists of definite peaks over a wide temperature range (as found by Kornelsen) corresponding to a larger variety of He binding possibilities than would otherwise be seen without predamaging. Would you comment on the merits of this technique as a method of studying He metal interactions.

WILSON: Kornelsen has done some predamaging with Ar ions. I think this technique is very valuable provided questions about surface effects (he only uses ~40 KeV implants) and the role of the Ar itself are answered. I don't think the method gives *direct* information about He-metal potentials, however, as, for example, scattering data can provide.

KULCINSKI: What evidence do you have that the experimentally measured release of helium atoms is due to single atom migration as opposed to the formation and migration of helium bubbles, especially at the higher temperatures?

WILSON: This, of course, is the heart of the matter. There is no direct experimental evidence that the gas release is due to single He atoms rather than bubbles. I should point out, however, that our calculated activation energies for single He atom motions are at least in qualitative agreement with experiment.

WYNBLATT: Do you have any idea about the contribution to the experimental results from diffusion along dislocations formed during irradiation?

WILSON: We have not taken dislocations into account in our analysis to date but recognize their possible importance.

TORRENS: What aspect of your calculation causes the point of inflection in the logarithmic plot of the potential at about half the lattice spacing?

WILSON: All of our potentials are repulsive at short separations and have an attractive well at intermediate separations. Any such potential must have a point of inflection when plotted on a semi-log scale. Even the "Wedepohl" potentials themselves (or perhaps Lenz-Jensen potentials as March has pointed out) have shallow minima in them and hence would show a point of inflection on a semi-log scale.

DE WETTE: Have you treated the helium atom quantum mechanically? Quantum effects are very important for helium; it is because of large zero-point motion that helium does not solidify under its own vapor pressure at absolute zero. If a helium atom is caged in a vacancy or interstitial position the zero-point energy is probably quite considerable and may have an important effect on the energy considerations. Can one legitimately neglect it?

WILSON: Using a typical vibration frequency, v, of 10^{13} sec^{-1}, the zero point energy, $E = \frac{1}{2}hv$, turns out to be ~0.02 eV. Our formation energies for He in interstitial positions are ~2 eV, two orders of magnitude larger than zero-point energies.

REFERENCES

1. P. Gombas, Handbuch, d. Phys., Vol. 36 (Springer Verlag, Berlin, 1956) p. 109.
2. See Ref. 8 in text.
3. K. Günther, Ann. Physik (7) **14**, 296 (1964), Kernergie, **7**, 443 (1964).
4. A. A. Abrahamson, Phys. Rev. **123**, 538 (1961); **130**, 693 (1963); **133**, A990 (1964).
5. See Ref. 4 in text.
6. See Ref. 3 in text.
7. See Ref. 5 in text.
8. E. V. Kornelsen, Proceedings of the 1971 International Conference on Solid Surfaces, Boston; Mass., Oct. 11-15, 1971, to be published in J. Vac. Sci. Tech.

ON PSEUDOPOTENTIAL CALCULATIONS OF POINT DEFECTS IN METALS

R. Chang

North American Rockwell Science Center
Thousand Oaks, California 91360

ABSTRACT

Results of pseudopotential calculations of the relaxational parameters and formation energy of point defects in aluminum and sodium are presented. The results are encouraging and suggest that the apprehension "a second-order perturbation pseudopotential calculation may break down on account of strong localized distortion near the point defect" is not justified. Future interests in this area are (i) refinement of the pseudopotential form factors suitable for all-purpose electronic and atomic property calculations in the simple metals and (ii) search of pseudopotential form factors and computational techniques for similar calculations in the transition metals.

INTRODUCTION

Recent literature is rich in information describing the application of pseudopotentials (PP) to studies of the band structure and electronic properties of simple metals and semiconductors.[1-4] The PP approach to band structure has been of paramount importance in the elucidation of the electronic properties of solids. "First principle" band calculations must deal with difficult questions surrounding the construction of self-consistent crystal potentials and with elaborate machine calculations that often obscure the underlying physical features. By contrast, the PP method largely avoids these difficulties at very little, if any, cost in accuracy. The resulting simplicity has led to the development of deep physical insights that indicate quite clearly how certain features of the band structure arise and that have made possible straightforward calculation of rather complex physical features such as optical properties, phonon spectra, and even the small structural contributions to the cohesive energy.

The treatment of atomic-like properties in solids by the PP method has not yet been fully developed, and the number of applications in this area is rather limited.[5] Such slow development is due to the inaccuracies inherent in the method of calculation, i.e., the large cancellation of direct and indirect interactions between ions makes the result of PP calculations very sensitive to the choice of PP form factors.

This paper presents a summary of efforts on applying the PP method to study relaxation effects close to a specific type of point imperfections, namely single vacancies and interstitials. The subject is timely since it is the present apprehension of many that a strong localized perturbation may be associated with a vacancy or interstitial to cause breakdown of the second-order perturbation theory.[5] We are particularly interested in studying the sensitivity of the results to the choice of PP, and the suitability of experimentally determined PP[3] to describe localized properties rather than extended properties of the solids. Aluminum is chosen as one of the metals for study because of the simplicity of its structure and because its PP form factor has been tested not only in band-structure calculations[1,2,6], but also in lattice vibration spectra[7-10], binding energies[11], X-ray[12], optical spectra[13] and superconducting properties[14]. Sodium is the second metal chosen for study for similar reasons[15-22] and for the additional reason that it has a bcc structure in contrast to the fcc structure of aluminum. Furthermore, the experimental vacancy formation energies in aluminum[23] and sodium[24] are available in the literature for comparison with the PP calculations. It has also become increasingly clear that atomic calculations of point defects in metals via the PP method can provide criteria important to the selection of the best PP form factors for other atomic calculations. This is because the formation energy of point defects (vacancies in particular) in metals is usually in the range of 0.01 to 0.1 Ry (0.14 to 1.4 eV) and is extremely sensitive to the slightest variation of the PP form factors.

2 PP CALCULATIONS OF POINT DEFECTS IN SIMPLE METALS

The classical way of treating point defects in solids as if they were embedded in an elastic continuum is well known.[25-27] Relaxation in the neighborhood of the point defect according to certain interatomic potentials was simultaneously introduced.[28-37] Only very recently have the concepts of PP been introduced to treat point defects in simple metals.[38-40] The literature is still lacking information on a careful PP calculation of point defects in simple metals. Whether the strong localized perturbation in the neighborhood of a point defect will lead to a breakdown of the PP method remains to be seen.

In the PP formalism, the energy of formation of a point defect is calculated as a function of the distortion parameters of the atoms in the neighborhood of the point defect and minimizing it with respect to those parameters. The formation energy of a point defect ΔE_d is the difference in energy between a system of N atoms, with a point defect, occupying a volume $N\Omega$ and a perfect crystal of N atoms at the equilibrium volume $N\Omega_o$

$$\Delta E_d = E_d(N,\Omega) - E_p(N,\Omega_o) \quad . \tag{1}$$

Both E_d and E_p are quantities proportional to N, while ΔE_d is of order one. Due to the fact that Ω_o is the equilibrium volume, in Eq. 1 $E_d(N,\Omega)$ can be replaced by $E_d(N,\Omega_o)$, committing only a second-order error, of order $1/N$ and therefore negligible. Thus

$$\Delta E_d = E_d(N,\Omega_o) - E_p(N,\Omega_o) \quad . \tag{2}$$

Following Harrison, the formation energy can be written[1]

$$\Delta E_d = E_{fe} + \Delta E_{es} + \Delta E_{bs} \quad , \tag{3}$$

where the three terms are, respectively, the free-electron contribution, the electrostatic contribution, and the band-structure term. The free-electron energy of a system is a function of the volume of the system, and since the defect and the perfect systems have the same volume in Eq. 2

$$\Delta E_{fe} = 0 \quad . \tag{4}$$

The problem is now restricted to the calculation of ΔE_{es} and ΔE_{bs}. For both terms[1] it is necessary to calculate the structure factors for both the perfect (p) and the defect (d) structures

$$S(\mathbf{q}) = \frac{1}{N} \sum_j \exp(-i\mathbf{q}\cdot\mathbf{R}_j) \quad , \tag{5}$$

where q is the lattice wave number and R_j the position vector of atom j. For the perfect crystal, the structure factor is

$$S_p(\mathbf{q}) = \sum_G \delta_{q,G} \quad , \tag{6}$$

where G is the reciprocal lattice vector, and δ the Kronecker delta. For the defect crystal, it is useful to imagine its structure as derived from a perfect crystal with (N+1) atoms (vacancy case) or N atoms (interstitial case), to which we have (i) removed the atom at the origin $\mathbf{R} = 0$ (vacancy case) or inserted the Nth atom at a given interstitial site (interstitial case) and (ii) displaced the position of other atoms j in the neighborhood from $\mathbf{R}_{oj}$ to $\mathbf{R}_j$. Therefore

$$S_d(\mathbf{q}) = \frac{N+1}{N} S_p(\mathbf{q}) + \frac{1}{N} \{T(\mathbf{q}) - 1\} \ldots \text{(vacancy)} \tag{7a}$$

$$S_d(\mathbf{q}) = \frac{N}{N-1} S_p(\mathbf{q}) + \frac{1}{N-1} \{T(\mathbf{q}) - 1\} \ldots \text{(interstitial)} \quad , \tag{7b}$$

where

$$T(\mathbf{q}) = \sum_j [\exp\{-i\mathbf{q}\cdot\mathbf{R}_j\} - \exp\{-i\mathbf{q}\cdot\mathbf{R}_{oj}\}] \ldots \quad . \tag{8}$$

In the summation of Eq. 8, all relaxation is assumed to be radial, i.e.

$$\mathbf{R}_j = (1+D_j)\mathbf{R}_{oj} \quad . \tag{9}$$

This assumption is reasonable since the noncentral component is negligibly small compared with the central component of the ion-ion forces for the elements discussed here.[41,42]

Chang and Falicov[41] show that, for small displacement approximation, keeping terms only up to the second power in the displacements, the electrostatic energy contribution ΔE_{es} consists of four terms:

Δ_0 The electrostatic energy difference for the point defect without relaxation

Δ_1 the electrostatic energy arising from the displacement of each ion in the defect lattice

Δ_2 (the negative of) the interaction energy between the missing ion (vacancy case) or (the positive of) the interaction energy between the interstitial ion (interstitial case) with the dipole generated by each displaced ion

Δ_3 the electrostatic energy due to the mutual interaction between the dipoles generated by the displaced ions.

Similarly, the band-structure contribution ΔE_{bs} consists of three terms:

Δ_4 A constant term independent of the distortions, arising from the fact that in Eq. 7 the structure factor $S_p(q)$ corresponds to (N+1) atoms instead of N (vacancy case) or N atoms instead of (N-1) (interstitial case)

Δ_5 contribution arising from the mixed term in the square of Eq. 7

Δ_6 contribution from the square of the last term of Eq. 7 averaged over all angles in q space.

The pertinent expressions for Δ_0 to Δ_6 are described in detail in Ref. 41 and 42 and are not reproduced here. Computations of the terms Δ_4, Δ_5 and Δ_6 involved the use of the PP form factors $W(q)$ and the screen parameter $Q(q)$ also discussed in detail in these references.

2.1 Vacancy in Aluminum

The details of the calculation of relaxation parameters and formation energy of a vacancy in aluminum were treated in Ref. 41 and are not repeated here. It is shown there that the energy of formation of a vacancy in aluminum and the relaxation parameters of first few shells in the neighborhood of the vacancy are very sensitive to the choice of PP. The PP form factor $W(q)$ finally chosen is empirical and is obtained in two steps:

(i) $W(q) = [0.4416q^2 - 0.830] \left[\exp\{0.935 (q^2 - 0.877)\} + 1 \right]^{-1}$ (10)

The PP reproduces the Ashcroft[6] empirical values:

$W(000) = -0.5736$ Ry (2/3 Fermi energy)
$W(111) = +0.0179$ Ry
$W(200) = +0.0562$ Ry

(ii) A horizontal shift of the $W(q)$ versus q curve by a very small amount, i.e., replacing q by (q - 0.02) in the equation for $W(q)$. This horizontal shift is found necessary in order to yield not only the vacancy-formation energy in agreement with experiment but also reasonable relaxation parameters and formation energy of an interstitial in aluminum to be described in the following section. The results for a vacancy in aluminum are summarized in Table I.

The small horizontal shift of $W(q)$ yields not only the vacancy formation energy in agreement with experiment but also displacements smaller than those previously calculated by Chang and Falicov.[41]

2.2 Interstitial in Aluminum

The calculation is started by inserting an aluminum atom at the center of the fcc unit cell. The same PP described above for the vacancy

Table I. Relaxation (5 shell) and energy of formation of a vacancy in aluminum.

Shell	Displacement, $\Delta d/d$	Formation Energy, Ry.
110	−0.013	+0.059 (calculated)
200	−0.033	
211	−0.011	+0.055 − 0.080 (experimental, Ref. 23)
220	−0.007	
310	−0.005	

calculation is used. The detailed calculations are published elsewhere. Results of the calculations are summarized in Table II.

Table II. Relaxation (5 shell) and energy of formation of an interstitial in aluminum

Shell	Displacement, $\Delta d/d$	Formation Energy, Ry.
100	+0.158	
111	−0.031	+0.138 (calculated)
210	+0.056	
221	+0.008	
300	+0.007	

The relaxation parameters and the formation energy of an interstitial in aluminum are quite reasonable. This suggests that PP calculations of point defects in simple metals are promising and that the apprehension "a strong localized perturbation near the point defect will lead to breakdown of the PP calculation" is not justified.

2.3 **Vacancy in Sodium**

Details of the calculation of the relaxation parameters and energy of formation of a vacancy in sodium have been reported recently by Chang.[42] The following PP, obtained in two steps, is used:
(i) The optimized model potential of Shaw and Pynn.[43]
(ii) A small horizontal shift of the Shaw-Pynn optimized model potential (replacing q by $q + 0.04$) in order to fit exactly the Fermi surface data of Lee.[17]
The results of the calculation are shown in Table III.

Table III. Relaxation (9 shell) and formation energy of a vacancy in sodium

Shell	Displacement, $\Delta d/d$	Formation Energy, Ry.
111	− 0.047	
200	+ 0.013	+ 0.035 (calculated)
220	− 0.014	+ 0.031 ± 0.002
		(experimental, Ref. 24)
222	− 0.011	
400	+ 0.006	
331	− 0.003	
420	+ 0.0003	
422	− 0.0017	

DISCUSSION

The horizontal shift of $W(q)$ along the q axis is arbitrary, and is, like the vertical shift proposed by Lin and Phillips[44], for the convenience of calculation. It would be more desirable, e.g., to plot $W(q)$ versus q^2 and shift $W(q)$ along the q^2 axis for the obvious reason that a linear term in q will not be generated in Eq. 10 for $W(q)$. Better yet, the constants in Eq. 10 should be redetermined to fit not only the Fermi-surface data, but also defect-formation energies and other atomic-property calculations. Since there is as yet no unique way of writing out the PP for simple metals, the small vertical or horizontal shifts are considered adequate for the present if such shift yields the calculated point-defect formation energy in agreement with experiment. This is because the formation energy, particularly for a vacancy, is small and extremely sensitive to the choice of the PP. It is hoped that calculations using the same PP can be extended to other studies and that further minor modification of the PP reported in this paper may be necessary in order to yield a generalized PP for all-purpose electronic and atomic property calculations. It is also desirable in future calculations to introduce noncentral displacements of the ions and to use Eq. 1 more rigorously in preference to Eq. 2 wherein a volume change of the defect crystal is neglected.

Harrison suggested that the volume-independent part of the structure-dependent band-structure energy may be thought as the indirect interaction between the ions and may be obtained from the PP form factors by Fourier integration.[1] This, added to the electrostatic part of the interaction, will provide a total effective interaction and is given as a two-body, central force interaction potential between the ions:[4]

$$V(R) = \frac{Z^2 e^2}{R} + \frac{2\Omega}{(2\pi)^3} \int \phi_{bs}(q) \exp(i\,\mathbf{q}\cdot\mathbf{R})\, d^3q \tag{11}$$

where Z is outer number of electrons of charge e; R is the interionic distance; Ω is the atomic volume; q is the lattice wavenumber; ϕ_{bs} is the energy-wavenumber (or band structure) characteristic. Eq. 11 has been used by several investigators to obtain the phonon dispersion and point-defect relaxation and formation energy of the alkali metals.[19,38,39] The displacements obtained by these investigators for a vacancy in sodium are much greater than those calculated by the PP method.[42] Since our calculations have shown that the large cancellation of the electrostatic and band-structure energies requires a proper choice of the PP and extremely careful formulation of the electrostatic and band-structure terms, the much simplified Eq. 11 does not appear to warrant any degree of accuracy or reliability.

Since there is presently no way of writing out the PP form factors that are unique, the only feasible means is to fit PP form factors to as many sets of reliable experimental data as possible. This is further complicated by the lack of fundamental understanding of (i) exchange and correlation effects of the electrons, (ii) nonlocality of PP and the on-Fermi-surface approximation, (iii) energy dependence of PP, (iv) structure dependence of PP (such as the k dependence of the diagonal matrix elements), (v) breakdown of the second-order perturbation theory, etc. These subjects have been discussed in considerable detail by Heine and Weaire[4] and are not repeated here.

PP calculations of electronic and atomic properties of transition metals are far from satisfactory. Several recent publications deal with the choice of PP for the transition metals.[45-51] The author has recently tried to calculate the formation energy and relaxation of a vacancy in copper based on the PP recently proposed by Moriaty[51] without much success. Since the development of PP for the calculation of point defects in transition metals is a difficult task at present and defect calculations in transition metals employing empirical two-body interatomic potentials[28-37] have met limited success, we might have to be satisfied with empirical two-body potential calculations for the transition metals at present.

REFERENCES

1. Harrison, W. A.: *Pseudopotentials in the Theory of Metals* (W. A. Benjamin, Inc., New York, 1966).
2. Falicov, L. M.: *Energy Bands in Metals and Alloys,* ed. by L. H. Bennett and J. T. Wafer (Gordon and Breach, New York, 1968) p. 73.

3. Cohen, M. L. and Heine, V.: *Solid State Physics* (Academic Press) **24**: 38, 1970.
4. Heine, V., and Weaire, D.: *Solid State Physics* (Academic Press) **24**: 250 (1970).
5. Heine, V., and Weaire, D.: *Solid State Physics* **24**: 345 (1970).
6. Ashcroft, N. W.: Phil. Mag. **8**: 2055 (1963).
7. Animalu, A.O.E. et al.: Nuovo Cimento **44B**: 159 (1966).
8. Coulthard, M.: J. Phys. **C3**: 820 (1970).
9. Gupta, H. C., and Tripathi, B. B.: Phys. Rev. **B2**: 248 (1970); J. Chem. Phys. **54**: 1883 (1971).
10. Wallace, D. C.: Phys. Rev. **187**: 991 (1969).
11. Kleinman, L.: Phys. Rev. **146**: 472 (1966).
12. Rooke, G. A.: J. Phys. **C1**: 776 (1968).
13. Hughes, A. J., et al.: J. Phys. **C2**: 102 (1969).
14. Carbotte, J. P., and Dynes, R. C.: Physics Letters **25A**: 532, 685 (1967).
15. Schneider, T., and Stoll, E.: *Neutron Inelastic Scattering,* Intern. Atomic Energy Agency, Vienna, 1968.
16. Rice, T. M.: Phys. Rev. **175**: 858 (1968).
17. Lee, M.J.G.: Proc. Roy. Soc. **A295**: 440 (1966).
18. Cowley, R. A., et al.: Phys. Rev. **150**: 487 (1966).
19. Ho, P. S.: Phys. Rev. **169**: 523 (1968).
20. Shyu, W. M., and Gaspari, G. D.: Phys. Rev. **163**: 667 (1967).
21. Kushwasha, S. S., and Rajput, J. S.: Phys. Rev. **B2**: 3943 (1970).
22. Price, D. L., et al.: Phys. Rev. **B2**: 2983 (1970).
23. Simmons, R. O., and Balluffi, R. W.: Phys. Rev. **117**: 52 (1969); ibid **119**: 600 (1960).
24. Feder, R., and Charbnau, H. P.: Phys. Rev. **149**: 464 (1966).
25. Eshelby, J. D.: Acta Met. **3**: 487 (1955); Solid State Physics **3**: 79 (1956).
26. Lie, K. C., and Kohler, J. S.: Advances in Physics **17**: 421 (1968).
27. Masumura, R. A., and Sines, G.: J. Appl. Phys. **41**: 3930 (1970).
28. Huntington, H. B.: Acta Met. **2**: 554 (1954).
29. Kanzaki, H.: J. Phys. Chem. Solids **2**: 24 (1957).
30. Tewordt, L.: Phys. Rev. **109**: 61 (1958).
31. Girifalco, L. A., and Weizer, V. G.: Phys. Rev. **114**: 687 (1959).
32. Gibson, J. B., et al.: Phys. Rev. **120**: 1229 (1960).
33. Seeger, A., et al.: J. Phys. Chem. Solids **23**: 639 (1962).
34. Johnson, R. A., and Brown, E.: Phys. Rev. **127**: 446 (1962).
35. Johnson, R. A.: Phys. Rev. **134**: A1329 (1964); Acta Met. **12**: 1215 (1964); Acta Met. **13**: 1259 (1965); also this conference.
36. Doyama, M., and Cotterill, R.M.J.: Phys. Rev. **137**: A994 (1965).
37. See also several papers in *Proceeding of Conference on the Calculation of the Properties of Vacancies and Interstitials,* U.S. National Bureau of Standards, Special Publication No. 287, 1966 (U.S. Government Printing Office).
38. Torrens, I. M., and Gerl, M.: Phys. Rev. **187**: 912 (1969).
39. Ho, P. S.: this conference.
40. Flocken, J. W., and Hardy, J. R.: Phys. Rev. **177**: 1054 (1969).
41. Chang, Roger, and Falicov, L. M.: J. Phys. Chem. Solids **32**: 465 (1971).
42. Chang, Roger: J. Phys. Chem. Solids **32**: 1409 (1971).
43. Shaw, R. W., Jr., and Pynn, R.: J. Phys. **C2**: 2071 (1969).
44. Lin, P. J., and Phillips, J. C.: Advances in Physics **14**: 257 (1965).
45. Hodges, L., and Ehrenreich, H.: Physics Letters **16**: 203 (1965).
46. Mueller, F. M.: Phys. Rev. **153**: 659 (1967).
47. Deegan, R. A.: J. Phys. **C1**: 763 (1968).
48. Pettifor, D. G.: J. Phys. **C3**: 366 (1970).
49. Harrison, W. A.: Phys. Rev. **181**: 1036 (1969).
50. Deegan, R. A.: Phys. Rev. **188**: 1170 (1969).
51. Moriaty, J. A.: Phys. Rev. **B1**: 1363 (1970).

DISCUSSION of paper by R. Chang

BULLOUGH: It seems essential to me to remove the Kanzaki forces so that the time equilibrium configuration is achieved.

VITEK: It is known that in metals the volume-dependent part of the crystal energy is generally much larger than the part corresponding to the pair interaction (in the simplest approximation it is given by the deviation from the Cauchy relations for the elastic constants). The point defects, in contrast to dislocations, for example, are essentially the defects changing the volume of the crystal (e.g., they may be considered as dilatation centers). In your calculations, however, you assumed that the volume-dependent part of the energy may be neglected.

DUESBERY: I am rather confused by the formulation of this problem in reciprocal space. In real space, the effect of the volume term is quite clear. A Taylor expansion of the lattice ion-ion potential energy in small displacements will yield first-order terms that are zero only if the elastic constants satisfy the Cauchy relations. Equilibrium is achieved by adding the volume-dependent contribution of the electron gas, which must contribute a term $1/2\ (C_{12} - C_{44})\ \delta V$, where δV is the absolute volume charge associated with the ionic displacements. A point defect in a discrete solid gives rise to a volume change of order one atomic volume. For a sufficiently large crystal, the associated strains and displacements of the ions tend to zero. Hence, the ion-ion energy, which depends only on relative displacements, or, equivalently, strains, does not change due to this volume change. The volume-dependent term, however, which depends on the absolute volume change, will contribute an energy of the order of $1/2\ (C_{12} - C_{44})\ V_a$, where V_a is the atomic volume. This term, which in real-space calculations, is frequently neglected, can be a large fraction of an electron volt for typical metals. While Dr. Ho's work seems to include this term, I think Dr. Chang may have neglected it.

JOHNSON: Just a comment on the effect of volume relaxation. If you allow no volume relaxation, a certain amount of energy is "stored" in bonds. If you allow volume expansion, the interior region does work on the elastic region. This can give rise to a large $P\Delta V$ energy, i.e., we "transfer" energy from the "bonds" to "elastic stored energy". Certainly the configuration is altered by this transfer, but the defect indulges a negligible change.

CHANG: The pseudopotential formalism presented here simply avoids the question of the effect of volume change on point-defect formation energy since it is shown that the effect is of second order, hence negligible. If one removes the Kanzaki forces as suggested by Dr. Bullough or considers the contribution $1/2 \, (C_{12} - C_{44}) \, \delta V$ as discussed by Dr. Duesbery, other corrective terms will come into play in the formulation which will cancel essentially most of the energy contributions arising from such a consideration. It is best to leave further discussions to Professor Hirth. In future pseudopotential calculations, however, it would be desirable to assess the effect of volume change on all three contributions to the defect formation energy ΔE_{fe}, ΔE_{es}, and ΔE_{bs}, even if it is a second-order effect.

WILSON: I noticed that the vacancy and interstitial formation energies which you calculate for increasing numbers of relaxed shells (up to 5) did not converge. How can you compare your value to experiment?

CHANG: They should converge very rapidly beyond the fifth shell, although actual calculations have not been carried out to prove the point.

COMPUTER SIMULATION OF THE SHORT-TERM ANNEALING OF DISPLACEMENT CASCADES

D. G. Doran

WADCO Corp., Hanford Engineering Development Laboratory
Richland, Washington

and

R. A. Burnett

Battelle, Pacific Northwest Laboratories
Richland, Washington

ABSTRACT

An important source of damage to a solid exposed to high-energy neutrons (or ions) is the displacement of atoms from normal lattice sites. In a fast reactor, energies of tens of keV may be transferred to an atom and thus initiate a displacement cascade consisting of a localized high density of interstitials and vacancies. These defects will subsequently interact with one another to form clusters and to reduce their density by mutual annihilation. This short-term annealing has been simulated with a small computer using an atomic model of γ-iron based on the work of

Johnson. The input cascades are due to Beeler. Results were obtained with both large (104 sites) and small (32 sites) annihilation regions. The former results in about one-half the residual defects of the latter, and a smaller fraction of clustered defects. Cluster size distributions and several examples of spatial distributions are given. Randomizing the spatial distribution of defects in a typical cascade geometry is found to diminish vacancy clustering and enhance interstitial clustering.

1 INTRODUCTION

An important source of damage to a solid exposed to high energy neutrons (or ions) is the displacement of atoms from normal lattice sites. In a fast reactor, energies of tens of keV may be transferred to the primary knock-on atom (PKA), producing in medium and high atomic weight materials a displacement cascade comprising a localized high density of vacancies and interstitials. These defects will subsequently interact with one another to produce clusters and to reduce their density by mutual annihilation – a process called short-term annealing in the present context.

The objective of this work was to determine by computer simulation the defect-clustering characteristics of γ-iron, a stand-in for the stainless steel used in fast reactors. Both high and low temperatures were studied, but only the high temperature results are reported here. Also investigated was the influence of the cascade formation process on the short-term annealing characteristics of a given density of defects. This was examined with a "randomized cascade" created by introducing defects at randomly selected positions within a typical cascade volume.

In a previous paper[1] the annealing simulation work initiated by Beeler[2] and Besco[3] was extended and applied to a description of short-term annealing of cascades in α-iron (bcc lattice). This paper is concerned with a new program to simulate short-term annealing in an fcc lattice, and its application to cascades produced by Beeler.[2] The new code employs a model for γ-iron based on the work of Johnson[4]. Beeler used a copper model in his fcc work, but there is no basis for distinguishing between copper and γ-iron at the present level of sophistication.

2 COMPUTATIONAL DETAILS

2:1 The Machine

A hybrid computer facility* consisting of a Beckman 2133 analog computer and a PDP-7 digital computer (8K memory) was used in this

---
*Operated by Computers and Control Section of Battelle-Northwest.

work. A random noise generator used to generate random numbers was the only contribution from the analog side. The digital program, HAPFCC (Hybrid Anneal Program – FCC), was written in assembly language.

The primary reason for using the small computer is the high level of man-machine interaction it provides – a significant consideration because the annealing runs are open-ended. Parameters can be changed through teletype entry at the beginning of a run, and several options can be exercised through console switches to control a run in progress. In addition, the program runs faster than an equivalent FORTRAN program on a UNIVAC 1108, and the cost is much lower with the small machine.

Initial defect distributions were read into the computer from punched cards, and intermediate and final distributions were stored on magnetic tape. Both tabular and graphic outputs were recorded with an electrostatic line printer.

HAPFCC can handle a maximum of 432 defect pairs within that portion of the 8K memory not required for program instructions. This was sufficient to handle the largest cascades presently available, but created a minor limitation in studying randomized cascades (as shown below).

2.2 Program Description and Operation

The annealing program, a correlated random walk on an fcc lattice, is defined by the following: the identity of mobile defects, jump vectors and jump probabilities of mobile defects, definition of correlated jumps and their relative jump probabilities, clustering criteria, and annihilation criteria.

During each time step, each member of each migrating cluster (of the appropriate type) is considered once, in random order, for a jump. A permissible jump vector is chosen at random, possible correlations sought, and the jump performed or not performed according to the assigned probability through selection of another random number. If a jump results, an examination is made for possible clustering or annihilation.

High-temperature (nominally 800 K) operation was characterized by two successive stages because of the large difference in mobilities of vacancies and interstitials. During the high-temperature-interstitial stage, the much slower vacancies were ignored except as they participated in annihilation of migrating interstitials. During the subsequent vacancy stage, interstitials were immobilized to conserve computer time. This procedure poses no problem at the beginning of the vacancy stage because only immobile interstitial complexes remain in the vicinity of the vacancies. Ultimately, however, some of the by-products of annihilation are mobile interstitials. An effort was made to determine the behavior of the newly created mobile interstitials (see Results section).

The time step in the high-temperature-interstitial stage was such that the uncorrelated jump probability of the most mobile defect (I_1)* was 0.5; the correlated probability was unity. The vacancy stage time step was chosen so that the highly mobile V_2 jumped once per step.**

It was found that 1000 to 2000 time steps in a given stage generally reduced the frequency of defect interactions to the order of one per 500 steps.

Ten 20-keV and ten 5-keV cascades were selected for processing. These cascades are a subset of those that provided the cluster size distributions of Ref. 2. Those distributions were obtained by Beeler from a computer program CLUSTER which applied clustering criteria to the output of the CASCADE program. For the present work, only the CASCADE outputs were utilized. Inherent in them is the recombination of all related interstitial-vacancy pairs (unrelated pairs are not recognized) separated by less than $\sqrt{5}$ half-lattice units (hlu); the interstitials are at octahedral sites. The initial operation of the HAPFCC program is to assign each octahedral interstitial to a randomly chosen neighboring fcc site and to assign to it an orientation along a randomly chosen major axis.

The spatial distribution of defects in a displacement cascade is rather special in that a vacancy-rich central region is surrounded by an interstitial-rich peripheral region. An effort was made to determine how strongly this configuration influences the cluster-size distribution. The geometrical configuration of one 20-keV cascade (2076) was determined by computer plotting the defects plane by plane; a simplified representation of the contribution from each plane was used (Fig. 1). A library of alternating interstitial and vacancy sites was prepared by randomly selecting sites within the geometrical figure that resulted from stacking the slabs (total volume of ~15,000 hlu^3). A "randomized cascade" was created by selecting 400+ defects of each type from this library, eliminating unlike defects that were first or second neighbors (as was done with 20-keV cascades), adding more defects, etc., until the desired number of defects was reached. This iterative procedure was necessary because of the program limitation of 432 defect pairs. Only two iterations were required to reach the goal of ~250 pairs, the number of defects in a 20-keV displacement cascade after first- and second-neighbor unlike defects have recombined.

3 THE γ-IRON MODEL

The model is based on Johnson's simulation of point defects and small clusters in γ-iron.[4] The activation energies found by him lead to the

---
*The symbols I_n and V_n designate interstitial and vacancy clusters, respectively, of size n.
**The relative jump probabilities should be reasonably representative of the temperature range, 600 to 1000 K.

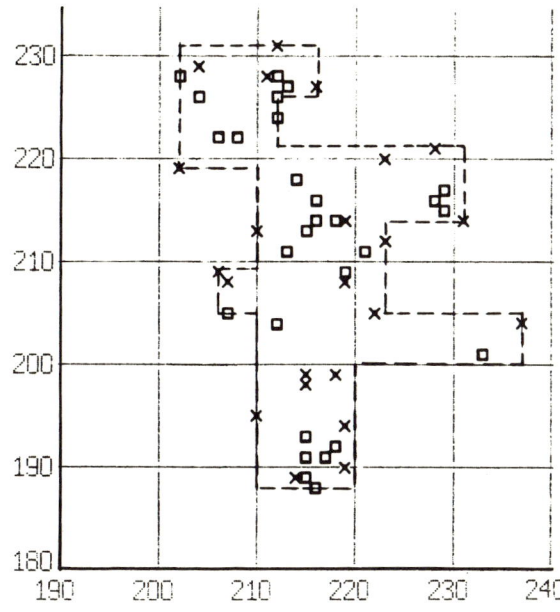

Fig. 1. A single plane through the pre-anneal configuration of a 20-keV cascade (2076), and its representation used in defining a randomized cascade.

following hierarchy of the mobile defects included in this work, starting with the most mobile: I_1, I_2, I_3, V_2, V_3, V_4, and V_1. The large gap in mobility between the slowest moving interstitial (I_3) and the fastest moving vacancy (V_2), even at high temperatures, led to separate interstitial and vacancy annealing stages in HAPFCC.

Johnson found interstitial behavior to be very complex, and his treatment is incomplete. While all of his results are not incorporated, their essence has been retained.

The stable I_1 is, following Johnson, the <100> split interstitial. In migrating, an x-oriented I_1 can jump to eight possible first-neighbor sites such that $\Delta x = \pm 1$; e.g., a jump in the x-z plane corresponds to $\Delta y = 0$, $\Delta z = \pm 1$ and a new (z) orientation. The motion of an I_1 that is within a 4th-neighbor separation of another I_1 is correlated such that the jump probability is unity toward the formation of the stable I_2 (see Fig. 2a). If relative orientations preclude stable I_2 formation, the jump probability for increasing the separation is unity. If an I_1 is within a 4th-neighbor separation of a clustered interstitial or two or more interstitials (clustered or single), a jump to decrease the total separation has a probability of unity, while a jump to increase the total separation is not permitted.

In addition, a longer-ranged correlation for single interstitials can be used; viz., a jump toward another interstitial has probability α if the proposed jump site is a 4th or nearer neighbor of the other interstitial (in the present work, $\alpha = 1$).

A first-neighbor bond constitutes clustering. The I_2 configuration is always two parallel interstitials, their common orientation being

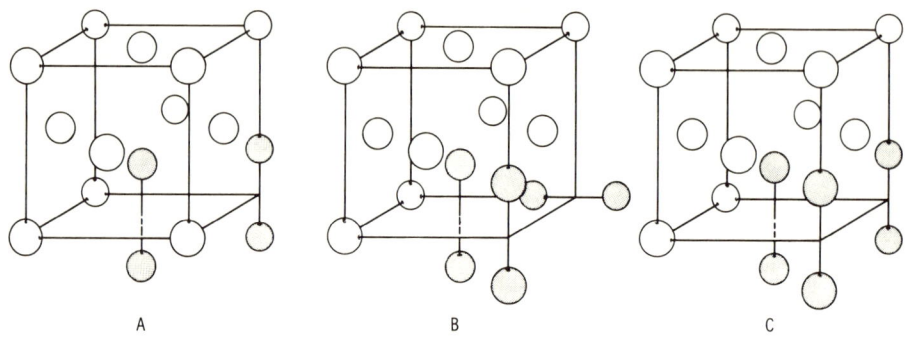

Fig. 2. Interstitial configurations studied by Johnson.[4] (a) The stable I_2. (b) The most stable I_3. (c) An alternative form of I_3.

perpendicular to a line joining their centers (Fig. 2a). The I_2 migrates only in the plane perpendicular to its orientation, with an uncorrelated jump probability of 0.12 relative to the I_1. The 4th-neighbor correlation scheme above applies to each member of an I_2.

The I_3 is permitted to exist in two I_3 (112)* configurations (see Fig. 2b, c). Since the I_2 is already in the most stable configuration, the third interstitial can take up one of four sites in the plane perpendicular to the orientation of the I_2. Furthermore, it can be oriented parallel to the I_2 or lie in the plane of the I_3; the latter (Fig. 2b) was found by Johnson to be the lowest-energy configuration. Migration of the "parallel" I_3 is confined to the plane of the defect, but the most stable I_3 can migrate in three dimensions. Jump probabilities relative to the I_1 are 0.011, but the stable form can reorient with a relative probability of 1.0.

An interstitial cluster of four or more is considered immobile, and no orientation information of its members is retained. Hence the morphology of the cluster has no significance. No dissociation of clusters is permitted.

Mobile species of vacancies are confined to V_1, V_2, V_3, and V_4. Either member of the V_2, a nearest-neighbor pair, can jump to one of the four first-neighbor sites in the plane bisecting their bond. Dissociation of the V_2, or any vacancy cluster, is not permitted.

The stable forms of the V_3 and V_4 are the triangle, V_3 (111), and the tetrahedron, V_4 (111111), and these configurations are immediately imposed when such clusters form. The V_3 can reorient by shifting the associated vacancy among the four sites of the tetrahedron; the reorientation frequency is equivalent to the V_2 jump frequency. The V_3 migrates with either a V_3 (112) or V_3 (113) intermediate state. The V_4 migrates

---

*Indicates 3 bonds, 2 of which are 1st-neighbor bonds and one of which is a 2nd-neighbor bond.

Table I. Vacancy correlations.

Defect	Jump Type	Activation Energy, eV	Jump Probability Relative to Isolated V_2
V_1	Isolated jump	1.32	0.002
$V_1^{(a)}$	One $2 \to 1$ [b]	1.28	0.004
	Two $2 \to 1$	1.24	0.007
	One $3 \to 1$	1.07	0.086
	One $4 \to 1$	1.07	0.086
	Two $3 \to 1$	0.82	1.0
	Two $3 \to 1$ + One $4 \to 1$	0.57	1.0
	Four $3 \to 1$	0.32	1.0
$V_2^{(c)}$	Isolated jump	0.9	1.0
$V_3^{(c)}$	Reorientation	0.9	1.0
	Isolated jump	1.02	0.165
$V_4^{(c)}$	Isolated jump	1.15	0.024

(a) May be single or member of a cluster.
(b) $m \to n$ designates jump from m^{th} neighbor separation to n^{th} neighbor separation.
(c) Data refer to one member of the cluster.

effectively one half-lattice unit in a <100> direction, undergoing simultaneously a 90-degree rotation. Higher-order vacancy clusters are considered immobile.

Vacancy correlations, limited to short range, are summarized in Table I. Some of the associated activation energies are based on work by the authors using Johnson's γ-iron potential in the code DEFECT.* Relative jump probabilities are, of course, dependent on the temperature.

An important but inadequately defined parameter, the annihilation region (AR), describes the attraction of a vacancy for an interstitial and the resulting annihilation of both defects. The AR is a set of sites relative to the position of the interstitial and dependent on its orientation, such that, if a vacancy occupies one of the sites, the interstitial and vacancy are considered to suffer mutual annihilation.

---

*The principal difference between DEFECT, developed by Beeler, and Johnson's program is that a rigid boundary in the former replaces the elastic continuum of the latter.

In a previous paper,[1] the AR was taken to increase with temperature, but this procedure was probably unjustified.[5] For temperatures so low that interstitial migration is rare, the effective AR would indeed increase with temperature, but, as interstitials become freely migrating, the vacancy sinks lose their effectiveness and the effective AR decreases.

The approach in the present work is to study the sensitivity of the results to the AR by using two regions (see Table II). The small AR is the 32-site spontaneous (0 K) annihilation region determined by Johnson; the large AR is a 104-site region obtained by moving one interstitial jump beyond the 32-site region.

Table II. Annihilation regions. Let split interstitial be located at origin and oriented in direction b.

Neighbor	Type	Exceptions	Number
Small Region (32 Sites)			
1	110	$\Delta b = 0$	8
3	211	$\Delta b = \pm 2$	16
4	220	$\Delta b = 0$	8
Large Region (104 Sites)			
1	110	$\Delta b = 0$	8
2	200	$\Delta b \neq 0$	4
3	211	$\Delta b = \pm 2$	16
4	220	None	12
5	310	$\Delta b = 0$	16
6	222	None	8
7	321	$\Delta b = \pm 2$	32
9	330	$\Delta b = 0$	8

4 RESULTS AND DISCUSSION

4:1 Displacement Cascades

4.1.1 Time History. As indicated above, computer runs were continued until defect interactions became infrequent – generally 1000 to 2000 time steps in each stage. One cascade was carried to 4000 time steps; in this case, no interactions occurred after 2000 steps.

Little annihilation took place in the vacancy stage relative to that in the interstitial stage, but there was a general increase in the fraction of

Table III. Average annihilation characteristics.

	Number of Defect Pairs					
	γ-Iron (Present Work)				α-Iron (Ref. 1)	
	20 keV		5 keV		20 keV	5 keV
Annealing Stage	Large AR[a]	Small AR[b]	Large AR[a]	Small AR[b]	(c)	(c)
CASCADE output	406	406	93	93		
CASCADE output, 1st and 2nd neighbors annihilated	253	253	—	—		
			High Temperature			
After initial annihilation	107	188	25	50	140	40
After interstitial stage	52[d]	102[d]	12[d]	22[e]	50	13
After vacancy stage	43[f]	79[f]	11[f]	19[g]	42	12

(a) 104-site AR.
(b) 32-site AR.
(c) 62-site AR.
(d) 2000 steps in interstitial stage.
(e) 1000 steps in interstitial stage.
(f) Additional 2000 steps in vacancy stage.
(g) Additional 1000 steps in vacancy stage.

mobile interstitials. This occurs when a migrating vacancy interacts with an immobile interstitial cluster so as to reduce the cluster (or a portion of it) to three or less. (Of course, there is general attrition of larger clusters also.) Several runs in which this phenomenon was prominent were continued in the interstitial mode to discover the fate of the newly created mobile defects. It was found that there is a strong tendency for these newly created defects to recluster with one another.

Decreasing the AR from 104 to 32 sites simply increased the rate of annihilation by the same factor that the defect population was increased; i.e., the percentage decrease per time step in the number of defects was the same in both cases.

4.1.2 Residual Annihilation and Clustering. The mutual annihilation of interstitials and vacancies that occurred during the short-term annealing simulation is summarized in Table III. The number of residual defects was approximately 2 pair/keV between 5 and 20 keV for the large AR and ~4 pair/keV for the small AR. The cluster-size distributions are given in Tables IV and V. With the large AR, 20-30 percent* of the surviving interstitials and 40-60 percent of the surviving vacancies are clustered. Most of the other defects migrate away and lose their identification with the cascade. The corresponding numbers for the small AR are 35-50 percent of the surviving interstitials and 60-70 percent of the surviving vacancies.

The spatial distribution of defects is illustrated in Fig. 3 to 5. In each postanneal configuration, the mobile interstitials have been removed to

---

*The percentage varied with cascade energy. In each case, the first number corresponds to 5 keV, the second to 20 keV.

Table IV. Interstitial cluster size distributions, high-temperature case

	Number of Interstitial Clusters of Size n per Cascade					
	γ-Iron				α-Iron (Ref. 1)	
	20 keV		5 keV			
n	Large AR[a]	Small AR[b]	Large AR[a]	Small AR[c]	20 keV	5 keV
1	11.7	12.5	6.0	5.9		
2	5.9	5.0	1.0	1.5		
3	2.6	6.25	0.3	1.2		
4	1.7	3.0	0.3	1.2		
5	0.3	2.5	0.2	0.1		
6	0.4	1.5	0	0		
7	0.1	1.0	0	0.1		
8	0.1	0	0	0.1		
9	0	0	0	0		
10	0	0.25	0	0		
≥3	5.2	14.5	0.8	2.7	6.6	1.9
≥4	2.6	8.25	0.5	1.5	2.5	0.4
≥5	0.9	5.25	0.2	0.3	0.8	0

(a) Average of 10 cascades; 2000 interstitial + 2000 vacancy time steps.
(b) Average of 4 cascades; 1000 interstitial + 1000 vacancy time steps.
(c) Average of 10 cascades; 1000 interstitial + 1000 vacancy time steps.

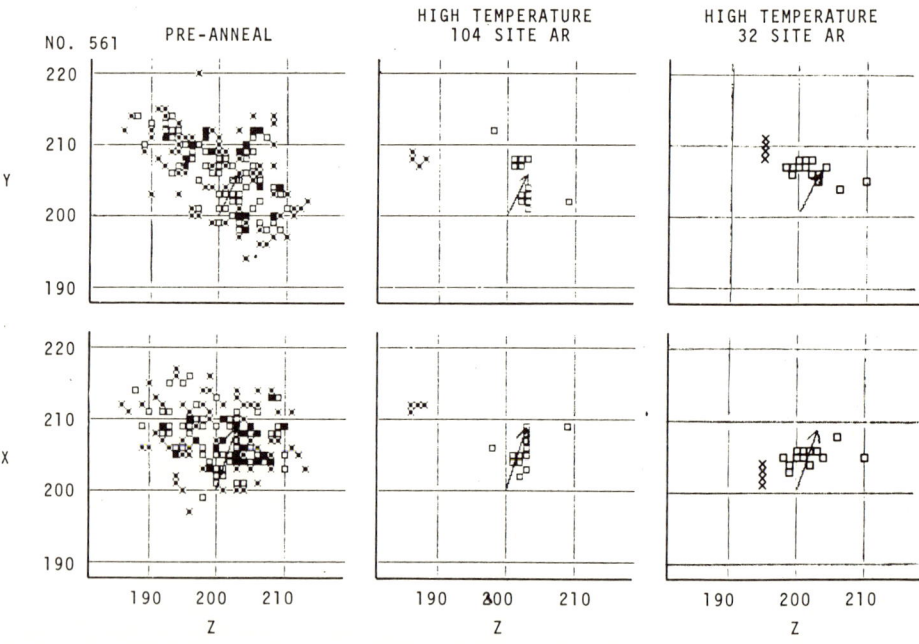

Fig. 3. A projection onto two planes of pre- and postanneal configurations of a 5-keV cascade for two annihilation regions. Squares and X's are vacancies and interstitials, respectively. Take-off point of PKA is 200, 200, 200; direction is shown by arrow. Dimension is half-lattice constant.

Table V. Vacancy cluster size distributions, high-temperature case

	Number of Vacancy Clusters of Size n per Cascade					
	γ-Iron				α-Iron (Ref. 1)	
	20 keV		5 keV			
n	Large AR[a]	Small AR[b]	Large AR[a]	Small AR[a]	20 keV	5 keV
1	10.2	15.2	3.8	4.8		
2	1.2	1.75	0.6	0.8		
3	0.6	0.25	0.3	0.4		
4	0.6	1.0	0.1	0.1		
5	1.3	1.5	0.2	0.3		
6	0.6	1.75	0	0.1		
7	0.9	0.5	0	0.2		
8	0.3	1.0	0.1	0.1		
9	0.4	0.5	0.1	0.1		
10	0.2	0.25	0.1	0		
> 10	0.2[c]	1.75[d]	0.1[e]	0.5[f]		
≥ 4	4.5	8.25	0.7	1.4	3.7	1.0
≥ 7	2.0	4.0	0.4	0.9	1.8	0.5
≥ 10	0.4	2.0	0.2	0.6	0.7	0.12

(a) Average of 10 cascades.
(b) Average of 4 cascades.
(c) $2V_{11}$ in 10 cascades.
(d) V_{11}, $3V_{12}$, V_{13}, V_{14}, V_{22} in 4 cascades.
(e) V_{11} in 10 cascades.
(f) $5V_{12}$ in 10 cascades.

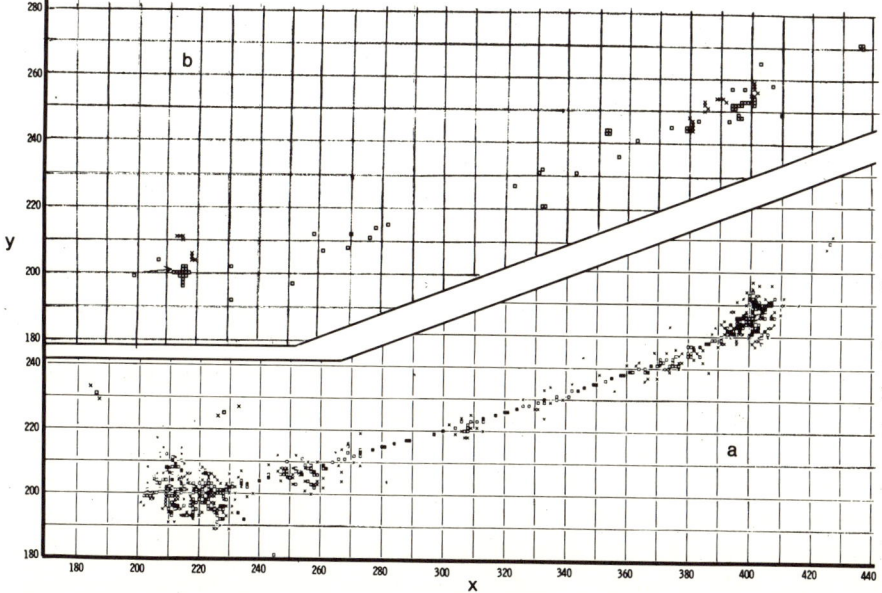

Fig. 4. A quasichanneled 20-keV cascade (2079). See caption Fig. 3. (a) Preanneal. (b) High-temperature postanneal, 32-site AR.

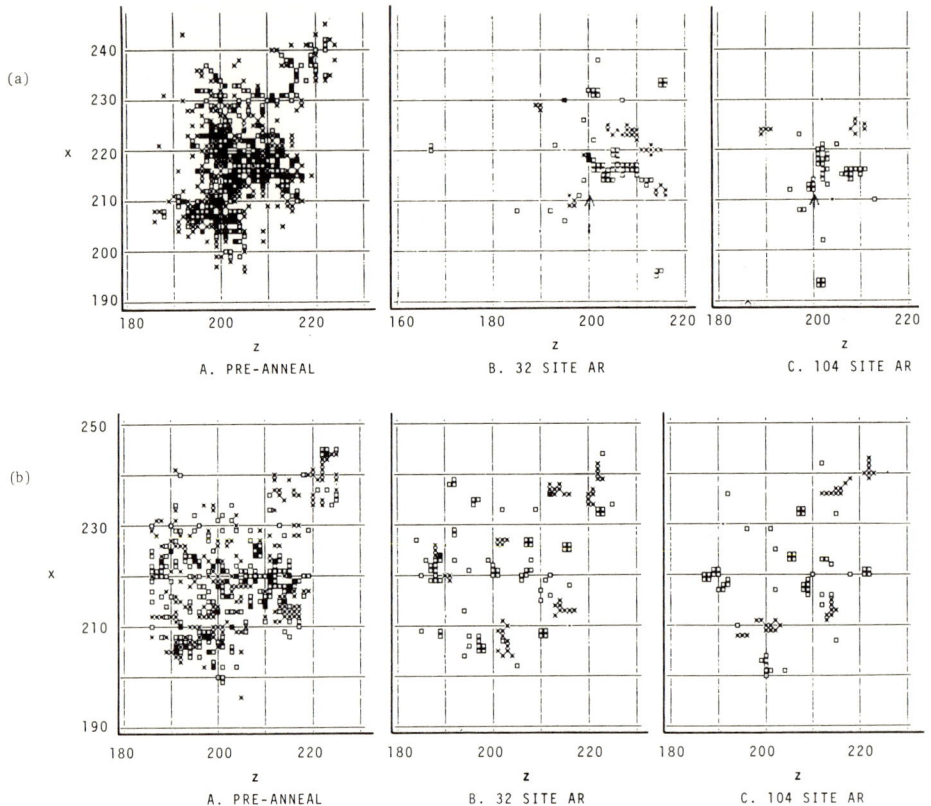

Fig. 5. (a) A compact 20 - keV cascade (2076). (b) A "randomized cascade" (77) in the geometry of cascade 2076. See caption Fig. 3.

account for the 10^6-10^8 jumps that each would have made, if permitted, during the vacancy stage. The quasichanneled cascade of Fig. 4 (in which no immobile interstitial clusters survived) was the only one of its type studied. More common was the rather compact cascade shown in Fig. 5a which served as the model for the randomized cascade. In regard to 5-keV cascades, it was not uncommon to have no residual immobile interstitial clusters (i.e., $I \geqslant 4$) or vacancy clusters ($V \geqslant 5$) when the large AR was used.

4.1.3 Effect of Annihilation Region. Results for large and small AR's are directly compared for 20-keV cascades in Fig. 6. Plotted are the fractional distributions of residual defects (i.e., each sums to unity) in clusters to show the influence of the AR on the relative clustering within a cascade. An absolute comparison is easily made using Tables III to V. The small AR promoted clustering of both interstitials and vacancies. The results for 5-keV cascades were similar, but shifted somewhat toward smaller cluster sizes.

Fig. 6. A comparison of the distribution of defects in clusters after simulated high-temperature anneals using small and large annihilation regions.

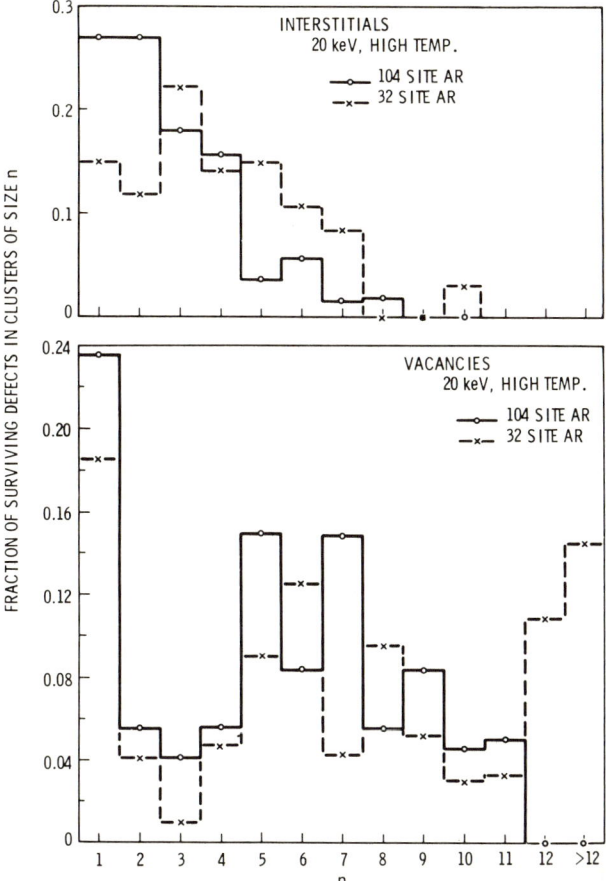

4.1.4 Energy Dependence. Fractional defect distributions for 20- and 5-keV cascades are compared in Tables IV and V. At 5 keV, there are relatively more single interstitials and fewer large clusters. Maximum interstitial cluster sizes were 8 and 5 for 20 and 5 keV, respectively, for the large AR, and 10 and 8, respectively, for the small AR. The number of interstitial clusters generally increased faster than the energy for the two energy values investigated; e.g., the number of clusters of size ⩾4 was proportional to $E^{1.2}$.

As to vacancies, 5-keV cascades had proportionately more small clusters than 20-keV cascades and also proportionately more large clusters. The latter was especially true for the large AR, for which the maximum cluster size was 11 at each energy. The small AR resulted in maximum cluster sizes of 22 and 12 for 20- and 5-keV cascades, respectively.

The number of vacancy clusters generally increased faster than the energy, as for interstitials. For clusters of size ⩾4 (or ⩾5) the number per cascade was proportional to $E^{1.3}$.

4.2 Randomized Cascades

4.2.1 Time History. Two randomized cascades, patterned after the same displacement cascade but with independent defect configurations, were run in high-temperature simulations with both large and small AR. The rate of annihilation was the same as for the corresponding displacement cascade runs. The decrease in the fraction of mobile interstitials with time during the interstitial stage, however, was surprisingly insensitive to the value of the AR, and the rate of decrease was higher than the average rate exhibited by the real displacement cascades for either value of the AR.

4.2.2 Residual Annihilation and Clustering. The results of mutual annihilation for the randomized cascades is summarized in Table VI. The small AR resulted in 70 percent more residual defects than did the large AR.

Clustering of defects in randomized cascades is summarized in Table VII. The creation of mobile interstitials in the vacancy stage was common in these runs, particularly for the small AR; hence the distributions are weighted somewhat toward small cluster sizes.

Cluster-size distributions for randomized cascades and actual cascades are compared in Fig. 7 and 8. There is a significant difference in sensitivity to the AR in the two cases. The randomized cascades were sensitive only at the large-cluster end of the distributions for both interstitials and vacancies. The maximum interstitial cluster size was decreased somewhat while the maximum vacancy cluster size was increased by decreasing the AR. Actual cascades, on the other hand, exhibited a decrease in small clusters and an increase in large clusters when the AR was decreased. It is this difference in behavior that gives rise to the AR effect that is apparent in Fig. 7 and 8.

The decreased influence of a change in the AR on the randomized cascades is probably a result of the more-uniform spatial distribution of defects. It appears that the mean distance between interstitials is significantly increased and, hence, clustering decreased when the AR was

Table VI. *Average annihilation characteristics of randomized cascades, high-temperature case*

Annealing Stage	Number of Defects	
	Large AR	Small AR
Initial loading — 1st- and 2nd-neighbor annihilation	257	257
After initial annihilation	118	186
After interstitial stage (1000 steps)	62	111
After vacancy stage (1000 steps)	54	91

Table VII. Cluster size distributions for randomized cascades, high-temperature case

	Number of Clusters of Size n per Randomized Cascade [a]				
	Interstitials			Vacancies	
n	Large AR	Small AR	n	Large AR	Small AR
1	5.0	9.5	1	19.0	35.5
2	4.0	5.0	2	1.0	3.5
3	2.0	7.5	3	0	1.5
4	3.0	2.0	4	3.0	2.5
5	1.5	1.5	5	2.0	1.0
6	1.0	3.5	6	0	1.0
7	0	1.0	7	1.0	1.5
8	0.5	0	8	0.5	0
9	0	0	9		0
10	0	0.5	10		0
>10 [b]	0.5		>10		1.0 [c]
$\geqslant 3$	8.5	16.0	$\geqslant 4$	6.5	7.0
$\geqslant 4$	6.5	8.5	$\geqslant 7$	1.5	2.5
$\geqslant 5$	3.5	6.5	$\geqslant 10$	0	1.0

(a) Average of two "cascades". (b) $1\text{-}I_{11}$ in two "cascades". (c) $1\text{-}V_{11}$ and $1\text{-}V_{13}$ in two "cascades".

increased in actual cascades, but not for randomized cascades. The same is true for vacancies, but the bunching of vacancies in actual cascades (as, for example, in Fig. 1) causes enhanced vacancy clustering for either AR relative to the randomized cascades.

The spatial distribution of defects in one of the randomized cascades is shown in Fig. 5b for direct comparison with Fig. 5a. The difference in interstitial clustering is particularly evident.

4.3 Comparison With α-Iron (bcc)

In previous work on short-term annealing, a model of α-iron (bcc) was used in which a 62-site AR was employed for high-temperature runs and a 30-site AR for low-temperature runs.[1]

The annihilation characteristics of the α-iron data are included in Table III for comparison with the present results. It would, of course, be expected that present results obtained with the large AR would be most comparable to the α-iron results. In fact, the effective difference between the 62-site bcc region and the 104-site fcc region is less than suggested by the number of sites involved. The bcc annihilation region is very

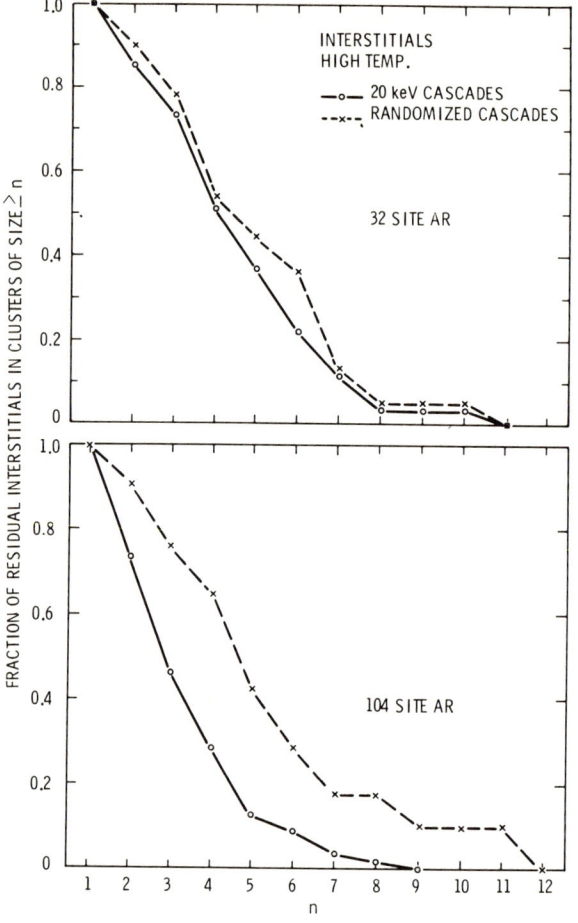

Fig. 7. A comparison of integral distributions of interstitials in clusters for displacement cascades and "randomized cascades."

anisotropic, extending farthest along the close-packed <111> directions, whereas the fcc annihilation region is more nearly isotropic. Unpublished work by the authors has shown that, in the bcc case, an isotropic region encompassing 112 sites produces annihilation approximately equivalent to the 62-site region.

The similarity in the high-temperature results, after the initial annihilation, is indeed striking and certainly somewhat coincidental* considering the numerous differences in the manner in which high- and low-temperature runs were made.

Some of the results on the cluster-size distributions for α-iron are included in Tables III to V. The large AR γ-iron results are quite similar

---

*Low-temperature runs were less similar.

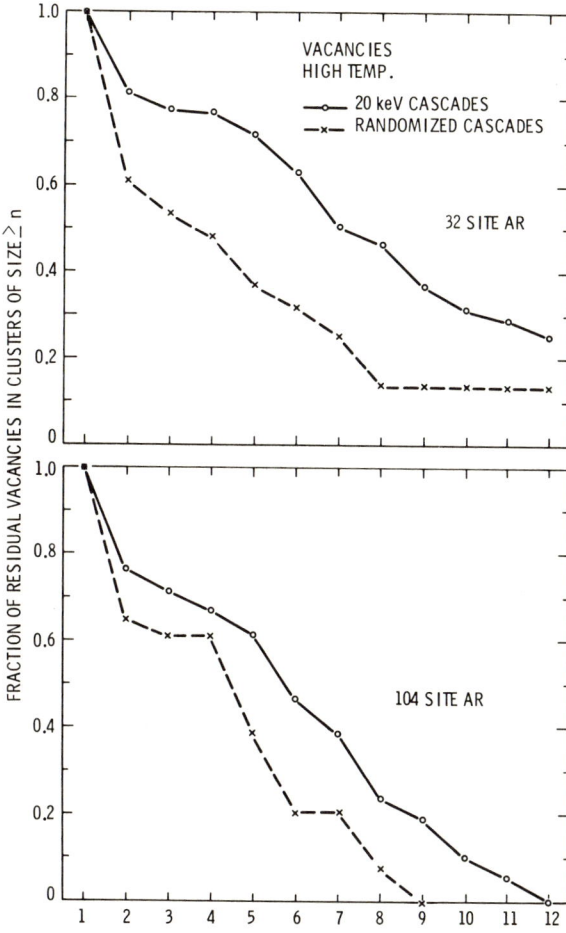

Fig. 8. A comparison of integral distributions of vacancies in clusters for displacement cascades and "randomized cascades."

to the α-iron results for both interstitials and vacancies. The major difference is the lower production of large vacancy clusters (≥ 10) by 20-keV cascades in the fcc case.

5 SUMMARY AND CONCLUSIONS

The displacement cascades simulated by Beeler[2] have been "temperature corrected" (to nominally 800 K) by means of a simulated short-term anneal. The annealing model was based on the simulation of γ-iron by Johnson.[4] Results were obtained with both large (104 sites) and small (32 sites) annihilation regions; the latter is considered the more realistic at the nominal temperature.

It was found that the 104-site annihilation region led to the survival of ~2 Frenkel pairs/keV between 5 and 20 keV; the 32-site region led to ~4 Frenkel pairs/keV. With the large annihilation region, 20-30 percent of the surviving interstitials and 40-60 percent of the surviving vacancies are clustered – the other defects migrate away and lose their identification with the cascade. The corresponding numbers for the small annihilation region are 35-50 percent of interstitials and 60-70 percent of vacancies. The number of clusters per cascade increased somewhat faster than the energy.

The principal effect of randomizing the spatial distribution of defects in a typical cascade geometry was to diminish vacancy clustering and enhance interstitial clustering somewhat.

ACKNOWLEDGMENT

This work was supported by the United States Atomic Energy Commission.

REFERENCES

1. Doran, D. G.: Rad. Effects, **2**: 249 (1970).
2. Beeler, Jr., J. R.: Phys. Rev., **150**: 470 (1966).
3. Besco, D. G.: General Electric Company Tech. Report GEMP-644, 1967.
4. Johnson, R. A.: Phys. Rev., **145**: 423, **152**: 629 (1966).
5. Merkle, K. L.: Private Communication.

DISCUSSION on paper by D. G. Doran and R. A. Burnett

DORAN: I would like to comment on a recent development. Torrens and Robinson have kindly furnished us several of their copper cascades. On running them through our annealing program, we find that the number of surviving defects is only about 50 percent of that found in the present work (Beeler's cascades) and clustering is considerably decreased relative to the present work. These results reflect the fact that their cascades have an appreciably lower density of defects than do Beeler's cascades. I hasten to add that the three cascades we have examined may not be completely typical and these results should be considered tentative.

BEELER: Our 20-keV cascades for copper were simulated using the Johnson II potential. Torrens used a different potential. I wanted to state this explicitly.

DORAN: Yes, the potentials are different but the difference seems small. Another obvious difference is in the displacement criteria used.

VINEYARD: What fraction of the defects escaped to infinity instead of clustering or recombining?

DORAN: Using the small annihilation region (AR), 41 of the interstitials and 23 of the vacancies were still mobile after the short-term anneal of a 20-keV cascade. The corresponding numbers were 12 and 8 at 5 keV. The values found with the large AR were just three-fourths of the above values.

There is some ambiguity in expressing these numbers as percentages for the following reasons. The number of defect pairs produced in the CASCADE program is ~400 and ~90 at 20 keV and 5 keV, respectively, but some close-pair recombinations (defined in text) have not yet been performed at this stage of the computation. Beeler subsequently applied a 42-site spontaneous AR (copper model) to get his final numbers of 175 and 46 for 20 and 5 keV cascades from which his displacement efficiencies were computed.

The present work is based on a nickel-annealing model applied to the same cascades. The spontaneous AR is just the small (32 site) AR. After its application (prior to any defect migration) the numbers of defect pairs are 188 and 50 at 20 and 5 keV. Percentages of residual mobile defects based on these numbers (percentages reported for α-iron in Ref. 1 were calculated analogously) are ~24% of the interstitials and ~14% of the vacancies for the small AR, and ~17% of the interstitials and ~11% of the vacancies for the large AR.

On the other hand, there may be more interest in basing the percentages on the numbers that would result from the usual Kinchin and Pease[1] treatment, viz., 400 and 100 defect pairs at 20 and 5 keV. This would, of course, reduce the percentages above by approximately one-half.

TORRENS: How relevant are the defect-correlated jump-activation energies calculated in a static lattice simulation to the annealing of a highly defected lattice region at elevated temperatures?

DORAN: I don't know, of course, how well static-lattice simulations apply at elevated temperatures. I don't know of a better way to estimate correlations. These estimates (at least some of them) were made in the presence of a high density of defects. The differences in activation energies that give rise to the correlations are relevant, I believe, in the range of about 600-1000 K.

TORRENS: Have you run any anneals with no correlated jumps at all and, if so, what is the effect of this on the clustering?

DORAN: We have not yet varied the correlations in the fcc case. We did, however, in the earlier bcc work and found that varying the correlation had a significant effect on clustering.

REFERENCES

1. Kinchin, G. H. and Pease, R. S. Rep. Prog. Phys. **18**, 1 (1955).

COMPUTER SIMULATION OF ATOMIC DISPLACEMENT CASCADES IN SOLIDS

I. M. Torrens

Centre d'Etudes Nucléaires de Saclay
France

and

M. T. Robinson

Solid State Division, Oak Ridge National Laboratory
Oak Ridge, Tennessee 37830, U.S.A.

ABSTRACT

Atomic displacement cascades originating from primary atoms displaced with energy up to 100 keV have been investigated by computer simulation. The model includes inelastic energy loss and finite temperature effects. Results obtained for crystals of copper, iron and gold indicate that long-range channeling and focusing are rare. The numbers of vacancies created and of those remaining are both directly proportional to the primary energy corrected for inelastic loss over the energy range up to 100 keV. The number of clusters of i vacancies is in constant ratio to the number of clusters of (i - 1) vacancies, this ratio being independent of primary energy.

1 INTRODUCTION

When a solid material is subjected to incident radiation, in the form of fast neutrons or high-energy charged particles, the ensuing degree of damage depends on a number of factors. The nature of the incident particle, its kinetic energy, and the type of interaction it may undergo with the atoms, atomic nuclei, or electrons of the solid will determine the *primary damage*. That is, the number and energy spectrum of atoms displaced by direct interaction with the incident particle (sometimes termed primary knock-on atoms or PKA's).

The primary displaced atom, upon leaving its lattice site, interacts with its neighbors in the solid, which are displaced in their turn if they receive sufficient kinetic energy. In this way a displacement cascade of energetic atoms builds up in the solid. The energy dissipation in the cascade is a combination of elastic and inelastic processes. The latter represents the interaction of a moving cascade atom with atomic electrons, resulting in energy loss through excitation and ionization, which is particularly important for high-energy collisions but does not contribute to damage production (at least in metals). The development of an atomic displacement cascade is a many-body problem, and analytical theories have so far been confined to cascades in amorphous solids. Some early results of a computer simulation investigation of displacement cascades in crystalline solids are reported here. The computer code permits the simulation of cascades originating from PKA's of energy up to several hundred kilovolts in a wide range of solids. Inelastic energy losses and the effects of finite temperature are included. Cascades may be started in an already-damaged crystal or in one containing a given concentration of defects. A surface in any specified orientation may be present. It is possible to use the code to study phenomena related to ion bombardment, such as ion injection along crystal channels and ejection of atoms from surfaces.

2 ANALYTICAL THEORIES OF CASCADE FORMATION

A review of the general principles underlying the analytical treatment of displacement cascades has been given by Robinson.[1] The most elementary theory is that of Kinchin and Pease.[2] Assuming an amorphous solid, two-body hard-sphere atomic collisions, a sharp displacement threshold E_d, and no inelastic energy losses, they obtain for the number of vacancies created in a cascade originating from a PKA of energy E:

$$\nu(E) = \frac{E}{2E_d} \quad . \tag{1}$$

If an energy E_{in} is lost during the cascade through inelastic processes, the total energy available for elastic collisions, or *damage energy*, is $(E - E_{in})$. The percentage of E which produces damage is defined as the *damage efficiency* $\mathcal{E}(E)$. When a more realistic interatomic potential is used instead of the hard-sphere model, the scattering cross section favors lower energy transfers, resulting in a smaller number of displaced atoms per cascade. Eq. (1) may then be rewritten:

$$\nu(E) = \frac{\kappa(E - E_{in})}{2E_d} \quad . \qquad (2)$$

The factor κ, the *displacement efficiency*, depends on the scattering law assumed to apply. An important question is whether it also depends on the energy E. Analytical theory predicts that since the vast majority of cascade atoms are generated from low-energy collisions, $\nu(E)$ should remain directly proportional to E and κ should be constant for cascades in an amorphous solid. If any lattice effects or other phenomena not taken into account in the analytical treatment introduce an energy dependence in κ, this should be revealed by computer simulation.

3 COMPUTER SIMULATION TECHNIQUES

With the computer, one can set up or simulate a situation in which an atomic displacement cascade is generated in a chosen crystal, the atoms interacting with an interatomic potential deemed realistic. It is highly important in interpreting computer simulation results to investigate to what extent these results depend upon the chosen model. The simulation technique may be applied to PKA energies in the kilovolt range at the expense of reducing the atomic collisions to two-body processes, although some allowance may be made for nearly simultaneous collisions. The two-body approximation improves rapidly as the energy increases. Apart from an early Monte Carlo simulation in an amorphous solid by Yoshida[3], the principal work in this field up to the present was initiated by Beeler and Besco[4], who simulated cascades generated by energetic primaries up to 30 keV in static crystal lattices of BeO, iron, and copper. The interatomic potentials used were a hard-sphere model with energy- and impact-parameter-dependent sphere radius, a Bohr screened Coulomb potential, a Thomas-Fermi potential, and a Born-Mayer form. Their model did not include inelastic collision losses, which would tend to be significant in metals even at 10 keV (see Results section).

4 THE SIMULATION MODEL

In the present simulation, the aim has been to develop as general and versatile a computer code as possible, applicable to many different types of solid, and permitting simple manipulation of different aspects of the model. The computation must be rapid and efficient, so that a sufficient number of cascades for statistical analysis may be produced in a computation time not prohibitive in cost.

4.1 The Crystal Lattice

The only limitation on the type of crystal arises from the requirement that atomic positions must be referred to an orthogonal coordinate system; thus, monoclinic and triclinic lattices are excluded. Up to five different types of atom may be included in the crystal – as low-density impurities or as an integral part of a crystalline compound or alloy.

A considerable waste of computer space would be incurred by storing in its memory the complete section of crystal lattice occupied by a cascade. A list of only 15-20 near-neighbor sites for one atom of each type in the crystal is retained. Together with the actual position of a cascade atom is stored a *reference origin,* a nearby lattice site. Before each collision the computer scans the neighbor list with respect to the reference origin to find the next target. This procedure removes any limit on the dimensions of the cascade other than the number of atoms it contains.

4.2 The Interatomic Potential

For the range of collision energies considered here, it is necessary to include nuclear repulsion in the interatomic potential, which then takes the general form:

$$V(r) = \frac{Z_1 Z_2 e^2}{r} f(r) \tag{3}$$

where $Z_1 e$ and $Z_2 e$ are the nuclear charges of the colliding atoms and $f(r)$ is a screening function to allow for shielding by the intervening electrons. Of the various expressions available for $f(r)$ we have chosen the Molière approximation[5] to the Thomas-Fermi screening function $\chi(r/a)$, given by

$$f(r) = 0.35 \exp\left(-0.3 \frac{r}{a}\right) + 0.55 \exp\left(-1.2 \frac{r}{a}\right) + 0.10 \exp\left(-6.0 \frac{r}{a}\right). \tag{4}$$

The constant a is a screening radius comparable to the Bohr radius a_0. In a potential given by Firsov as a reasonable approximation to the energy of interaction of two Thomas-Fermi atoms, the screening function was $\chi(r/a)$ and the screening radius was given by:

$$a = \frac{0.885 \, a_0}{\left(Z_1^{1/2} + Z_2^{1/2}\right)^{2/3}} \tag{5}$$

Since the Molière potential involves an approximation to $\chi(r/a)$, we are not necessarily justified in using the above expression for a. In the simulations, the screening radius has been determined by adjusting the Molière potential so that it has the same numerical value at the nearest-neighbor distance in the crystal as a Born-Mayer potential whose parameters were determined from elastic-constant data. The screening radii used for simulation in the three metals are compared with the corresponding Firsov values in Table I, and the potentials are compared with the Born-Mayer and Firsov potentials in Fig. 1.

Table 1. Comparison of modified screening radius a used in the present work with the Firsov value for copper, iron, and gold (in Å).

	Cu	Fe	Au
Firsov a	0.0960	0.0996	0.0688
Modified a	0.0738	0.0781	0.0752

4.3 The Collision Process

Classical two-body scattering is assumed to hold true for all collisions. Fig. 2 illustrates the collision of a projectile atom with an initially stationary target atom. The scattering integrals may be evaluated in closed

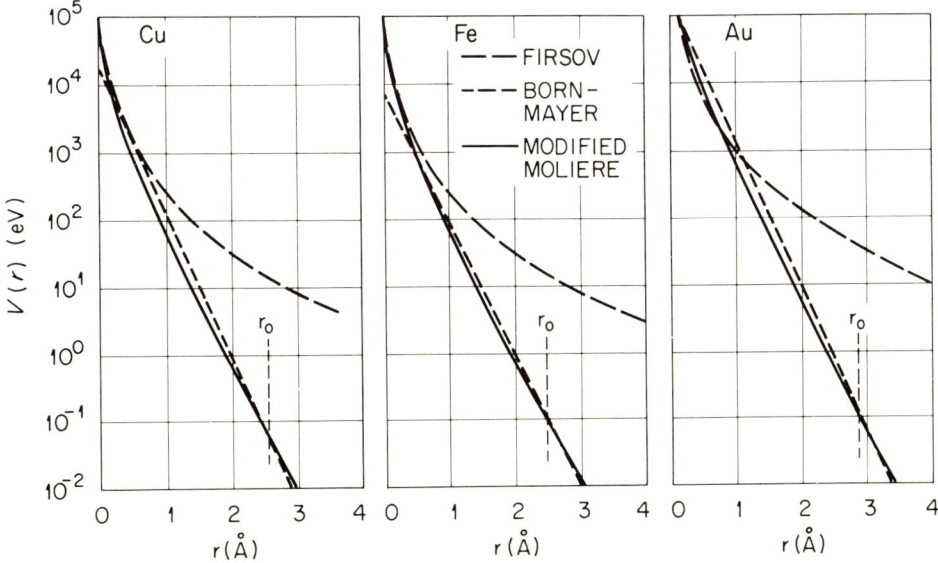

Fig. 1. Interatomic potential $V(r)$ used in the simulation (modified Molière) compared with the Firsov and Born-Mayer forms for copper, iron, and gold.

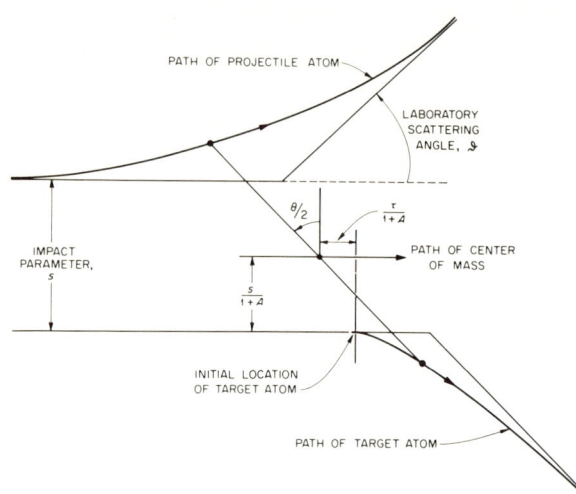

Fig. 2. Classical two-body scattering event. The figure indicates the trajectories of the two particles, their positions and that of the center of mass at the distance of closest approach. Also shown are the scattering angles in the barycentric and laboratory systems, the impact parameter s and the time integral τ.

form only for a small group of power potentials. Here, the energy transfer and projectile deflection in an atomic collision are computed by numerical integration.[1]

A moving-cascade atom, particularly in an ordered lattice, frequently passes almost directly between two or more lattice atoms so that it collides with them at almost the same time. The collision is then many-body and the two-body approximation is unsatisfactory, tending to overestimate the energy loss and deflection of the projectile. We allow for this by writing into the search procedure *simultaneity criteria,* which essentially require the impact parameters of the target atoms to be comparable and the bases of the impact parameters on the projectile trajectory to be close together. If these criteria are satisfied by two or more targets, we compute the results of a collision of the projectile with each target separately, then sum the projectile energy losses and vectorially sum the deflections. This procedure slightly underestimates the energy lost in the simultaneous collisions. Previous simulations included no provision for nearly simultaneous collisions.[3,4]

It would be unrealistic to follow the primary projectile until it stops, then to follow a secondary atom until it in turn stops, and so on. In the present simulation, the most energetic atom in the cascade is always followed.

4.4 The Target Search Procedure

The first step is to construct a cylinder about the motion vector of the projectile as axis, with initially specified dimensions. Then the list of neighbor sites with respect to the reference origin of the projectile is searched (effectively generating a small section of crystal) to find all neighboring atoms within the cylinder. If a target site is found to be vacant, it is ignored. Having found more than one possible target atom in the cylinder, the simultaneity test is applied. Any atom that fails this test when compared with the nearest target is rejected for the present collision, but may of course be re-selected for the next collision of the cascade atom.

4.5 The Atomic Displacement Model

The low-energy aspects of a displacement cascade are the most difficult to treat by simulation. At energies below about 100 eV many-body effects supervene. In our model a sharp isotropic displacement threshold energy E_d (~25 eV) is assumed. Atoms which receive energy greater than E_d in a collision join the cascade, but make no further collisions unless their energy is greater than $2E_d$. A cascade atom is followed until its energy drops below $2E_d$ in a collision, after which it is considered stopped. If, in this collision, its energy drops below E_d and a lattice atom is displaced, this is treated as a replacement collision and the projectile is put into the target site. Otherwise, stopped atoms are left where they are until the end of the cascade.

When no further cascade atoms remain with energy above $2E_d$, the cascade atoms that are not already in lattice sites as a result of replacement collisions are subjected to a recombination test. If a cascade atom is within a specified *recombination radius* of a vacancy, it is put into the vacancy. If not, it becomes an interstitial. The recombination radius used in the simulation was normally 1.5 lattice parameters, but has been varied down to zero. In future studies, the recombination will be separated from the cascade creation in order to study it in more detail.

In an earlier version of the code the recombination was performed during the cascade and stopped atoms were either annealed into vacancies or made interstitial immediately after the collision in which they stopped. They could then be restruck later in the cascade as either substitutional atoms or interstitials. This was found to occur very rarely. In the present code no "redisplacements" are permitted, since an atom whose energy drops below $2E_d$ will not have sufficient time to settle into either a substitutional or interstitial configuration during the time necessary for the cascade (~10^{-14} sec). It is at least as accurate, and conceptually more realistic, to ignore this atom with respect to later collisions in its neighborhood as it is to put it into any chosen position.

Thus in the present code all sites once vacated (even in a replacement collision) are considered vacant until the end of the cascade. No interaction between cascade atoms of any energy is permitted, and the only interstitials which may become collision targets are preexisting interstitials inserted in the crystal before the beginning of the cascade. At the end of the cascade, atoms that become interstitial are left precisely where they were at the end of their last collision, since the time-scale threshold between cascade creation and thermal annealing becomes confused at this point and we are at least as justified in leaving interstitial motion to the annealing stage.

In the work of Beeler et al., the displacement threshold above which atoms joined the cascade was twice the sublimation energy of the metal. An amount E_d of kinetic energy was then subtracted from the initial energy of the displaced atom as it joined the cascade. This resembled the model proposed by Snyder and Neufeld.[7] Each cascade atom was allowed to make at least one collision, and was brought to rest as an interstitial if its energy fell below E_d. It was then immediately put into one of the possible split interstitial configurations and tested for Frenkel pair stability. If it failed this test, it was recombined with the appropriate vacancy and was available as a target for later collisions. The same criticism that caused us to leave any recombination until the end of the cascade may be applied to this procedure.

4.6 Inelastic Collision Energy Losses

We have used the results of a theory due to Firsov[8], who calculates the electron excitation energy in a single two-body collision of impact parameter s:

$$\Delta E_{ex} = \frac{\alpha}{(1 + \beta s)^5} , \qquad (6)$$

where

$$\alpha = 0.05941 \, (Z_1 + Z_2)^{5/3} \, [E/M_1]^{1/2} \text{ eV} \qquad (7)$$

and

$$\beta = 0.3042 \, (Z_1 + Z_2)^{1/3} \, \text{Å}^{-1} . \qquad (8)$$

Here the projectile energy E is expressed in eV and M_1, the mass of the projectile, in atomic mass units.

For the impact parameter s we substitute the distance of closest approach in the collision, which seems more reasonable at low energies since it is the degree of overlap of the electronic shells of the colliding atoms which contributes to the energy loss. The inelastic loss calculated in a collision is subtracted from the kinetic energy of the projectile at the end of the collision.

4.7 Finite Temperature Effects and Thermal Vibrations

Thermal vibrations are simulated in the adiabatic approximation[9], which is reasonable for the time interval involved in an atomic collision at these energies. The target atom is given a small thermal displacement from its lattice site and considered stationary for the duration of the collision. The rms vibration amplitude is a function of the lattice and of the crystal temperature, and is calculated using the Debye model.[10] Then the thermal displacement for any given target atom is selected using a random table-scan procedure based on the inverse probability function for a Gaussian distribution of this amplitude.

4.8 PKA Selection and Cascade Computation

The numerical results are based on statistical analysis of sets of 10 cascades. In each set, only the PKA direction of motion is varied — all other aspects of the model being fixed. The direction is randomly selected within a solid angle of 1/48 of the total sphere.

The computation was carried out in double precision on IBM System/360, Model 75 or Model 91 machines. Typical computation times for events run in copper are 2 sec for a 10-keV cascade and 70 sec for a 100-keV cascade, on the Model 91. Although so far 100 keV has not been exceeded, this is by no means an upper limit. We estimate that cascades from primaries of energy up to or greater than 1 MeV could be simulated before present machine capacity is exceeded.

The vacancy and interstitial spatial distribution at the end of the cascade has been analyzed by a separate program[11] that determines the number and characteristics of clusters of vacancies or interstitials. This information is useful from the point of view of displacement spike formation and the nucleation of voids in irradiated metals. It is also used in a recently developed code to simulate the thermal annealing of displacement cascades[12].

5 RESULTS AND DISCUSSION

Cascades have been simulated in fcc copper and gold and in bcc iron, with PKA energy varying from 0.1 keV up to 100 keV, in the case of a static lattice and with thermal vibrations included. The value of displacement threshold energy unless specifically mentioned was 25 eV for copper and iron and 35 eV for gold.

We have not yet made any detailed analysis of the spatial distribution of point defects in a cascade. In qualitative terms the higher-energy PKA's initiate cascades which leave a column of defects (with perhaps some subsidiary branches) ending in one or more vacancy-rich zones, with most interstitials scattered around the outside. Branching of cascades following a close collision of a high-energy cascade atom is quite common.

In Fig. 3 the damage energy $(E - E_{in})$ is plotted against the PKA energy E for cascades in copper with and without thermal vibrations included. The dashed line shows the corresponding analytical estimate[1] based on the Lindhard theory of inelastic losses[13]. Also plotted in the *damage efficiency*, as shown in Table II for the three metals.

Table II. Results of atomic displacement cascades in copper, iron, and gold, over the PKA energy range 10 keV to 100 keV, for static lattice (SL) with thermal vibrations included and for crystal temperature 600 K.

		Copper	Iron	Gold
Damage Efficiency[a] $\mathcal{E}(E)$, percent	SL	72	70	91
	600 K	64	66	90
Displacement Efficiency, κ	SL	0.869 ± 0.004	0.882 ± 0.007	0.860 ± 0.006
	600 K	0.809 ± 0.004	0.817 ± 0.004	0.812 ± 0.004
Survival Fraction, F_s[b]	SL	0.251 ± 0.006	0.267 ± 0.005	0.208 ± 0.005
	600 K	0.228 ± 0.003	0.249 ± 0.003	0.187 ± 0.003
Cluster Factor, f_c[b]	SL	0.276 ± 0.018	0.207 ± 0.013	0.424 ± 0.010
	600 K	0.262 ± 0.010	0.181 ± 0.010	0.408 ± 0.003

(a) For 100-keV cascades.
(b) Recombination radius = 1.5 lattice parameters.

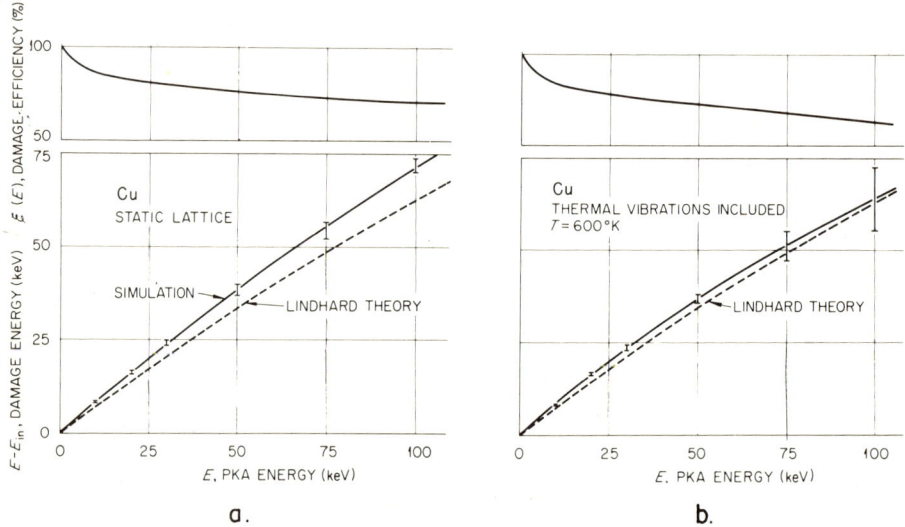

Fig. 3. Variation of damage energy $(E - E_{in})$ and damage efficiency $\mathcal{E}(E)$ (%) with PKA energy E for cascades in copper, (a) static lattice, (b) thermal vibrations included, $T = 600$ K. The dashed lines is an analytical estimate based on the Lindhard theory. Error bars in this and later figures indicate the statistical spread over samples of 10 cascades.

Fig. 4 compares the number of vacancies created in cascades simulated in copper (before any recombination) with that predicted by the Kinchin-Pease ratio based on the damage energy available. Similar curves have been obtained for iron and gold. The two curves are related by the displacement efficiency κ, which (plotted against PKA energy for copper and iron in Fig. 5) shows no tendency to vary with E in any systematic fashion. The introduction of thermal vibrations causes κ to decrease by some 7 percent from the static-lattice value, but again it remains constant over the PKA energy range. Preliminary results of cascades at 0 K (with zero-point vibrations) indicate a further reduction in κ from the 600 K value. The same tendencies are observed for cascades in gold; the weighted mean values of κ over the energy range are given in Table II for copper, iron, and gold. Contrary to the effect reported by Beeler[14], no evidence has been found for a decrease of κ with increasing E. Variation of the interatomic potential caused a change in κ, but it retained its energy invariance. Variation of the displacement threshold and removal of the inelastic energy loss option produced no noticeable effect on κ. When thermal vibrations were introduced, the removal of the simultaneous collision possibility caused κ to increase rather than decrease in copper and iron and to remain the same in gold. The precise reason for the sensitivity of κ to the presence of thermal vibrations is not at present fully understood, but its energy invariance seems to be nonetheless well established. This agrees with the conclusions of analytical cascade theory.

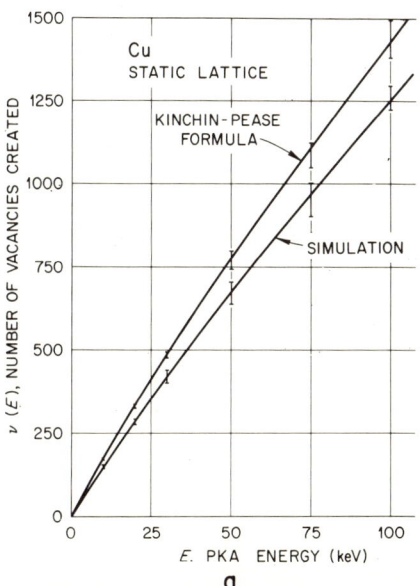
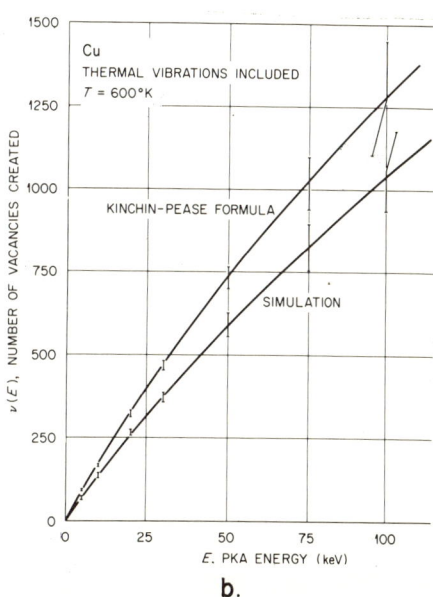

Fig. 4. Variation of $\nu(E)$ with E for cascades in copper, compared with that predicted by the Kinchin-Pease ratio based on the same damage energy, (a) static lattice, (b) T = 600 K.

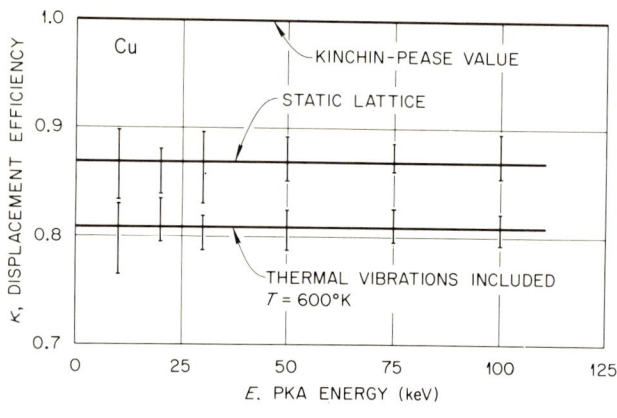

a. Copper

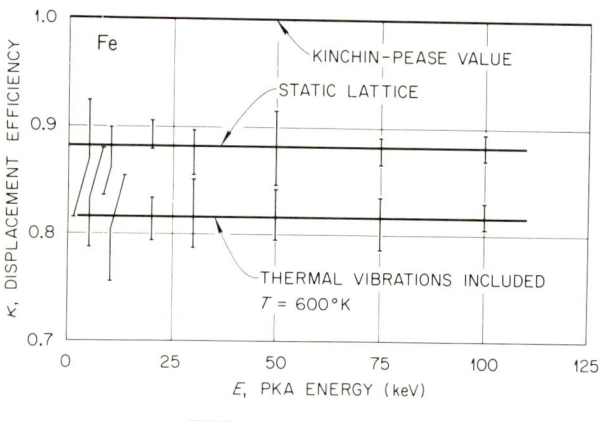

b. Iron

Fig. 5. Displacement efficiency κ plotted against PKA energy for (a) copper and (b) iron.

The ratio of the number of vacancies surviving after interstitial-vacancy recombination within a radius of 1.5 lattice parameters to the number created, which we term the *survival fraction* F_S, is shown in Table II for the three metals. F_S depends on the spatial extent of the cascade and the model recombination criteria. Again we find that the introduction of thermal vibrations causes a small drop in F_S, but there is no significant variation with PKA energy. This indicates that there is a negligible degree of overlap of different branches of the cascade at higher energies, which would tend to increase the amount of recombination and decrease F_S. [It was in fact the product $F_S\kappa$ that was stated by Beeler to decrease as E increases.]

The analysis of defect clusters showed that while clusters of up to 15 vacancies are present, only one cluster of four interstitials has been observed in more than 1000 cascades run so far, and even tri-interstitials are a rare occurrence. The condition used to define a cluster is that adjacent point defects must be separated by distances not greater than the lattice nearest-neighbor distance.

A more detailed analysis produced the interesting empirical result that to a good approximation if N_i is the number of clusters of i vacancies in a cascade there exists a constant *cluster factor* f_c, such that

$$f_c = \frac{N_{i+1}}{N_i} \quad . \tag{9}$$

Thus if there are N_1 single vacancies there will be $f_c N_1$ divacancies, $f_c^2 N_1$ trivacancies, and so on, f_c being, in fact, the probability of clustering. Table II shows the numerical values of the cluster factor for the three metals. This quantity also is independent of PKA energy. Since the statistics are necessarily worse in 10-cascade samples for clusters than for single-defect properties, we are unable to conclude for certain at present whether the introduction of thermal vibrations affects f_c. This empirical rule is very well obeyed in the cascades run in copper and iron, but less well in gold due to the presence of more large clusters than the rule predicts. We cannot presently explain the phenomenon in any satisfactory way, but further investigation of the whole recombination and clustering problem is planned. The asymmetry between interstitial and vacancy clustering, together with the existence of this cluster factor, indicates a marked correlation in the vacancy creation during the cascade.

Only rare instances of long-range channeling have been observed. In copper and gold, the phenomenon has not been observed in the static-lattice simulation, and once (100-keV cascade in copper) in finite-temperature simulations. In iron, some long-range channeling has occurred. Medium- and short-range channeling of cascade atoms is more common, characterized by a high inelastic energy loss and a sharp drop in $\nu(E)$ compared with the average, again most common in iron and practically nonexistent in gold. Channeling is more frequent when thermal vibrations are present, probably a consequence of small thermal displacements increasing the instantaneous acceptance angle of a channel to an atom displaced from a lattice site, and decreasing the blocking effect. These observations are consistent with the energy invariance of κ, since if long-range channeling were a frequent occurrence at high energy it would tend to reduce κ owing to the large number of subthreshold energy losses in such events. Medium- and short-range channeling worsens our statistics but does not alter significantly the value of κ.

6 CONCLUSION

Many of our numerical results are model dependent and at this early stage of the work it is not possible to say to what extent they reproduce reality. But there is convincing evidence for some qualitative tendencies. One is that long-range crystallographic effects such as focusing and channeling (for particles ejected from a lattice site) are unimportant, at least in cubic lattices, in influencing cascade development. Second, despite considerable variation of the model, the results indicate strongly that the displacement efficiency κ does not vary with PKA energy. Neither does the survival fraction F_s after close-pair recombination. This we believe is a real effect. Finally, the empirical rule which leads to a cluster factor f_c is closely followed in copper and iron, although more large clusters appear in gold. This rule could be very useful in predicting the frequency of large clusters that might be nucleation centers for voids, under given irradiation conditions.

The simulation project is in its early stages and will be extended to other materials and higher energies. It is hoped to make a more detailed analysis of cascade spatial characteristics and close-pair recombination, and to study the thermal annealing of cascades under different conditions.

ACKNOWLEDGMENTS

We are grateful to D. K. Holmes of Oak Ridge National Laboratory and to Y. Adda of Centre d'Etudes Nucléaires de Saclay for many valuable discussions during the course of this work. This research sponsored in part by the U.S. Atomic Energy Commission under contract with Union Carbide Corporation and in part by the Commissariat a l'Energie Atomique, France.

REFERENCES

1. Robinson, M. T.: in *Proceedings of the 1971 International Conference on Radiation-Induced Voids in Metals,* U.S.A.E.C., to be published.
2. Kinchin, G. H., and R. S. Pease: Rep. Prog. Phys., **18**: 1 (1955).
3. Yoshida, M.: J. Phys. Soc. Japan, **16**: 44 (1961).
4. Beeler, J. R., Jr., and D. G. Besco: J. Appl. Phys., **34**: 2873 (1963).
5. Molière, G.: Z. Naturforsch., **2a**: 133 (1947).
6. Firsov, O. B.: Zh. Eksp. Teor. Fiz., **24**: 79 (1953) [English translation: U.S.A.E.C. Report ORNL-tr-1635 (1967)].
7. Snyder, W. S., and J. Neufeld: Phys. Rev., **97**: 1636 (1955).
8. Firsov, O. B.: Zh. Eksp. Teor. Fiz., **36**: 1517 (1959) [English translation: Soviet Phys.-JETP, **36**: 1076 (1959)].
9. Robinson, M. T., and O. S. Oen: Phys. Rev., **132**: 2385 (1963).
10. Blackman, M.: *Handbuch der Physik,* Vol. 7, Part I, p. 377 (1958).
11. Barré, B.: private communication.
12. Lanore, J. M.: to be published.
13. Lindhard, J., V. Nielsen, M. Scharff, and P. V. Thomsen: Kgl. Danske Videnskab. Selskab, Mat.-fys. Medd., **33**: No. 10 (1963).
14. Beeler, J. R., Jr.: Phys. Rev., **134**: A530 (1964); ibid., **150**: 470 (1966).

DISCUSSION on paper by I. M. Torrens and M. T. Robinson

BULLOUGH: Do you think there is any significant transient potential-energy loss during the formation of these high-energy corrodes? Is this loss only important for low-energy trajectories?

TORRENS: Transient potential-energy losses are probably only important for collisions below the threshold energy. The transient losses are small compared to the total subthreshold energy loss. To investigate their importance would require a many-body dynamic simulation of collisions at the low-energy end of the cascade. Such an investigation is planned as a check on this and other aspects of the atomic-displacement model.

KULCINSKI: When you include the directionality of displacement energy I would expect that the displacement efficiency would decrease; do you have any idea at thy present time how this would alter the spatial distribution of defects?

ROBINSON: (i) I would expect the directional dependence of E_d to be really influential only for PKA energies in the threshold region. Analytical theory shows that alterations in the threshold details do not change the linear dependence of the number of defects on the PKA energy above some value E_d^*: This may be defined as the energy above which the displacement probability is unity regardless of the direction of displacement.

(ii) The same should be true for ion bombardment as well, in spite of the low mean transferred energy mentioned by Kulcinski, since most of the damage will come from those few collisions producing high-energy recoils.

JACKSON: Did you see any focussons and if so how long were they?

TORRENS: We have not so far observed any long-range focussons. However, by stopping a cascade atom when its energy drops below $2E_d$ we have biased the model slightly against focussons or replacement sequences with energies between E_d and $2E_d$. Atoms in this energy range are assumed to be incapable of increasing the number of vacancies created. They could, however, propagate replacement sequences, and we intend to investigate this by lowering the stopping-energy criterion. We would not anticipate any increase in the amount of damage created, but details of its spatial distribution could change.

PARSONS: When your computer code is used to simulate a cascade in germanium or silicon, does it predict that the resultant damaged region will have an amorphous structure rather than a defect-rich crystalline region?

TORRENS: This question may not be adequately treated at the present time since the simulation program has not been used as yet for any diamond cubic crystals. The amorphous nature of the damaged region would tend to be established during the thermal annealing of the damaged region created by the cascade.

WILSON: You began by stating that you are interested in the potentials only in the region >10 eV but then you fit your potential to a Born-Mayer form at first-neighbor separations. Exactly what interatomic potential at various separations are you using?

TORRENS: The Molière potential is used in our calculations at all separations. The Born-Mayer potential has been widely used at energies in the intermediate range. Our choice of the nearest-neighbor separation for matching our Molière potential to the Born-Mayer form is somewhat arbitrary, but it gives a potential that appears reasonable over the complete range of separation involved in the simulation. We note that variation of the potential screening radius over a 20 percent range produced little change in the number of vacancies created in the cascade, but altered their spatial distribution.

WILSON: Your method, including the vibrational modes, seems particularly applicable to the study of diffusion processes. As you no doubt know, one could view the motion of an atom passing through a saddle-point configuration while various local modes are hindering or aiding the motion. Have you looked at this problem with your method?

TORRENS: Our adiabatic approximation assumes that the incoming-atom velocity is high compared to that of the vibrating-lattice atom, which is true for a cascade atom but certainly not for a diffusing lattice atom passing through the saddle point. A molecular-dynamics treatment would be more relevant to the diffusion problem.

HO: I do not understand the energy independence of the efficiency of producing displaced atoms. Could you explain?

TORRENS: The displacement efficiency K is defined with reference to the damage energy $(E-E_{in})$ after subtracting the total inelastic loss [see Eq. 2]. The ratio $\nu(E)/E$ will of course decrease as the primary energy E increases, since an increasing fraction of E disappears as inelastic energy loss during the cascade.

TEMPERATURE DEPENDENCE OF THE VACANCY FORMATION ENERGY IN KRYPTON BY MOLECULAR DYNAMICS

R. M. J. Cotterill and L. B. Pedersen

Dept. of Structural Properties of Materials
The Technical University of Denmark
2800 Lyngby - Denmark

ABSTRACT

A molecular dynamics study of the lattice vacancy in krypton shows that the formation energy of this defect decreases by about 20 percent with increasing temperature, in the range 0 to 96 K.

1 INTRODUCTION

It is a well-established fact that the temperature dependence of the concentration of lattice vacancies in a crystalline lattice is described by the equation

$$n_V/N = \exp\left(S^F_{Vv}/k\right) \exp\left(E^F_V/kT\right) \tag{1}$$

where n_V and N are, respectively, the number of vacancies and the number of atoms. S_{Vv}^F is the change in vibrational entropy, and E_V^F the change in potential energy when one vacancy is formed. The magnitudes of these latter two parameters can be determined if n_V can be measured as a function of T. It is usual to assume that S_{Vv}^F and E_V^F are not themselves functions of T, and the experimental evidence so far accumulated seems to indicate that they are constant. The experimentally accessible range of temperature is, however, limited by the necessity of having n_V large enough to allow accurate measurement. With the presently available techniques, this range is approximately 0.6 T_m - T_m, where T_m is the absolute melting point. It is not yet possible to determine experimentally whether or not E_V^F and S_{Vv}^F are constant over the entire solid-state existence range.

Recent papers by Flynn[1] and Friedel[2] have stressed the importance of anharmonic effects in processes involving point defects. The magnitude of E_V^F is determined by the atomic configuration around the vacancy and the interatomic potential, and S_{Vv}^F is related to the change in the frequency spectrum when one atom is removed from the interior of the crystal and placed on the surface. Anharmonicity affects the details of these factors, and since the degree of anharmonicity is a function of temperature, the net result could be temperature-dependent values of E_V^F and S_{Vv}^F.[3,4]

There have been several attempts to calculate the magnitude of E_V^F and S_{Vv}^F. The traditional method of obtaining E_V^F has been via molecular statics, in which the interactions between discrete atoms are simulated by an analytical potential (usually two-body and central). The formation energy is found from the difference in potential energy between a perfect crystal and one containing a vacancy, both being in the equilibrium configuration.[5] Calculations of this type give values that are strictly valid only at 0 K, and, for the pure elements, the derived value of E_V^F is often (within a few percent) equal to the sublimation energy.[6] Glyde[7] has extended this type of approach in an attempt to examine the temperature dependence of E_V^F and S_{Vv}^F for solid argon (using the experimentally determined variation of lattice parameter with temperature). He found that E_V^F and S_{Vv}^F, respectively, decrease and increase with increasing temperature. Glyde was also able to demonstrate that many-body forces need not be invoked.

This paper describes calculations of the temperature dependence of E_V^F using the technique of molecular dynamics in which the thermal motions of individual atoms are followed by solving the classical equations of motion. The changes in potential energy that give rise to E_V^F are then found from suitable time averages. It is found that E_V^F decreases with increasing temperature, in agreement with Glyde's quasimolecular-statics calculations.

2 BASIC RELATIONSHIPS

A crystal is in equilibrium when its Gibbs free energy G is a minimum, G being given by

$$G = U + PV - TS \tag{2}$$

where U, P, V, T, and S are, respectively, the internal energy, pressure, volume, temperature, and entropy. When n_V vacancies are present, and the N atoms are now arranged on $N + n_V$ sites, the Gibbs free energy is changed by an amount

$$\delta G = n_V (\delta U_V^F + P \delta V_V^F) - T \delta S_V^F \quad , \tag{3}$$

where the term in parentheses is the change in enthalpy per vacancy. Glyde shows that the second term in the enthalpy is small and can be neglected.[7] The entropy change can be divided into configurational and vibrational components thus

$$\delta S_V^F = \delta S_{Vc}^F + n_V S_{Vv}^F \quad , \tag{4}$$

where

$$\delta S_{Vc}^F = K \ln \left\{ N!/(N - n_V)! \, n_V! \right\} \tag{5}$$

and where S_{Vv}^F is the vibrational entropy change per vacancy. The internal energy has potential and kinetic components, but the latter will be unchanged by the introduction of vacancies at constant temperature. We therefore write

$$\delta U_V^F = E_V^F \quad , \tag{6}$$

where E_V^F, the familiar "vacancy formation energy", is a change in potential energy. Minimizing G with respect to changes in n_V then leads to Eq. 1.

3 COMPUTATIONAL PROCEDURE

As has been described previously, the molecular-dynamics technique permits one to calculate the positions of all atoms in a model assembly as a function of time by solving the classical equations of motion.[8-11] The model crystal is made effectively infinite by use of periodic boundary conditions. The number of atoms M in the irreducible cell determines both the long wavelength cut-off and the defect concentration in such an "experiment", and it is usually made as large as the computer memory size will allow; in the present case it was 500. The temperature, T, of the system is related to the velocities, v_i, by the relation

$$3MkT/2 = \sum_{i=1}^{M} m v_i^2/2 \quad , \quad (7)$$

where m is the atomic mass. The pressure P follows from the virial theorem.

$$P = \rho k T - (\rho/3M) < \sum_{i=1}^{M} \sum_{j>i}^{J} r_{ij} \, \delta V(r_{ij})/\delta r_{ij}> \quad , \quad (8)$$

where ρ is the number density and r_{ij} is the distance between the ith and jth atoms in the assembly. In the calculations described here for krypton the interactions were represented by a Lennard-Jones function so that the potential energy between the ith and jth atoms was

$$V(r_{ij}) = \epsilon \left\{ (r_o/r_{ij})^{12} - 2(r_o/r_{ij})^6 \right\} \quad . \quad (9)$$

The constants ϵ and r_o are a function of the truncation number J. They were calculated by the self-consistent method described by Girifalco and Weizer[12] using the known extrapolated 0 K values of E_S (=1.86 x 10^{-13} ergs) and lattice parameter (=5.646 x 10^{-8} cms) as given in Ref. 13 and 14, respectively. It was assumed that these constants were temperature independent. For a truncation at 54 neighboring atoms, the constants had the values, $\epsilon = 2.364 \times 10^{-14}$ ergs, and $r_o = 4.080 \times 10^{-8}$ cms. In a classical system of finite size there will be instantaneous (and antiphase) fluctuations in the total potential and kinetic energies. It can be shown that the mean square fluctuation in temperature will be that given by the relation[15]

$$<(\delta T)^2>/<T>^2 = (1 - 3k/2C_v)/M \quad , \quad (10)$$

where C_v is the specific heat at constant volume. This is yet another reason for making M as large as possible. One sees, therefore, that the calculation of the change in potential energy due to the introduction of a defect is essentially a signal-to-noise problem, and it is found that time averages over at least 1000 time steps of calculation (i.e., over 10^{-11} seconds) are necessary to sufficiently reduce the computational errors.

In a typical calculation of E_V^F by this technique the motions of the atoms in a perfect crystal, at a chosen pair of values of nominal pressure and nominal temperature, are followed for several thousand time steps (each lasting 10^{-14} sec). The potential energy of the crystal U_p, the instantaneous value of which is given by

$$U_p = \frac{1}{2} \sum_{i=1}^{M} \sum_{j=i}^{J} V(r_{ij}) \quad , \tag{11}$$

is then time averaged. A second run is then carried out, at the same pressure and temperature, for a crystal containing a vacancy, and the time-averaged potential energy of a crystal containing M atoms (i.e., $M + 1$ sites) is determined. E_V^F is then the difference between these two mean potential energies. In practice, the temperature and pressure can be controlled by incrementally changing the velocities of the atoms and varying the volume of the irreducible cell, respectively. Such control is invariably necessary because the departure of either of these parameters from their nominal values can only be determined with sufficient accuracy after averages have been taken over about one thousand cycles.

4 RESULTS

The results of a check on the basic properties of the model are shown in Fig. 1. The equilibrium lattice parameter was calculated as a function of temperature for zero pressure. It can be seen from the figure that the molecular-dynamics model shows a greater expansion over the entire solid range, compared with the experimental curve[16]. Fig. 2 shows the calculated temperature dependence of the vacancy formation energy at zero pressure. The upper curve on this figure shows the dependence calculated by molecular statics. In this calculation the atoms were all located on the lattice sites and the change in energy results from the change in lattice parameter. For this calculation the lattice-parameter values were taken from the model molecular-dynamics calculations referred to above. The lower curve represents a smooth line drawn through the actual molecular dynamics results which were as follows:

$E_V^F (0 \text{ K}) = 1.82 \times 10^{-13}$ ergs

$E_V^F (61 \text{ K}) = 1.71 \times 10^{-13}$ ergs

$E_V^F (96 \text{ K}) = 1.44 \times 10^{-13}$ ergs ,

the calculational errors increasing with temperature and reaching $\pm 0.15 \times 10^{-13}$ ergs at 96 K. The experimental value of the vacancy-formation

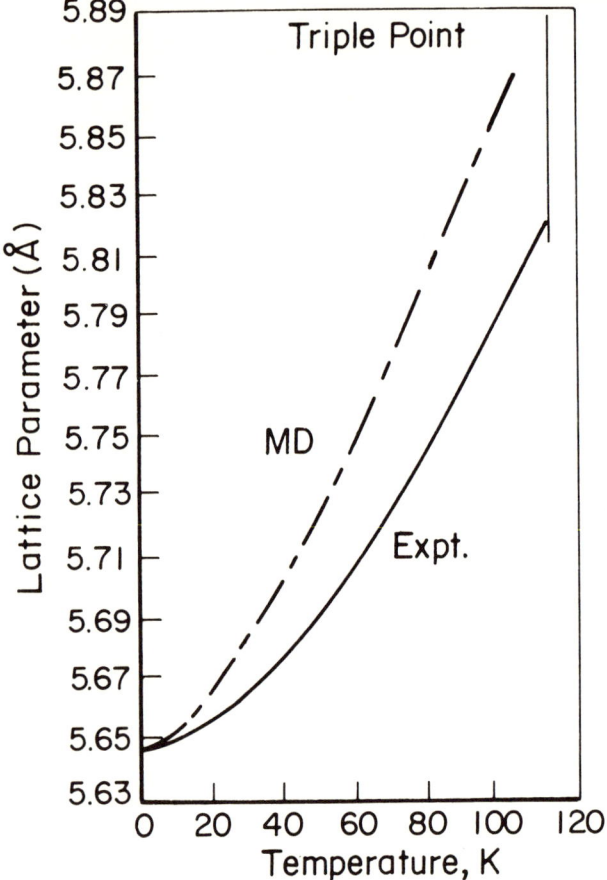

Fig. 1. Lattice parameter versus temperature. ———— Experimental curve. —·—·—·—·—·— Molecular Dynamics curve.

energy in krypton, measured in the vicinity of T_m, is 1.24 x 10^{-13} ergs.[17] If similar calculations are carried out for temperatures much closer to the melting point (116 K at 1 atm), trouble is encountered because of incipient vacancy migration.

5 DISCUSSION

It is interesting to note that the temperature dependence of E_V^F becomes particularly marked around the experimental Debye temperature (66 K). This tends to reinforce the belief that the effect is due to the actual thermal vibrations rather than to a secondary effect such as volume change of the crystal upon formation of the vacancy. Indeed, the present calculation gives a formation volume which does not differ from one atomic volume by more than the uncertainty of the volume caused by the

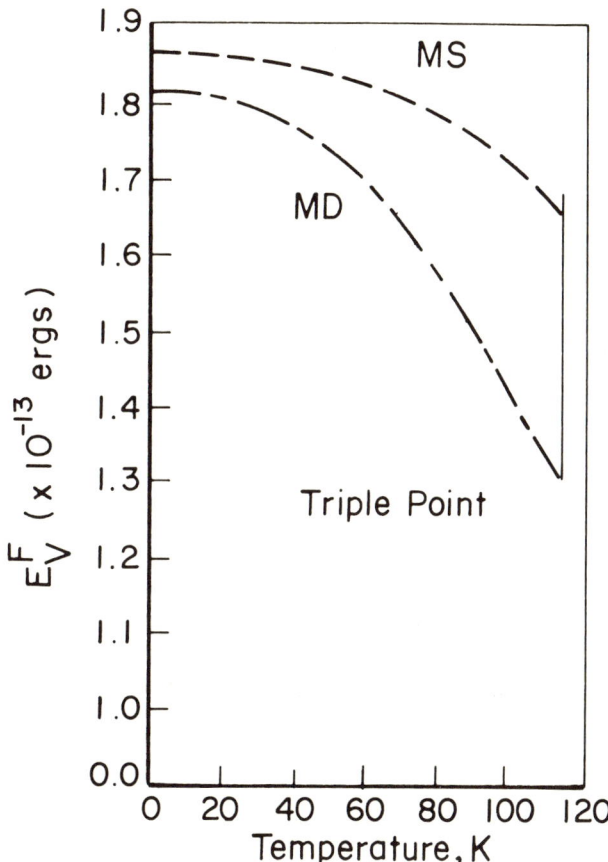

Fig. 2. Vacancy formation energy versus temperature. ------------ Molecular Statics. _._._._. Molecular Dynamics. Corresponding points at a given temperature have the same lattice parameter.

noise fluctuations in the temperature and pressure. Moreover, although thermal expansion contributes to the decrease of the vacancy formation energy, this contribution is only about 20 percent of the total observed change. In this connection it should be noted that the molecular-statics curve of Fig. 2, being based on the molecular dynamics model results for the temperature dependence of the lattice parameter, cannot be directly compared with the results of Glyde. To obtain a direct comparison it would have been necessary to carry out a molecular-statics calculation based on the experimental expansion curve. This clearly would have given an even weaker temperature dependence of the vacancy-formation energy. The difference between the molecular-statics and molecular-dynamics curves shown in Fig. 2 is due to the fact that, at a finite temperature, the atoms are not located on their lattice sites. The mechanism by which the actual vibrations make the major contribution is not yet clear, but it is possibly due to subtle many-body effects. (The many-body effects referred to here should not be confused with the many-body forces referred to in the Introduction. The molecular-dynamics model in the present study examines the properties of what might be referred to as a "many-two-body system".)

Fig. 3 shows a comparison between the temperature dependence of the vacancy concentration in krypton obtained experimentally by Losee and Simmons[17] and theoretical values calculated from the determination of E_V^F. The experimental results have been drawn as the straight line given by Losee and Simmons[17] together with a linear extrapolation of their results. The theoretical curves have been plotted using two different approximations. One was a linear approximation based on the relationship

$$E_V^F = E_{10} + A_1 (T - T_{10}) \qquad T > T_{10} \qquad (12)$$

where T_{10} and E_{10} are, respectively, 61 K and the corresponding vacancy formation energy, and A_1 is a constant. The second approximation was based on the power series relationship

$$E_V^F = E_{20} + A_2 T^2 + B_2 T^3 \quad , \qquad (13)$$

which relates to a basis of 0 K where E_{20} is the vacancy-formation energy at that temperature. Using the relation given by Zener[18]

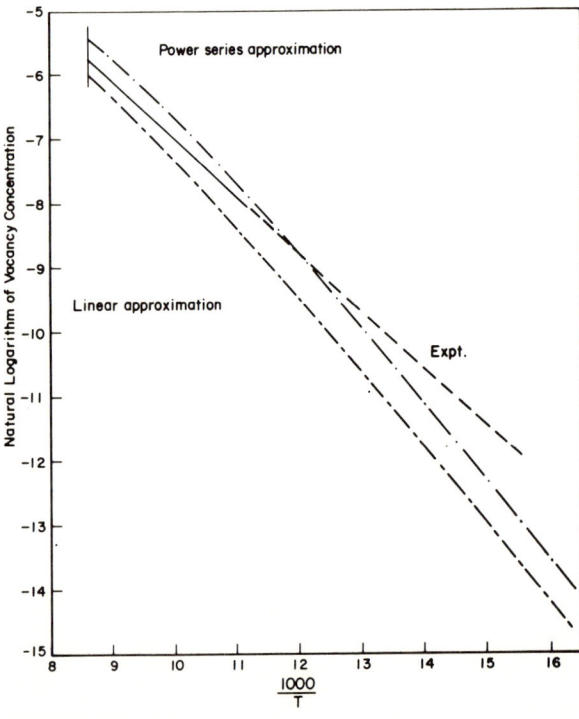

Fig. 3. Natural logarithm of vacancy concentration versus reciprocal temperature. ——— Experimental curve. —..—..—.. Linear Approximation $\ln(n_V/N) = S_{10}/k - E_{10}/kT + A_1/k (\ln(T/T_0) - (T - T_{10})/T)$ —.—.—.. Power series approximation $\ln(n_V/N) = S_{20}/k - E_{20}/kT + A_2T/k + B_2T^2/(2k)$.

$$\delta S_{Vv}/\delta T = (1/T)\delta E_V^F/\delta T \quad , \tag{14}$$

one obtains

$$S_{Vv}^F = S_{10} + A_1 \log T/T_{10} \tag{15}$$

for the linear relation and

$$S_{Vv}^F = S_{20} + 2A_2 T + (3/2) B_2 T^2 \tag{16}$$

for the power-series relation.

The values of the constants were determined to be

$A_1 = -7.71 \times 10^{-16}$ ergs/K

$A_2 = -9.22 \times 10^{-19}$ ergs/K/K

$B_2 = -3.33 \times 10^{-20}$ ergs/K/K/K

and the values of S_{10} and S_{20} were fixed so that the vibrational entropies were 2 k at 116 K (the triple point).[17]

$S_{10} = 7.77 \times 10^{-16}$ ergs/K

$S_{20} = 11.63 \times 10^{-16}$ ergs/K

Considering the error latitudes quoted by Losee and Simmons[17] and the agreement of the curves shown in Fig. 3, it seems reasonable to conclude that the experimental results do not rule out a temperature dependence of E_V^F.

ACKNOWLEDGMENTS

It is a pleasure to acknowledge helpful discussions with F. W. deWette, J. W. Martin and A. Rahman during the course of this work.

REFERENCES

1. C. P. Flynn: Zeits. für Naturforschung, **26**, 99 (1971).
2. J. Friedel in *Vacancies and Interstitials in Metals*, ed. A. Seeger et al. (North-Holland, 1970) p 787.
3. A. Seeger and H. Mehrer, p 1 in Ref. 2.
4. J. Frenkel: *Kinetic Theory of Liquids* (Dover, 1955) p 14.
5. M. Doyama and R.M.J. Cotterill in *Lattice Defects and Their Interactions*, ed. R. R. Hasiguti (Gordon & Breach, New York, 1967) p 79.
6. Tabulated values are given in A. C. Damask and G. J. Dienes, *Point Defects in Metals* (Gordon and Breach, New York, 1963); *Lattice Defects in Quenched Metals*, ed. R.M.J. Cotterill et al. (Academic Press, 1965); and in Ref. 2.

7. H. R. Glyde: J. Phys. Chem. Solids, **27** 1659 (1966).
8. B. J. Alder and T. Wainwright: J. Chem. Phys., **31**, 459 (1959).
9. J. B. Gibson, A. N. Goland, M. Milgram and G. H. Vineyard: Phys. Rev., **120**, 1229 (1960).
10. A. Rahman: Phys. Rev., **136**, A405 (1964).
11. L. Verlet: Phys. Rev., **159**, 98 (1967).
12. L. A. Girifalco and V. G. Weizer: Phys. Rev., **114**, 687 (1959).
13. R. H. Beaumont, H. Chihara and J. A. Morrison: Proc. Phys. Soc. (London) **78**, 1462 (1961).
14. D. N. Batchelder, D. L. Losee and R. O. Simmons: J. Phys. Chem. Sol., **Suppl. L. 10**, 843 (1967).
15. J. L. Lebowitz, J. K. Percus and L. Verlet: Phys. Rev. **153**, 250 (1967).
16. Tabulated values are given in E. R. Dobbs and K. Luszczynski, Proc. Internat. Conf. Low Temp. Phys., Paris, 1955, pp. 439-440.
    G. H. Cheesman and C. M. Soane, Proc. Phys. Soc. (London), **70B**, 700 (1957).
    B. F. Figgins and B. L. Smith, Phil Mag. **5**, 186 (1960).
17. D. L. Losee and R. O. Simmons, Phys. Rev. Letters **18**, 451 (1967).
18. C. Zener: *Imperfections in Nearly Perfect Crystals,* eds. W. Shockley, J. H. Hollomon, R. Maurer and F. Seitz (J. Wiley and Sons, Inc., New York, 1952).

DISCUSSION on paper by R.M.J. Cotterill and L. B. Pedersen

SEEGER: Would a low migration energy, i.e., a rapid movement of the vacancy at finite temperature, interfere with your calculation of the energy of formation?

PEDERSEN: Not unless it were very low and/or the temperature very high, neither of which is the case. If it is assumed that our vacancy crystal is activated toward diffusion and we wanted to make a correction for that, we would get lower energies of formation than the two reported here, because the potential energy would be higher.

SEEGER: There have been quite a number of analytical calculations on the temperature dependence of the formation energy, including one on vacancies in the rare gases by Fischer[1]. These calculations agree on what is also instinctively clear, namely that anharmonic effects make the formation of a vacancy easier at high temperatures. This means that both the entropy and the energy of formation of vacancies increase with increasing temperature, and I believe that this should be true not only for all simple crystals but also for realistic models. Could you comment on why you think your model behaves differently?

PEDERSEN: The fact that we have assumed the sublimation energy to be independent on temperature could be of significance.

de WETTE: The size of the box and the number of particles in it determine the density of the system. In your studies as a function of temperature, have you taken the density change (thermal expansion) into account?

PEDERSEN: The volume was fixed so as to make the pressure essentially equal to zero atmospheres.

TORRENS: Have you looked at vacancy migration in your model? Preliminary molecular-dynamical calculations by Levesque at Orsay and myself indicate that some complicated many-body processes play a significant part in the vacancy jumps.

PEDERSEN: We have not yet looked at the migration of vacancies. Our model was built to handle only fixed neighbor interactions. It is highly improbable that we would be able to calculate, for example, migration energies because of the "signal-to-noise" problem.

TYSON: Have you taken into account the temperature dependence of the sublimation energy in arriving at your estimate of the vacancy formation energy?

PEDERSEN: No. The reason is that we do not know the temperature variation of the sublimation energy. Experimentally it turns out, however, that there is no significant variation of the sublimation energy with temperature.

REFERENCE

1. K. Fischer, Z. Phys. **155**, 59 (1959).

AGENDA DISCUSSION: POINT DEFECTS

L. A. Girifalco [*]

School of Metallurgy and Materials Science
University of Pennsylvania
Philadelphia, Pennsylvania 19104

H. H. Yoshikawa [**]

Irradiation Analysis Section
WADCO Corporation
P. O. Box 1970
Richland, Washington 99352

The agenda discussion covered five major topics. First, Professor J. P. Hirth provided an exposition on volume effects on vacancy formation. Second was the suitability and applicability of pseudopotential theory. Third was a discussion on the effect of inclusion of zero point energies and their quantum effects. Fourth was an extended discussion on

---

[*]Chairman.
[**]Secretary.

migration energy calculations. Fifth was a discussion on possible experimental checks on the reliability of potentials.

On Volume Effects

The expose by Professor Hirth is given in Appendix A to this Discussion. Hirth concluded by pointing out that the discussion applies only to situations where the initial and final thermodynamic states are well defined, as is the free energy change. A problem can arise, however, in certain types of atomic calculations where the interatomic potentials do not yield the complete free energy of a system. In these cases, one must superpose an "effective" pressure on the system and must include work done against this pressure to give the total free energy of a defect in systems which undergo an overall volume change. This effect is discussed in Appendix B which Dr. M. Duesbery submitted after the conference.

Pseudopotentials

It is clear that detailed results using pseudopotentials can be misleading. There is general agreement both among opponents and proponents of the use of pseudopotential theory that one cannot assign reliability to all of the quantities calculable from theoretical principles.

As an example, Roger Chang cited his results using the Heine-Abarenkov, the Shaw-Pynn, and modified Shaw-Pynn potentials[1]. The results for sodium, shown below, indicate the sensitivity of the energy of vacancy formation, ΔE_d, and of various strain coefficients, Z, to the choice of potential. For some coefficients, not only is there a change in magnitude, there is even a change in sign.

The definition of the quantities tabulated is obtained from the equations relating the change in energy to the strains.

	Heine-Abarenkov	Shaw-Pynn	Modified Shaw-Pynn
ΔE_d	0.06	0.049	0.031
Z_1	-1.92	-2.28	-2.50
Z_{37}	-1.35	+0.47	+3.81
Z_{15}	+0.87	+1.93	+3.51

$$\Delta E_d = \Delta_0 + \Delta_1 + \Delta_2 + \Delta_3 + \Delta_4 + \Delta_5 + \Delta_6$$

$$\Delta_6 = Z_0 + \sum_{i=1}^{9} Z_i e_i + \sum_{i=1}^{9} Z_{ii} e_i^2 + \sum_{i=1}^{9} \sum_{j=1}^{9} Z_{ij} e_i e_j$$

Seeger felt that in connection with the point defect problem the pseudopotential methods were not applicable except for alkali metals. Second-order perturbation theory is applicable only for alkali metals, and it is clear that the details of the energy build-up for different configurations is highly dependent on the model.

In Chang's opinion, despite the deficiencies of the pseudopotential model, the modified Shaw-Pynn potential should be quite useful, particularly if the parameters of the potential are fitted to agree with experimental data.

Harrison felt the pseudopotentials were quite useful in giving an indication of the important processes that are taking place even though the detailed values of quantities calculated might be questionable. The sensitivity of results, such as vacancy scattering, to the choice of pseudopotential reflects directly the limited validity of the perturbation theory. In this respect Harrison agreed with Seeger but he did not agree that the difficulty is necessarily more serious in the polyvalent metals than in the alkalis. An important feature of the method is that it systematically addresses the entire problem, but the alternative of modeling a calculation entails the danger of possible neglect of important factors. For example, in scattering by stacking faults in monovalent metals, the correct calculation using pseudopotentials indicates scattering only in fourth order, whereas several earlier models gave second-order contributions that would have cancelled out any systematic approach.

In Ashcroft's opinion, useful pseudopotentials exist. While it is true that different structures are very sensitive to the pseudopotential, one must be circumspect about the quantities to be calculated rather than to condemn the entire technique.

Ho felt that volume-dependent properties are probably poorly described by using pseudopotentials, but that those properties dependent on structure should be reasonably well described.

As an example, he cited calculations of displacements around a defect. In this case, using pseudopotentials guarantees that the minimum configuration energy is calculated.

In the calculation of the minimum configuration energy, the harmonic part in the change of the structure factor comes from the form of the dispersion curves and hence is not dependent on the pseudopotential.

Bullough reported that the computation of the quantity, G_α, gives a consistent value for the formation volume.

Quantum Effects

Professor Girifalco posed the question, "What happens to equilibrium-state calculations when the zero point energy is added?" When this is done, the condition that the potential energy be a minimum, no longer has to be satisfied.

Professor Harrison responded that the Born-Oppenheimer approximation should be adequate and thus would offer a means for coping with this situation. Professor Maradudin felt the change would simply be to minimize the Helmholtz free energy.

Professor Seeger felt that in the calculation of the energy of formation of defects the answer at high temperature was readily obtained. In this instance, the effect of the zero point energy would be confined to the lower temperature region and would have little effect for results intended for application at higher temperature.

Migration Energies

Dr. Torrens noted that migration energies were generally calculated with a static lattice approximation; because this represents an unrealistic condition, one must have reservations about the results. The question then arises, "How much faith should one have in a static treatment of migration phenomena?"

Dr. Wilson reported that he has used six different potentials in calculating helium migration in a copper lattice. Although the energy of formation of the impurity atom changes drastically, the migration energy proves to be relatively insensitive to the potential form. Also, the atomic displacements around the impurity atom were found not to be highly sensitive to the potential form.

Dr. Wynblatt felt also that the migration energy should not be highly sensitive since most potentials depend on input derived from the observed elastic properties of the material. In terms of the energy surface traversed by the migrating atom, the problem simply becomes that of treating the distortion of the saddle point between equilibrium configurations for the migrating atom.

Professor March proposed an experimental method to decide whether noncentral forces must be included. In considering a single crystal of a bcc material such as iron, if the 411 and 330 reflexions in X-ray scattering have different intensities, then spherical pseudoatoms are insufficient to describe the charge density and a noncentral potential such as the one proposed by R. A. Johnson must be included.

In response to a question from Shewmon as to the possibility of potentials being extended inappropriately, Johnson pointed out that the potentials should give reasonable information for physical mechanisms involving defects and should provide guidelines for interpreting experiments even though the detailed numbers resulting from the calculations cannot be fully trusted. For example, it is found that divacancies are more mobile than single vacancies in the fcc lattice. In calculations for bcc transition metals, second-neighbor pairs of vacancies are found to be most stable, and partial dissociation with a migration energy slightly greater than that for single vacancies is required for divacancy migration. As another example, the extension of the work in analyzing carbon clustering should have some validity when the clustering is associated with the effect of the strain field. As the direct carbon-carbon interaction is questionable, results depending on the strength of that bond would be more questionable.

Beeler reported that, in the fcc lattice, the single interstitial collects a carbon atom whereas in the bcc lattice the second neighbor divacancy collects carbon atoms.

Professor Weiner commented on results of migration-energy calculations based on potentials fitted to elastic constants and phonon dispersion relations. He found that the migration energy tended to zero for specific values of the critical shear stress for a modified Frenkel-Kontorova model based on a series of piecewise-linear spring joining the atoms of a solid.[3] The zeros indicate points of incipient instability. From this result, it is clear that the use of elastic and phonon behavior data are insufficient in themselves to guarantee the validity of migration-energy calculations.

Possible Experimental Checks

There was a general call for experimentalists to provide data to check the reliability of potentials. Possible suggestions such as field microscopy and low energy electron diffraction have the deficiency that both provide information only on surface atoms which experience a potential field not typical of the bulk of the material. Seeger indicated that experiments are being conducted by Schilling at Julich on diffuse X-ray scattering in aluminum. The experiment is intended to determine the distortion around interstitial atoms introduced in aluminum by irradiation at liquid-helium temperatures. The density of defects is estimated at 10^{-3}.

In all probability, difficulties are associated with the interpretation of this experiment, but it appears to offer the best hope for experimental study of the distortions around interstitials.

Beeler suggested that the Mossbauer effect in iron might be used to develop potentials.

It is clear that there is a major deficiency of experimental methods for providing information which could directly settle many of the controversies associated with the different potential forms proposed for the calculation of defect properties.

APPENDIX A

presented by

J. P. Hirth
Metallurgical Engineering Department
Ohio State University

In the discussion of volume effects in connection with vacancy formation in terms of potential theories, a number of apparently conflicting views were presented. As these effects can be represented thermodynamically, they are qualitatively model insensitive. Therefore, for convenience, they are discussed in terms of a continuum, linear, elastic model. All of the seemingly conflicting views are shown to be correct in certain cases.

Fig. 1 shows several reversible work cycles representing vacancy formation. Because the free energy and volume are state variables, the cycles perforce give the same net free energy and volume changes for the same initial and final states. The first case is that of constant pressure (A, B, C) with the external pressure P equal to zero. In Step B, an atom is removed from the interior and placed at a kink site on the surface, the system being constrained not to relax. This step involves a work term nominally equivalent to the heat of sublimation and a volume expansion v_1, equal to the atomic volume. As the surface structure no longer has a role in further considerations, the expanded system is represented as a

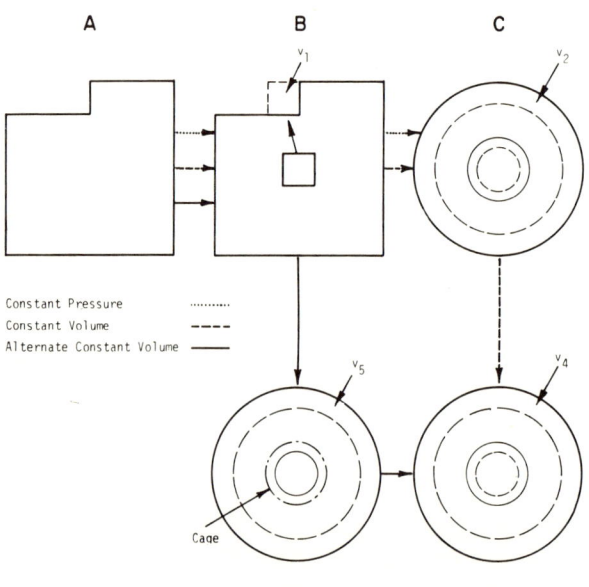

Fig. 1. Representation of formation of a vacancy by removal of an atom to a kink site (A, B). Reversible work cycles for vacancy formation are: constant pressure (A, B, C), constant volume (A, B, C, D), and alternative constant volume (A, B, E, D).

sphere. Further, to allow in the continuum model for volume changes associated with nonlinear relaxations near the vacancy, we follow J. Friedel[4] and model the vacancy as a ball coherently glued to a spherical hole initially larger than the hole and with smaller bulk modulus K than the matrix.

Relaxation occurs in Step C. While as discussed by W. Harrison, the volume change v_2 can in principle be of either sign, it is generally expected to be negative as shown in Fig. 1. The sum of the reversible work terms in Steps B and C equals G, the Gibbs free energy of formation of the vacancy, appropriate for the isothermal constant pressure process, and the volume change $v_3 = v_1 + v_2$ is the formation volume. G and v_3 are the quantities measured experimentally.

Now consider the constant volume process (A, B, C, D). The constant-volume state is achieved by compressing configuration C by an amount $v_4 = -v_3$. The elastic work equals the reversible work in this step and is given by

$$W = Ke^2 V/2 = K(v_4/V)^2 V/2 = (Kv_4/2)(v_4/V) \qquad (1)$$

where e is the dilatation and V is the total volume of the system. The first term in parenthesis in the result of Eq. 1 is of the order of the formation energy G, while the second term in parentheses is of the order 10^{-22} for a 1-cc crystal. Thus W is completely negligible compared to G. Since the Helmholtz free energy of formation $F = G-W$ it follows that $F = G$. Thus, as stated by W. Harrison and R. Bullough, and in apparent conflict with A. Seeger and M. Duesbery, the volume work term, W, is negligible in comparing vacancy formation energies in the two above cases. Moreover, Step D illustrates the remarks made by R. Johnson: a substantial energy change occurs in the atomic bonds near the core in his atomic calculations, represented here by a change in energy stored in the ball, but a nearly compensating energy change occurs in the matrix so that the net-work term is very small. This point is discussed in detail in the context of continuum elastic crossterms in Ref. 5.

As discussed by G. Ehrlich, it appears that the first term in P. Ho's work corresponds to Step B. In that case, Ho's energies appear to be about twice the heat of sublimation and hence a factor of two too large. However, Ho's calculation corresponds to a cycle equivalent to (A, B, E, D) with his first energy term that of Steps A, B, E. To achieve State E, Configuration B is compressed by an amount $v_5 = -v_1$. In addition, Ho does not relax the boundary of his Region II, and here this is represented by a rigid cage in Fig. 1 which prevents relaxation at the vacancy. Thus, relative to Configuration C or D, Configuration E has a much higher free energy, F, corresponding to the stored elastic energy associated with the cage constraint: this accounts for the factor of two discussed above. Configuration E is then relaxed to Configuration D by removing the cage and allowing an *inner* volume relaxation v_6. The added strain energy giving the factor of two flows out of the system in this step and the net free energy change is the same as for cycle (A, B, C, D) as, of course, already indicated by Ho's final result.

Finally, we reconsider Case (A, B, C) in the presence of a finite external pressure, P. Now in Steps B and C additional work is done by the system against external pressure in an amount Pv_3, which contributes to the free energy of formation. On the other hand, no such work is done in the constant-volume process. Hence the free energies of formation in the two cases now differ by Pv_3 and the concentration of vacancies will differ by a factor $\exp -(Pv_3/KT)$, in agreement in this case with the suggestions of Seeger and Duesbery mentioned above. With $P = 10^{10}$ dyne/cm^2, the Pv_3 contribution will be about equal to the free energy of formation in the absence of pressure, so the effect can be large.

APPENDIX B

presented by

M. S. Duesbery
National Research Council of Canada

For a hypothetical elastic body comprised of material points interacting via interatomic potentials, there may be linear, volume-dependent terms contributing to the point defect formation energy. In particular, if the material points are arranged in a primitive centrosymmetric array and if the binding forces are central, we have the specialized elastic continuum of Cauchy and St. Venant. If the elastic constants of this continuum do not satisfy the Cauchy relations, an external pressure

$$P' = 1/2\,(C_{12} - C_{44}) \quad,$$

where C_{ij} are elastic constants in Voigt notation, is necessary to maintain the body in equilibrium. For such a body, a term $P'dv$, where dv is the formation volume, must be added to the point defect formation energy.

The Cauchy relations are not necessary conditions on the elastic continuum, but the Cauchy theory provides a framework for the representation of real metallic solids. To a first approximation, the cohesive energy of simple metals can be separated into a central ion-ion potential that depends only on the relative separation of ion pairs, and an auxiliary energy that depends only on the volume of the body. The problem of the defect in a simple metal is thus analogous to that of an imperfect Cauchy body. For the typical metal, the term $P'dv$ is several tenths of an electron volt, and thus is of the same order as measured point defect energies.

This means that the computation of defect energies in metals by consideration only of central pair-interactions cannot be justified to any approximation, unless the pair potential is constructed deliverately to satisfy the Cauchy relations. Since the equilibrium pressure P' is given in terms of the elastic constants, and since the formation volume is obtainable directly from the defect strength, there seems no reason for the frequent neglect of this important term.

References

1. R. W. Shaw, Jr. and R. Pynn, J. Phys., **C2**, 2071 (1969).
2. R. Chang, J. Phys. and Chem. of Sol., **32**, 1409 (1971).
3. J. H. Weiner and W. T. Sanders, Phys. Rev., **134**, A1007 (1964).
4. J. Friedel, *Dislocations*, Pergamon Press Ltd. (1964).
5. J. P. Hirth and J. Lothe, *Theory of Dislocations*, McGraw-Hill, New York (1968), Chap. 2.

Part Four

DISLOCATIONS AND STACKING FAULTS

INFLUENCE OF DISLOCATIONS ON ELECTRON MICROSCOPE CRYSTAL LATTICE IMAGES

J. R. Parsons

Atomic Energy of Canada Limited
Chalk River Nuclear Laboratories
Chalk River, Ontario, Canada

ABSTRACT

A brief description is given of (1) the electron-optics of lattice image formation, (2) the electron-microscope facility used to obtain these images, and (3) some unexpected oscillations in lattice plane spacing observed near end-on dislocations in germanium and aluminum.

1 INTRODUCTION

An electron microscope can be operated so that the spacing of atomic planes in a crystalline material can be observed. This has led to the possibility of obtaining such images from regions of a crystal containing crystalline defects — e.g., dislocations, precipitates, damaged regions produced by fast-particle bombardment, etc. These images contain information about the perturbing influence of the defects at an atomic level, in that they show terminating planes, bending of the resolved planes, or both, depending on the character of the defect. Further, lattice images show fringes of good contrast right into regions of the defect where the strain field is known to be high — e.g., at the core of a dislocation as in Fig. 3. One would like to use these images to differentiate between the various lattice models used to describe these regions of high strain in which Hooke's law is not applicable.

Before a detailed analysis can be performed on a lattice image in which the perturbing influence of a lattice defect is evident, the relationship between the imaged position of a lattice plane and its actual position in the specimen must be known. Ideally one would hope for a one-to-one correspondence.

In a recent paper, Cockayne et al. have summarized the results of both a theoretical and experimental investigation of this relationship.[1] In this work, lattice images at inclined dislocations in germanium were used. The effect on the number of terminating lattice planes and their bending at these dislocations was examined when either the diffraction geometry or the objective lens focussing current was changed. The hoped for one-to-one correspondence was not found to exist except in the special case of a dislocation oriented so that it was viewed end-on.

In this paper a brief description is given of (i) the electron-optics of lattice image formation, (ii) the electron-microscope facility used to obtain these images, and (iii) some unexpected oscillations in lattice plane spacing observed near end-on dislocation in germanium and aluminum.

2 ELECTRON-OPTICS OF LATTICE IMAGE FORMATION

2.1 Illumination, Contrast, Specimen Thickness

If, as in Fig. 1, the transmitted and one diffracted beam of amplitudes ϕ_o and ϕ_g and phases θ_o and θ_g are recombined then, from Ref. 1, the intensity at any point x on the image plane is

$$I(x) = |\phi_o|^2 + |\phi_g|^2 + 2|\phi_o||\phi_g| \cos\left(\frac{2\pi x}{dM} + \theta_g - \theta_o + \Gamma_g - \Gamma_o\right) , \qquad (1)$$

where M is the magnification, d the spacing to be resolved and

$$\Gamma_g - \Gamma_o = \frac{-2\pi}{\lambda} C_o^{\cdot\cdot} \beta^4 + \frac{2\pi}{\lambda} \frac{\Delta f}{2} \beta^2 \qquad (2)$$

is the phase change introduced into the diffracted beam ϕ_g at angle β by the defocus Δf and the spherical aberration $C_o^{\cdot\cdot}$ of the objective lens. From Eq. 1 the lattice image *contrast* is a maximum when

$$\frac{\Delta I}{I} = \frac{4|\phi_o||\phi_g|}{|\phi_o|^2+|\phi_g|^2} \cos\left(\frac{2\pi x}{dM} + \theta_g - \theta_o + \Gamma_g - \Gamma_o\right) = \text{maximum} , \quad (3)$$

which occurs when $|\phi_o| = |\phi_g|$ and when the argument of the cosine function is an integral multiple of π.

From the dynamical theory of electron diffraction[2] for the two-beam case,

$$|\phi_o| = \cos\frac{\pi z}{\xi_g} \quad \text{and} \quad |\phi_g| = \sin\frac{\pi z}{\xi_g} , \quad (4)$$

so that $|\phi_o|$ is first equal to $|\phi_g|$ when the *specimen thickness* $z = \xi_g/4$; 108 Å for germanium (111) reflection; where ξ_g is the two-beam extinction distance.

For a perfect crystal, $\theta_g - \theta_o = \pi/2$ in Eq. 3. Thus, if the microscopist operates the objective lens at focus so that $\Delta f = 0$ in Eq. 2 and uses *tilted illumination* so that both ϕ_o and ϕ_g make equal angles (i.e., $\beta/2$) with the optic axis, then, as Dowell[3] has demonstrated, the effect of spherical aberration is nullified and the argument of the cosine function in Eq. 3 is first equal to π when $x = dM/4$.

Thus, the lattice image of a perfect crystal will show an image shift of $d/4$ and will have maximum contrast when the specimen is $\xi_g/4$ thick and the microscope is operated with tilted illumination and a focussed objective lens.

2.2 Defocussing

Experience has shown that it is not always possible to recombine ϕ_o and ϕ_g with a focussed objective lens. Astigmatism, lens alignment, and

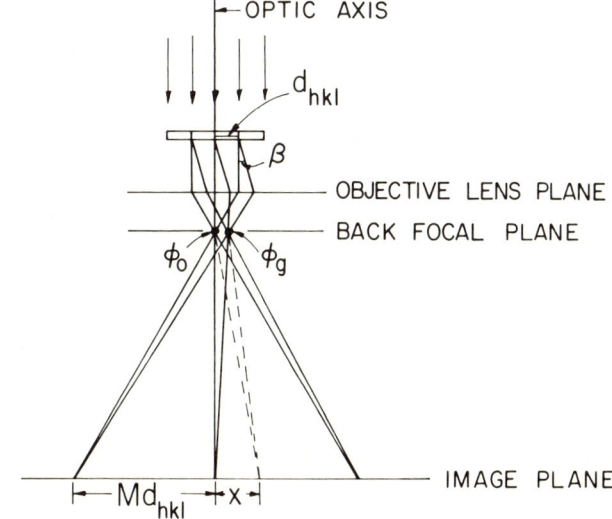

Fig. 1. Electron-optics of lattice image formation.

operator eye defects work against this ideal condition. The influence of a defocussed objective lens on the imaged position of a lattice plane in the specimen can be determined as follows. From Eq. 1 a lattice plane at an origin in the specimen is imaged at a position

$$x = \left[(\theta_g - \theta_o) + (\Gamma_g - \Gamma_o)\right] \frac{dM}{2\pi} \quad . \tag{5}$$

Thus, if an incremental focal series of images is taken of the same specimen area and from the first to the last image in this series the objective lens focussing current is changed from Δf to $\Delta f + \delta \Delta f$, then (from Eq. 2 and 5) the imaged position of this plane and all others will shift by an amount

$$\Delta x = \frac{[\delta \Delta f] \beta^2 \, dM}{2\lambda} \quad . \tag{6}$$

For (111) planes in germanium, good lattice image contrast still exists when $\delta \Delta f = 500$ Å. Thus for $\beta \sim 10^{-2}$ rads., $M = 250,000$X, $d \sim 3$ Å, and $\lambda = 0.037$ A, Δx in Eq. 6 is $\sim 5 \times 10^{-3}$ cm. This is too small to detect in Fig. 3 and other lattice images as it is less than the random variation in the imaged position of a lattice plane. Thus, there is no gross shift in the imaged position of a lattice plane due to changes in objective-lens focussing current. This has been confirmed experimentally.[1]

2.3 Diffraction Geometry

The diffraction geometry is changed by altering the direction of beam incidence on the specimen. In Fig. 2 two different directions of beam incidence have been used. Both images show the same set of resolved planes and the same inclined dislocation in germanium. Gross changes in the imaged position of the lattice planes and in the number of terminating planes (arrowed in Fig. 2) appear. These changes were also observed in the calculated images.[1]

Lattice-image calculations by Mannami[4] and Cockayne[5] show that this lack of a one-to-one correspondence can be avoided if the dislocation is viewed in an end-on orientation. In this orientation the lattice displacement produced by the dislocation is not a function of foil thickness as it is for an inclined dislocation. Thus, at an end-on dislocation the crystal diffracts as a perfect crystal with a rigid-body translation and a one-to-one correspondence between imaged and actual position is expected. Experimentally, with modern two-axis tilt stages, it is a fairly straightforward procedure to orient a dislocation in the electron microscope so that it is being viewed end-on.

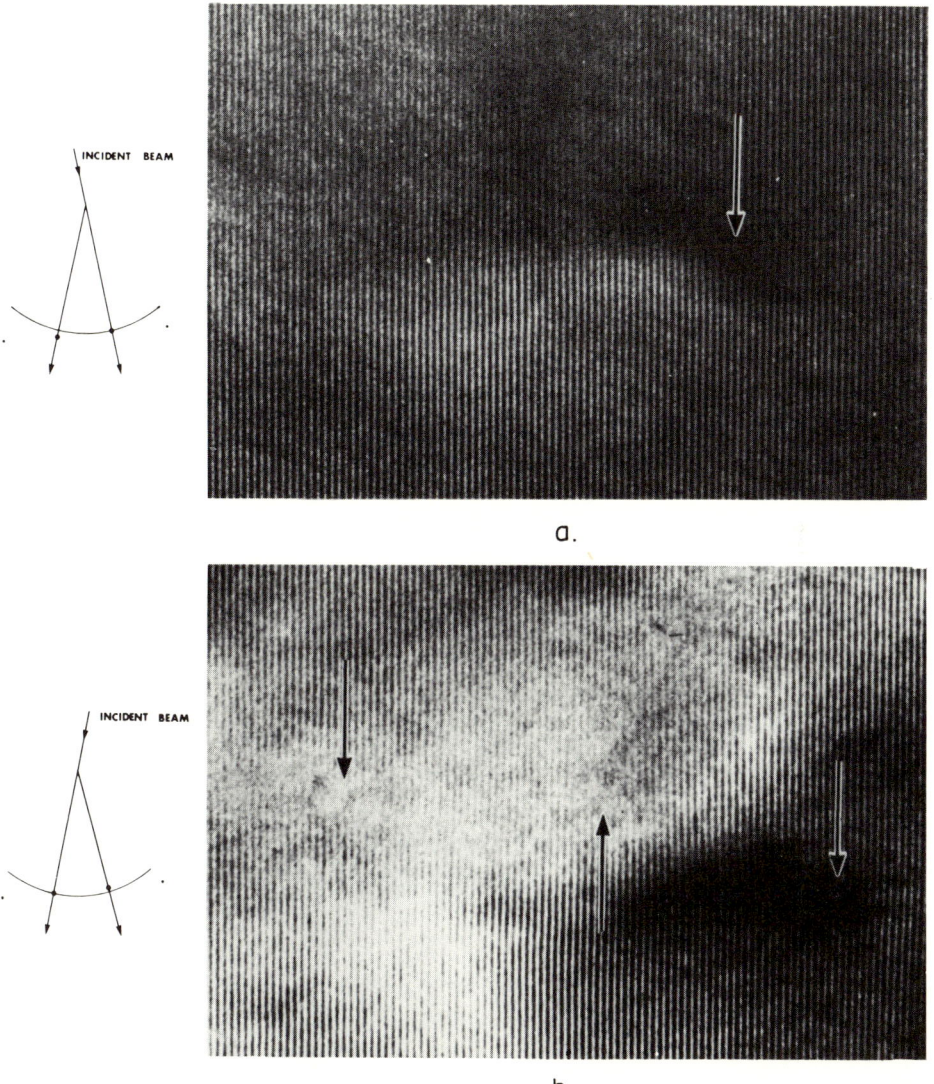

Fig. 2. The germanium (111) lattice images in (a) and (b) contain the same inclined dislocation. Note the changes in diffraction contrast, number of terminating fringes (arrowed) and fringe distribution that occur when the transmitted beam is combined with the diffracted beam + g in (a) and - g in (b).

3 MICROSCOPE FACILITY

A separate, air-conditioned building has been constructed to house the microscope and associated equipment. The building is well removed from large sources of a-c disturbance. The microscope sits on a 5-ton concrete block that is spring supported on two concrete piers on a rock foundation. The operator sits on a floor cantilevered over the block. The microscope is separately grounded to copper-mesh plates in the Ottawa River. Valuable discussions with Dr. F. Moran, University of Chicago, and Dr. T. Komoda, Kokubunji, Japan, have gone into the design of this facility. For lattice-image work, thin film, self-cleaning objective apertures, and pointed filaments are used. It is possible with this facility to visually observe a lattice image with a 3 Å spacing on the fluorescent screen at any time of the day.

4 DISLOCATION LATTICE IMAGES

As discussed by Bassett et al. the effect of a dislocation line on a directly resolved image of the $(hk\ell)$ planes is the occurrence of N terminating planes.[6] For a dislocation of Burgers vector $[uvw]$, the number of terminating $(hk\ell)$ planes is

$$N = hu + kv + \ell w \quad . \tag{7}$$

Note that N is to be taken as the algebraic sum of the number of terminating planes so that when Eq. 7 is applied to the dislocation in Fig. 2, N is conserved, i.e., $N = +1$ in both Figs. 2a and 2b.

4.1 End-On Dislocation in Germanium

Fig. 3 shows the perturbing influence of an end-on 60-degree dislocation on the regular 3.27 Å spacing of $(\bar{1}11)$ planes in germanium. This description of the dislocation in Fig. 3 was determined by Parsons et al.[7] from the following: $N = 1$ and $(hk\ell) \equiv (\bar{1}11)$ in Eq. 7, foil normal $\sim$ [110]; and an asymmetry exists in that the imaged lattice planes to the left of the arrowed extra plane in Fig. 3 are displaced more than those to its right.

The magnitude and extent of the strain field produced by the dislocation in Fig. 3 was determined by making microdensitometer tracings. The tracings were made coincident with the end of the arrowed terminating plane and in a direction perpendicular to the resolved planes. The above mentioned one-to-one correspondence allowed the strain-field component ϵ_{xx} of the dislocation in Fig. 3 to be obtained directly from measurements made on these tracings. The results have been plotted in Fig. 4. Strains were accurate to ± 0.03. This error, shown as $\Delta\epsilon$ in Fig. 4, associated with each data point in the figure, represents the average spacing deviation from the mean spacing of 100 planes recorded in a specimen area not containing a dislocation.

4.2 End-On Dislocations in Aluminum

The lattice image in Fig. 5 shows how a set of $\{111\}$ planes in aluminum is affected by a low-angle tilt boundary formed by eight end-on edge dislocations.

LATTICE IMAGES 469

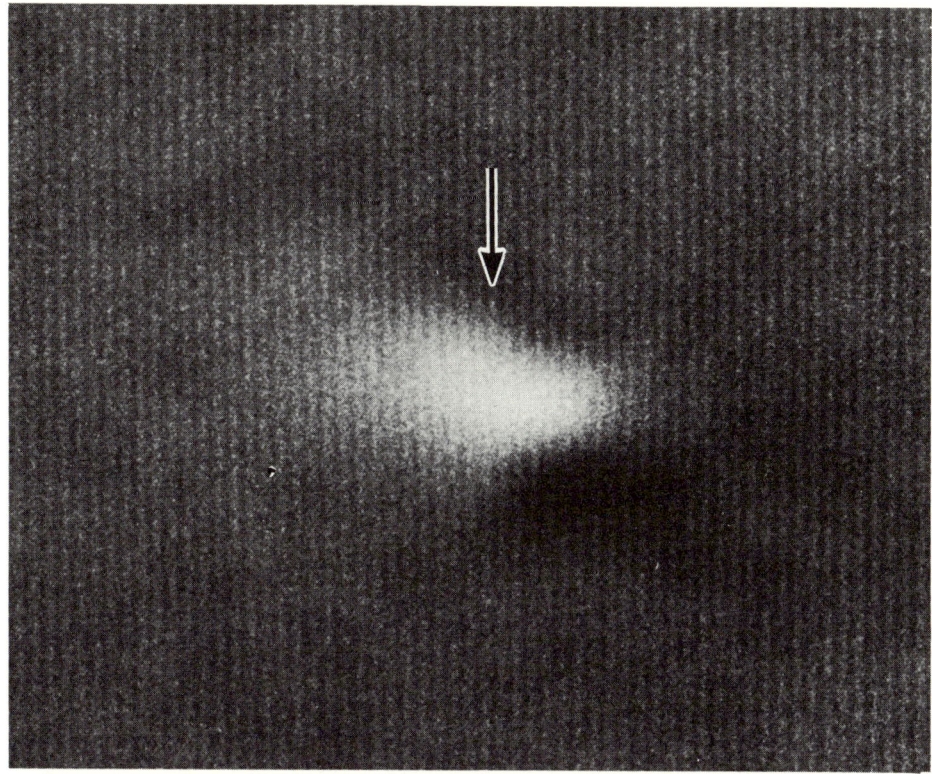

Fig. 3. Lattice image showing how a 60-degree dislocation, viewed end-on, perturbs the regular 3.27-Å spacing of (111) planes in germanium.

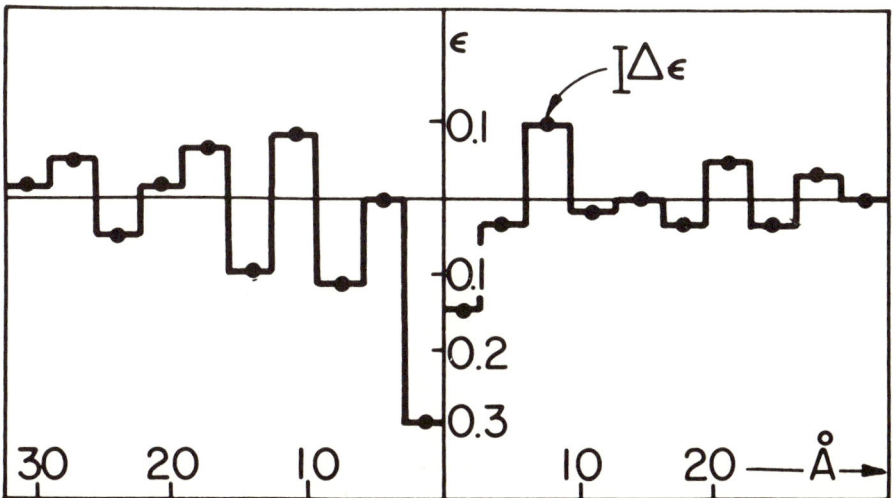

Fig. 4. Strain-field profile perpendicular to and coincident with the end of the arrowed terminating plane in Fig. 3. All data points have an error $\Delta\epsilon$.

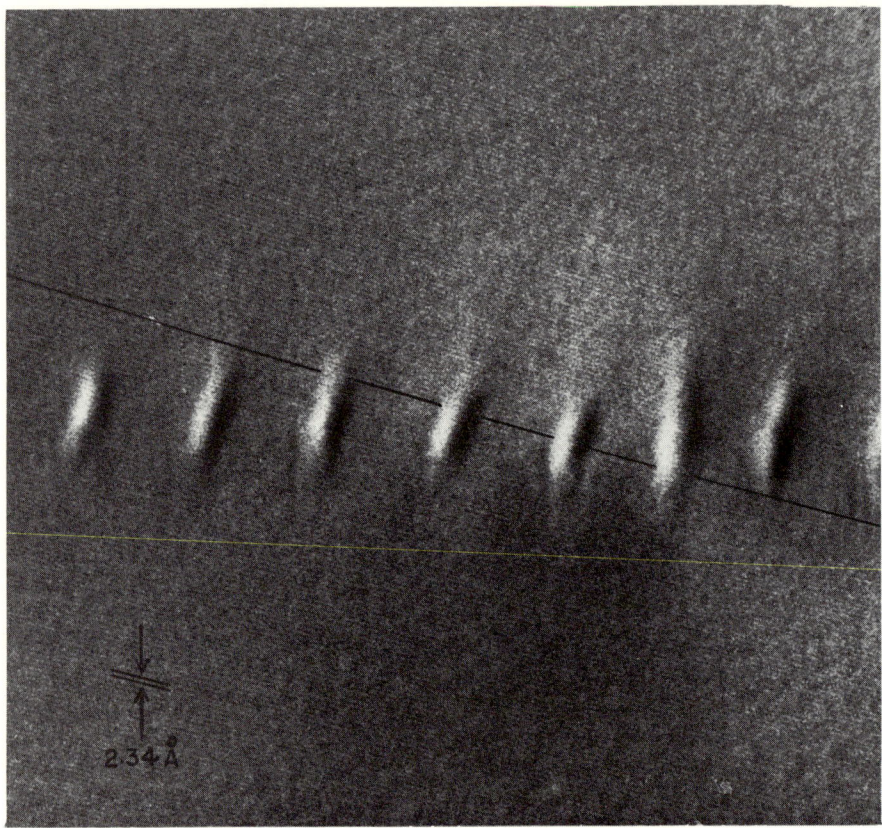

Fig. 5. Low-angle tilt boundary in aluminum. The ~3 degree orientation change can best be seen by viewing the figure at a glancing angle in a direction parallel to the inked line.

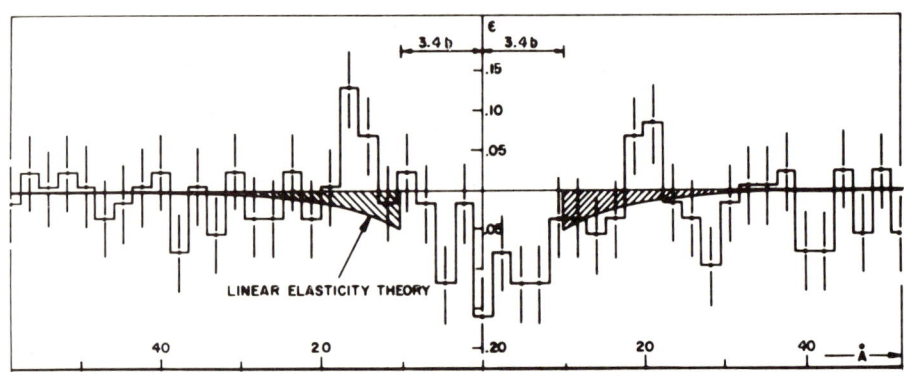

Fig. 6. Strain-field profile perpendicular to and coincident with the end of the terminating plane at one of the dislocations in Fig. 5. Strains, ϵ, are accurate to ±0.045 as indicated by vertical lines at each data point.

As a regular array of edge dislocations, their cumulative effect is to produce an orientation change, i.e., they form a grain boundary. This change in orientation can be most readily seen by viewing Fig. 5 at a glancing angle in a direction parallel to the inked line. From measurements made on the figure, the boundary is a {221} plane and the change in orientation across it is 2.8 ± 0.1 degree. A property of this type of boundary is that the mean separation between dislocations that is required to produce an orientation change of θ is expected to be $D = 0.5\ b\ \sin(\theta/2)$. Assuming a Burgers vector $b = 0.5\ <110>$ for the dislocations in the boundary, D is calculated to be 60 ± 2 A. This is very comparable to the measured mean separation of 55 ± 4 A. This analysis is based on the assumption of a one-to-one correspondence between imaged and actual lattice plane positions.[8] Once again, because the dislocations are end-on, the assumption appears justified as it has provided an accurate measure of the spacing between dislocation cores in the boundary.

The magnitude and extent of the strain field produced by one of the dislocations in the boundary of Fig. 5 was determined as described in Section 4.1. The results have been plotted in Fig. 6. Strains, in this case, were accurate to ± 0.045. This error is shown by the vertical lines at each data point. For comparison purposes, the same figure contains a plot of the strain-field profile expected, from the concepts of linear elasticity theory, for a single infinitely long edge dislocation in aluminum. Calculations for this comparison plot were made up to the dislocation core region (6.8 b wide) and for points 2.5 A above the slip plane.

5 DISCUSSION

It is clear that end-on dislocation lattice images present a powerful means for investigating dislocation properties since a one-to-one correspondence is expected between imaged and actual lattice plane positions. Fig. 3 and 5 have been analyzed on this basis with the following results based on a comparison of Fig. 4 and 6.

In the core region of a dislocation the strains are both nonuniform and high with a maximum of ~0.30 (Fig. 4) for a single, isolated 60-degree dislocation in germanium, and a maximum of ~0.15 (Fig. 6) for an edge dislocation in an aluminum low-angle tilt boundary.

Completely unexpected in these strain field plots is the existence of small-scale expansions and contractions of the lattice which appear in Fig. 4 for a distance of ~15 Å from the center of the dislocation. Similar small-scale oscillations in lattice plane spacing appear in Fig. 6 for a distance of 20 to 30 Å from the center of the dislocation.

The reason for these expansions and contractions is not understood. It would be most interesting to ascertain whether they are observed when a perturbed interatomic potential is used to similate the effect of a dislocation in a crystal lattice.

REFERENCES

1. Cockayne, D.J.H., J. R. Parsons, and C. W. Hoelke: Atomic Energy of Canada Ltd., report CRNL-539 (1970), also Phil. Mag. (in press).
2. Howie, A., M. J. Whelan: Proc. Roy. Soc. (London), **Ser. A, 263**: 217 (1961).
3. Dowell, W.C.T.: J. Phys. Soc. Japan, **17** (Suppl. B-11), 175 (1962).
4. Mannami, M.: J. Phys. Soc. Japan, **17**: 1160 (1962).
5. Cockayne, D.J.H.: (private communication).
6. Bassett, G. A., J. W. Menter and D. W. Pashley: Proc. R. Soc. **A, 246**: 345 (1958).
7. Parsons, J. R., M. Rainville and C. W. Hoelke: Phil. Mag. **21**: 1105 (1970).
8. Parsons, J. R., C. W. Hoelke: J. Appl. Phys. **40**: 866 (1969).

DISCUSSION on the paper by J. R. Parsons

BASINSKI: What is the accuracy of determination of the lattice positions in the strained region near the core of the dislocation? Atomistic calculations, at least for a screw dislocation in a model sodium lattice, show fractional Å deviations from the predictions of elasticity. If we therefore wanted to use the direct lattice imaging technique to test such a calculation we would require a resolution of at least 0.1 Å.

PARSONS: The residual diffraction contrast present at the dislocations in Fig. 3 and 5 changes the background on which the microdensitometer traces of lattice plane positions appear. Lattice plane positions are not more difficult to determine from these traces in the strained region near the core of the dislocation than they are in regions more removed from the dislocation. Hence the error bars shown in Fig. 4 and 6 are the same in both regions.

Your other comment is interesting in that although we cannot determine the position of a single lattice plane to an accuracy of 0.1 Å, the quality of the lattice images is good enough to determine its position with an *average* accuracy of $\pm$ 0.045 x 2.34 $\sim$ $\pm$ 0.1 Å for aluminum and $\pm$ 0.03 x 3.26 $\sim$ $\pm$ 0.1 Å for germanium.

ROBINSON: Would you comment on the effects of chromatic aberration on the images you observe?

PARSONS: Chromatic aberrations increase the width of the fringes in a lattice image, and their deteriorating effects are minimized by using titled illumination.

ROBINSON: Why does the correspondence between the real lattice planes and those in the observed image fail when the dislocations are not end-on?

PARSONS: When the dislocation is inclined, lattice displacements vary with depth in the foil, whereas for an end-on dislocation the displacements are constant through the foil. The phase shift θ_g-θ_O determines lattice plane positions in a 2-beam lattice image. This phase shift is constant for an end-on dislocation. Hence a one-to-one correspondence is expected in this case.

COWLEY: The interpretation of your electron microscope images in terms of positions of lattice planes depends on the use of the "column approximation" with the dynamical theory of diffraction in the crystal. For end-on dislocation images, a rough estimate of the diameter of the column that may be used (using the usual criteria and the specimen thicknesses involved) suggests that, for the resolutions of your images, the column approximation is on the verge of breaking down. Also the detailed study of fringe spacings you have made involves a much more severe test of the approximations than is normally contemplated in electron microscopy. Hence I suggest that the oscillations in fringe spacing observed may be a result of n-beam dynamical diffraction processes not included in the normal theoretical treatment.

A complete calculation of contrast for a nonperiodic object such as a dislocation can be made by use of an artificially induced large periodicity in the method of Grinton and Cowley (in press) that was developed for nonperiodic biological specimens. Such calculations on model objects would suggest that oscillations in fringe spacings can be induced by diffraction effects.

PARSONS: I agree with you that such calculations would indeed be valuable.

VINEYARD: Have you considered the effects of jogs or kinks and the possibility that they might be responsible for the variations in planar spacings you observe?

PARSONS: No, I have not, but it is an interesting suggestion if Dr. Cowley's suggested calculations do not account for the observed oscillations.

ON THE MOTION OF THE $\frac{a}{2}$ <111> SCREW DISLOCATION IN α-IRON

P. C. Gehlen
Battelle, Columbus Laboratories
Columbus, Ohio 43201

ABSTRACT

The atomic configuration of a microkink along a $\frac{a}{2}$ <111> screw dislocation in α-iron is described. The core in the kink is 2 to 3 Burgers vectors wide and 4 to 5 Burgers vectors long. The energy of the kink is 0.08 eV and the atomic positions in the core are not very sensitive to the size of the model and to the choice of interatomic force law. It is shown that infinitely long straight dislocations move only under Peierls stresses considerably larger than expected experimentally. Kinked dislocations on the other hand move under more reasonable stresses.

1 INTRODUCTION

Deformation and fracture of metals are ultimately controlled by phenomena happening on the atomic level, particularly by the generation and motion of defects such as dislocations. Since local atomic positions at the core of a dislocation, cannot usually be resolved experimentally, they must be simulated, using high-speed computers. The $\frac{a}{2}$ <111> screw dislocation, which appears to play a predominant role in the slip behavior of bcc metals, has been studied extensively by lattice simulation. Most authors[1-4,6,7] using a variety of interatomic potentials and isotropic or anisotropic elasticity (to position the boundary atoms) find that the dislocation is made up of three narrow faults related by the ternary screw axis of the undeformed lattice. Bullough and Perrin[4] reported an asymmetric core configuration, but Vitek et al.[3] have reported that if the core is viewed on the correct {112} planes, the symmetry is restored.

Recently, using an interatomic potential derived by Bullough and Perrin[5], I have shown that at least two different core configurations for the $\frac{a}{2}$ <111> screw dislocation have similar energies and thus should coexist in this model of α-iron: one retains the threefold symmetry of the undeformed lattice, the other doesn't[6]. Diener et al.[7] also have reported two different configurations; however theirs are related by the symmetry elements of the lattice while mine are not.

Except for Chang[1], none of the above authors have reported a value for the Peierls stress of the $\frac{a}{2}$ <111> screw dislocation. Vitek[8], however, is currently studying the subject. In my opinion Chang's method does not realistically reproduce a physical phenomenon: he arbitrarily displaces atoms in one-half of the crystal with respect to the other half and monitors the changes in potential energy until a maximum occurs. At that point he feels that he has overcome the Peierls barrier ($\frac{\mu}{150}$ in his calculation, μ being the shear modulus). Vitek uses a static calculation (see Ref. 9) and finds that his dislocation moves when a stress of about $\frac{\mu}{20}$ is applied to the crystallite. My own findings using a dynamic method show that no dislocation motion occurs for stresses as large as $\frac{\mu}{10}$ to $\frac{\mu}{15}$. The difference between Vitek's and my results may arise either from a difference in relaxation procedure or from the fact that Vitek uses a much larger model (2500 versus 250 atoms) which may attenuate the effects of the boundary on the behavior of the core. If the experimentally obtained critical resolved shear stress is extrapolated to 0 K, one obtains a value of about $\frac{\mu}{300}$ for the resistance of the lattice to dislocation motion (see Ref. 10). The purpose of the present research is to elucidate the discrepancy beetween this value and the calculated ones.

All the above models, using periodic boundaries along the dislocation line, implicitly assume that the dislocation moves as an infinitely long straight line and hence neglect the role of kinks which is central in current theories of yielding of bcc metals. Accordingly, a 3-dimensional simulation of a kinked dislocation has been made. As a first step, a 3-dimensional unkinked dislocation was simulated. No kinks were spontaneously nucleated and the dislocation was still sessile. The second step was to attach a small kink onto the dislocation line. Preliminary results show that in this model the dislocation will move under reasonable stresses. This paper describes the kink model and some of the preliminary results on the resistance of the lattice to kink motion.

2 PROCEDURE

As the computational procedure has been described in detail in Ref. 6 and 11, only a brief summary and modifications are considered here.

2.1 Computational Method

The method used is based on the GRAPE computer programs[12] and consists in solving the Newtonian equations of motion of a set of N atoms that interact according to a given interatomic potential. After suitable reduction, these equations are

$$\dot{x}^i_j \left(t + \frac{\Delta t}{2} \right) = \dot{x}^i_j \left(t - \frac{\Delta t}{2} \right) + (\Delta t/m) F^i_j(t) \tag{1}$$

and

$$x^i_j(t + \Delta t) = x^i_j(t) + \Delta t \left\{ \dot{x}^i_j \left(t + \frac{\Delta t}{2} \right) \right\}, \tag{2}$$

where $x^i_j(t)$ is the ith component of the position vector of atom j (with respect to an arbitrary origin) at time t, $\dot{x}^i_j(t)$ the derivative of $x^i_j(t)$ with respect to time, Δt is the time step and

$$F^i_j(t) = \sum_{\substack{k \\ (k \neq j)}} \partial \phi [x_j(t)] / \partial x^i , \tag{3}$$

where ϕ is the interatomic potential. Since in this particular instance it is assumed that the forces are axial — that is to say, that they are independent of the spatial orientation of the bond — the vectors in Eq. 3 may be replaced by a scalar. The potentials used in this study are those derived by Johnson for α-iron[13].

To determine the equilibrium configurations of various defects, Eq. 1 and 2 are solved alternately until a minimum in the potential energy is found. To reduce computer time, the energy-quench method was used: whenever the kinetic energy reaches a maximum, all velocities are set equal to zero before relaxation is continued. At the beginning of relaxation it is assumed that all atoms are at rest. However, when temperature effects are to be simulated, the initial velocity are computed according to the formalism given in Part Six of these Proceedings.

This shortcut is justified only when the ultimate aim of the calculation is a final equilibrium configuration. If, however, the transition between two such configurations is to be studied, a fully dynamic method should be used. The kinetic energy is not quenched but is allowed to oscillate freely.

In view of the large number of atoms required by 3-dimensional models, a high-speed mass storage device was used (capacity: 2.5×10^5 60-bit words; maximum transfer rate to/from central core: one word per 2×10^{-7} second).

With a time step Δt of about 3×10^{-15} seconds, about 30 minutes of computer time are required to generate a core configuration.

2.2 The Boundary Problem

To provide the computer with a framework within which to relax the core atoms, the so-called boundary atoms are usually held in fixed positions. This scheme is based on the assumption that the boundary atoms are at a sufficiently large distance from the core so that linear elasticity holds. If this assumption is not valid, it is necessary to have additional forces on the boundary atoms to simulate the effects of the continuum in which the crystallite is contained. The treatment of Gibson et al.[14] has been modified by assuming that the crystallite is a cylinder. If a is the spacing between atoms in the surface and b is the spacing between the two surface layers that are affected by the continuum, it can be shown that radial forces acting on the first and second surface layers are

$$F_1 = \left(\frac{R-b}{R}\right)^2 pa^2 \qquad (4)$$

$$F_2 = \left[1 - \left(\frac{R-b}{R}\right)^2\right] pa^2 \qquad (5)$$

and

$$p = \frac{2\mu}{R}\delta R \qquad (6)$$

where R is the radius of the boundary, δR the radial expansion, and μ the shear modulus. If p and δR are associated with the tangential force and displacement, the same equations for F_1 and F_2 will result.

This formalism was used with a few of the kink models and whenever dislocation motion was attempted.

The original atomic positions for straight dislocations were calculated using linear anisotropic elasticity.[15] Yoffe's formalism was used for the kink models.[16] Fig. 1 shows how the superposition of three angular dislocations results in a kink. If the Burgers vector, **b**, is in the z-direction, the displacements are given by the following equations*:

$$u_x = \frac{b}{8\pi(1-\nu)} \left(\frac{x^2}{r(r-y)} - \frac{x^2}{r_A(r_A-y+Y)} + (1-2\nu) \log \frac{r_A-y+Y}{r-y} \right) \quad (7)$$

$$u_y = \frac{bx}{8\pi(1-\nu)} \left(\frac{y}{r(r-y)} - \frac{y-Y}{r_A(r_A-y+Y)} - \frac{1}{r-y} + \frac{1}{r_A-y+Y} \right) \quad (8)$$

$$u_z = b\alpha + \frac{bx}{8\pi(1-\nu)} \left(\frac{z}{r(r-y)} - \frac{z}{r_A(r_A-y+Y)} \right) \quad (9)$$

where ν is Poisson's ratio, Y the width of the kink,
$$r^2 = x^2 + y^2 + z^2, \quad r_A = x^2 + (y-Y)^2 + z^2 \quad ,$$

and

$$\alpha = \frac{1}{4\pi} \left[\tan^{-1}\left(\frac{y-Y}{x}\right) - \tan^{-1}\left(\frac{y}{x}\right) + \tan^{-1}\left(-\frac{xr_A}{(y-Y)z}\right) - \tan^{-1}\left(-\frac{zr}{yz}\right) \right]. \quad (10)$$

The kink was introduced in the lattice as shown schematically in Fig. 2. If Eq. 7, 8, and 9 are used as such, only the lower part of the kink will have the expected twofold symmetry around a [$\bar{1}$01] axis through A. The upper half is symmetric with respect to an axis that is shifted by $\frac{b}{2}$. This anomaly introduces an undesirable stress on the model which makes it impossible to retain the defect inside the crystallite. To circumvent this difficulty, the origin of the coordinate system of the upper half was shifted by $\frac{b}{2}$. Only in this case could a kink with symmetry compatible with that of the remainder of the lattice be generated.

3 RESULTS

3.1 Kink Configuration

Three consecutive planes constituting a repeat distance along the [111] direction are shown schematically in Fig. 3. The largest

---
*In Yoffe's paper the α term in Eq. 9 is given with the wrong sign.

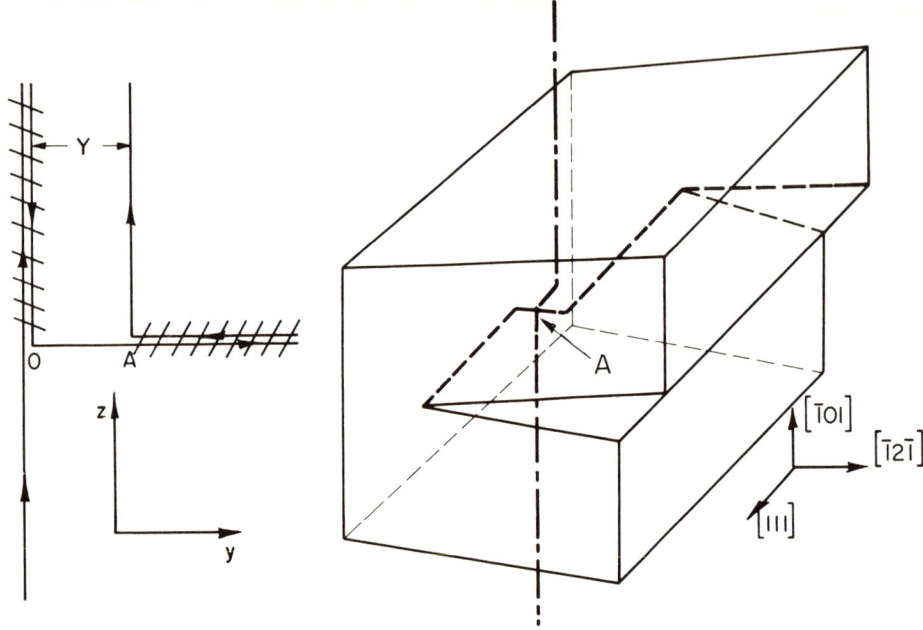

Fig. 1. A kink is made up of two angular dislocations and a straight screw dislocation. The cross hatched regions are those parts of the dislocations that cancel each other out.

Fig. 2. Schematic view of how the kink is introduced in the lattice.

3-dimensional model is made up of ten such repeat distances and contains a total of about 1200 movable and 700 boundary atoms. Although a variety of kink models was generated, attention is focused on the following model (see also Fig. 4): the first straight screw part of the kink is along the [111] direction and intersects the (111) plane at S_1; the pure edge part along the [$\bar{1}2\bar{1}$] direction joins points S_1 to S_2 in the plane of the figure, and, finally, the second straight screw part lies parallel to the [111] direction starting from S_2. The glide plane for this particular kink is thus the ($\bar{1}01$) plane. Both screw parts are left-handed and are positioned at the centers of gravity of triangles such that the lowest possible core energy results.[17] Since triangles that have one side in common have alternating high and low core energies, the shortest possible length for the edge part of the kink is one-third the repeat distance along the [$\bar{1}2\bar{1}$] crystallographic direction, or 2.335 Å for a lattice parameter of 2.86 Å. Fig. 4 shows two consecutive ($\bar{1}01$) planes in an undeformed bcc lattice. The lattice has twofold symmetry around each lattice site and also around points midway between lattice sites (such as A). From Eq. 9 it results

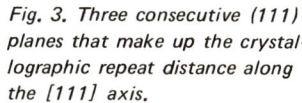

Fig. 3. Three consecutive (111) planes that make up the crystallographic repeat distance along the [111] axis.

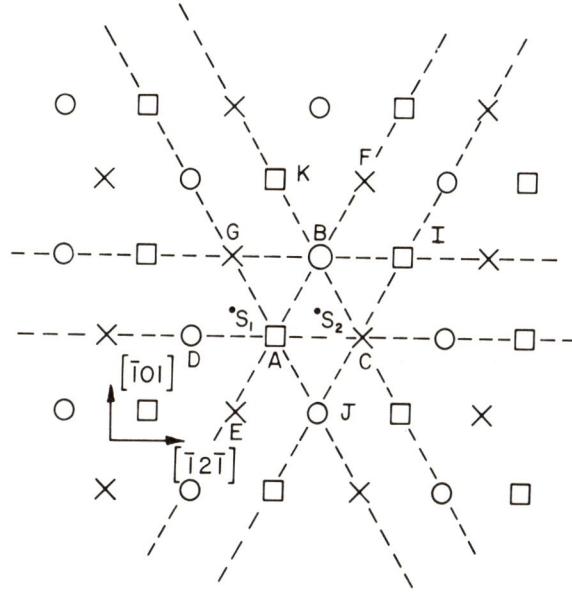

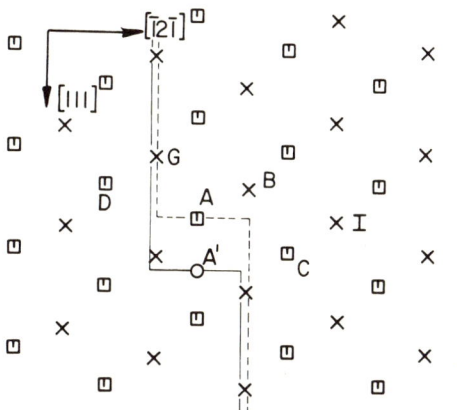

Fig. 4. Two consecutive [$\bar{1}$01] planes in an undeformed bcc lattice. The labels of lattice sites are as in Fig. 3. Solid and dashed lines show the positions of the kink before deformation. After deformation they will be superimposed.

that after deformation A and A' will be superimposed. Therefore, if the twofold symmetry of the entire lattice is to be preserved after deformation, the position of the origin of the coordinate system must be shifted by $b/2$ when the ($\bar{1}$01) plane containing Points S_1 and S_2 is crossed.

Using this procedure has very little effect on the displacements; typically atomic positions calculated with and without the shift in origin differ by less than 10^{-3} Å in the boundary and by less than 10^{-1} Å in the immediate vicinity of the core. It may be worthwhile to point out that a similar situation was encountered with the <100> edge dislocation. Fig. 2

in Ref. 11 shows how the core becomes asymmetric when the coordinate system is not shifted when crossing the slip plane. In that case little difference in the two core configuration resulted once relaxation was terminated; however, in this case — since the kink is much more mobile than the edge dislocation — it was not possible to retain the defect in the crystallite due to the extraneous strain resulting from the asymmetry.

After relaxation the atomic configuration in the immediate vicinity of the kink is shown in Fig. 5 and 6. Fig. 5 shows two consecutive ($\bar{1}01$) planes immediately above and below the pure edge part. It can be seen that the kink width is only about 2-3 Burgers vectors in the [$\bar{1}2\bar{1}$] direction. Parallel to the Burgers vector, the kink stretches out over about 4-5 b.

Fig. 6 shows part of the [$\bar{1}2\bar{1}$] plane containing atoms A and K. The trace of the kink is shown by the broken line; the X represents the projection of the pure edge part. The edge part of the dislocation is well visible in the figure with the traction region to the left of the kink line and the compression region to its right.

Whenever Johnson's potentials are used, the simulated core structure of the $\frac{a}{2}$<111> straight screw dislocation retains the threefold symmetry of

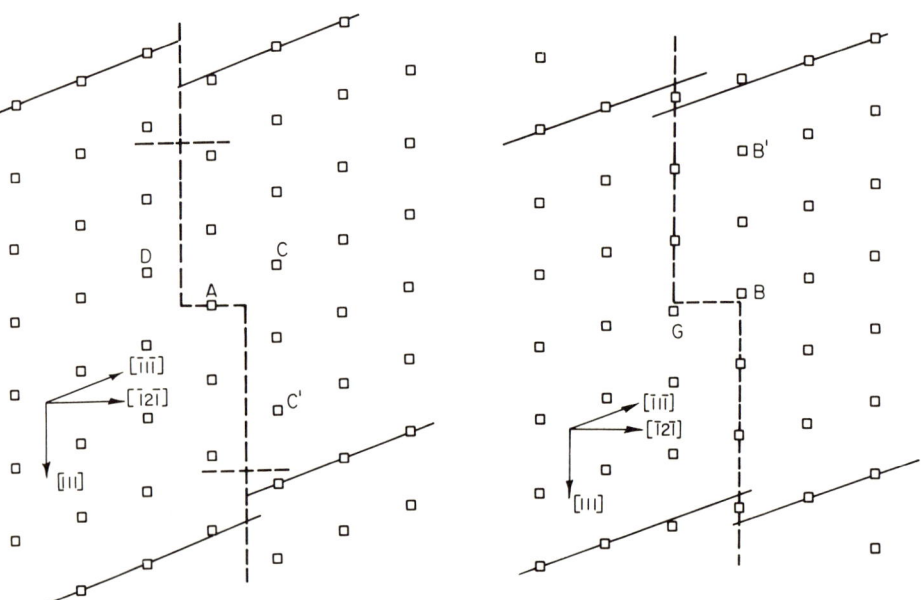

Fig. 5. Two consecutive ($\bar{1}01$) planes in the vicinity of the kink core. The labels of atoms are as in Fig. 3.

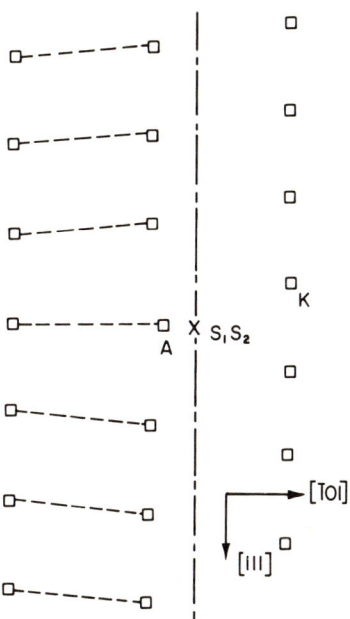

Fig. 6. (121) plane crossing the kink in the center of the pure edge segment. The labeling of lattice sites is as in Fig. 4. The broken line represents the trace of the kink. X is the projection of the pure edge segment.

the bcc lattice. By using Bullough and Perrin's potential[5], it was possible to generate asymmetric cores[6]. I have attempted to "force" asymmetric cores into a Johnson lattice containing a straight screw dislocation, but without success. With the presence of the kink, however, the threefold symmetry is lost and the faults on the {110} planes making up the core are no longer related by that symmetry element. However, the asymmetry is not nearly as well defined as in Ref. 6 where one atom in the core was shifted by $b/4$ when going from one plane to the other. Here the transition is smooth and becomes barely detectable at some distance from the core. This confirms the fact that in models of α-iron using the Johnson potential, asymmetric configurations are metastable.

3.2 The Energy of the Kink

The total energy of the crystallite can be represented by two terms, one is volume dependent and is associated with electronic effects; the other one is only structure dependent.

Potentials such as the Johnson potential are meant to account only for the volume-independent term. For most simple defects such as straight edge and screw dislocations the volume is conserved and the stored strain energy versus the logarithm of distance from the core can be calculated in a straightforward manner. From such plots, core energy, effective whole radius, etc., may then be obtained.[1,2,11]

For the kink model no such plot could be obtained. This was attributed to small volume changes that occur along the dislocation line. To obtain an approximate value for the kink energy the following scheme was used. First the stored energy was calculated in a small cylinder about $2.5b$ away from the center of the kink. This cylinder had its axis parallel to the axis of the boundary cylinder (and thus to the Burgers vector), the same radius and a thickness of $1b$ (one repeat distance). Then the same energy was calculated in a similar cylinder, but containing five repeat distances and centered on the pure edge part. Assuming that the small cylinder is sufficiently far away from the kink, the energy stored within it can be taken as the strain energy of a pure screw dislocation. Therefore, if multiplied by five and subtracted from the energy in the large cylinder, the kink energy should result. This procedure yields a value of 0.08 eV which is in reasonable agreement with results reported elsewhere.[8]

As pointed out above this value is only approximate since the reference cylinder may not be far enough from the kink. However, in order to minimize the effects of inhomogeneities in volume variations along the dislocation line this compromise had to be made.

3.3 The Effects of Model Size, Flexible Boundaries, and Choice of Potential

One of the basic assumptions in computer simulation is that the boundary atoms are sufficiently far away from the defect so that they may be positioned according to linear elasticity. This problem is only partially solved by the use of flexible boundary conditions since the action of the continuum on the crystallite is only crudely approximated.

The size of the model must thus be gradually increased until no major changes occur in the configuration to be simulated. Table I shows the lengths of a few representative bonds as a function of model size. Bonds AC, AC', and AD are in the $(\bar{1}0\bar{1})$ plane just below the kink (see Fig. 5); GB and GB' in the plane just above and AB and AB' are bonds linking the two planes. Models containing from 523 to 1220 movable atoms were tried. When rigid boundaries were used, bond lengths varied by only about 1 percent when going from the small model to the large one. The largest differences between the 817- and 1120-atom models were about 0.003 Å and were considered small enough not to justify a further increase in model size. Finally, the large model was run with flexible boundaries and again the variations are of the order of 1 percent. Since

Table I. Bond Lengths as a Function of Model Size[1] (in Å)

Radius[2]:	10.8[3]	13.5[3]	10.8[3]	13.5[3]	13.5[4]
Length:	16.0	16.0	25.0	25.0	25.0
Number of Atoms:	523	782	817	1220	1220
AC	2.6715	2.6742	2.6852	2.6865	2.7021
AC'	4.3228	4.3192	4.2944	4.2924	4.2577
AD	2.5788	2.5767	2.5732	2.5722	2.5662
GB	2.4626	2.4622	2.4583	2.4597	2.4633
GB'	5.9082	5.9148	5.9748	5.9780	6.0377
AB	2.4664	2.4687	2.4615	2.4634	2.4613
AB'	5.6745	5.6797	5.7171	5.7201	5.7566

(1) All models use the Johnson Unmodified Potential.
(2) The boundary of the model is cylindric.
(3) With rigid boundary.
(4) With flexible boundary.

the flexible boundaries are designed to effectively make the crystallite infinite, the corresponding bond lengths should change in the same direction as when going from the smallest rigid boundary model to the larger one. From the table one can see that this is the case, except for bond GB.

Another way to approximate the sensitivity of the configuration to various inputs into the calculation is to alter the interatomic force law. Johnson has derived two potentials[13], one fits the experimental elastic constants of iron (referred to as the unmodified potential), the other fits the elastic constant after some electronic contributions have been removed from them (modified). The large model was run with both potentials and the corresponding bond lengths are shown in Table II. It can be seen that the configuration is no more sensitive to changes in the force law than to changes in the model size.

Table II. Bond Lengths for Kink in Large Model (in Å)

Bond	Modified Johnson Potential	Unmodified Johnson Potential
AC	2.6917	2.6865
AC'	4.2824	4.2924
AD	2.5736	2.5722
GB	2.4659	2.4597
GB'	5.9866	5.9780
AB	2.4596	2.4634
AB'	5.7170	5.7201

3.4 Preliminary Results on the Motion of Kinks Under Stress

As pointed out above, Peierls stresses for straight dislocations obtained with computer simulation are considerably larger than the value of $\frac{\mu}{300}$ suggested by experimental observation.[10] In the present study, two different models have been used to displace the dislocation. In the first, two parallel dislocations of the same sign were positioned next to each other in the lattice so that a repulsive force resulted between them. In the second, a strain corresponding to the desired stress level was applied to the crystallite by displacing the atoms from their relaxed positions. The latter method was used to study kink motion. To allow for positional fluctuations along the dislocation line (which is not possible in the previous models where periodic boundaries were used) similar experiments were conducted with 3-dimensional models. In all cases, no discernible dislocation motion was observed for stresses as high as $\frac{\mu}{10}$ to $\frac{\mu}{15}$. Since no thermal vibrations and no impurities or vacancies were included in the model, microkink generation was not expected and indeed did not occur. There is, however, hope that this might occur when the asymmetric and symmetric configurations obtained with the Bullough and Perrin potential[5] will be juxtaposed in a 3-dimensional crystal.

Although this study of kink motion is not complete, preliminary results indicate clearly that kinked dislocations move much more readily than straight ones. Starting with the configurations of Fig. 5 and 6, a shear stress of $\frac{\mu}{200}$ was applied in the $(\bar{1}01)$ plane by displacing the boundary atoms accordingly. In Fig. 7, it can be seen that the dislocation has straightened out. Particularly, notice that in Fig. 7b the edge part of the kink is no longer visible. Similar runs have been made for lower stresses and, at present, it is estimated that at 0 K the Peierls stress for kink motion is about $\frac{\mu}{500}$. This result is an upper limit and a careful investigation is currently underway to see if changing the time step in Eq. 1 and 3 and/or increasing the relaxation time will lower the value reported here.

Preliminary tests also show, as expected, that the critical stress for kink motion is temperature dependent. For instance, while at 0 K no apparent kink motion occurs for a shear stress of $\frac{\mu}{500}$, the kink moves readily when the "temperature" is raised to 25 K. The temperature is raised by adding a given amount of kinetic energy to all movable atoms. The accuracy of this method in reproducing realistic atomic motions at a given temperature is currently being tested and will then be applied to a study of the Peierls stress as a function of temperature.

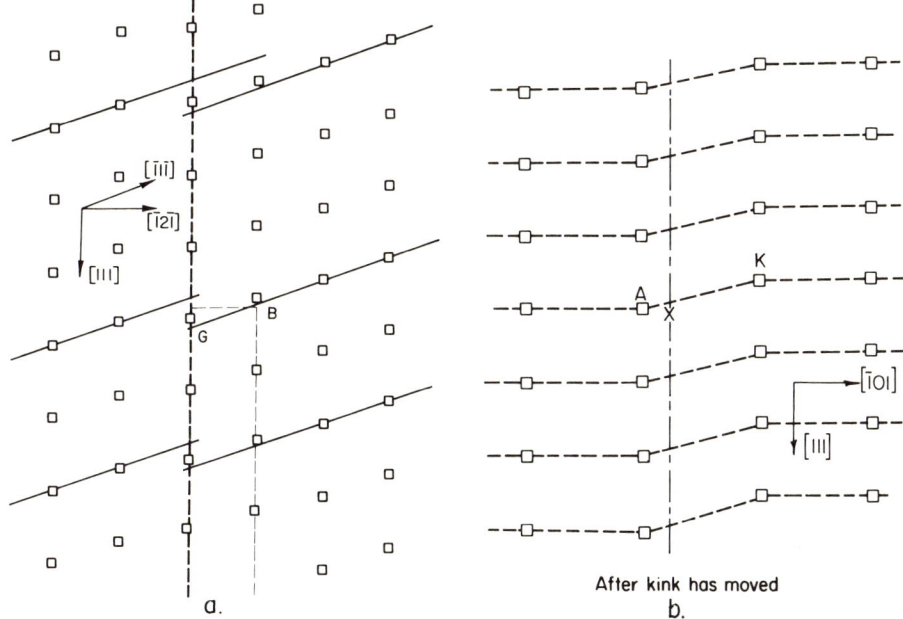

Fig. 7. (a) *Same as Fig. 5 after kink has moved out of crystal.* (b) *Same as Fig. 6 after kink motion.*

All the runs performed so far indicate that the speed at which the kink moves is about the speed of sound in iron. This is to be expected since the crystallite contains no obstacles to slow down the kink once it has accumulated enough energy to start moving.

It should be pointed out that, even though the value found for the Peierls stress is lower than the experimental value, it may not be in complete agreement with the experimental facts since kink motion must be preceded by double kink nucleation. This process probably requires a higher stress than kink motion; however, in the presence of lattice defects such as vacancies and interstitials and positional fluctuation along otherwise straight dislocations, kink nucleation may be less difficult than the motion of perfectly straight dislocations.

4 CONCLUSIONS

- If Yoffee's formalism is used to introduce a kink into the lattice, the origin of the coordinate system must be shifted by $b/2$ to maintain the overall symmetry of the lattice.
- The kink width is found to be 2-3 Burgers vectors and its length, 4-5 Burgers vectors.
- The energy of the kink was found to be 0.08 eV. This result is only approximate but compares well with other data.

- The core configuration is not very sensitive to model size and even less sensitive to the force law.
- Straight $\frac{a}{2}$ <111> screw dislocations have a Peierls force, at least an order of magnitude larger than the one experimentally observed.
- Preliminary studies of kink motion, show that kinds move under a more realistic stress of $\frac{\mu}{500}$. Kinks travel at speeds near the speed of sound in iron.
- The Peierls stress for kinks is temperature sensitive.

REFERENCES

1. R. Chang, Phil. Mag., **6**, 1021 (1967).
2. P. C. Gehlen, G. T. Hahn, and A. R. Rosenfield, *Fundamental Aspects of Dislocation Theory*, J. A. Simmons, R. de Wit and R. Bullough, Eds., NBS Spec. Publ. 317 (1970), p. 305.
3. V. Vitek, R. C. Perrin, and D. K. Bowen, Phil. Mag., **21**, 1049 (1970).
4. R. Bullough and R. C. Perrin, *Dislocation Dynamics*, A. R. Rosenfield, G. T. Hahn, A. L. Bement, and R. I. Jaffee, eds. (McGraw-Hill, New York, 1968), p. 175.
5. R. Bullough and R. C. Perrin, *Radiation Damage in Reactor Materials* (IAEA, Vienna, 1969), Vol. II, p. 233.
6. P. C. Gehlen, J. Appl. Phys., **41**, 5165 (1970).
7. G. Diener, R. Heinrich, and W. Schellenberger, Phys. Stat. Sol. (b), **44**, 403 (1971).
8. V. Vitek, private communication.
9. Z. S. Basinski, M. S. Duesberry, and R. Taylor, this conference.
10. W. A. Spitzig and A. S. Keh, Acta Met., **18**, 1021 (1970).
11. P. C. Gehlen, A. R. Rosenfield, and G. T. Hahn, J. Appl. Phys., **39**, 5246 (1968).
12. A. Larsen, GRAPE — A computer Program for Classical Many-Body Problems in Radiation Damage (Brookhaven Nat. Lab. Assoc. U., Inc. 1964).
13. R. A. Johnson, Phys. Rev., **145**, 423 (1966).
14. J. B. Gibson, A. N. Goland, M. Milgram, and G. H. Vineyard, Phys. Rev., **120**, 1229 (1960).
15. J. P. Hirth and P. C. Gehlen, J. Appl. Physics., **40**, 2177 (1969).
16. E. H. Yoffe, Phil. Mag., **5**, 161 (1960).
17. H. Suzuki, *Dislocation Dynamics*, A. R. Rosenfield, G. T. Hahn, A. L. Bement, and R. I. Jaffee, eds. (McGraw-Hill, New York, 1968), p. 679.
18. J. P. Hirth and J. Lothe, *Theory of Dislocations* (McGraw-Hill, New York, 1968).

DISCUSSION on paper by P. C. Gehlen

WILSON: The inclusion of a larger number of relaxed atoms is equivalent to using an anisotropic continuum as opposed to the more usual method of including a radial continuum. Perhaps this explains why the Vitek et al. calculation (work in progress), which uses 2500 atoms, differs from Gehlen's which uses 250 atoms with a radial continuum.

DUESBERY: Dr. Vitek and I have calculated the Peierls stress of an infinite screw dislocation in the bcc lattice, using the same Johnson potential, and find a value of about 0.04 of the shear modulus. We used a static method of calculation, and a block size of 45 x 45 Burgers vectors. Do you think the discrepancy is due to your use of a smaller block size?

GEHLEN: It is difficult for me to estimate the effect of having a smaller crystallite, since I have never used a model with variable dimension to calculate the Peierls barrier. In calculating the properties of static dislocation cores, however, I have used a large range of model sizes without appreciable effects on the results.

WILSON: In reply to Dr. Duesbery's comment: In every case in which we performed our calculations by both the direct search and gradient search methods we obtained exactly the same answer. The gradient search is faster but the direct search is, in my opinion, more reliable.

BULLOUGH: I think it is very important to set the boundary conditions using anisotropic elasticity. Also the boundary conditions should, if possible, be consistent with the core relaxations as they proceed. This can be achieved by including the higher order (linear) elastic terms in a variational procedure to maintain accurate consistency with the core.

GEHLEN: It is my experience that in the case of the static core configurations of straight edge[1] and screw[2] dislocations, the use of isotropic or anisotropic elasticity to position the boundary atoms made very little difference as far as the configuration itself was concerned. I agree with you that, if dislocation motion is occurring, the boundary should be consistent with the core at all times. Dr. Kanninen is working on such boundary conditions and will talk about them tomorrow.

CHANG: Don't you think that it is more appropriate to obtain the kink energy by comparing the energy of a large enough crystal containing a straight screw dislocation with that containing a screw dislocation and a pair of kinks having the same boundary conditions?

GEHLEN: As I pointed out in my presentation, the value of the kink energy is only approximate. To do this correctly one should obtain a plot of the stored strain energy versus $\ln R$. Since there are small variations in volume along the dislocation line, this procedure doesn't work in this case. I agree that it would probably be more realistic to use the method that you suggest, however, since kink pairs in real crystals are expected to be about 50 A apart, our model is too small to accommodate them.

CHANG: Can you generate such a pair of kinks in a crystal containing a straight screw dislocation by applying an external stress to the crystallite?

GEHLEN: At this time we have not attempted to generate kinks by applying a strain to the crystal. That is, however, the next step that we want to undertake.

SHEWMON: Empirically the low temperature yield stress in iron is strongly influenced by temperature and interstitial impurities. Have you tried to simulate these?

GEHLEN: So far we have only made one run at temperature, namely when the stress was $\frac{\mu}{500}$ we raised the temperature to 25 K. At 0 K the kink didn't move; at 25 K, it did. A systematic study of the Peierls stress as a function of temperature will be undertaken.

WEINER: Have you computed the critical shear stress for a perfect crystal and how this stress depends upon the details of the potential employed?

GEHLEN: No, I have not attempted to do that.

BASINSKI: Since the potential extends for an appreciable distance, how do you define the energy of a dislocation for very small radii? Do you define it as the energy associated with the atoms included within the radius, but allowing the interactions to be extended outside the region considered. I wonder whether the comparison with elasticity calculations is meaningful for very small radii.

GEHLEN: In the plots showing the stored strain energy versus $\ln R$[1] where R is the radius of a cylindrical annulus centered on the dislocation line, the stored strain energy in a given annulus is obtained by calculating the total potential energy of all the atoms within the annulus and then subtracting the corresponding energy that these atoms would have if they were in an undeformed crystal. This method does not put any restrictions on the validity of the result which is as good as the potential itself is.

DUESBERY: Dr. Vitek and I think that the study of the motion of the kinks which have been introduced ab initia onto the screw dislocation does not say anything about the Peierls stress for the motion of these dislocations. This type of kink will obviously be removed quickly when an external stress is applied and we are again left with the straight screw dislocation which has to be moved in order to continue the movement. The kink motion described may thus be relevant to, for example, microdeformation, but not to the macrodeformation of bcc metals.

GEHLEN: I agree with Dr. Vitek that double kink nucleation must precede and is a more difficult process than kink motion in the macrodeformation range. We are not convinced that kink nucleation is necessarily as difficult as moving screw dislocations as rigid lines. Defects in the lattice, vacancies, impurities, atoms or defects along the screw dislocation line connected with positional fluctuations along an otherwise straight dislocation[2] may serve to reduce the kink nucleation stress. This view derives some support from both Dr. Vitek's and our own simulations of the α-iron lattice with the Johnson potential. According to Dr. Vitek (private communication) the stress to move the screw as a straight, rigid line (in the absence of thermal vibrations) is 0.04 μ. Our stress is about 0.1 μ. Both of these values are an order of magnitude larger than the values obtained by extrapolating critical resolved shear stress measurements iron single crystals to 0 K[3] (about 0.003 μ).

REFERENCES

1. See Ref. 11.
2. See Ref. 6.
3. See Ref. 10.

ON THE FACTORS CONTROLLING THE STRUCTURE OF DISLOCATION CORES IN B.C.C. CRYSTALS

V. Vitek,
Department of Metallurgy
University of Oxford, Oxford, England

L. Lejček
Institute of Physics
Czechoslovak Academy of Sciences, Prague, Czechoslovakia

D. K. Bowen
School of Engineering Science
University of Warwick, Coventry, England

ABSTRACT

It is argued that the restoring forces acting between two parts of a crystal displaced along a plane comprise the main influence upon the structure of a planar dislocation core. This hypothesis has been investigated using a Peierls-type model in which the sinusoidal restoring force law was replaced by force laws calculated using three different interatomic potentials. The structure of an $\frac{a}{2}$<111> edge dislocation on a {112} plane was calculated in detail using this model and compared with a computer simulation of the dislocation inside a finite crystallite. The cores were in both cases described as a continuous distribution of partial dislocations of density $\rho_x(x)$ the shape of which is the same for both types of calculations. The physical meaning of maxima, minima, and inflexions in ρ_x, their relation to the restoring forces, and their interpretation in terms of generalized splitting are discussed.

1 INTRODUCTION

Dislocation core structures have recently been calculated, using central forces, in fcc[1,2], hcp[3], and bcc[3,4,5] crystals. In fcc and hcp crystals stable stacking faults can always be formed on the close-packed planes. The basic feature of the dislocation in these structures was, therefore, supposed to be a dissociation into two partials with large inelastic displacements confined mainly to the fault plane. This has been confirmed by computer calculations.[1,2,3] Originally, it was also assumed that a similar type of splitting can exist on {110} and {112} planes in bcc crystals.[6] However, atomistic calculations[7,8] show that the existence of stable stacking faults in these crystals is unlikely, and for the central-force potentials used, e.g. in Ref. 3 and 4, they certainly do not exist. Thus, splitting in the same sense as in close-packed crystals cannot take place.

Nevertheless, the atomistic calculations show that large core displacements are still confined to one (e.g., edge dislocation: this paper) or several nonparallel (screw dislocation[4]) crystallographic planes. This suggests that the core structure will be controlled mainly by the restoring forces acting between two halves of the crystal when displaced with respect to each other along these crystallographic planes. The structure of a total or partial dislocation whose core is confined to a crystallographic plane can then be investigated by the Peierls model[9] when the simple sinusoidal law is replaced by the force law corresponding to this crystallographic plane.

In the present paper the restoring forces for {112} planes in bcc crystals are first presented. They have been calculated using the central-force potentials which were used to calculate the core structure of screw dislocations.[3,4] The core structure of an $\frac{a}{2}$<111> edge dislocation lying on a {112} plane is then calculated both on the Peierls model and by the computer simulation of a finite crystallite with the dislocation in the middle. The results of both calculations are compared, and the manner in which the core structure is determined by the functional dependence of the restoring forces on the displacement vector is discussed.

2 RESTORING FORCES FOR {112} PLANES

The restoring forces for {110} and {112} planes in bcc and {111} planes in fcc crystals were first calculated when the stability of single layer stacking faults on these planes was investigated.[7] A generalized stacking fault was formed by cutting the crystal along a crystallographic plane and shifting one part of the crystal with respect to the other by a vector f. The stacking fault energy γ was then calculated as a function of the vector $\mathbf{f} \equiv (f_x, f_y)$ and a surface $\gamma(\mathbf{f})$ thus obtained. Obviously, the components of the restoring force are: $-\frac{\partial \gamma}{\partial f_x}$, $-\frac{\partial \gamma}{\partial f_y}$. The calculations in Ref. 7 were made using various central-force potentials and allowing relaxations perpendicular to the fault plane in several layers adjacent to the fault.

For a (112) plane, axes with x in the $[\bar{1}\bar{1}1]$ and y in the $[1\bar{1}0]$ directions were chosen. Due to the crystal periodicity the limits of the displacements are: $0 \leq f_x < b = a\sqrt{3}/2$, $0 \leq f_y < a\sqrt{2}/2$, where a is the lattice constant. The results show that the surface $\gamma(\mathbf{f})$ generally possesses a deep valley in the $[\bar{1}\bar{1}1]$ direction for $f_y = 0$ and increases very steeply in the $[1\bar{1}0]$ direction.[7] Thus the restoring forces increase very rapidly in this direction. As the Burgers vector of a dislocation lying on a (112) plane is $b = \frac{a}{2}[\bar{1}\bar{1}1]$, most of the displacements will be in the $[\bar{1}\bar{1}1]$ direction. Since the y component of the restoring force becomes very large for small displacements in the $[1\bar{1}0]$ direction it is very unlikely that displacements perpendicular to the Burgers vector will exist in the dislocation core. This has been confirmed by the relaxation calculations for the edge dislocation (no elastic displacements are in this direction even in the anisotropic case). We shall consider therefore, only the x component of the restoring force for $f_y = 0$. The restoring force $-\frac{\partial \gamma}{\partial f_x}$ has been calculated by the same method of Ref. 7 using the potentials J_0 and J_2 and an effective potential for sodium[10]. J_0 and J_2 are empirical potentials adjusted so as to obtain modified elastic constants for α-Fe and to assure the stability of the bcc lattice; they are truncated between second- and third-nearest neighbors (as first suggested by Johnson[11]). Potential J_2 gives the magnitude of γ approximately three times lower than potential J_0. The interatomic potential for sodium has been derived from first principles; during the calculations it was truncated after the sixth-nearest neighbors. A significant relaxation perpendicular to the fault occurs in six layers of atoms. It always corresponds to an expansion of the lattice and is biggest for $f_x \approx b/2$.

The calculated dependences of $\dfrac{\partial \gamma}{\partial f_x}$ on f_x are shown in Fig. 1. The restoring force becomes zero only once (besides at $f_x = 0, b$). This corresponds to a maximum in $\gamma(f_x)$. For all three force laws, the positive and

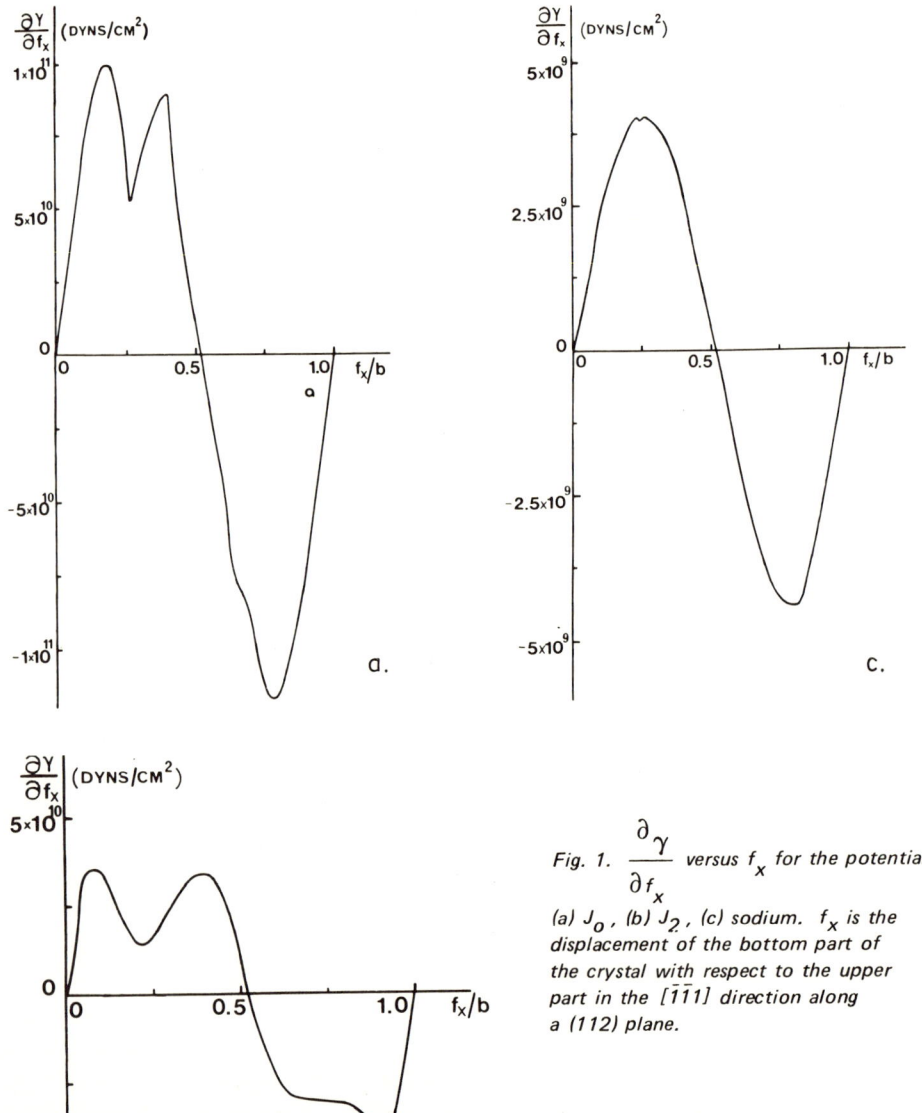

Fig. 1. $\dfrac{\partial \gamma}{\partial f_x}$ versus f_x for the potentials (a) J_0, (b) J_2, (c) sodium. f_x is the displacement of the bottom part of the crystal with respect to the upper part in the $[\bar{1}\bar{1}1]$ direction along a (112) plane.

negative parts of the curve are different. This corresponds to the asymmetry of the displacements in the twinning ($[\bar{1}\bar{1}1]$ when the bottom part of the crystal is displaced with respect to the upper part) and antitwinning directions, respectively. Generally, the magnitude of the restoring forces is larger for displacements in the antitwinning sense. The asymmetry is much more pronounced for J_0 and J_2 than for sodium; deep minima occur in the dependence $\frac{\partial \gamma}{\partial f_x}$ on f_x on the positive (twinning) side for potentials J_0 and J_2 whilst the minimum of the same type is very shallow for sodium. This will be reflected in the differences of the dislocation core structure when calculated using different potentials.

3 PEIERLS TYPE MODEL OF THE CORE

In the Peierls model, a dislocation is obtained by displacing two halves of the crystal along a plane of cut, xz, (axis z is in the direction of the dislocation line) by the vector $\mathbf{f}(x)$ which fulfills the conditions

$$\mathbf{f}(-\infty) = 0$$
$$\mathbf{f}(+\infty) = \mathbf{b} \quad . \tag{1}$$

We shall call $\mathbf{f}(x)$ the vector of disregistry. The model is obviously suitable for a total or partial dislocation whose core is confined to a single crystallographic plane; this is very likely for an edge dislocation. It is further assumed that the two halves of the crystal behave as isotropic elastic continua and that the atomistic structure is taken into account only by a periodic restoring force which acts between them. The core structure is then calculated from the condition that the force due to the elastic stresses between the two halves of the crystal must be equal to the corresponding restoring force in the plane of the cut. This leads to the well-known Peierls integro-differential equation for the vector of disregistry, $\mathbf{f}(x)$. The dislocation core structure can then be interpreted in terms of dissociation as a continuous distribution of partial dislocations of density

$$\rho_\alpha(x) = \frac{d f_\alpha(x)}{dx} \quad (\alpha = x, y) \quad . \tag{2}$$

The Peierls equation represents, then, a necessary condition for the minimization of the total energy of the dislocation. If we consider that displacements exist only in the x direction, the elastic energy of the continuous distribution of the partials of density $\rho_x(x)$ is for an edge dislocation

$$E_{el} = -\frac{\mu}{4\pi(1-\nu)} \int_{-\infty}^{+\infty} \int_{-\infty}^{+\infty} \rho_x(x_1) \rho_x(x) \ln(|x_1-x|) \, dx_1 dx \quad , \tag{3}$$

where μ is the shear modulus and ν the Poisson ratio of the material. The stacking fault energy of this 'dissociation' is

$$E_\gamma = \int_{-\infty}^{+\infty} \gamma(f_x(x)) \, dx \tag{4}$$

where $\gamma(f_x)$ is the γ-surface for the plane of the cut. The necessary condition for the functional minimization of the total energy

$$E_{tot} = E_{el} + E_\gamma$$

with respect to $f_x(x)$ is

$$\delta E_{tot}(f_x) = 0 \quad .$$

Simple variational calculation shows that it is equivalent to the equation

$$\frac{\mu}{2\pi(1-\nu)} \int_{-\infty}^{+\infty} \frac{1}{x-x_1} \left(\frac{df_x(x_1)}{dx_1} \right) dx_1 = -\frac{\partial \gamma}{\partial f_x(x)} \quad , \tag{5}$$

which is the Peierls integro-differential equation for $f_x(x)$ when the simple sinusoidal force law[9] is replaced by $-\frac{\partial \gamma}{\partial f_x(x)}$. This has to be solved together with Eq. 1.

It can easily be shown, however, that for any f_x satisfying Eq. 5.

$$\delta^2 E_{tot}(f_x) = 0$$

and, therefore, Eq. 5 is not a sufficient condition for the minimization of the total energy, i.e., there can be more than one solution, but only one of them corresponds to minimizing the total energy. This non-uniqueness of the solution occurs, for example, if stable stacking faults can exist and the dislocation splits into partials separated by the fault ribbons. Along the stacking-fault ribbon $\frac{df_x}{dx} = 0$ and $\frac{\partial \gamma}{\partial f_x} = 0$ so that the width of the splitting cannot be found from Eq. 5. It only determines the structure of the partials whilst the width of splitting has to be found from the minimization of the total energy. Since the physical meaning of Eq. 5 is that it is the condition of the equilibrium between the elastic and restoring forces, it is plausible to assume that the core structure will be determined uniquely by Eq. 5 if no stable stacking faults can exist, i.e., when no ribbons of a constant displacement f_x may be formed.

The core structure of the $\frac{a}{2}[\bar{1}\bar{1}1]$ edge dislocation on a (112) plane has been calculated on the Peierls model for the restoring forces presented in the previous section. Numerical solution of Eq. 5 has been performed by an iteration method described in the Appendix. The constants used in the calculation are: $\mu = 5.77 \times 10^{11}$ dynes/cm^2, $\nu = 1/3$, and $a = 2.86$ Å for J_0 and J_2; and $\mu = 9.23 \times 10^9$ dynes/cm^2, $\nu = 0.457$, and $a = 4.234$ Å for sodium; the shear modulae are those corresponding to shear in $\langle 111 \rangle$ direction on a $\{112\}$ plane. The results of these calculations are shown in Fig. 2.

The width of the core varies with the potential and is controlled by the relative strength of the restoring forces which can be roughly assessed by $\frac{1}{\mu}\left|\frac{\partial \gamma}{\partial f_x}\right|$ at local maxima; the width decreases with increasing

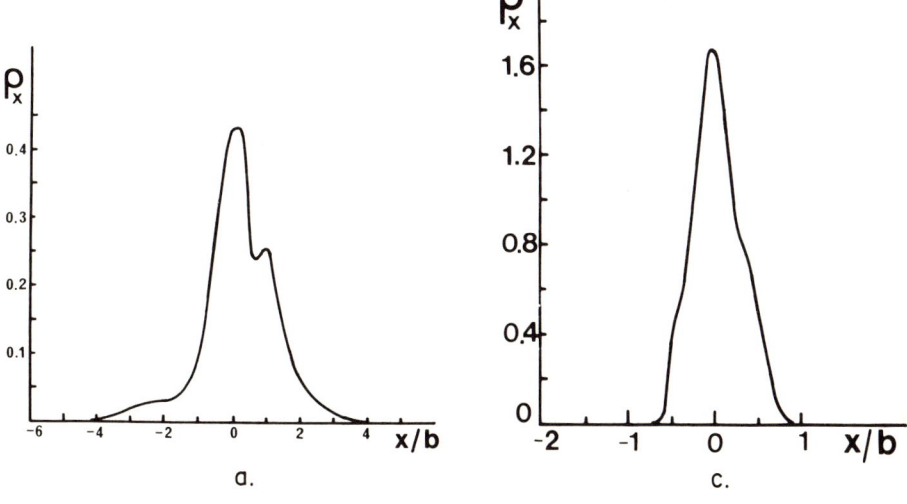

Fig. 2. The core structure calculated on the Peierls model depicted as a dependence of the density, ρ_x, of the continuous distribution of partial dislocation on x. Corresponding restoring forces were calculated using a) potential J_0; b) potential J_2; c) potential for sodium.

strength. For the highest maximum, $\frac{1}{\mu}\left|\frac{\partial\gamma}{\partial f_x}\right|$ = 0.07, 0.2, 0.49 for J_2, J_0, and sodium, respectively. It is seen that, for all three, the density is asymmetrical and more widely spread on the twinning side than on the antitwinning side of the dislocation.* For the potentials J_0 and J_2, the density ρ_x possesses two maxima, the smaller one on the twinning side of the dislocation; whilst, for sodium only an inflexion appears. This obviously reflects the differences in the functional dependence of the restoring forces in these three cases.

4 RELAXATION CALCULATION OF THE CORE STRUCTURE

The calculation of the core structure of the $\frac{a}{2}[1\bar{1}1]$ edge dislocation lying on a (112) plane has been made by a computer simulation of a block of the discrete lattice with the dislocation in the middle. Central forces corresponding to the potentials J_0 and J_2 and to the sodium potential, respectively, have been assumed to act between the atoms of the lattice. The dislocation was introduced into the crystal by applying to all the atoms a displacement corresponding to the anisotropic elastic displacement due to a dislocation in the middle of the block. The boundary atoms were then kept fixed in their displaced positions whilst all the other atoms of the block were relaxed so as to have zero total force on each moving atom. The number of relaxed atoms was 1152 for J_0 and J_2, and 1296 for sodium. For this size of block the core structure was certainly not influenced by the boundary conditions. This was confirmed by gradually increasing the size of the crystallite and following the corresponding changes in the core; they were already negligible for blocks half the size of that used for the final calculations.

In all three cases no displacements parallel to the dislocation line exist, which is in agreement with the considerations of Section 2. Most of the large inelastic displacements parallel with the Burgers vector are concentrated between two of the (112) planes. Of course, this concentration of the displacements is not total. Appreciable, although, smaller inelastic displacements occur between another two adjacent layers of atoms in the case of potentials J_0 and J_2 (they are larger for J_0) and between another four layers in the case of sodium.

In Fig. 3, the positions of the atoms in the dislocated lattice, relaxed for the potential J_2 are shown in the $[1\bar{1}0]$ projection, i.e., projected along the dislocation line. The planes between which most of the displacement is concentrated are marked A and B. It is seen that the core is widely spread on the twinning side of the dislocation (i.e., in the $[\bar{1}\bar{1}1]$

---

*On the twinning and antitwinning sides, the displacements across the (112) plane due to the dislocation are in the twinning or antitwinning sense, respectively (see in Fig. 3 relative displacement between the planes A and B).

Fig. 3. *[110] projection of the atoms of the bcc lattice with ½ [111] edge dislocation on a (112) plane, relaxed using the potential J_2. The dislocation core is concentrated between the planes marked A and B.*

direction) but no ribbons of a stacking fault occur. It is not, however, easy to make any detailed assessment of the core structure from this type of picture. All the results of the relaxation calculations are, therefore, presented in Fig. 4 as plots of $\rho_x(x) = \dfrac{\mathrm{d}f_x(x)}{\mathrm{d}x}$ versus x, where the x coordinate is in the $[\bar{1}\bar{1}1]$ direction. These plots have been constructed from the coordinates of the atoms in the two planes between which most of the displacement is concentrated, as follows: for each plane, the x components of the displacements of the atoms from their positions in the ideal lattice are plotted against x (the x-coordinate of the atoms in the ideal lattice). This gives two sets of points; that for the upper plane decreasing with x, that for the lower increasing with x. Smooth curves were drawn through these two sets of points; the separation of the curves

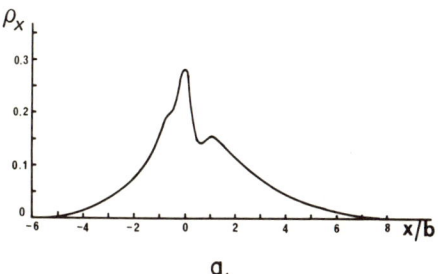

a.

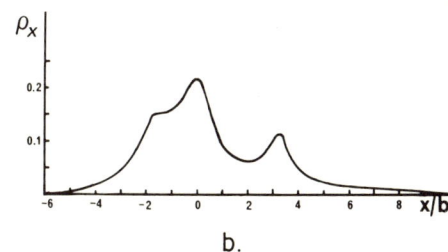

b.

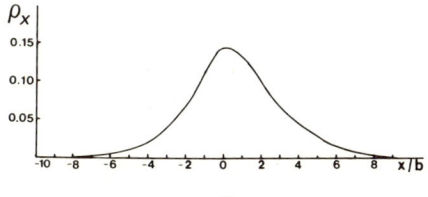

c.

Fig. 4. *The core structure calculated by the relaxation method depicted as a dependence of the density, ρ_x, of the continuous distribution on partial dislocations on x. (a) Potential J_0. (b) Potential J_2. (c) sodium.*

at any given value of x is the disregistry f_x of the lower part of the crystal relative to the upper part. Thus f_x may be measured at any x and the derivative $\dfrac{df_x(x)}{dx}$ obtained by numerical differentiation. The results of the relaxation calculations represented in this way can obviously be compared directly with those of the Peierls-type calculations.

5 DISCUSSION

The core structure of the edge dislocation $\frac{a}{2}[\bar{1}\bar{1}1]$ on a (112) plane has been described in terms of a continuous distribution of partial dislocations of the density $\rho_x(x)$. The principle qualitative features of the density ρ_x are:

(i) The overall symmetry, showing, e.g., a slower decrease on the twinning than on the antitwinning side of the dislocation.

(ii) The number and positions of peaks and inflexions which show the nonuniform distribution of the Burgers vector in the core.

It is seen when comparing Fig. 2 and 4 that these features are the same for the cores calculated both on the Peierls model and by computer simulation. However, the results of the two methods differ, particularly in the widths of the cores and, therefore, in the height of the peaks since

$$\int_{-\infty}^{+\infty} \rho_x(x)\, dx = b \ .$$

In the Peierls model the width of the core depends on the strength of the restoring forces which can be assessed by $\dfrac{1}{\mu}\left|\dfrac{\partial \gamma}{\partial f_x}\right|$ at its peak values; the width decreases with increasing strength. During the calculation of the restoring forces a uniform relative displacement $\mathbf{f}$ of the two parts of an infinite crystal is assumed. The relaxation perpendicular to the fault is then also uniform along the plane of cut. On the other hand, the displacement varies continuously in the dislocation core; thus the relaxation perpendicular to the fault will also vary along the plane of cut. Since the dilatation obtained during calculation of the restoring forces minimizes γ, the nonuniform dilatation of a core will effectively increase both γ and $\left|\dfrac{\partial \gamma}{\partial f_x}\right|$. However, in calculating the restoring forces, the displacement parallel to the fault is considered only in the plane of the cut whilst a relaxation also exists in several adjacent layers of a dislocation core (see

Section 4). This obviously leads to an effective decrease of γ and $\left|\frac{\partial \gamma}{\partial f_x}\right|$. A balance of these two effects will then determine the differences between Peierls type and relaxation calculations. In the case of J_2, as the relaxation in the planes other than the plane of the cut is very small, the effect of the dilational relaxation should predominate; the core calculated on the Peierls model is, indeed, wider than that calculated by relaxation. On the other hand, in the case of J_0 the parallel relaxation is larger and, thus, the Peierls-type calculation gives a narrower core than does the relaxation calculation. For sodium, as relaxations spread into four layers parallel with the plane of the cut, the core calculated on the Peierls model is much narrower than that calculated by computer simulation.

Since the two methods used reveal the same principal features of cores confined to a single crystallographic plane, we can conclude that they are basically determined by the functional dependence of the corresponding restoring forces. When no stable stacking faults can exist, the following consideration is plausible: when one part of the crystal is gradually being shifted with respect to the other to form a dislocation, the disregistry will vary more rapidly when it corresponds to a local maximum in $\left|\frac{\partial \gamma}{\partial f_x}\right|$ than when it corresponds to a local minimum or inflexion. This variation then corresponds to a local maximum/minimum or inflexion in $\rho_x = -\frac{df_x(x)}{dx}$. Therefore, if no stable stacking faults exist, any local minimum or inflexion in $\left|\frac{\partial \gamma}{\partial f_x}\right|$ implies a minimum or inflexion in the density of the continuous distribution of partials describing the core structure; the absolute minimum $\left|\frac{\partial \gamma}{\partial f_x}\right| = 0$ must obviously not be considered. This rule is observed in all three cases studied in the present paper.

If stable stacking faults may be formed, then the restoring force changes sign more than once and $\frac{\partial \gamma}{\partial f_x}$ may be divided into two or more parts, each of which possesses only one sign change. The solution of the Peierls equation then corresponds to two or more discrete partial dislocations separated by ribbons of stacking faults. The structure of each of the partials is determined by the corresponding part of the restoring-force law, and the above considerations apply to each of them. This has been checked by both Peierls-type and relaxation calculations using a potential cut off close enough to the second-nearest neighbor so that a stable stacking fault with displacement vector $\frac{a}{12}[\bar{1}\bar{1}1]$ may exist on the (112) plane[8]. Both calculations show the formation of two distinct partials $\frac{a}{12}[1\bar{1}\bar{1}]$ and $\frac{5a}{12}[\bar{1}1\bar{1}]$ separated by a stacking fault ribbon. In the Peierls model, however, the width of the splitting cannot be uniquely determined (see Section 3).

Generally, if the dislocation density ρ_x possesses several peaks or inflexions this might be interpreted as a dissociation into several discrete overlapping partial dislocations. This is identical with the suggestion of the generalized splitting of dislocations[12] where partial dislocations are separated by ribbons of unstable stacking faults. However, there is an essential difference between the generalized splitting, and splitting with stable stacking faults: if a ribbon of an unstable stacking fault is formed, the restoring force along this ribbon, $-\frac{\partial \gamma}{\partial f_x}$, is nonzero and, therefore, a shear stress will act in the fault plane; for a stable stacking fault $\frac{\partial \gamma}{\partial f_x} = 0$. The shear stress in the fault has to be compensated for by a distribution of the Burgers vector in the fault ribbon. This obviously implies that $\rho_x(x)$ will have to be nonzero there; this means that the cores of the partials of a generalized splitting must largely overlap. The width of a generalized splitting will thus always be small and it will generally not behave in the same way as a splitting with stable stacking faults. For example, an applied external stress which does not change the width of the normal splitting may also partly balance the shear stress in the fault and, thus, change the width of splitting together with the Burgers vectors of the partials.

The calculation of the restoring forces for any crystallographic plane, when an interatomic interaction is assumed, is relatively simple and generally much less time consuming than the calculation of the core structure. Knowledge of the restoring forces, however, can be favorably used to assess and/or interpret the main features of the cores of dislocations confined to these crystallographic planes. It could be used even if the dislocation core were distributed in several nonparallel crystallographic planes (e.g., screw dislocation in bcc crystals[3,4]) each of which contains a well-defined part of the Burgers vector. The determination of the dislocation core by the restoring forces then applies to the individual partial dislocations.

The structure of the dislocation core has been discussed so far in an unstressed crystal. However, in order to study dislocation motion, the investigation of the influence of an external stress on the core structure is of great importance. Usually, we are interested in a lattice friction stress, often called the Peierls-Nabarro stress, which has to be applied to start the dislocation moving in the otherwise ideal lattice. A simple method to calculate this stress was suggested by Nabarro.[13] However, the friction stress depends very strongly on the width of the core (e.g., exponentially for the case calculated by Nabarro) which, as discussed above, is not reliably known from the Peierls-type calculations. Furthermore, in Nabarro's model it is supposed that the functional dependence of the density $\rho_x(x)$ does not change under stress, but that it moves rigidly through the lattice. This important assumption need not be valid since most of the influence of the stress may arise through changing the density ρ_x. This can be studied either on Peierls model by including external stress in Eq. 5 or by atomistic relaxation calculations. Some preliminary relaxation calculations indicate that the dislocation moves by sudden jumps accompanied by changes in ρ_x rather than by a rigid motion of ρ_x through the lattice. Investigation of these problems is now in progress.

ACKNOWLEDGMENTS

The authors would like to thank Professors P. B. Hirsch, F.R.S., and T. W. Christian, and Dr. F. Kroupa for their interest. V. V. would like to thank I.C.I. for a research fellowship.

APPENDIX

SOLUTION OF THE PEIERLS EQUATION

Equation 5 is a nonlinear singular integro-differential equation and can be rewritten as[14]:

$$f_x(x) = -\frac{2(1-\nu)}{\mu\pi} \int_{-\infty}^{x} \left[\int_{-\infty}^{+\infty} \frac{\partial \gamma}{\partial f_x(t)} \frac{dt}{t-k} \right] dk , \qquad (6)$$

which has to be solved together with the boundary conditions (Eq. 1).

The numerical solution of Eq. 6 has been obtained by an iteration method using the following iterative formula:

$$f_x^{(n+1)}(x) = -\frac{2(1-\nu)}{\mu\pi} \lambda_n \int_{-\infty}^{x} \left[\int_{-\infty}^{+\infty} \frac{1}{t-k} \left\{ \frac{\partial \gamma}{\partial f_x^{(n)}(t)} - \frac{\partial \gamma}{\partial f_x^{(n)}(k)} \right\} dt \right] dk \qquad (7)$$

where

$$\lambda_n = \frac{b}{K_n} \quad (8)$$

and

$$K_n = -\frac{2(1-\nu)}{\mu\pi} \int_{-\infty}^{+\infty} \left[\int_{-\infty}^{+\infty} \frac{1}{t-k} \left\{ \frac{\partial\gamma}{\partial f_x^{(n)}(t)} - \frac{\partial\gamma}{\partial f_x^{(n)}(k)} \right\} dt \right] dk \quad (9)$$

Integrations in Eq. 9 must not be interchanged. The iteration process (Eq. 7) converges to a function $f_x^{(c)}(x)$; simultaneously λ_n converges to a value λ. $f_x^{(c)}(x)$ fulfills the boundary condition (Eq. 1) and is the solution of the Peierls equation which corresponds, however, to the restoring force $-\lambda \frac{\partial\gamma}{\partial f_x^{(c)}}$. The solution, $f_x(x)$, of Eq. 5 is then obviously

$$f_x(x) = f_x^{(c)}\left(\frac{x}{\lambda}\right) . \quad (10)$$

During the numerical calculation the interval $(-\infty, +\infty)$ has to be replaced by an interval $(-x_0, x_0)$ in the Eq. 7 and 9. The value of $f_x(x)$ has to be fixed at these points so that the boundary conditions (Eq. 1) are replaced by

$$\int_{-x_0}^{x_0} \frac{d f_x(x)}{dx} dx = b - \Delta b \quad , \quad (11)$$

where Δb is a reasonably small quantity which determines the amount of the displacement outside the interval $(-x_0, x_0)$. However, the solution is not sensitive to the choice of x_0 since the parameter λ will automatically change the chosen interval into a correct one, with respect to Δb, $(-\lambda x_0, \lambda x_0)$.

REFERENCES

1. Cotterill, R.M.J., and Doyama, M.: Phys. Rev., 145: 465 (1966).
2. Doyama, M., and Cotterill, R.M.J.: Phys. Rev., 150: 448 (1966).
3. Basinski. Z. S., Duesbery, M. S., and Taylor, R.: Phil. Mag., 21: 1201 (1970).

4. Vitek, V., Perrin, R. C., and Bowen, D. K.: Phil. Mag., **21**: 1049 (1970).
5. Gehlen, P. C.: J. Appl. Phys., **41**: 5165 (1970).
6. Kroupa, F., and Vitek, V.: Can. J. Phys., **45**: 945 (1967).
7. Vitek, V.: Phil. Mag., **18**: 773 (1968).
8. Vitek, V.: Phil. Mag., **21**: 1275 (1970).
9. Peierls, R.: Proc. Phys. Soc., **52**: 34 (1940).
10. Basinski, Z. S., Duesbery, M. S., Pogany, A. P., Taylor, R., and Varshni, Y.P.: Can. J. Phys., **48**: 1480 (1970).
11. Johnson, R. A.: Phys. Rev. A., **134**: 1329 (1964).
12. Vitek, V., and Kroupa, F.: Phil. Mag., **19**: 265 (1969).
13. Nabarro, F.R.N.: Proc. Phys. Soc., **59**: 256 (1947).
14. Leibfried, G., and Lücke, K.: Z. Phys., **126**: 450 (1949).

DISCUSSION on paper by V. Vitek, L. Lejček and D. K. Bowen

DUESBERY: You have shown that the difference between the Peierls and fully discrete calculations is largest for the sodium potential. Do you think that this is due to the long-range nature of the sodium potential?

VITEK: The large relaxation for sodium extends over several planes parallel to the slip plane of the dislocation. It is very probably due to the long-range nature of the potential but a more detailed investigation is necessary for a definite answer.

TYSON: One should be careful in applying the method of continuum distribution of dislocations to the determination of atomic displacements, as some important anharmonic effects may be missed in this way. For an edge dislocation, the region above the slip plane is in compression and that below is in tension; since we expect any reasonable interatomic force to be stiffer in compression than tension, it then follows that relaxations from the harmonic lattice displacements will be greater below than above the slip plane. In fact, recent calculations have shown that a crack actually appears below the extra half plane of a cube edge dislocation in iron.[1,2] The approach you have outlined will miss this feature, which may be very important in some problems.

VITEK: The Peierls model will certainly miss the anharmonic effects outside the plane of the cut; thus, any detailed study of a particular dislocation has to be made by fully atomistic calculations. The method outlined in this paper shows, however, very clearly the relation between the main features of the dislocation core (i.e., distribution of the Burgers vector) and the restoring forces in the plane of the cut. This can be very useful, for example, when studying the variation of the core structure with the interatomic forces. The restoring forces can always be calculated easily and their relation to the core structure is then very straightforward.

BULLOUGH: It is not clear to me why you consider the involvement of the Peierls analysis and model to be useful. In any event, for such narrow dislocations I would have thought that the "vertical cut model" "of van der Merwe"[3] would be much more appropriate.

VITEK: The object of this study has been to show the relation between the restoring forces in the slip plane of a dislocation and the structure of its core (distribution of the Burgers vector). For this purpose the Peierls model with the cut along the slip plane is the most appropriate. Furthermore, the calculated cores are not extremely narrow.

HIRTH: In connection with nonlinear effects, particularly with respect to the smearing out of sharp peaks in the calculated density of continuous dislocations, I wonder whether you have considered an intermediate case between the linear continuum calculation and the fully atomistic calculation. Your stacking-fault energies are calculated in the presence of a continuous shear. A second-order effect would be introduced by considering the stacking-fault energy as a function of both shear displacement and a gradient in the shear displacement.

VITEK: We have not considered directly the dependence of the stacking-fault energy and restoring forces on the gradient in the displacements. This problem is discussed, however, when comparing the Peierls-type and fully atomistic calculations.

REFERENCES

1. Bullough, R., and Perrin, R. C., *Dislocation Dynamics,* eds. A. R. Rosenfield, G. T. Hahn, A. L. Bement and R. I. Jaffee (McGraw-Hill, New York, 1968), p. 175.
2. Gehlen, P. C., Rosenfield, A. R., and Hahn, G. T.: J. Appl. Phys., **39**, 5246 (1968).
3. van der Merwe, J. H.: Proc. Phys. Soc. (London), **63**, 616 (1950).

EXTENDED DEFECTS IN COPPER AND THEIR INTERACTIONS WITH POINT DEFECTS

R. C. Perrin[*], *A. Englert*[**], *and R. Bullough*[*]

[*] *Theoretical Physics Division*
Atomic Energy Research Establishment
Harwell, Berkshire, England

[**] *Université Libre de Bruxelles*
Bruxelles, Belgium

ABSTRACT

The atomic configuration of the glissile edge dislocation and the intrinsic and extrinsic sessile Frank dislocations in copper have been obtained using an atomic model of the copper lattice with an interatomic pair potential extending through second neighbors. The extent of dissociation into Shockley partials of the glissile edge dislocation is obtained by an iterative procedure involving continual readjustments to the boundary displacements. The atomic model has also been used to estimate the interaction energy between intrinsic point defects and the three dislocation types when the point defects are in proximity with the dislocations.

1 INTRODUCTION

Many important physical and technological phenomena occur as a direct consequence of the interaction between point defects and dislocations. Such interactions have been shown to have a particularly significant effect upon the kinetics of recovery processes in deformed crystalline solids[1] and are a direct cause of the well-known metallurgical phenomenon of strain aging. Strain aging usually involves the long-range migration of interstitial impurities to the dislocations where they either segregate as a precipitate phase or form a (dilute) Boltzmann impurity atmosphere. In either case the kinetics of impurity depletion is significantly affected by the elastic interactions between the dislocation stress field and the impurity atoms. The precise effect on the kinetics can be fully understood only when the detailed spatial variation of the point defect-dislocation interaction is known in the neighborhood of the dislocation. Intrinsic point defects such as vacancies and self-interstitials will also interact with the dislocations. For example, the formation and growth of voids in certain neutron-irradiated metals is observed to be very sensitive to changes in the dislocation density[2,3]; it is suggested that such voids can grow only because the interstitials interact more strongly with the dislocations than do the vacancies. Again a complete understanding of this process requires a knowledge of the dislocation-point defect interactions.

Unfortunately, it is not easy to calculate the interaction when the point defect is a foreign atom (substitutional or interstitial); clearly, the strength of such a point defect cannot be deduced directly from the intrinsic properties of the host lattice. For the intrinsic point defect, however, Bullough and Hardy[4] have shown how the appropriate Kanzaki forces (which define the strength of such defects) can be deduced in terms of the properties of the perfect lattice when the harmonic lattice model is assumed. Their treatment of the vacancy and the vacancy-vacancy interaction in copper assumed an axially symmetric harmonic model which enables the force constants to be related to derivatives of an effective two-body potential between the ions. To obtain the vacancy-vacancy interaction, they used a Fourier-transform procedure in which the relaxation process was effectively achieved by direct summation over points of the reduced Brillouin zone. While this method of lattice statics is certainly convenient for single or pairs (or a superlattice[5]) of point defects, it is not very convenient for studies involving topological defects such as dislocations; it is also particularly awkward to include the possibility of dislocation dissociation in such a harmonic model.* To discuss such a dislocation and its interaction with intrinsic point defects in copper, the authors have constructed an effective two-body potential using the original Bullough

---
*No doubt it could be done with a somewhat artificial change of force constants over a planar region and a subsequent quasiharmonic treatment.

and Hardy force constants to define its derivatives at first- and second-neighbor separations. The relaxation process is then performed in direct space by successive relaxation of an appropriate crystallite of freely interacting atoms. The construction of this potential with its adjustments to fit various other physical properties of copper and an outline of the relaxation procedure are described. Results are given for three physically important edge dislocations: the pure edge dislocation dissociated into a pair of Shockley partial dislocations separated by an intrinsic stacking fault, the pure edge Frank extrinsic dislocation that represents a planar aggregate of interstitials, and the pure edge Frank intrinsic dislocation that represents a planar aggregate of vacancies. Detailed atomic core configurations and the spatial variations of the intrinsic point defect-dislocation interactions are discussed. Finally, the significance of the results are briefly discussed with particular reference both to the interstitial bias conjecture in the theories of void growth and to the general problem of the short-ranged interaction between point defects and dislocations.

2 THE POTENTIAL AND RELAXATION PROCEDURE

The interatomic potential used in the present study, originally constructed by Englert, et al.[6], consists of two parts: a volume-dependent potential and a pairwise potential; the former simulates the cohesive energy arising from the free electron gas, and the latter describes the ion-ion interactions that arise via polarization of the electron gas. The pair potential, conveniently represented by a spline function, extends up to the third-neighbor separation in the perfect copper lattice at which point (for computational convenience) it is arbitrarily set to zero with zero slope and curvature. At interatomic distances less than the first-neighbor separation in the perfect lattice the potential was matched to the Born-Mayer repulsive potential for copper with parameters that ensure consistency with observed displacement energies. The spline function has the form

$$V(r) = A_k(r - r_k)^3 + B_k(r - r_k)^2 + C_k(r - r_k) + D_k , \qquad (1)$$

where the ranges and parameter values are given in Table I. These parameter values ensure that the theoretical lattice is a stable fcc lattice at the observed lattice spacing for copper of 3.608 A; they are also consistent with the axially symmetric force constant values obtained by Bullough and Hardy[4] by fitting to the observed elastic constants and phonon-dispersion data for copper. These harmonic response properties impose constraints only on the derivatives of $V(r)$ and are supplemented by two additional physical constraints that remove some of the arbitrary nature of $V(r)$; the

Table I. The Coefficients of the Pair Potential for Copper[a]

K	r_k(A)	A_k	B_k	C_k	D_k
1	1.	-667.9458	1081.8838	-628.5649	138.1100
2	1.5	-49.0449	79.9655	-47.6408	10.8050
3	2.0	-3.2382	6.3981	-4.4591	0.8453
4	2.551	-0.258148	1.045285	-0.357854	-0.210930
5	3.061199	-2.221407	0.650164	0.507164	-0.155699
6	3.341810	1.507669	-1.219882	0.347295	-0.011272
7	3.607658	-0.080144	-0.017445	0.018353	0.023168
8	4.209149	2.186182	-0.162063	-0.089620	0.010455
9	4.311190	-1.575972	0.507171	-0.054405	0.001945
10	4.418461	0.0	0.0	0.0	0.0

(a) As given by a spline function consisting of 10 cubic equations

$$V(r) = A_k(r-r_k)^3 + B_k(r-r_k)^2 + C_k(r-r_k) + D_k$$

each valid for $r_k \leq r \leq r_{k+1}$. V is in eV and r_k in A.

potential is adjusted to be consistent with both the observed intrinsic stacking-fault energy of copper (70 erg/cm^2) and with the vacancy-formation energy of approximately 1 eV. The potential thus defines a model that is precisely identical to the harmonic lattice model[4] over the harmonic range but has been supplemented to permit the sensible existence of stacking faults and to ensure that the absolute energies of defects will be reasonable.*

To obtain the atomic configurations associated with the straight edge dislocations, a rectangular parallelepiped of freely interacting atoms was set up such that the atoms formed a perfect fcc lattice with the copper lattice spacing subject to the interatomic potential given by Eq. 1. The faces of the parallelepiped were appropriate {110}, {111}, and {112} crystallographic planes, and the dislocation was arranged, in all cases, to lie through the center of the assembly and orthogonal to the two {112} faces. In the direction of the dislocation line, specifically the [11$\bar{2}$] (z-axis) direction, the assembly was only six lattice planes thick and periodic boundary conditions were imposed across the two (11$\bar{2}$) faces. In the other two orthogonal directions ([$\bar{1}$10] - the x direction and [111] - the y direction) the assembly was made up of 50 {110} lattice planes and 18 {111} planes. The parallelepiped was deliberately made extensive in the x direction to accommodate the dissociation into Shockley partials of the glissile edge dislocation. A total of 900 atoms was thereby involved. The required defect configuration was achieved by first imposing the appropriate anisotropic elastic displacements on the boundary layer of atoms

---

*It is recognized that this potential approach cannot be used to accurately estimate absolute formation energies of defects since important electronic processes have been ignored; this paper deliberately restricts its application to defect-interaction problems since such interactions depend directly on the response function of the lattice that is well defined at least within the harmonic approximation[4].

and then dynamically relaxing[7] the inner atoms under the potential (Eq. 1). In fact, the central-atom region that was allowed to relax consisted of 36 x 14 x 6 lattice planes — a total of 504 atoms. The next section shows that the equilibrium separation of the partials in the glissile edge dislocation was found to be almost exactly $9b$, where b is the magnitude of the total Burgers vector of the edge dislocation ($b = \frac{a}{2}$ [$\bar{1}10$] where a is the lattice spacing). To study the sessile Frank dislocations, it was again convenient to put the dislocation lines parallel to the z axis ([$\bar{1}12$] direction). The two sessile partial dislocations associated with the two Frank dislocations were taken, purely for convenience, to be separated by the same distance of $9b$; this necessitated the insertion of 17 interstitials for the extrinsic Frank dislocation and the removal of 17 lattice atoms for the intrinsic dislocation.

To obtain the interaction energies between these dislocations and intrinsic point defects, it was necessary to extend the parallelepiped in the z direction and then drop the periodic boundary conditions on these faces. As the excessive size made it infeasible to include the entire dislocation within such a three dimensional block, small sections of 18 x 12 lattice planes in the x and y directions were selected and made into three-dimensional assemblies by adding together five atomic repeat layers (30 lattice planes) in the z direction. Of this assembly, the central region of 10 x 8 x 14 lattice planes (185 atoms) was free to relax. The boundary atoms were held in their previous relaxed dislocation positions and the point defect was located near the center of the assembly; the position of the point defect relative to the dislocation was then varied by simply changing the location of this small atomic assembly relative to the dislocation itself.

3 THE EDGE DISLOCATIONS AND THEIR INTERACTIONS WITH POINT DEFECTS

Three physically significant edge dislocations can occur in a metal such as copper with a fcc lattice. These are: (i) the usual glissile pure edge dislocation which can, and does, dissociate into a pair of Shockley partial dislocations separated by an intrinsic stacking fault on the (111) plane, and (ii) the two sessile pure edge dislocations that are associated with planar aggregates of vacancies or interstitials. These latter dislocations usually exist as closed dislocation loops, but for the sake of computational convenience they are represented as long, straight dislocations parallel with a <112> direction. For the purpose of studying the short-range point-defect interactions with such dislocations and their core configurations, the simplification to a straight dislocation should not cause any real loss of generality.

3.1 The Glissile Edge Dislocation

This dislocation, lying along the [112] direction with a total Burgers vector $\mathbf{b} = \frac{a}{2}[\bar{1}10]$, can dissociate into a pair of Shockley partial dislocations

$$\frac{a}{2}[\bar{1}10] = \frac{a}{6}[\bar{1}2\bar{1}] + \frac{a}{6}[\bar{2}11] \quad .$$

To determine the equilibrium separation of these partials, proceed as follows.

(i) From the general anisotropic elastic theory of dislocations obtain the displacement field for each of the above partial dislocations.

(ii) Superimpose these elastic solutions to correspond to the above dissociated dislocation situated midway between two (111) planes near the center of the complete atomic assembly with a separation between the *elastic* partials of d_0.

(iii) Calculate the relative displacements across the slip plane by calculating the deviation of the atoms in the upper (111) slip plane from the centroids of the corresponding triangles of atoms in the lower (111) slip plane. The crystallographic positions of the partials are then defined as the values of x at which this relative displacement has a value equal to exactly ½ of its extreme value; similarly, the range of x over which the relative displacement is bounded by ¼ of its extreme values is a measure of the widths of the separate partials. This crystallographic definition of the partial positions in general yields a separation very slightly greater than the original (elastic continuum) setting.

(iv) Allow the central region to relax under the potential (Eq. 1) and determine the separation of the partials by (iii) above. This relaxation will cause the partials to change their separation to d_1. Now recalculate the elastic displacements of the boundary atoms to correspond to two partials separated by d_1 and allow the central region to again relax.

(v) Continue this iteration until the separation of the partials defined by (iii) is the same for both the central region after relaxation and for the elastic setting of the boundaries.

With an initial choice of $d_0 = 8b$ this iterative process eventually converges to give a final separation* between the partials of

$$d = 9.2b \tag{2}$$

A projection on to the $(11\bar{2})$ plane of the final atomic configuration is shown in Fig. 1. For clarity the figure includes only two of the six

---

*In previous work on this dislocation Ref. 6 reported an inability to obtain a precise equilibrium separation. Now, it is believed that the difficulty was probably due to incomplete relaxation under the potential.

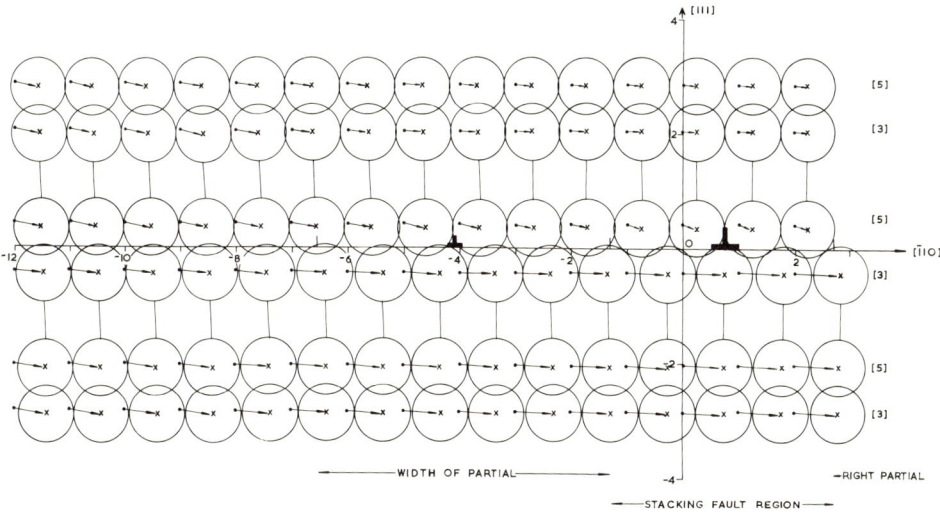

Fig. 1. The relaxed atomic configuration associated with the left hand partial of the dissociated edge dislocation. This is a (11$\bar{2}$) projection and, for clarity, only the atoms of two of the six (11$\bar{2}$) atom planes are displayed. The precise position of the partial is indicated by a small dislocation symbol; the larger symbol indicates the center of the total edge dislocation. The arrows show the displacements from the perfect lattice configuration caused by the introduction of the dislocation.

(11$\bar{2}$) layers; the 3rd and 5th layers were chosen so as to include the atoms adjacent to the actual slip plane. The left partial is shown completely and its center at $x = -4.1$ is indicated by the small dislocation symbol. The center of the complete dislocation (at $x = 0.75$) is indicated by the larger dislocation symbol. The figure also includes the perfect lattice points and the motion of these lattice points as caused by the introduction of the dislocation is shown by the arrows. The center of this partial and its width were obtained from the relative displacement prescription in (iii) above. The relative displacement across the slip plane is shown more clearly in the (111) projection given in Fig. 2, which shows the complete dissociated dislocation.

It is clear from these figures that by allowing this dislocation to achieve equilibrium under the potential (Eq. 1) the partials have increased their separation somewhat from the separation expected on the basis of continuum elasticity and a constant stacking fault energy of 70 erg/cm^2*
and also, perhaps what is more important, the individual partials have increased their "Peierls" width from the elastic width of about $2b$ to just over $5b$; this transition of the relative displacement variation from the elastic to the relaxed atomic configuration is shown in Fig. 3. The elastic displacements appropriate to the $9.2b$ separation have been deliberately

---

*From the anisotropic calculations of Teutonico[8] and Bullough[9] d(elastic) = $7.9b$.

516   R. C. PERRIN, A. ENGLERT, AND R. BULLOUGH

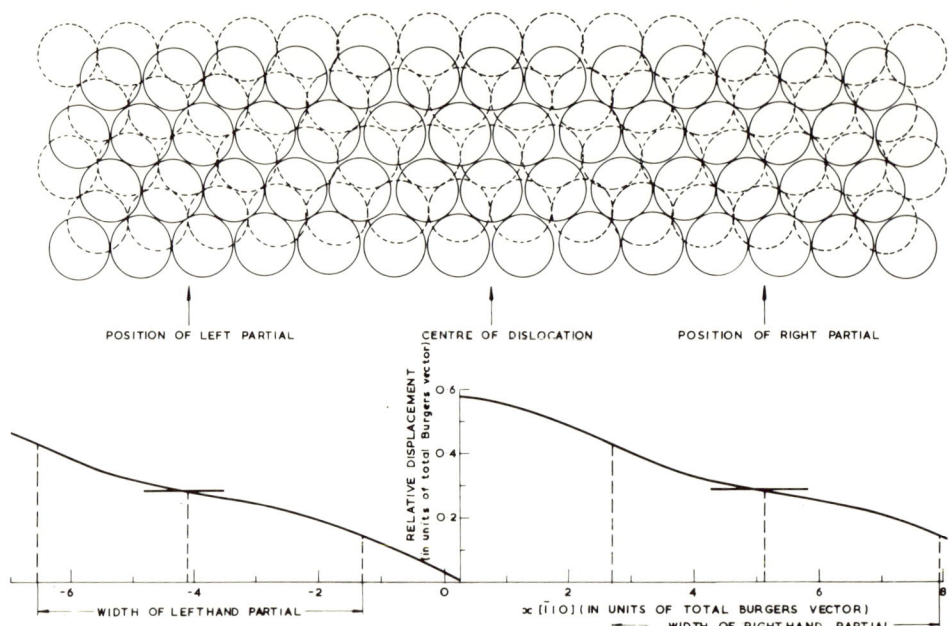

Fig. 2. A (111) projection of the atoms above and below the slip plane for the relaxed dissociated edge dislocation and the variation of relative displacement across this slip plane. The relative displacement curves were obtained by calculating the deviation of the atoms in the upper (111) slip plane from the centroids of the corresponding triangles of atoms in the lower (111) slip plane; the arrows are included in the projection to clarify this correspondence. The precise positions and widths of the two partials are indicated. − − − upper (111) plane; ———— lower (111) plane.

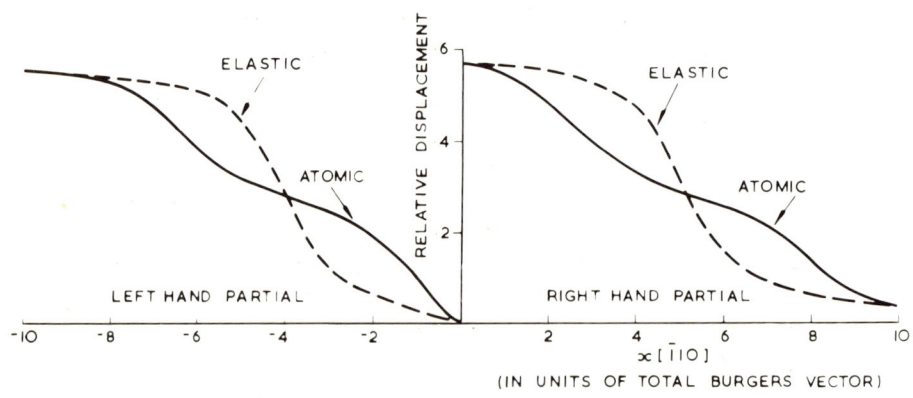

Fig. 3. A comparison between the (anisotropic) elastic and fully relaxed relative displacements across the slip plane of the dissociated edge dislocation. In both cases the partials are separated by 9.2b. The elastic partial dislocations have a width of about 2b whereas the relaxed partials have each increased their widths to just over 5b.

imposed over *all* the atoms to expose the width changes due to relaxation. The large increase in width of the partials, of course, reduces the region between the partials over which the faulting is at all perfect. In fact, Fig. 1 and 2 show that out of a total partial separation of $9.2b$ only about $4b$ can be considered as resembling a stacking fault in the strictly perfect sense. This spread of misfit into the faulted region (and on either side of it) leads to the rather interesting point-defect interactions referred to below. However, it is interesting to further explore the relation between the elastic and relaxed configurations in the immediate slip-plane region. To do so, we have calculated the e_{xx} tensile and compressive strains between the atoms in the atomic layers on either side of the slip plane; these tensile strains beneath the dislocation give a direct indication of such phenomena as microcracking or possible rapid-diffusion channels. The elastic situation and the corresponding fully relaxed situation are shown in Fig. 4a and b, respectively. The difference is quite striking. These strains are highly localized in the neighborhood of the partials in the elastic model; whereas, in the atomic model, they are distributed in an oscillatory manner across the entire dissociated dislocation. It is particularly interesting to see that the maximum tensile strains that were directly beneath each partial in the elastic model have now moved to positions near the center of the fault. This result certainly suggests that such faulted dislocations could well contribute to pipe diffusion by substantial transport down the fault itself.

The interaction energies between this dislocation and vacancies and interstitials have been calculated in the way described in Section 2. The results for both point defects are displayed in Fig. 5a, in which the numbers are the interaction energies in electron volts and a positive value indicates attraction. In this $(11\bar{2})$ projection of the left partial, the depth of the various atoms is indicated by using a different atom symbol for each $(11\bar{2})$ plane. The associated number in ( ) brackets refers to the vacancy created by removing that particular lattice atom. The number in [ ] brackets refers to the interstitial and is associated with a lattice atom when the local relaxed interstitial configuration is clearly itself associated with the position of that lattice atom (usually a split dumb-bell configuration of some type); when the interstitial adopts a definite new site this is shown with its associated [ ] bracket (this happens when interstitials are placed immediately above the dislocation, as shown by the O symbol in the figure).

At once, the striking difference between the vacancy and interstitial energies is seen. Above the slip plane the vacancy is always attracted; whereas, below the dislocation, it is weakly attracted in the slip plane near the partial with weak repulsion near the center of the fault. At only a lattice spacing away beneath the dislocation, it is always repelled. The

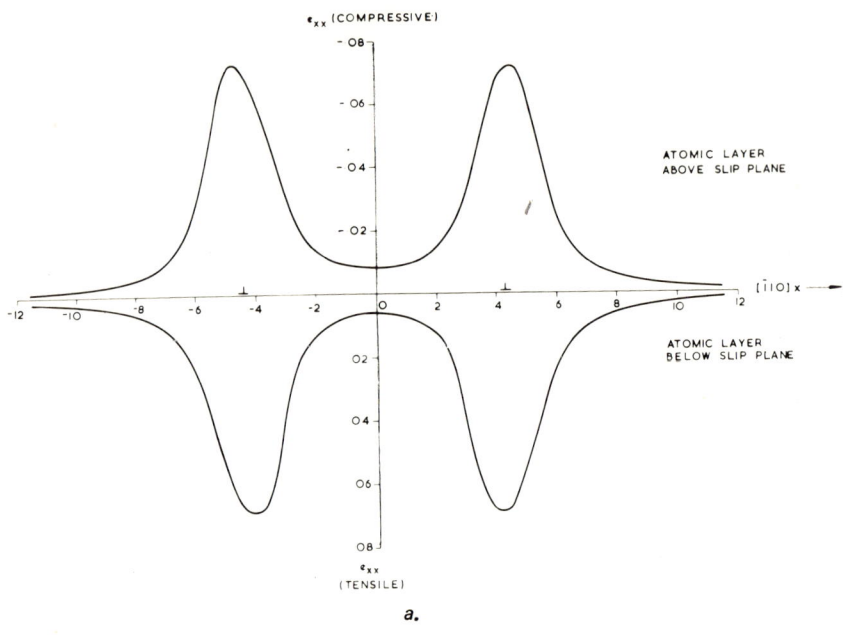

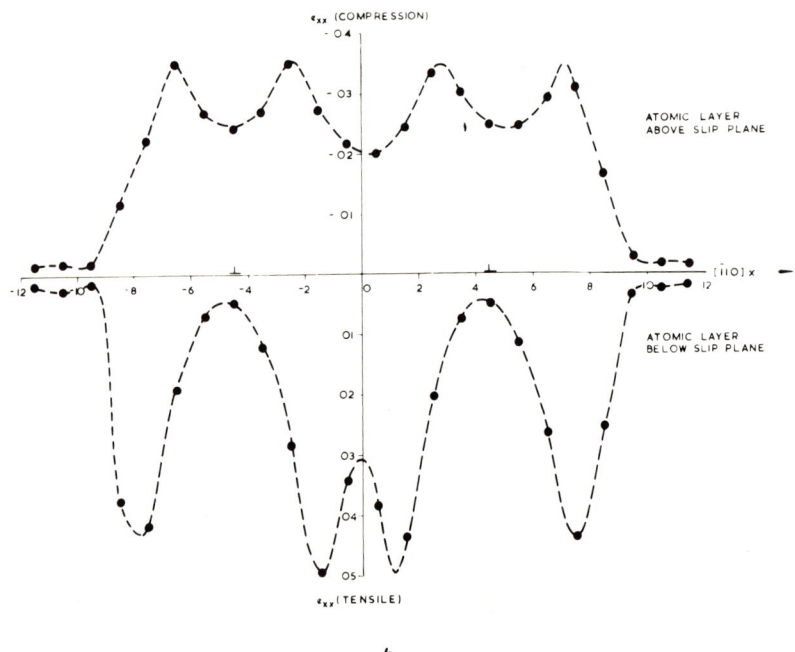

Fig. 4. The e_{xx} strains in the (111) atomic layers immediately above and below the slip plane for (a) the "elastic" dissociated edge dislocation, and (b) the "fully relaxed" dissociated edge dislocation.

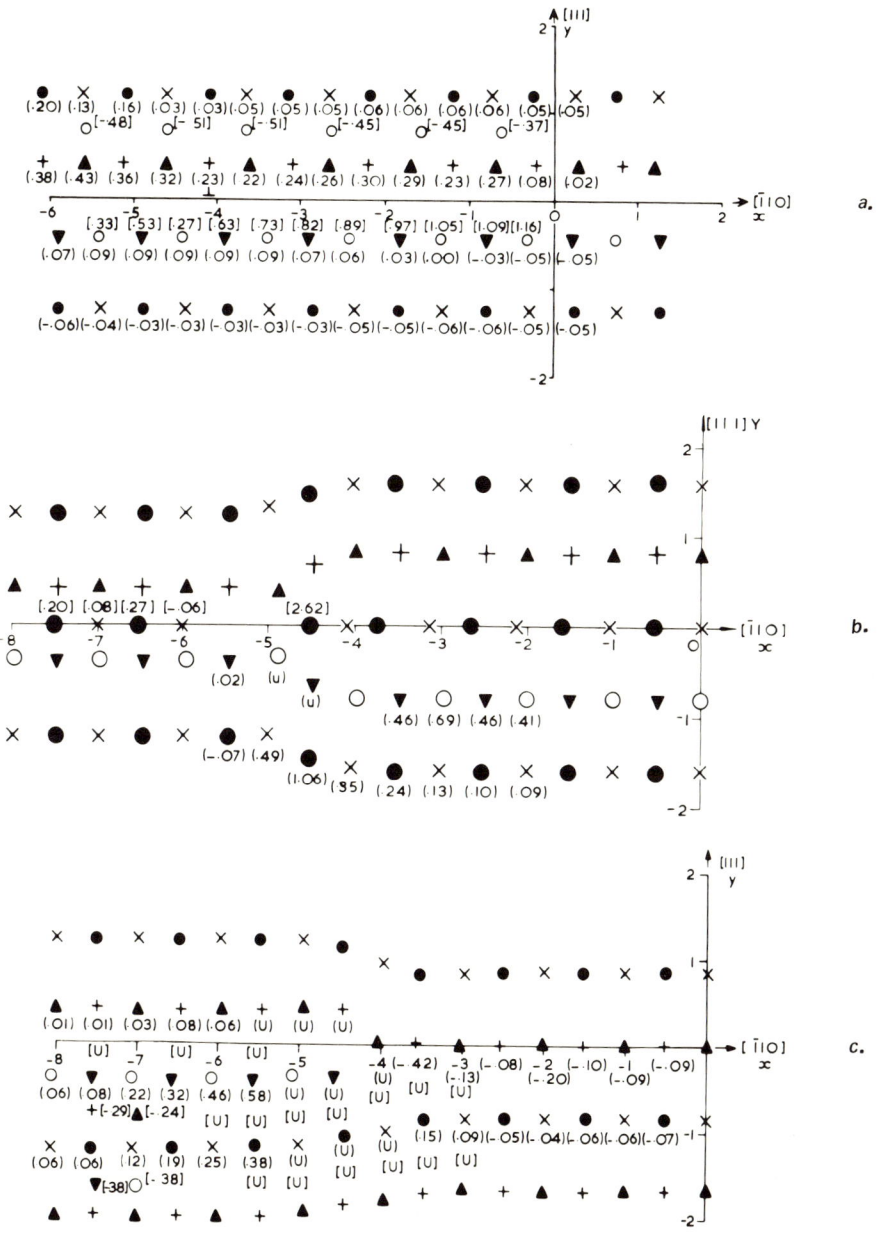

Fig. 5. (a) The point-defect Shockley partial-dislocation interaction energies in electron volts. When the number is positive the interaction is attractive and vice versa. If the interstitial is not clearly associated with a lattice atom after relaxation its separate position is indicated. (b) The Frank partial dislocation (extrinsic) relaxed configuration displayed as a projection in the [11$\bar{2}$] direction. (c) The Frank partial dislocation (intrinsic) relaxed configuration displayed as a projection in the [11$\bar{2}$] direction.

attractive interaction all around the immediate vicinity of the partial would seem to be a clear indication of a significant inhomogeneity interaction that should be only weakly angular dependent, whereas the longer ranged asymmetry is consistent with the eventual complete dominance of the size-effect interaction with its strong angular dependence. By contrast, the interstitial is strongly attracted beneath the dislocation and strongly repelled above it — there is no indication of a significant inhomogeneity effect. The well-distributed nature of these energies across the dissociated dislocation (consistent with Fig. 4b) should lead to point-defect-atmosphere formation that is not localized on the partials and thus one can reasonably envisage the entire dissociated dislocation absorbing such defects by climb without the necessity of constriction processes.

3.2 The Frank Sessile Dislocations ($b = \frac{a}{3}[111]$)

The relaxed configurations for the extrinsic and intrinsic dislocations are shown in Fig. 5b and 5c, respectively. (Because such dislocations are sessile, the previous iterative relaxation procedure was not necessary.) Both projections in these figures are in the $[11\bar{2}]$ direction and all the $(11\bar{2})$ atom layers are displayed by using various atom symbols. One interesting consequence of the relaxation for the extrinsic partial dislocation is the large inward displacement of the two atoms located at $x = -5$ in Fig. 5b. Again the interaction energies between these dislocations and interstitials and vacancies are shown. When the interstitial does not associate with a lattice atom its position is indicated with an appropriate extra atom symbol plus the interaction energy in [ ] brackets. For several positions near the center of these dislocations, the point defects did not have a metastable configuration and inevitably drifted into the center of the dislocation under the relaxation; such positions are labelled by a [u] for unstable interstitial or by a (u) for unstable vacancy.

4 DISCUSSION

With the potential used in this work, the equilibrium configuration of the interstitial in the otherwise perfect lattice is not the <100> split configuration suggested by alternative potentials for the face-centered lattice but is the "true" symmetrical interstitial configuration. In the compressed region above the slip plane of the Shockley partial the interstitial relaxes into these symmetrical sites as indicated by the O symbols in Fig. 5a. Below the slip plane, however, the interstitial adopts a <110> split configuration in most cases; the exceptions are the atoms at $x \sim -4.5$ and $x = -5.5$ (very near the partial), where the interstitial atom adopts a position slightly out of the (111) plane. It is particularly interesting to see that the maximum interstitial attraction is very large (1.16 eV) and that it occurs near the center of the fault rather than in the immediate neighborhood of the partials; this is consistent with the well-distributed nature of the tensile strains as exemplified by Fig. 4a, where the relaxation has clearly shifted the strain maxima from beneath the partials toward the center of the faulted region.

There are a large number of sites around the Frank partials from which the point defects are instantaneously captured by the dislocation. Around the intrinsic partial, nine sites indicated in Fig. 5c by (u) have been found on which the vacancy is unstable; the final figure is probably twelve if the behavior in the plane above the fault plane is the same as in the plane below. From all these positions the vacancy moves to the center of the dislocation and extends the fault. Interstitials are strongly repelled from the plane of the intrinsic fault on the compressive side ($y=0$, $x<-4$ in Fig. 5c) and are displaced spontaneously through the (111) planes. At a large number of sites marked [u], the interstitial is unstable and moves either directly or by displacing an adjacent lattice atom into the center of the partial. At least thirty such unstable sites are estimated in the vicinity of these partials from which an interstitial will instantaneously move into the dislocation core.

The results on all three dislocations are quite consistent with the basic hypothesis suggested as an explanation of void growth in irradiated metals at temperatures of about one-third of the melting point. That is, the interstitials suffer a stronger attraction to the radiation-produced and deformation-produced dislocations than do the vacancies. These preferential interstitial interactions not only provide the reason for the slight excess steady-state vacancy concentration with the consequent possibility of void growth, but also explain why the vacancy dislocation loops (Frank intrinsic partials) cannot exist in such irradiation conditions since they interact just as strongly with the interstitials (which destroy them) as do the Frank extrinsic partials (which grow by adsorbing interstitials).

REFERENCES

1. Bullough, R. and Newman, R. C.: Rep. on Prog. in Physics, **33**: 101 (1970).
2. Nelson, R. S.: Proceedings of International Conference on Radiation-Induced Voids in Metals; Albany, New York (June 1971).
3. Bullough, R. and Perrin, R. C.: Proceedings of Conference on Radiation-Induced Voids in Metals; Albany, New York (June 1971).
4. Bullough, R. and Hardy, J. R.: Phil. Mag. **17**: 833 (1968).
5. Stoneham, A. M.: Jnl. Physics F (Metal Physics): to appear (1971); also Harwell Rep. TP.446 (1971).
6. Englert, A., Tompa, H. and Bullough, R.: Fundamental Aspects of Dislocation Theory. Nat. Bureau of Standards. Special Publication 317; Washington, D.C. (1970).

7. Bullough, R. and Perrin, R. C.: Proc. Roy. Soc.: **A305**: 541 (1968).
8. Teutonico, L. J.: Acta. Met. **11**: 1283 (1963).
9. Bullough, R.: *Dislocations,* Harwell Rep. A.E.R.E. PGEC/L.33 (1964).

DISCUSSION of paper by R. C. Perrin, A. Englert and R. Bullough

JOHNSON: What is the magnitude of the bump in your potential near the second neighbor compared to the depth of the well near the first neighbor?

PERRIN: The depth of the potential well is -0.24 eV and the value of the potential at the small maximum near the second neighbor is about 0.025 eV.

JOHNSON: The more structure in a potential, the more likely it is for a calculation to converge to a local minimum rather than the true minimum. Joe Beeler has given an excellent example of the difficulties you can get into in this regard. Did you have any problems from this effect?

PERRIN: First, we used a dynamic relaxation procedure that is less likely to converge to a local minimum than a static procedure and, second, two initial positions for the interstitial were tried.

MARCH: The change in ionic density due to the dislocation presumably has to be screened by the conduction electrons. There would then be an "electrostatic" interaction with a charged defect like a vacancy.

Is this comparable in magnitude with the values of the interaction energy that you calculate?

SEEGER: For the simplest possible model (free electrons, edge dislocations) this problem was treated by Cottrell et al.[1] The result is that in this case the electric interaction has the same position dependence as the elastic interaction. Save for exceptional cases (very small electric interaction) the electric interaction is the smaller of the two. Refinements have been attempted in recent years, but they have not changed this picture qualitatively. The electric interaction is expected to be fairly important in determining the position of lowest energy that a vacancy or an impurity atom may take up.

SHEWMON: In treating the kinetics of point defects diffusing to sinks one often uses a capture radius to describe the defect-dislocation interaction. Can you comment on the relative values of the radii for vacancies and for interstitials from your work?

PERRIN: We are not yet in a position to give values of the relative capture radii for vacancies and interstitials, although that is ultimately one of the objects of the work. Certainly the capture radius for interstitials will be appreciably greater than that for vacancies, particularly for the Frank partial dislocations where we have found that the interstitials are attracted athermally into the core from a large number of surrounding sites.

DAHL: What configuration did the interstitials assume in your calculations in the compressive and tensile regions?

PERRIN: On the compressive side of the Shockly dislocation, the interstitial adopted the same configuration as in the perfect lattice (i.e., the octahedral configuration). On the tensile side of the dislocation it was found that the interstitial was absorbed into the plane immediately below the slip plane and adopted a <110> split configuration due presumably to the large strains in this direction.

DAHL: What procedure was employed to assure that the cell boundary did not affect the computational results?

PERRIN: In calculating the configuration of the Shockley dislocation the anisotropic elastic displacements on the boundary were continually adjusted to correspond to the splitting of the partial dislocations in the central region.

WYNBLATT: In your studies of the interaction of vacancies and interstitials with a dislocation split into Shockley partials, do you observe any effects that could be interpreted as climb of the dislocation?

PERRIN: Because the Shockley dislocation has a strong attraction for point defects over the full width of the dislocation, it would seem likely that it should climb relatively easily.

DORAN: My concern is with the application of your potential to consideration of interstitials — for which atomic separations are far from normal. It appears to me that one test of applicability is to determine if the potential reproduces the known pressure-volume behavior of the solid. Has any such test of your potential been made?

PERRIN: We have not yet studied the pressure-volume behavior predicted by this potential.

PARSONS: I am most intrigued by the oscillations you show in your calculated displacement field near the dislocation core because of their similarity to those that appear in end-on dislocation lattice images. Thus, are the oscillations you showed real or are they a consequence of the potential you used? If they are real, what are they due to?

PERRIN: As we have not investigated the dislocation configuration with any other potential, I cannot say how sensitive they are to the details of the potential, but we believe they are real. On a hard-sphere model, it is easy to imagine that the e_{xx} strain is small at the symmetrical positions corresponding to the center of the partial and the center of the fault and larger when in intermediate positions.

TYSON: Do you feel that the unexpected strain distribution you find above and below the slip plane is a general feature, or does it depend on the potential? Have you examined other calculations, for example, those of Cotterill and Doyama[2] using Morse potentials, to check this point?

PERRIN: We have not examined the calculations of other authors in this respect as it needs a detailed knowledge of the atomic configurations.

REFERENCES

1. Cottrell, A. H., Hunter, S. C. and Nabarro, F.R.N.: Phil. Mag. **44**, 1064 (1953).
2. Cotterill, R.M.J. and Doyama, M.: Phys. Rev. **145**, 465 (1966).

PARTIAL DISLOCATION INTERACTIONS IN A FACE-CENTRED CUBIC MODEL SODIUM LATTICE

Z. S. Basinski, M. S. Duesbery and R. Taylor

Division of Physics
National Research Council of Canada, Ottawa 7, Canada

ABSTRACT

Preliminary results presented deal with the mutual interaction and individual core structures of Shockley partial dislocations in a fcc model sodium lattice. Deviations from the predictions of linear elasticity are apparent for separations greater than about 10 Burgers vectors, due to boundary conditions, and also for separations less than about 8 Burgers vectors. The nonlinearity at small separations is of an unexpected sense, such that the partial separation is greater than the linear elastic value. The possible sources for nonlinearity are discussed in the light of the results.

1 INTRODUCTION

It is established that in many fcc metals, the stable screw dislocation is dissociated into two Shockley partials separated by an intrinsic stacking fault.

In pure metals the partial separation, as calculated from linear elasticity, is usually less than about 15 Burgers vectors. For example, in pure copper, with a stacking-fault energy of about 80 erg/sq cm, the width of the stacking fault is about 2.5 Burgers vectors. For these small separations, the use of linear continuum elasticity is seriously in question.

For many properties, the behavior of the dissociated dislocation is different from that of the perfect dislocation; for example, all short-range interactions and the ease of cross-slip will depend strongly on the extension of the dislocation core. In the absence of any better model, it has been conventional to treat these properties by means of linear elasticity, even for very small partial dislocation separations.

Here is presented a preliminary report on work intended to investigate the validity of linear elasticity for the treatment of dissociated dislocations in the limit of high stacking-fault energy. In the discrete dislocation model used, the ions interact via a two-body effective ion-ion potential appropriate for sodium. The derivation of and the justification for the potential have been discussed elsewhere (Basinski, Duesbery, Pogany, Taylor and Varshni[1]; Basinski, Duesbery and Taylor[2,3] to be referred to as BDT1, BDT2, respectively).

2 METHOD OF CALCULATION

Only those aspects of the calculation which differ from the methods used in BDT1 and BDT2 are discussed. The effective ion-ion potential represents an fcc lattice that is stable to small inhomogenous displacements, but metastable to the large homogeneous shear corresponding to the fcc-hcp phase transformation. In fact, the energy of an intrinsic stacking fault in the fcc structure is -3.5 erg/sq cm. Thus, to allow introduction to the fcc lattice of a dissociated dislocation with a stacking fault of finite width, it is necessary to apply an external shear stress to the lattice. The effect of this stress, which is of such a sense as to force together partials bounding an intrinsic fault, is to increase the stacking-fault energy by an amount $|\sigma b_e|$ where σ is the applied stress and b_e is the edge component of the partial Burgers vector. By varying the stress, the magnitude of the stacking-fault energy, and hence the separation of the partials, can be varied.

In the present work, only screw dislocations are treated. As in previous calculations (BDT1, BDT2), the relaxation of an assembly of ions is considered using periodic boundary conditions along the dislocation line. The remaining boundaries are fixed at the positions dictated by linear anisotropic elasticity. The adequacy of these boundary conditions for calculation of the dislocation core structure, and for treatment of the movement of total dislocations, has been discussed in some detail in

BDT1, BDT2, respectively. In these cases, interpretation could be restricted to regions remote from the fixed boundaries. In the present calculation, the ion block used is of size and shape such that the maximum linear extent of both primary and cross-slip planes is $70b$, where b is the magnitude of the total Burgers vector, and dislocations dissociated by up to $30b$ are considered. Since this involves calculating interactions having a range up to $1/2$ the width of the ion block, it is clear that a more-careful appraisal of the boundary conditions may be needed. This point is considered later.

The core structures are interpreted in terms of the stress field around the dislocation, which can be defined on a discrete basis (BDT2). If the x_3 axis is taken parallel with the dislocation line, with the x_2 axis normal to the slip plane, the stress tensor will have the general form

$$\sigma = \begin{pmatrix} \sigma_{11} & E_{12} & S_{13} \\ E_{12} & \sigma_{22} & S_{23} \\ S_{13} & S_{23} & \sigma_{33} \end{pmatrix},$$

where the components denoted by E and S are due primarily to edge and screw Burgers vectors, respectively. In the present treatment the diagonal terms are ignored, and the edge and screw fields, E and S, are treated separately. At any ion, the screw field can be reduced to a single shear stress S'_{23} by a rotation $\theta = -\tan^{-1}(S_{13}/S_{23})$ about the x_3-axis, and thus can be represented by a line of magnitude S'_{23}, drawn at an angle θ to the x_1-axis (i.e., parallel with the local shear plane). The edge field E at an ion has only a single component. It is represented by a line of magnitude E_{12}, drawn normal to the slip plane.

We shall also make use of the misfit across the slip plane, i.e., the difference in displacement between ions above and below the slip plane (this is essentially the displacement gradient method used by Vitek et al.[4]), to define most easily the width and separation of partials.

3 THE PARTIAL DISLOCATION CORE STRUCTURE

We consider the core structure of a screw dislocation, which should approximate to the regime of linear elasticity (see Section 4), with a partial separation of $10.5b$. The dislocation has a total Burgers vector and line direction $a/2[101]$, and lies on the (111) slip plane. All diagrams show projections normal to [101]. The ion heights are omitted in general for clarity, but for reference purposes are included on the diagram in Fig. 1a, in units of the $\langle 101 \rangle$ repeat distance.

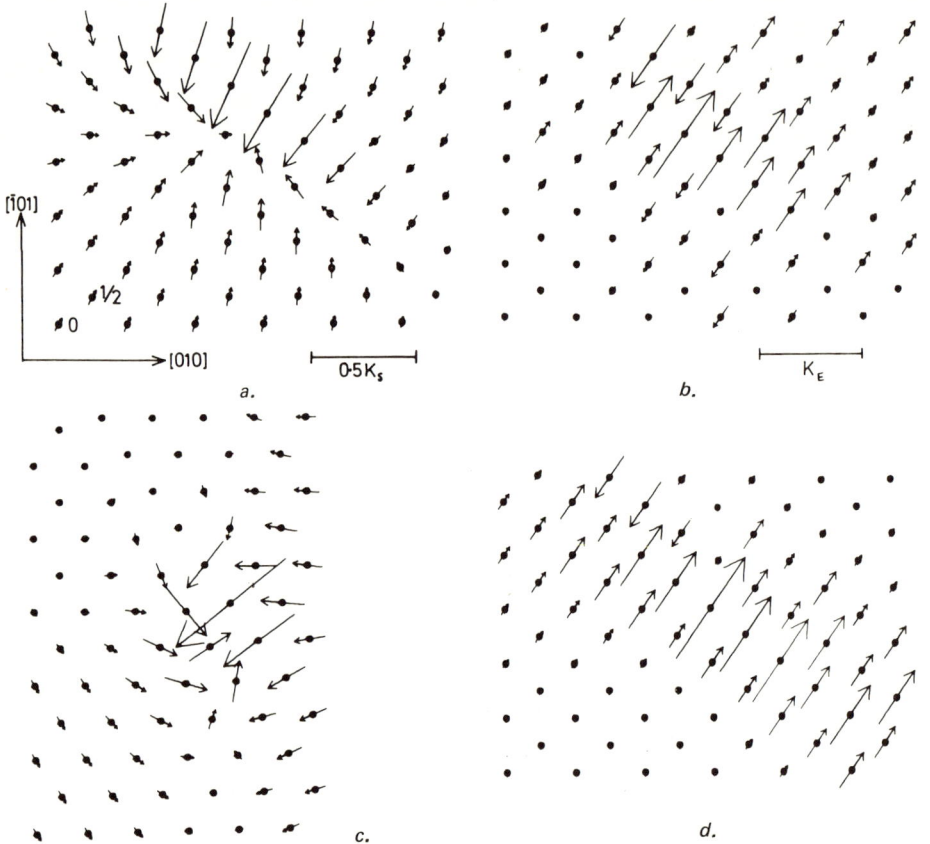

Fig. 1. The edge and screw stress fields of the partial dislocation core for a dissociation in the linear regime; (a) The screw stress field of the relaxed core; (b) The edge stress field of the relaxed core; (c) The screw stress field of the linear elastic core; (d) The edge stress field of the linear elastic core.

3.1 The Stress Field

In Fig. 1a and 1b are shown the screw and edge stress fields, respectively, of the left-hand partial of the relaxed dislocation. The corresponding linear elastic fields are shown for comparison in Fig 1c and 1d. The edge fields (Fig. 1b and d) are similar in character, and are extended in the slip plane. The relaxed screw field (Fig. 1a), however, shows distinct differences from the elastic solution (Fig. 1c). The stresses in the relaxed core are confined more strongly to the slip plane than suggested by linear elasticity.

It should be noted, however, that even for a separation of $10.5b$, the edge stresses in the stacking fault region are nonzero, indicating incipient nonlinearity.

3.2 The Displacement Misfit

In Fig. 2a and b the screw and edge misfit across the slip plane are shown as a function of distance in the slip plane normal to the dislocation line. The edge misfit varies smoothly through the dislocation cores, with a functional form reminiscent of the Peierls-Orowan dislocation. The separation of the partials, i.e., the distance between the points at which the edge misfit is exactly one-half the edge component of the intrinsic fault vector, is clearly defined. The width of the core, defined as the extent of the region around the core in which the misfit is less than one-half of the asymptotic value, is $\sim 5b_p$ (b_p is the magnitude of the partial Burgers vector).

The screw misfit (Fig. 2a), in contrast, is not at all Peierls-like, and shows strong oscillations in the neighborhood of the dislocation cores. This is presumably because the screw misfit accommodation is not restricted to a single plane and serves to emphasize the doubtful application of the discrete planar-misfit dislocation models to screw dislocations.

4 ELASTIC INTERACTION OF PARTIAL DISLOCATIONS

4.1 Influence of Boundary Conditions

The effect of fixed boundary conditions can be understood by reference to the diagrams in Fig. 3, which show the misfit across the slip

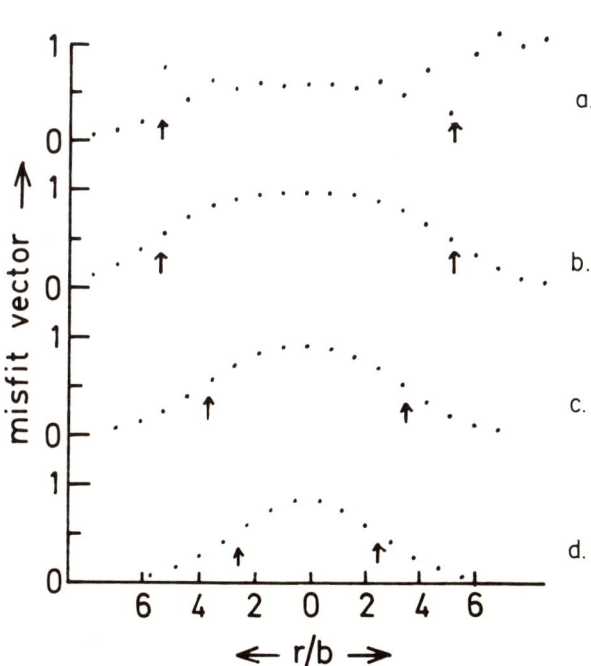

Fig. 2. The screw and edge misfit across the slip plane as a function of distance in the slip plane. The vertical ordinate is in units of the perfect misfit vector, i.e., $1/2a \langle 110 \rangle$ for the screw misfit, and $1/12a \langle 211 \rangle$ for the edge misfit; (a) The screw misfit for a separation of 10.5b; (b) The edge misfit for a separation of 10.5b; (c) The edge misfit for a separation of 7.6b; (d) The edge misfit for a separation of 5.0b.

plane. The vertical lines B, B' represent the fixed boundaries; the regions outside these boundaries are prevented from relaxing. Fig. 3a shows the misfit for a dislocation that has not yet moved, i.e., the boundary displacements for this dislocation are exact within the limits of linear elasticity. Fig. 3b shows the same dislocation after movement (solid line), with the original misfit field in broken lines. Two effects are apparent. First, the area swept by the dislocation is reduced by the presence of the boundaries, since the regions outside B, B' remain fixed. The second effect arises within the boundaries because the misfit discontinuities at the boundaries in the idealized model (Fig. 3b) will be smoothed out in a real calculation; this effect also will tend to reduce the area swept by the dislocation. For a virtual movement under a stress σ which would, in the absence of fixed boundaries, sweep an area ΔA, the work done is

$$\Delta W = \sigma b (\Delta A - \delta A) \quad ,$$

where δA is the reduction in swept area due to the boundary conditions. Thus the effect of fixed boundaries is to reduce the applied stress by an amount $\sigma \delta A/\Delta A$. It is clear from Fig. 3 that the effect will be small as long as the misfit gradient at the boundaries is small; for movement close to the boundaries, however, the retarding stress should increase sharply.

For the idealized case in Fig. 3, the retarding force can be estimated. The boundary force is just the gradient of the energy lying outside the fixed boundaries, which, for a screw dislocation in an isotropic elastic medium, reduces to a set of repulsive image dislocations.

4.2 The Partial Dislocation Separation

Fig. 4 shows the variation of the reciprocal partial dislocation separation, b/r, with the applied stress, in units of dynes/sq cm. The broken straight line indicates the variation predicted by linear anisotropic

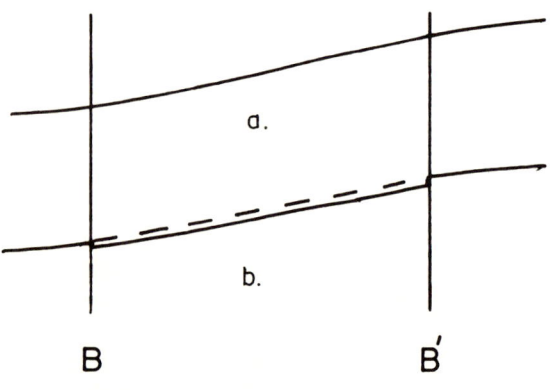

Fig. 3. Schematic diagram of the effects of fixed boundary conditions. The lines, B,B' represent the fixed boundaries; Line (a) The unperturbed displacement field; Line (b) The displacement field after a core translation. The original field is shown in dashed lines.

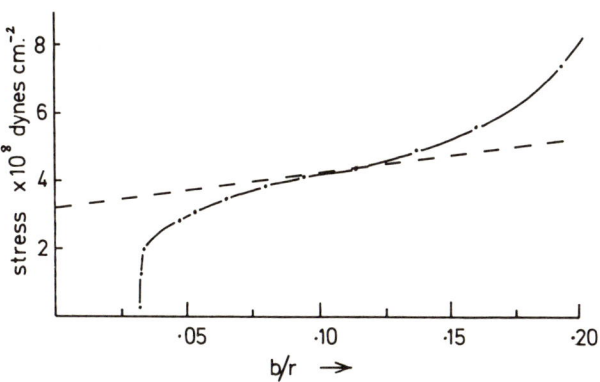

Fig. 4. The functional dependence of the reciprocal partial dislocation separation in units of b/2 on the applied stress, in units of dyne/sq cm. The dashed line shows the linear elastic predictions.

elasticity, and the solid curve shows the calculated variation for boundary conditions appropriate to a separation of $10.0b$. For separations $>10.5b$, the calculated values become increasingly smaller than the elastic result, until, at $\sim 30b$, the partials are no longer free to move apart. In the range $8b$-$10.5b$, the behavior is as predicted by linear elasticity. Thus it is likely that the discrepancy at larger separations arises from boundary effects. The approximate image correction discussed above can account for half of the difference out to a separation of $\sim 15b$, indicating that the smoothing of the misfit discontinuity at the boundaries also has a significant effect.

For separations $<8b$, the interaction shows strong nonlinear elastic effects, of such a sense that the calculated separation is larger than predicted by elasticity. This sense of nonlinearity is unexpected, and will be considered in detail later.

Unfortunately, it was not possible to constrict the partials closer than $5b$, due to the onset of a phase transformation to the hcp structure. It seems probable that this collapse of the lattice is due to the basic instability of the fcc model lattice to large deformation. Work is in progress to examine ways of suppressing the transformation, in order to allow extension of the calculations to the interesting region of complete constriction and cross-slip.

The sense of the nonlinearity in the present results, however, suggests that the stress required for constriction may be considerably larger than the elastic value. For example, at a separation of $5b$, the stress is already large enough to give an elastic separation of $2b$.

4.3 The Effect of Core Overlap on the Stress Field

Fig. 5 a and b show the screw and edge stress fields, respectively, of the core configuration for a partial separation of $5b$. The overlap of the core stresses is quite evident, and there is no stress-free stacking-fault region.

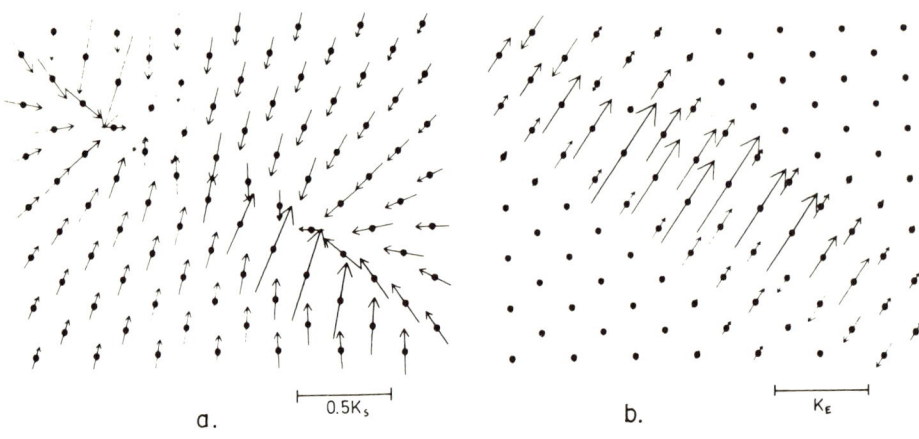

Fig. 5. The stress field of the partial dislocations in the nonlinear region, for a separation 5.0b; (a) The screw field; (b) The edge field.

Comparison with the linear core of Fig. 1a and b shows that the edge field is not greatly distorted. The screw field, on the other hand, is severely distorted; the shear planes in the partial cores are rotated, and the range of the stress field out of the slip plane is increased.

5 DISCUSSION

The results given above, representing a preliminary report, do not, at present, allow drawing any firm conclusions regarding the effects of nonlinearity on partial dislocation separation. Therefore, the discussion below concentrates more on the validity of the assumed model.

5.1 Nonlinear Effects in Dissociated Dislocations

The partial-dislocation-separation suggests a strong nonlinearity for small stacking-fault widths, of such a sense that the necessary stress to maintain a given partial separation is larger than predicted by linear elasticity. This sense of nonlinearity is unexpected and is in fact the reverse of that in the associated hcp lattice (BDT1). Thus it is important to consider the possible sources of this effect in more detail.

(i) As the partial separation decreases, the overlap of strain fields with the stacking fault will decrease the effective fault vector, and thereby the edge components of the partial dislocation Burgers vectors. The extent of this change can be seen from Fig. 2c and d, which show the edge-misfit maps for separations of $7.6b$ and $5.0b$, respectively; the maximum fault vector is reduced by 10 percent for Fig. 2c and by 15 percent for Fig. 2d, from the perfect intrinsic fault vector. Since the intrinsic fault is a metastable configuration, any small change in fault vector must increase the total fault energy, and will tend to reduce the partial separation below the elastic value. The associated change in the edge component of the Burgers vectors will reduce the attractive part of the partial interaction, while leaving the repulsive part (i.e., that due to the screw components) unchanged. The net force on the partials due to the applied stress is also changed, since the applied stress interacts only with the edge components of the Burgers vectors. The change in Burgers vector, tending to increase the elastic separation, thus acts in opposition to the stacking-fault-energy change.

We can estimate the magnitude of these effects from the expression for the elastic separation r,

$$r = (K_s b_s^2 - K_e b_e^2)/2\pi (\gamma + \sigma b_e)$$

where K_s, K_e are the screw and edge energy factors, and b_s, b_e are the screw and edge Burgers vector components. Using values for this fcc model lattice of $K_s = 4.0 \times 10^{10}$ dynes sq cm, $K_e = 2.0 \times 10^{10}$ dynes sq cm, and $\gamma = -3.5$ ergs sq cm, then for an elastic separation of $2b$, the effects due to fault-energy change and Burgers-vector change are, respectively,

$$\Delta r/b \sim -1.5 \Delta \gamma / \gamma$$

$$\Delta r/b \sim -8.5 \Delta b_e / b_e$$

For $\Delta b_e / b_e = -0.15$ (appropriate to the discrete core for a separation of $5.0b$) and for a stacking fault energy change of about 20 percent, the net effect gives an increase of the dislocation separation by about 40 percent. Thus this effect alone cannot explain the calculated nonlinearity.

(ii) The elastic interaction of the partials will deviate from the linear $1/r$ dependence for close separations, due both to overlap of the cores and to the finite range of interatomic forces. For very small separations, the interaction must be finite (i.e., smaller than the linear elastic interaction). As shown in BDT1, for a dissociated screw dislocation in the hcp lattice, the equilibrium separation of $3.5b$ can be obtained by relaxation either from the undissociated elastic solution, or from the elastically dissociated dislocation (with a separation of $7.1b$). Provided that the dislocation in the fcc phase behaves in a similar manner, the interaction energy for infinitesimal separation is positive (i.e., repulsive) and decreases monotonically with increasing separation.

Thus, for sufficiently small separations, the interaction force will be smaller than the elastic value, and the expected nonlinearity would show a continuous and monotonic decrease in the linear interaction. While the above arguments do not preclude the possibility of an enhanced interaction over intermediate ranges, there seems to be no physical reason for this to be so. It may be noted that differences in behavior are possible for a large Peierls stress. However, in the present case, the Peierls stress is likely to be of a magnitude similar to that in the model hcp lattice (i.e., less than 5×10^{-4} of the shear modulus (BDT2)), and will thus be negligible in comparison with the elastic forces.

We expect the elastic nonlinearity in the partial interaction to decrease the partial separation, and to have an effect opposite to the calculated nonlinearity.

(iii) The present calculation uses applied stress to stabilize the lattice and to modify the effective stacking-fault energy. Since the stresses are large (up to about $0.05 K_e$, where K_e is the edge energy factor), and are quite close to the theoretical shear strength of the perfect fcc lattice ($0.023 K_e$ (Basinski et al.[3])) for opposite shear in the sense of the fcc to hcp phase transformation, we should consider possible changes under stress of the elastic constants. Using the expressions in Ref. 3, the elastic constants in a state of homogeneous applied stress can be calculated. The calculation of the anisotropic energy factors from the conventional expressions is not possible, since the stressed material is noncubic. If we ignore the weak noncubic terms, and estimate the energy factors from the cubic elements only, then for the applied stress of 7×10^8 dynes cm^{-2}, both K_s and K_e are increased by about 50 percent. This effect will increase the elastic separation of $2b$ by 50 percent, i.e., in the sense of the calculated nonlinearity.

(iv) It is possible also that truncation of the effective ion-ion potential may affect the partial dislocation separation, particularly for a long-range cutoff, as in the present case. Since the surface density of neighbors increases as the radius of the sphere of interacting neighbors increases, steadily larger numbers of ions can leave and enter the interaction range under the influence of the partial strain fields. A sufficiently large change in the number of neighbors can change the effective energy of the stacking fault. In the present calculation, as in BDT1 and BDT2, the potential is cut off at 8.38 Å, at which point the energy is zero, and the derivatives are small. For the fcc structure, this cut-off corresponds to 1.57 lattice parameters, just inside the large (310) shell of neighbors, but well outside the smaller (220) shell. Thus perturbation of the lattice is likely to cause a net increase in the number of ions at the extreme limit of the interaction range. Since the potential is negative in this region, the effective stacking-fault energy will be decreased, and, thus, the partial separation will tend to be larger than the value for the unperturbed stacking fault. This effect will be more pronounced for larger perturbations of the ideal fault structure, i.e., at small partial separations.

The discussion above has considered both physical and numerical sources for nonlinearity in the apparent partial dislocation interaction. The effects of stress on the elastic constants and numerical effects due to potential cutoff can account for most of the nonlinearity, and thus it is impossible to ascribe any definite sense or magnitude to the elastic nonlinearity in dislocation interaction. It is clear that the isolation of the elastic effect requires a more-careful treatment of the other effects discussed above. In particular, it seems important to use an effective ion-ion potential with appropriate truncation that describes an fcc lattice which is stable to both small inhomogeneous deformations and to large homogeneous strains.

5.2 Boundary Effects

As shown in Section 4, the use of fixed boundary conditions has a strong influence on the partial dislocation interaction. For dislocation separations of ~1/2 the width of the discrete block, the boundary effects dominate. For smaller separations, the boundaries give rise to appreciable deviations from linearity in the partial separation − applied stress relation. It seems clear that for this particular calculation, careful treatment of the boundary conditions is mandatory.

ACKNOWLEDGMENTS

The authors are grateful to Mr. G. Champion for modifications to and maintenance of the computing equipment.

REFERENCES

1. Z. S. Basinski, M. S. Duesbery, A. Pogany, R. Taylor and Y. P. Varshni, 1970, Can. J. Phys., **48**, 1480.
2. Z. S. Basinski, M. S. Duesbery and R. Taylor, 1970, Phil. Mag., **21**, 1049.
3. Z. S. Basinski, M. S. Duesbery and R. Taylor, 1971, Can. J. Phys., in the press.
4. V. Vitek, R. C. Perrin and D. K. Bowen, 1970, Phil. Mag., **21**, 1049.

DISCUSSION on paper by Z. S. Basinski, M. S. Duesbery, and R. Taylor

BULLOUGH: It seems strange to use a potential to study defects in a face-centered lattice when that potential is inappropriate for such a lattice.

DUESBERY: The object of this paper has been to investigate the properties of dislocations in different structures using the same effective ion-ion potential, in an effort to eliminate the possible effects of using different force laws. The potential is consistent with the structural properties of sodium, which transforms at low temperatures from the bcc structure to a mixture of hcp and fcc phases.

However, our results indicate that calculations with a negative stacking-fault energy and a stabilizing applied stress hinders the analysis of the elastic nonlinearity in the partial dislocation interactions.

BULLOUGH: If you apply a shear stress to a dissociated dislocations both partials should glide in such a way that the whole dislocation moves. Your stress appears to drive the two partials together.

DUESBERY: Consider the dislocation referred to a Cartesian set of axes X_i, with the X_2 axis normal to the slip plane and the X_3 axis along the positive sense of the dislocation line. Then for a Shockley partial, the screw and edge components of the Burgers vectors are b_3 and b_1, respectively, and $b_2 = 0$. For an applied shear stress tensor with nonzero components σ_{12}, σ_{21} only, it can be verified from the Peach-Koehler[1] equation that the only force on the partial dislocation is $F_1 = \sigma_{21} b_1$.

For the two Shockley partials forming a dissociated screw dislocation, the edge components of the Burgers vectors are equal and opposite. Thus, from the above equation, there is no total force on the dislocation, but only an expansion or contraction of the partials, depending on the sign of the stress.

For an applied stress tensor with nonzero components σ_{23}, σ_{32} only, there is only a total force, of course, since the screw components of the Burgers vector, b_3, have the same sign in both partials.

HARRISON: In many simulations periodic boundary conditions are used. With dislocations it could seem more appropriate to use a sort of antiperiodic boundary conditions corresponding to dislocations of alternate signs. Has that been done?

DUESBERY: Periodic boundary conditions have been used in a lattice-statics calculation of the screw dislocation core structure by J. R. Hardy (Phil. Mag., in the press). Neither periodic nor antiperiodic boundary conditions are appropriate for calculation of most dislocation properties, however, since the leading term in the dislocation-dislocation interaction force varies inversely with the separation, and thus has too large a range.

REFERENCE

1. M. O. Peach and J. S. Koehler, Phys. Rev., **80**, 436 (1950).

THE MOTION OF SCREW DISLOCATIONS IN A MODEL B.C.C. SODIUM LATTICE

Z. S. Basinski, M. S. Duesbery and R. Taylor

Division of Physics
National Research Council of Canada, Ottawa 7, Canada

ABSTRACT

The behavior of the screw dislocation core in the presence of an external uniaxial stress of varying sense and orientation has been examined for a bcc model lattice, using an effective ion-ion potential for sodium developed from first principles. The Peierls stress is strongly orientation-dependent, and has a minimum value of $0.0076G$, where G is the shear modulus. The mechanism for dislocation movement can be planar or nonplanar, depending on the orientation of the applied stress, and can give rise to crystallographic slip on {110} or {112} planes or to noncrystallographic slip.

1 INTRODUCTION

1.1 General

It has long been recognized that the use of continuum elasticity in the description of properties of dislocated lattice is justified only for long-range interactions that are insensitive to the discrete structure. Because of this, theories of the microscopic behavior of isolated dislocations have relied largely on induction from experiment, supported by the predictions of the quasi-discrete Peierls-Nabarro & Frenkel-Kontorova dislocation models.[1]

In recent years, with the aid of high-speed computing devices, it has become possible to treat the microscopic properties of lattice defects from a more-basic point of view, and to explain defect properties from a more-reliable theoretical basis.

One problem of particular interest, which has received much attention in the literature and to which this paper is devoted, is the microscopic structure and behavior of the screw dislocation core in bcc metals. The belief is widespread that the properties of the bcc screw dislocation are responsible for the large Peierls stress in bcc metals and for the unusual slip geometry of bcc metals (for a review of these topics, see Ref. 2), but the application of continuum methods fails to give a definitive answer. The early calculations of the discrete core structure in the bcc lattice (Chang[3]; Bullough and Perrin[4]; Gehlen, Hahn, and Rosenfield[5]; Vitek, Perrin, and Bowen[6] (referred to hereafter as VPB); Basinski, Duesbery, and Taylor[7] (referred to hereafter as BDT1)), while differing in method of interpretation and conclusions, agree in principle on the nonplanar nature of the misfit accommodation in the core region. It was shown further that this nonplanarity is absent for extended screw dislocations in hcp (BDT1) and fcc structures (Basinski, Duesbery, Taylor[8] to be referred to as BDT2); thus it appears that the nonplanar screw core in bcc metals may be a property peculiar to the bcc lattice.

This nonplanar core structure is somewhat reminiscent of the threefold dissociation model of bcc screw dislocations advanced by Hirsch[9], whose suggestion that such a nonplanar split could be responsible, on a continuum model, for a large Peierls stress was confirmed by later workers (see Christian and Vitek[2]). While the discrete core cannot be considered to be dissociated in the continuum sense, the behavior is in some ways analogous to that of the Hirsch model. It was shown by Basinski et al.[10] (to be referred to as BDT3), using an effective ion-ion potential valid for both bcc and hcp sodium, that the Peierls stress for the screw dislocation in this model bcc lattice is at least 25 times greater than in the hcp lattice. BDT3 considered also the motion of screw dislocations in the bcc lattice, and found that movement takes place by means of successive slip steps on

{110} planes, with the selection rule that no two consecutive steps can occur on the same {110} plane. The work in BDT3, however, applies only to the case of an applied pure shear stress. To provide a more direct comparison with experimental results, and to give a more complete picture of core behavior under stress, the work described here repeats much of the work in BDT3, but for an applied uniaxial stress.

1.2 The Dislocation Core in a Model bcc Sodium Lattice

Those parts of BDT1,2,3 that have a direct bearing on the current work are reviewed briefly. The method of calculation is as used and described in BDT2; the matrix of ions used is periodic along the dislocation line, with lateral dimensions 45 x 45b, where b is the magnitude of the Burgers vector. Lateral boundaries are fixed at positions dictated by anisotropic elasticity; this approximation has been considered in detail in BDT2 and BDT3. The ions are assumed to interact via a two-body effective ion-ion potential calculated from first principles; the derivation of and justification for the form of the interaction are discussed in detail in Ref. 11.

The core is represented by means of the screw stress field (for derivation, see BDT2 and BDT3); thus, for example, in Fig. 1, the lines have magnitude proportional to the shear stress on the ion acting in the direction of the Burgers vector, and are drawn parallel with the local plane of shear (see BDT2 and BDT3). All diagrams are drawn on a [111] lattice projection. For clarity, the ion coordinates in the [111] direction are omitted; for reference purposes, the stacking order is shown in the bottom left corner of Fig. 1a-1d, in units of the $<111>$ repeat distance. In all cases, the sense of the dislocation line is [111], and the Burgers vector is $a/2[111]$, defined by the FS/RH convention.

Fig. 1a shows the screw stress field of the anisotropic elastic screw dislocation core. It should be noted that this elastic solution has both the $<111>$ three-fold and $<101>$ two-fold symmetry of the lattice. Fig. 1b shows the relaxed core configuration in the absence of applied stress, obtained in BDT1, BDT3. The relaxed core has [111] triad symmetry, but is not invariant to the [10$\bar{1}$] diad; therefore the configuration of Fig. 1c, which is obtained from Fig. 1b by a diad rotation about [101], exists also (VPB, BDT3). In BDT3, the cores of Fig. 1b and c, were called S1, S2, respectively; for the present work, it is appropriate to relabel the cores as R and L, respectively, denoting the right and left-handed senses of rotation of the stresses on the ion rows immediately adjacent to the core, looking along the positive [111] sense of the dislocation line. For the purposes of this paper, the notation is unique.

The cores R and L are equivalent, stable configurations. In the bcc lattice, however, a second, metastable, dislocation site is possible; in this

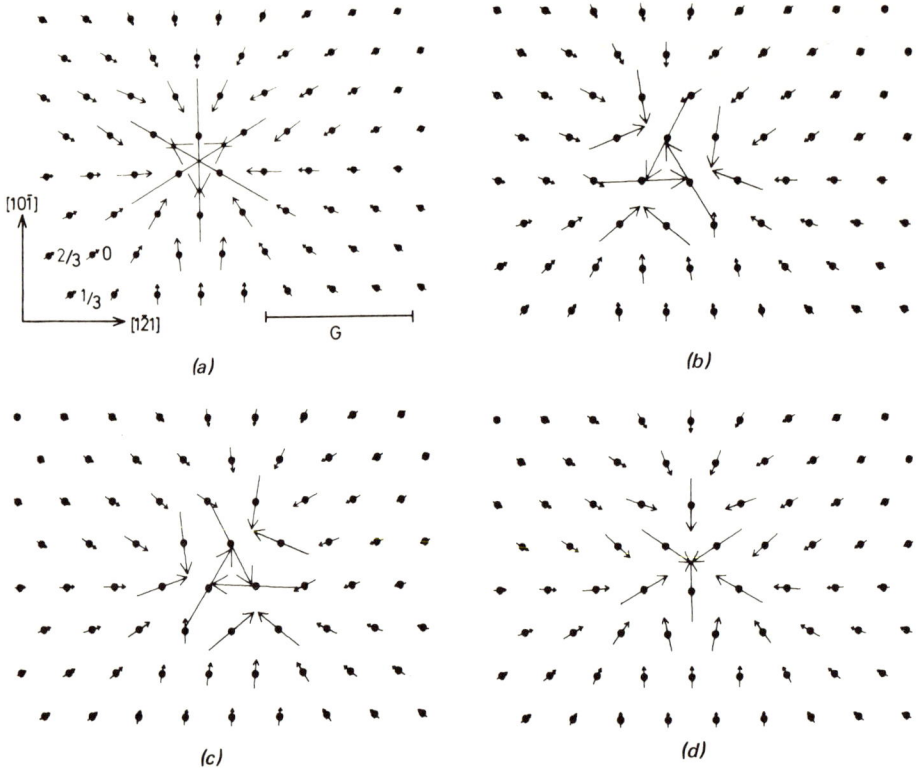

Fig. 1. The shear stress field of the discrete screw dislocation core in a bcc lattice. (a) The anisotropic elastic non-equilibrium core; (b) The stable relaxed R-type core; (c) The stable relaxed L-type core; (d) The metastable relaxed M-type core.

core the spiral stacking sequence of the lattice at the dislocation core is destroyed by screw dislocation. Fig. 1d shows the metastable core calculated in BDT1; this core is referred to as the M-core. The M-core has the same six-fold symmetry as the anisotropic dislocation, and thus exists in only one configuration. Note that the M-core is only weakly metastable; it collapses to one or other of the stable configurations at a stress ~1/6 of the minimum Peierls stress for the stable dislocation (BDT3).

2 A SIMPLIFIED MODEL FOR THE SCREW DISLOCATION CORE

This section introduces a simplified representation for the screw dislocation core structure, which will be helpful in dealing with the complex core configurations that can develop in the presence of an external stress. It must be emphasized that the representation is not intended as a formal model, but only as a device to simplify visualization of the core behavior.

2.1 Representation of the Stress-Free Core

Consider the elastic dislocation core of Fig. 1a. The local planes of shear on the three nearest-neighbor ion rows intersect in the dislocation site, and are directed uniformly inwards. This indicates the presence of a shear stress source, and suggests a simple representation for the discrete core fields. The screw stresses, as pointed out in BDT3, conform for vector addition, and subject to the approximation of linear superposition, we can reconstruct the essential features of the dislocation core by a linear combination of "shear stress sources" of varying strength. The diagram of Fig. 2a shows the shear stress source arbitrarily defined to have a magnitude of 2; a stress source of strength n is represented by a triangle with an inscribed figure "n". In this way the salient features of the discrete cores R and L can be represented as shown in Fig. 2b. It can be verified easily that the representations of Fig. 2b reproduce quite closely the core configurations of Fig. 1 out to the third-nearest-neighbor ion rows; beyond this range the stress field differs little from the elastic solution.

We can pursue the representation further to estimate the relative ease of possible dislocation movements. The external work necessary for any core translation can be expressed as a function of the changes in the internal stress field and the interionic force constants. In the absence of a formal treatment, the dominant term is assumed to be quadratic in the stress changes. Thus a rough estimate of the ease of dislocation movement can be obtained by considering the sum of the squares of the changes in strength of the composite shear stress sources.

2.2 Motion of the Screw Dislocation Core in Pure Shear

The selection rules for dislocation motion in pure shear obtained in BDT3 are now discussed. Consider first the hypothetical case of yield under zero stress. It was shown in BDT3 that the basic translation vector for screw dislocations in the model bcc sodium lattice is $a/3<211>$; i.e., the repeat distance on $\{110\}$ planes. Since there are two equivalent core configurations, R and L, there are four possible translation modes, i.e., the translation $+a/3[1\bar{2}1]$ in the notation of Fig. 1 can take place from and to either the L or R core configurations. In subsequent discussion, a translation of the dislocation core from an initial core configuration N1 to a final core N2 is denoted by N1/N2. Thus the four possible $a/3[1\bar{2}1]$

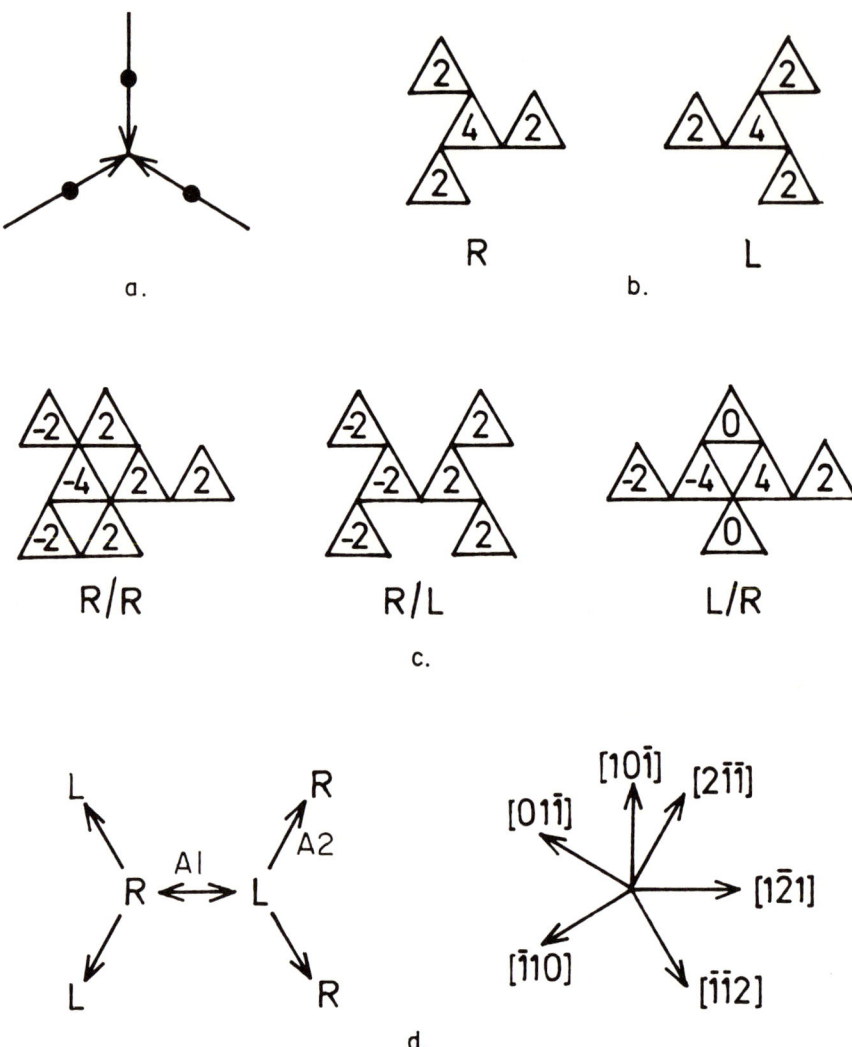

Fig. 2. (a) The discrete shear stress source of strength 2; (b) Stress-source representation of the stable cores R and L; (c) Stress-source representation of core movement; (d) The basic slip mechanism of the stress-free core.

translations are R/R, R/L, L/R and L/L; the cases R/R and L/L are obviously degenerate. The diagram of Fig. 2c illustrates the source-strength changes necessary for the three movements R/R, R/L, and L/R; it may be verified that the sum-square changes are 40, 24, and 40 respectively. Thus, on this simplified model, the $+a/3[1\bar{2}1]$ translation is expected to take place most easily by a R/L movement; i.e., a unit translation on the $(10\bar{1})$ plane, together with a conversion of the R-type core to the L configuration. It can easily be verified that the most likely step for the L core on this model involves a reconversion to the R-type core, and that the directions of these likely steps are opposite to those for movement of the R core. To avoid confusion, the most likely steps for the L core are referred to as R/L movements, to emphasize the similarity with the R core. A further $+a/3[1\bar{2}1]$ translation now requires a L/R-type movement, which by the above arguments is expected to be a relatively more difficult step; however, the L-type core can move by a R/L-type step with a translation vector of $a/3[\bar{1}\bar{1}2]$ or $a/3[2\bar{1}\bar{1}]$, i.e. of either of the {110} planes nonparallel with the initial $(10\bar{1})$ slip plane. These arguments have been developed in more detail in BDT3, in which the R/L type movement is called a *preferred* step, while the L/R and R/R-type translations are referred to as *unlikely* steps. On the assumption that only preferred movements can take place, the fundamental slip geometry of the L and R-type screw cores is as shown in Fig. 2d, in which the double-headed arrows indicate the possible reversible preferred translations. These arguments apply only in the limit of vanishing applied stress. For example, if the initial preferred translation is caused by a driving pure shear stress on the $(10\bar{1})$ plane, then the stress acting in the nonparallel {110} planes is only half as much. Thus the translation vector for the second slip step above will be determined by whether the preferred slip step (out of the $(10\bar{1})$ plane) can take place at less than half the stress needed for the unlikely slip step (continued movement in the $(10\bar{1})$ plane). For the model sodium lattice, as shown by the calculations of BDT3, movement for an applied pure shear stress in the direction of the Burgers vector takes place exclusively by the preferred translations. In general, different stresses will be necessary for the two steps, because of the nonplanarity, and because the cores L and R will not be degenerate in energy for nonzero applied stresses. Following BDT3, we define the smallest stress needed to cause a repeat translation of the core as the *first critical shear stress*; the accompanying movement, which involves a change in core configuration, is defined as an *equivalence translation*. For example, the R/L movement for a translation vector $+a/3[1\bar{2}1]$ (labelled A1 in Fig. 2d) is an equivalence translation, since the final core state is not the same as the starting configuration.

It can be seen that two consecutive equivalence translations (e.g., A1 and A2 in Fig. 2d) reproduce the initial core configuration exactly. The resultant path defines an *identity translation*, and the stress required for this step, which is the stress required for large-scale dislocation motion, is referred to as the *second critical shear stress*.

An important consequence of this intrinsically nonplanar dislocation movement is that the concept of cross-slip loses its meaning. Each identity translation is an independent slip event. In the presence of an applied stress, the activation energies for the possible identity translations will, in general, be different, and the microscopic slip surface will be traced by a statistical combination of identity translations, taking into account the influence of thermal fluctuations and internal stresses. Thus the wavy slip lines and noncrystallographic slip planes observed frequently in the deformation of bcc metals are easily explained on this model. The occurrence of crystallographic slip on $\{112\}$ planes (i.e., the plane of the identity translation) is possible in cases where the activation energy for one particular identity translation is much smaller than any others, or at very low temperatures. Crystallographic slip on $\{110\}$ planes, however, would require the cooperation of two different identity translations, and is therefore unlikely, since even when the activation energies are the same, the independence of the slip events will result in waviness of the slip surface.

BDT3 extended the above arguments to include the relationship between the activation energies of the possible preferred translations in the presence of a general applied pure shear stress and showed that an orientation dependence of the critical shear stresses is expected. In particular, an asymmetry for a shear stress on $\{112\}$ planes is predicted. Direct calculation showed this asymmetry to be of the sense observed experimentally, i.e., such that the stress required for slip in the twinning sense is less than that for slip in the antitwinning sense.

3 MOTION OF THE SCREW DISLOCATION CORE IN UNIAXIAL STRESS

This section considers motion of a screw dislocation under the influence of an external uniaxial tension and compression of varying orientation. The core configurations in uniaxial stress are discussed, then the slip geometry and yield stress are treated.

3.1 The Core Configurations in Uniaxial Stress

For an applied pure shear stress, the core configuration in the stable equilibrium sites can be distorted for stresses close to the second critical shear stress, but is always recognizably of L or R type. For a uniaxial stress, however, distinctly different core configurations can be stabilized by tensile stresses close to <100>. The form of these configurations, and their relationship to the basic L and R cores, can be seen most easily by analysis in terms of the stress-source model of Section 2. Fig. 3a and b show the additional core configurations stable under uniaxial stress, together with the stress-source representations; as for the L and R cores, the stress-source method gives a fair description of the essential features of the cores.

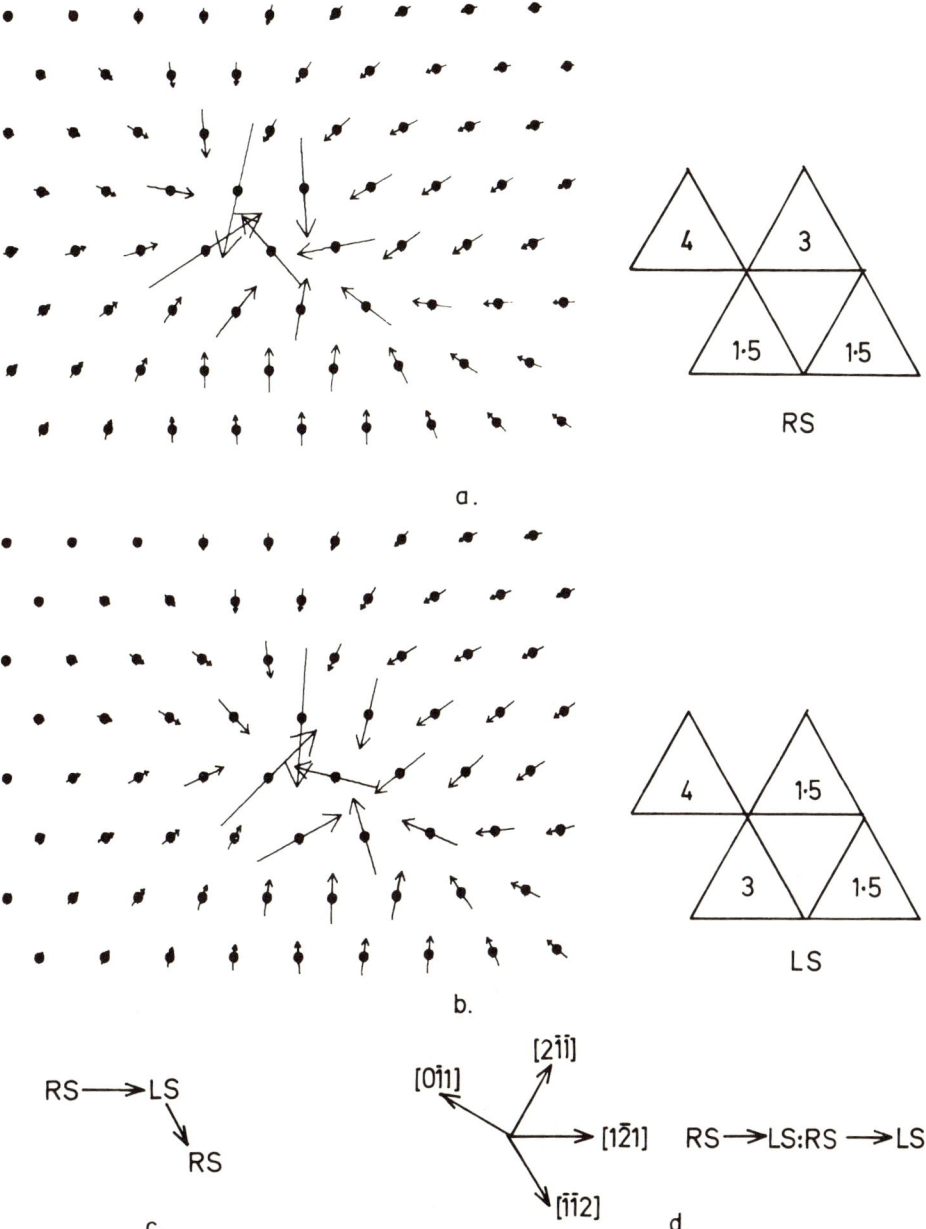

Fig. 3. The core configuration and slip geometry of the stress-stabilized cores RS and LS. (a) The stress field and source representation of the RS-type core; (b) The stress field and source representation of the LS-type core; (c) The non-planar slip mechanism of the cores, RS, LS; (d) The planar slip mechanism of the cores RS, LS.

Using the arguments of Section 2, we can investigate the possible slip geometry of the LS, RS cores of Fig. 3a and b. For an equivalence translation with vector $+a/3[1\bar{2}1]$, the sum of the squared stress changes for RS/RS, RS/LS, LS/RS and LS/LS, respectively, are 30.5, 21.5, 42.5, and 38.0. On this model we expect a slip geometry as shown in Fig. 3d. The RS core has only one stress-aided preferred movement, with a translation vector $a/3[1\bar{2}1]$. The resulting LS-type core has three possible stress-aided movements, in the directions $[2\bar{1}\bar{1}]$, $[1\bar{2}1]$ and $[\bar{1}\bar{1}2]$, but only one of these (the $[\bar{1}\bar{1}2]$ direction) is a preferred translation. Thus, the only identity translation is a [011], and the only possible slip mode is crystallographic. To allow for the possibility that the core configuration may change without translation an operation, such as between cores N1 and N2, is represented by the expression N1:N2 (note that the operation is symmetric N2:N1=N1:N2).

For the four cores R, L, RS, and LS, there are six distinct operations, for which the sums of the squared stress changes are shown in Table I. For the R:L core change, the squared stress change is the same as for R/L; however, in the presence of an applied stress, the latter will be preferred, since the accompanying translation absorbs more work. The RS:LS change on the other hand, requires a very small stress change and may occur in preference to the RS/LS translation. This suggests the possibility of the alternative slip geometry shown in Fig. 3d, in which the initial RS/LS equivalence translation with vector $+a/3[1\bar{2}1]$, and the only possible slip mode for this geometry is crystallographic, on the $(10\bar{1})$ plane.

Thus the possible geometries of Fig. 3d can give rise to crystallographic slip on {112} planes if the activation energy for the RS/LS equivalence translation is much smaller than that for the LS:RS core change, and on {110} planes in the opposite limit. For comparable energies, a noncrystallographic mixture of the mechanisms is expected.

3.2 The Slip Geometry in Uniaxial Stress

Fig. 4 indicates the orientations of uniaxial tension and compression, which have been treated. For all orientations, in compression and for most of the reference stereographic triangle in tension, the only stable dislocation cores are of R and L type. Movement takes place, as in the case of pure shear, by the geometry of Fig. 2d (for greater detail, see BDT3).

Table I. The Sum of the Squared Stress Changes for Core Configuration Change Without Translation

	R	L	RS	LS
R	*	24	13.5	19.5
L	24	*	19.5	13.5
RS	13.5	19.5	*	4.5
LS	19.5	13.5	4.5	*

Fig. 4. Stereographic projection of uniaxial stress orientations. (a) Tension; (b) Compression.

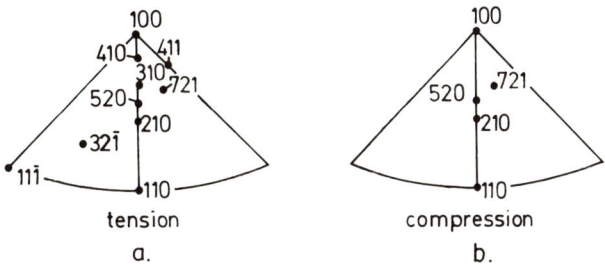

For orientations close to [100], for tensile stresses only, the cores RS, LS are stable under stress. For the [100], [410], [310], and [520] orientations, dislocation movement is by the mechanism of Fig. 3d, with identity translation vector $a/3[1\bar{2}1]$. For the [411] orientation, the core moves by the mechanism of Fig. 3c, with identity translation vector $a[01\bar{1}]$. A mixture of the two mechanisms operates for the [721] orientation, for which case the core begins moving with the mechanism of Fig. 3c, but then switches to that of Fig. 3d. Since both these mechanisms operate in different orientation ranges, it seems likely that the activation energies are comparable, so that long-range crystallographic slip either on {110} or {112} planes would be unlikely at finite temperatures. It is possible, however, that for other metals, with a similar screw dislocation core structure, crystallographic slip may be preferred.

3.3 The Critical Shear Stresses in Uniaxial Stress

This section considers the influence of orientation and sense of stress on the second critical shear stress. As noted by BDT3, in the absence of effects due to zero-point energy of the ions and to quantum-mechanical tunnelling, the second critical shear stress can be identified with the Peierls stress at $0°K$; the present work considers only the second critical shear stress. Resolution of the stresses in accordance with the Schmid Law must be made on the operative equivalence translation plane which bears the least stress (BDT3), for the slip geometry of Fig. 2d. The same resolution rules apply for the nonplanar geometry of Fig. 3c; for the planar geometry of Fig. 3d however, the equivalence translation plane is unique, and resolution must be made on the {110} slip plane.

Table II summarizes the second critical shear stresses, resolved according to the above convention, for several orientations of tension and compression. The nature of the slip geometry is shown also; the symbols N and X (Table II) refer to the nonplanar (Fig. 2d, 3c), and planar (Fig. 3c) geometries, respectively, with the expression NX denoting a mixture of the two.

The yield stresses are in units of $.01G$, where G is the shear modulus in the $\langle 111 \rangle$ direction, and are uncertain to about $\pm.001G$ in this

Table II. The Second Critical Shear Stress in Units of 0.01G

Axis	Compression	Slip Mode	Tension	Slip Mode
100	1.06	N	1.23	X
411	*	*	1.20	N
721	1.23	N	1.50	NX
520	1.40	N	1.48	X
210	0.93	N	0.93	N
110	0.90	N	0.90	N
111	*	*	0.76	N
321	*	*	1.82	N

preliminary work; an asterisk in Table II indicates that the result is not available.

It may be seen from Table II that, for orientations in which the slip geometry is the same in tension and compression (i.e., [210] and [110]), there is no detectable sense-dependence of the yield stress (it should be noted, however, that the uncertainty limits are coarse, and would not detect, for example, the 10 percent asymmetry on {112} planes found by BDT3). When the slip geometry is different (i.e., for orientations close to [100]), there is a difference between tensile and compressive yield stresses for the same orientation. The magnitude of the difference varies with orientation, reaching a maximum of at [100] of about 20 percent of the smaller, compressive yield stress.

The orientation dependence for a specified sense of stress has been studied in more detail for tensile stresses. The tensile yield stress varies by a factor of two over the range of orientations considered, and thus this source of plastic anisotropy is much larger than those due either to asymmetry effects (BDT3) or to sense-dependent effects.

The minimum yield stress occurs for the [111] orientation, and has a magnitude of $0.0076G$, which is significantly smaller than the minimum Peierls stress in pure shear of $0.0105G$ (BDT3). The largest yield stress of $0.0154G$ occurs for an orientation [321], in the central region of the standard stereographic triangle; at the corners of even symmetry, the yield stress drops to $0.0123G$ and $0.009G$ for the [100] and [110] axes, respectively.

The range of values available in compression is too limited to permit a full comparison with the orientation dependence in tension. The results in Table II, however, suggest that the yield stress differential between the [100] and [110] orientations is smaller than in tension.

The results above show that the Peierls stress for screw dislocations in the model sodium lattice is sensitively dependent on the nature and orientation of the applied stress, and suggest that the Schmid law of constant critical shear stresses may be inappropriate for bcc metals in general. Unfortunately, it is not yet possible to determine the exact nature of the interaction between the dislocation core and the applied stress, but it is hoped that continued work may clarify the situation.

4 DISCUSSION

In this section are discussed the more important deductions that can be made from the preliminary results presented in Section 3, and from earlier calculations made on the same model dislocation core (BDT1, BDT3). As mentioned in Section 1, the fundamental effective ion-ion potential used gives good agreement with the available experimental measurements on the bulk and microscopic properties of sodium. Thus, within the limits of the calculations, it is possible that the results reported above are indicative of the properties of real alkali metals. A wider application of the results to the bcc structure in general is more questionable. It seems probable that some properties of the core, such as the nonplanarity of shear accommodation in the core region, and the order of magnitude of the Peierls stress, may be more sensitive to the lattice structure than to the effective ion-ion potential. However, the exact magnitude of the Peierls stress and the details of the core structure (and hence the possible slip mechanisms) may be sensitive to the assumed potential. Clearly it is necessary to investigate the effect of the assumed potential before any assessment of the general applicability of a single set of results can be made.

4.1 The Peierls Stress for Screw Dislocations

It was shown by Basinski et al.[12], using the same sodium potential, that the theoretical shear strength of a perfect bcc model lattice in uniaxial stress never exceeds that in pure shear, but may be up to a factor of two smaller. The results of Section 3.3 show a similar phenomenon for the Peierls stress. The calculated Peierls stress is smallest for the <111> orientation, and is about 25 percent smaller than the minimum value in pure shear. The calculated minimum of $0.0076G$, however, is still a factor of two higher than suggested by experimental measurements on alkali metals (see BDT3) and is more in agreement with the values obtained for bcc transition metals. However, as pointed out in BDT3, the calculated Peierls stresses are upper limits, and so the discrepancy may not be too serious.

As mentioned above, the applicability of the results for sodium to all bcc metals, even in a qualitative manner, is doubtful. Thus the calculated orientation dependence should not be compared critically with experimental measurements in different metals. The available experimental data on the orientation dependence of the yield stress of bcc metals cover the behavior of bcc transition metals only. The details of the dependence differ from metal to metal, but some inequalities appear to be valid for all transition metals (Duesbery[13]). For example, the yield stress in tension for a <100> orientation is always larger than the yield stress in compression for a <100> axis. The reverse is true for the present bcc model lattice.

However, the presence of a large orientation dependence in the model sodium lattice indicates the feasibility of an orientation dependence in more complex metals.

The results of Section 3.3 suggest that there are at least two sources of orientation dependence. The first, having a character similar to the dependence in pure shear (BDT3), seems to arise from direct interaction between the applied stress and the core, without any great distortion of the core structure; for this effect there is little or no sense-dependence of the Peierls stress for the same orientation (but note that the present calculations are too coarse to detect the asymmetry on {112}). A second source comes from the stabilization under stress of new dislocation core structures and slip geometry for orientations close to <100> and for tensile stresses only; here, the primary effect is a difference between yield stresses in tension and compression for the same orientation.

4.2 The Slip Geometry

One of the more interesting consequences of the application of uniaxial stress to the dislocation core is the orientation dependence of the basic slip geometry, due to the formation of stress-stabilized core structures. Earlier theories of the slip geometry in bcc metals (Vitek and Kroupa[14], Vitek[15,16], Duesbery and Hirsch[17], Duesbery[13]) enjoyed fair success in explaining the observed occurrence of wavy slip lines and noncrystallographic slip planes in bcc metals. To explain the observation in certain cases of crystallographic slip on {110} planes, however, it was necessary to postulate a different slip mechanism (Vitek and Kroupa[14]). The results of Sections 3.1 and 3.2 show that such a change in mechanism is indeed possible for the bcc screw dislocation core, and that the mechanism change can be caused simply by a suitable orientation of the applied stress.

It seems unlikely that the present study has uncovered all the possible slip mechanisms of the screw dislocation in the bcc lattice. In view of the consequences of these abnormal slip mechanisms for interpretation of experimental results (e.g., the correct resolution of stresses), it will be of interest to examine the effects of different potentials and of different forms of applied stress.

ACKNOWLEDGMENTS

The authors wish to acknowledge the assistance of Mr. G. Champion in the implementation of necessary hardware modifications to the computing equipment.

REFERENCES

1. *Dislocation Dynamics,* e.g., 1968, eds, A. R. Rosenfield, G. T. Hahn, A. L. Bement and R. I. Jaffee (McGraw-Hill).
2. J. W. Christian and V. Vitek, 1971.
3. R. Chang, 1967, Phil. Mag., **19**, 501.
4. R. Bullough and R. C. Perrin, 1968, *Dislocation Dynamics* (McGraw-Hill).
5. P. C. Gehlen, G. T. Hahn and A. R. Rosenfield, 1969, Conference on Fundamental Aspects of Dislocation Theory, (N.B.S., Washington, D.C.).
6. V. Vitek, R. C. Perrin and D. K. Bowen, 1970, Phil. Mag., **21**, 1049.
7. Z. S. Basinski, M. S. Duesbery and R. Taylor, 1970, Phil. Mag., **21**, 1201.
8. Z. S. Basinski, M. S. Duesbery and R. Taylor, this conference.
9. P. B. Hirsch, 1960, Fifth International Congress of Crystallography, Cambridge, U.K.
10. Z. S. Basinski, M. S. Duesbery and R. Taylor, 1971, Can. J. Phys., in the press.
11. Z. S. Basinski, M. S. Duesbery, A. Pogany, R. Taylor and Y. P. Varshni, 1970, Can. J. Phys., **48**, 1480.
12. Z. S. Basinski, M. S. Duesbery and R. Taylor, 1970, Second International Conference on the Strength of Metals and Alloys (Am. Soc. for Metals).
13. M. S. Duesbery, 1969, Phil. Mag., **19**, 501.
14. V. Vitek and F. Kroupa, 1966, Phys. Stat. Sol., **18**, 703.
15. V. Vitek, 1967, Phys. Stat. Sol., **18**, 687.
16. V. Vitek, 1967, Phys. Stat. Sol., **22**, 453.
17. M. S. Duesbery and P. B. Hirsch, 1968, *Dislocation Dynamics* (McGraw-Hill).

DISCUSSION on the paper by Z. S. Basinski, M. S. Duesbery and R. Taylor

VITEK: I would like to point out that the methods used for depicting the core structure are very essential for understanding the results. The most important requirement of these methods is to show the localization of the core. This is very well satisfied by the method of local stresses used by the authors. Another method that possesses the same quality suggested by Vitek et al.[1] uses the relative displacements of the atoms due to the dislocation. Each of these methods provides, however, slightly different information and they should be used to complement each other.

VINEYARD: I surmise that your method of calculating corresponds, in finite differences approximation, to solving the equations of motion for a set of atoms with zero masses, acted on by the forces of your model plus viscous damping. Is this indeed what you do?

DUESBERY: We have used the method of successive overcorrections to solve the system of nonlinear simultaneous equations defining equilibrium of an assembly of interacting ions. Essentially, this involves displacing each ion in turn through a distance proportional to the force, the proportionality constant being the overcorrection factor.

The analogue in the dynamic method would indeed be the assumption of zero mass and an infinitely viscous medium, but with the ratio of the time step to the damping constant remaining finite. This ratio would then correspond to the overcorrection factor.

SEEGER: I have two questions: (1) If you put an increasing stress on your crystal until the screw dislocation moves, what mechanism prevents it from slipping repeatedly? (2) What mechanism determines the time scale of the movement from one stable position to the next?

DUESBERY: (1) None; (2) The time scale has no real meaning in a static calculation. The total number of iterative steps needed depends upon the overcorrection factor.

GEHLEN: You said that during relaxation (while the dislocation is moving) you calculate the force acting on each atom, then displace the atom accordingly "as far as you can get away with". It seems to me that on the contrary this displacement should be as small as possible since you want to avoid to accidentally skip over metastable positions. Another way of looking at what you are doing is to say that by increasing the displacement you are effectively raising the temperature of your crystal.

DUESBERY: The concept temperature is inappropriate to a static calculation. The overcorrection factor defines the rate of convergence only. The incremental applied stress determines the difference between initial and final solutions. For large stress increments when there is a large difference between initial and final states, it is possible to miss metastable solutions; by the use of sufficiently small increments, these numerical problems can be avoided. We do in fact observe this effect. For example, the minimum shear stress required for an equivalence translation is between $0.008G$ and $0.009G$, provided that stress increments of $\leqslant 0.001G$ are used. For a single stress increment of $0.007G$, the dislocation does not move, but for an increment of $0.008G$, movement takes place; in this way the use of excessively large stress increments causes an underestimate of the Peierls stress, as suggested by Dr. Gehlen.

REFERENCE

1. See Ref. 6.

ATOMISTIC CALCULATION OF PEIERLS-NABARRO STRESS IN A PLANAR SQUARE LATTICE

W. R. Tyson

Trent University, Peterborough, Canada

ABSTRACT

Fully atomistic calculations for an edge dislocation have been performed using an empirical pairwise potential bonding atoms in a planar square array. The lattice friction stress for motion of a <10> dislocation is much larger than the Peierls-Nabarro result and consistent with more recent theoretical approximations.

1 INTRODUCTION

A detailed knowledge of the structure and energy of the highly distorted region at the dislocation core is vital for a proper quantitative understanding of lattice friction stress, dislocation-solute interactions, cross-slip processes, and a host of other mechanisms thought to control the flow stress of crystals. The first attempt to calculate atomic configuration in this region was due to Peierls[1], and the resulting "Peierls-Nabarro model" of the dislocation core has received widespread attention. As well as providing a realistic picture of the core structure, the model leads[2] to a lattice friction stress

$$\tau_p = \frac{2\mu}{1-\nu} \exp\frac{-4\pi}{b} \zeta \quad , \tag{1}$$

where μ and ν are the shear modulus and Poisson's ration, b is the Burgers vector, and $2\zeta = d/(1-\nu)$ is the dislocation width where d is the spacing between slip planes.

Eq. 1 predicts values of the friction stress $\tau_p \sim 10^{-4} \mu$. Working within the framework of the same model, Foreman et al.[3] have shown that wider cores and much lower values of τ_p/μ result if more realistic relations are used for the variation of the force acting between slip planes as they are sheared relative to each other. More recently, the method of calculating τ_p has been severely criticized, and Kuhlmann-Wilsdorf[4] has argued that Eq. 1 underestimates τ_p by several orders of magnitude. Moreover, the Peierls-Nabarro model predicts that the two symmetric configurations for a <100> dislocation in a simple cubic lattice are unstable and of equal energy which means that the dislocation energy has a periodicity of one-half the lattice spacing. These features and the various modifications of the model have been reviewed by Nabarro[5] and Hirth and Lothe[6].

The unexpected results of the Peierls-Nabarro model may be due to the fact that it is semiatomistic only, in that the lattice on both sides of the slip plane is approximated by semi-infinite continua and that no account is taken of the position of other than first-neighbor rows of atoms across the slip plane. In this paper, the results of a truly atomistic calculation using a phenomenological interatomic potential to describe an edge dislocation in a two-dimensional lattice are compared with the Peierls-Nabarro model.

2 CHOICE OF POTENTIALS

As the elastic displacement field of a <100> edge dislocation in a simple cubic lattice is two-dimensional, all atoms lying in a plane perpendicular to the dislocation line remain in that plane. Hence, all essential features of the dislocation may be displayed in a two-dimensional lattice — the structure chosen for study in this work. Besides saving a considerable amount of computer time, this selection allowed atomic configurations to be easily visualized.

To further simplify the treatment, an interatomic potential was chosen so that the lattice would be elastically isotropic; this imposed conditions on the slope and curvature of the potential. The shape of the potential was chosen so that a square lattice would have the same cohesive energy as a close-packed (hexagonal) lattice, which has a considerably smaller atomic volume at equilibrium. No volume-dependent term was considered. The potential is shown as Curve A in Fig. 1. Some calculations were also performed with potential B, which displays oscillations as would be expected for a "pseudopotential"; potential B produces a markedly anisotropic square lattice. Both potentials give very nearly sinusoidal force-displacement relations for shear of a perfect crystal (Fig. 2).

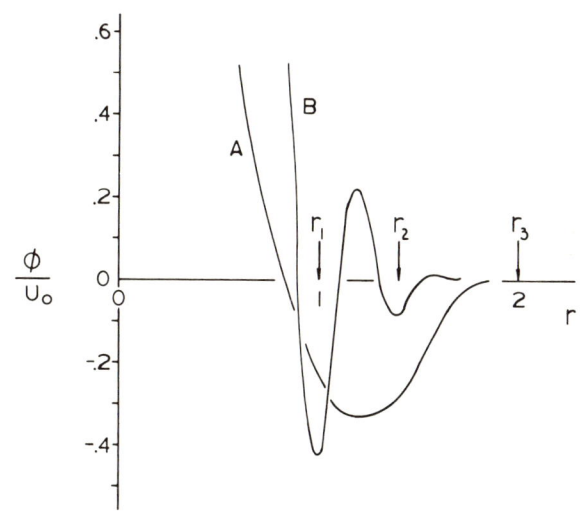

Fig. 1. Interatomic potentials for the study of two-dimensional square lattice; r_1, r_2, and r_3 represent the positions of the first, second, and third nearest neighbors in this lattice.

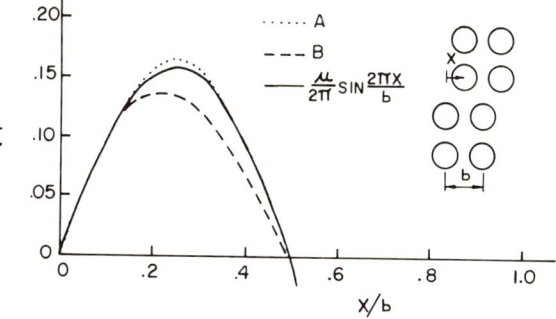

Fig. 2. Force-displacement relations for shear of a perfect crystal bonded by potentials A and B of Fig. 1.

3 CORE STRUCTURE AND ENERGY

Dislocations with $b = a[10]$ (a is the lattice parameter) were formed by displacing atoms to positions given by the elastic strain field of a continuum (anisotropic for potential B) from solutions presented by Hirth and Lothe[6], and the lattice was then allowed to relax to equilibrium using a dynamic "quench" method (see, for example, Gehlen et al.[7]). Using potential A, the relaxed configuration is not very different from the elastic solution, as shown in Fig. 3. However, a single function is not sufficient to describe displacements above and below the slip plane, and atoms above the slip plane were found to relax from the elastic positions considerably more than those below (Fig. 3). Thus, it is difficult to define a "width", although Fig. 3 indicates that the half-width is roughly $\zeta \sim 1.3b$ which is somewhat larger than the Peierls estimate $\zeta = b/2(1-\nu) = 0.67b$. Atoms were relaxed within a radius of 10 lattice spacings of the dislocation; as a check on convergence, results using a relaxation radius of 8 lattice spacings were found to be almost identical.

The total strain energy E within a radius r of the dislocation core for potential A is shown in Fig. 4 for the configuration labelled I (inset, Fig. 4). Points represent data for individual atoms, which contribute an amount to the strain energy $\Delta E = 1/2\ (\Sigma\phi - \Sigma\phi_0)$, where the summation is taken over all neighbors within range of the potential; ϕ_0 denotes the energy for the atomic positions in a perfect unstrained lattice. The strain energy predicted from continuum elasticity may be expressed (Cotterill and Doyama[8]) as

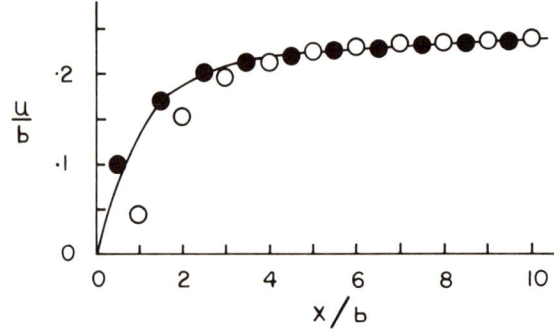

Fig. 3. Displacements parallel with the Burgers vector for a dislocation in configuration I (see Fig. 4), potential A. Solid line shows results from linear elasticity; points show relaxed displacements for atoms above the slip plane (open) and below (filled).

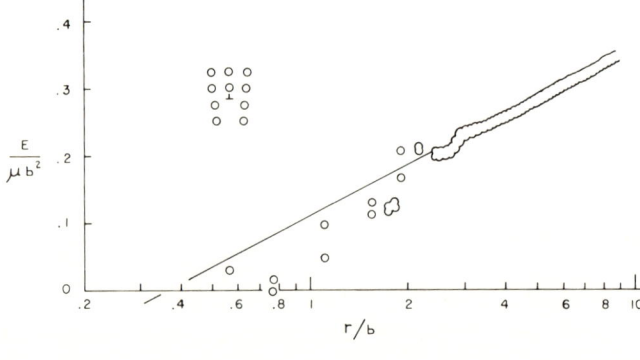

Fig. 4. Strain energy for configuration I (see inset), potential A, within radius r of the dislocation.

$$E = \frac{\mu b^2}{4\pi(1-\nu)} \ln \frac{r}{r_{eh}},$$

where the "equivalent hole radius" $r_{eh} = b/\alpha_p$ with $\alpha_p \sim 1$ to 4 (Hirth and Lothe[6]); from Fig. 4, $r_{eh} = .35\,b$ and so $\alpha_p = 2.8$. Further, the "core radius" inside which nonlinear effects are important is $r_c \sim 3b$, and the core energy within this radius is $U_c \sim 0.6\,U_0$ where the cohesive energy per atom $U_0 = 1/2\Sigma\phi_0$ is 0.367 μb^2 for potential A. With the dislocation in the other symmetrical configuration II midway between two rows in the upper half-plane, the corresponding results are $r_{eh} = 0.33\,b$ and $\alpha_p = 3.1$, with the core radius and energy indistinguishable from configuration I. Core relaxations were found to be much more significant for potential B, and α_p was found to be 0.56 and 1.2 for configurations I and II, respectively.

4 PEIERLS BARRIER

Two properties are of interest regarding lattice resistance to dislocation motion: the uniform shear stress required to move the dislocation (Peierls stress τ_p), and the energy that must be provided locally to move the dislocation to the next equilibrium site (Peierls energy U_p). They may be separately determined by imposing different conditions during relaxation.

To find τ_p, a uniform shear strain γ was imposed on the whole lattice with the dislocation in configuration I. The atoms were then allowed to relax to their equilibrium positions. At a critical strain γ_p, the dislocation was found to move toward the boundary of the relaxed region, defining the Peierls stress $\tau_p = \mu\gamma_p$; it was found that $\tau_p/\mu = 0.021 \pm 0.001$ for potential A.

To evaluate U_p, the dislocation must be held in positions corresponding to core displacements between configurations I and II. This was achieved in the present case by displacing the dislocation to $x = \alpha b$ with $0 \leqslant \alpha \leqslant 0.5$ and imposing elastic displacements corresponding to that position. Finally, the two atoms nearest the center of the dislocation (one above, one below the slip plane) were displaced to positions given by a linear interpolation between the corresponding fully relaxed positions at $\alpha = 0$ and $\alpha = 0.5$ (configurations I and II), and held there during relaxation. The atomic configurations are shown in Fig. 5, and the increase in energy within the relaxed region above the value for $\alpha = 0$ is plotted in Fig. 6. The Peierls energy is $U_p \sim 0.0038\mu b^2$, i.e., $U_p \sim 0.017 U_c$. Also, configuration II appears to be metastable, a possibility foreseen by Hirth and Lothe[6]. The Peierls stress may be derived from the energy profile;

$$\tau_p = \frac{1}{b}\frac{dU}{dx} = 0.019\mu$$

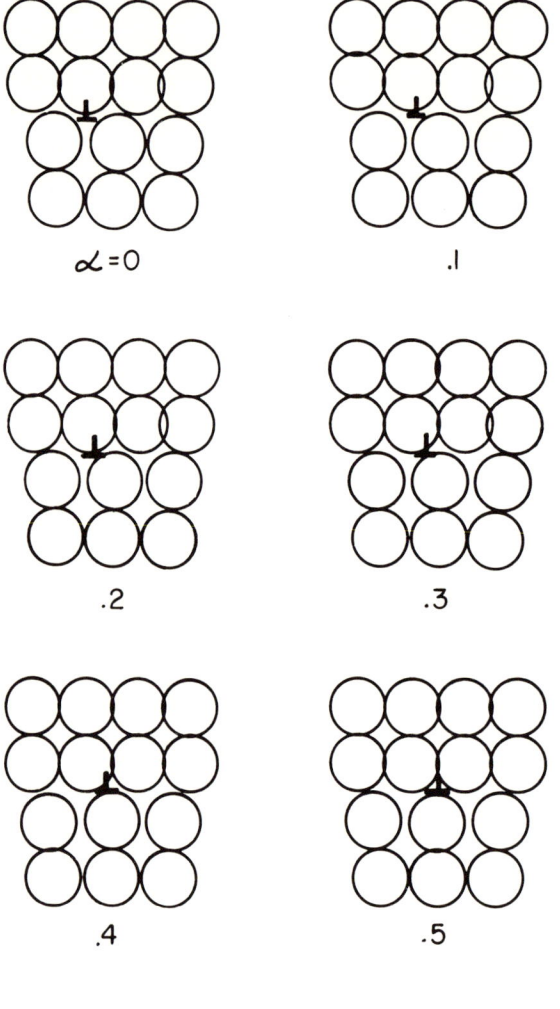

Fig. 5. Atomic arrangement in the core of a dislocation moving from configuration I ($\alpha = 0$) to configuration II ($\alpha = 0.5$) using potential A.

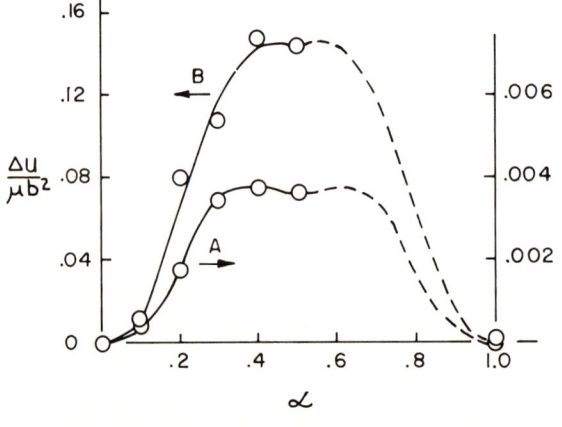

Fig. 6. Peierls barrier for motion of dislocations using potentials A and B. Right-hand scale refers to potential A.

for potential A (in reasonable agreement with the result found above). The derivative is taken at the inflection point. For comparison, τ_p calculated from Eq. 1 for this case is 0.00061μ, a factor of 30 smaller than the fully atomistic result. Also shown in Fig. 6 is the Peierls barrier for potential B. In this case, $\tau_p = 0.54\mu$ where μ corresponds to (01)[10] shear; this enormous stress, a consequence of the marked relaxation in the core region, is discussed below.

The atomic configurations during glide of a [10] dislocation using potential B are shown in Fig. 7. At $\alpha = 0$, the atom directly above the dislocation line relaxes considerably in a direction normal to the Burgers vector **b**, and as α increases to 0.5 this atom moves much further in this direction than it does parallel to **b**. The energy difference between $\alpha = 0$ and $\alpha = 0.5$ is about one-quarter of the core energy for this potential. Since the theoretical strength for (01) [10] shear is 0.14μ which is

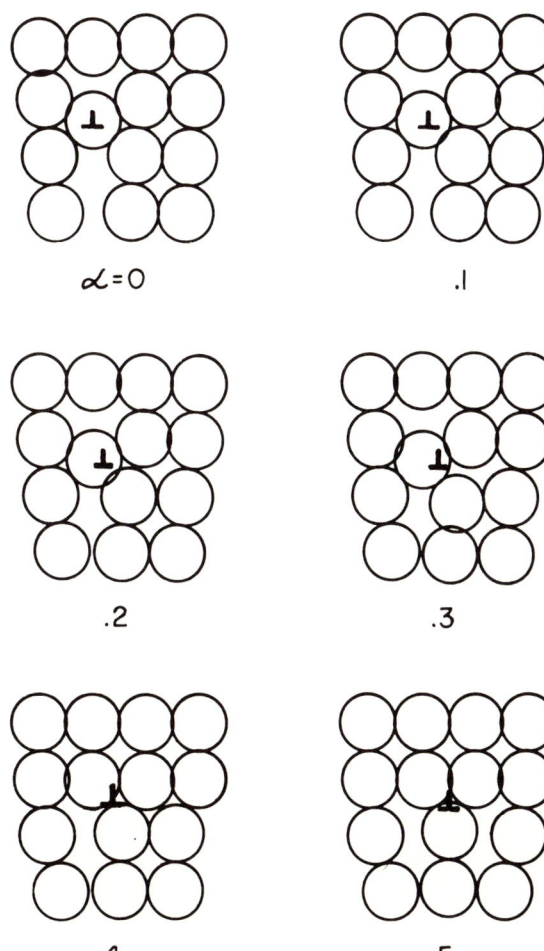

Fig. 7. Atomic arrangement in the core of a dislocation moving from configuration I ($\alpha = 0$) to configuration II ($\alpha = 0.5$) using potential B.

considerably less than τ_p, the lattice will fail under an applied shear stress before the dislocation will move. The dislocation is effectively sessile.

To study the effect of different atomic arrangements across the slip plane, a dislocation with **b** = [11] was studied using potential A. It was found that each of the two symmetric positions of the dislocation is an unstable equilibrium state; the instability may correspond to the dissociation [11] → [10] + [01].

5 SUMMARY

For potential A, the relaxed-core configuration is not greatly different from that given by linear elasticity, although for potential B considerable relaxation was observed. For both potentials, the lattice friction stress is much larger than that predicted by the Peierls-Nabarro result and more in line with more recent theoretical estimates (see, for example, Ref. 4).

Although quantitative predictions for any specific material require a knowledge of the appropriate interatomic potential and a sound theoretical justification for its form, the results presented here demonstrate that the friction stress for motion of an edge dislocation in a lattice held together by two-body central forces can be quite large.

REFERENCES

1. Peierls, R. E.: Proc. Phys. Soc. London **52**: 34 (1940).
2. Nabarro, F.R.N.: Proc. Phys. Soc. **59**: 256 (1947).
3. Foreman, A. J., et al.: Proc. Phys. Soc. **64A**: 156 (1951).
4. Kuhlmann-Wilsdorf, D.: Phys. Rev. **120**: 773 (1960).
5. Nabarro, F.R.N.: *Theory of Crystal Dislocations,* Oxford (1967).
6. Hirth, J. P. and J. Lothe: *Theory of Dislocations,* McGraw-Hill (1968).
7. Gehlen, P. C. et al.: J.A.P. **39**: 5246 (1968).
8. Cotterill, R.M.J. and M. Doyama: Phys. Rev. **145**: 465 (1966).

AGENDA DISCUSSION: DISLOCATIONS AND STACKING FAULTS

J. P. Hirth [*]
The Ohio State University
Columbus, Ohio 43210

G. T. Hahn [**]
Battelle, Columbus Laboratories
Columbus, Ohio 43201

The discussion gained momentum rapidly as *A. Seeger* examined special problems connected with the boundary conditions for small atomic arrays in an elastic continuum. *A. Seeger* offered a method of coupling near-boundary atoms to the continuum when the atomic interactions are long ranged. He also discussed an approach for treating the nonlinear response of the continuum and attending lattice dilatations. The conferees then engaged in an increasingly heated discussion of potentials: Johnson's potential and the "humps and bumps" of the Bullough-Perrin potential, Englert's potential and the pseudopotentials.

Calm returned as *J. Weiner* moved to a new theme, describing a simplified atomic model for treating dislocation motion. *W. Tyson* followed with a report of atomic dislocation models, one of which had an

---
[*]Chairman.
[**]Secretary.

astoundingly large Peierls stress as well as a highly unconventional core structure. His paper is published in full in this volume. The question of simulating temperature, raised by *P. Shewmon,* was touched briefly. Then it was time to reconcile the computer simulations of bcc dislocation movement with experiments. *A. Seeger* quickly reviewed six low-temperature, dislocation, relaxation processes along with the attending peaks and mechanisms. Here the energy of the conferees ebbed and was finally spent with *M. Duesbery's* remarks on the mounting of the Peierls barrier. These are the highlights; more details are given in the following paragraphs.

Potentials

The discussions of empirical pair potentials reflected the participants' concerns with reliability of potentials and the methods used to construct them. When these potentials are fitted accurately to the elastic constants they usually reproduce phonon-dispersion curves approximately. The reverse is also true, but both sets of data usually cannot be fitted simultaneously. *R. Bullough* believes the elastic constants should be fitted since these are more accurately known than are the dispersion curves. *Huntington* on the other hand believes that if there appears to be a real difference between the elastic constants and the neutron-dispersion curves, neutron data should be preferred on the basis that short-wavelength information is more significant for defect behavior than for $k = 0$.

P. Ho recommended that potentials be determined by fitting to the second- as well as the third-order elastic constants to account for the anharmonic effects. Such a procedure has already been employed by Chang and Graham[1]. In any event, since the elasticity measurements only characterize derivatives of the potential close to the lattice points, absolute values and the depth of potential wells can be adjusted independently. It was generally agreed that the vacancy formation energy tests the potential close to the lattice spacing, but that extended defects, such as crowdions and stacking faults were sensitive to the entire potential, possibly to the Friedel oscillations, and certainly to the potential cutoff. Results for interstitials are most sensitive to the repulsive part of the potential. Calculations of the relative stability of the tetrahedral versus the octahedral interstitial site have not shed much light on the validity of potentials; particularly since the initial conditions may influence the result. It was *R. A. Johnson's* experience that defect formation energies were model sensitive, whereas migration energies were relatively insensitive to the choice of potential. *Seeger* suggested that the stacking fault energy should be sensitive to phase shifts of the Friedel oscillations relative to the fault plane, but *W. Harrison* took exception to this view with respect to second-order perturbation theory. *Harrison* argued that there are no phase shifts of the oscillations in the second-order calculation and that there is no more reason to expect the second-order calculation to fail for this problem than for others. In fact, it would seem somewhat more reliable here since no long wavelength components of the pseudopotential (which are larger in amplitude than the short-wavelength components) enter. He argued that, even if one did believe that these phase shifts mentioned by *Seeger* were important, the best approach would still be to do the calculation in wavenumber space and proceed to higher order. This would then include not only the phase shifts but other corrections, formally and probably numerically of the same magnitude, that reflect the breakdown of the two-body interaction picture. *Harrison* recalled that in his early calculation of the stacking-fault energy in magnesium[2], he had more confidence in his prediction of 50 ergs/cm[2] than in the other calculations and was therefore very much disappointed that there were no experimental values for comparison.

There was considerable discussion of the Bullough-Perrin potential for α-iron (see Fig. 1a).[3] The most noteworthy features of this potential are the sharp rise in energy between second- and third-neighbor distances and the fact that its cutoff distance is larger than that for the Johnson potentials. When matched to the phonon-dispersion data, it gave a surface energy of 1000 ergs/cm[2], gave stacking-fault energies of 700 and 800 ergs/cm[2], and was consequently regarded as an improvement over the Johnson version. *P. Gehlen* reported that since the equilibrium positions of the near neighbors in the perfect bcc lattice are very close to the bottom of the potential wells, this potential is very anharmonic. For instance, when this potential was used to simulate a $\frac{a}{2}$ <111> screw dislocation in α-iron a core radius of about 70 Å resulted.[4] *R. Bullough* noted that as the nonlinear behavior of the potential was not an "input", it could be changed without modifying any of the harmonic properties. *V. Vitek* maintained that this potential yields lower "stacking-fault energies" (in the sense of partial displacements in his paper in this volume) in the antitwinning direction than in the twinning direction on (112) planes — a result inconsistent with the lower resolved shear stress for dislocation motion in the latter sense.

A. Englert gave a more complete account of the potential for copper[5] employed in her paper with Bullough and Perrin. The potential, shown in Fig. 1b, is consistent with the stacking-fault energy of 70 ergs/cm[2]. With the cutoff distance at the third-neighbor distance r_3, the stacking-fault energy is determined by the value of the potential at a distance $(8/3)^{1/2}r_1$. The amount of relaxation depends on the slope of the

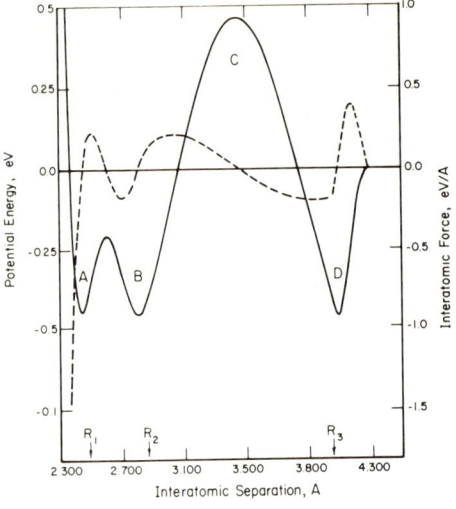

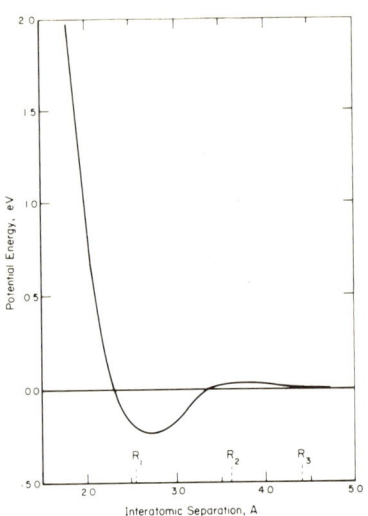

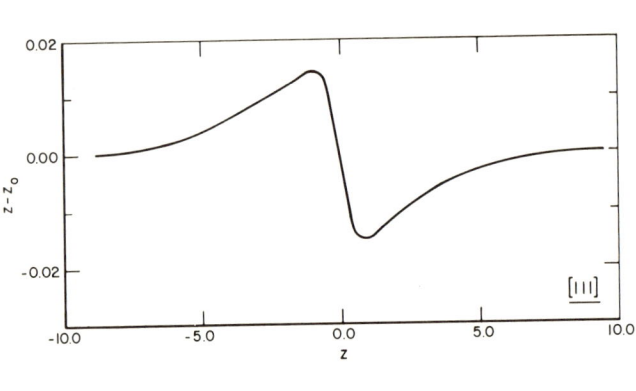

Fig. 1. (a) The interatomic potential as given by Bullough and Perrin. The broken line shows the resulting interatomic forces, (b) The pair potential for copper in electron volt units, r_1, r_2, and r_3 are the first-, second- and third-neighbor spacings. (c) the displacements of the adjacent close-packed planes parallel to an intrinsic fault, in the [111] direction orthogonal to the fault. The displacements u_z are all away from the fault (at $z = 0$) and are given in units of r_1, the 1st neighbor distance. The relaxed stacking fault energy is 70.5 ergs/cm^2.

potential at this distance and at the distances r_1 and r_2. The displacement of the (111) planes around the fault is represented in Fig. 1c.

Vitek agreed with the point of view that the simple potentials adjusted to a few parameters (e.g., elastic constants) cannot be adequate when one wishes to calculate a quantitative value of a physical quantity for a particular material. For example when the potential is adjusted to a set of elastic constants it can still be varied so as to change stacking-fault energy.[6] On the other hand if one is interested in qualitative properties of defects in the metals crystallizing in a given structure, those potentials can be very useful. An example is the structure of the core of screw dislocations in bcc metals, the basic features of which have been supposed to be the same in all bcc materials. Here, the potential is used only to make a model of a mechanically stable bcc lattice (assuming central forces between the atoms); calculations using different potentials indeed show principally the same core structure.[4,6,7] Further, the possibility of varying the stacking-fault energy (or restoring forces, see Vitek's paper in this volume) enables us to investigate cores of different widths, asymmetry, etc.

In summary of the discourse on potentials, equilibrium was not reached among the participants. Users of the simple potentials based on parametric fits to physical phenomena were satisfied that simulations employing such potentials were useful in modeling defect behavior and felt, moreover, that their potentials reasonably reflected major anharmonic effects. Those espousing potentials based on fundamental principles (such as the inclusion of long-range electronic effects) took the position that the simple potentials were not useful in describing real materials. However, the more-refined potentials were highly model sensitive, and appeared to require considerable development to unequivocally represent material properties. Pending such development, both approaches seem justified for an interim period.

Boundary Conditions

Following an introduction of the topic by *Hirth, A. Seeger* emphasized the importance of imposing the correct boundary conditions on discrete atom arrays in defect-simulation studies. Special problems are encountered with calculations involving long-range potentials, such as for ionic crystals, and for defects with long-range strain fields, such as for dislocations.

Seeger advocated a method previously used both by Johnson et al.[8] and by Seeger et al[9], in simulations of point-defect properties for accommodating the long-range potentials of atoms at and near the boundary of the array. The method employs a transition region (Region II) which is inserted between the discrete crystallite (Region I) and the continuum (Region III). The position of atoms in Region II were derived from the continuum solution, but the energy is calculated by means of the potentials used in Region I. All the "bonds" extending out from the array are accommodated by making the width of Region II comparable to the range of the potential. Finally, the energy is minimized with respect to all the displacements in Region I.

Hirth noted that complications are also introduced by third-order elastic effects — primarily that of crystals being harder in compression than in tension — that are contained in the potential but are suppressed outside the discrete array by the usual linear elastic treatment of the boundary conditions. These effects are important for defects with long-range strain fields, such as the dislocation where the nonlinearities produce a crystal volume expansion, ~ 1 atomic volume/atom plane. Simple models can be constructed to physically reproduce such effects. For example, in a linear, isotropic continuum approximation, at large distances from the core the elastic field associated with such a dilatation can be modeled by the linear elastic field of a line of dilatation along the core. More realistically, the long-range field should be modeled by the linear-elastic field of two rows of orthogonal, point-force couples of different magnitudes lying along the core for dislocations with inversion crystal symmetry along their axes or by more general point-force arrays in other cases. An analog of the latter case is the replacement of the field of carbon in iron by that of a tetragonal point-force array, and that for the former case of a ball in a hole for the substitutional solute atom analog. In particular, for the nondilatational portion of the field, long-range interactions between dislocations will be introduced and the line tension of dislocations will be modified by a factor of 1 to 2 by such terms. Thus, the nonlinear terms could markedly affect the theory of dislocation interactions. Of course, for other than qualitative implications of such terms, a full third-order anisotropic treatment is required as discussed below.

In the context of atomic calculations, large errors in both defect configuration and energy can be introduced if only the linear defect elastic displacement field is imposed on the boundary: the system will be artificially constrained if the long-range fields arising from nonlinear effects near the defect are not also superposed on the boundary. The influence of nonlinear terms has emerged clearly from the analysis of the work of Granzer and Eisenblätter on dislocations in alkali halides.[10] The expansion in the core was compensated by a shell of compression just inside the region where the computer solution was joined to the linear elastic solution. This artifact is avoided if the elastic solution included the third-order elastic effects.

The nonlinear problem is particularly difficult in the anisotropic elastic case (no analytical solution is even available for a point force in the anisotropic case) relevant to the long-range field. A general method for treating the anisotropic elastic case has been presented by Teodosiu and Seeger.[11] Teodosiu has since added a general technique for including the angular-dependent terms of linear-elasticity theory that are of the same order of magnitude as the nonlinear terms included. In a Frankfurt-Stuttgart collaboration, the method is now being employed for calculations on dislocations in alkali and silver halides.

R. Bullough agreed that the nonlinear solution should be included in the boundary region around the dislocation, but emphasized the magnitude of the task involved. He felt that the nonlinear solution has to be included in the full anisotropic form. In addition, he noted that the Teodosiu-Seeger formulation of the third-order solution is derived by a perturbation procedure in which the source of the perturbation arises from the core configuration (and strength). Unfortunately, the nonlinear solution is everywhere sensitive to the core region and there are convergence problems, so that the solution can only be obtained by a very careful, self-consistent procedure. *Bullough* emphasized a position that he has long taken that in addition to the nonlinear effects, one should include the full **linear** elastic displacement field of the dislocation, including, for example, terms associated with surface tractions at free surfaces. *J. Weiner* made the suggestion that trends in the effects of boundary conditions might be examined by simple models that can be treated analytically as well as discretely. He cited as an example Sanders treatment of the Peierls stress of the Rosenstock-Newell model: a simple cubic, 2-dimensional lattice held together by linear springs.[12]

Temperature Simulation

P. Shewmon raised the question of finite temperature effects in computer simulations. *R. Bullough* commented on the difficulties of conducting simulations at finite temperatures from the viewpoint of computer time. Temperature has been included by Tsai as part of a defect migration study.[13] Atoms in the array were given a Boltzman[11] distribution of velocities and permitted to vibrate. Equipartition of the energy was obtained after one thermal cycle. However, the procedure proved to be very time consuming, requiring 30 iterations per vibrational period. Calculations of this kind were also reported by *P. Gehlen,* who together with *J. Beeler* is studying the influence of temperature on the stress for kink motion in α-iron (Johnson potential). Preliminary results are given below:

Temp., K	σ_K-Stress for Kink Motion
0	$\sigma_K > \frac{\mu}{500}$
25	$\sigma_K < \frac{\mu}{500}$

Another type of problem arises with respect to temperature effects. For dislocation glide or climb or for crack propagation, a possible mechanism is nucleation and growth of double kinks or jogs.

568 J. P. HIRTH AND G. T. HAHN

Dislocation Simulations Versus Experiments

A. Seeger presented a brief review of the mechanisms of deformation of bcc metals at low temperature and the connections with the simulations of dislocation movement reported in Part IV of this volume. The interpretation of bcc deformation is complicated because there is a multiplicity of possible mechanisms. Citing the work of Chambers[14], *Seeger* identified four mechanisms and the corresponding internal friction peaks. In Fig. 2 these mechanisms are reflected by the relaxation strains connected with the modulus defect associated with the different mechanisms. The mechanisms are tabulated below and illustrated in Fig. 3 (numerical values referring to α-iron).

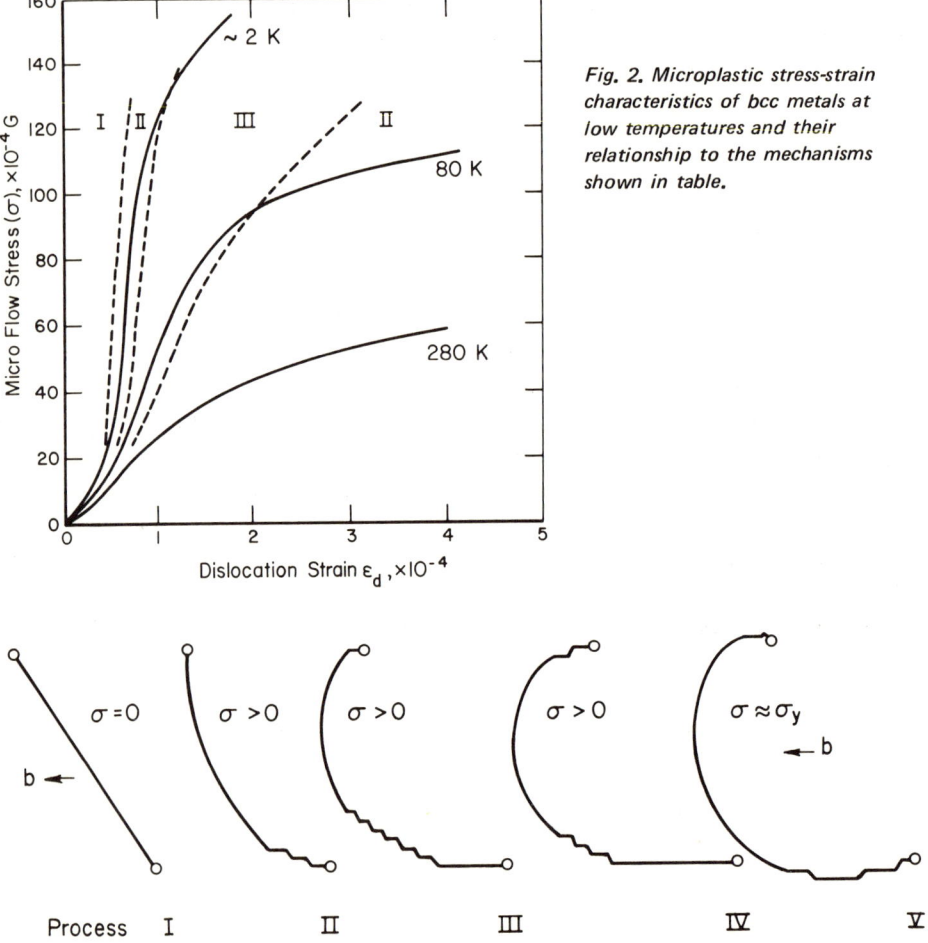

Fig. 2. Microplastic stress-strain characteristics of bcc metals at low temperatures and their relationship to the mechanisms shown in table.

Fig. 3. Mechanisms corresponding to stress-strain curves shown in Figure 2.

Process	Controlling Mechanism	Corresponding Internal Friction Peak
I Low stress	Motion of edges[a]	Not established[a]
II	Motion of preexisting, geometric kinks (0.1 eV) in screw dislocations	α
III High stress	Motion of kinks in presence of H_2 in screw dislocations	β
IV Macroscopic	Double-kink generation (0.7 eV) in screw dislocations	γ, γ'

(a) Newer, so-called, α' and δ peaks have been proposed to be associated with this process.

According to *Seeger,* the kink calculations reported by Gehlen at this conference support the identification of the α-peak with kink motion in screw dislocations. Macroscopic flow was clearly controlled by double-kink generation, a process whose stress requirements were difficult to establish from first principles. *Seeger* mentioned a possibility raised by Gehlen's study of the screw core[3], that an as-yet nameless defect along the screw dislocation line where symmetric and asymmetric segments meet could be a favorable site for double-kink nucleation. *Duesbery* disagreed with the latter point, arguing that such defects, if they indeed exist, would annihilate under stress. He stated that since the activation volume for double-kink nucleation is $\sim 50\ b^3$, the generation of double kinks at 0° K must involve pushing a significant length of line over the Peierls barrier. He believes, therefore, that the stress for double-kink nucleation is best represented by the resistance to the motion of a straight dislocation rather than by the resistance to kink motion. Hirth feels that both of these questions remain open. He has shown that there is a transition stress, at which control by kink motion[16,17] goes over to control by kink nucleation[18,19] at a fraction of the Peierls stress. Moreover, the activation energy for screws and edges would differ for either type of rate control. Thus, both mechanisms are possible in the present case depending on the actual magnitude of the Peierls stress. A similar possibility would exist for Gehlen's defects; they would tend to annihilate under stress, but new pairs of defects would tend to nucleate, analogously to the double-kink case.

References

1. R. Chang and L. T. Graham, in *Int. Conf. on Point Defect Calculations,* NBS, p. 53 (1963).
2. W. A. Harrison, *Pseudopotentials in the Theory of Metals,* Benjamin, N.Y., 1966, p. 208.
3. R. Bullough and R. C. Perrin, in *Radiation Damage in Reactor Materials* (IAEA, Vienna, 1969), Vol. II, p. 233.
4. P. C. Gehlen, J. Appl. Phys., **41**, 5165 (1970).

5. A. Englert, H. Tompa, and R. Bullough, in *Fundamental Aspects of Dislocation Theory*, NBS Special Publication 317, p. 273.
6. V. Vitek, R. C. Perrin, and D. K. Bowen, Phil. Mag., **21**, 1049 (1970).
7. Z. S. Basinski, M. S. Duesbery, and R. Taylor, Phil. Mag., **21**, 1201 (1970).
8. R. A. Johnson and E. Brown, Phys. Rev., **127**, 446 (1962).
9. A. Seeger, E. Mann, and R. von Jan, J. Phys. Chem. Sol., **23**, 639 (1962).
10. F. Granzer, G. Wagner, and J. Eisenblätter, Phys. Stat. Sol., **30**, 587 (1968); J. Eisenblätter, ibid., **31**, 71 (1969).
11. C. Teodosiu and A. Seeger, in *Fundamental Aspects of Dislocation Theory*, NBS Special Publication 317 (1970) p. 877.
12. W. T. Sanders, Phys. Rev., **128**, 1540 (1962).
13. D. S. Tsai, R. Bullough, and R. C. Perrin, J. Phys. C. Solid State Phys., **3**, 2022 (1970).
14. R. H. Chambers, *Dislocation Relaxations in BCC Transition Metals*, Phys. Acoustics, W. P. Mason, ed., IIIA, Academic Press, New York and London (1966), p. 123.
15. J. P. Hirth, in *Inelastic Behavior of Solids,* Eds. M. F. Kanninen, W. F. Adler, A. R. Rosenfield, and R. I. Jaffee (McGraw-Hill, Inc., 1968).
16. J. P. Hirth and J. Lothe, *Theory of Dislocations* (McGraw-Hill, Inc., 1968).
17. A. Seeger and P. Schiller, Acta Met., **10**, 348 (1962).
18. V. Celli, M. Kabler, T. Ninomiya, and R. Thomson, Phys. Rev., **131**, 58 (1963).
19. J. E. Dorn and S. Rajnak, Trans. Met. Soc. AIME, **230**, 1052 (1964).

Part Five

SURFACES AND INTERFACES

EXPERIMENTAL STUDIES OF ATOMIC BEHAVIOR AT CRYSTAL SURFACES

Gert Ehrlich
Coordinated Science Laboratory and
Department of Metallurgy
University of Illinois at Urbana-Champaign
Urbana, Illinois

1 INTRODUCTION

Possibly the simplest surface problems are those dealing with the behavior of elemental solids. Such important phenomena as crystal and thin-film growth, evaporation, surface diffusion, and sintering all involve atomic events at the solid-vacuum interface. However, the state of understanding of even these simple problems mirrors that in more complex areas of technological interest: although at present measurements are available on the overall event, only little has been done to establish the atomic mechanisms directly.

To a truly surprising degree, our notions about crystal surfaces are based on simple models of interatomic forces, not on experiment. The structure and stability of crystal planes, the nature of the sites at which atoms are held at a crystal surface, as well as the energetics involved in transporting material from one site to another, are largely known only through theoretical estimates. In fact, it is difficult to divorce a formulation of the kinetics of surface processes from the specifics of the pair-binding pictures popular in the 1930's.

The actual behavior of atoms at a crystal surface, as well as the structure of the surface itself, not only are of interest for their vast technological importance but also present a problem and challenge to our understanding of the cohesion of solids. Within a crystal, the atomic geometries of experimental interest are limited. By now the structure of the perfect crystal (for elemental solids at least) is taken for granted, having become routinely accessible through standard techniques. Of primary concern are the energy of an atom at a normal site and its variation with temperature, the lattice energy during the motion of an atom, together with the energetics and attendant configuration of vacancies and line imperfections.

In contrast, crystal surfaces offer a varied array of atomic environments. Each crystal plane presents a different geometry. An understanding of the relation between this atomic arrangement and binding would be important in predicting the equilibrium form; for small crystals, this is dictated by the surface tension as a function of orientation. Once the structure of the surface is established, the properties of atoms on the different surfaces are important for their contributions to the kinetics of various transport processes. Even without considering defects, such as dislocations or grain boundaries, it is apparent that at the vacuum-crystal interface we are confronted with a much wider range of atomic sites than in the bulk. Reliable predictions of their structure and energetics would be extremely helpful in the description of any practical surface process.

Till now the bewildering variety of surface geometries which may contribute to a real crystal, as well as the absence of reliable theoretical guidelines, have slowed progress in understanding surfaces on the atomic level. In a sense this very complexity may be an advantage, however. It is at the surface that we can subject our notions of atomic forces to a critical test. By placing atoms on different crystal planes we sample entirely new and different atomic configurations, which, in contrast to the bulk, are likely to play a part in real processes. The all-important prerequisite for taking advantage of this richness is the means of obtaining experimental information on the atomic structure of surfaces, and on the properties of atoms held there. Until a few years ago, such techniques were not available; it is for this reason that the atomic behavior of surfaces remained unexplored by experiment.

In surveying recent efforts to probe the crystal surface on the atomic level, it is convenient to proceed in two parts. Studies of the properties and atomic arrangement of the surface layer are examined first. The remainder of this review is devoted to the behavior of individual atoms held at the surface. The former has become quite a fashionable field, with a vast literature and life of its own. Much of the effort here has been adequately reviewed and this presentation is therefore brief. By comparison, examination of individual atoms is only in its infancy. It is, moreover, a field of great promise frought with considerable difficulties. As such, the present state of this subject is examined in some detail.

2 STRUCTURE AND PROPERTIES OF CLEAN SURFACES

2.2 Energetics

The thermodynamic quantity of interest in the description of surfaces is the surface tension γ; this is defined as the surface excess of the Landau function L per unit area, where $L \equiv F-G$, the difference between the Helmholtz and Gibbs free energies.[1] For a crystal, γ is a function of the orientation of the surface. For small crystals, it is this variation of surface tension with orientation that dictates the form to be expected at equilibrium through the famous construction of Wulff.[2] On larger crystals, for which surface processes no longer dominate the overall shape, the surface tension dictates the stability of a macroscopic plane to faceting[3], as well as the grooving found at twin boundaries[4]. Authoritative discussions of the formalism underlying these phenomena are available.[1,2,5,6]

All of these effects have been studied in attempts to derive information on the dependence of surface tension on atomic arrangement. Although the information so obtained is thermodynamic (and, as such, properly requires a statistical treatment), at low temperatures it is the energy contribution that dominates. With a knowledge of the surface tension it should therefore be possible to test the applicability of any given interatomic potential to surface phenomena. It is this aim that has motivated much of the recent work.

The experimental problem appears deceptively simple. All that needs to be done is to establish small particles in their equilibrium shape by annealing at a high temperature. The distance h_i of crystal face i from the center will then be directly proportional to the surface tension γ_i of that face, and the sequence of surface tensions is immediately accessible. The execution of such experiments is not easy, however. Equilibration must be done at high temperatures ($T/T_m > 0.8$), close to the melting point, where contamination by diffusion of impurities from the bulk is difficult to avoid. The support for such particles is especially troublesome. Ideally it should be nonwetting; this, however, generally means that the support surface is heavily contaminated and will in turn act as a source of contamination for the surface studied[7]. On the other hand, if the support interacts strongly with the crystallites of interest, it is likely to affect the overall shape.

Another problem is posed by evaporation, which is significant at the high equilibration temperatures. Unless suppressed, the crystal shape may be dominated by the kinetics of this process, rather than by the surface tension. Attempts to establish an equilibrium environment are generally made by enclosing the sample in a container of the same material and isothermal with it. However, the vapor pressure, p, of a small crystal differs from that of a large surface, p_∞, according to the thermodynamic relation

$$\ln p/p_\infty = 2 \frac{v}{kT} \frac{\gamma_i}{h_i},$$

where v is the atomic volume. Small particles may therefore still be affected by evaporation. A final uncertainty is introduced by the need to quench in the equilibrium form established at high temperatures. This is generally not a rapid process, and the shape finally examined may, in fact, reflect a much lower temperature ($\sim T_m/2$) than that corresponding to the annealing condition[6].

Examination of twin-boundary grooving does not suffer from several of these difficulties. It is still subject to contamination. The only studies in which this has been reasonably well suppressed are those relying on the field-emission or field-ion microscope[8,9,10] to examine the shape of a metal tip after equilibration at high temperatures. Although doubts must still remain about the role of the support and about evaporation, the studies in the microscope can be done under ultrahigh-vacuum conditions, under continuous surveillance of surface cleanliness.

Data on the surface tension of metals, in its dependence upon orientation, are listed in Table I. The discord for fcc metals is clear. Although differences in the techniques of analyses may be responsible in part, a likely and important source of difficulty appears to be contamination. This need not necessarily come from the ambient. For the fcc metals especially, it is extremely difficult to remove impurities from the bulk; at high temperatures they may then diffuse to the surface.[11] Even the analysis of bubbles in metals, which has also been used to derive information on the structural anisotropy of surface tensions, is subject to this problem. Work in which contamination does not intrude has been concentrated on the refractory body-centered cubic metals. For these, Drechsler and Nicholas[9] have succeeded in fitting the observed variations from one plane to the next with values calculated assuming a Mie potential

$$V(R) = \frac{\epsilon\, nm}{m-n} \left[\frac{1}{m} \left(\frac{R_o}{R} \right)^m - \frac{1}{n} \left(\frac{R_o}{R} \right)^n \right],$$

Table I. Variation of surface tension with orientation.

Plane	FCC Crystals: $\gamma_{hkl}/\gamma_{111}$			
	Au, Cu, Ag, Ni [a]	Ni [b]	Ni [c]	Au [d]
(111)	1	1	1	1
(100)	1.045	.948	.999	1.071
(211)	1.10	1.018	1.016	1.049
(311)	1.115	1.009	1.014	1.062
(331)	1.14		1.015	1.043
(310)	1.14	.998	1.013	1.060
(210)	1.16	1.008	1.012	1.055
(110)	1.15	1.008	1.008	1.048

Plane	BCC Crystals: $\gamma_{hkl}/\gamma_{110}$			
	Ta [e] 1400 K	W [e] 2400 K	Mo, W [c] 2400 K	W [f] 2600 K
(110)	1	1	1	1
(211)	1.068	1.021	1.022	1.022
(100)	1.050	1.025	.992	1
(111)	1.083	1.027	1.023	1.026
(210)			1.017	1.017
(310)	1.065	1.027	1.014	1.012
(221)	1.080	1.029		1.027
(331)			1.018	1.023
(411)	1.070	1.039		1.023
(311)			1.022	1.029

Crystal Versus Bubble Shape

	$\gamma_{100}/\gamma_{110}$		$\gamma_{100}/\gamma_{111}$
Bubbles in Cu [g]	1.2	Bubbles in Al [g]	1.03
Cu Shape [a]	.91	Cu Shape [a]	1.045
Ni Shape [c]	.99	Ni Shape [c]	.999
Bubbles in Mo [g]	1.14		
Mo Shape [c]	.992		

(a) B. E. Sundquist, Acta Met. **12**, 67 (1964).
(b) H. Mykura, Acta Met. **9**, 570 (1961).
(c) J. F. Nicholas, Austral. J. Phys. **21**, 21 (1968); also Ref. 9.
(d) W. L. Winterbottom and N. A. Gjostein, Acta Met. **14**, 1041 (1966).
(e) Ref. 8.
(f) Ref. 10.
(g) R. S. Nelson, D. J. Mazey and R. S. Barnes, Phil. Mag. **11**, 91 (1965).

in which ϵ marks the depth of the well, R_o is the pair separation at the minimum, and m and n are the exponents of the attractive and repulsive parts, respectively. The uniqueness of this result remains to be tested. Attempts have also been made to use Morse potentials to evaluate surface tensions of the (100), (111) and (110) planes for a variety of fcc metals.[12] The absolute values are quite far off the mark, however. For silver, as an example, the surface tension is estimated at 2508 erg/cm^2 on the (100), compared with a value of 1140 derived from experiment[13] (at $T \sim 1200$ K). Similar disparities found for copper suggest that these attempts to transfer potentials derived from bulk criteria to the surface are ill-founded.

As yet no effort has been made to examining the surface tension of nonmetals and its variation with structure, even though such information would be of great interest, especially for covalent solids and for rare gas crystals.

2.2 Atomic Arrangement of Clean Surfaces

The commercial availability of ultrahigh-vacuum equipment for low- and high-energy electron diffraction has brought about a dramatic upsurge in structural studies of solid surfaces. The technique, as well as many of the results have been thoroughly reviewed.[14,15,16] Therefore only a few general comments are in order.

The information most directly deduced from diffraction studies is the symmetry and spacing of the unit cell. In work with low-energy electrons, inelastic events rapidly diminish the intensity of the incident beam as it penetrates into the bulk, and only the outermost lattice planes are sampled.[17] The diffraction pattern therefore reflects only the two Laue conditions appropriate to the lattice vectors **a** and **b** in the surface plane:

$$\mathbf{S} \cdot \mathbf{a} = 2\pi h \qquad \mathbf{S} \cdot \mathbf{b} = 2\pi k \quad .$$

Here **S** is the diffraction vector, given by $\kappa - \kappa_o$, the difference between the wave vector of the scattered and incident radiation, and h and k are integers. The third condition, involving the projection of the diffraction vector **S** on the lattice vector not in the surface plane, is not satisfied. In studies with high-energy electrons the experiment is arranged so that the electron beam strikes the surface at grazing incidence, and again samples only the outermost region. The atomic grid constituting the surface layer is therefore immediately accessible.

A survey of structural information, derived from low-energy electron-diffraction experiments on surfaces, is provided in Table II. For metals, it appears that the atomic spacing of the surface layer is generally the same as in the bulk of the crystal. Surface cells with multiple spacing are

Table II. Atomic arrangement of surfaces from LEED studies.

Surface	Structure	Investigator
Al (110), (111), (110)	(1×1)	F. Jona, J. Phys. Chem. Solids **28**, 2155 (1967).
Cr (100)	(1×1)	C. A. Haque and H. E. Farnsworth, Surface Sci. **1**, 378 (1964).
Fe (110)	(1×1)	A. J. Pignocco and G. E. Pellissier, Surface Sci. **7**, 261 (1967).
(100)	(1×1)	A. J. Pignocco and G. E. Pellissier, J. Electrochem. Soc. **112**, 1188 (1964).
(111)	(1×1)	H. Kobayashi and S. Kato, Surface Sci. **18**, 34 (1969).
Ni (111), (110)	(1×1)	R. L. Park and H. E. Farnsworth, Surface Sci. **2**, 527 (1964).
Cu (100)	(1×1)	R. N. Lee and H. E. Farnsworth, Surface Sci. **3**, 461 (1965).
(111)	(1×1)	C. A. Haque and H. E. Farnsworth, Surface Sci. **4**, 195 (1966).
(110)	(1×1)	G. Ertl, Surface Sci. **6**, 208 (1967).
Nb (110)	(1×1)	T. W. Haas, Surface Sci. **5**, 345 (1966).
Mo (110)	(1×1)	T. W. Haas and A. G. Jackson, J. Chem. Phys. **44**, 2921 (1966).
(100)	(1×1)	K. Hayek, H. E. Farnsworth, and R. L. Park, Surface Sci. **10**, 429 (1968).
(111)	(1×1)	H. E. Farnsworth, C. A. Haque, D. M. Zehner, and G. Barton, Surface Sci. **17**, 1 (1969).
(310)	(1×1)	
Rh (100), (210)	(1×1)	C. W. Tucker, Jr., Acta Met. **15**, 1465 (1967).
(110)	(1×1)	C. W. Tucker, Jr., J. Appl. Phys. **37**, 4147 (1966).
Pd (100)	(1×1)	J. C. Tracy and P. W. Palmberg, J. Chem. Phys. **51**, 4852 (1969).
(110)	(1×1)	G. Ertl and P. Rau, Surface Sci. **15**, 443 (1969).
(100)	(2×1), (2×2), c(2×2)	A. M. Mattera, R. M. Goodman, and G. A. Somorjai, Surface Sci. **7**, 26 (1967).
Ag (100)	c(2×2), ring	A. M. Mattera, R. M. Goodman, and G. A. Somorjai, Surface Sci. **7**, 26 (1967).
(100)	(1×1)	P. A. Palmberg, T. N. Rhodin, and C. J. Todd, Appl. Phys. Letters **10**, 122 (1967).
Ta (110)	(1×1)	J. E. Boggio and H. E. Farnsworth, Surface Sci. **1**, 399 (1964).
W (110)	(1×1)	R. M. Stern, Appl. Phys. Letters **5**, 218 (1964).
(111)	(1×1)	N. J. Taylor, Surface Sci. **2**, 544 (1964).
(100)	(1×1)	J. Anderson and W. E. Danforth, J. Franklin Inst. **279**, 160 (1965).
(211)	(1×1)	C. C. Chang and L. H. Germer, Surface Sci. **8**, 115 (1967).
Re (0001)	(1×1)	H. E. Farnsworth et al., Surface Sci. **17**, 1 (1969).
	(1×1)	G. J. Dooley, III, and T. W. Haas, Surface Sci. **19**, 1 (1970).
Ir (100)	(1×1)	J. T. Grant, Surface Sci. **18**, 228 (1969).
Pt (100), (110), (111)	(1×1)	C. W. Tucker, Jr., J. Appl. Phys. **35**, 1897 (1964).
Pt (100)	(2×1), (5×1), ring	H. B. Lyon and G. A. Somorjai, J. Chem. Phys. **46**, 2539 (1967).
Pt (110)	Ring	
(111)	(2×2), (3×3), ring	
Pt (100)	(1×1)	J. T. Grant and T. W. Haas, Surface Sci. **18**, 457 (1969).
Au (100)	(5×1), (6×6), ring	A. M. Mattera, R. M. Goodman, and G. A. Somorjai, Surface Sci. **7**, 26 (1967).
	(5×1)	P. A. Palmberg, T. N. Rhodin, and C. J. Todd, Appl. Phys. Letters **11**, 33 (1967).
(100)	(5×1), (1×1), ring	D. G. Fedak and N. A. Gjostein, Acta Met. **15**, 827 (1967).
(111)	(1×1), ring	
(110)	(1×2), (1×1), ring	
Si, Ge (111), cleaved	√3×1	J. J. Lander, G. W. Gobeli, and J. Morrison, J. Appl. Phys. **34**, 2498 (1963).
Sb, Bi (0001)	(1×1)	F. Jona, Surface Sci. **8**, 57 (1967).
(01$\bar{1}$2)	(1×1)	
(11$\bar{2}$0)	Reconstructed	
GaAs, GaSb ($\underline{111}$) A	(2×2)	A. U. MacRae, Surface Sci. **4**, 247 (1966).
($\overline{111}$) B	(3×3)	
GaAs, GaSb InAs, InSb (110)	(1×1)	A. U. MacRae and G. W. Gobeli, J. Appl. Phys. **35**, 1629 (1964).
LiF (100)	(1×1)	E. G. McRae and C. W. Caldwell, Surface Sci. **2**, 509 (1964).
NaF (100)	(1×1)	E. G. McRae and C. W. Caldwell, Surface Sci. **7**, 41 (1967).
Graphite (0001)	(1×1)	J. J. Lander and J. Morrison, J. Appl. Phys. **35**, 3593 (1964).

reported for the noble fcc metals. This is a problem on which there is still not unanimity in the literature. It is significant that with metals such as palladium, platinum and silver, for which multiple spacings have been reported, studies on stringently clean surfaces (listed in Table II) have yielded ordinary spacings. This should not be surprising in view of the well-established problem of really cleaning the fcc metals. Gold is the one material for which quite a careful study has revealed anomalous spacings. For this alone, such effects may prove to be typical of the clean surface. It will certainly be desirable to pursue even this more carefully.

The situation is quite different for the semiconductors — there restructuring of the surface layer from the arrangement typical of the bulk appears to be the rule rather than the exception. Here again the possibility has been raised that some of the structures may be stabilized by impurities.[18] At the moment it appears, however, that such structures are found even under very clean conditions and are therefore characteristic of the material itself.

A question that aroused some interest in the early renaissance of surface diffraction is the lattice spacing perpendicular to the vacuum-crystal interface.[19] Since the third Laue condition is no longer applicable in LEED studies, such information is not immediately obvious from the geometry of the diffraction patterns. Instead, attempts were made to infer the spacing from the details of the intensity-versus-wavelength relations measured experimentally. Unfortunately, it has been clearly demonstrated that these features are sensitively dependent upon the details of the model used to describe the interaction of the incident electrons with the lattice.[20] Reliable information concerning the spacing perpendicular to the interface therefore does not appear accessible through such an analysis.

In assessing the structural information derived by low-energy electron diffraction, it is important to keep in mind several important experimental limitations. LEED, like any other diffraction technique, is responsive to long-range order. When diffraction patterns are observed, they only establish that somewhere in the area covered by the electron beam there are regions with an ordered structure. Such a pattern may not be seriously affected by contamination, nor does the structure observed necessarily represent more than a small fraction of the total area. These details can often be deduced from an analysis of the intensity or from an examination of inelastic events. The observation of sharp diffraction patterns with a geometry characteristic of that expected from the bulk by itself is not reliable as a technique for determining the cleanliness of the crystal.

There are many illustrations of this. Well-developed patterns have been found for copper surfaces; under examination in the optical microscope, or with high energy electrons at grazing incidence, oxide scale was definitely established.[21] On (111) GaSb, a surface populated with droplets of liquid gallium still gave an unusually good pattern.[22] On copper again, quantitative studies of faceting by Mykura[23] suggest that as little as 15 percent of the surface may suffice to give a sharp pattern. Finally, on a Pt (111) it has been found that after a good electron-diffraction pattern is achieved, further cleaning results in no pattern changes; however, the scattering characteristics of the surface toward a molecular beam of deuterium continue to indicate further important changes in the surface.[24]

Much of the early LEED work depended entirely upon this criterion of sharpness to establish the cleanliness of the surface, even though there were qualitative indications aplenty that this was not sufficient. The early studies must be evaluated very carefully. Only recently have additional techniques been applied to ascertain the chemical composition of the surface layer, independent of the diffraction effects. The most popular of these is the energy analysis of Auger electrons emitted when the surface is bombarded by electrons with energies of a few kV.[25] This technique has revealed very significant contamination levels, surprisingly by sulfur.[26] In the future, the routine application of this and other analytical methods, such as determination of the appearance potential for x-rays[27], should eliminate this uncertainty.

2.3 Atomic Vibrations

In any attempt to determine the dynamics of surface processes on the atomic level, we need information on the average spacings of atoms in the ideal surface layer: it is also vital to understand the dynamics of surface atoms. These excursions are of interest as well in understanding the thermodynamic properties of the interfacial layer at finite temperatures.[28]

In fact, heat-capacity studies have been used to derive information on atomic vibrations at the surface. These should contribute a term proportional to the area, varying as T^2, instead of the usual volume dependent T^3 term given by Debye. However, the experiments are hard to interpret unequivocally. What is measured are deviations of the temperature dependence from the expected Debye curve when the material studied is present in a finely divided form with a high surface-to-volume ratio. This makes it exceedingly difficult to specify accurately the nature of the surfaces studied.[29] Work on single-crystal planes is required to allow reasonable interpretations. Such measurements probably could be done by studying very thin crystal foils that, when heated in a vacuum, would constitute their own adiabatic calorimeter. The difficulty of such work is obvious.

With the advent of diffraction studies on surfaces, information on the excursions of atoms on single-crystal planes has been obtained from the intensity of the diffraction peak as a function of the temperature.[30] As is known from X-ray work, this dependence is given by the Debye factor exp-2M. For a lattice of independent oscillators and radiation of wavelength λ

$$M = 8\pi^2 \langle u^2 \rangle \left(\frac{\sin \theta}{\lambda} \right)^2 \quad;$$

here $\langle u^2 \rangle$ is the mean-square displacement of atoms from their average positions, parallel to the diffraction vector **S**, and θ is the Bragg angle.[31] If the lattice can be adequately represented by a Debye spectrum with a characteristic temperature Θ, then in the range of high temperatures $T > \Theta$,

$$\langle u^2 \rangle = 3\hbar^2 \, T/mk\Theta^2 \quad.$$

From the slope of a semilogarithmic plot of the peak intensity against T we can therefore derive the characteristic temperature Θ, and also the mean square displacement $\langle u^2 \rangle$. It is customary to present results in terms of Debye temperatures as a matter of convenience; this should not obscure the fact that the temperature factor involves a mean-square displacement whether or not a Debye analysis is justified.

A low-energy electron beam will sample a region of the crystal near the surface, the depth of which increases with electron energy. To explore the excursions of atoms in the outermost atom layer, measurements of the temperature dependence must be carried down to low energies (40 eV and less).[17] At energies of 300 eV and above, effects typical of the bulk are obtained. Early studies of this nature were actually done with high-energy electrons at a glancing angle to the plane of interest to sample the surface layer.[32] Most recent work has been in LEED experiments on the temperature variation of diffraction peaks.

The data obtained under ultrahigh-vacuum conditions are summarized in Table III, and are limited to metals. At the surface, it appears that vibrational amplitudes exceed those in the bulk by roughly a factor of 2. Qualitatively this is the expected result. An atom at the surface is subject to the restoring force of neighbors on only one side. The results are in fair accord with model calculations based on first- and second-neighbor interactions.[33,34] The data on the (110) are especially interesting. Atoms on these planes are arranged in protruding close-packed rows along the [$\bar{1}$10]. Intuitively we would expect that fluctuations would occur more readily in the (110) at right angles to the close-packed rows, that is along [001], than in the rows themselves. This is indeed observed, with a characteristic temperature of 310 degrees along the [$\bar{1}$10], compared to 220 degrees both along the [001] and normal to the surface. In contrast, on the (111) plane, which in the fcc lattice is close packed, no anisotropies in the vibrational amplitudes are found to within the experimental error. This is

Table III. Atomic excursions at surfaces.

	$\dfrac{u_\perp}{u_{Bulk}}$	$\Theta_\perp$	Θ_D	Investigator
Ag (111)	1.46	155	225	E. R. Jones, J. T. McKinney, and M. B. Webb, Phys. Rev. **151**, 476 (1966).
(110)	1.48	152	225	J. M. Morabito, R. F. Steiger, and G. A. Somorjai,
(100)	2.16	104	225	Phys. Rev. **179**, 638 (1969).
Pd (100), (111)	1.95	140	273	R. M. Goodman, H. H. Farrell, and G. A. Somorjai,
Pb (111)	1.64	55	90	J. Chem. Phys. **48**, 1046 (1968).
Pt (100), (110), (111)	2.12	110	234	H. B. Lyon and G. A. Somorjai, J. Chem. Phys. **44**, 3707 (1966).
Ni (110)	1.77	220	390	A. U. McRae, Surface Sci. **2**, 522 (1964).

surprising; vibrations normal to the surface would be expected to have larger amplitudes than within the (111).

There are two cautions to remember. All measurements to date have been on fcc metals. In none of these studies has the surface composition been established by an independent technique such as Auger spectroscopy. This would certainly prove most desirable in order to remove any doubt about the interpretation. A question of a different nature has been raised concerning the theoretical framework underlying these analyses. Laramore and Duke[35] have carried out calculations of the elastic scattering of electrons from an fcc lattice modeled on aluminum, assuming s-wave scattering only, and a Debye spectrum for the phonons. They find that the Debye temperature deduced from the temperature dependence of the scattering intensity depends markedly on the scattering power of the ion cores at the surface relative to those in the bulk, as well as on the penetration of the electrons into the lattice. Such effects will limit the ultimate accuracy with which the vibrations of the outermost atom layer can be disentangled from those of the bulk. However, with care in the analysis of the measurements it should be possible to approach within 10 percent of the actual surface parameters.

3 OBSERVATION OF INDIVIDUAL ATOMS

3.1 The Experimental Problem

In contrast to the wealth of material now available on the structure of individual crystal planes, little is known about the properties of atoms self-adsorbed on such surfaces. This gap arises largely because of the experimental difficulties involved in characterizing atomic behavior.

Severe requirements are imposed on any technique that is to be useful in this area:
(i) It has to be capable of operating under ultrahigh-vacuum conditions in order to avoid contamination of the crystal surface.
(ii) It has to resolve the atomic structure of the surface.
(iii) The act of observation must not perturb the specimen.
(iv) The behavior of individual atoms must be discernable.

Although no single technique incorporates all of these desirable features, considerable information has become available over the years concerning the properties of gases and vapors adsorbed on crystals. In principle more complex than the surface layer of a crystal holding its own atoms, adsorbed gases have one great practical advantage — they differ from the substrate. Given the state of experimental techniques of the past decade it has been a reasonably straightforward problem to characterize the energetics and kinetics of such adsorbed layers.

Consider the behavior of alkali metals on a refractory metal, such as tungsten. The number of alkali atoms on the surface at any time can be measured by rapidly heating the crystal (flashing) and counting the atoms and ions evolved, either with a mass spectrometer or ionization gauge; from the kinetics of desorption, the energetics of binding can be deduced in the usual way from an Arrhenius plot.[36] Information about the crystal surface can be inferred from LEED. Inasmuch as the number of alkali-metal atoms held on unit area of the substrate is comparable to the number of available surface sites of the lattice, the contribution of defects such as lattice steps can be readily isolated. The arrangement of concentrated layers of the alkali atoms is again accessible through LEED studies, as scattering from this layer differs significantly from that of the bare metal itself.[37] Finally, a cesium atom held at a crystal surface will suffer an extensive rearrangement of its electron cloud, approaching in some ways the properties of an ion. This polarization manifests itself as a considerable change in the surface dipole and, therefore, in the electron work function of the surface. Taking advantage of this and of the high spatial resolution of the field-emission microscope, which is responsive to changes in work function, it is a straightforward matter to document the surface mobility of such atoms.[38]

The situation is entirely different in dealing with atoms at their own lattice. In principle, we should be able to draw upon the store of information available about the properties of the parent lattice. However, the experimental problems become formidable. There is no longer a simple way of determining the location of the adatoms. The equilibrium concentration of atoms on the flats is very small — they are held entirely at growth steps. An atom added to the surface at temperatures high enough for mobility will be incorporated into the steps. Under these conditions, the kinetics of evaporation yield information only about the overall energy change involved in removing an atom from a growth step into the vapor (the heat of vaporization of the crystal), not about the energetics of binding on individual planes. Only if the dislocation density is so low that diffusion between steps is small compared to evaporation (a difficult criterion to realize in practice) could any useful binding information be obtained in such experiments.

Calorimetric measurements of the heat released when atoms are allowed to adsorb on their own lattice are subject to the same difficulty. Only if atomic mobility is small would the measured energy reflect the interaction with a plane. Such studies are within the realm of the possible, but would be extremely difficult to execute. Calorimetric measurements can be made on very thin foils with a large surface-to-volume ratio. The primary problem becomes one of preparing the extremely thin specimen necessary to minimize conductivity losses, while maximizing surface effects, at the same time maintaining a high degree of surface perfection. Additional obstacles, such as minimizing the radiation input from the source of metal atoms, remain to be overcome. The difficulties of such measurements have been so obvious as to keep anyone from even attempting them.

The same factors intrude in any attempt to measure the properties of atoms self-adsorbed on planes of different geometries. The mobility is now no longer readily accessible to observation in the field-emission microscope. The work-function changes produced by putting an atom on its own lattice are small compared to the effects due to chemically different vapors. Radio-tracer experiments might still yield mobility information. This would relate to motion over atomically flat regions only if a detection technique were available with a resolution smaller than the distance between lattice steps. So severe is this requirement that no attempts of the sort have been reported.

This dismal situation has changed dramatically, however, with the invention of the field ion microscope (FIM) by Müller.[39]. This instrument obviously satisfies several of the important criteria listed earlier – it can resolve individual atoms at the surface and is simple to operate under ultrahigh-vacuum conditions. What little we presently know about properties of atoms held on their own lattices has been learned through the application of this technique. Our emphasis will therefore be on this method – its strengths and weaknesses, and on the analysis of the information about atomic behavior of solids gained by it.

3.2 Field Ion Microscopy

Although extremely powerful, the field ion microscope is quite simple. The surface is prepared in the form of a sharply pointed tip with a radius of a few hundred A. As indicated in Fig. 1 this is mounted in the center of a tube equipped with a fluorescent screen deposited on a transparent electrode. Fields on the order of 4-5 V/A are easily achieved at the tip surface when a negative voltage of 10-20 kV is applied at the screen. Any gas admitted to the tube will ionize, as the potential confronting an electron in the vicinity of the ion core is thinned down by the high field. This phenomenon was already known in the 1920's, and led Oppenheimer[40] to formulate his description of tunneling. In the limit of low fields,[41] the probability of field ionization is given by

$$k_F = 0.94 \times 10^{15} \frac{F}{\sqrt{I}} \exp{-.684 \frac{I^{3/2}}{F}} \text{ sec}^{-1}, \tag{1}$$

where I is the ionization potential of the gas in eV, and F the field in V/A. More elaborate expressions are necessary for a quantitative description of the ionization at finite rates, but Eq. 1 contains the important physical parameters — the barrier height, in this case $I^{3/2}$, divided by the field.

Close the surface, the electron potential is reduced through the image effect, as shown in Fig. 2, and field ionization occurs preferentially at the more-protruding atom sites. The ions created at the surface are accelerated by the applied field, replicating the location of atom sites on the fluorescent screen. The magnification is equal to the ratio of the screen radius R, divided by the radius of the emitter r times a constant β (with a value between 1.5 and 2) to account for image compression by the supports. An estimate of the resolution δ of the microscope, dictated largely by the thermal energy of the ions formed at the tip, has been given by Müller[39,42] as

$$\delta = 2\left[\left(\frac{\beta^2 r \hbar^2}{2mecF}\right)^{1/2} + \frac{4\beta^2 rkT}{ecF}\right]^{1/2};$$

where c is the conversion factor between the applied voltage V and the field F, and T the temperature of the surface.* If the finite size of the image gas atom is also included, the limiting resolution appears to be ~2A. For the highest resolution it is desirable to operate at high fields, using image gases with a high ionization potential preferably helium. Microscopy with helium, at a field of 4.5 V/A, has the added advantage of allowing operations at low temperatures without building up obscuring gas

---
*Henceforth in this article, energies are measured in eV, distances in A, and potentials in volts.

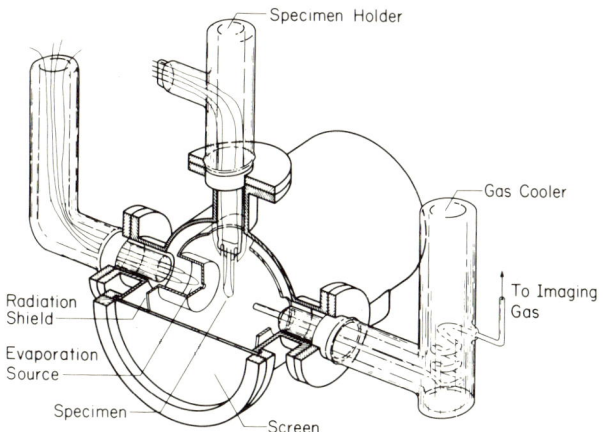

Fig. 1. Field-ion-microscope tube, for study of surface phenomena under ultrahigh-vacuum conditions.

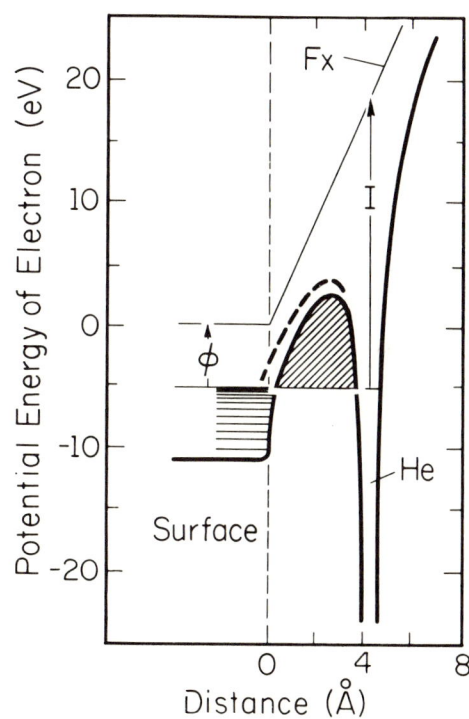

Fig. 2. Ionization of helium in an electric field F. ϕ = work function, I = ionization potential in eV. Dashed line indicates electron potential in free space; heavy solid line gives potential at the surface.

deposits on the surface. In practice, the limiting resolution is achieved only under special circumstances. As is apparent from the image in Fig. 3, low-index planes are generally not resolved; ionization takes place preferentially at the edges of the planes where the field is higher.

The images produced in this fashion are faint. Gas pressures are limited to 10^{-3} mm by the requirement that the path of the ion created at the surface not be disturbed by collisions with gas atoms. In view of the

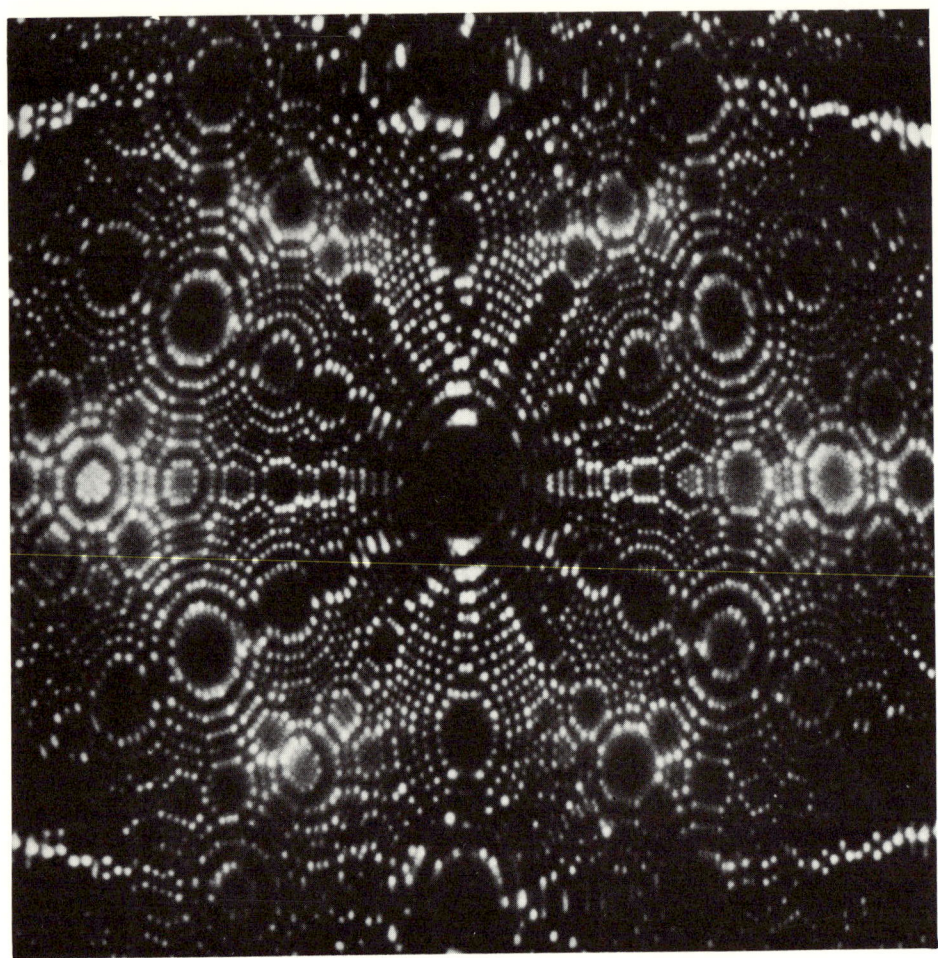

Fig. 3. Field-ion image of clean tungsten. [110] oriented.

inefficiency of the optical coupling involved in photographing the image, 10^6 or more ions may be necessary to depict an atom, requiring exposures on the order of minutes. A variety of intensifying devices has been adapted to cut down the exposure. The full details of these instruments, as well as the general theory and operation of the FIM has been presented in authoritative reviews and books.[39,42,43] The aim here, however, is to emphasize one very important point — the great resolving power of the FIM is obtained only under severe operating conditions. It is therefore vital in any investigation to ensure that the act of observation has not seriously perturbed the specimen.

3.3 Field Desorption

The full power of the FIM can be realized only because of the extreme perfection of the surface. This is achieved by maintaining the surface at a low temperature (usually ~20 K) and raising the field to a value at which evaporation occurs. For the same applied voltage, the more-protruding parts of the tip, being at a higher field, evaporate more rapidly. The surface so produced approaches the same degree of perfection as the bulk. Above all, it does not suffer from the thermal disorder typical of samples annealed at high temperatures, at which an occasional atom may be displaced from its normal site.

How does the high field bring about evaporation? This is important not only for the formation of the surface, but also in understanding the application of the FIM to the study of binding energies at the surface. Consider a schematic potential curve for an atom, as a function of the distance from the surface atoms, as in Fig. 4. For anything except some alkali metals, it is cheaper to remove a neutral at a desorption energy χ_0^o rather than an ion of charge state n, requiring an energy χ_0^n; the difference amounts to $I_n - n\phi$, I_n being the energy required to create such an ion in the gas phase, and ϕ the work function of the surface. For the atom, the actual potential is not known. For the ion it is assumed that the energy can be adequately represented by an image term, $-3.6n^2/x$ eV, where x is the distance from the image plane in A.

The presence of a high field changes the potential curves significantly. The metal atom is the less seriously affected. The field causes some polarization, lowering the energy by $P_0 = \frac{1}{2}\alpha_0 F^2$ in the familiar second-order Stark effect. The energy of the ion is not only reduced by $\frac{1}{2}\alpha_+ F^2$ through polarization, a small effect by comparison with the neutral atom, but by the direct interaction between the charge with the field, amounting to $-nFx$. It is important to realize the distinction between the two distance scales introduced here. For the ion, the distance x is conveniently measured from the image plane, for the atom, the cores at the surface serve as a more natural origin. Unlike a classical conductor, the metal allows fields to penetrate a small distance, on the order of the screening length Δ.[44] For the ion, it is therefore appropriate to reckon distances starting from a screening length behind the classical image plane.[45]

Crucial to any quantitative prediction is an assumption that has become traditional: the high field pushes back the electron cloud to such an extent that in effect, an ion remains on the surface. For this ion the superposition of the image potential $-3.6n^2/x$ upon the applied field $-Fnx$ produces a peak in the energy-distance relation, known as the Schottky barrier. Ignoring polarization, this peak is $3.795\ n^{3/2}F^{1/2}$ eV below the vacuum level, and occurs at $x_s = \left(\frac{3.6n}{F}\right)^{1/2}$ A. Provided the field and image terms are based on the same distance scale, field penetration does not alter the Schottky barrier.

As the field at the surface is increased, the height of the barrier localizing the ion at the surface is reduced. Classically, the ion evaporates

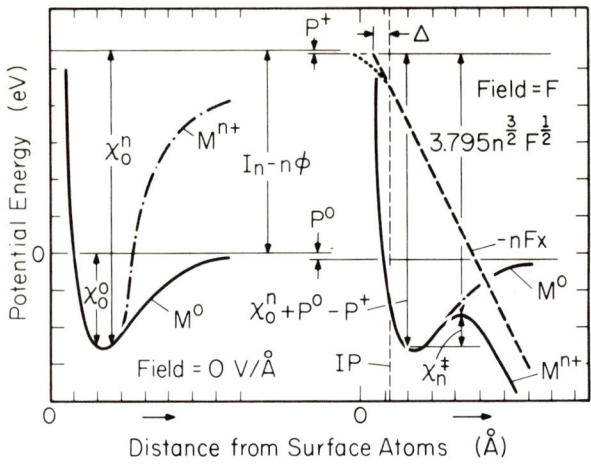

Fig. 4. Field evaporation of metal on image force model model. $\chi_n^{\ddagger}$ = activation energy for desorption in high field; I P = image plane; Δ = screening length.

over the barrier once its height has been reduced so that the available thermal energy suffices to propel the ion away from the surface. The rate of this process is

$$k_E = \nu \exp\left(-\frac{\chi_n^{\ddagger}}{kT}\right)$$

$$\chi_n^{\ddagger} = \chi_o^n + P - 3.795 n^{3/2} F^{1/2} \quad \text{and} \quad \chi_o^n = \chi_o^o + I_n - n\phi \,, \tag{2}$$

where ν is an effective vibrational frequency, and P a term accounting for all polarization effects. Tunneling of the atoms through the barrier, rather than evaporation over it, is also a possibility.[46,47,48] For the picture just presented this does not significantly alter the formalism – tunneling can be accommodated by replacing the actual temperature, T, with an effective tunneling temperature, T_t, at which the rate of barrier penetration from the ground state is just equalled by the rate of evaporation over the barrier.

On a classical picture and in the limit of $T = 0$ K, the applied field must be high enough to cause the barrier to vanish. This field value, calculated on the assumption that the polarization corrections are negligible, is given by

$$F = \frac{1}{n^3}\left(\frac{\chi_o^o + I_n - n\phi}{3.795}\right)^2 . \tag{3}$$

For shaping the surface it is necessary to remove many atom layers. The energy χ_o^o in this process is the heat of evaporation at $T = 0$ K, that is the energy to remove an atom from a kink site. Typical values predicted by Eq. 3 for the fields to produce a well-ordered surface are listed in Table IV.

It is apparent that the fields (4.5 V/A) for helium-ion imaging are comparable to those at which evaporation of the surface may occur. In some of the more interesting materials, rapid evaporation makes helium-ion imaging impossible. However, ionization can be accomplished at lower fields by working with neon (I = 21.6 eV) or argon (I = 15.8 eV). For these, adequate ion images are obtained at 3.5 and 2.2 V/A, respectively. Unusually detailed images are obtained at voltages 2/3 those required with pure helium when small amounts (5 percent) of hydrogen are added to helium.[49] Although the mechanism of this promoting effect is still not well understood, it is exceedingly useful for depicting surfaces of the softer metals.

Another factor initially limited the application of the FIM to the more-refractory metals. The fields required for evaporation are dictated by the energy for removing an ion; the cohesion of the material, however, depends upon the interatomic forces. The former may be high, the latter low, so that relatively high fields may be needed to shape a sample of low

Table IV. Field evaporation conditions.

Element	Charge n	Evaporation Field, V/A Calc.	Expt.[a]	Polarization Energy, eV	Polarization Correction, %
Si	1	4.72			
	2	3.40	3.0	−1.20	−24.5
	3	6.03			
Ti	1	3.91			
	2	2.48	2.5	.07	1.4
	3	4.16			
Fe	1	4.26			
	2	3.42	3.6	.53	12.3
	3	5.51			
Cu	1	3.09	3.0	−.10	−2.9
	2	4.36			
	3	7.69			
Nb	1	6.59			
	2	3.49	4.0	1.42	20.7
	3	4.34			
Mo	1	5.56			
	2	3.76	4.5	1.97	32
	3	4.9			

(a) Data from Ref. (42).

mechanical strength. This is generally unsuccessful, resulting in slipping and even complete destruction of the tip. Müller found a way around this problem.[49] Small additions of hydrogen enhance the rate of evaporation in a high field, so that delicate materials, such as iron, can be successfully formed to give surfaces of high perfection without mechanical damage.

The conditions for FIM are quite severe. However, using one trick or another, observation of elemental solids has been successfully extended to materials as weak as gold.[50] For the higher-index planes at least, the FIM has the capability of revealing the arrangement of surface atoms. A quantitative evaluation of atomic spacing is not feasible, as the factors affecting the imaging process are too many and interrelated. However, such spacings have been adequately documented in LEED studies.

Of much greater concern at the moment is the crudeness of the theory on which the interpretation of the evaporation process is based. How evaporation can provide us with quantitative information on the forces acting on an atom at a surface is examined next.

4 DIRECT DETERMINATION OF ATOMIC BINDING ENERGIES

Perhaps the most tempting feature of the FIM is that the desorption energy χ_o^o of atoms on well characterized sites appears to be accessible through the relation

$$\chi_o^o = n\phi - I_n - P + 3.795 n^{3/2} F^{1/2} + kT \ln(\nu/k_E) \quad , \qquad (4)$$

which follows immediately from Eq. 2. However, this is a hope, founded on many assumptions and oversimplifications which need most careful examination. The following discussion not only distinguishes the areas of doubt, but also emphasizes the degree of uncertainty likely to enter.

4.1 Image Force Model

First in the list of assumptions on which the usual picture of field evaporation is based is the premise that the high field strips the evaporating atom of electrons prior to evaporation. Evaporation can then be viewed as the escape of an ion over a Schottky hump. This assumption provides an explicit form for the potential curves involved in the evaporation. Without it, determination of the field for evaporation would not immediately yield quantitative information on binding energies.

The image potential is valid for a charged particle at some distance from a semi-infinite metal surface. In fact, in discussing field evaporation, Tsong and Müller[51] argue that the expression for the image potential maintains its exact form, without any need to account for field penetration in real metals. Recently, however, there has been considerable theoretical effort to establish the interaction energy between a point charge and a semi-infinite metal.[52,53] There now appears general agreement that this potential can be expressed as $-3.6n^2/x$, provided x denotes the distance to the classical image plane *plus* the screening length Δ, as has been the convention in this paper. The screening length, however, is not the usual Thomas-Fermi value $1/\lambda$, but has instead been approximated as

$$\Delta = \lambda^{-1} + \pi(4k_F)^{-1} \quad ,$$

where k_F is the wave vector at the Fermi surface.

At smaller separations, the surface structure may no longer be represented by a perfect plane; the shape of the sample will then affect the interaction with the ion, as pointed out by Lewis[54]. Closer in still, at distances on the order of the atomic spacing, we can expect quantum mechanical effects to cause deviations from the image potential.[55] At fields typical for the evaporation of metals, the Schottky hump is within an angstrom of the equipotential surface. At such distances the simple image model must break down, seriously impairing the a priori validity of the basic field-evaporation expression.

To make useful predictions, or to deduce binding energies from experimentally observed evaporation fields, one requires information on the polarizability term. This is an extremely uncertain matter. What enters here is α, the difference between the polarizability α_o of the atom on the surface in the high field in which desorption occurs, and the polarizability of the evaporating ion α_+. The only way to determine this correction appears to be through desorption experiments in high fields.[56] If we accept the validity of the model, then the difference between the experimentally observed field and that calculated ignoring P gives us the correction due to polarization.

An alternative, more soundly based method is to measure the field dependence of the rate of evaporation. From Eq. 2 we find that

$$\left(\frac{\partial \ln k_E}{\partial \ln F}\right)_T = \frac{1}{kT}\left(\frac{3.795}{2} n^{3/2} F^{1/2} - \alpha F^2\right) \tag{5}$$

so that the polarization term is directly accessible to experiment.

The activation energy for field evaporation, $\chi_n^{\ddagger}$, can be derived as usual from the temperature dependence of the rate. Since

$$\chi_n^{\ddagger} = \chi_o^o + I_n - n\phi + \frac{1}{2}\alpha F^2 - 3.795 n^{3/2} F^{1/2} \quad , \tag{6}$$

and we know the polarization term from the field dependence, the binding energy of the atom can be derived from these measurements. It should be clear, however, that at the low temperatures (T < 200 K) at which these experiments must be done, the activation energy $\chi_n^{\ddagger}$ is very small ($\chi_n^{\ddagger} < \frac{1}{2}$ eV). The value of χ_o, the atomic binding energy, is therefore dependent entirely upon the field terms. Measurements at different fields have been carried out for tungsten, molybdenum, and platinum, yielding $\alpha = 3.5$ A^3, 5 A^3, and 8.3 A^3, respectively.[46] At the evaporation field this amounts to a correction on the order of 4 eV, compared to values less than half that obtained from the difference between the actual and predicted evaporation fields (in Table IV).

It is clear that the polarization correction can be significant. The discord between estimates derived by different methods suggests some deficiency in the approach. Note furthermore that the polarization is likely to be site-specific.[57] To obtain binding energies for atoms at different planes, knowledge of the correction at the specific sites is necessary. The field dependence of the evaporation rate of atoms in that particular configuration could again be used to determine the polarizability term. This is a much more difficult procedure, however, than for evaporation of the surface as a whole.

Granting for the moment the validity of the model, as well as a knowledge of the polarization, it is still necessary to know both the ionization potentials I_n and the work function ϕ of the material studied. The latter have been quite well documented, and average work functions for a variety of clean surfaces are known.[58] Again the work function varies from one plane to the next. Any attempt to determine binding energies rests upon a knowledge of the work function for the plane on which an atom is held. So far such data are available only for silicon, germanium and a limited number of refractory metals: tungsten, tantalum, molybdenum and rhenium. Unfortunately it is precisely for these metals that information on ionization potentials is sparse. Moore's[59] authoritative summary on ionization potentials can sometimes be misleading, as many of the values listed for the refractory metals have been obtained by extrapolation schemes, and are by now outdated.

Finally, the binding energy deduced from field desorption depends upon the charge state n of the ion formed. This has been predicted by calculating the field for evaporation of differently charged entities using Eq. 3; that charge is assumed to evolve for which the evaporation field is lowest. This procedure is certainly consistent, but robs us of another independent check of the validity of the basic field evaporation model.

Measurements of the charge state of ions have recently become available from studies in the atom probe.[60] In this instrument, shown schematically in Fig. 5, ions created by field evaporation are allowed to drift

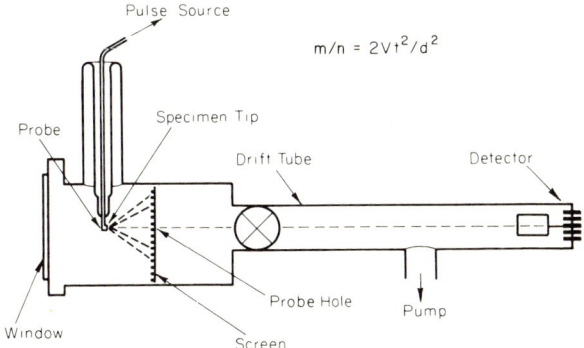

Fig. 5. Schematic of atom probe field ion microscope.

down a field free space; the time of flight t to a detector a known distance d from the emitter is then accurately measured. From a knowledge of this time and of the kinetic energy of the particles (that is, of the potential V through which they have been accelerated), the charge-to-mass ratio n/m can be immediately determined according to $n/m = \dfrac{d^2}{2Vt^2}$.

The earliest studies of this sort established an anomaly.[60,61] On the image force model, formation of charge state +2 is predicted for most metals. There is an occasional element that we can expect to desorb with charge +1 in a high field; none are predicted to evolve in charge state +3.[46] The experiments revealed quite a different behavior – triply and often quadruply charged ions have been reported for many elements. A partial qualitative survey of this phenomenon is given in Table V.

Several factors may be involved in the formation of these highly charged ions. As most of the work reported has been done under moderate vacuum conditions, the possibility exists that adsorbed layers may have exerted a significant effect.[62] Many of the charge-state predictions are based on values of the ionization potentials with little experimental foundation. The possibility has also been raised that such highly charged entities may not be the product of field evaporation. They could instead be formed away from the surface, by further ionization in the high fields surrounding the emitter, but after the evaporation process is completed.[62]

Table V. Ion Abundance in Atom Probe

	Charge State				Ref.
	+1	+2	+3	+4	
Fe	16	60	2	–	(62)
Rh	–	37	14	–	(62)
W	1	3	275	2	(61)
Ir	–	7	1	–	(61)

Another uncertainty enters because in atom-probe experiments, in which the ion is removed by a very rapid pulse (with a rise time of nanoseconds), the instantaneous field operating on the tip is difficult to define. It is therefore not inconceivable that ions could be desorbed at fields above the minimum value necessary for evaporation.[63] Last, the standard picture of the field evaporation process may just be seriously lacking – which, in view of the many assumptions already mentioned, would not be surprising.

The full implications of atom-probe studies for the mechanism of field evaporation are not yet clear. However, it appears that surface contamination is not the sole culprit. Studies in an ultrahigh-vacuum atom probe, by Chambers et al.[63] have revealed Mo^{+3} as well as Mo^{+2}. This is the only system for which meaningful conclusions can be expected: of the more refractory elements, it is the only one for which both the work function and the higher ionization potentials have been adequately documented. The efficiency of postionization has also been examined carefully by these investigators; from their work it emerges that this is insufficient to account for the high charge states observed. The possibility exists that some of the highly charged ions may be created due to local field enhancement at the surface. A closer study of these phenomena is in order. However, the findings in the atom probe undermine confidence in the validity of the classical view of field evaporation.

4.2 Charge-Exchange Model

An alternative to the traditional view of field evaporation in metals is to consider the evaporation along the lines useful in dealing with atoms covalently held at a solid.[45,56] Just as before, evaporation at high fields now involves a transition from an atomic to an ionic state. The limiting step is not evaporation of the ion, however, but the crossing from the atomic to the ionic potential curve, as indicated in Fig. 6. The activation energy for field evaporation, $\chi_n^{\ddagger}$, is now just the difference between the potential energy of the atomic ground state

$$V_a(F) = -\chi_o^o - \frac{\alpha_o F^2}{2}$$

and the potential of the atomic or ionic curve at the crossing point x_c, where

$$V_i(F,x) = I_n - n\phi - nFx - \frac{1}{2}\alpha_+ F^2 + V_i(0,x) \quad . \tag{7}$$

The first two terms relate the potential energy of the ion to that of the vacuum level of the atom; the third term gives the lowering of the

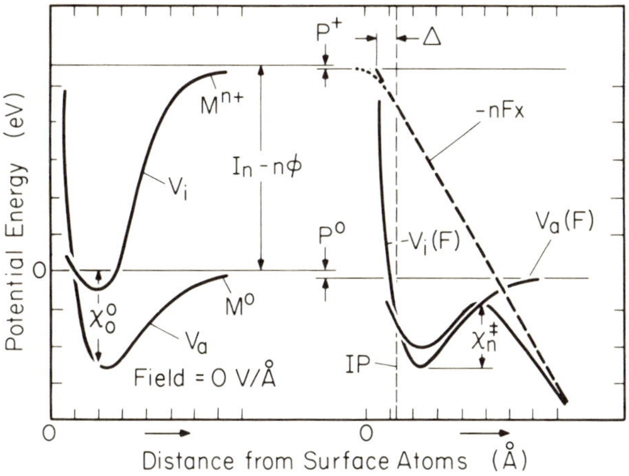

Fig. 6. Potential diagram for field evaporation in charge-exchange model. Schottky hump is to the left of intersection between atomic and ionic curves.

potential energy of the point charge by the field (with x including again the screening distance Δ). The usual polarization term, together with the interaction energy of the ion with the surface, complete the relation for the ionic potential. Note that effects such as level broadening have been neglected here.

The activation energy is now given as

$$x_n^{\ddagger} = V_i(F, x_c) - V_a(F)$$

$$= x_o^o + I_n - n\phi - nFx_c + \frac{1}{2}(\alpha_0 - \alpha_+)F^2 + V_i(0, x_c) \quad . \tag{8}$$

To make any immediate predictions, we need all the usual parameters, together with information on the crossing point x_c as well as on the value of the ionic interaction energy $V_i(0, x_c)$. For the latter we again have to assume the image potential. Since we do not know the course of the potential for the atoms, some arbitrary assumption has to be made to fix the crossing point x_c. It appears that in this formulation of the desorption process, it is necessary to make even more assumptions than in our previous model to relate the desorption field to the binding energy of the atom, x_o^o.

A direct experimental evaluation of many of the significant parameters involved is possible,[56] however, by following the procedures already outlined for the image force model.

From the variation of evaporation rate with field we obtain

$$kT\left(\frac{\partial \ln k_E}{\partial F}\right)_T = nx_c + \alpha F \quad ,$$

$$\alpha = (\alpha_0 - \alpha_+) \quad , \tag{9}$$

assuming that the crossing point x_c does not depend sensitively upon the field. Both nx_c and the polarization correction are therefore immediately available. The temperature dependence of the evaporation as usual yields $\chi_n^{\ddagger}$. Provided the ionic interaction term $V_i(0,x_c)$ is known (together with I_n and ϕ), we have available all the parameters necessary to fix χ_o^o.

Tsong and Müller[56] have recently carried out this analysis for the field evaporation of a tungsten tip, and find $\alpha = 3.44$ Å^3. Assuming that tungsten evaporates with a charge $n = +3$, they also derive a value of 0.55 Å for x_c, the location of the crossing point. The polarizability term is in excellent agreement with the value previously found by Brandon[46] assuming an image force model for evaporation. An activation energy $\chi_n^{\ddagger}$ of 0.16 eV was determined at a field of 5.7 V/Å. Since the energy of vaporization is known, it is of interest to check the validity of the image potential. With the values of the various constants given in Ref. 48, this $\chi_n^{\ddagger}$ leads to an ionic potential of -23.5 eV, compared to -26.2 eV predicted by the image potential. Note that this amounts to a disparity of only 0.6 Å in x_c.

It should be apparent that the procedures for deriving information from field evaporation experiments is much the same for both models of the evaporation process. In principle, the appropriate model could be selected if the potential curves and other parameters (α, x_c, I_n, ϕ) were available. This is not the case at the moment, nor are such data likely in the near future. In lieu of this, a critical comparison of the field dependence of the evaporation process predicted by the two models might be instructive. If evaporation over a Schottky hump is limiting, then $(kT \ln k_E - 3.795 n^{3/2} F^{1/2})$ should vary as F^2. If charge exchange is the slow process, however, the log of the evaporation rate should best be represented by a 2nd order polynomial.

The results of analyzing the data of Tsong and Müller[56] in both ways are shown in Fig. 7. The image potential model gives a considerably better fit of the data, yielding a polarizability of 4.7 Å^3. The scatter of the data is too large, however, to allow a convincing decision. In fact, within the last few months Tsong[64] has presented new measurements of field evaporation for tungsten; from these he deduces a value of $\alpha = 4.6 \pm 0.6$ Å^3 — quite different from his earlier results.

Both approaches to field evaporation suffer from the same difficulties. Both require a knowledge of the higher ionization potentials often not available for the more interesting metals; both rely on a knowledge of the potential energy of the ion in proximity to the surface, and this is the real limitation in determining the binding energies. In principle it is possible to avoid this problem. Studies of field evaporation at higher

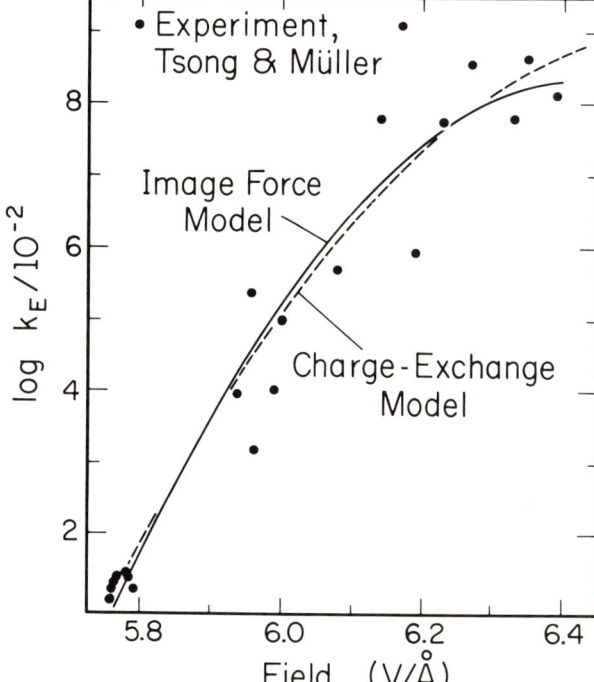

Fig. 7. Predictions of image force and charge exchange model for field evaporation of tungsten. Experimental data from Ref. 56.

temperatures, that is, at lower fields, would place the peak of the barrier further from the surface, insuring the validity of the image potential for the ion. Unfortunately there are fundamental limitations on the temperature range that can be explored. When determining the binding energy of atoms on different sites, the temperature must be low enough to prevent significant diffusion while the field is on. Using short-duration pulses it may be feasible to carry out measurements up to 200 K. For tungsten atoms on the (211), we estimate that at this temperature evaporation would occur with a rate constant of 1 atom/sec at ~5.2 V/A. The peak of the barrier on the image force model would still be only 1.2 A from the image plane – not a significant improvement over low-temperature measurements. At the moment, progress in this particular area appears to depend upon further theoretical effort.

4.3 Experimental Studies of Atomic Binding

Attempts to measure the binding of metal atoms on different planes of a tungsten crystal have been made by two groups: studies on tungsten atoms were carried out by Ehrlich and Kirk[48]; these were confirmed and extended to a series of 5-d atoms, such as rhenium, tantalum, and platinum, by Plummer and Rhodin[65]. Both rest on the application of the image force model to the interpretation of the experimentally determined desorption fields.

The procedure in the first of these studies was to deposit atoms on a field evaporated emitter from a resistively heated source. As is apparent from Fig. 8, individual atoms are clearly revealed on the different planes. The voltage necessary for desorption was then determined by increasing the d-c potential on the emitter in small increments, imaging the surface between each voltage application to establish the location of the atoms removed. Only during the ion microscopy was image gas present; desorption was carried out in an ultrahigh-vacuum after the image gas had been pumped out.

For a given applied potential, the field may vary by as much as 30 percent from one plane to another.[66] The correction required to account for this can be determined from the variation in voltage necessary to maintain optimal focus of the ion image.[67] The connection between voltage and field, $F = cV$, is made through measurements of the current-versus-voltage characteristics for the field emission of electrons; since the work function of the tungsten surface is known, the Fowler-Nordheim relation, which establishes the electron emission as a function of the field, yields the necessary proportionality constant c.[68]

As already pointed out, there is serious doubt about the validity of the image force model. To check the applicability of the model, measurements were made of the field for evaporation of the tip as a whole, a process for which the desorption energy is just the tabulated value of the heat of vaporization. Unfortunately, spectroscopic values of the higher ionization potentials for tungsten are not available. The values listed by Moore are extrapolations from elements in the neighboring rows, in part based on outdated material.[48] New extrapolations were used to arrive at $I_2 = 26.2$ eV and $I_3 \sim 57$ eV. With these values it appears that the image

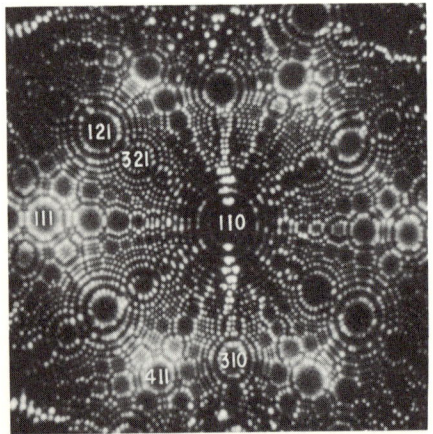 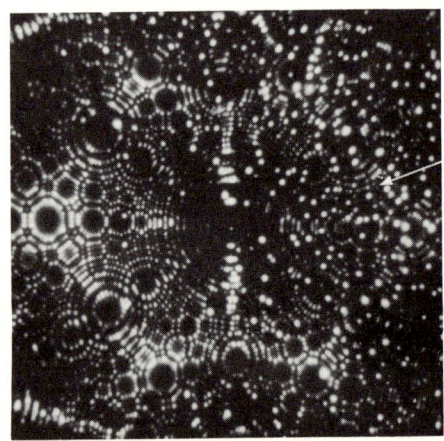

Fig. 8. Deposition of tungsten atoms on a tungsten surface at 20 K. Left: clean tungsten surface produced by field evaporation. Right: same surface, after exposure to W atoms. Source direction indicated by arrow.

force model gives a reasonably good account of the field evaporation, the polarization term amounting to only 6 percent of the heat of vaporization. The polarizability derived from this correction was then used to correct the atomic binding energies on different planes. As already pointed out, this agreement is no assurance of the validity of the approach. A correction amounting to more than 5 eV is derived from the field dependence of the evaporation rate. For molybdenum, which should behave quite similarly, the necessary physical constants are well established. For this metal the data in Table IV indicate a significant disparity between the experimental desorption field and the value predicted by ignoring polarization.

The procedures in the only other binding study presently available, that of Plummer and Rhodin,[65] are generally much the same. There are, however, some significant differences, the most important being that desorption was carried out in the presence of the image gas, using a pulsed field. Pulses of less than 100-microsecond duration were found to suppress any effect due to the presence of the gas. This assertion has recently come under some suspicion.[56] In their atom probe work, Müller et al[69] were able to identify products such as WHe^{+++} and WHe_2^{+++} when tungsten was field desorbed at a helium pressure of 10^{-4}mm. In a field of 4.5 V/A, the image gas is held to the surface with an energy of 0.14 eV. It is this adsorbed layer that appears to be involved in the reduction of the evaporation field by the image gas which has been observed in the past. Even when the field is relaxed, this layer disappears only slowly, so that pulsed fields are unlikely to avoid the effects of the image gas. The details of the mechanism by which metal-helium ions are formed are certainly not yet clear. It does appear, however, that at temperatures up to 77 K it is a helium-metal complex rather than just an atom that evolves during field evaporation in the presence of helium gas. The extent to which the models previously described are applicable is therefore placed in even more serious doubt, and measurements done in the presence of helium must be treated with caution.

The results of the two studies on tungsten atoms are listed in Table VI. It is of interest that the desorption fields found for the various planes are in reasonably good agreement. The desorption energies differ, but these disparities arise from different choices of the work functions and ionization potentials. The most striking feature of the measurements is the relationship between binding energy and surface structure revealed here. It has generally been believed, largely on the basis of nearest neighbor models, that binding on atomically smooth planes would be weaker than on rough surfaces. This is not at all the trend apparent from the field desorption studies. In fact, on very rough planes such as the (411) (an atomic model of which is shown in Fig. 9) the binding energy found from field desorption is smaller than on a relatively close-packed surface such as (211). These trends are examined more closely later.

Table VI. Binding Energy of Tungsten Adatoms[a]

Plane	ϕ, eV		Desorption Field, V/Å		Binding Energy, eV	
	E&K	P&R	E&K	P&R	E&K	P&R
(110)	5.5	8.0	3.68	3.45	5.3	8.2
(100)		4.80	–	5.01		8.0
(211)	4.88	4.90	4.88	4.52	7.0	6.9
(310)	4.34	–	5.21	–	6.7	–
(111)	4.40	4.45	4.88	4.79	6.0	6.7
(321)	4.54	4.60	5.02	4.98	6.7	7.4
(411)	4.40	–	4.96	–	6.2	–

(a) E&K, Ref. 48; P&R, Ref. 65.

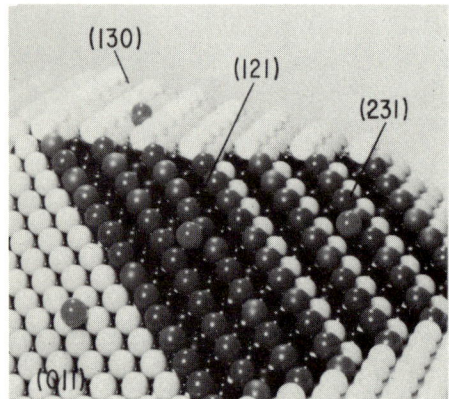
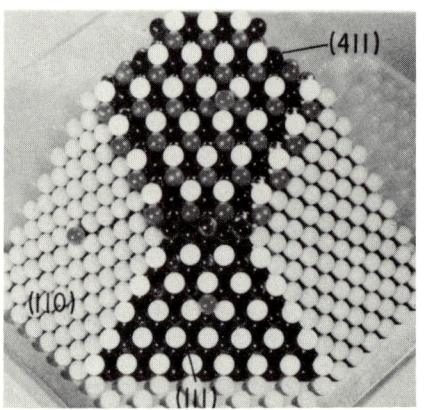

Fig. 9. Hard-sphere model of body-centered-cubic surfaces.

Attempts have recently been made to measure the relative binding of atoms at various lattice steps and corners by determining the relative rates of field evaporation. These measurements have not yet been checked in the absence of image gas and may therefore be difficult to interpret. The field evaporation of tungsten atoms adsorbed on the (110) has also been determined as a function of the field. From these studies, Tsong[64] deduces a polarizability $\alpha = 6.8 \pm 1$ Å^3. No attempt was made to examine the binding.

5 DIRECT DETERMINATION OF ATOMIC MIGRATION ENERGIES

It should be apparent that the full power of field desorption studies will be realized only once the theory of these simple experiments is placed on a sounder footing. At the moment the information available is titillating, but hardly conclusive.

Somewhat less direct information on binding energies is available, again through the FIM, by methods that are experimentally more complicated, but rigorous. This is achieved through studies of atomic motion on single crystal planes.[70] As we have already seen (Fig. 8) the field ion microscope is able to locate individual atoms adsorbed on a surface, even though it may not always be able to resolve the arrangement of the lattice atoms underneath. It is therefore for the first time possible to observe Brownian motion directly.

5.1 Random Walks and Surface Diffusion

In surface-diffusion experiments on macroscopic samples, only the overall mass transport[71] has been observed, presumably involving the transfer of atoms from lattice edges onto the flats, as suggested in Fig. 10. The individual steps in this overall process have not been accessible to direct study in the past. The closest anyone has come to this is in the growth of whiskers. There the absence of dislocations insures that the limiting step in the growth is actually the transport of material over the atomically smooth sides. However, the analysis of these experiments is based upon approximations to the diffusion equation which limit the validity of the energetics deduced.[72]

Information on the migration of atoms is of considerable interest, as this constitutes one of the significant events in the kinetics of growth and evaporation, as well as in shape changes of crystals. In our context, however, the migration can be viewed primarily as an experimental probe of the interatomic forces at the surface. In diffusing over a plane, at temperatures at which an atom hops from one equilibrium adsorption site to another, the rate of jumping to any one of the nearest-neighbor sites is given in the usual approximation of absolute rate theory by

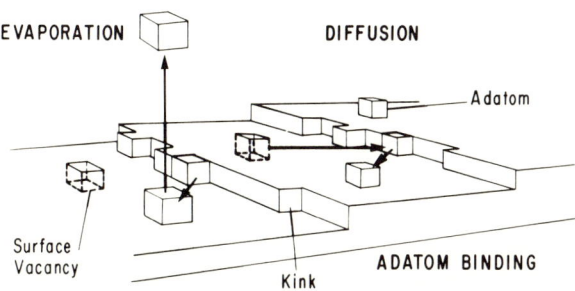

Fig. 10. Schematic of atomic processes participating in mass transfer over a surface.

$$w = \nu \exp\frac{\Delta S_m^{\ddagger}}{k} \exp -\frac{V_m^{\ddagger}}{kT}. \tag{10}$$

Here, ν is the vibrational frequency of the atom in the direction of the motion, and $\Delta S_m^{\ddagger}$ and $V_m^{\ddagger}$ are the entropy and energy of activation. The latter two quantities can be visualized, in analogy with a semiempirical description of diffusion in the bulk[73], as indicating the changes of the entropy and energy of the system when an atom is moved from the equilibrium site to the saddle point, while confined to a plane perpendicular to the direction of motion.

The mean-square distance $\langle R^2 \rangle$ covered by an atom executing a random walk of many steps during an interval τ, is just

$$\langle R^2 \rangle = K w \tau \ell^2, \tag{11}$$

where K is a numerical constant that accounts for the geometry, and ℓ^2 is the mean-square jump distance. For a (100) plane in the bcc lattice, c is 4 and $\ell = a_0$, where a_0 is the lattice spacing. For all practical purposes these quantities are temperature independent, the variation in the distance covered arising entirely through the jump rate w. The mean-square distance for unit time interval is simply related to the activation energy $V_m^{\ddagger}$ through

$$\ln \frac{\langle R^2 \rangle}{\tau} = \ln K\ell^2 \nu + \frac{\Delta S_m^{\ddagger}}{k} - \frac{V_m^{\ddagger}}{kT}; \tag{12}$$

the latter can be obtained as usual from a semilogarithmic plot against $1/T$. There are really only two assumptions inherent in this analysis.
(i) The atom describes an uncorrelated random walk.
(ii) Transfer from an equilibrium site to another occurs by jumping over the barrier rather than by tunnelling through it.
Jumping is readily shown as the dominant process by an order of magnitude calculation. That the atom actually moves at random from site to site, rather than covering long distances once activated, is certainly in accord with our present understanding of energy accommodation by a lattice. Model calculations suggest that an atom colliding with its own crystal is thermalized rapidly, within a period measured in tens of atomic vibrations. As shown later, this thermalization is also directly susceptible to observation in the FIM. On colliding with a site, the jumping atom should be trapped very efficiently; when it is activated to go on to its next jump, an atom will do so without any knowledge of its prior history.

The assumption that the jump frequency ν, the length ℓ, and the entropy of activation are temperature insensitive is not crucial to the empirical determination of the barrier to migration. Deviations from constancy would just manifest themselves as curvature in the Arrhenius plot if taken over a sufficient temperature interval.[74]

The activation energy is not as informative as the binding energy of an atom to the lattice; it is after all a difference – between the energy of an atom at the saddle point for migration and the energy at a normal site. From this it may never be possible to deduce unequivocally the forces acting on an atom. At the very least, the migration energy should, however, provide a rigorous test for any proposed atomic potential.

It is important to realize that in moving over a crystal the atom does not just sample the potential field of a rigid array of cores – the lattice itself will be deformed in the vicinity of the migrating atom. In the interpretation of the experimental quantities, the configuration of the saddle point as well as of the normal sites must be kept in mind.

5.2 Diffusion Measurements in the Field Ion Microscope

At the temperature at which a metal surface can be viewed without introducing thermal disorder ($T < 600$ K), the concentration of atoms adsorbed on a smooth lattice plane is negligible. For example, on surfaces warmed to room temperature no transfer of atoms from edges to flats is observed. However, it is a rather simple matter to place atoms on various surfaces by evaporation from a source aimed at the sample, just as in measurements of the binding energy of atoms (see Fig. 8).

Determination of the diffusion parameters rests entirely on measurements of the mean-square displacement $<R^2>$. In principle, this determination[70] proceeds as follows:

An atom is deposited on the plane of interest, with the surface maintained at a low temperature. Its location is determined by FIM at 20 K. The lattice is heated to the temperature of the diffusion experiment; migration is allowed to proceed for a time chosen so that the interval during which the temperature is changing makes only negligible contributions to the diffusion. The surface is then cooled. No fields are applied to the surface during this period, and the sample is maintained in an ultrahigh-vacuum environment to ensure freedom from any extraneous effects. Only when the lattice has been restored to a low temperature, at which atom mobility can be expected to be negligible, is the imaging gas introduced. The field necesssary for imaging the surface, to establish the location of the atom after diffusion, is applied only for the briefest possible interval. What the FIM affords, therefore, is a measure of the displacement of an atom during a known time interval τ, at a temperature T. Such a diffusion sequence is shown in Fig. 11.

Can such measurements be made without disturbing the location of the atom during imaging? This can be checked by taking a series of field

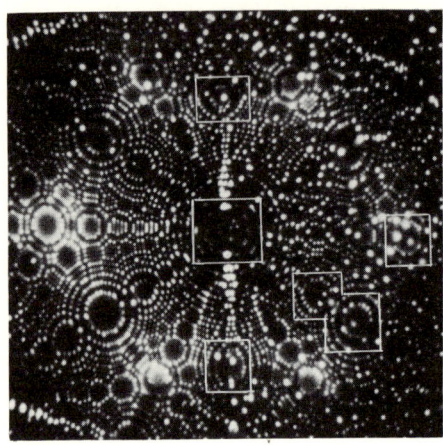 

Fig. 11. Motion of W adatoms at ~320 K. Left: Atoms deposited on clean tungsten surface. Right: Diffusion has not taken place on (11$\bar{1}$), (310), and (130); atoms on other planes outlined by rectangles have moved after 300 sec at 320 K.

ion micrographs at 20 K. In measurements on tungsten atoms adsorbed on various surfaces, no displacement has been observed during such a series, indicating that the position of an atom is not significantly affected by the act of observation.

The mean-square displacement of atoms on a given plane, the quantity of interest for the diffusion studies, is obtained by observing a sequence of displacements, at a fixed temperature, squaring each and taking the average. There are at least four difficulties in this procedure.

Distance determination. This is difficult to carry out quantitatively in the FIM, as the magnification varies considerably over a curved surface.[75] However, on planes for which atomic structure is resolved, this itself serves as a calibration. Semiquantitative estimates are possible for smoother surfaces, based on the average radius of the emitter tip which can be determined by direct observation. Finally, we know that the minimum distance an atom can jump is equal to the spacing of the surface. This can just be used to calibrate the excursions observed on a photograph of uncertain magnification. In the last analysis, a quantitative knowledge of the distance is not important for a determination of the energetics. As long as measurements at different temperatures are made on the same surface, the distance scale used never affects the activation energy.

Correlation effects. Two effects may vitiate the assumption that the atoms execute a random walk on the surface. One is the possibility that atoms once excited require many collisions with the lattice sites for deactivation. Under these circumstances correlation could be expected between jumps to subsequent sites. Apart from the calculations[76] of energy transfer, which suggest that this possibility is unlikely, the rate of localization can be observed directly[77]. Consider what happens when an emitter is illuminated from one side by a beam of atoms. If atoms are localized at the site of initial impact, then the boundary of the deposit should be identical with that obtained if the surface had just been exposed to a light source. In the other extreme, when atoms remain excited even after many collisions with the lattice, the tip would be uniformly covered by atoms, without a strong preference for the side in direct line with the source. Such observations have been made by Ehrlich[77], and also by Young et al.[78]; both indicate only a very limited transgression of atoms across the shadow line.

Correlation effects may also enter because of the limited size of the surfaces on a field ion source. For an atom at a boundary, the assumption of randomness is not maintained. An atom coming to the edge of a lattice plane may just roll over the side and then be held permanently at a step. In this event the random walk ceases once the atom reaches an edge. Alternatively, we may imagine that the lower coordination of edge atoms increases the potential energy of an atom adsorbed there, compared to an atom held on a large flat. The lattice edge then would act as a reflecting boundary. Once an atom strikes a boundary, the next step is no longer taken at random – the atom must reverse its direction.

Corrections for such boundary effects can be made in the expression for the mean-square distance covered by an atom.[79] The important point that emerges from these is that $<R^2>$ no longer varies linearly with the number of jumps or the time, but involves higher powers as well. It is reasonably simple to restrict measurements to time intervals for which these corrections are negligible. That the atom behaves in accord with Eq. 11 can also be established by changing the interval over which diffusion is observed. Doubling the interval, for example, should double the mean-square displacement. This direct experimental check should serve as a final validation of the observations.

Atom identification. In principle all observations could be made on a single atom adsorbed on a crystal plane. The interpretation of the measurements then becomes quite unequivocal, but the measurements would be tedious in the extreme. The general procedure, therefore, is to deposit many atoms on the surface, so that statistics on the distances covered during a diffusion interval can be obtained rapidly. One of the drawbacks of this technique is that identification of the displacement of any one atom is no longer possible. We have no certain method for deciding which atom moved where. This difficulty is not overwhelming, however. For a random walk, small excursions are preferred; provided the atoms are not too close together initially, this rule makes it readily possible to determine diffusion distances. Consistency of particular assignments can be checked by analyzing the same micrographs, but merely increasing the interval over which diffusion is allowed to occur. If the assignments are valid, the mean-square distance should increase linearly with the length of the diffusion interval.

Temperature measurements. Only a limited range of temperatures is accessible in surface-diffusion studies. An upper limit is set (at ~ 600 K) by the instability of the highly perfect surface[80]; a lower limit is determined by the cessation of diffusion at ~250 K. The actual range varies from one system to the next, as well as with the particular plane examined, and is affected by extraneous factors such as diffusion of impurities from the sides of the emitter. Activation energies good to approximately 5 percent can be determined, however, provided the tip temperature is calibrated and controlled by what are now standard procedures.

None of the problems in determining mean-square displacements are critical — they can all be overcome by reasonably routine experimental procedures. Although tedious and time-consuming, such measurements should yield sound values for migration energies as well as other parameters of interest.

5.3 Experimental Studies of Surface Mobility

Observations of atomic mobility by FIM have been quite limited. Müller indicated the possibility of such studies long ago.[81] Quantitative measurements were not attempted until the work of Ehrlich and Hudda, who examined the diffusion of tungsten atoms on three planes of a tungsten crystal.[70] Their results are listed in Table VII. The structures of the three planes are rather different, as is apparent from the atomic models in Fig. 9. The (211) and (321) planes are characterized by protruding rows in the [111] direction. On the (110) plane, these rows are adjacent to one another, forming a smooth and reasonably closely packed plane. In conformity with these structures, the diffusion is observed to occur preferentially along the [111] on both the (211) and (321).

Only a limited number of planes is accessible to quantitative observation — the size of the planes diminishes as the Miller indices become higher. Furthermore, those planes located away from the apex are more susceptible to contamination. For these reasons, the only other study of atomic mobility, by Bassett and Parsley,[82] concentrated on the same planes. Their measurements for tungsten atoms quantitatively confirmed the earlier studies, as indicated in Table VII. These investigators also examined tantalum and rhenium atoms. It is of interest that the general trends on the three planes are much the same for the different atoms as well.

Table VII. Surface migration on tungsten.

	W[a]		W[b]		T_a[b]		Re[b]	
	D_o, cm²/sec	V_m, eV	D_o, cm²/sec	V_m, eV	D_o, cm²/sec	V_m, eV	D_o, cm²/sec	V_m, eV
(110)	$3 \cdot 10^{-2}$	.96	$2.1 \cdot 10^{-3}$	.87	$4.4 \cdot 10^{-2}$	.78	$1.5 \cdot 10^{-2}$	1.04
(211)	$2 \cdot 10^{-7}$	.57	$3.8 \cdot 10^{-7}$	.58	$9 \cdot 10^{-8}$	.49	$1.1 \cdot 10^{-2}$	.88
(321)	$1 \cdot 10^{-3}$	.87	$1.2 \cdot 10^{-3}$	.84	$1.9 \cdot 10^{-5}$	.67	$4.8 \cdot 10^{-4}$	.89

(a) Ref. 70. (b) Ref. 82.

In principle, additional information should be available through analysis of the distribution curves governing individual displacements. The standard Rayleigh distribution is applicable only if all jump directions in a plane are equally likely or, equivalently, if jumps occur with equal probability along the Cartesian coordinates. In either situation, the number of jumps during any time interval must be large compared to unity.[79] Neither condition is met in the experiments so far available. A detailed evaluation of the distribution observed for atomic displacements is not easy, however. Many more experimental observations are required for this than for the mean square displacement. Furthermore, if these data are obtained as usual by examining the motion of many atoms on one plane, difficulties in the interpretation arise — the atom may associate into dimers and higher upon collision.

The only attempt of this sort indicated a behavior quite different from a Rayleigh distribution.[70] Most of the atoms were actually concentrated at the origin, where the Rayleigh distribution predicts none. This of course would be the behavior if atom clusters are present and are immobile compared with the monatomic species. However, this could also be expected for a random walk in which the number of jumps is very small, and therefore subject to severe fluctuations.

Observations of atom association and dissociation have been reported.[83] As yet, these are not really quantitative. Only the dissociation of weakly held clusters of iridium on the (110) of tungsten has been examined; for dimers the bond strength is as low as 0.07 eV. While this is a very interesting phenomenon in its own right, it makes more difficult the determination of jump statistics — these will have to be measured in single-atom experiments. Difficult as these may be, they should be worthwhile as a means of probing the atomic mechanism of jumping.

6 ANALYSIS OF ATOMIC MEASUREMENTS

The results just outlined, tenuous though they are, for the first time provide data against which to compare our present notions of atomic forces. A few words about the implications of these results appear appropriate here, in order to focus more clearly upon the unexpected as well as upon those aspects that will require further study.

In the past, predictions about atomic behavior at interfaces have been based largely on the concept of nearest-neighbor bonding. Although recognized as an oversimplification, the ease of making predictions (especially with the aid of recently developed counting techniques[84]) has done much to perpetuate this point of view. On any such model the rough surfaces, on which an atom is most effectively surrounded by lattice atoms, should show the highest binding energies. Also on such rough surfaces, atomic movement from one equilibrium site to another must occur over highly protruding sites, at which the coordination of the migrating atoms is low. The barrier to diffusion should likewise be high on atomically rough planes and low on smooth planes. Neither of these expectations is satisfied by the measurements just presented.

To pinpoint the disparities more clearly, it is useful immediately to examine the next stage of sophistication in simulating atomic behavior, in which longer range interactions are allowed. Most of the calculations on defect properties in solids have been done with Morse potentials. The three parameters defining the potential have been evaluated for a number of solids by Girifalco and Weizer[85], using the heat of sublimation, the equilibrium lattice spacing, together with the compressibility. In Table VIII are listed the binding energy for tungsten atoms, as well as the energies of migration, obtained by summing the interaction between the adatom and nearby lattice atoms. In these estimates the atoms are allowed to sit at a normal lattice site, and the crystal is assumed to terminate at the appropriate surface without any change in the spacing. This of course is quite unrealistic, as the atomic spacing will not be maintained at the bulk value, but will instead assume the value that minimizes the total energy, at least at low temperatures. However, this relaxation causes only small quantitative changes in the binding energies[48,86] and does not alter the fundamental disagreement between the experimental values for atomic binding and those calculated from Morse potentials. The calculated values increase as the roughness increases; the experimental values seem to be relatively constant with variations in the structure of the surface. If anything, the experiments suggest a decrease in the binding energy.

There is a particularly startling behavior in the diffusion of atoms on their own lattice. On the atomically smooth (110) the barrier to migration is as high as on the rough (321). On the (211), a plane with a structure similar to that of the (321), the barrier amounts to only 60 percent of that on the (321). It is this contrast that is important. One possible way of interpreting the difference between (211) and (321) is to invoke differences in the binding energy of atoms on the two planes caused by differences in the second-nearest-neighbor environment. These would not be adequately described by the Morse potential calculations.

It has long been recognized that the use of pair potentials in cohesive energy calculations can be justified for a metal only under limiting assumptions about constancy of volume[87], since the displacement of one ion core will exert a general effect upon its neighbors through a redistribution of the screening electrons. The difference in the atomic arrangement of the (211) and (321) leads to a different screening of cores at the surface, and it has been suggested that this could account for the big difference in the diffusion barriers observed on the two planes.[70] Unreliable though the absolute values of the binding energies measured by field evaporation are, the close agreement found on the (211) and (321) suggests that this is not the appropriate explanation, even though differences in electron distribution may well be accountable for the remarkable behavior of the (110) plane.

Instead it appears appropriate to assign the differences in the behavior of the (211) and (321) to the detailed dynamics of atom motion on the two planes.[48] The correct explanation is already suggested by the remarkably low value of the frequency factor on the (211), which is orders of magnitudes smaller than on the (321). It should be recalled that the activation energy for surface migration mirrors variations of the atomic potentials only if the lattice is assumed to be rigid and unaffected by the migrating atom. On the (211) these assumptions are unlikely. This plane is

Table VIII. Morse potential estimates for tungsten adatoms on tungsten, eV.

	Binding Energies[a]		
(110)	5.82	(111)	8.10
(211)	7.76	(321)	8.18
(310)	7.67	(411)	8.42

	Migration Energy		
		Relaxed Lattice	
	Rigid[b]	E&K[b]	W&G[c]
(110)	.47	.44	.54
(211)	.95	.78	.43
(321)	.96	.91	—

(a) All atoms in normal sites; sum out to three lattice spacings.
(b) Ref. 48.
(c) Ref. 86.

made up of [111] rows of tungsten atoms protruding above the surface. In migrating over the plane, an atom must jump over a considerable barrier in moving along a rigid [111] channel of atoms. Alternatively, the atoms in the [111] rows may participate in the diffusion process, moving aside to allow the migrating atoms to roll over a much reduced barrier. Now the overall activation energy for the process must include the energy required to displace the lattice atoms from their normal position. If this is a likely process on the (211), we would not expect it to be significant on the (321). On this plane a migrating atom is in contact with only one protruding atom row in the lattice; the other side of the diffusion channel is made up of lattice atoms embedded in a flat. On the (321), therefore, the atoms forming the diffusion channel are not easily able to move aside and we might expect a higher activation energy.

These entirely qualitative notions have been buttressed by calculations of the overall activation energy by Ehrlich and Kirk[48] assuming interatomic forces obeying a Morse potential. More extensive estimates (also listed in Table VIII), which more consistently account for possible lattice deformation around the migrating atom, have been made by Wynblatt.[86] Both confirm that fluctuations in the position of lattice atoms on the (211) will open up a lower-energy path for migration. This view is also consistent with the low frequency factor for the migration process which is observed experimentally. For diffusion to occur with a low activation energy over the (211), the lattice atoms in the [111] rows must displace outward just as the migrating atom is activated to move by them. The probability of this simultaneous event involving three atoms should be low, and this has been confirmed by quantitative estimates.

Our ability to understand qualitatively the migration on the (211) compared to that on the (321), and the almost quantitative agreement between the barrier on the (211) as calculated using Morse potentials and experimental results, only points up the failure of these attempts to account even qualitatively for atomic behavior elsewhere. This is especially so for migration on the (110). Here the experimental results are by far the most extensive; the interpretation of the experiments is also quite unequivocal. At the least, we can conclude that Morse potentials will not serve to adequately characterize surface behavior.

As has been pointed out previously, the experimental results on binding energies are not really trustworthy. As they are the only values available, however, they probably should not be rejected out of hand. The experimental results cannot be fitted with either Morse or Lennard-Jones potentials. However, if the experimental data on atomic binding energies are accepted as valid, we can use these data to derive an interatomic force law.

Proceeding entirely heuristically, we postulate a trial potential in the form of a polynomial

$$V(R) = \sum_{n=0}^{N} C_n R^n \tag{13}$$

The binding energies are not much affected when the surface is allowed to relax to its minimum-energy configuration. As a first effort, therefore, all atoms are assumed to be fixed at the normal lattice positions. The potentials in Eq. 13 are summed for atoms at a kink site as well as on the (211), (310), and (411) planes. Together with the condition that at $T = 0$ K the potential energy of the lattice must be a minimum with respect to changes in the lattice spacing, and a knowledge of the compressibility (the second derivative of the potential with respect to the lattice spacing), this gives enough relations to define the first six coefficients in the polynomial trial function.

The potential obtained when these summations are carried out to three lattice spacings is shown in Fig. 12. It is of interest that it has the oscillatory form recently found for potentials in the bulk.[88] Even this interatomic potential does not satisfactorily predict surface energetics. As is apparent from the estimates in Table IX, the value for the (321) is much in excess of that observed. Similarly, the activation energy predicted for migration over the (110) is much the same as that obtained from Morse potentials, and still too low compared with experiment.

It is too early to tell if such an empirical approach will have any utility. The problem right now is largely experimental — the determination of reliable surface energetics. Field evaporation has not yet provided

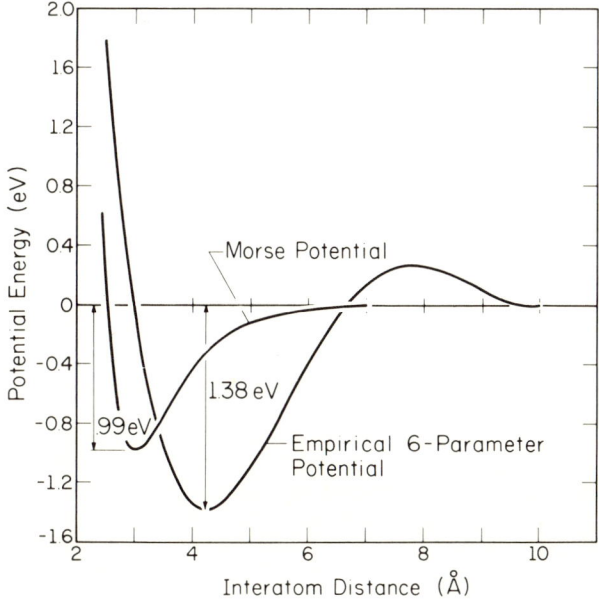

Fig. 12. Comparison of Morse potential with empirical polynomial potential derived from surface data for tungsten.

Table IX. Predictions of polynomial potential.

		Energies (eV)	
		Calculated	Experimental [a]
Binding			
	(110)	7.01	5.3-6.3
	(111)	5.95	6.0
	(321)	10.2	6.7
	(100)	5.91	7.0
Diffusion			
	(110)	.38	.96

(a) Data from Refs. 48, 65, 70.

soundly based binding energies of atoms. The hope is, however, that increased effort will be devoted to the determination of migration barriers. With values available for a larger number of low-index planes, it should be feasible to test more rigorously the possibility of simulating behavior at surfaces with empirical potential curves.

ACKNOWLEDGMENTS

Thanks are due to Marian Vesely for providing as yet unpublished analyses, and for help with some of the calculations. This paper, as well as current studies of atoms at crystal surfaces, have been made possible through support from the National Science Foundation under Grant GK-16593.

The Coordinated Science Laboratory is supported by the Joint Services Electronics Program (U. S. Army, U. S. Navy, and U. S. Air Force) under Contract DAAB-07-67-C-0199.

REFERENCES

1. C. Herring, in *Structure and Properties of Solid Surfaces,* ed. by R. Gomer and C. S. Smith (Univ. of Chicago Press, Chicago 1953) pp. 5-81.
2. W. W. Mullins, in *Metal Surfaces: Structure, Energetics and Kinetics* (American Society for Metals, Metals Park, Ohio 1963) pp. 17-98.
3. C. Herring, Phys. Rev. **82**, 87 (1949).
4. H. Mykura, Acta Met. **5**, 346 (1957).
5. W. L. Winterbottom, in *Surfaces and Interfaces I, Chemical and Physical Characteristics*, ed. by J. J. Burke, N. L. Reed, and V. Weiss (Syracuse Univ. Press, Syracuse, N. Y. 1967) pp. 133-165.

6. H. Mykura, in *Molecular Processes on Solid Surfaces*, ed. by E. Drauglis, R. D. Gretz, and R. I. Jaffee (McGraw-Hill Book Co., New York 1969) pp. 129-145.
7. W. L. Winterbottom, Acta Met. **15**, 303 (1967).
8. M. Drechsler and H. Liepack, *Adsorption et Croissance Cristalline* (Editions du CNRS, Paris 1965) pp. 49-79.
9. M. Drechsler and J. F. Nicholas, J. Phys. Chem. Solids **28**, 2609 (1967).
10. A. Müller and M. Drechsler, Surface Sci. **13**, 471 (1969).
11. A. J. Melmed, J. Appl. Phys. **38**, 1885 (1967).
12. J. J. Burton and G. Jura, J. Phys. Chem. **71**, 1937 (1967).
13. H. Udin, *Metal Interfaces* (American Society for Metals, Cleveland, Ohio 1952) pp. 114-133.
14. J. W. May, Advances in Catalysis **21**, 151 (1970).
15. C. B. Duke, Ann. Rev. Mat. Sci. **1**, 165 (1971); P. J. Estrup and E. G. McRae, Surface Sci. **25**, 1 (1971).
16. E. Bauer, in *Techniques of Metals Research*, ed. by R. F. Bunshah (Interscience Publishers, N. Y. 1969) Vol II. Part 2, Chapt. 15, 16.
17. J. T. McKinney, E. R. Jones, and M. B. Webb, Phys. Rev. **160**, 523 (1967).
18. A. J. Van Bommel and F. Meyer, Surface Sci. **12**, 391 (1968).
19. L. H. Germer, A. U. MacRae, and C. D. Hartman, J. Appl. Phys. **32**, 2432 (1961); A. U. MacRae and L. H. Germer, Annals N.Y. Acad. Sci. **101**, 627 (1963).
20. M. P. Seah, Surface Sci. **17**, 181 (1969).
21. G. W. Simmons, D. F. Mitchell, and K. R. Lawless, Surface Sci. **8**, 130 (1967).
22. A. U. MacRae, in *Surfaces and Interfaces I, Chemical and Physical Characteristics*, ed. by J. J. Burke, N. L. Reed, and V. Weiss (Syracuse Univ. Press, Syracuse, N. Y. 1967) pp. 29-52.
23. R. J. Reid and H. Mykura, J. Physics **D2**, 145 (1969).
24. D. L. Smith and R. P. Merrill, J. Chem. Phys. **52**, 5861 (1970).
25. L. A. Harris, J. Appl. Phys. **39**, 1419, 1428 (1968).
26. For a review, see E. N. Sickafus and H. P. Bonzel, Recent Progr. Surface Sci. **4**, 115 (1971).
27. R. L. Park, J. E. Houston, and D. G. Schreiner, Rev. Sci. Inst. **41**, 1810 (1970).
28. G. C. Benson and K. S. Yun, in *The Solid-Gas Interface*, ed. by E. A. Flood (Marcel Dekker, N. Y. 1967) Vol. 1, pp. 203-269.
29. D. Patterson, J. A. Morrison, and F. W. Thompson, Can. J. Chem. **33**, 240 (1955). J. A. Morrison and D. Patterson, Trans. Faraday Soc. **52**, 764 (1956).
30. P. J. Estrup, in *The Structure and Chemistry of Solid Surfaces*, ed. by G. A. Somorjai (J. Wiley & Sons, N. Y. 1969) 19-1.
31. R. W. James, *The Optical Principles of the Diffraction of X-Rays* (G. Bell & Sons, London 1948).
32. C. Menzel-Kopp and E. Menzel, Z. Physik. **144**, 538 (1956).
33. F. O. Goodman, Surface Sci. **3**, 386 (1965).
34. R. F. Wallis, in *The Structure and Chemistry of Solid Surfaces*, ed. by G. A. Somorjai (J. Wiley & Sons, N. Y. 1969) 11-1.
35. G. E. Laramore and C. B. Duke, Phys. Rev. **2B**, 4783 (1970).
36. J. B. Taylor and I. Langmuir, Phys. Rev. **44**, 423 (1933).
37. A. U. MacRae, Surface Sci. **13**, 130 (1969).
38. L. W. Swanson, R. W. Strayer, and F. M. Charbonnier, Surface Sci. **2**, 177 (1964).
39. E. W. Müller, Advances Electronics Electron. Phys. **13**, 83 (1960).
40. J. R. Oppenheimer, Phys. Rev. **13**, 66 (1928).
41. M. H. Rice and R. H. Good, Jr., J. Opt. Soc. Am. **52**, 239 (1962).
42. E. W. Müller and T. T. Tsong, *Field Ion Microscopy* (American Elsevier, N. Y. 1969).
43. K. M. Bowkett and D. A. Smith, *Field Ion Microscopy* (North-Holland Publishing Co., Amsterdam 1970).
44. E. W. Müller, Phys. Rev. **102**, 618 (1956).
45. R. Gomer and L. W. Swanson, J. Chem. Phys. **38**, 1613 (1963).
46. D. G. Brandon, Phil. Mag. **14**, 803 (1966).
47. T. T. Tsong, Surface Sci. **10**, 102 (1968).
48. G. Ehrlich and C. F. Kirk, J. Chem. Phys. **48**, 1465 (1968).
49. E. W. Müller, S. Nakamura, O. Nishikawa and S. B. McLane, J. Appl. Phys. **36**, 2496 (1965).

50. D. G. Ast and D. N. Seidman, Appl. Phys. Letters **13**, 348 (1968).
51. T. T. Tsong and E. W. Müller, Phys. Rev. **181**, 530 (1969).
52. D. M. Newns, Phys. Rev. **B1**, 3304 (1970).
53. D. E. Beck and V. Celli, Phys. Rev. **B2**, 2955 (1970).
54. T. J. Lewis, Proc. Phys. Soc. (London) **67B**, 187 (1954).
55. C. Herring, in *Metal Surfaces* (American Society for Metals, Cleveland, Ohio 1952) pp. 1-19.
56. T. T. Tsong and E. W. Müller, Phys. Stat. Sol. **1a**, 513 (1970).
57. E. W. Müller, in *Molecular Processes on Solid Surfaces,* ed. by E. Drauglis, R. D. Gretz, and R. I. Jaffee (McGraw-Hill Book Co., New York 1969) p. 400.
58. J. C. Rivière, Solid State Surface Sci. **1**, 180 (1969).
59. C. E. Moore, *Atomic Energy Levels*, Nat. Bur. Stand. (U. S.), Circ. 467, Vol. III (1958), Table 34.
60. E. W. Müller, J. A. Panitz, and S. B. McLane, Rev. Sci. Instr. **39**, 83 (1968); E. W. Müller, Naturwiss. **57**, 222 (1970).
61. S. S. Brenner and J. T. McKinney, Appl. Phys. Letters **13**, 29 (1968).
62. E. W. Müller, Quarterly Rev. **23**, 177 (1969); *Structure et Proprietés des Surfaces des Solides* (Editions CNRS; Paris 1970) p. 81.
63. R. S. Chambers, G. Ehrlich, and M. Vesely, 17th Field Emission Symposium, New Haven, August 1970.
64. T. T. Tsong, 31st Phys. Electronics Conf., NBS, Gaithersburg, Maryland, March 1971.
65. E. W. Plummer and T. N. Rhodin, J. Chem. Phys. **49**, 3479 (1968).
66. S. Nakamura, J. Electron Microscopy (Tokyo) **15**, 279 (1966).
67. E. W. Müller, Acta Met. **6**, 620 (1958).
68. A. van Oostrom, Philips Res. Rept. Suppl. 1 (1966).
69. E. W. Müller, S. B. McLane and J. A. Panitz, Surface Sci. **17**, 430 (1969).
70. G. Ehrlich and F. G. Hudda, J. Chem. Phys. **44**, 1039 (1966).
71. N. A. Gjostein, in *Surfaces and Interfaces I, Chemical and Physical Characteristics*, ed. by J. J. Burke, N. L. Reed, and V. Weiss (Syracuse Univ. Press, Syracuse, N. Y. 1967) pp. 271-304.
72. J. A. Simmons, R. L. Parker, and R. E. Howard, J. Appl. Phys. **35**, 2271 (1964).
73. A. LeClaire, Progress in Metal Phys. **4**, 265 (1953).
74. G. S. Rushbrooke, *Introduction to Statistical Mechanics* (Clarendon Press, Oxford 1949) pp. 316-319.
75. M. Drechsler and P. Wolf, Proc. IV Intern. Congr. Electron Microscopy, Berlin 1958 (Springer-Verlag, Berlin, 1960) Vol. II, p. 835.
76. G. Ehrlich, Ann. Rev. Phys. Chem. **17**, 295 (1966).
77. G. Ehrlich, in *Metal Surfaces: Structure, Energetics, and Kinetics* (American Society for Metals, Metals Park, Ohio) pp. 221-258.
78. T. Gurney, Jr., F. Hutchinson, and R. D. Young, J. Chem. Phys. **42**, 3939 (1965); R. D. Young and D. C. Schubert, ibid., 3943.
79. G. Ehrlich, J. Chem. Phys. **44**, 1050 (1966).
80. C. W. Frank and L. D. Schmidt, Surface Sci. **10**, 275 (1968).
81. E. W. Müller, Z. Elektrochem. **61**, 43 (1957).
82. D. W. Bassett and M. J. Parsley, J. Phys. **D3**, 707 (1970).
83. D. W. Bassett, Surface Sci. **23**, 240 (1970).
84. J. K. Mackenzie, A.J.W. Moore, and J. F. Nicholas, J. Phys. Chem. Solids **23**, 185 (1962).
85. L. A. Girifalco and V. G. Weizer, Phys. Rev. **114**, 687 (1959).
86. P. Wynblatt and N. A. Gjostein, Surface Sci. **22**, 125 (1970).
87. W. A. Harrison, *Pseudopotentials in the Theory of Metals* (Benjamin, New York, 1966).
88. A. Englert, H. Tompa, and R. Bullough, *Proc. Conf. Fund. Aspects Disloc. Theory*, NBS, Washington, April 1969.

DISCUSSION on paper by G. Ehrlich

WILSON: Do you assume in the calculations that the lattice spacing normal to the surface is the same as in the bulk?

EHRLICH: In fitting our polynomial potential all atoms were assumed to be in their normal lattice sites even at the surface; that is, the lattice spacing was not allowed to relax. We have made estimates of binding energies in which the outermost layer is relaxed, and calculations for completely relaxed surfaces have been made by Dr. Wynblatt. These estimates show that relaxation has only a small effect (a few percent) upon the value of the binding energy.

WYNBLATT: In reference to this, I would like to say that in the calculation of adatom binding energies although there may be a significant relaxation energy associated with the defect, most of that is lost when the adatom is pulled off the surface because the surface must then be restored to the perfect surface configuration. This "unrelaxation" process cancels most of the adatom relaxation energy. This however is not the case where adatom migration is concerned. Here, the relaxation effects can be very important in certain instances.

TORRENS: Do you think it would be possible, by increasing the number of parameters if necessary, to fit a polynomial potential to all the experimental surface binding and migration energies which you have obtained, or are the data for different surface orientations incapable of explanation using a single pair potential?

EHRLICH: Our efforts to construct a pair potential by calling primarily on surface data must really be considered quite exploratory. I suspect that all the existing data could be accommodated by a suitably adjusted polynomial potential, but we have not yet done this. I have serious doubts, however, that this potential would adequately reproduce binding energies on planes of geometries different from those used to derive the coefficients in the potential.

ASHCROFT: I wonder how sensitive the binding energy results are to the assumption of an *image term* in the potential energy of a surface atom. It would seem that the simple classical $1/x$ form would require amputation at very small distances x from the surface. Apart from the 'one-dimensionality' implied by this model, how are the physical and geometrical differences between planes of low rational index incorporated in the image potential cut off?

EHRLICH: The assumption of a simple image term of the form $1/x$ is vital in the derivation of binding energies from field evaporation measurements. However, termination of the potential at some cutoff distance would not seriously affect the interpretation of the desorption fields, provided the cutoff occurs to the left of the barrier.

It is generally recognized (as discussed in the paper itself) that the classical image approximation will fail close to the surface, in the general region in which the barrier is located. Quantum effects on the screening have been evaluated for a semi-infinite electron gas. Fortunately, these can be included by adding a penetration length Δ to the distance x in the image potential; this leaves the formalism of field evaporation intact. However, no serious attempt has yet been made to account for the effects of the atomic structure of the surface on the form of the ionic potential.

VINEYARD: Can you be sure that the atom does not escape by first migrating to the edge of a face, where the field will be higher and the binding lower?

EHRLICH: Evaporation measurements are made by raising the field in small increments, and observing the location of the adsorbed atoms after each increase. At the low temperatures at which these experiments are performed, no motion of the atoms is observed prior to removal from the surface. The field at which removal occurs should therefore be characteristic of evaporation from a particular site. It is, in fact, found that the voltages for desorption from the center of a plane are considerably higher than close to the edges, due to changes in the curvature over the emitter surface.

ROBINSON: Could the high-charge states observed in the field evaporated ions have been produced by Auger processes occurring *after* the ions have left the surface? This could have implications for the interpretation of the binding energies.

EHRLICH: Postionization after field evaporation has occurred is indeed one of the processes that may possibly contribute to the creation of the high-charge states found in the atom probe. We have made estimates of the efficiency of this process, assuming a 1-dimensional system[1]. For molybdenum at a field of 5 V/Å, the 1-dimensional calculations indicate that essentially all +2 ions will postionize to +3, but only 3 in 10^5 will convert to +4. It turns out, however, that the 1-dimensional model is a poor representation of the actual system. Far away from the surface, field ionization has been worked out in 3 dimensions[2]. In a field of 5 V/Å, the rate at which Mo^{+2} ionizes to Mo^{+3} in free space is $\sim 5 \cdot 10^{10}$ sec^{-1}; the 1-dimensional model, however, predicts a rate on the order of 10^{16} sec^{-1}. To obtain a reliable guide to the role of postionization in the appearance of high-charge states, a 3-dimensional calculation close to the surface appears necessary. Such a calculation is underway, and preliminary estimates suggest that postionization will make only minor contributions.

HUNTINGTON: It seems that there is one sort of many-body force that may favor mobility down grooves on atomic serrated planes over that on close packed planes. It is the interaction between the adatom and the electron gas. Such an interaction would tend to clamp the adatom to the atomistically close-packed surface and slow its motion; on the serrated surface this effect would be less because the overlapping electron distribution in the groove would tend to lubricate motion down it.

EHRLICH: It will be interesting to evaluate this interaction quantitatively, inasmuch as the variation of atomic mobility from one plane to the next is not even qualitatively understood. The experimental values of the barrier to migration are about the same on a rough surface [the (321)] as on a smooth one [the (110)], and this may possibly indicate a smoothing of the diffusion channels on the former.

KULCINSKI: Your present work is concerned with the migration of atoms identical to the base metal. Could you comment on the possibilities of such studies with specific impurities such as interstitial atoms?

EHRLICH: Surface studies of gases have been attempted with the FIM. They are difficult, however, inasmuch as the act of observation seriously perturbs the adsorbed layer. With self-adsorbed metal atoms this is not so significant a problem. Just as for adsorbed gases, studies of interstitials are likely to be tricky. To observe them they must be at or close to the surface. There the severe conditions necessary to achieve atomic resolution are likely to perturb the atoms under examination. However, at Cambridge, Fortes and Ralph[3] have been quite successful in looking at oxygen interstitials.

REFERENCES

1. See Ref. 63.
2. C. Lanczos, Z. Physik, **68**, 204 (1931).
3. M. A. Fortes and B. Ralph, Acta Met. **15**, 707 (1967).

SIMULATING SURFACES BY THE SUMMATION OF PAIRWISE INTERATOMIC POTENTIALS

D. P. Jackson

Chemistry and Materials Division
Atomic Energy of Canada Limited
Chalk River, Ontario, Canada

ABSTRACT

Workers in surface science are often led, for lack of other more precise methods, to simulate surfaces by the computer summation of pairwise interatomic potentials, usually of very simple forms. In this paper, this approach is considered for a variety of surface problems. The surface collision problems: atom surface scattering and surface atom ejection are discussed in detail, and a new simulation of the latter is reported. The planar relaxation of surface layers is considered as a possible source of information on surface potentials.

1 INTRODUCTION

Many problems in the physics and chemistry of solid surfaces require an accurate knowledge of the interatomic potentials acting at a surface.

The distribution of electrons in the solid changes rapidly near the surface, and the ion cores at the surface relax somewhat due to the asymmetrical bonding present there. Hence, an interatomic-potential scheme adequate for describing bulk phenomena may require modification before it can be used in surface problems. The extent of the differences between "bulk" and "surface" potentials is not known at present. Preceding chapters of this volume show significant progress in developing bulk potentials; however, very few interatomic potential models are designed particularly for application to surfaces. Therefore, surface simulations have been confined largely to the summation of pairwise interatomic potentials (SPP) technique. Here we briefly review some classes of surface problems simulated by SPP methods, making no attempt at a comprehensive literature survey and concentrating in particular on three of them.

The following remarks on the SPP technique apply to all of the cases discussed below. The SPP method involves several assumptions. The most important is that the interatomic interactions can be described by a sum of pairwise-acting potentials. This point, discussed exhaustively in the preceding chapter on potentials, is strictly justified only for special cases. Past work on surfaces has been done mainly with very simple empirical potential expressions such as the Lennard-Jones and Morse formulae applied to metals; therefore, large approximations are introduced. Further, the parameters for these empirical potentials are derived from fitting experimental data: for the Lennard-Jones potential the properties of gases are used and for the Morse potential the bulk properties of metals. Hence surface data is not used to obtain the potentials introducing a further approximation. In spite of these limitations to the SPP technique, at the moment it seems to be the only feasible treatment for many surface problems. If the pragmatic attitude is taken that SPP is a reasonable tool for calculations, realizing its essentially empirical nature, then the obvious strategy for improving the situation for surfaces is to base the SPP parameters on *surface* experimental data. In this way one could derive effective surface SPP parameters that hopefully would result in better quantitative models for surface interactions. Some possibilities towards this goal are mentioned below.

The classes of surface problems for which SPP simulations have been employed may be defined as follows:

Physical adsorption – binding energy, migration energies
Surface structure – relaxations, rearrangements, gross structure
Surface lattice vibrations
Defects – extended, point
Collision phenomena – scattering, ejection.

Examples of most of these problems are discussed in this volume with the notable exception of "collision phenomena" which forms the main topic of this paper. In addition, "surface relaxations" are briefly discussed as a method of obtaining information on surface potentials. There is, at present, a real need for an accurate interatomic potential description for these surface problems. Until this is available one must still rely on the SPP approximation.

2 THE SCATTERING OF ATOMS FROM SURFACES

The problem considered here is the following: a neutral atom is scattered from a "perfect" crystal surface – what energy and momentum are exchanged and what are the final coordinates and velocity of the scattered atom? Such problems were first studied many years ago to demonstrate atomic diffraction[1] and more recently have become of crucial importance in rarefied gas dynamics as applied to vacuum systems and particularly to the aerodynamics of space vehicles in high orbits and planetary atmospheres[2]. Experimental work in this field has been stimulated by the development of well-collimated molecular beams of comparatively high energy.

Generally speaking the choice of a theoretical treatment depends on E_0 the energy of the incident beam.[3] If E_0 is small and the deBroglie wave length is comparable to the lattice spacing, quantum-mechanical theories must be used to describe the atomic-diffraction and phonon-exchange processes that occur at the surface. With increasing E_0, classical theories become feasible and the scattering interaction is described in terms of three physical concepts: the inertial collision with a scattering center on the surface, the restituitive properties of the lattice to which the scatterer is bound, and the long-range collective field of the surface. For higher E_0, the second effect is of less importance since the characteristic time for a collision becomes small in comparison with the natural frequencies of the lattice – the collision occurs so quickly that the lattice atom "doesn't have time" to react until after the scattered atom has left the surface region. One can then approximate the surface as consisting of *free* (i.e., decoupled) scatterers; often one may assume that these scatterers are harmonic oscillators in order to include surface-temperature effects. For these cases, the details of the surface field are important, the structure of the field being the major influence in the scattering process. Both SPP[4,5] and hard sphere[6,7] models have been used under these conditions.

The so-called hard-cube model[8] has had the most success in comparison with experiment. This treatment assumes a free harmonic oscillator, with frequency corresponding to the surface temperature, and with a *smooth* surface in the sense that the scattering is described by an inertial process in which no momentum parallel to the surface is exchanged. The relative success of this theory leads one to believe that the effective surface field is smoother than had hitherto been postulated. However, from other kinds of experiments (e.g., field-ion microscopy), we know that there is definite structure in the surface field. The question now is why does the surface field appear to be so smooth to the incoming atoms? Goodman has suggested a quantum mechanical basis for this phenomenon at lower energies involving the details of surface phonon exchange.[9] However it may be that, for some reason, the beam energies involved in the experiments are insufficient to allow significant penetration of the incoming atoms into the regions of the field where structural variations are significant; the inertia and temperature effects described by the model probably are very much more important than any field-structure effect for the experimental energies treated. One may picture the surface field seen by an incoming atom as consisting of two parts: a long-range attractive force that acts on the atom before and after collision, and a short-range repulsive force whose effect is felt from at most a few atoms at the collision site. In effect the hard-cube model says that this repulsive force field is smooth.

Even in relatively simple models[10] the attractive field may give rise to some interesting effects. Assume that the surface has an attractive potential well of depth E^* and that the forces due to this potential are solely in the direction normal to the surface. Then an incoming atom will experience an acceleration and a change in effective incidence angle due to this potential thereby modifying the scattering parameters at the surface. After collision, denoting the atom's energy by E' with component $E'_\perp$ normal to the surface, the atom's subsequent trajectory will belong to one of the following categories[10]

(i) $E' > E^*, E'_\perp > E^*$
(ii) $E' < E^*$
(iii) $E' > E^*, E'_\perp < E^*$

In case (i) the atom *escapes* the influence of the surface but suffers a deceleration and change in angle to the surface as it does. In case (ii) the atom cannot escape the attractive field and it is *absolutely trapped* although it may require many further collisions with the surface to dissipate its energy[11]. Case (iii) is called *pseudo (or semi) trapping*.[10,11] Here the trajectory curves back toward the surface and undergoes one or more additional repulsive collisions. Eventually a case (iii) atom will either escape (i) or suffer absolute trapping (ii). Surface scattering models originally indicated that the proportion of atoms trapped was unrealistically high; this was a result of a failure to distinguish between cases (ii) and (iii). To accurately calculate the accommodation coefficients (based on the energy and momentum exchanges in surface scattering) and the resulting density and velocity distributions of the scattered atoms, one must include the effects of pseudotrapping.

Atom surface scattering is very instructive in illustrating the important collective effects interatomic potentials exhibit at surfaces. It is remarkable that models with very simple initial assumptions can lead to a very complex system of interactions mainly due to the influence of the long-range attractive forces of the surface. Information about effective surface potentials is difficult to extract from experimental atom surface scattering data due to the difficulties of disentangling the various scattering phenomena taking place.

3 THE EJECTION OF ATOMS FROM SURFACES

Surface ejection may be regarded as the "reciprocal" of the surface scattering problem. The question here is: what is the subsequent trajectory of a surface atom after a knock-on from below? This problem is of interest in radiation damage and particularly in sputtering. Many sputtering theories have been proposed, but they are mainly concerned with the mechanisms of energy transfer from the sputtering ion beam to the solid atoms and the subsequent propagation of this energy through the solid.[12,13] Little attention has been given to ejection at the surface which might be considered as the terminal-boundary problem of the sputtering process. It is found experimentally that, even for relatively high energy incident ion beams, the atoms eventually ejected from the surface may have very much lower energies and, hence under these circumstances, a detailed knowledge of the ejection process becomes imperative.

Recently we have been simulating ejection from cubic metal surfaces using an SPP technique with Morse potentials. The SPP parameters used have not been modified for surfaces but have been taken from Girifalco and Weizer[14]; because atomic trajectories starting deep in the crystal will eventually be incorporated into the model, it seems preferable to retain the bulk parameters throughout. Fig. 1 is a schematic of a typical configuration used in the calculations for a bcc (100) surface. The atom to be ejected has been placed in field represented by the SPP sum of atoms relatively near it plus a contribution due to a "smeared-out" approximation for the field due to atoms relatively remote from it. This approximation consisted of summing the contributions of an infinite "stack" of disks with planar atomic density equal to that of the atomic layers[15]; Crowell[16] has used a similar method with the Lennard-Jones potential. Fig. 2 compares this approximation, represented by Eq. 2, with the usually made assumption that the lattice points are smeared out through the volume of the solid and represented by Eq. 1. The case considered is the evaluation of the semi-infinite lattice sum

$$\sum_{\text{½-lattice}} e^{-kr}$$

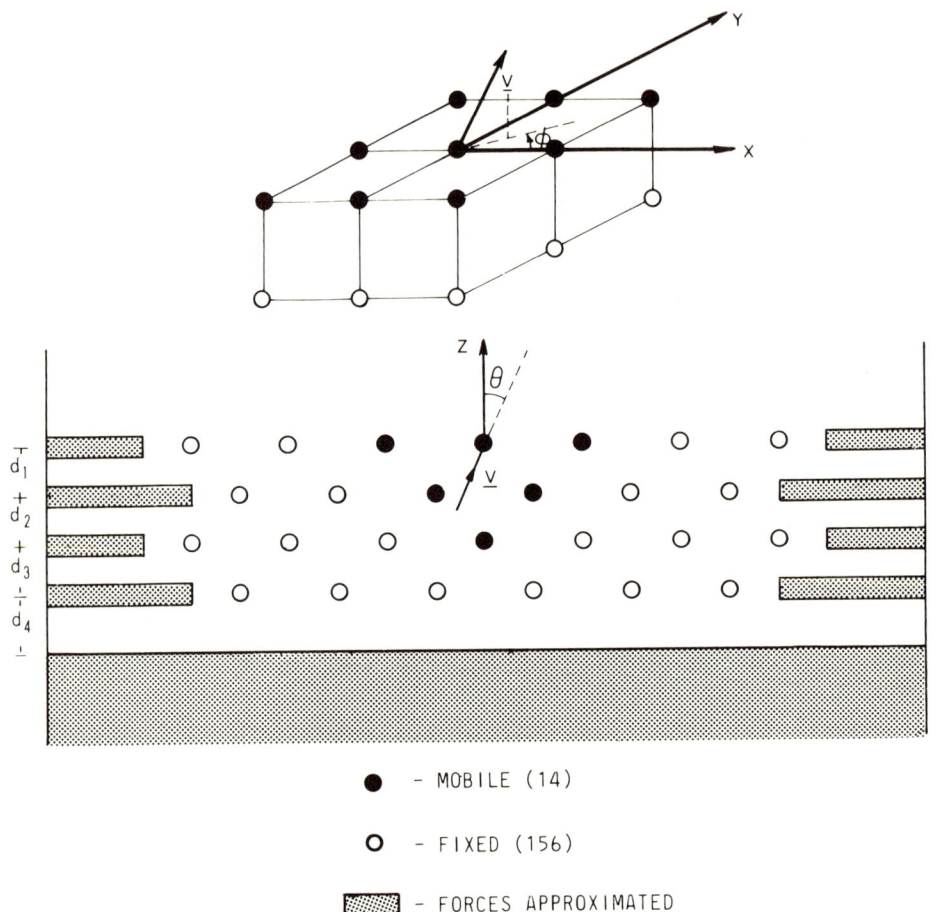

*Fig. 1. Ejection from a bcc (100) surface showing the configuration used to simulate the surface field. The PKA vector **v** makes angles θ and φ with the principle axes. The numbers are typical of those used in the calculations.*

above a bcc (100) surface with $k = 5$. S_A and S_X are the actual computed values of the lattice sums above the two external points on the surface denoted "A" and "X" on the inset. Eq. 2 allows a good approximation to the field at a distance of a few lattice spacings and is clearly superior to Eq. 1; it also results in substantial savings in computer time. Having thus built up the field, it is necessary then to relax the surface layers so that there are no forces present before the ejected atom is set in motion. This process is considered in somewhat more detail in Part 4 below.

A knock-on represented by the vector **v** in Fig. 1 is applied to the atom to be ejected. Clearly it is necessary that the immediate neighbors of the struck atom be allowed to move so that energy can be adsorbed by the lattice. Hence one must simultaneously integrate the equations of

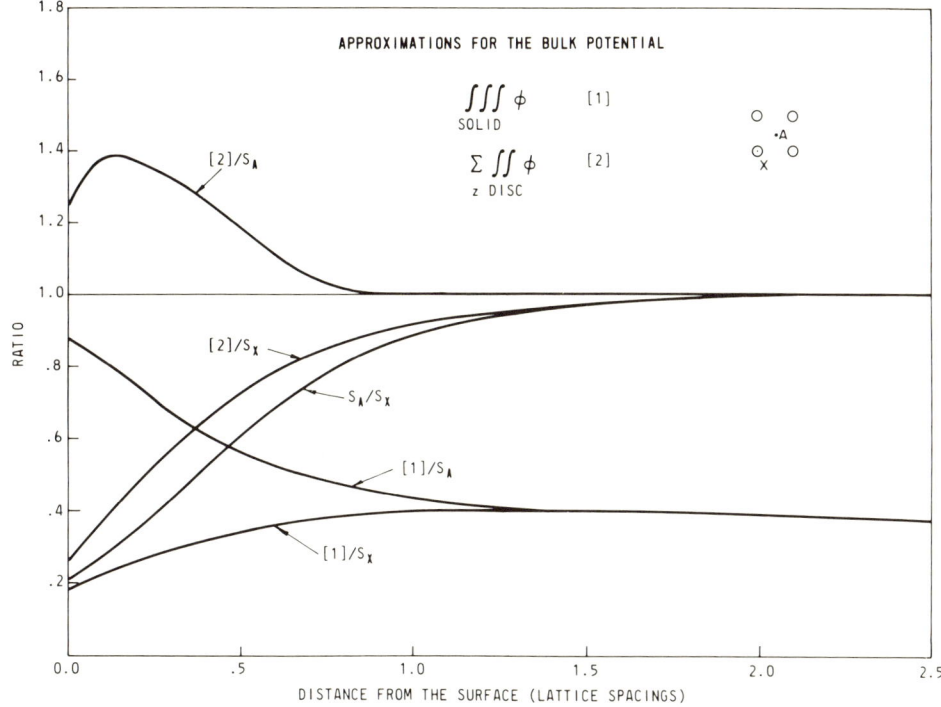

Fig. 2. Comparisons of the approximations used to model the potential above the surface of a semi-infinite solid (bcc (100)). Eq. [1] and [2] represent two methods of approximating the potential sum given in the text. S_A and S_X are the actual lattice sums above the surface Points A and X of the insert.

motion for several atoms. In the example given in Fig. 1 this involves the simultaneous solution of 84 first-order differential equations. Therefore one must balance the complexity of the field model against computation time. This is a particular consideration in collision models since one usually requires a large number of trajectory solutions to simulate an ejection pattern or scattered flux distribution. Configurations such as the one illustrated in Fig. 1 are compromises between these factors.

Fig. 3 shows the motion of the primary knock-on atom (PKA) of Fig. 1 when it is given a knock-on normal to the surface. When the incident energy E_O is less than a threshold energy (~11.5 eV for the *computer* tungsten calculated here), the energy of the knock-on is dissipated by the motions of the surrounding atoms. This occurs with greater initial excursions as E_O is increased until the atom finally breaks free from the surface. The threshold energy has been found to comprise the energy necessary to overcome the surface attractive field plus the energy represented in the relaxation of the remaining surface atoms

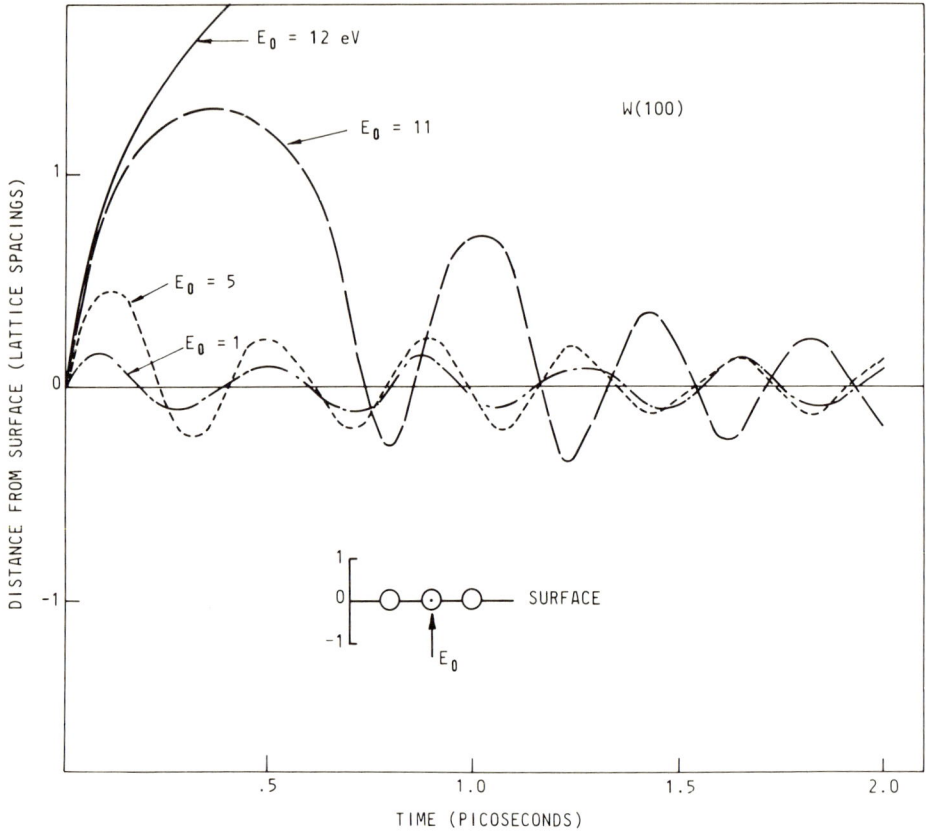

Fig. 3. Motion of the PKA as a function of initial energy E_0 with $\theta = o$ for a tungsten (100) surface, potential parameters from Ref. 14.

around the *hole* it leaves. The latter is a small fraction of the former for tungsten, but it may become a significant component in other systems. These results are generally what one would expect, however the situation rapidly becomes very complicated when angle θ (see Fig. 1) is much different from zero. A few of the possible trajectory types that can occur are shown schematically in Fig. 4. In addition to the trapping and pseudotrapping effects mentioned in connection with scattering, re-entrant atoms and, particularly, secondary atoms are possible. The total picture then becomes very complex since, in general, a trajectory's history may comprise a combination of the above possibilities.

The eventual aim of this program, now in a preliminary stage, is to generate ejection patterns, i.e., the pattern of sputtered particle impacts on a collector held above the target and the corresponding velocity distribution. This objective requires the inclusion in the model of several effects

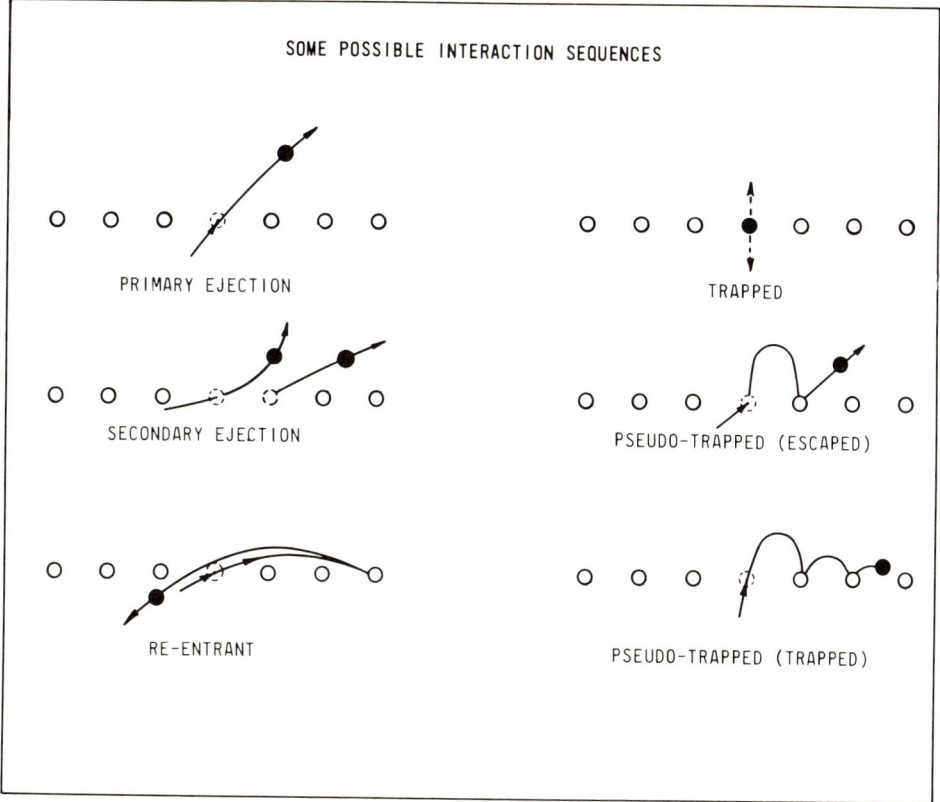

Fig. 4. Schematic of some possible trajectory events in ejection. (The trajectory paths are drawn for illustrative clarity and are not intended to be physically accurate.)

that are not mentioned here, however initial examples of *surface* ejection patterns show variations in the velocity and distribution of the ejected atoms with PKA energy which imply that surface effects are significant in determining the final form of simulated experimental ejection patterns.

4 SURFACE RELAXATIONS

One may consider a surface as a defect and inquire then what are the relaxations of the ion cores around this defect? These surface relaxations have been treated for some time both theoretically and experimentally. The effect is well established in semiconductors[17], particularly in germanium[18], by low-energy-electron-diffraction (LEED) measurements; however, LEED results for metals are sparse[19] and apparently subject to problems in interpretation[20]. Theoretical treatments have been mainly incidental to investigations of adsorption, surface vibrations, and surface structure; some of these results are mentioned below.

We have recently used an SPP method to calculate the surface relaxations of the low-index planes of seven fcc and nine bcc metals[21] using the Morse potentials of Girifalco and Weizer[14]. The total potential of the semi-infinite solid was calculated by the SPP technique. The relaxations were obtained by minimizing this total potential in terms of the interplanar spacings of the atomic layers near the surface. In view of what has been said above about SPP calculations, it is unlikely that the results will be quantitatively correct but some useful qualitative conclusions may be drawn from them.

The results for fcc metals are shown in Fig. 5. Here the relaxations of the first surface layer (as a percentage of the bulk interplanar spacing) are plotted against the parameter group e^{ar_0} of the Morse potential:

$$\varphi = D \{ e^{-2a(r-r_0)} - 2 e^{-a(r-r_0)} \} .$$

The variation of these first layer relaxations for the three low index planes is in the order $(110) > (100) > (111)$. This ordering agrees with the previous work of Wynblatt and Gjostein[22] for copper and Cotterill and Doyama[23] for copper, gold, and aluminum. It does not agree with the previously published fcc-metals calculations of Burton and Jura[24], however this disagreement is probably due to a factor error.[27] Similarly for bcc metals the ordering is $(111) > (100) > (110)$. Plausibility arguments based on the number of bonds an atom has in the various plane types show that these orderings are reasonable.

The most interesting feature of Fig. 5 is the sensitivity of the relaxations to the potential parameters, that is they are model sensitive.

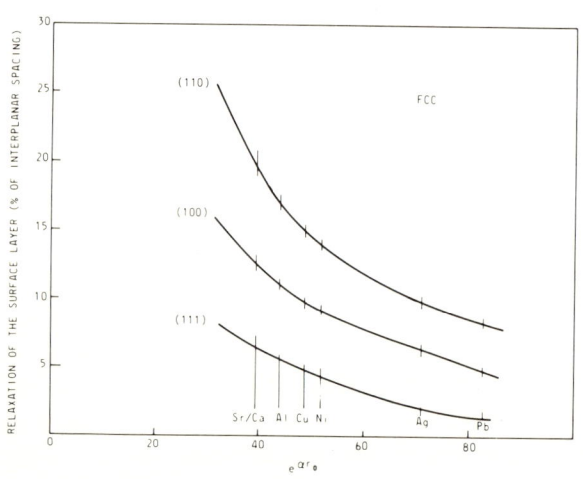

Fig. 5. The first surface layer relaxations of seven fcc metals, expressed as a percentage of the bulk interplanar spacing and plotted against a parameter group of the Morse potential.

For instance it would be very interesting to see the experimental relaxations plotted on a similar graph as a test of the possibility of using simple potentials and as a method of selecting between proposed potential expressions. In conjunction with other data one could then fit parameters that would be more valid than those presently used. Surface relaxation experiments therefore are potentially very useful as a source of information about interatomic potentials. Compared with other possible means of obtaining potential data, use of this phenomena seems to be the most clear cut as to the relationship between the experimental results and the theory. This of course is not true in collision processes because of their complexity. Problems in LEED-pattern interpretation have made relaxation data difficult to obtain up to now; however the use of Rutherford backscattering[25,26] with multiple alignment techniques in ultrahigh vacuum is an interesting possibility for obtaining further relaxation results in future.

REFERENCES

1. Estermann, I. and Stern, O. Z. Phyik **61** (1930) 95.
2. French, J. B. NATO AGARDograph 112 (1966).
3. Goodman, F. O. Surf. Sci. **26** (1971) 327.
4. Oman, R. A. *Proc. 5th Int. Symp. on Rarefied Gas Dynamics,* Vol. 1, p. 83.
5. Lorenzen, J. and Raff, L. M. J. Chem. Phys. **49** (1968) 1165.
6. Goodman, F. O. Surf. Sci. **7** (1967) 391.
7. Jackson, D. P. and French, J. B. *Proc. 6th Int. Symp. on Rarefied Gas Dynamics,* Vol. 2, p. 1119.
8. Logan, R. M. and Stickney, R. E. J. Chem. Phys. **44** (1966) 195.
9. Goodman, F. O. J. Chem. Phys. **53** (1970) 2281.
10. Jackson, D. P. University of Toronto UTIAS Report 134 (1968).
11. Goodman, F. O. ref. 7, p. 1105.
12. Kaminsky, M. *Atomic and Ionic Impact Phenomena on Metal Surfaces,* Chapter 10, Berlin (1965).
13. Carter, G. and Colligon, J. S. *Ion Bombardment of Solids,* Chapter 7, London (1968).
14. Girifalco, L. A. and Weizer, V. G. Phys. Rev. **114** (1959) 687.
15. Jackson, D. P. to be published.
16. Crowell, A. D. J. Chem. Phys. **22** (1954) 1397.
17. Lander, J. J. and Morrison, J. J. Chem. Phys. **37** (1962) 729.
18. Hansen, N. R. and Haneman, D. Surf. Sci. **2** (1964) 566 see also Taloni, A. and Haneman, D. Surf. Sci. **10** (1968) for an SPP calculation of these relaxations.
19. MacRae, A. U. and Germer, L. H. Phys. Rev. Lett. **8** (1962) 489.
20. Park, R. L. and Farnsworth, H. E. Surf. Sci. **2** (1964) 527.
21. Jackson, D. P. Can. J. Phys. **49** (1971) 2093.
22. Wynblatt, P. and Gjostein, N. A. Surf. Sci. **12** (1968) 109.
23. Cotterill, R.M.J. and Doyama, M. *Lattice Defects and Their Interactions,* New York (1967) 1.
24. Burton, J. J. and Jura, G. J. Phys. Chem. **71** (1967) 1937.
25. MacRae, A. U. Surf. Sci. **13** (1969) 130.
26. Davies, J. A. J. Vac. Sci. and Tech. **8** (1971) 487.
27. Note added in proof: Private communications with Dr. J. J. Burton have cleared up this disagreement, which was indeed due to factor problems. The relaxations given here and in Ref. 21 are presumably correct.

DISCUSSION on paper by D. P. Jackson

ROBINSON: In actual sputtering experiments, the lattice atoms are usually displaced from their normal sites several times before they can be ejected from the target and there might also be consequences of the inclusion of the ion material into the target. It would seem that these effects would need to be included in your model to obtain agreement with observed ejection patterns. Furthermore, it would be necessary to include the energy spectrum of the source atoms in your calculation.

JACKSON: Your comments are well taken and the effects you mentioned must be included in a "full" comparison with experiment. However, I feel that there is a significant amount of surface content in the sputtering patterns observed experimentally and at the moment I have restricted myself to trying to understand these purely surface effects.

INTERACTION ENERGY AND CONFIGURATION OF LEDGES ON (001) COPPER SURFACES

P. Wynblatt
Scientific Research Staff
Ford Motor Company
Dearborn, Michigan 48121

ABSTRACT

A computer model of the surface, consisting of a crystal with (001) terraces and either one or two ledges, has been constructed. Interaction between the atoms of the model has been described by means of a Morse potential. Given this force law and a previously developed zero-force atomic relaxation technique, it has been possible to establish the configuration of an isolated ledge, the attendant displacement field as well as the increase in energy which results upon introduction of a ledge on an otherwise perfect surface.

Consideration of surfaces with two ledges has allowed computation of the interaction between all three types of ledge pairs, as a function of ledge spacing. The results obtained show that the interaction between the displacement fields of a pair of ledges is short range and repulsive in all cases.

Estimates of the anisotropy of surface energy, on the basis of the calculated interaction, are found to agree qualitatively with the trends observed experimentally at high temperatures. When the present results are combined with a suitable high-temperature model, they are found to provide a correction in the proper direction.

1 INTRODUCTION

The atomistic interpretation of surface phenomena relies heavily on the Terrace-Ledge-Kink (TLK) model of surface structure. Better understanding of the phenomena therefore demands a detailed knowledge of both configuration and interaction of the various components of the TLK model. The present study represents a further attempt in that direction.

Both theoretical analyses and interpretation of measurements of the anisotropy of surface energy have been couched in terms of ledge energy and the energy of interaction between ledges.[1,2,3,4] The latter has generally been ascribed to one of two possible sources[2,3,4]: interaction between the displacement fields arising from local distortion at ledges and/or an effective high-temperature interaction stemming from entropy effects associated with the wandering of ledges about their mean positions. The purpose of this study is to estimate the ledge energy as well as the displacement field interaction by determining the local atomic displacements in the vicinity of a ledge. In particular, it is considered of interest to establish the range of the displacement field interaction in view of a recent phenomenological analysis[5] that predicts a long range interaction between ledges.

Thus far, most attempts at calculating atomic displacements, both in the vicinity of a perfect surface[6,7,8] and around point defects at the surface[8,9,10], have employed a pairwise interaction model. The shortcomings of this type of model for metals are well known, and the failure of the approach to predict surface defect properties quantitatively is well documented.[8,9] In spite of these difficulties, it is felt that the pairwise interaction model still provides a reasonable means of obtaining qualitatively meaningful information about the issues under investigation here, namely: the configuration of surface ledges and the nature and range of the displacement field interaction between ledges.

2 METHOD OF CALCULATION AND RESULTS

2.1 The Model

An atomic model of the surface, consisting of a crystal with (001) terraces and either one or two ledges lying along the [110] direction, was employed in the calculations. The geometry of the model is illustrated in Fig. 1a and 1b. All three possible types of ledge pairs were considered and these were labeled sequential (S), face to face (F) and back to back (B), as illustrated in Fig. 1c. Interaction between the atoms of the array was described by means of a Morse potential[11]. In this formulation, the energy of a pair of atoms i and j is expressed as

$$\varphi(r_{ij}) = D \left[\exp \left\{ - 2\alpha(r_{ij} - r_o) \right\} - 2 \exp \left\{ - \alpha(r_{ij} - r_o) \right\} \right] \quad , \tag{1}$$

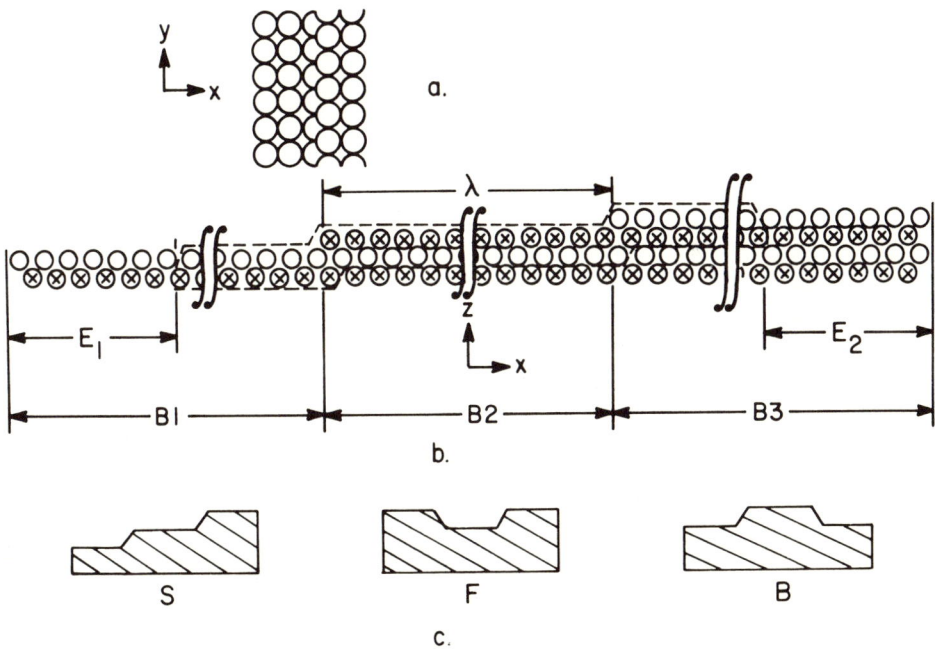

Fig. 1. (a): Top view of (001) terrace with [110] ledge. (b): Cross-sectional view of (001) surface with two ledges. Open circles depict atoms lying in the plane of the figure, crossed circles represent atoms half an atomic diameter below the plane of the figure. The coordinate system used, given in (a) and (b), does not lie along the cube edges, but rather, x and y lie along close packed directions in the (001) surface and z lies parallel to the surface normal. The array is divided into variable-length blocks B1, B2 and B3 (see text) and two ends of fixed length E1 and E2. Dashed line encloses atoms allowed to relax in computations. (c): Cross-sectional view of the three types of ledge pairs, sequential (S), face to face (F) and back to back (B).

where r_{ij} is the separation of the pair and D, α and r_o are empirical constants that take on values of 0.3429 (eV), 1.3588 (A^{-1}) and 2.866 (A) for copper. The energy of the ith atom in the array can then be expressed as:

$$\Phi_i = \sum_{j \neq i} \varphi(r_{ij}) \quad ,$$

and the total energy of the array as: $N\Phi_i/2$. For convenience, the range of the potential was limited so that only atoms within a distance of 4.5 lattice parameters (a_o) of a given atom were allowed to contribute to its energy. The introduction of such a cut-off decreases the energy of an atom in a perfect lattice by an insignificant amount ($\sim 10^{-6}$ eV) but appreciably reduces computation time. Furthermore, because of this cut-off, the dimensions of the model crystal were effectively infinite in the y and z directions. The dimension of the array in the x direction could be changed depending on the requirements of a particular computation, but was limited to a maximum of 50 interatomic spacings ($b_o = a_o/\sqrt{2}$).

Previous computations for perfect (001) copper surfaces[6,10] have shown that the surface energy is minimized by a normal displacement of the atoms near the surface, in such a way as to increase the interplanar spacing. Consequently, before starting any relaxation computations involving ledges, the first three surface layers were displaced outwards by 0.128878, 0.032082, and 0.006970 interplanar spacings, respectively. The ensuing configuration is henceforth referred to as the *perfect surface* configuration. The surface displacements were applied independently to the first layers of each of the three blocks labeled B1, B2, and B3 (or each of two blocks for cases involving a single ledge) in Fig. 1b. This constituted the starting configuration for the ledge-relaxation calculations.

To minimize end effects, the two ends of the array, labeled E1 and E2 in Fig. 1b, were held at their perfect surface configuration and not allowed to relax during the subsequent relaxation process for a length equivalent to the range of the potential ($4.5a_o \cong 6.4b_o$). This procedure does not entirely eliminate end effects since it imposes artificial constraints on the ends; however, it avoids the complication of accounting for the additional corners and surfaces that would result if the model crystal were allowed to have freely relaxing ends. Furthermore, it will be shown that this device does not affect the results of the computation, within the limits of the approximations used.

The equilibrium configuration of the array was determined by an atomic relaxation process. Only those atoms enclosed within the dashed line in Fig. 1b were allowed to relax. The relaxed position of any given atom in the array was determined by computing the coordinates of the atom corresponding to a zero net force on it. This zero-force technique has been previously described in detail[10] and will not be reiterated here. A representative atom for a given row was selected and allowed to relax, thus determining an appropriate displacement vector; all other atoms in that row were then displaced by the same vector. This process was applied to each row in turn and iterated until the energy change on successive iterations was less than 10^{-4} eV. In all computations, the energy difference between the initial and final states (i.e., the relaxation energy) was then assumed to be correct to within 10^{-4} eV. The symmetry of the array dictates that the net force on any atom, in the y direction, must be identically zero; therefore, only the x and z components of the displacement vector, $\mathbf{u}_x$ and $\mathbf{u}_z$, were computed.

To maximize the interledge spacing for which computations could be performed, it was necessary to determine the closest approach of a ledge to the end of the array, consistent with the approximations of the model. At a distance of 13.25 interatomic spacings from the end of the array, a ledge will not interact atomically with the constrained portion (E1 or E2 of Fig. 1b) of the array. Also, at that distance from the end, calculation shows that the ledge-relaxation energy differs by less than 10^{-4} eV from the ledge-relaxation energy computed for an isolated ledge at the center of the array. This distance was therefore taken as the closest permissible approach to the edge. Furthermore, this limitation ensured that constrained ends did not materially affect the results.

2.2 Ledge Energy and Configuration

The energy of an isolated ledge may be obtained from the following thought experiment. Consider first a perfect surface of infinite extent, a portion of which is illustrated schematically in Fig. 2a. Next, remove that part of the uppermost layer of atoms to the left of Plane A and replace it elsewhere on the surface, thus forming two ledges as shown in Fig. 2b. If the two ledges are sufficiently widely separated there will be no sensible atomic interaction between them. (Consistency with the present approximations requires only that the ledges be separated by more than $4.5a_0$, for the atomic interaction to vanish.) Furthermore, before relaxation is allowed, there can be no displacement field interaction between the two ledges. The total energy change for the crystal (i.e., for the formation of two ledges) therefore simply amounts to the energy of the *"bonds"* that are broken across Plane A in Fig. 2a. Computation shows that this results in an energy change per atom diameter along the ledge of 0.8219 eV per pair or 0.4110 eV per ledge. A second energy contribution arises from atomic relaxation. When an array of maximum size (50 b_0 in length), with a single ledge at the center, is allowed to relax, the energy of the array decreases by 0.0857 eV per atom diameter along the ledge. This is the

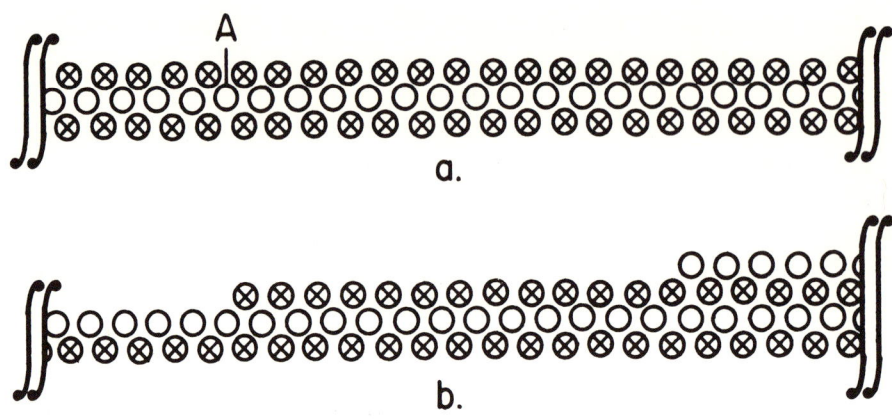

Fig. 2. (a): Section through perfect terrace. Open and crossed circles as in Fig. 1. (b): Section through same terrace after atoms to the left of Plane A have been transferred to the top right, thus forming two ledges.

ledge-relaxation energy. Thus, the net energy change associated with the formation of a ledge is found to be 0.3253 eV per atom.

The computation of the ledge-relaxation energy also yields the equilibrium configuration of an isolated ledge. For the purpose of reference, it is useful to distinguish between the upper and lower terraces associated with a ledge, as indicated in Fig. 3a. The tangential component of the displacement field, u_x, is found to be always directed toward the ledge and, for the most part, is considerably larger than the normal component, u_z. A plot of log ($|u_x|/b_0$) versus x for the atoms in the first surface layer, on either side of the ledge, is given in Fig. 3a. The figure shows that the magnitude of u_x drops off exponentially with x, with the exception of the "core" of the ledge where displacements are less well behaved. Furthermore, it is interesting to note that the virtual center of the distortion, given by the point of intersection of the extrapolated linear portions of the plot, is somewhat displaced toward the lower terrace of the ledge. Thus, the distortion is not symmetric about the geometric midpoint of the ledge and, as shown later, this result causes the various possible ledge pairs to interact in different ways.

With the exception of the core region, the normal component of the displacement field is positive (i.e., directed away from the crystal) and is given in Fig. 3b for atoms in the first surface layer. (It should be recalled that these displacements represent the excess of displacement over and above the perfect surface displacements initially imposed on the surface which are also directed away from the crystal.) These displacements fall off more rapidly than exponentially and are also asymmetrically disposed

LEDGES ON (001) COPPER SURFACES 639

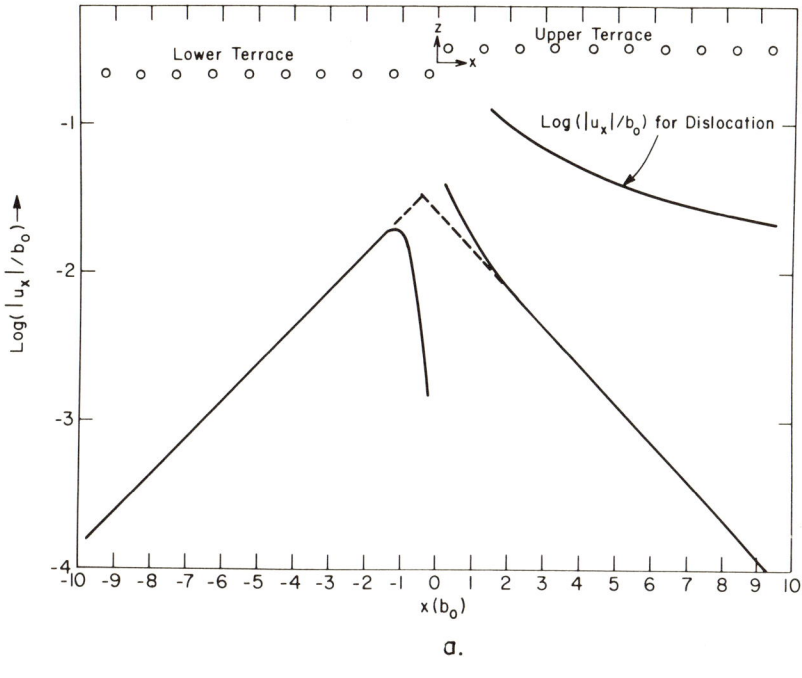

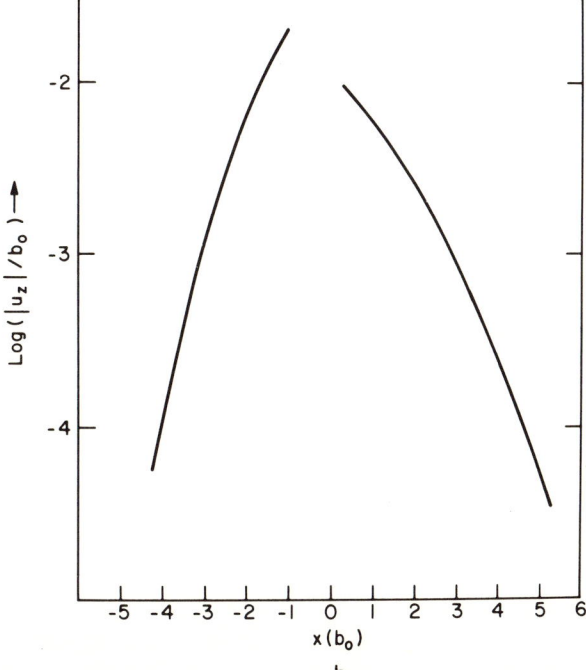

Fig. 3. (a): Tangential component of displacements, u_x, about isolated ledge, as a function of x for atoms in the first surface layer. Displacement field for an edge dislocation is also shown (see text). (b): Normal component of displacements, u_z, as a function of x for atoms in the first surface layer.

about the ledge. The general aspect of an isolated ledge is given schematically in Fig. 4.

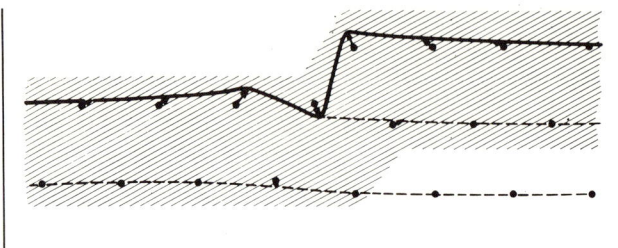

Fig. 4. Schematic of atom displacements in vicinity of ledge. Shading shows region allowed to relax.

2.3 Interaction Energy and Ledge Pair Configuration

For the present purposes, the interaction energy of a pair of ledges separated by a distance λ is defined as the energy change of the pair that results when the spacing of the pair is decreased from infinity to λ. It is convenient to consider the interaction energy as being made up of two contributions, a short-range atomic interatomic and an interaction arising from the overlap of the displacement fields that result from atomic relaxation. Consider first a surface consisting of a pair of ledges separated by a distance large compared with the range of atomic forces, say 20 b_0, such that only the interaction between displacement fields prevails. The relaxation energy for the pair, ΔE_{R2}, may be computed from the difference in energy between the relaxed and unrelaxed configurations. ΔE_{R2} is then compared with twice the relaxation energy for a single isolated ledge, ΔE_{R1}, giving the interaction energy as $E_I = \Delta E_{R2} - 2\Delta E_{R1}$, where ΔE_{R2} and ΔE_{R1} are defined as negative quantities. A positive interaction energy implies repulsion between the ledges, conversely a negative E_I implies attraction. The computations show that this contribution to the interaction leads to repulsion for all three types of ledge pairs. As the interledge spacing is decreased, a point is reached when atomic interaction comes into play. That contribution may be obtained from the following thought experiment. Imagine that a third ledge is present on the surface, a reference ledge, and that it is arbitrarily far removed from the pair under consideration. The interledge spacing is then decreased by either transferring rows of atoms from the reference ledge to the pair or vice versa.* For example, for face-to-face pair of ledges, the interledge spacing is decreased by transferring atom rows from the reference ledge to either ledge in the pair. When the transferred row first interacts, its energy is lowered by the additional bonds it forms with its partner. The interaction is thus attractive. Similarly, it is found that the atomic interaction is attractive for back-to-back ledges but repulsive for sequential ledges.

---

*This device has the advantage that it permits the energy changes to be followed in a system where the total number of atoms is conserved and so avoids certain conceptual difficulties that arise when nonconservative systems are employed.

The dependence of the total interaction energy on the interledge spacing, λ, is shown in Fig. 5. It can be seen that, at relatively large pair separations, the interaction is repulsive in all three cases, but is definitely short range, dropping below 10^{-4} eV for interledge spacings larger than about 10 or 12 b_0. For spacings less than about 4 b_0, the displacement field interaction is overwhelmed by the shorter-range atomic interaction, the total interaction becoming attractive for face-to-face and back-to-back ledges and more strongly repulsive for sequential ledges. It is interesting to note that the displacement field interaction is largest for face-to-face and smallest for back-to-back ledge pairs. This must be a consequence of the asymmetry of the displacement field which was pointed out earlier in connection with Fig. 3a. Recall that the virtual center of the distortion was found to be located on the lower terrace of the ledge, thus causing the displacements to be greater on that terrace for a given value of $|x|$. Since face-to-face ledges interact along the lower terraces, the largest interaction energy for that case is consistent with the displacement field asymmetry. By the same reasoning one might expect back-to-back pairs to display the smallest interaction, since they interact along the upper terraces, and sequential ledge pairs to interact in some intermediate fashion, since the upper terrace of one of the ledges interacts with the lower terrace of its partner. This is the behavior observed in Fig. 5.

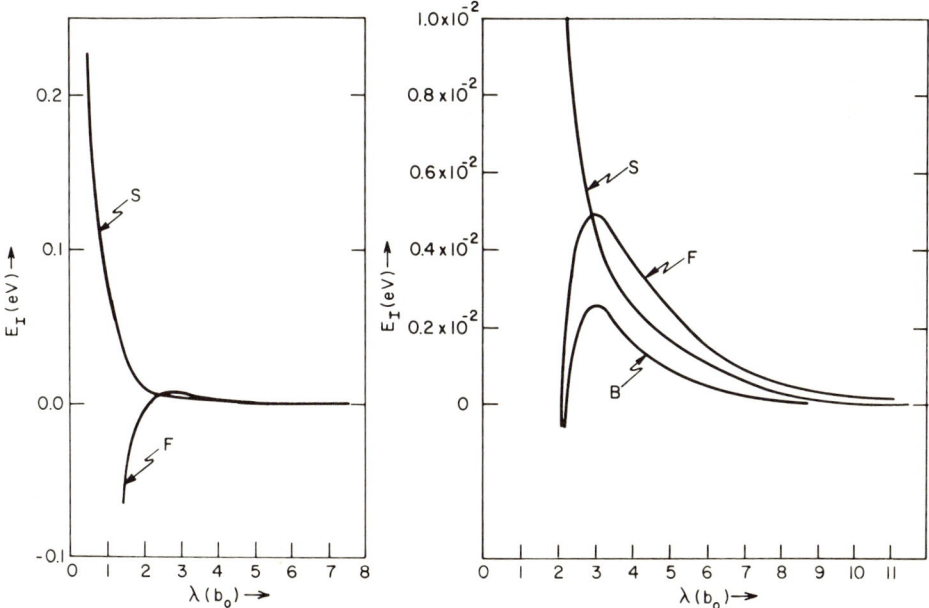

Fig. 5. Interaction energy, E_I, as a function of ledge spacing, λ, for sequential (S), face-to-face (F) and back-to-back (B) pairs. Left, general aspect; right, detail.

642 P. WYNBLATT

The displacement fields for sequential ledge pairs are illustrated in Fig. 6. The figure shows that the mutual effect on the displacement fields is only significant over a region, 5 to $6b_o$ wide, half way between the two ledges. Pair configurations for the other two types of pairs are similar but they do not display the slight asymmetry of Fig. 6.

3 DISCUSSION

As mentioned earlier, the anisotropy of surface energy has often been interpreted in terms of some of the quantities calculated here. A means is therefore offered for comparing the present results with experiment. In general, the surface energy, γ, of a flat crystalline surface depends on the crystallographic orientation of the plane of the surface. The orientation dependence of the surface energy is generally represented in the form of a three-dimensional polar plot, where the energy of a surface oriented normal to the radius vector is given by the magnitude of the vector. This representation is known as a γ-plot. For simplicity, a cut through the three dimensional plot, i.e., a two-dimensional γ-plot, is considered.

A convenient way of relating the variation of surface energy with orientation to the structure of a TLK surface has been put forward by Cabrera[2]. Consider a surface of energy γ, rotated by an angle θ from some reference low-index terrace, as illustrated in Fig. 7. The energy of such a surface may be expanded as a power series:

$$\gamma(\theta)/\cos\theta = a_o + a_1 \tan\theta + a_2 \tan^2\theta + \ldots , \qquad (2)$$

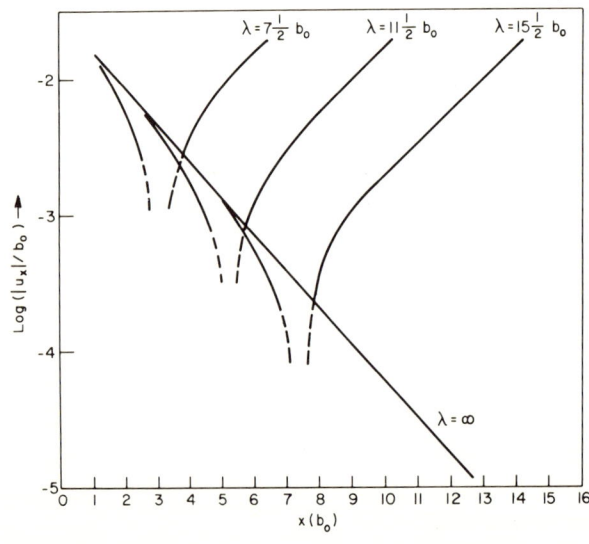

Fig. 6. Tangential component of displacements, u_x, as a function of x for atoms in the first layer of the surface between a pair of sequential ledges. Left ledge of pair is always situated at $x = 0$. Displacements are shown for pair spacings of $\lambda = 7\frac{1}{2}\,b_o$, $11\frac{1}{2}\,b_o$, $15\frac{1}{2}\,b_o$, and ∞.

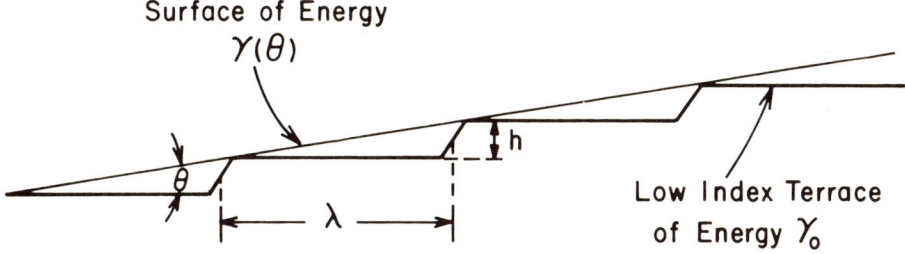

Fig. 7. Schematic of surface inclined at an angle θ to a low-index reference terrace of surface energy γ_0. h is the height of monoatomic ledge.

where $a_0 = \gamma_0$, the surface energy of the reference surface, a_1 is proportional to the energy of an isolated ledge, and a_2 and higher order coefficients are related to the interaction between ledges. A similar approach is adopted here where $\gamma(\theta)$ is expressed as:

$$\gamma(\theta)/\cos\theta = \gamma_0 + [E_0 + E_I(\lambda)]/\lambda \quad , \tag{3}$$

where E_0 is the energy per atom diameter of an isolated ledge, $E_I(\lambda)$ is the interaction energy of a pair of ledges per atom diameter of ledge and $1/\lambda$ is the ledge density.* Since $\tan\theta = h/\lambda$, Eq. 3 may be rewritten as:

$$\frac{\gamma(\theta)}{\gamma_0} = \cos\theta + \frac{[E_0 + E_I(\theta)]}{h\,\gamma_0}\sin\theta \quad . \tag{4}$$

Eq. 3 and 4 are predicated on the following assumptions. First, recall that the quantity $E_I(\lambda)$, or $E_I(\theta)$, as calculated here, represents the energy change of a pair of ledges that results when the spacing of the pair is decreased from infinity to λ. It is assumed: that a ledge on a surface of given orientation interacts only with its first neighbors, and that the interaction of the ledge with its two neighbors is simply the sum of its interaction with each neighbor. Thus, the energy change of a single ledge on a surface due to its interaction with its two neighbors is taken to be equal to the interaction energy of an isolated pair, i.e., $E_I(\lambda)$. These two assumptions are realistic for large λ (small θ) but must break down for small λ. Eq. 4, therefore, may be regarded as equivalent to Eq. 2 in which the higher order interaction terms have been neglected.

---
*The formulations of Eq. 2 and 3 differ somewhat in that the latter combines ledge energy and interaction into a single term. This is convenient because of the present definition of the interaction energy; however, Eq. 3 could, in principle, be modified to conform more closely to Eq. 2.

Although the anisotropy of surface energy is generally presented in the form of a γ-plot, the information is often gathered from measurements of normalized surface "torques", $1/\bar{\gamma}\, d\gamma/d\theta$, where $\bar{\gamma}$ is some mean value of the surface energy. For comparison with experiment, Eq. 4 can be differentiated w.r.t. θ, giving

$$\frac{1}{\gamma_0}\frac{d\gamma}{d\theta} = \frac{[E_0 + E_I(\theta)]}{h\gamma_0}\cos\theta + \left[\frac{1}{h\gamma_0}\frac{dE_I(\theta)}{d\theta} - 1\right]\sin\theta, \quad (5)$$

where the torque has been normalized with respect to γ_0 rather than $\bar{\gamma}$. This should only have a negligible effect on the comparison with experiment since measurements show[1,12] that $(\gamma_{max} - \gamma_0)/\gamma_0$ lies between 0.01 and 0.03. In the present case, $\gamma_0 = \gamma(001)$. This quantity has been computed by Burton and Jura[6] (using the same potential energy function and including relaxation effects) to be 1.640 eV/atom; also, for the terraces considered here and since λ has units of b_0, h simply takes on the value $1/\sqrt{2}$. The normalized γ-plot and torque plot can now be calculated from Eq. 4 and 5 and are given in Fig. 8.

Qualitatively, the features of Fig. 8 are in agreement with the measurements of Robertson and Shewmon[1] and Hondros and McLean.[12] There are however significant discrepancies. The value of the ratio γ_{max}/γ_0 is calculated to be 1.040; whereas, the experimental values are 1.016 and 1.012 for Ref. 1 and 12, respectively. Also, the maximum energy (i.e., zero torque) is calculated to occur at 16 degrees but measured to lie at 19 degrees[12] and 27 degrees[1]. Finally, the experimental intercept on the torque plot, $1/\gamma_0\, d\gamma/d\theta|_{\theta=0} = E_0/h\gamma_0$, is 0.07 or 0.09 for Ref. 1 and 12, respectively, which is considerably lower than the calculated value of 0.28. However, the present computations are strictly valid only at the absolute zero of temperature whereas the experimental values are obtained at temperatures of about 1300°K where other effects, mentioned in the introduction, are known to intervene. Those effects are considered shortly, but first it is interesting to compare the present calculation with another recent theoretical estimate of the interaction between ledges.

Blakely and Schwoebel[5] have used a phenomenological approach to estimate the interaction between an isolated pair of ledges. They assign the capillarity forces acting on a curved surface to the interaction between sequential ledges and show that the force on a ledge due to the presence of a second ledge on the surface is,

$$F = \frac{\gamma\Omega}{2ca_0}\frac{1}{\lambda^2}, \quad (6)$$

where Ω is the atomic volume and c is a constant with a value of about 3. The change in energy of the pair resulting from a decrease of the

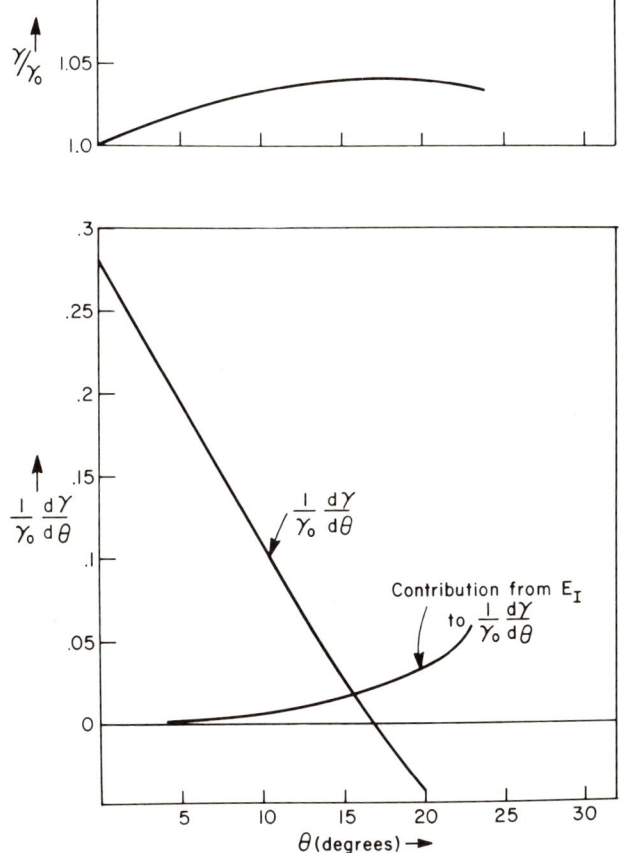

Fig. 8. Plots of γ/γ_o versus θ as calculated from Eq. 4 and $1/\gamma_o \, d\gamma/d\theta$ from ledge interaction is also shown.

interledge spacing from infinity to λ, i.e., the interaction energy as defined here, is therefore given by

$$E_I = \int_\infty^\lambda F d\lambda = \frac{\gamma \Omega}{2ca_o} \frac{1}{\lambda} \quad . \tag{7}$$

This interaction is repulsive and definitely long range. Blakely and Schwoebel propose that this interaction could arise from either or both of two possible sources: either from interaction of the displacement fields of the two ledges which, as suggested by Cabrera[2], might be similar to the interaction between dislocations or from an effective interaction at high temperature stemming from entropy effects associated with the wandering of ledges. The present investigation has shown that the interaction of ledges through their displacement fields, unlike the interaction of dislocations, is short range. For comparison with a ledge, the displacement field of an edge dislocation (in the direction of the Burgers vector and at a distance $b_0/2$ from the plane defined by the dislocation line and the Burgers vector) is shown in Fig. 3a. It can be seen that the displacements associated with a dislocation are both larger and longer range than those of a ledge. Having ruled out the displacement field interaction as the source of the long-range interaction between ledges, it is interesting to consider now the other possible factor so as to determine whether it may suitably complement the present approach.

Gruber and Mullins[13] have investigated the entropy contribution to the ledge free energy which arises from wandering (or kink density) of the ledge and the influence of interledge spacing in restricting the kink density, thus decreasing this entropy contribution.[13] The free energy of a ledge, and therefore the surface free energy, calculated in this manner becomes a function of the interledge spacing, and the result may be viewed as an effective interaction between ledges. Gruber and Mullins write the surface free energy as

$$\frac{\gamma}{\gamma_0} = \cos\theta + \varphi(\theta,\beta) \frac{\beta kT}{h\gamma_0} \sin\theta \quad , \tag{8}$$

where the function $\varphi(\theta, \beta)$ is determined from a statistical mechanical analysis, β is defined as E_0/kT and where it is assumed that straight ledges at absolute zero do not interact. The function $\varphi(\theta,\beta)$ can vary from negative values to 1; however, negative values need not concern us here as they correspond to surface melting. For a given value of β, φ increases with θ from some positive value, approaching 1 in the limit of large θ where the ledge free energy cannot be decreased as wandering is restricted by the close bunching of the ledges. For a given value of θ, φ increases with β and approaches 1 in the limit of large β where wandering (or kink formation) becomes energetically unfavorable.

To introduce the effect of temperature into the present calculations, the interaction of ledges at absolute zero is simply grafted onto Eq. 8 by substituting $[\beta kT + E_I(\theta)]$ for βkT, thus

$$\frac{\gamma}{\gamma_0} = \cos\theta + \varphi(\theta,\beta) \frac{[\beta kT + E_I(\theta)]}{h\gamma_0} \sin\theta \quad . \tag{9}$$

Clearly, this expedient can serve only as a first-order approximation since any interaction between ledges will affect the derivation of $\varphi(\theta,\beta)$; however, it provides certain insights into the effect of coupling the interactions. For proper comparison with the analysis of Gruber and Mullins, their values of $T = 1000$ C and $\gamma_0 = 1670$ erg/cm^2 (0.677 eV/atom) are used. Fig. 9 gives the torque plots as calculated from Eq. 8 and 9 for $\beta =$

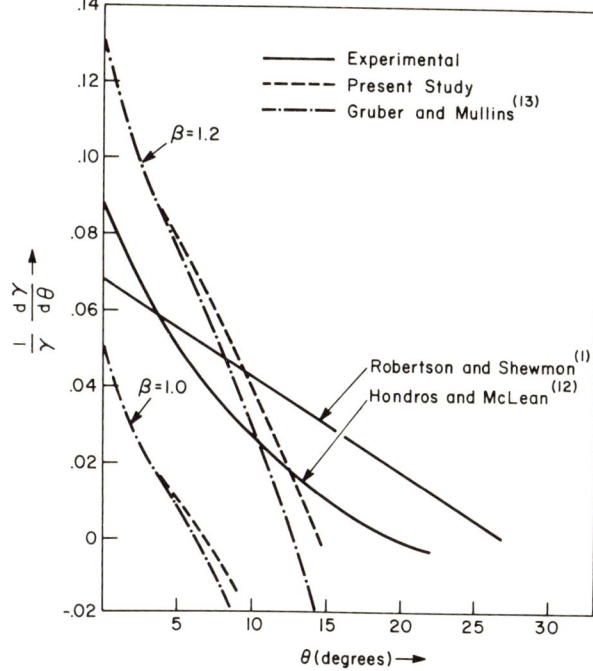

Fig. 9. Plots of $1/\gamma_o \, d\gamma/d\theta$, obtained by differentiating Eq. 8 (dot-dash) and Eq. 9 (broken line), versus θ. Also given are the corresponding plots from measurements (solid lines).

1.0 and 1.2. The corresponding plots from the measurements of Robertson and Shewmon[1] and Hondros and McLean[12] are also given. It is clear that the high temperature effects formulated by Gruber and Mullins have a significant effect on the present analysis (Fig. 8) and conversely, that the interaction between straight ledges at absolute zero provides a small but significant improvement of the Gruber and Mullins analysis for $\theta > 4$. The more recent measurements of Hondros and McLean are fairly well bracketed by the two theoretical plots and it is clear that the data at low angles could be fitted reasonably well by using some intermediate value of β. Deviations at higher misorientation may be partly due to the neglect of higher order interaction effects.

SUMMARY AND CONCLUSIONS

The atom displacements associated with a surface ledge are found to decrease sharply with increasing distance from the ledge, the tangential component of the displacement field dropping off exponentially and the normal component, even more rapidly.

As a result, the displacement field interaction between all three possible types of ledge pairs is short range. Furthermore, this interaction is found to be repulsive in all cases and is significant only over a range of ledge pair spacings from about 11 to 4 interatomic distances. The displacement field interaction cannot therefore be used to rationalize the long-range ledge pair interaction predicted by phenomenological models.

For ledge pair separations smaller than about 4 interatomic distances, the displacement field interaction is overwhelmed by the shorter-range interatomic forces that reinforce the repulsive interaction between sequential ledge pairs but reverse the interaction between face-to-face and back-to-back pairs, making it attractive.

The calculated anisotropy of surface energy at 0 K is found to agree qualitatively with the trends observed experimentally at high temperatures. Furthermore, when the present results are combined with a suitable high-temperature model they are found to provide a correction in the proper direction.

Finally, it is worth noting that in spite of the well-known shortcomings of the pairwise-interaction model, the present study demonstrates once again that the model can provide a simple means to useful insights which may be difficult, if not impossible, to obtain from other available techniques.

ACKNOWLEDGMENTS

I am greatly indebted to Professors J. M. Blakely and R. L. Schwoebel for making their work available to me and for permission to reference the work before publication. I wish to thank Dr. N. A. Gjostein for constant encouragement and valuable discussions throughout the course of this study. Thanks are also due to Drs. C. L. Magee and W. L. Winterbottom for their critical review of the manuscript.

REFERENCES

1. Robertson, W. M., and Shewmon, P. G., Trans. Met. Soc. AIME, **224**: 804 (1962).
2. Cabrera, N., Surface Sci., **2**, 320 (1964).
3. Hirth, J. P., *"Energetics in Metallurgical Phenomena"*, W. M. Mueller (ed.), Vol II, p. 1, Gordon and Breach, New York, 1965.
4. Mullins, W. W., ASM Seminar on *"Metal Surfaces"*, p. 17, ASM, Metals Park, 1962.
5. Blakely, J. M., and Schwoebel, R. L., to be published in Surface Science.
6. Burton, J. J., and Jura, G., J. Phys. Chem., **71**, 1937 (1967).
7. Bonneton, F., and Drechsler, M., Surface Sci., **22**, 426 (1970).
8. Wynblatt, P., Surface Sci., **22**, 125 (1970).
9. Ehrlich, G., and Kirk, C. F., J. Chem. Phys., **48**, 1465 (1968).
10. Wynblatt, P., and Gjostein, N. A., Surface Sci., **12**, 109 (1968).
11. Girifalco, L. A., and Weizer, V. G., Phys. Rev., **114**, 687 (1959).
12. Hondros, E. D., and McLean, M., *"Structure et Proprietes des Surfaces Solides"*, p. 219, Editions du Centre National de la Recherche Scientifique, Paris, 1970.
13. Gruber, E. E., and Mullins, W. W., J. Phys. Chem. Solids, **28**, 875 (1967).

DISCUSSION on paper by P. Wynblatt

GIRIFALCO: A class of experiments exists that might be used to obtain some information on potentials at surfaces. These involve heat of wetting and heat of adsorption measurements. When coupled with preadsorption techniques of the type proposed by W. D. Haskins, they can be used to determine the range of surface forces and the distribution of energy over the surface. If the wetting (or adsorbing) material is chemically similar to the substrate, geometric mean laws can be used to relate AA, AB, and BB potentials.

As Dr. Ehrlich has pointed out, such experiments (and their interpretation) are difficult. However, I believe that the primary difficulty is in surface preparation and contamination. The ability to map regions of the surface that are energetically different by preadsorption techniques should be useful.

EHRLICH: To derive information that will be useful in the construction of potential functions for the gas-solid interface it is necessary to measure binding energies on different crystal planes. This is not at all easy. Attempts to determine the heat of adsorption of rare gases have been made in the field emission microscope by Engel and Gower[1], as well as in our laboratory, but these measurements are still very crude.

WYNBLATT: Any information pertaining to surface interactions can, in principle, be used to formulate interatomic potentials near the surface. Another experiment that is similar to those mentioned and that may be useful in this connection was performed by Domange and Oudar[2]; they related changes in the structure of fcc metal surfaces to the presence of specific concentrations of adsorbed sulfur. However, properties that depend on the presence of a second element are generally more difficult to translate into useful information.

HUNTINGTON: I'm confused by your statement that the distortion about the ledge doesn't tend to smooth it out.

WYNBLATT: If one decomposes the vector connecting the two rows of atoms that constitute the ledge into a normal and a tangential component, then one finds that the normal component is increased by the relaxation process, thus increasing the height of the step.

HARRISON: We should keep in mind that surface properties of the type you have discussed contain the ambiguity as to the relaxation of neighbors to a vacancy that I discussed. For the potentials you used the last layer of atoms moved from the surface and the ledges were sharpened rather than smoothed. I suspect that these conclusions are correct, but it could very well be that they are all backwards and that the reverse conclusions are true. It depends upon the details of the potentials and we are very uncertain about these at the surfaces.

WYNBLATT: I agree. Any potential formulation that predicts attraction between first neighbors will presumably lead to the reverse of the picture I presented.

SHEWMON: What physical properties go into determining the potential you use? Can you adjust any of the constants by comparison with surface properties? Or, how sensitive are your results to varying the potential?

WYNBLATT: The physical properties that go into the potential are the compressibility, sublimation energy, and lattice parameter. None of the constants were adjusted to surface properties, and since I have only used this particular potential, I do not know how sensitive the present results might be to the form of the potential. However, in his earlier talk, Dr. Ehrlich showed that calculations of other surface properties do not appear to be any more sensitive to variation of the potential than bulk properties.

DORAN: Where did you truncate your potential?

WYNBLATT: At 4.5 lattice parameters.

WEINS: What type if boundary conditions did you use? If they were fixed or periodic what was the size of the unit used for the calculation?

WYNBLATT: Because of the truncation of the potential, the atomic array used was effectively infinite in two of the three dimensions. The third dimension was changed from calculation to calculation, as required, and the end portions of the array in that dimension were constrained as explained in detail in the manuscript.

HIRTH: Your figure showing the relaxation at a ledge was of interest with respect to F. C. Frank's suggestion of nucleation of an edge dislocation by punching-in of a ledge. The core-scale configuration at the ledge resembled an incipient dislocation.

If one regarded this configuration as such a dislocation, it would have an image dislocation near the surface and the resultant pair would have the displacement field of a dislocation dipole. It might be of interest to determine whether such a dipole field would fit the observed displacements. If so, this would provide an easy method of introducing many-layer relaxations normal to the surface by continuum methods.

BULLOUGH: You commented that the interaction between the ledges was not like the interaction between dislocations. It seems to me they were in part rather like the interaction between any two dislocations. Have you considered this possible interpretation?

WYNBLATT: Because of the difference in the nature of the displacement fields of an isolated ledge and of a dislocation it would probably require a special array of dislocations to produce the observed interaction.

REFERENCES

1. T. E. Engel and R. Gomer, J. Chem. Phys. **52**, 5572 (1970).
2. J. L. Domange and J. Oudar, Surface Sci., **11** 1244 (1968).

COMPUTER CALCULATIONS OF DYNAMICAL SURFACE PROPERTIES OF CRYSTALS

F. W. de Wette

The University of Texas at Austin
Austin, Texas 78712

1 INTRODUCTION

This paper reviews the results of numerical studies of surface properties of crystals based on the use of realistic interatomic potentials. These studies involve detailed calculations of static, dynamical, thermodynamic and scattering properties of noble gas and ionic crystals, systems for which rather realistic interatomic (interionic) potentials are available. The results show that, although in a sense one studies model systems, a wealth of new and detailed information can be obtained, which goes significantly beyond what can be found with analytic methods. Moreover, it is precisely this detailed information that is needed for detailed comparison with experimental results. In a sense, calculations of this type might be called *computer experiments:* If one relies on the theory to give an adequate description of the phenomena, if the interatomic potential that is used is reasonably reliable, and if unnecessary approximations are avoided, then one can expect the calculations, in principle, to yield all the details of the phenomena. A comparison with experiment will then provide a test, both for the underlying theory and for the details of the interaction potential.

This review is not intended to be exhaustive. Its aim is to discuss a number of selected results and indicate their significance. For details of the calculations, and for results of other workers in the field, we refer to more-detailed publications.

The calculations described here have been carried out for slab-shaped crystals with two flat surfaces. Such a geometry permits feasible calculations, while at the same time yielding all the interesting surface effects (except the extremely long wavelength vibrational modes). Two methods of calculation have been used: lattice dynamics and molecular dynamics. *Lattice dynamics* in the quasiharmonic approximation is probably valid for surface vibrations of noble-gas crystals below about one sixth of the melting temperature (as compared to about $1/3$ T_M for the bulk) due to the larger amplitudes of vibration at the surface. *Molecular dynamics* on the other hand, is a classical method that takes anharmonic effects into account automatically; it is valid above about one-half the Debye temperature (it neglects zeropoint motions). Thus lattice dynamics and molecular dynamics are complementary methods.

The methods by which the surface dynamical calculations have been carried out are reviewed. A number of selected results for the noble gas crystals are described, and recent results for ionic crystals, including the specific heat which is compared with experimental results, are reviewed.

2 METHODS OF CALCULATION

2.1 Lattice Dynamics

The calculations for the noble-gas crystals have been performed for crystalline slabs (films) of fcc structure with (100), (111), and (110) orientations. The particles are taken to interact through a Lennard-Jones (LJ) 12-6 potential

$$\phi(r) = 4\epsilon \left[(\sigma/r)^{12} - (\sigma/r)^6 \right] . \qquad (1)$$

Here ϵ and σ are the LJ potential parameters and r is the distance between the interacting particles.

The first step in performing a lattice-dynamics calculation for thin films is to determine the static displacements (surface relaxation) of the crystalline planes parallel with the surface. This is done by minimizing the static energy of the slab with respect to the interplanar distances. Physically this procedure corresponds to allowing the planes to relax until the force on every atom is equal to zero. The main effect of the static displacements is to change the force constants near the surface.

The calculations of the dynamical properties are based on the quasi-harmonic approximation. In this approximation the frequencies $\omega_p(\bar{q})$ are determined by numerically solving the matrix eigenvalue equation[1]

$$\sum_{\beta \ell_3'} D_{\alpha\beta}(\ell_3 \ell_3'; \bar{q}) \xi_\beta(\ell_3'; \bar{q}p) = \omega_p^2(\bar{q}) \xi_\alpha(\ell_3; \bar{q}p) \quad , \qquad (2)$$

where $D_{\alpha\beta}(\ell_3\ell_3';\bar{q})$ is the dynamical matrix and the $\xi_\alpha(\ell_3;\bar{q}p)$ are its eigenvectors. Here $\alpha = x, y$ or z, ℓ_3 labels the planes in the slab, $\bar{q}$ is a two-dimensional wave vector and p labels the modes associated with given $\bar{q}$ ($p = 1,2, \ldots 3N_3$, where N_3 is the number of planes in the slab).

In the quasiharmonic approximation, the surface mean-square amplitudes of vibration are given by

$$<u_\alpha^2(\ell_3)> = \frac{\hbar}{2NM} \sum_{\bar{q},p} |\xi_\alpha(\ell_3;\bar{q}p)|^2 \frac{\coth[\hbar \omega_p(\bar{q})/2k_BT]}{\omega_p(\bar{q})} \quad , \qquad (3)$$

where N is the particle number, M the particle mass, k_B Boltzmann's constant, and T the temperature.

2.2 Molecular Dynamics (MD)

Molecular-dynamics calculations solve Newton's equations of motion for a system of interacting particles. The same systems are considered here as were studied with lattice dynamics, i.e., slab-shaped fcc crystals with (100), (111), and (110) orientations in which the particles interact through a LJ potential. The advantage of MD calculations is two-fold: (i) the particles seek the proper mean positions corresponding to the temperature of the system, so that thermal expansion is automatically taken into account, and (ii) the particles carry out vibrations governed by the classical equations of motion, so that anharmonicity is automatically included.

The method of calculation has been described elsewhere.[2,3] The equation of motion of a particle, with a damping force included, is

$$\frac{d^2 s^\ell}{d\tau^2} = 24 \sum_{\ell' \neq \ell} \left(\frac{1}{s^{\ell\ell'}}\right)^8 \left[2\left(\frac{1}{s^{\ell\ell'}}\right)^6 - 1\right] s^{\ell\ell'} - \gamma \frac{ds^\ell}{d\tau} \quad . \qquad (4)$$

Here $s^\ell = r^\ell/\sigma$ where r^ℓ is the position vector of particle ℓ; $s^{\ell\ell'} = s^\ell - s^{\ell'}$ and $s^{\ell\ell'} = |s^{\ell\ell'}|$. Further, γ is the dimensionless damping constant and τ is the dimensionless time defined by $\tau = (\epsilon/M\sigma^2)^{1/2} t$; ϵ and σ are the LJ potential constants.

In a MD experiment, initial kinetic energy can be given to the system by giving the particles initial velocities or initial deviations from their mean positions. During the experiment the kinetic energy (temperature) can be changed by manipulating the damping constant; in this fashion kinetic energy can be put in or taken out of the system.

3 RESULTS FOR MONATOMIC FCC CRYSTAL SLABS

In the following are presented selected results for the fcc crystal slabs in which the particles interact through a LJ potential. Since the LJ interaction gives a reasonable description of the pair interaction of noble-gas atoms, the results may be taken to adequately represent surface phenomena in noble gas crystals. In addition, qualitatively, the results apply to all fcc crystals in which the particles interact through short-range central forces.

3.1 Phonon Dispersion Curves – (100) Surface*

The two-dimensional wave vectors $\bar{q}$ are chosen in the irreducible element (IE) of the two-dimensional Brillouin zone (BZ) corresponding to the surface under consideration, in this case the (100) surface. In Fig. 1, the BZ for the (100) surface of an fcc crystal is shown, with the IE surrounded by heavy lines. In Fig. 2, are shown the dispersion relations $\omega_p(\bar{q})$ for a 21-layer slab with (100) orientation, for values of $\bar{q}$ lying along the boundaries of the IE. Two cases are shown: (a) without, and (b) with static relaxation taken into account. Notice the following features: (i) There exist "bands" of dispersion curves which correspond to the (acoustic) bulk bands. (ii) There exist gaps between these bulk bands. (iii) Surface modes can exist below the bulk bands (S_1, S_2, S_4, S_5) or in these gaps (S_6, S_7, S_8, S_9, S_{10}), the latter generally at short wavelengths. (iv) Surface modes can exist within the bulk subbands (S_3, S_4) for wave vectors $\bar{q}$ along symmetry directions, as a result of symmetry decoupling of these modes from the surrounding bulk bands. (v) Comparing Fig. 2a and 2b it is evident that the qualitative features of the surface phonon spectrum are sensitive to changes in the surface force constants. (e.g., the modes S_2, S_9, and S_{10} are not present in Fig. 2a). (vi) Of all the surface modes mentioned, only S_1 and S_4 persist in the long-wavelength limit. Along $\overline{\Gamma M}$ S_1 has the character of a so-called "generalized Rayleigh wave": its polarization is SV (shear vertical) and its amplitude shows an approximately exponential decay with increasing distance from the surface. (vii)

---

*For a complete discussion of surface modes in slabs with fcc orientation, see Ref. 4.

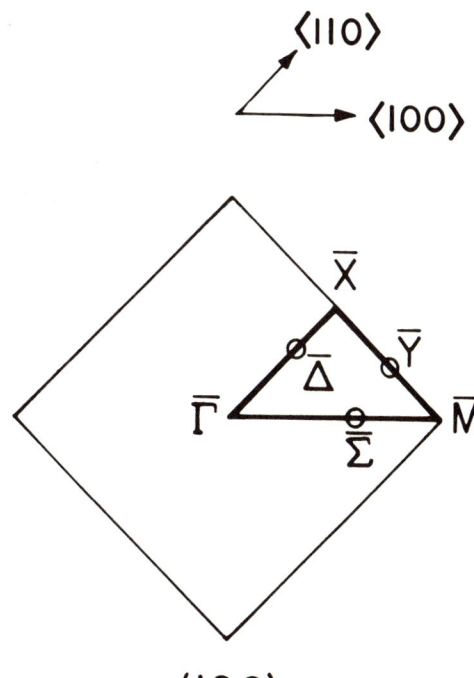

Fig. 1. Two-dimensional Brillouin zone for the (100) surface of a fcc crystal. The irreducible element is outlined by heavy lines.

Along $\overline{XM}$ the modes S_1 and S_4 show "hybridization": they approach each other closely but instead of crossing there is a sudden interchange in character: going from left to right, S_1 changes from SH (shear horizontal) to SV, and S_4 from SV to SH.

An examination of the squared amplitudes $|\xi(\ell_3;\bar{q}\,p)|^2$ as a function of layer number ℓ_3 reveals the variation of the amplitude of a given mode away from the surface. For details of the results refer to Ref. 4. Here it is just mentioned that the attenuation of surface modes with distance from the surface is generally rather complex: the magnitude of $|\xi(\ell_3)|^2$ need not fall off monotonically, or even regularly. It is possible for a mode localized near the surface to have its maximum amplitude in some layer beneath the surface layer. In fact, as a result of the "perturbation" caused by the surface, surface modes of the same polarization character are "peeled off" from a given bulk band in a series, with a first-layer mode peeled off first, then a second-layer mode, etc.

In Fig. 3 are shown dispersion curves of a (100)-oriented, 21-layer slab of which the particles of the two outer layers have a mass (M_L) different from that of the substrate particles (M_S); for mass ratios $M_L/M_S = 0.25$ and $M_L/M_S = 4.0$, respectively.[5] Note that in both cases three surface mode branches are associated with the presence of such an "adsorbed monolayer", which persist into the small $\bar{q}$ region (long wavelengths). For

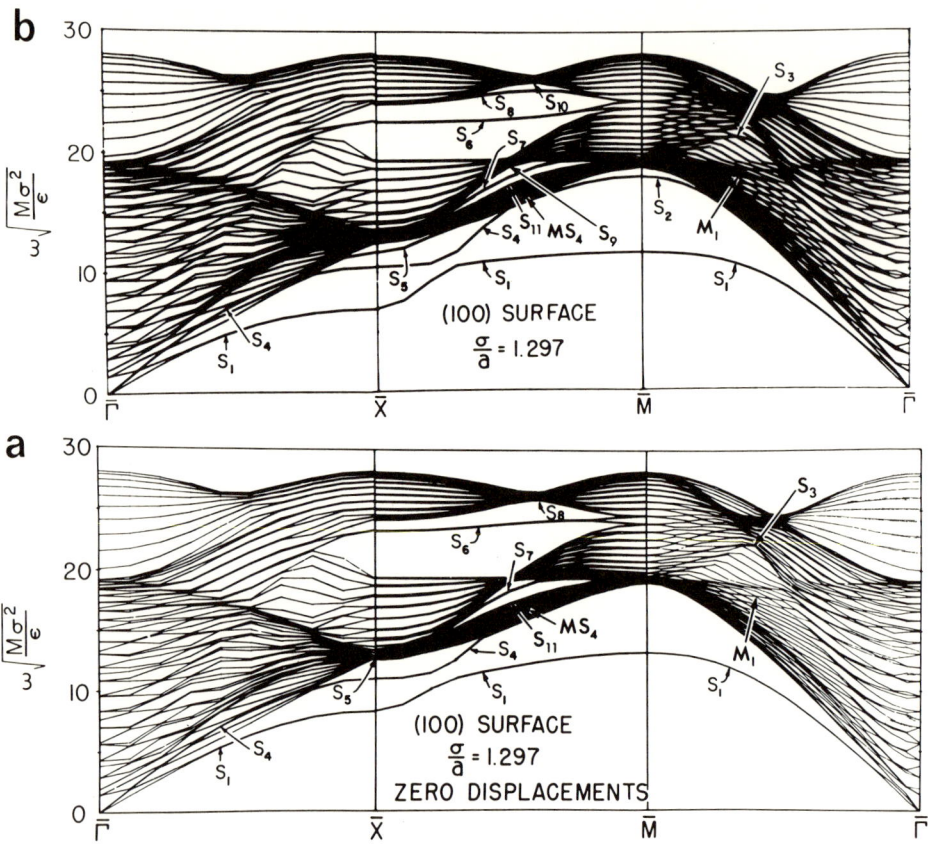

Fig. 2. Dispersion relations $\omega_p(\bar{q})$ for a 21-layer slab with (100) surfaces; (a) without, and (b) with surface relaxation taken into account.

light atoms they lie above the bulk bands (A_1, A_2, and A_3) and are "optical" in character: the light particles move in the field of the "static" heavy particles. For heavy atoms the surface modes (S_1, A_2', and A_3') lie below the bulk bands and are "acoustical" in character. Finally, a heavy adsorbed layer enhances the localization of the generalized Rayleigh waves at the surface while a light layer has the opposite effect.

These results for the phonon dispersion curves have been presented in sufficient detail to demonstrate the rich variety of results that is obtained in numerical studies of this kind.

3.2 Surface Phonon Frequency Distribution

In Fig. 4, a histogram is shown of the frequency distribution function $f(\omega)$ for a 11-layer slab with (111) surfaces. The distinct, narrow peak at

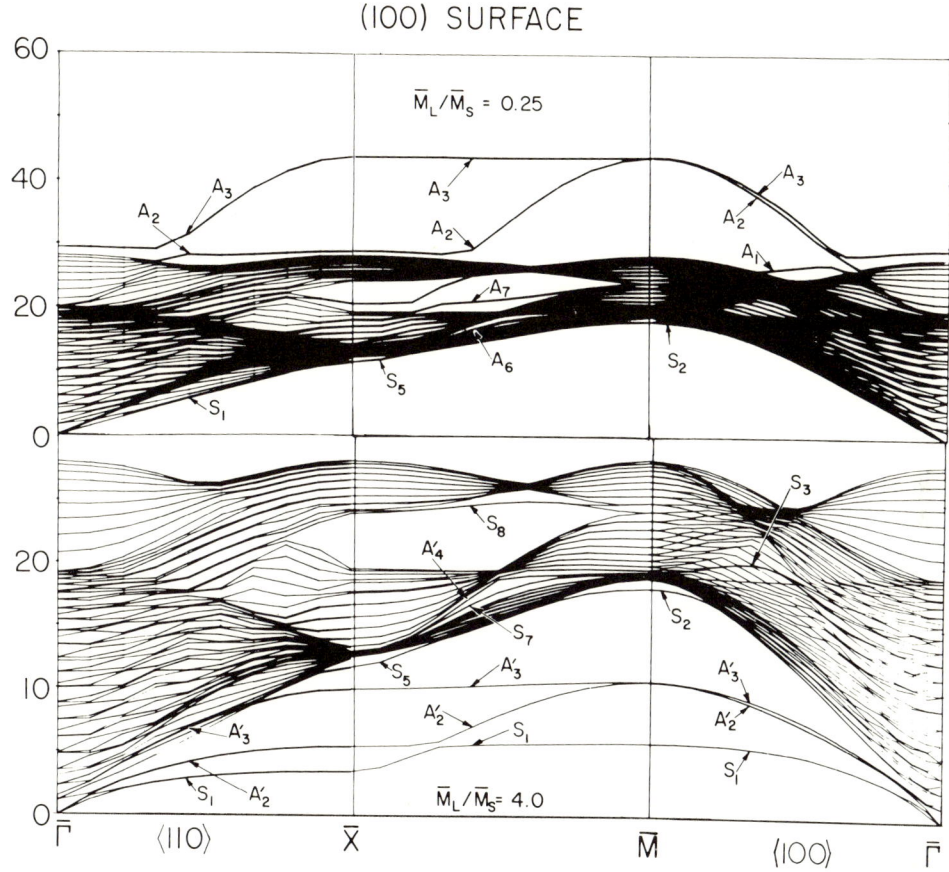

Fig. 3. Dispersion relations $\omega_p(\bar{q})$ for a 21-layer slab with (100) surfaces, of which the particles of the outer layers have a mass (M_L) different from that of the substrate particles (M_S): for mass ratios $M_L/M_S = 0.25$ and 4.0.

$(M\sigma^2/\epsilon)^{1/2}\omega \approx 10$ is associated predominantly with the Rayleigh modes. It is the excitation of these surface modes that gives rise to a peak in the surface specific heat.[6] The surface frequency distribution $f^s(\omega)$ (pictured in the lower half of Fig. 4) is obtained by subtracting out the bulk frequency distribution. It is this function that can be used for calculating the surface thermodynamic properties. For instance, the surface specific heat is given by[6]

$$C_V^s = \frac{3k_B}{A_0} \int_0^{\omega_{max}} d\omega\, f^s(\omega)\, \frac{x^2 e^x}{(e^x - 1)^2}, \qquad (5)$$

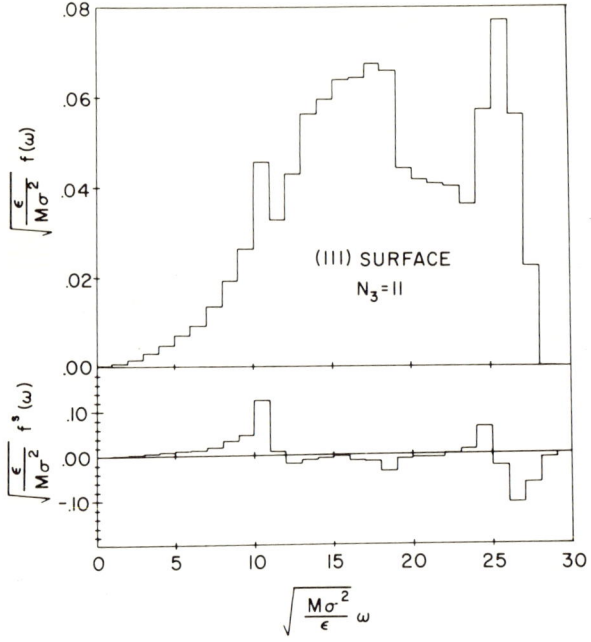

Fig. 4. Frequency distribution function $f(\omega)$ and surface frequency distribution function $f^s(\omega)$ for an 11-layer slab with (111) surfaces. The density is that of the static crystal ($\sigma/a = 1.297$, where $\sqrt{2}a$ is the nearest-neighbor distance).

where $x = \hbar\omega/k_B T$, and A_0 is the area per surface particle.

3.4 Mean-Square Amplitudes of Vibration[7]

The measured quantities which so far have provided the most direct evidence of the existence of surface vibrations are the mean-square amplitudes of vibration (cf. Eq. 3), which, through the Debye-Waller factor, determine the intensity of the Bragg peaks in diffraction experiments. As a consequence, the surface mean-square amplitudes of vibration can be determined from the temperature dependence of the Debye-Waller factor in low energy electron diffraction (LEED) experiments.[8]

A substantial discrepancy has been noted between the measured mean-square amplitudes of vibration in nickel, silver, platinum, palladium, and lead, and those calculated with simple force constant models.[8d] It has been suggested that decreases in the surface force constants due to static relaxation effects might account for this decrepancy.[8d,9] The present calculations on noble-gas crystals suggest that, for a complete explanation of this discrepancy, surface anharmonicity and differential thermal expansion also have to be taken into account.[7]

Fig. 5 shows the dependence of the mean-square amplitudes of vibration $\langle u_\alpha^2 \rangle$ on the distance from the surface[3]; m labels a plane of atoms parallel with the surface plane, with $m = 1$ at the surface. The

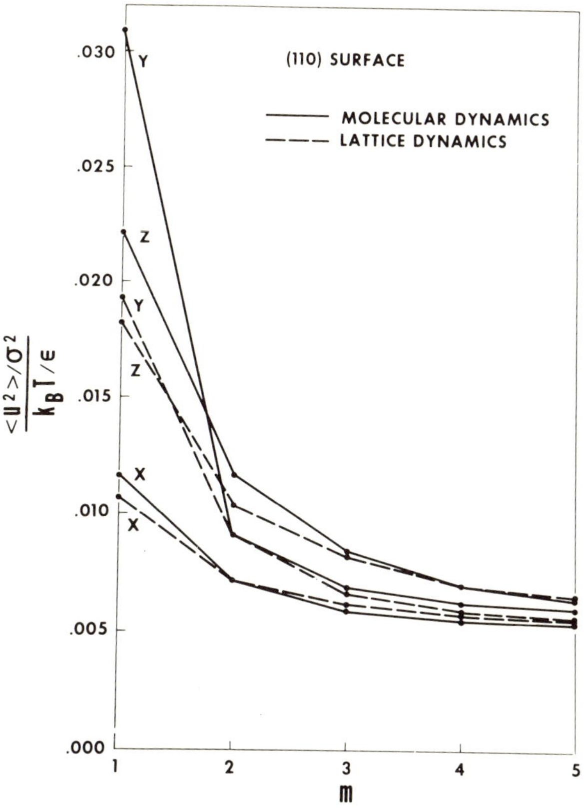

Fig. 5. Mean-square amplitudes $\langle u_x^2 \rangle$, $\langle u_y^2 \rangle$, and $\langle u_z^2 \rangle$ for a (110) surface at about half the melting temperature ($\sigma/a = 1.28$).

results shown are for a (110) oriented slab at about half the melting temperature. The lattice dynamics results (dashed lines) show the increase in $\langle u_\alpha^2(m) \rangle$, and increasing anistropy toward the surface. (In all cases considered in this paper the z axis is chosen perpendicular to the surface. In particular, for the (110) oriented slab the x and y axes are taken to point in the [$\bar{1}$10] and [001] directions, respectively[7]). These well known results follow from the greater vibrational freedom of the surface particles, and from the loss of isotropy at the surface. The solid lines show the same quantities as obtained from a MD calculation at the same density and temperature. It can be seen that, in general, the MD and lattice-dynamics results are in good agreement at the center of the crystal, but that there are considerable differences at the surface, particularly for $\langle u_y^2 \rangle$. These differences must be due to anharmonicity, since they are too large to be attributed to statistical fluctuations in the molecular dynamics results. These results therefore indicate that at about half the melting temperature, anharmonic effects are small in the bulk, but cause substantial increases in the mean-square amplitudes at the surface. Thus, lattice dynamics in the quasiharmonic approximation will lose its validity at lower temperatures at the surface than in the bulk.

Fig. 6 shows the ratio $<u_\alpha^2>_{surface}/<u_\alpha^2>_{bulk}$ as a function of T/Θ_D (Θ_D is the Debye temperature) for slabs of different orientation. It is found in the calculation that this ratio is practically independent of density (within 10 percent for the density range of the noble-gas solids) which makes it in practice a universal function of T/Θ_D. This result follows from Eq. 3 in view of the fact that the Grüneisen constant for the surface is essentially equal to that for the bulk. The main features of the temperature dependence of $<u_\alpha^2>_s/<u_\alpha^2>_b$ are a rapid increase up to about half the Debye temperature, and a leveling off to a constant asymptotic value at higher temperatures. These features follow from the fact that the contributions to $<u_\alpha^2>$ due to the low-frequency modes increase with temperature, and since surface modes are predominantly low-frequency modes, the ratio $<u_\alpha^2>_s/<u_\alpha^2>_b$ increases. At high temperatures, both $<u_\alpha^2>_s$ and $<u_\alpha^2>_b$ are proportional to T, and their

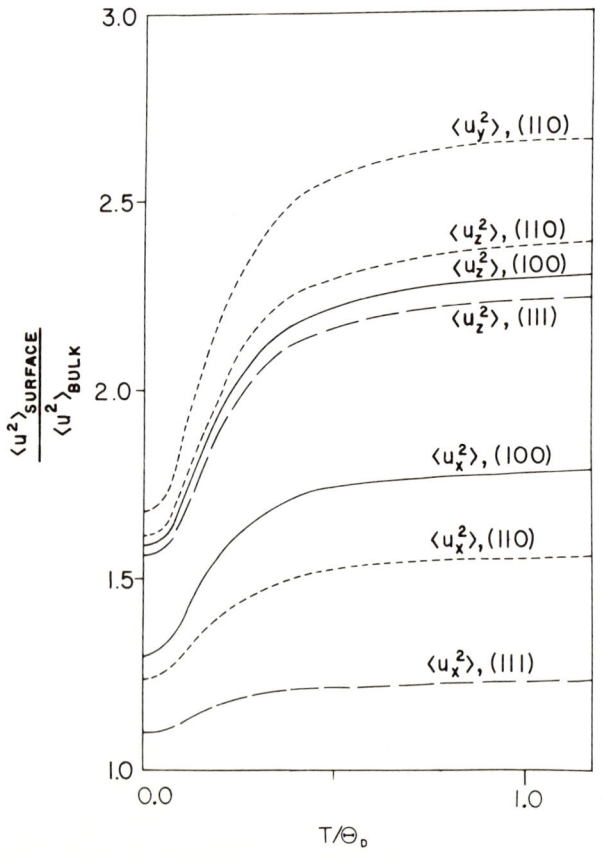

Fig. 6. Temperature dependence of $<u_\alpha^2>_{surface}/<u_\alpha^2>_{bulk}$ for (100), (111), and (110) surfaces.

ratio thus becomes constant. The rise in $<u_\alpha^2>_s/<u_\alpha^2>_b$ has recently been observed in measurements on the (110) surface of chromium[10].

3.5 Surface Thermal Expansion

The introduction mentioned the static displacement, i.e., the outward relaxation of the crystalline planes near the surface. This effect is calculated by minimizing the static energy of the crystal with respect to the interplanar distances. At finite temperatures, additional displacements, the so-called *dynamical displacements* are caused by the fact that a vibrating atom at the surface encounters neighbors in the next plane, upon moving toward the center of the crystal, but encounters no other atoms upon moving outward. On the average, therefore, an atom will be farther from the center of the crystal when vibrating than when stationary. With lattice dynamics, this effect (which is temperature dependent) is calculated by minimizing the free energy (at each temperature) as a function of the interplanar distances. With MD the effect is found from the mean positions of the crystalline planes as functions of temperature. These displacements may be expressed in terms of δ_m, the fractional change in the distance between the mth and the $(m + 1)$th planes.*

Fig. 7 shows the displacements δ_m calculated with lattice dynamics (dashed curves) and molecular dynamics (solid curves). The dashed curve marked $T = 0$ represents the static displacements at absolute zero, obtained from minimizing the static energy. The dashed curve, marked $T^{**} = 0.356$, is obtained from minimizing the free energy at about half the melting temperature ($\frac{1}{2}T_M$). Since these curves express the *fractional* increases in the interplanar spacing, the $T = 0$-curve would be valid for all temperatures if the rate of thermal expansion were the same at the surface as in the bulk. The fact that the fractional displacements increase with temperature means that the rate of thermal expansion is larger at the surface than in the bulk. The solid curves in Fig. 7 indicate the MD results. Notice that at $\frac{1}{2}T_M$ the lattice dynamics and MD results are in complete agreement; this indicates that at $\frac{1}{2}T_M$ the effect of anharmonicity on the displacements is negligible. The curve indicated by $T^{**} = 0.547$ gives the MD displacements at $\frac{3}{4}T_M$. Although valid for a somewhat lower density, the further increase in fractional displacements δ_m with T indicates that the differential thermal expansion at the surface persists at temperatures close to the melting temperature.

---

*The fractional changes δ_m are defined by: $d_m = d(1 + \delta_m)$, where d_m is the distance between the mth and the $(m + 1)^{th}$ plane, and d the distance in the bulk between corresponding planes.

664 F. W. de WETTE

4 RESULTS FOR IONIC CRYSTALS

Optical surface modes in ionic crystals have been treated both in the continuum approximation[11] and the rigid-ion approximation.[12,13] A number of apparent discrepancies between the results of these earlier treatments have recently been resolved in the course of a detailed study of the dynamical properties of slab-shaped NaCl crystals with (100) orientation.[14,15] A subsequent study of the surface thermodynamic properties of NaCl has resolved a long standing discrepancy between the measured and calculated specific heat of NaCl.[16]

4.1 Surface and Pseudo-Surface Modes of NaCl

The study of the surface and pseudo-surface modes of vibration for the (100) surface of an NaCl crystal, was based on the same rigid-ion model as used by Tong and Maradudin[13], as well as on the shell model. The latter are the first shell-model calculations of surface modes of vibration. The calculations in both models have been carried out for a slab-shaped crystal in which the ions are taken to interact through the long-range Coulomb potential, and the short range repulsive Born-Mayer potential, truncated at nearest neighbors.

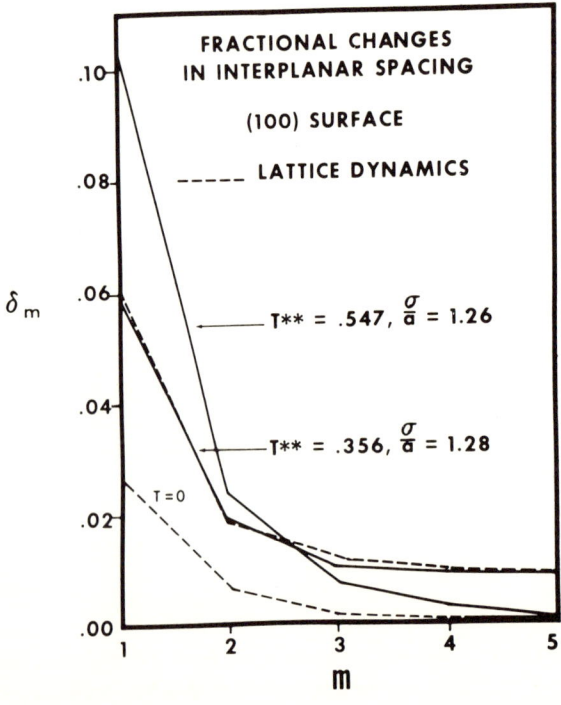

Fig. 7. Dynamic displacements δ_m for a (100) surface at $T = 0$ and $\frac{1}{2}T_M$ as determined with lattice dynamics (dashed lines), and at $T = \frac{1}{2}T_M$ and $\frac{3}{4}T_M$ as determined with molecular dynamics (solid lines).

Fig. 8 shows the phonon-dispersion curves $\omega_p(\bar{q})$ for the (100) surface of NaCl for $\bar{q}$-values taken around the edges of the irreducible element of the *BZ* (see Fig. 1). Notice the significant difference in overall appearance of these curves between the rigid-ion model results (Fig. 8a) and the shell-model results (Fig. 8b).

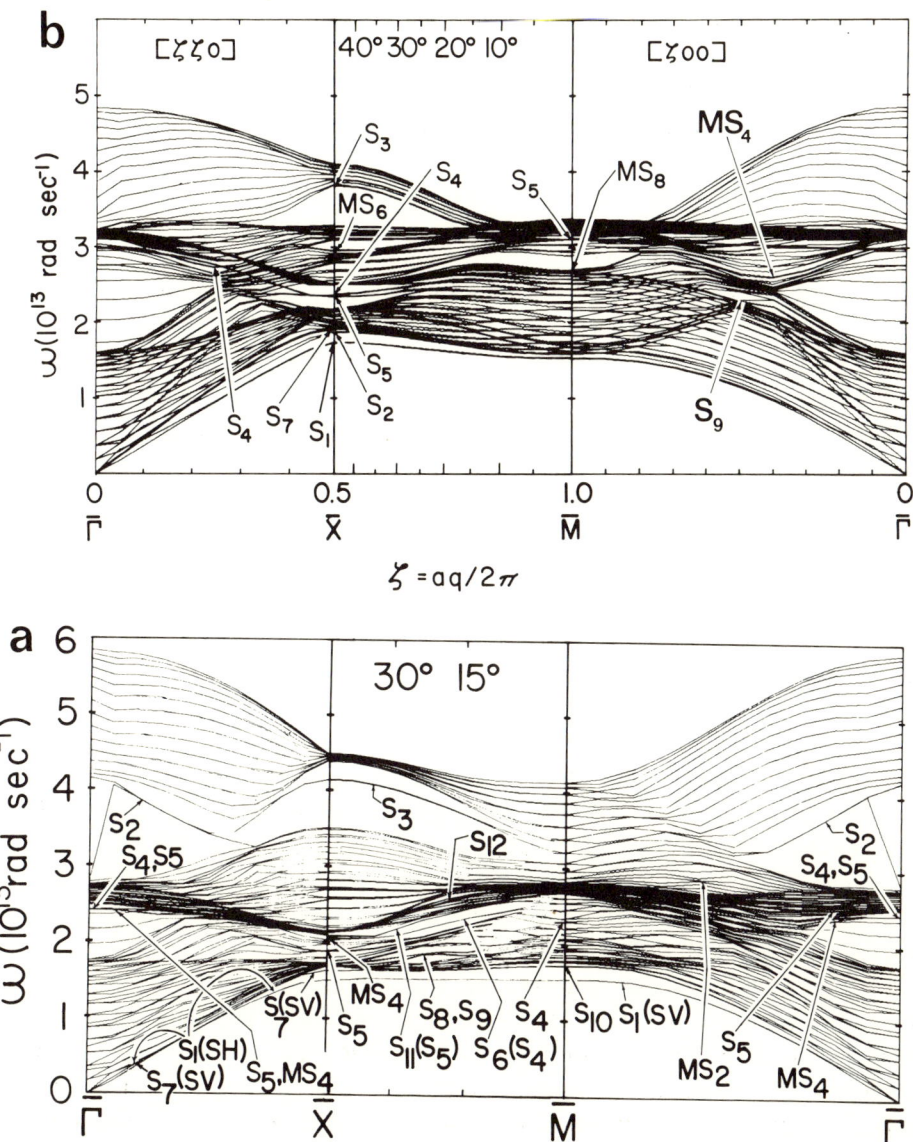

Fig. 8. Dispersion curves for a 15-layer slab of NaCl with (100) surfaces, calculated with: (a) the rigid-ion model of Ref. 13, and (b) the shell model.

In the rigid-ion results, note to the existence of a very large number of surface modes. Of these, only the long wavelength optical modes S_2 (Fuchs-Kliewer modes[11]), and S_4 and S_5 (Lucas modes[12]) and the acoustical Rayleigh mode S_1 (SV) have been described earlier.[13] All the other surface modes, existing mainly in the gaps in the bulk bands at short wavelengths have been discovered quite recently.[14,15] Modes labeled with MS_i are pseudo-surface modes; they are continuations into the bulk bands of the corresponding surface modes S_i existing in gaps at larger wave lengths. Surface modes "penetrating" the bulk bands, unless they are symmetry decoupled from the latter, lose their strictly surface character and become pseudo-surface modes.

Fig. 8b shows the $\omega_p(\bar{q})$-curves as given by the shell model.* This model, as opposed to the rigid-ion model takes the polarizabilities of the ions into account. Besides the difference in overall appearance of the $\omega_p(\bar{q})$-curves, the main difference with the rigid-ion model results is that, in the shell-model results, the FK modes, which are deeply penetrating and optically active modes, never appear as pure surface modes. Instead, they are always buried within the bulk bands as pseudo-surface modes, whose existence is indicated in the figures by a disturbance near $\bar{\Gamma}$ in the longitudinal optical bulk band (upper band). In contrast to the deeply penetrating FK modes, the Lucas modes consists of two pairs of surface modes with a very small penetration depth at $\bar{q} = 0$. The position of the FK and Lucas modes with respect to the bulk bands, in the neighborhood of $\bar{q} = 0$, is schematically depicted in Fig. 9.

4.2 Surface Thermodynamic Functions for NaCl

The surface thermodynamic functions (energy, free energy, entropy, specific heat) have been calculated as functions of temperature for the (100) surface of NaCl.[16] The calculations were based on the rigid-ion calculation of the dynamical properties.[15]

The present discussion is limited to a comparison of the results for the specific heat, with the experimental results of Morrison and Patterson[18], and Barkman et al.[19]

Fig. 10 shows the results for the surface entropy S^s and surface specific heat C_v^s for the (100) surface of NaCl. The softening of the lattice vibrations in the presence of a surface causes S^s and C_v^s to be positive. The peak in C_v^s is due, predominently, to the excitation of the low frequency surface modes. The points plotted in Fig. 10 represent the experimental data of Morrison and Patterson[18] for the excess heat capacity of finely divided NaCl. The position of the theoretical peak (at 45 K)

---

*An 11-parameter shell model was used, with values of the parameters as determined by Ref. 17.

DYNAMICAL SURFACE PROPERTIES 667

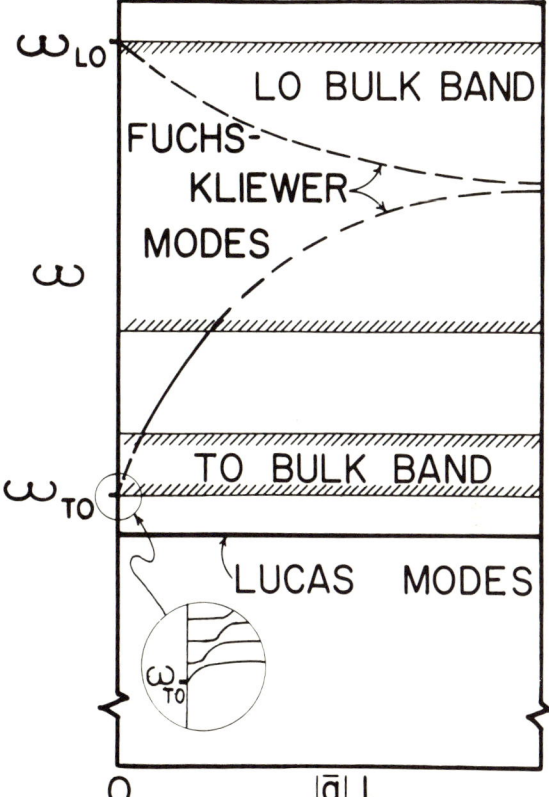

Fig. 9. Schematic of the LO (longitudinal optical) and TO (transverse optical) bulk bands, and of the Fuchs-Kliewer, (FK) and Lucas modes, in a thick slab at long wavelengths. The dashed lines indicate that the FK modes are pseudo-surface modes within the bulk bands. As indicated in the circular inset, these dashed lines actually represent a locus of interaction among hybridizing branches.

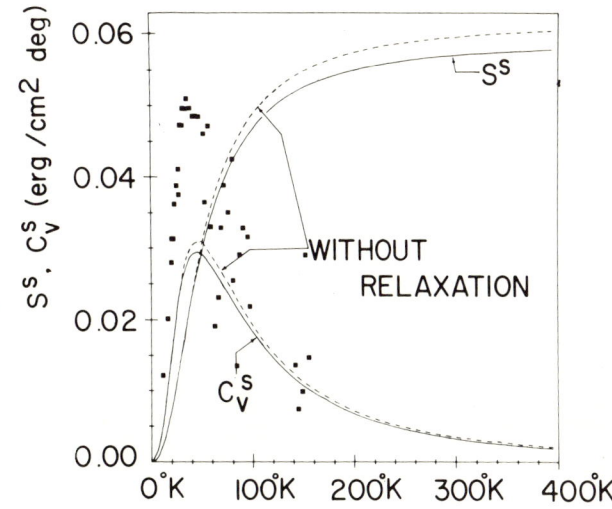

Fig. 10. Surface entropy S^s and surface specific heat at constant volume C_v^s for the (100) surface of NaCl. The experimental points are the excess molar heat capacity (ΔC_p) data for Sample B of Ref. 18.

is in remarkably good agreement with the experimental data. The discrepancy in the peak height may be due to uncertainties in determining the surface area of the experimental sample.

In Fig. 11, the calculated results for C_v^S are compared with experimental data for C_p^S in the low temperature region from 5 to 20 K, where the excess heat capacity should be due predominantly to surface contributions and vary quadratically with temperature. It is seen that present calculated results (solid line), which exhibit a strictly quadratic temperature dependence, agree quite well with the experimental results of Barkman et al.

5 CONCLUSION

This paper has reviewed results of numerical studies of surface dynamical and thermodynamic properties on the basis of realistic interatomic potentials. It has been shown that large-scale numerical calculations of this kind, based on reliable theories and avoiding unnecessary approximations, can yield a wealth of detailed information. A comparison of the results of such calculations with accurate experimental data can provide tests for the basic interactions involved, as well as for the theories or models underlying the calculations. It appears that these types of "computer experiments", as distinct from more analytical theories, can play an increasingly important role in our understanding of physical phenomena.

ACKNOWLEDGMENTS

This work was supported in part by the Air Force Office of Scientific Research, Office of Aerospace Research, USAF, under Grant No. AFOSR-71-1973. The author also wishes to thank the Robert A. Welch Foundation for support.

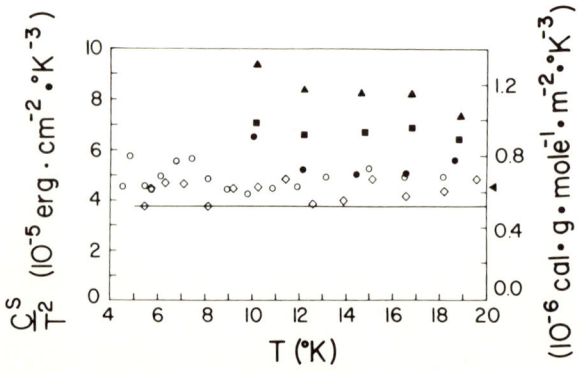

Fig. 11. Surface specific heat divided by T^2. Experimental points: solid points, Morrison et al.[18]; open points, Barkman et al.[19] The present results are given by the straight line from 5 to 20 K at 3.80×10^{-5} erg cm^{-2} deg^{-3}.

REFERENCES

1. R. E. Allen, G. P. Alldredge, and F. W. deWette, "Studies of Vibrational Surface Modes. I. General Formulation." Phys. Rev. B **4**, 1648 (1971).
2. A. Rahman, Phys. Rev. **136**, A405 (1964).
3. R. E. Allen, F. W. deWette and A. Rahman, Phys. Rev. **179**, 887 (1969).
4. R. E. Allen, G. P. Alldredge, and F. W. deWette, "Studies of Vibrational Surface Modes. II. Monatomic fcc Crystals", Phys. Rev. B **4**, 1661 (1971).
5. G. P. Alldredge, R. E. Allen, and F. W. deWette, "Studies of Vibrational Surface Modes. III. Effect of an Adsorbed Layer". Phys. Rev. B **4**, 1682 (1971).
6. R. E. Allen and F. W. deWette, J. Chem. Phys. **51**, 4820 (1969).
7. R. E. Allen and F. W. deWette, Phys. Rev. **179**, 873 (1969); ibid **188**, 1320 (1969).
8. Recent measurements of the Debye-Weller factor have been carried out by:
   (a) E. R. Jones, J. T. McKinney, and M. B. Webb, Phys. Rev. **151**, 476 (1966), (b) H. B. Lyon and G. A. Somorjai, J. Chem. Phys. **44**, 3707 (1966), (c) R. M. Goodman, H. H. Farrell, and G. A. Somorjai, J. Chem. Phys. **48**, 1046 (1960), (d) J. M. Morabito, R. F. Steiger, and G. A. Somorjai, Phys. Rev. **179**, 638 (1969), (e) D. Tabor and J. Wilson, Surface Science **20**, 203 (1970).
9. R. F. Wallis, in *The Structure and Chemistry of Solid Surfaces*, edited by G. A. Somorjai (John Wiley, New York 1969).
10. R. Kaplan and G. A. Somorjai, "Low Temperature LEED Study of the (110) Surface of Chromium", Solid State Comm. **9**, 505 (1971).
11. R. Fuchs and K. L. Kliewer, Phys. Rev. **140**, A 2076 (1965).
12. A. A. Lucas, J. Chem. Phys. **48**, 3156 (1968).
13. S. Y. Tong and A. A. Maradudin, Phys. Rev. **181**, 1318 (1969).
14. T. S. Chen, R. E. Allen, G. P. Alldredge, and F. W. deWette, Solid State Comm. **8**, 2105 (1970).
15. T. S. Chen, G. P. Alldredge, F. W. deWette, and R. E. Allen, "Surface and Pseudo-Surface Modes in Ionic Crystals", Phys. Rev. Letters **26**, 1543 (1971).
16. T. S. Chen, G. P. Alldredge, F. W. DeWette, and R. E. Allen, "Surface Thermodynamic Functions for NaCl", J. Chem. Phys. **55**, 3121 (1971).
17. R. E. Schmunk and D. R. Winder, J. Phys. Chem. Solids, **31**, 131 (1970).
18. J. A. Morrison and D. Patterson, Trans. Farad. Soc. **52**, 764 (1956).
19. J. H. Barkman, R. L. Anderson, and T. E. Brackett, J. Chem. Phys. **42**, 1112 (1965).

DISCUSSION on paper by F. W. de Wette

ASHCROFT: For a certain class of interatomic potentials it has been suggested[1] that long-range order in two dimensions may be absent. Possibly this is only appropriate in the thermodynamic limit, but in answer to a similar question on the equivalent spin system Dr. Vineyard has indicated that peculiarities are observed in the spin-spin correlation functions. Can these theorems be tested with your method?

deWETTE: The long-range order in particle positions is related to the mean-square amplitude of vibrations. It has been known for a long time that the mean-square amplitudes in a free film of infinite extent, like those in an infinite two-dimensional crystal, diverge at nonzero temperatures. Although the divergence is slow (logarithmic) it can have a sizeable effect in films of sufficiently small thickness. We have reported

670   F. W. de WETTE

calculations of this effect for thin films of Lennard-Jones material some years ago.[2] So we have indeed tested the theorem you mentioned in one particular case. It is amusing to note that a two-dimensional harmonic crystal would have to be roughly of the size of the solar system, for long-range order to be lost.

SHEWMON: You've given a comparison between surface specific heat and your calculations. Are there any similar checks available for the surface modes you described?

deWETTE: There are transmission experiments of 25-KeV electrons through LiF foils by Boersch, Geiger and Stickel.[3] The energy losses are attributed to the excitation of the transverse optical Fuchs-Kliewer modes. This interpretation has been confirmed by work of Fujiwara and Ohtaka[4] and by Chase and Kliewer[5]. Ibach[6], has performed low energy electron inelastic scattering experiments on ZnO surfaces. The loss spectrum has been interpreted as resulting from multiple excitations of surface phonons. Finally, Rieder and Hörl[7], have performed neutron-scattering experiments on crystallites of MnO of various grain sizes. A surface phonon spectrum is derived which is suggestive of the existence of the Rayleigh, the Lucas and the Fuchs-Kliewer modes. However, the use of the so-called *incoherent approximation* casts some doubt on the reliability of the obtained surface vibrational spectrum.

One of the most direct evidences of the existence of surface vibrations are the surface mean-square amplitudes of vibration, which are larger than those in the bulk, and which can be derived from the temperature dependence of the Debye-Waller factor as measured in LEED experiments.

HO: I don't think the surface specific heat at low temperatures gives too much information about surface interatomic forces or surface lattice vibrations (particularly the optical modes). It seems to me that almost any reasonable two-dimensional atomic model will yield a T^2 dependence at low temperatures. However, the coefficient of this dependence depends more on the model. Can you comment on this?

deWETTE: I agree; thermodynamic properties are integrated properties and do not give direct information about the details of the atomic interactions. The fact that we get the coefficient of the T^2 dependence in agreement with the measurements of Barkman et al.[8] indicates, however, that the long-wavelength acoustical modes are represented fairly well by the model used.

REFERENCES

1. N. D. Mermin, Phys. Rev. **176**, 250 (1968).
2. F. W. de Wette and R. E. Allen, Phys. Rev. **187**, 878 (1969).
3. H. Boersch, J. Geiger, and W. Stickel, Phys. Rev. Lett., **17** 379 (1966).
4. T. Fujiwara and K. Ohtaka, J. Phys. Soc. Japan, **24**, 1326 (1968).
5. J. B. Chase and K. L. Kliewer, Phys. Rev. **B2**, 4389 (1970).
6. H. Ibach, Phys. Rev. Lett., **24**, 1416 (1970).
7. K. H. Rieder and E. M. Hörl, Phys. Rev. Lett., **20** 209 (1968).
8. J. H. Barkman, R. L. Anderson, and T. E. Brackett, J. Chem. Phys. **42**, 1112 (1965).

A COMPUTER SIMULATION STUDY OF GRAIN BOUNDARIES IN FCC GAMMA-IRON AND THEIR INTERACTIONS WITH POINT DEFECTS

R. E. Dahl, Jr.,
WADCO Corporation
Richland, Washington

J. R. Beeler, Jr.
The Ohio State University
Columbus, Ohio

and

R. D. Bourquin
Battelle-Northwest
Richland, Washington

ABSTRACT

The atomic structure near tilt grain boundaries in gamma-iron was determined through the discrete lattice studies. Regions of good and bad fit were apparent. The extent of the strain fields was determined. Calculated energies of high-angle, low-angle, and coincidence-site grain boundaries agree with measured values. The influence of grain boundaries on the formation and migration energies of vacancies was found to be very directional. The effects of carbon atoms on grain-boundary structure were studied. A "healing" of the grain-boundary structure was found to occur with one carbon atom in a misfit region. Migration energies of carbon impurities in and to a boundary were calculated.

The GRAINS code in quasidynamic and fully dynamic modes using Johnson's potentials produced these results. Graphical computer output provided useful visual descriptions of grain-boundary morphology.

1 INTRODUCTION

The scope of this report is limited to a discussion of grain boundaries, their structure, energies, and their interaction with vacancies and carbon impurity atoms. This is an attempt to describe the application of computer simulation to these areas of concern for polycrystalline metallurgy with brevity, yet with sufficient depth and references for the more interested reader. Each topic is discussed in a separate section.

1.1 The Structure and Energy of Simple Tilt Boundaries in γ-Iron

The structure of grain boundaries is of fundamental interest in understanding the ability of some grain boundaries to drive recovery, recrystallization, and other processes to transform a material with a high dislocation density into nearly perfect crystals. The structure on an atomistic level of such boundaries has been largely unknown. The structure of low-angle boundaries of a simple configuration is perhaps as well known as that of dislocations. However, even in this area, the somewhat qualitative models make it difficult to propose mechanisms for the important metallurgical phenomena that depend largely on grain-boundary behavior.

This study was restricted to asymmetric tilt boundaries in an fcc lattice because there is some experimental and theoretical information that can be used to test and corroborate the results. The orientation studied was that of a grain boundary lying in the (010) plane having an axis of rotation <100>.

1.2 Construction and Relaxation of Cells

The computer experiments described in this report were conducted using GRAINS, a quasi-dynamic code described in detail elsewhere.[1] Atoms were moved (according to the classical physics equations of motion) to equilibrium positions. The integration scheme and numerical analysis procedures developed by Gibson et al.[2] were followed. The iron-iron interaction function was the Johnson I[3] function for α and γ-iron.

Iron-carbon interactions were described by the function developed by Johnson[4], which has an attractive and a repulsive range.

The carbon-carbon interactions are described by a simple repulsive function which treats the interacting carbon atoms as hard spheres. The interatomic function was a cubic set equal to zero at 1.40 A, twice the atomic radius for carbon given by Slater[5]. The approach to the cut-off is asymptotic. The structure of carbon clusters was of less interest in this study than the accommodation of the grain-boundary region to the clusters. Atoms bonded together in a graphitic- or diamond-type structure would occupy less volume than that of contiguous spheres, thus the hard-sphere case would cause the maximum strain.

The computational cells used in this study had to be sufficiently large so that the rigid cell boundaries would not influence the grain boundaries. However, only one repeat distance was required of the depth in the [100] direction of the cells used to determine grain-boundary structure and energy since no [100] motion could occur. The tilt grain boundaries were in an (010) plane. The grain boundaries extended 28 half-lattice constants in the [001] direction and one-half lattice constant in the [100] direction. Simulation of an infinite grain boundary was achieved through imposition of periodic boundary conditions simulating image planes. Angles less than 3 degrees could not be simulated in this computational cell because the dislocation spacing was greater than the length of the cell. The length of the cell provides several repeats of the grain-boundary structure for most of the grain boundaries studied. The rigid boundaries were located approximately 15 A from the grain boundary, and were found not to perturb the results of the calculations.

Approximately four minutes of computational time on the UNIVAC-1108 computer were required for each of the grain-boundary determinations for computational cells that contained approximately 500 atoms. The computations were terminated when the variation in potential energy from successive iterations was less than 0.003 eV.

1.3 Grain-Boundary Structure

Determination and understanding of grain-boundary structure is a requisite for understanding the mechanisms of diffusion, quenching, and the interaction of defects with grain-boundary surfaces.

The structure of a 20-degree boundary is presented in Fig. 1. This figure, which is a direct computer output, indicates very exactly the

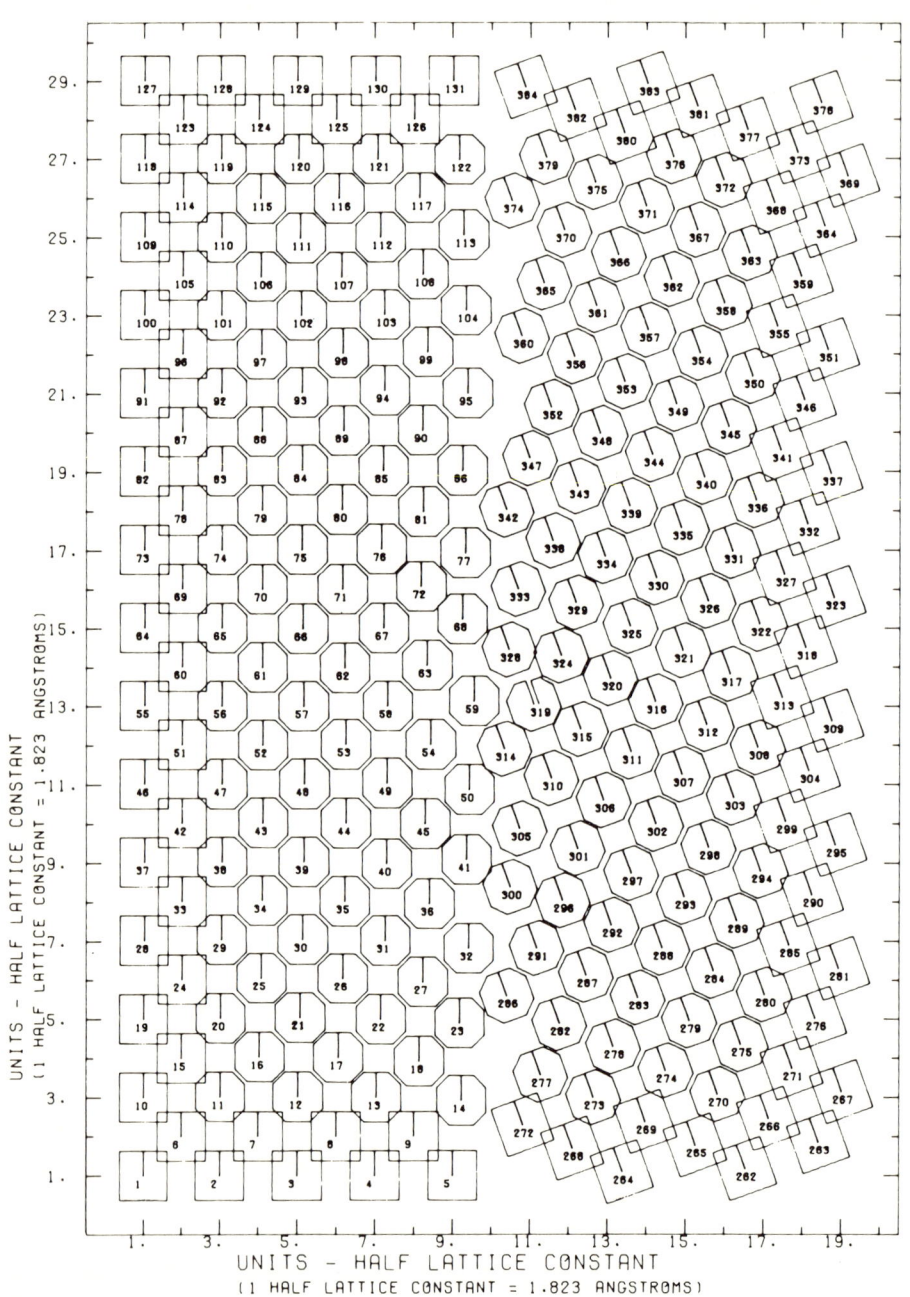

Fig. 1. Structure of 20.0-degree-tilt grain boundary.

computed positions of the atoms at the ends of the fiducial marks within the symbols. Hexagons represent movable atoms and squares represent immovable cell-boundary atoms. The symbols representing the atoms in the grain on the right side of the grain boundary have been rotated through the grain-boundary angle so that the grains can be easily distinguished. The angle of the grain boundary and the lattice constant appear in the title and along the axes.

The structure of the grain boundaries is seen to include regions of *good fit* and *bad fit*. The shapes of these regions are somewhat complex since the bad-fit regions do not coincide from one plane to another. The bad-fit regions are of extreme interest since inclusions or precipitates could be accommodated there.

The width of a grain boundary can be defined as the extent of its strain field. Assigning the width of this region is a somewhat arbitrary choice which depends upon the magnitude of the displacement considered to be significant. Displacements of atoms in the vicinity of a 6-degree boundary are shown in Fig. 2. Displacements in high-angle and coincidence-site boundaries were found to be very similar in magnitude to

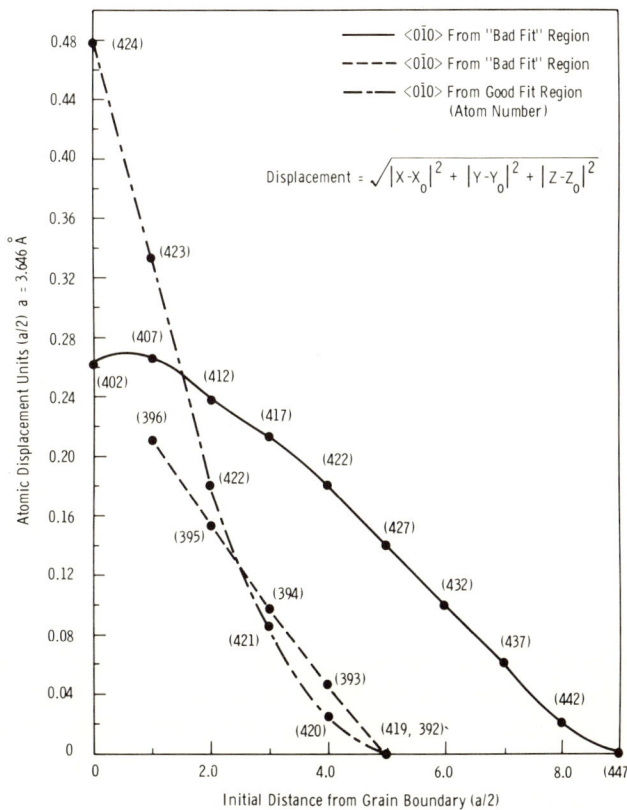

Fig. 2. Strain field in vicinity of a 6-degree-tilt boundary in γ-iron.

those shown for the 6-degree boundary. Atomic displacements toward the boundary (i.e., grain-boundary width) are greatest along close-packed lines that terminate in misfit regions of the boundary. Thus, grain boundaries influence defects at a greater distance than would be postulated from an average grain-boundary width.

Hirth and Lothe have calculated that no long-range stress fields arise for simple tilt boundaries.[6] According to their results, the stress is, at a distance of twice the dislocation spacing, 1 percent of that at an equivalent distance from a single dislocation. This is in general agreement with the results of these computer experiments. The energy of atoms within one atomic distance from the boundary was very nearly the same as the *bulk* value. Thus, the stresses from the boundary were short ranged. However, directional effects are apparent in analyses of stress and of strain data.

1.4 Grain-Boundary Energy

The energy of low-angle grain boundaries can be calculated according to the relationship

$$E = E_O \theta (A - \ln \theta) ,$$

which was derived by Read and Shockley.[7] The term E_O depends upon the elastic distortion and can be calculated directly from the elastic constants of the material or measured according to

$$E_O = \frac{bG}{4\pi} ,$$

where G is the shear modulus of the material. Its value for γ-iron is about 9×10^6 psi (6.2×10^{11} dynes/cm^2).[8] Thus, E_O is about 1790 ergs/cm^2. The parameter A, which depends upon the atom misfit along the boundary (the nonelastic deformation), has been very difficult to estimate or predict. Read suggests that, to estimate "A", grain-boundary energy data be presented in a plot of E/θ versus $\ln \theta$, where $(-E_O)$ is the slope and A is the zero-energy intercept.[9] The range over which a linear relationship applies will indicate the region of applicability of the dislocation model.

Grain-boundary energies as a function of the angle of rotation are presented in Fig. 3. The method for calculation of grain-boundary energies from the computer-experiment results is explained in detail in Ref. 10. Energies predicted with the Read-Shockley relation also appear as a curve in this figure. The values used for the arbitrary constants E_O and A in constructing this curve were 1790 ergs/cm^2 and 0.3, respectively. Computer-simulation energies calculated for 1 and 2-degree grain

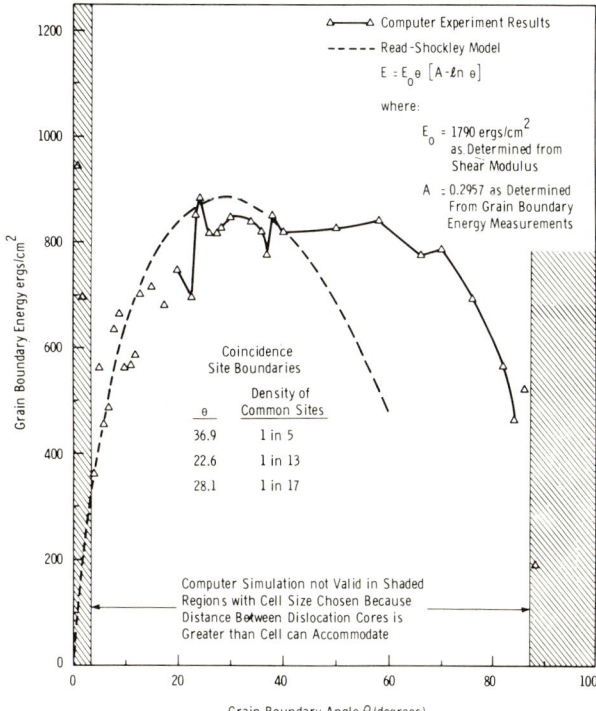

Fig. 3. Energy of tilt boundaries in γ-iron as a function of angle

boundaries were very high but, as explained previously, the computational cell was not adequate to contain the dislocation spacing required to construct these boundaries. Results for angles up to about 15 degrees are distributed about the Read-Shockley line, and the general shape of the curve through the computer-experiment data coincides reasonably well with the Read-Shockley line up to about 40 degrees. Both have an asymmetric distribution about a rotation of 45 degrees.

Three angles of rotation between 0 and 90 degrees produce coincident-site boundaries in fcc lattice for tilt boundaries having an axis of rotation [100]. These are 36.9, 22.6, and 28.1 degrees, and they are designated on Fig. 3. The density of common sites for these boundaries are one in five, one in thirteen, and one in seventeen, respectively.[11] Note that the energies calculated for these particular grain boundaries appear as cusps or low-energy configurations in the plot. The depths of the cusps are in proportion to the density of common sites for these boundaries, indicating the relative fit. The cusps are narrow, indicating that the coincident-site boundaries are shown to be unique configurations.

The agreement in the magnitude of grain boundary energies obtained through the computer experiments and those measured experimentally is good. The maximum grain-boundary interfacial energy determined in the computer experiments was about 950 erg/cm^2. Values cited in the literature for experimental measurements range from 700 ergs/cm^2 (Ref 12, 13) to 1100 ergs/cm^2 (Ref. 14). Grain-boundary angles were not specified in these reports.

A plot of E/θ versus $\ln \theta$ was made from the computer experiment results (see Fig. 4) and statistically analyzed. The value of E_O obtained from the slope of the linear least squares curve is 1678 ergs/cm^2, and the value of A derived from the intercept of this curve is 0.413 which is in agreement with those proposed by Read. Thus, the form of the Read-Shockley equation could be used to predict the energy of any low- or high-angle boundary up to about 70 degrees with reasonable accuracy.

The scatter of data encountered in the low-energy region is much more extreme than that for high-angle boundaries, making it difficult to ascertain where the dislocation model fails. A graphical analysis of the data in Fig. 4 indicates that a departure from a linear curve through the

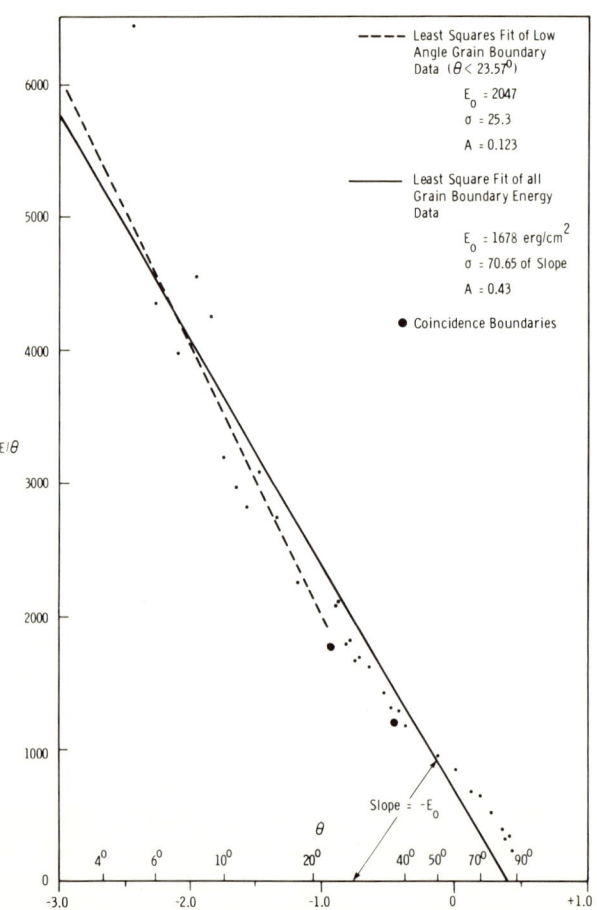

Fig. 4. Analysis of Read-Shockley grain-boundary-energy equation $E = E_o \theta [A - \ln \theta]$

very low-angle data occurs at about 12 degrees. However, the variance of the slope in this region obviates any precise conclusion.

The conclusions from grain-boundary-energy data are:

(i) The dislocation theory and computer experiment results are in quantitative agreement with experimentally measured physical and mechanical property data.

(ii) The dislocation model for tilt boundaries is strictly applicable only up to about 12 degrees. However, predictions of grain-boundary energies up to 40 degrees made with the Read-Shockley equations agree with the simulation results within about 18 percent.

(iii) The coincident-site boundaries are proven to be definite low-energy configurations that occur at discrete grain-boundary angles.

2 VACANCY FORMATION AND MIGRATION IN AND NEAR GRAIN BOUNDARIES

The importance of vacancies in diffusion, quenching, annealing, and radiation damage in metals motivated investigation of the influence of grain boundaries on formation and migration of vacancies. Strength and activation of grain-boundary sources and sinks are difficult to estimate because of the complexity of grain-boundary morphology and its influence on defects. Computer experimentation can contribute in this area since discrete lattice calculations make these problems tractable.

2.1 Method

The vacancy studies were conducted in a cell containing an asymmetric 6-degree boundary. This boundary angle was chosen since it was near the center of the range in which the dislocation model should be valid, and in which computer experimentation with a practical-size computational cell was feasible. Concurrent studies were made in perfect crystals to compare with the results of experimental measurements and other computer experiments.

Vacancy-migration energies were calculated by moving an atom toward a vacant site incrementally and calculating the configurational energy of the cell at each step. The moved atom was allowed two degrees of freedom so it could go to the lowest-energy position at each incremental step. Thus, the lowest-energy path between two positions, often a circuitous route, was followed.

These studies were conducted in a computer cell comprising seven (100) plane layers. The dimensions in the [010] and [001] (X and Y) directions were 19 and 21 half-lattice constants, respectively. A computer-drawn plot of the center (100) layer into which vacancies were introduced appears in Fig. 5. The Johnson[3] *I* iron-iron potential was used for both the bcc α-iron and the fcc γ-iron experiments. In α-iron, the range of the potential (3.44 A) is midway between the second- and third-neighbor distance, while in γ-iron it does not extend to second neighbors.

Formation energies for single vacancies and atomic displacements, determined in experiments in which vacancies were introduced into perfect crystals, are presented in Table I.

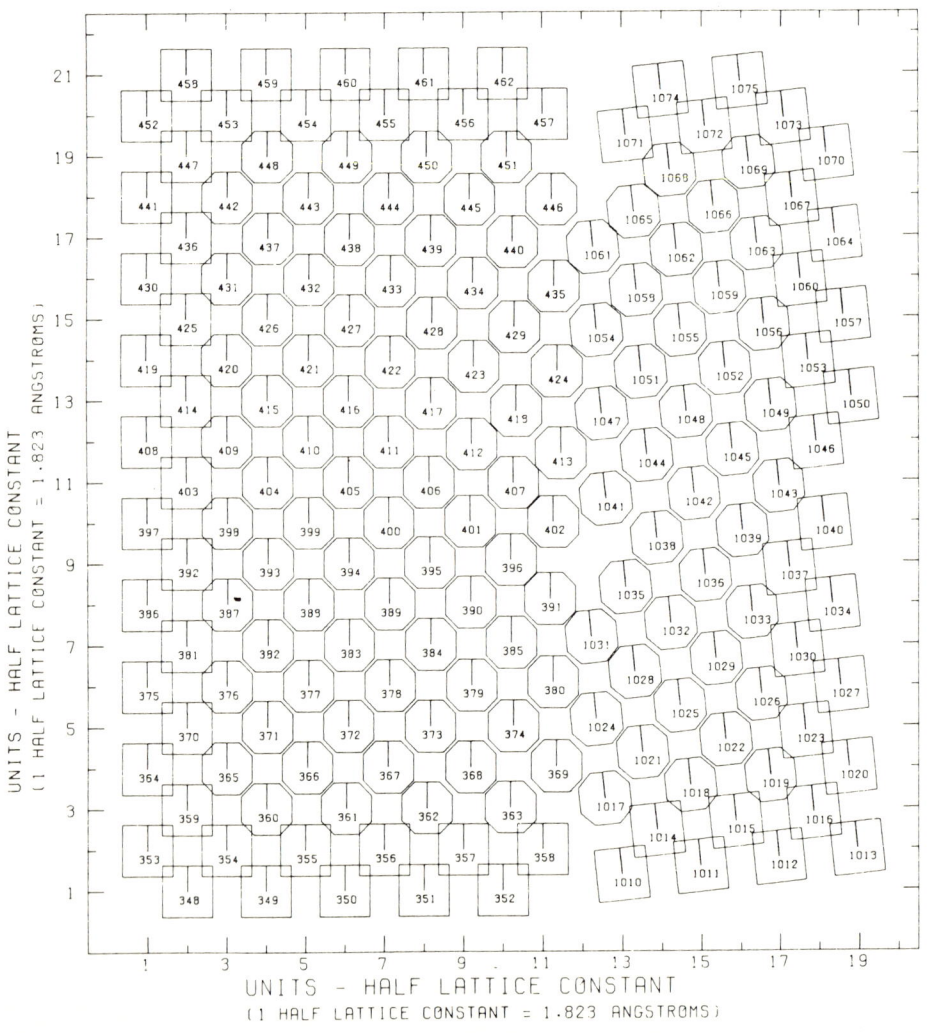

Fig. 5. A 6-degree-tilt grain boundary in γ-iron [center (100) plane].

Table I. Vacancy Formation Energies and Atomic Displacements in Perfect Crystals Calculated Using Grains

α-Iron (bcc) a = 2.86 A

Movement of atoms in vicinity of vacancy (units % $\frac{a}{2}$)

1st Neighbors	2nd	3rd	4th	5th
5.12	−5.18	0.79	−0.54	1.54

(Negative displacements indicate movement away from vacant lattice site)

Vacancy Formation Energy = 1.369 eV

γ-Iron (fcc) a = 3.646 A

Atomic movement toward vacant lattice site

1st Neighbors	2nd Neighbors
0.981	0.007

Vacancy Formation Energy = 1.50 eV

The migration energy in a perfect crystal of γ-iron was found by Johnson[3] to be 1.32 eV. Thus, the activation energy for self-diffusion, obtained by summing the formation (1.5 eV) and migration energies of a single vacancy, is 2.82 eV. This result is in agreement with the value of 2.8 eV obtained experimentally by Buffington et al.[15] It is also close to the activation energy of 2.91 eV measured for tracer diffusion of ^{59}Fe in iron by Heumann et al.[16]

2.2 Vacancy Formation

The results for vacancy-formation energies and atomic displacements appearing in Table I agree exactly with results obtained by Johnson[3,4] and by Doran,* who used the same potential but with the variational method of computer experimentation.

Vacancy formation energies determined for 6-, 36.9-, and 50-degree tilt boundaries (i.e., low, coincidence-site, and high-angle boundaries)

---
*D. G. Doran, Unpublished work, Battelle-Northwest, Richland, Washington (1969).

ranged from 0.08 to 1.5 eV depending on their position. That all formation energies for vacancies in these unstressed lattices were positive indicates that the systems are physically realistic and stable and that these boundaries would not be sources of interstitials unless stresses were imposed. The atoms removed to determine vacancy-formation energy included, in each case, the high-energy atoms along the boundaries.

The energy required to create vacancies differed greatly along the 6-degree grain boundary (Fig. 6). Because the energy required near the misfit regions was almost zero, atoms could be removed very easily at these positions. The formation energy at positions relatively deep into the grain was exactly equal to that determined in the perfect-crystal experiments. This value, 1.50 eV, is referred to as the *bulk* value in the following discussions.

The vacancy formation energies at sites lying along close-packed [0$\bar{1}$0] lines from misfit regions of a 6-degree boundary are much lower than the bulk value for about eight interatomic distances into the grain. Conversely, formation energies at positions along directions [$\bar{1}$00] agree with the perfect-crystal value within one atomic distance of the grain

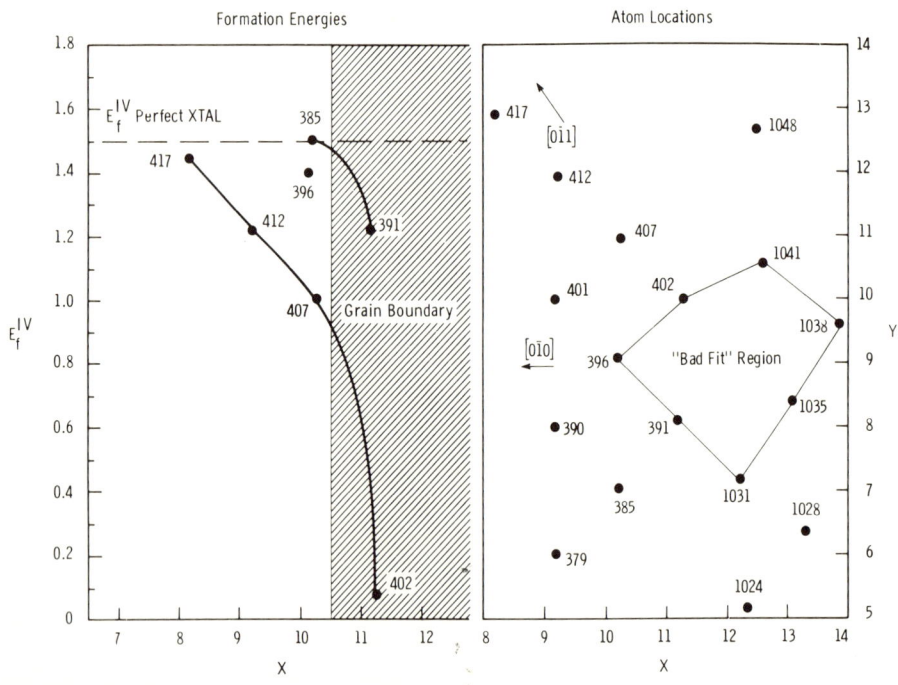

Fig. 6. Formation energies of single vacancies in and near a grain boundary.

boundary. The low vacancy-formation energies mean that the activation energy for diffusion will also be low. Therefore, there are paths of *easy* diffusion extending from the misfit regions of the boundary for some distance into the adjacent grains. As a result, the influence of a grain boundary extends significantly farther into a grain than the distance normally considered to be the grain-boundary width. However, this occurs along particular unique paths. The interaction zone, or zone of influence determined in these experiments does not extend to the same magnitude as the zone observed, in microscopy studies of radiation damage[17], to be denuded of loops and impurities. However, it is probable that through the vacancy concentration gradient established by the grain boundary and through the correlated motion of vacancies with other defects, diffusion to grain-boundary-misfit regions contributes to the cleanness of the zone near grain boundaries. Conclusions drawn for the low-angle boundary were found to apply to coincident-site and high-angle boundaries in a less extensive series of experiments conducted on several boundaries of each type.

2.3 Vacancy Migration

The energy required for vacancy migration into and away from the misfit region of a 6-degree boundary was determined by the series of experiments whose results are presented in Fig. 7. In these experiments, atom 407 was removed. Atom 402 (Fig. 5) was moved from its position at the edge of the grain-boundary hole along the $[0\bar{1}1]$ direction into the grain, thus simulating vacancy movement in the opposite direction. Only very little energy (about 0.01 eV) is required to move the vacancy to the hole, while 1.5 eV is required to move it away. Thus, the grain-boundary hole is definitely a sink for vacancies and a possible nucleation site for a large vacancy cluster.

A vacancy deficiency near internal surfaces formed by grain boundary *holes* is in agreement with the free-surface work, where easy vacancy migration toward the misfit regions is observed.

2.4 Conclusions

Perhaps the most important conclusions and demonstrations from the studies on vacancy-grain boundary interactions are:
(i) Determination of values for migration and formation energies of vacancies in the vicinity of a grain boundary so that diffusion can be more accurately calculated.
(ii) The extent of grain boundary-vacancy interaction zones. The easy diffusional paths extend farther from misfit regions of the grain boundary into the adjacent grain than would be anticipated from the grain-boundary width.

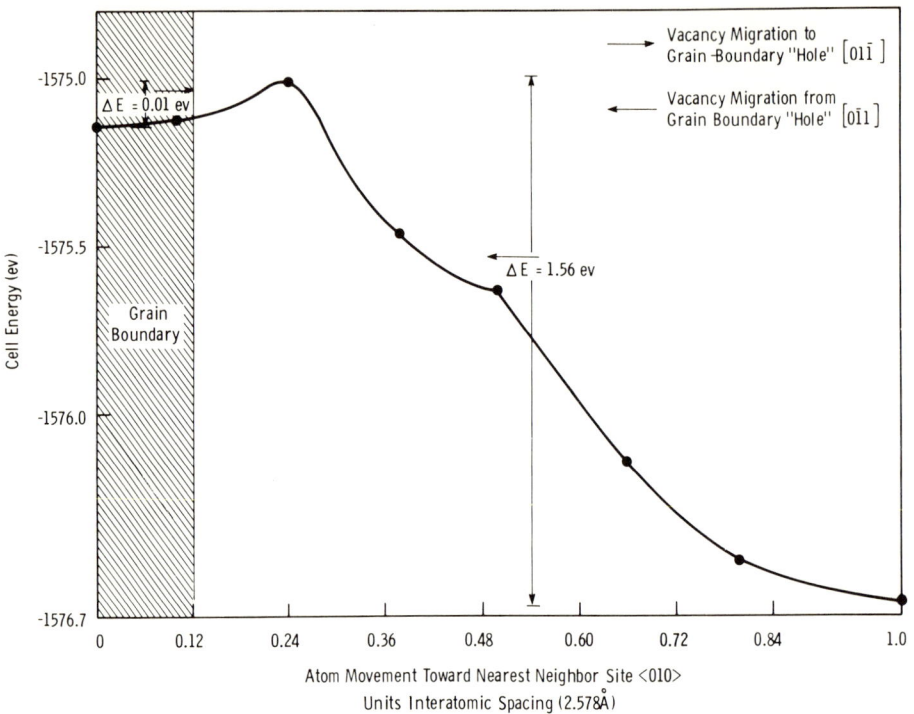

Fig. 7. Vacancy migration energy to a bad-fit region of a 6-degree-tilt boundary in γ-iron.

3 INTERACTIONS OF CARBON IMPURITY ATOMS WITH A LOW-ANGLE GRAIN BOUNDARY

Small concentrations of interstitial impurity atoms can have a large effect on the properties of metals. Carbon is perhaps the most important of these impurities, for the iron-based alloys. This study was initiated to obtain a qualitative understanding of the effects that carbon atoms would have on a grain boundary and to obtain some data on the relative energy required to place and move carbon impurity atoms in the vicinity of a grain boundary. Conclusions reached should also be pertinent to some other interstitial impurities.

Specific objectives were to:
(i) Determine the probable region of impurity-atom segregation
(ii) Determine the effects upon grain-boundary structure of impurity atoms
(iii) Estimate the energy required for migration of an impurity atom in a grain boundary.

3.1 Method

An asymmetric 6-degree-tilt grain boundary was also chosen for these studies. Perturbations appeared to be very local. Since the experiments were all conducted in or very near a misfit region, the results should be typical of high-angle grain boundaries.

The computational cells used to study grain boundaries and carbon impurity atoms were composed of 1275 iron atoms in seven (100) planes. Impurity atoms were added in the center layer so that strain fields could be easily contained within the cell. The grain boundaries were in an (010) plane and were essentially infinite in two directions [001], [100].

Calculations were continued, iterated, until the sum of the potential energies of all atoms in the cell differed by less than 0.005 eV from one iteration to the next. The surface energy of the boundary atoms was monitored during each iteration. No change was observed upon introducing any of the carbon clusters. Thus it was concluded that the rigid cell boundaries did not influence the computations.

3.2 Results

Single Carbon Interstitials. Carbon atoms were inserted individually at several locations in and near the 6-degree grain boundary to determine relative effects on the grain-boundary structure, and the extent of the region in which there was interaction between grain boundaries and impurity atoms. Effects of impurity atoms on a grain boundary are of interest because ordering or disordering of the boundary could affect the strength of the grain boundary and, therefore, polycrystalline materials. Structural perturbations could also affect grain-boundary migration and sliding by pinning.

Data compiled included the changes in cell energy, initial and final positions of the carbon atoms, and the energy of the carbon atoms for locations in or near the misfit region in the 6-degree boundary hole which, in plane $Z = 4$, is surrounded by atoms 391, 402, 1041, and 1035 (see Fig. 5). The center of this misfit region or *hole* is approximately at the coordinates $X = 12$, $Y = 9.5$, and $Z = 4$. Carbon atoms were added into the hole, at the edge of the hole, at 1 and 2 octahedral positions from the hole along a close-packed direction [$0\bar{1}1$], and directly into the cell normal to the boundary [$0\bar{1}0$].

Changes in the structure and energy of the grain boundary caused by the addition of an impurity atom near the center of the hole were most striking. The structure of the grain boundary (compare Fig. 5 and 8) became very well ordered, apparently because carbon atoms bonded with the neighboring iron atoms and drew them together. Iron-iron bonds were

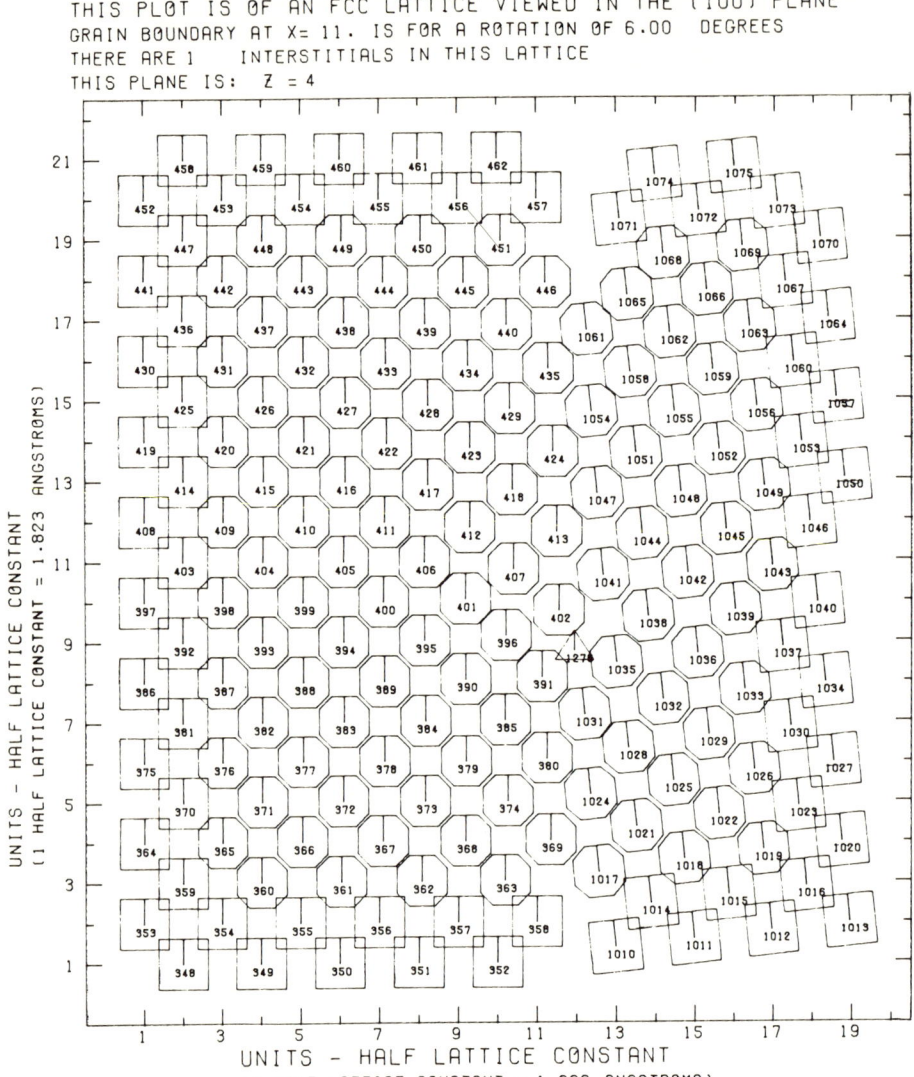

Fig. 8. A 6-degree-tilt grain boundary with a carbon interstitial atom in a misfit region plane Z = 4.

formed and strengthened, so that considerable relaxation took place in the entire configuration as shown by the absence of misfit regions in Fig. 8 as well as in the planes adjacent to that containing the carbon atom. The energy of the cell was reduced some 2.6 eV by the addition of the carbon atom in the grain-boundary hole. Approximately 1.6 eV of this energy change can be attributed to the relaxation of the lattice into a more stable configuration. The ordering or *healing* of the grain boundary, as shown in Fig. 8, would certainly strengthen the grain boundary and facilitate transmission of stresses and dislocation movement from grain to grain.

The results obtained in this investigation should be pertinent for other small interstitial impurity atoms such as boron or nitrogen. Small concentrations of boron located in grain-boundary regions are known to increase low-temperature ductility. The healing observed in these studies would cause such an effect.

The energy and structural changes accompanying the addition of carbon atoms in 1 and 2 octahedral positions away from the boundary in a close-packed [0$\bar{1}$1] direction (i.e., at coordinates 10, 10, 4, respectively) were almost identical to those observed when a carbon atom was added to a perfect crystal. The strains caused by carbon atoms located in a normal or near-normal octahedral site are so small that they do not influence the structure. Thus, carbon atoms located a distance of only one-half lattice unit from the boundary did not interact with the grain boundary.

Di- and Tri-Interstitials in a Grain-Boundary Hole. Experiments were conducted in which a second and then a third carbon impurity atom were added into the grain-boundary hole. Each experiment was begun with the strain field of the preceding experiment. Each subsequent carbon atom was added at exactly the same location and forced to seek equilibrium positions. Results of these experiments are found in Fig. 9. A large change in energy or structural ordering occurs upon the addition of the first carbon atom to the misfit region. The energy changes accompanying the addition of the second and third carbon atoms were approximately equal, and slightly less than that observed in the perfect crystal experiment. Thus, it seems certain that the tri-interstitial did not cause disordering or an extensive strain field extending into the grains. Apparently, the tri-interstitial, which assumed the shape of a platelet (Fig. 10), was easily accommodated in the misfit region of the grain boundary.

3.3 Summary

Perhaps the most significant observation is the ordering of a grain boundary by an impurity atom. A single carbon atom added to the misfit region of the grain boundary essentially caused complete local ordering of the structure.

Relatively large carbon clusters can be accommodated in misfit regions of a grain boundary without stressing the adjacent grain and, thus, their formation is energetically favorable. These clusters could function as pins to impede grain-boundary motion or transmission of strains from one grain to the next and, thus, would harden or strengthen a metal.

The interaction region where the formation and migration of interstitial impurities is influenced by the presence of grain boundary is seen to be quite small and certainly less than the interaction region observed between vacancies and grain boundaries. Migration of carbon atoms to a

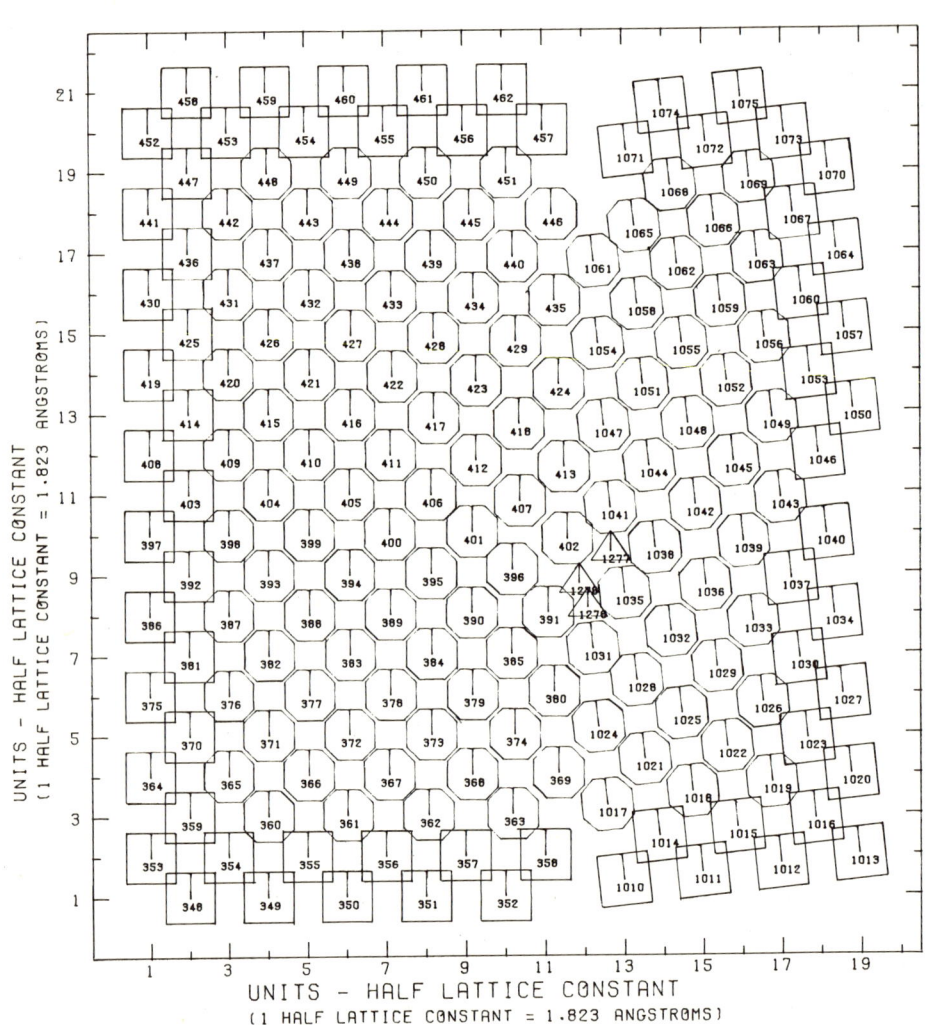

Fig. 9. A 6-degree grain boundary with 3 carbon interstitials in bad-fit region.

grain boundary would seemingly require correlated motion with a vacancy; since, if a carbon atom is located more than one-half lattice unit from the grain boundary, it is, in essence, in solution in a perfect crystal.

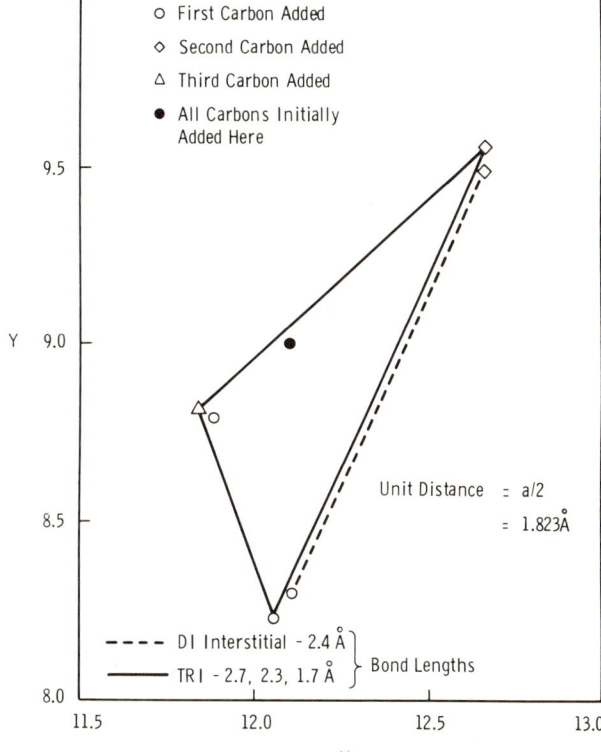

Fig. 10. Configuration of di- and tri-interstitial clusters in grain-boundary hole.

ACKNOWLEDGMENTS

This paper is based on work performed by Battelle-Northwest under U. S. Atomic Energy Commission Contract AT(45-1)-1830 [Dahl and Bourquin]; by the Hanford Engineering Development Laboratory, Richland, Washington, operated by WADCO Corporation, a subsidiary of Westinghouse Electric Corporation, under U. S. Atomic Energy Commission Contract AT(45-1)-2170 [Dahl]; and by the North Carolina State University under U. S. Atomic Energy Commission Contract AT(40-1)-3912 and Air Force Materials Laboratory Contract F33615-68-C-1012 [Beeler].

REFERENCES

1. Dahl, R. E., Jr., J. R. Beeler, Jr., and R. D. Bourquin: USAEC Report HEDL-SA-171, submitted for publication to Computer Physics Communications, 1971.
2. Gibson, J. B., A. N. Goland, M. Milgram, and G. H. Vineyard: Phys. Rev., **120** (4): 1229 (1960).

3. Johnson, R. A.: Phys. Rev., **145** (2): 423 (1966).
4. Johnson, R. A.: Acta Met., **13**: 1259 (1965).
5. Slater, J. C.: J. Chem. Phys., **41** (10): 3199 (1964).
6. Hirth, J. P., and J. Lothe: in *Theory of Dislocations*, p. 637, McGraw-Hill Book Co., New York, 1968.
7. Read, W. T., and W. Shockley: Phys. Rev., **78**: 275 (1950).
8. American Society for Metals: *Metals Handbook*, p. 1211, Vol. 1, 8th Edition, Metals Park, Ohio, 1965.
9. Read, W. T.: in *Dislocations in Crystals*, p. 155, McGraw-Hill Book Co., New York, 1953.
10. Dahl, R. E., Jr.: *A Computer Simulation Study of Tilt Grain Boundaries in Gamma-Iron and Their Interactions with Point Defects*, PhD Thesis, North Carolina State University, 1970.
11. Brandon, D. G., B. Ralph, S. Ranganthan, and M. S. Wald: A Field Ion Microscopy Study of Atomic Configuration at Grain Boundaries, Acta Met., **12**: 813 (1964).
12. Shewmon, P. G.: in *Recrystallization, Grain Growth, and Texture*, American Society for Metals, Metals Park, Ohio, 1965.
13. Inman, M. C., and H. R. Tipler: Interfacial Energy and Composition in Metals and Alloys, Met. Review, **8**: 105 (1963).
14. Friedel, J: *Dislocations*, p. 275, Pergamon Press, Oxford, 1964.
15. Buffington, F. W., K. Hirano, and M. Cohen: Self-Diffusion in Iron, Acta Met., **9**: 434 (1961).
16. Heumann, T. H., and R. Imm: Self-Diffusion and Isotope Effect in γ-Iron, J. Phys. Chem. Solids, **29**: 1613 (1968).
17. Carlander, R., S. P. Harkness, and F. L. Yaggee: Fast Neutron Effects on Type-304 Stainless Steel, Nucl. Appl., **7** (1): 67 (1969).

DISCUSSION on paper by R. E. Dahl, Jr., J. R. Beeler, Jr., and R. D. Bourquin.

WEINS: In attempting to grow straight off-coincidence symmetric tilt boundaries we found that the boundaries had kinks; the figure shows such a boundary in a silver bicrystal. The kinks could be explained on the basis of forming assymetric boundary segments with short periods. The fact that your energies for short period assymetric boundaries are reasonable is supported by the kinks observed. The work on grain-boundary kinking was presented at the 1971 Spring AIME Meeting and is to be published (M. J. Weins and J. J. Weins).

SHEWMON: Did you examine the energy cusps at symmetric coincident boundaries or only the assymetric boundary shown in your paper?

DAHL: All of the boundaries studied were assymetric tilt boundaries. Some 39 boundaries with differing angles of rotation were constructed to determine the dependence of grain-boundary energy upon the angle. Each of the coincidence boundaries within this range (36.9, 22.6, and 28.1 degrees) were constructed. Definite cusps found in the curve indicate that these are low-energy configurations. The cusps were quite narrow supporting the theory that they are unique configurations.

WYNBLATT: I have two questions:—
(1) Could you give some more details about the procedure for rejecting atoms from the boundary region?
(2) This type of model has recently been criticized on the basis that it predicts grain-boundary regions that have lower density than that of a liquid. This, of course, is physically unappealing. Do you have any comment on this? I also wonder whether Dr. Weins might like to comment on this issue.

DAHL: (1) The procedure for rejecting atoms in the interfacial region between the two grains is a critical step in the construction of a grain boundary. The problem occurs in a very narrow region but in that region the atoms cannot be placed in positions too close or too distant and retain a physically realistic model. The method employed was to superimpose the grains with a chosen distance of minimum allowable interatomic separation. If atoms in the rotated (superimposed) grain were closer than this distance to an atom in the other grain they were rejected. The distance of approach was chosen in a series of experiments in which this parameter was varied during construction. The energies of the computational cells containing the resulting boundaries were then computed to determine the value of the parameter which would give the lowest average energy per atom and therefore most realistic configuration. The first neighbor distance $\sqrt{2}a/2$ was predictably found to be the optimal value and was employed in the construction of all other grain boundaries reported.

(2) Estimation of atomic density in a grain boundary depends upon very arbitrary choices of the volume of boundary regions which are quite indistinct in many regions. However, the problem is real and was the basis for the rejection parameter study which was described. At values of separation too great or too small the effective atomic density was unrealistic and was reflected in high configurational energies. Thus by selecting an optimal configurational energy the density problem was resolved in a sensitive and objective manner. However, the accuracy and degree of realism is dependent upon the choice of the interatomic potential in the disordered region. The same potential was used in all regions during these studies.

WEINS: The comment was made that the *free volume* in the boundary is too high since it is greater than that of a liquid. Yet if one tried to constrain a liquid by two periodic boundaries and allowed the liquid to be only a few atom planes thick then one should not expect the liquid to have the same free volume as a liquid. Furthermore it is questionable if one could obtain a random structure with these boundary conditions.

The apparent free volume of the boundary will be a function of the number of planes considered; as more planes are considered, the calculated density differences will decrease. This means that the density difference is not necessarily greater than the liquid.

HUNTINGTON: What is the effect of the added carbon on self-diffusion in the boundary?

DAHL: The experiments to determine self-diffusion in a grain boundary with carbon impurity atoms have not yet been conducted. However, the closed structure that was caused by the addition of a carbon atom should reduce self-diffusion in the misfit region to rates comparable to those that would be observed in the bulk material (i.e., in a perfect crystal).

COMPUTER SIMULATION OF THE STRUCTURE OF HIGH ANGLE GRAIN BOUNDARIES

M. J. Weins
University of Illinois at Chicago Circle
Department of Materials Engineering
Chicago, Illinois

ABSTRACT

The structure of a 36.8 degree [001] tilt boundary with the boundary symmetrically located between equivalent (110) planes was determined using various pairwise potentials. The calculation was done in a two-stage process. First one crystal was allowed to translate with respect to the other, and then the atoms were allowed to move from their lattice sites. The resulting structures were consistent with experimental evidence.

1 INTRODUCTION

Previous models proposed for the structure of grain boundaries have been based on arguments of geometric fit and not on an energetic calculation. The present work generates a minimum-energy model for the structure of a high-angle-coincidence grain boundary. Unlike previous models the crystals are allowed to move relative to each other and the atoms are allowed to relax from lattice sites to sites that are energetically more favorable.

A grain boundary is the junction between two crystals. A grain boundary therefore has five degrees of freedom that must be specified for the grain boundary to be totally defined. Three degrees of freedom define the relative rotation of the crystals with respect to each other, and the two additional constraints are required to define the plane of the grain boundary. The grain boundaries that are discussed here have a common tilt axis and the grain-boundary plane is located symmetrically between equivalent planes in the crystals. This means that one can think of generating the relative orientations of the crystals from a single crystal by rotating one portion of the crystal with respect to the other portion by a specified angle around a designated tilt axis. Thus, for a symmetric tilt boundary where the crystals have a common [001] tilt axis and the (110) planes are rotated by 36.8 degrees, the relative orientation of the two crystals can be generated from a perfect crystal by rotating the (110) planes around a [001] axis until the (110) planes have an angular separation of 36.8 degrees. The grain-boundary plane is thus defined to be the plane that contains the tilt axis and bisects the angle by which the (110) planes are separated.

Early models for the structure of grain boundaries assumed that an amorphous liquid layer connected the crystals.[1,2,3,4] Recrystallization studies showed that certain angular relations between the crystals were more prevalent.[5] The amorphous-layer theory was inconsistent with this experimental evidence that showed that certain grain boundaries had special properties and with other experimental evidence that indicated the boundary region was only a few atom-diameters thick.[6]

The grain-boundary orientations that were dominant after recrystallization became known as coincidence orientations. This name resulted from the fact that rotation of one of the crystals with respect to the other about a common axis resulted in a superlattice that contained some of the atoms of both crystals.[7,8,9] The superlattice became the basis of a grain-boundary structural model known as the coincidence model. A coincidence lattice is shown in Fig. 1.

The coincidence model attributed the special properties of boundaries of these orientations to the fact that some of the boundary atoms were part of a superlattice common to both crystals. The coincidence model however, was unable to explain the experimental observation that boundaries in the vicinity of the coincidence orientations, as well as at exact coincidence, showed an increase in grain-boundary mobility.[10]

Fig. 1. Lattice coincidence model for a 36.8-degree [001] symmetric tilt boundary showing only one (001) plane. The small circles indicate the lattice sites of the coincidence lattice.

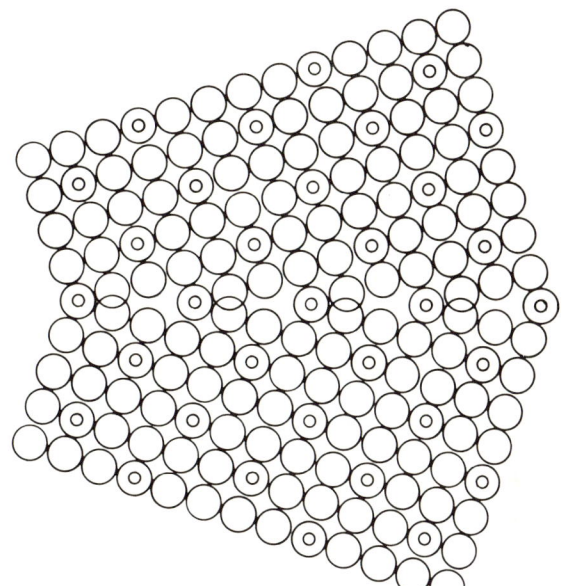

Bishop and Chalmers developed a model based on boundary coincidence. At the coincidence orientations this model corresponds to the coincidence model; however, as the crystals were rotated slightly off the coincidence orientations the lattice coincidence was lost, but boundary coincidence was maintained by adding a small stress field.[11] Bishop and Chalmers assumed that the boundary atoms were mutually shared by the two crystals and that these shared atoms could account for the special properties of the boundaries at and near the coincidence orientations. This model also allowed for a continuous change in grain-boundary energy as the orientation was moved from the coincidence orientation. Fig. 2 shows the Bishop-Chalmers representation for a 36.8-degree symmetric tilt boundary with a common [001] tilt axis and the (110) planes symmetrically disposed.

Fig. 2. The Bishop-Chalmers model for a 36.8-degree {001} symmetric tilt boundary. The period of the boundary is shown by the solid line beneath the figure.

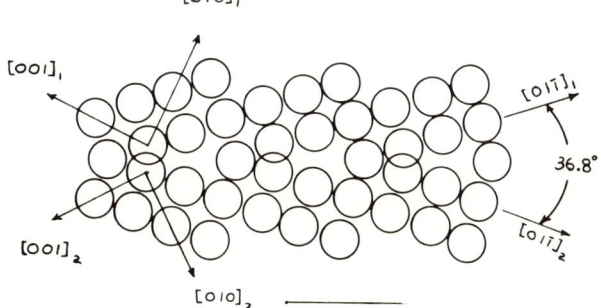

The short periodicity that was predicted by both the coincidence model and the Bishop-Chalmers model has been experimentally verified by Schober and Baluffi[12]. They used transmission electron microscopy to study boundaries in the vicinity of the coincidence orientations and demonstrated that the stress field for off-coincidence boundaries were periodic with a period consistent with the periodicity predicted by the coincidence lattice.

The confirmation of a short period being associated with the coincidence boundaries and the availability of high speed computers made it possible to calculate the structure of minimum energy for boundaries of the coincidence orientations. The short periodicity was essential because it meant the boundary region could be represented by a small number of atoms periodically extended. Less than a hundred atoms were required to represent the boundary in this study.

2 POTENTIALS

For this work three different forms of pairwise interactions were used, the Morse, the Lennard-Jones six-twelve, and the four-seven potentials. The constants for the potentials were selected to represent two different materials; gold and aluminum. These potentials are functions with a single minimum and they represent orientation independent central forces. Table I shows the form of the potentials used, as well as, the constants used to define them.

The four-seven and the six-twelve require two constants. These were the equilibrium separation of the atoms in a perfect crystal and some characteristic energy. The curvature of these potentials in the vicinity of the minimum is a function of the power law selected, and is not necessarily consistent with the observed compressibility of the material. The Morse potential, a three-parameter potential, requires three characteristics of the material for definition. For this study they were a characteristic energy and the constraints that the slope of the energy function be set equal to zero at the minimum and that the curvature of the energy function be consistent with the compressibility. Unlike the four-seven and the six-twelve potentials, the Morse potential takes into account the compressibility of a material and, thus (for problems in solids where the compressibility may be important) may be slightly more realistic.

As can be seen from examination of Table I a four-seven and a six-twelve potential based on the sublimation energy for both gold and aluminum were used[13]. The structure constants for these potentials were derived by Lennard-Jones.[14] For the Morse potential both the energy of sublimation and the vacancy energy were used. The Morse potential

Table I. Forms for Potentials Used in This Work

(1) Morse Potential

$$\psi(r_{ij}) = D\left\{\exp[-2\alpha(r_{ij}-r_0)] - 2\exp[-\alpha(r_{ij}-r_0)]\right\},$$

where $\psi(r_{ij})$ is the potential energy between atoms i and j at a separation of r_{ij}.

α, D, and r_0 are constants with the dimensions of reciprocal distance, energy, and distance, respectively.

The constants are as follow

Aluminum (Equilibrium atomic separation = 2.847547)

	D eV	α A$^{0-1}$	r_0 Å
Vacancy energy basis			
Nearest neighbors	0.14	2.277514	2.847800
Nearest neighbors + Second nearest neighbors	0.131147	2.266157	2.867348
Sublimation energy basis			
Nearest neighbors	0.526107	1.174971	2.847547
Nearest neighbors + Second nearest neighbors	0.426571	1.143819	2.954159

Gold (Equilibrium atomic separation = 2.874126)

	D eV	α A$^{0-1}$	r_0 Å
Vacancy energy basis			
Nearest neighbors	0.180000	2.969985	2.874127
Nearest neighbors + Second nearest neighbors	0.1749150	2.964278	2.880848
Sublimation energy basis			
Nearest neighbors	0.6033919	1.622149	2.847126
Nearest neighbors + Second nearest neighbors	0.528904	1.601566	2.92683

(2) Four-Seven Potential

$$\psi(r_{ij}) = \frac{-4.0*E*R_0^4}{C4*r_{ij}^4} + \frac{2.0*E*R_0^7}{C7*r_{ij}^7}$$

where C4 and C7 are crystal constants for an fcc material, 25.3383 and 13.3593, respectively. E is a characteristic energy, and R_0 is the equilibrium separation.

(3) Six-Twelve Potential

$$\psi(r_{ij}) = \frac{-4.0*E*R_0^6}{C6*r_{ij}^6} + \frac{2.0*E*R_0^{12}}{C12*r_{ij}^{12}}$$

where C6 and C12 are crystal constants for an fcc material, 14.4589 and 12.1313, respectively.

	Aluminum	Gold
Sublimation Energy, E $\frac{\text{kcal}}{\text{g-atom}}$	77.5	87.3
R_0, Å	2.85	2.878

constants were those determined by the procedure outlined by Cotterill and Doyama[15,16]. They constructed the Morse potentials in such a manner that the energy would be consistent with the energy of a perfect crystal if the interactions were summed in one case over only nearest neighbors, and in the second case over nearest- and second-nearest neighbors. The structure was determined employing both nearest- and second-nearest-neighbor potentials.

Fig. 3 and 4 are plots of the potential energy versus the normalized separation. The normalized separation is defined as the atom separation divided by the equilibrium separation. These figures show that the potentials differ in form although all of the functions have a single minimum, approach zero as the separation of the atoms becomes very large, and approach infinity as the separation of the atoms approaches zero. Fig. 3 plots selected Morse potentials. These potentials were chosen to be representative of the Morse potentials used in this work. The potential based on the sublimation energy of gold, plot A, has the deepest

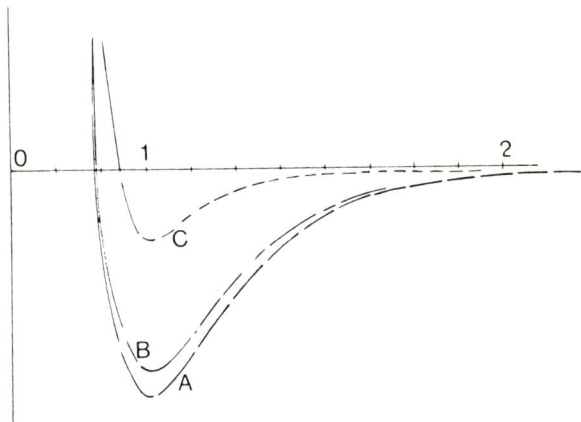

Fig. 3. Plot of various Morse potentials. The abscissa is the potential. The ordinate is the normalized separation. A and B are Morse potentials based on the sublimation energy of gold considering respectively nearest neighbors, and nearest neighbors plus second nearest neighbors. C is a Morse potential based on the vacancy energy for aluminum considering nearest neighbors.

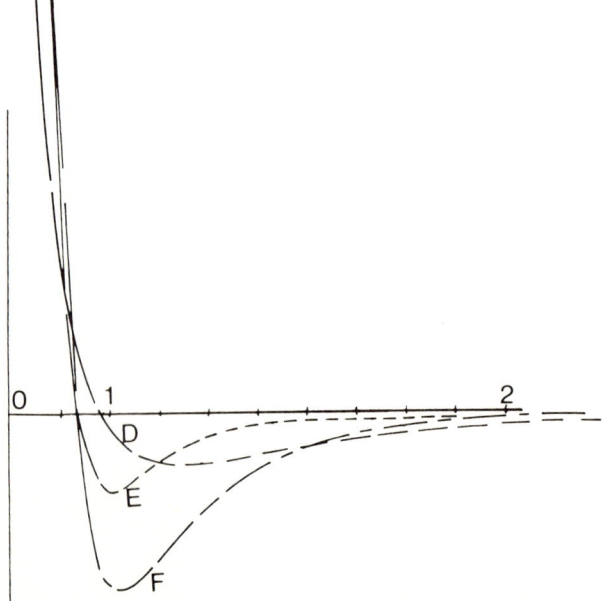

Fig. 4. Plot of various potentials. The abscissa in the potential and the ordinate is the normalized separation. D and F are respectively the four-seven and the six-twelve potentials based on the sublimation energy of gold. E is a Morse potential based on the vacancy energy of aluminum.

minimum. The depth of the minimum decreases as the number of neighbors considered in the calculation is increased. This change also resulted in a decrease in the curvature of the function in the vicinity of the equilibrium separation. This upward shift of the minimum is essential if the energy of a perfect crystal is to be obtained in both cases. Shallower minimums were obtained when the vacancy energy was used as the characteristic energy. This can be seen from examination of the plot, C, in Fig. 3. Plot C is for the Morse potential based on the aluminum vacancy energy and is contrived to give the energy of a vacancy in a perfect crystal if the interactions are summed over first-nearest neighbors. The potentials plotted in Fig. 3 represent the extremes in the Morse potentials.

Fig. 4 compares the Morse potential based on the aluminum vacancy energy, plot E, to the four-seven and the six-twelve potentials based on the sublimation energy of gold, plots D and F. These potentials differ greatly in form. The four-seven potential has the shallowest minimum and the minimum occurs at the greatest separation. The six-twelve potential has its minimum shifted to a separation slightly greater than the equilibrium separation and this potential approaches zero more rapidly than either the four-seven or the Morse potentials.

3 CALCULATION

One grain-boundary orientation was used to determine the effect of different potentials on the resultant minimum-energy structure. On the basis of earlier work using a Morse potential to determine the minimum-energy structure of a variety of coincidence boundaries it was decided to examine a 36.8-degree symmetric tilt boundary with a [001] tilt axis and the boundary located symmetrically between (110) planes.[17,18]

The calculations were performed using the same algorithm with only the potential being changed for each minimum-structure determination. The algorithm was a two-step procedure, the first step being a rigid translation that determined the relative translations of the crystals which resulted in minimum energy. The second step, the atomistic movement, allowed the atoms to move from their lattice sites to sites associated with lower energy.

Both portions of the algorithm required calculating the energy of the array of atoms that represented the boundary region. The interaction was assumed to be described by the potentials given in Table I, and the energy of the array of atoms was calculated from the following expression:

$$E = \frac{1}{2} \sum_i \sum_j \psi(r_{ij}) \quad ,$$

where

$\psi(r_{ij})$ is the interaction potential between the ith and the jth atoms

i and j are atomic indices, i is summed over all atoms in the array, j is summed only over those atoms that are at a separation of less than second-nearest neighbors, or less than third-nearest neighbors

$\frac{1}{2}$ corrects for the double counting of the interactions.

The number of atoms considered in the array will be a function of the periodicity of the boundary. The array of atoms needed to represent the boundary is periodic in two directions. Parallel with the tilt axis, the z direction, the structure can be represented by the stacking inherent in the crystal structure, ABABAB. Therefore in this direction only two layers are needed to represent the structure. In the plane perpendicular to the boundary and in the direction of the boundary plane, the x direction, there is periodicity of the structure that is a function of the angle by which the (110) planes are symmetrically disposed from the boundary plane. For a 36.8-degree boundary, this periodicity is 3.162 times the diameter of the atom. The periodicity is shown in Fig. 2. In the third orthogonal direction, the y direction, the structure is not periodic; however, it can be assumed that, sufficiently far from the boundary plane, the structure will be that of a perfect crystal. The number of atoms required in the y direction was thus selected to be sufficient that the addition or deletion of a plane of atoms in this direction did not alter the structure of the boundary.

Translational periodic boundary conditions were employed in both the x and z directions, and the crystals were extended in the y direction to insure that all of the atoms had proper coordination. The atoms in the y direction that were farthest from the boundary were rigidly constrained and, thus, remained on lattice sites and were not able to relax as atoms would at a free surface. This constraint insured that the atoms in the y direction farthest from the boundary plane that were included in the energetic calculation saw the configuration of a perfect crystal.

The first step of the calculation, the rigid translation, was done by calculating the energy of the array of atoms for various translational configurations. For this, the crystals were systematically moved in steps of 1/20th the period in the x direction, and 1/10th of the period in the z direction. At each of the x-z translations the y coordinate was allowed to vary in increments of 1/25th the atomic diameter until the minimum was found. The relative translations that resulted in lowest energy were used as the starting configurations for the atomistic relaxations.

The second step in the algorithm, the atomistic relaxation, was an iterative procedure that allowed the atoms to move from their lattice sites to sites of lower energy. Since the negative gradient of the energy is the force on each of the atoms, the atoms were first allowed to move in the direction of the negative gradient an amount proportional to the magnitude of the gradient. For successive iterations the direction of movement was defined to be the negative gradient modified by the gradient of the proceeding step. This method of determining the direction of search, known as the method of conjugate gradients, can be expressed as follows:

$$S^{n+1} = -\nabla E(x^{n+1}) + (wf)^n S^n ,$$

where:

S^{n+1} is the new direction of search

$\nabla E(x^{n+1})$ is the gradient of the energy function for the current value of the vector x

S^n is the previous direction of search

$(wf)^n$ is the weighting factor for the present determination and is defined as follows:

$$(wf)^n = \frac{\nabla E(x^{n+1}) \cdot \nabla E(x^{n+1})}{\nabla E(x^n) \cdot \nabla E(x^n)}$$

$\nabla E(x^n)$ is the gradient of the energy function for the previous value of the vector x

x is a vector array of the coordinates of the atoms and is 3N long, where N is the number of atoms selected to represent the boundary region.

The above procedure for selecting the direction of search gives more rapid convergence than would be obtained if only the gradient were considered.

The calculation was stopped when the change in energy after successive calculations was less than 0.1×10^{-3} eV for each of three successive steps.

4 RESULTS

Since the calculation was performed in two steps, the results after rigid translation are discussed separately from the results of the atomistic translation. Tables II through V give the rigid translations associated with the minimum grain-boundary energy. Tables II, III, and IV report the

Table II. Four-Seven Potential Translations Resulting in Minimum Energy

Aluminium

x period = 9.04 Å   z period = 4.04 Å

for consideration of atoms at a separation of less than the second-nearest-neighbor separation in a perfect crystal

x translation	y translation	z translation
0.904 Å	1.03 Å	2.02 Å
4.52	1.03	2.02
5.42	1.03	0
0	1.03	0

Gold

x period = 9.10 Å   z period = 4.07 Å

for consideration of atoms at a separation of less than the second-nearest-neighbor separation in a perfect crystal

x translation	y translation	z translation
0.910 Å	1.04 Å	2.04 Å
4.55	1.04	2.04
5.46	1.04	0
0	1.04	0

for consideration of atoms at a separation of less than the third-nearest-neighbor separation in a perfect crystal

x translation	y translation	z translation
3.64 Å	0.921 Å	2.04 Å
6.37	0.921	0

minimum translations for the four-seven, the six-twelve, and the Morse potentials where the energy was determined by summing the interactions between atoms that had a separation of less than the second-nearest-neighbor separation in a perfect crystal. Examination of these tables shows that, except for the Morse potential based on the gold vacancy energy, four translations result in minimum energy. All four of the minimums have the same y displacement and these minimums occur in pairs; one pair is associated with a small x translation, the other, with an x translation of approximately one half of the periodicity. For the small x translation, one of the minimums is associated with no z translation and, therefore, the AABB stacking across the boundary is maintained; the other minimum is associated with a translation of one half of the z period resulting in an ABAB stacking across the boundary. For the pair with the larger x translation the situation was similar with one minimum having no z

Table III. Six-Twelve Potential Translations Resulting in Minimum Energy

Aluminium

x period = 9.04 Å   z period = 4.04 Å

for consideration of atoms at a separation of less than the second-nearest neighbor separation in a perfect crystal

x translation	y translation	z translation
1.35 Å	0.797 Å	2.02 Å
4.05	0.797	2.02
5.85	0.797	0
8.56	0.797	0

Gold

x period = 9.10 Å   z period = 4.07 Å

for consideration of atoms at a separation of less than the second-nearest-neighbor separation in a perfect crystal

x translation	y translation	z translation
1.36 Å	0.805 Å	2.03 Å
4.10	0.805	2.03
5.91	0.805	0
8.65	0.805	0

translation and the other minimum having a z translation of one half of the period. This is shown for the four-seven potential on the basis of the sublimation energy of aluminum from in Fig. 5. The four minimums are shown in this figure. The left portions of Fig. 5a through 5d are the x-y projection, the right portions are y-z projections. The centers of the atoms are indicated and the boxes indicate the unit of structure that represents the boundary region and which is periodically extended in the x and z directions for the calculation. From examination of these figures it can be seen that, although there are four minimums, the minimums have an equivalence since each of the translations results in the boundary atoms having similar coordination.

The y translation, while it remained constant for each potential, varied with the potential. For the Morse potential, the more severe the potential the greater the y translation. The most severe Morse potential, that based on the gold vacancy energy had the greatest y translation, y=0.866A, and the Morse potential based on the sublimation energy of aluminum had the smallest y translation, y=0.114A. The four-seven and the six-twelve potentials had greater y translations than the Morse potentials. This is attributable to the fact that the separation of the minimum energy is slightly greater than the equilibrium separation in a

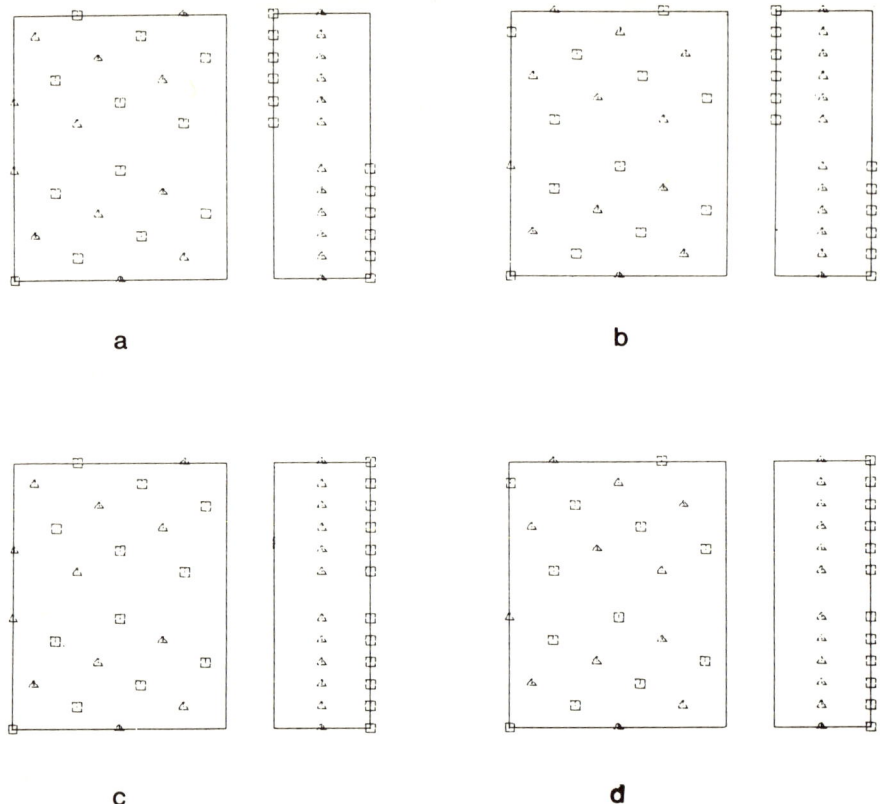

Fig. 5. The four equivalent minimum energy structures resulting from rigid translation are given in a, b, c, and d. These structures were determined employing a four-seven potential based on the sublimation energy of aluminum and considering atoms at a separation of less than second-nearest neighbors.

perfect crystal. Table II gives the minimum translations associated with the four-seven potentials for the calculation that considered separations of less than second-nearest-neighbor distances and for the calculation that considered separations of less than third-nearest neighbors. In the latter the y separation is reduced by the additional-neighbor consideration. This would be expected from the form of the potential.

For the Morse potential, based on the vacancy energy of gold, only two minimums were found, again one of these had a small x translation and the other had an x translation of approximately one half of the periodicity in the x direction. For both of these minimums the z translation was such that the crystals shifted one quarter the periodicity in the z direction, or one half of the separation of the (001) planes. This meant that the (001) planes of the two crystals no longer remained coplanar. The structure of this boundary is shown in Fig. 6.

Table IV. Morse Potential Translations Resulting in Minimum Energy [a]

Aluminium

x period = 9.01 Å   z period = 4.03 Å

Based on sublimation energy

x translation	y translation	z translation
2.25 Å	0.114 Å	2.01 Å
3.15	0.114	2.01
6.75	0.114	0
7.65	0.114	0

Based on vacancy energy

x translation	y translation	z translation
1.35	0.797 Å	2.01 Å
4.05	0.797	2.01
5.85	0.797	0
8.55	0.797	0

Gold

x period = 9.10 Å   z period = 4.06 Å

Based on the sublimation energy

x translation	y translation	z translation
0.91 Å	0.805 Å	2.03 Å
4.54	0.805	2.03
5.45	0.805	0
0	0.805	0

Based on the vacancy energy

x translation	y translation	z translation
0.45 Å	0.806 Å	1.016 Å
4.99	0.806	1.016

(a) For consideration of atoms at a separation of less than the second-nearest-neighbor separation in a perfect crystal.

Table V gives the translations associated with minimum energy for the Morse potential when atoms at a separation of less than third-nearest neighbors were considered in the calculation. As can be seen, the consideration of second-nearest neighbors decreases the y separation of the two crystals. This was previously observed for the four-seven potential. All Morse potentials produced a translation in the z direction when the calculation considered all atoms at a separation of less than third-nearest neighbors. The z translation was such that the planes perpendicular to the z direction were not coplanar, but rather staggered. For the gold Morse

708 M. J. WEINS

Table V. Morse Potential[a]

Aluminium

x period = 9.01 Å   z period = 4.03 Å

Based on the sublimation energy

x translation	y translation	z translation
0.901 Å	0.228 Å	1.61 Å
4.50	0.228	1.61
5.40	0.228	0.403
0	0.228	0.403

Based on the vacancy energy

x translation	y translation	z translation
0.45 Å	0.686 Å	0.805 Å
0.45	0.686	1.208
4.95	0.686	0.805
4.95	0.686	1.208

Gold

x period = 9.10 Å   z period = 4.06 Å

Based on the sublimation energy

x translation	y translation	z translation
0.454 Å	0.6906 Å	1.016 Å
4.999	0.6906	1.016

Based on the vacancy energy

x translation	y translation	z translation
0.454 Å	0.806 Å	1.016 Å
5.91	1.148	1.016

(a) For consideration of atoms at a separation of less than the third nearest neighbor separation in a perfect crystal.

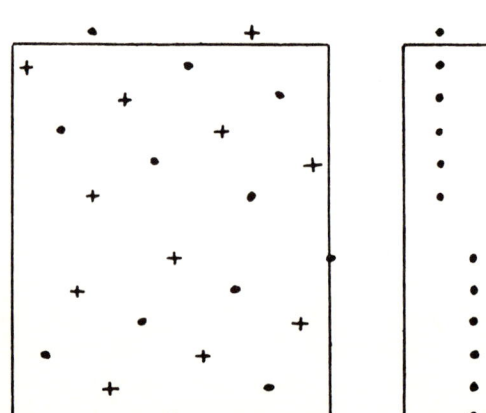
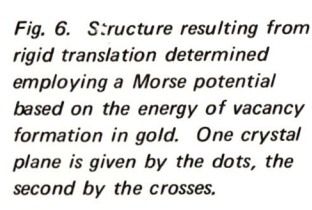

Fig. 6. Structure resulting from rigid translation determined employing a Morse potential based on the energy of vacancy formation in gold. One crystal plane is given by the dots, the second by the crosses.

potentials, the (001) planes were shifted to a position that was a quarter of the period in the z direction, and therefore the (001) planes of one crystal were located at a point midway between the (001) planes of the other crystal. For aluminum, the planes shifted just slightly off the coplanar position.

Second-nearest neighbors again had four equivalent minimums, as was true for the four minimums associated with the other Morse potential calculations.

The atomistic relaxation resulted in only slight movement of the atoms from the lattice sites. Fig. 7 and 8 show the structure after rigid translation and after atomistic relaxation when a Morse potential, based on the aluminum vacancy energy summed over atoms at a separation of less than third-nearest neighbors, was used to determine the structure. Examination of these figures shows that the atomistic relaxation results in very little shift in the atom positions. Fig. 9 shows the structure after atomistic relaxation for the four-seven potential based on the gold sublimation energy. Again the interactions were summed over all atoms at a separation of less than third-nearest neighbors. The shifts in the atomic positions was very slight relative to the atom positions after rigid translation.

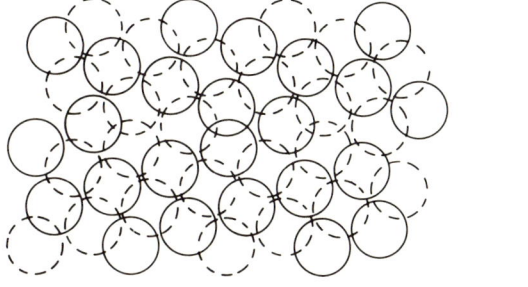

Fig. 7. Structure after rigid relaxation using a Morse potential based on the vacancy energy of aluminum and considering nearest neighbors and second-nearest neighbors. One set of planes in the x-y projection is given by the circles with solid lines, the second set of planes is given by the circles with broken lines. The y-z projection of the centers is to the right.

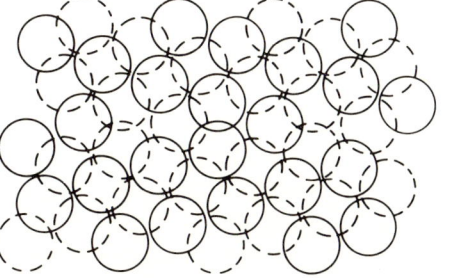

Fig. 8. Structure after atomistic relaxation determined using a Morse potential based on the vacancy energy of aluminum with the interactions summed over atoms at a separation of less than the third nearest neighbor separation. One set of planes in the x-y projection is given by the circles with solid lines, the second set of planes is given by the circles with broken lines. The y-z projection of the atom centers is on the right.

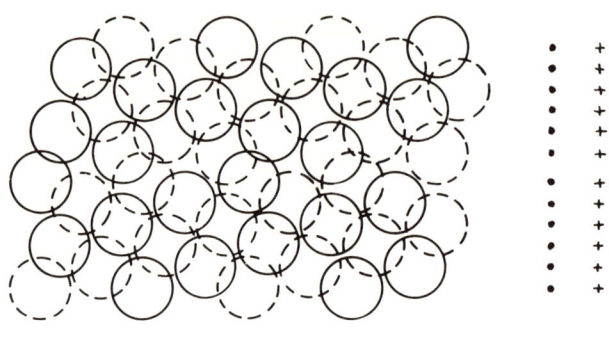

Fig. 9. Structure after atomistic relaxation determined using a four-seven potential based on the sublimation energy of gold with the interactions summed over atoms at a separation of less than the third-nearest-neighbor separation. One set of planes in the x-y projection is given by the circles with solid lines, the second set of planes is given by the circles with broken lines. The y-z projection of the atom centers is on the right.

5 DISCUSSION

While differences in the structure resulted from the use of the various potentials, certain common features were in all the structures. All of the minimum-energy structures generated had, relative to the Bishop-Chalmers and coincidence models, a translation in the y direction. The magnitude of the translation varied with the potential and the number of neighbors considered. With the exception of the aluminum Morse potential based on the sublimation energy, the y translation was in the vicinity of one half of an atomic diameter. Because of the soft nature of the aluminum Morse potential the smallest expansion was found in the y direction when the boundary was determined using this potential.

The greatest differences in the structure were in the z translations. In the calculation considering neighbors of a separation of less than that of second-nearest neighbors for all cases except for the Morse potential based on the vacancy energy of gold, the (001) planes of the two crystals remained coplanar. When neighbors at a separation less than third-nearest neighbors were considered, the structures generated by the Morse potentials did not show this coplanar feature. The lattice planes assumed this compromise position so that the atoms in the vicinity of the boundary could bring a maximum number of neighbors into their radius of consideration.

These minimum-energy structures were independent of the starting configuration. This was confirmed by using different starting configurations for the rigid translation and by randomly perturbating the atoms prior to atomistic relaxation.

This work indicates that the structure of grain boundaries is nontrival and is dependent on the material. Other models have assumed that the boundaries were material invariant.

The structures generated would predict anisotropic grain-boundary diffusion. The boundaries have channels of low atomic density parallel with the tilt axis, however, these channels are not found perpendicular to the tilt axis. This would lead one to predict more rapid diffusion parallel to the tilt axis than perpendicular. This is in agreement with experimental observations on grain boundary diffusion[19,20]. Furthermore, the structure can also be interpreted as consisting of sets of interpenetrating ledges. This steplike character of the boundary has been observed.[21] Thus, while differences in structure result from the use of various potentials, all the structures are in accord with experimental observations.

REFERENCES

1. M. Brillouin, Ann. Chem. Phys. **13**, 77 (1898).
2. G. Quincke, Proc. Roy. Soc. **A76**, 431 (1905).
3. W. Rosenhain, and J.C.W. Humphrey, J. Iron and Steel Inst. **87**, 219 (1913).
4. W. Rosenhain and D. Ewin, J. Inst. Met. **10**, 125 (1913).
5. M. L. Kronberg and F. H. Wilson, Trans. AIME **185** (1949).
6. M. J. Attardo and J. M. Galligar, Acta Met. **15**, 395 (1967).
7. W. Bollman, Phil. Mag. **16**, 363 (1967).
8. D. G. Brandon, B. Ralph, S. Ranganathan and M. S. Wald, Acta Met. **12**, 813 (1964).
9. D. G. Brandon, Acta Met. **14**, 1479 (1966).
10. K. T. Aust, *Recovery and Recrystallization of Metals*, Ed. L. Himmel, Interscience Publ. p. 131 (1962).
11. G. Bishop and B. Chalmers, Scripta Met. **2**, 133 (1968).
12. R. Schober and R. W. Baluffi, Phil. Mag. **31**, 846 (1970).
13. P. Tick Thesis MIT (1969).
14. J. E. Jones and A. E. Ingham, Proc. Roy. Soc. **A107**, 636 (1925).
15. R.M.J. Cotterill and M. Doyama, Phy. Rev. **145**, 465 (1966).
16. R.M.J. Cotterill and M. J. Doyama, Argon National Lab Report (1965).
17. M. J. Weins, H. Gleiter, and B. Chalmers, J. Appl. Phys. (to be published June 1971).
18. M. J. Weins, H. Gleiter, and B. Chalmers, Scripta Met. **4**, 235 (1970).
19. R. E. Hoffman, Acta Met. **4**, 97 (1956).
20. M. J. Weins and B. Chalmers, to be published.
21. H. Gleiter, Acta Met. **17**, 565 (1969).

DISCUSSION on paper by M. J. Weins

WILSON: Are all the potentials you chose devoid of a volume-dependent term?

WEINS: Yes.

HARRISON: It seems to me that the idea of coincidence boundaries is either a genuinely interesting effect of the long-range oscillatory forces extending across the boundary of an uninteresting artifact arising from the advantages of short-range periodicity in the plane of the boundary. If it is the former, one could get at it by focusing on the long-range interactions that you have left out completely.

WEINS: The coincidence boundaries can be thought of as being higher-order twins. Experiments have shown an increase in the planar separation of the planes parallel with the boundary across the boundary. This would eliminate the long range periodicity.

GIRIFALCO: There is an interesting difference between inverse power potentials and potentials of other types. When the pair-separation difference is expressed in reduced form (using the lattice parameter or nearest-neighbor distance), lattice sums for the inverse power potential factor into two parts. One part is a purely geometric sum containing no information on material properties. The other part contains material properties, but no structure information. In effect, the crystal structure and material properties are decoupled. Could this be the origin of the differences you obtained between results for the inverse power and Morse potentials? If so, the qualitative aspects of the results for the inverse power potential are more suspect than those for the Morse potentials.

WEINS: That is an interesting comment and might result in the Morse potentials being more reliable than the inverse power laws. I also feel that the Morse potentials are better because they are fit to the elastic properties of the material, and the elastic properties might be important in determining the structure of the boundary.

DAHL: Our experiments also show that the coincidence site boundaries are unique. The grain boundary energy for the 36.9-degree boundary, for which the density of common sites is 1 in 5, lies in a cusp which is very narrow in θ. Cusps are distinguishable for the other coincidence boundaries but are much less pronounced. The density of common sites is also much less than for the 36.9-degree boundary; therefore, the results appear to support the coincidence-site hypothesis.

JACKSON: When one calculates Morse potential parameters in the manner of Girifalco and Weizer[1] one obtains a_o – the equilibrium lattice spacing at 0 K. Of course, for any further calculations involving these calculated Morse parameters a_o must be used as the lattice spacing for consistency. This was noted by Cotterill and Doyama[2] in their derivation of truncated Morse potential parameters. How much did your a_o vary from the usually quoted "experimental" lattice spacing?

WEINS: I used the values of Cotterill and Doyama, which for gold is 4.06A and for aluminum, 4.03A.

REFERENCES

1. L. A. Girifalco and V. G. Weizer, Phys. Rev., **114**, 487 (1959).
2. See Ref. 15.

A STUDY OF CRACK PROPAGATION IN ALPHA-IRON

M. F. Kanninen
Applied Mathematics and Mechanics Division

and

P. C. Gehlen
Metal Science Group

Battelle, Columbus Laboratories
Columbus, Ohio 43201

ABSTRACT

In previous attempts to simulate crack extension on a (100) plane in α-iron, it was not possible to induce the rupture of atomic bonds at the crack tip. The work reported here overcomes this deficiency by considering the crack front to be jogged. It is then found that an existing crack will heal or extend, depending on whether the stress-intensity factor is less or greater than a critical value. The results are in quantitative agreement with the critical stress-intensity factor determined indirectly in previous work and are consistent with the Griffith criterion for quasibrittle crack growth.

1 INTRODUCTION

A previous paper[1] presented preliminary results of a computer simulation of a "Griffith Crack" on a (100) plane in α-iron. While appropriate for an initial attempt, the computation model used in that work was inadequate in certain respects. As a result, it was not possible to observe crack extension by the rupturing of the atomic bonds at the crack tip. The most obvious faults of the model are in the arbitrary positioning of the peripheral or "boundary" atoms and the use of periodic boundaries along the crack line. For periodic boundaries, only a single crystallographic repeat distance is included in the simulation – bordering planes are accounted for by imposing conditions representing an infinite number of identical repetitions of the planes under consideration. Hence, if bond rupture and bond shear are to occur, they must occur simultaneously along the entire crack front.

In retrospect, it was decided that this process is too difficult, at least in the absence of thermal effects. Concurrent work on dislocation motion[2] and a suggestion of Hirth[3] led to considering the effect of introducing a jog into an otherwise straight crack front in a crystallite which is extensive in all three dimensions. The crack front was therefore allowed to have local variations along its length. This paper presents preliminary conclusions of the study.

2 CONTINUUM FRACTURE MECHANICS

A brief review of continuum fracture mechanics is useful to introduce the terminology used in this work and to establish a frame of reference for the results of the simulation study. Such work is based upon the fundamental contribution of Griffith[4] in which, using current notation, a crack will extend, provided that

$$G \geqslant G_c = 2\gamma \qquad (1)$$

Here G, called the strain energy release rate, reflects the summation of the change in the elastic strain energy in a body containing a crack (if the crack were to extend by an infinitesimal amount) and the work done on the body during such a process. The "Griffith principle" states that crack extension occurs when G takes on a critical value G_c which is just equal to the energy required to form the new surfaces. For an elastic-perfectly brittle material, as assumed in Eq. 1, the energy required is twice the surface energy γ.

While direct application of the Griffith principle is always possible, it is usually much more convenient to utilize the alternative point of view developed by Irwin[5]. Irwin has demonstrated an equivalence between the use of the stress-intensity-factor parameter k and Griffith's energy-balance approach which can be expressed for an orthotropic crack system as[6]

$$G = \pi k^2 \left(\frac{a_{11}a_{22}}{2}\right)^{1/2} \left[\left(\frac{a_{22}}{a_{11}}\right)^{1/2} + \frac{2a_{12} + a_{66}}{2a_{11}}\right]^{1/2}, \quad (2)$$

where the a_{ij}'s denote the elastic compliances of the material*. Eq. 2 shows that a critical value k_c must correspond to G_c. Consequently, because G_c is a property of the material, k_c is also a property of the material. In particular, combining Eq. 1 and 2 gives

$$k_c^2 = \frac{2\sqrt{2}}{\pi} \frac{\gamma}{\sqrt{a_{11}a_{12}}} \left[\left(\frac{a_{22}}{a_{11}}\right)^{1/2} + \frac{2a_{12} + a_{66}}{2a_{11}}\right]^{-1/2}. \quad (3)$$

In the approach known as linear elastic fracture mechanics (or as Griffith-Irwin fracture mechanics) the criterion for crack extension is obtained by setting k — a function of the applied loads and geometry but independent of the material — equal to k_c. With this method it is not necessary to consider a specific geometry and applied loading in the following. Rather, a result applicable to a wide class of problems is being determined. To illustrate, in the simple case of an isolated crack in an infinite sheet under uniform tension σ acting in a direction normal to the crack plane, $k = \sigma\sqrt{c}$ where c is half the crack length. The fracture stress σ_f is obtained for these conditions by simply setting

$$k_c = \sigma_f \sqrt{c}$$

For an isotropic material and plane-stress conditions, $a_{11} = a_{22} = 1/E$, $a_{12} = -\nu/E$ and $a_{66} = 2(1 + \nu)/E$, where E is Young's Modulus and ν is Poisson's Ratio. Substituting these values into Eq. 3 and using the result to replace k_c in the above gives

---

*Conventionally, a subscript is attached to both G and k to indicate that the applied load is restricted to one of three different modes. In this work, only loadings that tend to open the crack are considered so that a distinction is not necessary between this state — Mode I — and loadings that tend to slide one face of the crack past the other — Modes II and III.

$$\sigma_f = \sqrt{\frac{2}{\pi}\frac{E\gamma}{c}},$$

which is the familiar Griffith equation for plane stress.

As the continuum theory does not provide any way of determining k_c, resort to experiment is always necessary. However, in computer simulation, a direct evaluation is possible. To set the stage for such a calculation in α iron, it is appropriate to specialize Eq. 3 for a cubic system. The result is

$$k_c = \left(\frac{2}{\pi}\frac{\gamma}{a_{11}}\right)^{1/2}\left[\frac{1}{2} + \frac{2a_{12} + a_{44}}{4a_{11}}\right]^{-1/4}. \quad (4)$$

The values consistent with those of the interatomic potential to be used in the atomic simulation — the modified potential of Johnson[7] — for separation on a (100) plane are calculated in Ref. 1. They are $a_{11} = 1.111$, $a_{12} = -0.556$, $a_{44} = 1.667$ (units of A^3/eV) and $\gamma = 0.081$ eV/A^2. Thus

$$k_c = 0.242 \text{ eV/A}^{5/2}, \quad (5)$$

which can be taken as arising from the Griffith theory. Because it is based completely on a continuum treatment, it is given for comparison purposes only.

The stress-intensity factor enters the simulation experiment through the equations for the linear elastic displacement field by which the boundary atoms are positioned. As given by Sih and Liebowitz[6], for a cubic system, the Cartesian displacements u and v in terms of a polar coordinate system located at the crack tip are

$$u = k\sqrt{\frac{r}{2}}\left\{(a_{11} + a_{12})\left[(1 - \alpha \sin 2\theta)^{1/4}\cos\varphi_2 + \right.\right.$$
$$+ (1 + \alpha \sin 2\theta)^{1/4}\cos\varphi_1\right] + (a_{11} - a_{12})\frac{\beta}{\alpha}\left[(1 - \alpha \sin 2\theta)^{1/4}\sin\varphi_2 - \right.$$
$$\left.\left. - (1 + \alpha \sin 2\theta)^{1/4}\sin\varphi_1\right]\right\} + \ldots \quad (6)$$

and

$$v = k\sqrt{\frac{r}{2}}\left\{\left(\frac{ba_{11} - a_{12}}{\alpha}\right)\left[(1 - \alpha \sin 2\theta)^{1/4}\cos\varphi_2 - \right.\right.$$
$$- (1 + \alpha \sin 2\theta)^{1/4}\cos\varphi_1 + 2a_{11}\beta\left[(1 - \alpha \sin 2\theta)^{1/4}\sin\varphi_2 + \right.$$
$$\left.\left. + (1 + \alpha \sin 2\theta)^{1/4}\sin\varphi_1\right]\right\} + \ldots, \quad (7)$$

where

$$b = \frac{2a_{12} + a_{44}}{2a_{11}}$$

$$\alpha = \sqrt{\frac{1-b}{2}}$$

$$\beta = \sqrt{\frac{1+b}{2}}$$

$$\varphi_1 = \frac{1}{2} \tan^{-1} \frac{\beta \sin \theta}{\cos \theta + \alpha \sin \theta}$$

$$\varphi_2 = \frac{1}{2} \tan^{-1} \frac{\beta \sin \theta}{\cos \theta - \alpha \sin \theta}$$

The terms indicated by ... in Eq. 6 and 7 are of order $r^{3/2}$ and can be neglected at atomic-scale distances from the tip of a macroscopic length crack, as considered here. Note that, in treating macroscopic length cracks, the present model differs significantly from those employed by Chang[8] and by Tyson and Alfred[9].

3 ATOMIC FRACTURE MECHANICS

It was expected that as the k level was increased (with the configuration allowed to relax at each level), some particular level would eventually be reached at which atomic bonds would be broken. In this way a direct evaluation of k_c could be made. Contrary to this expectation, however, no bond rupture was observed even when k was increased well beyond the value predicted by continuum theory. As an alternative, an indirect evaluation was made and, as reported in Ref. 1, the value $k_c = 0.274$ eV/A$^{5/2}$ determined.

Guided by concurrent attempts to simulate dislocation movement, which proved to be possible only when the process was allowed to proceed via a kink-propagation mechanism, a study of a jogged crack front was initiated. Just as in the previous work, the starting configuration was achieved by computing two equilibrium configurations for a given k level, each with crack tips at different positions. As in Ref. 1, the modified Johnson potential[7] and the quasidynamic method were used. The only essential differences were that more than the minimum number of planes were directly simulated and that the crystallite was taken as a circular cylinder rather than as a parallelepiped. Typical planes from two such configurations are shown in Fig. 1. Notice that in Fig. 1a, bond AB is broken, so that AD is the crack tip pair, while in Fig. 1b, AB is the crack tip pair.

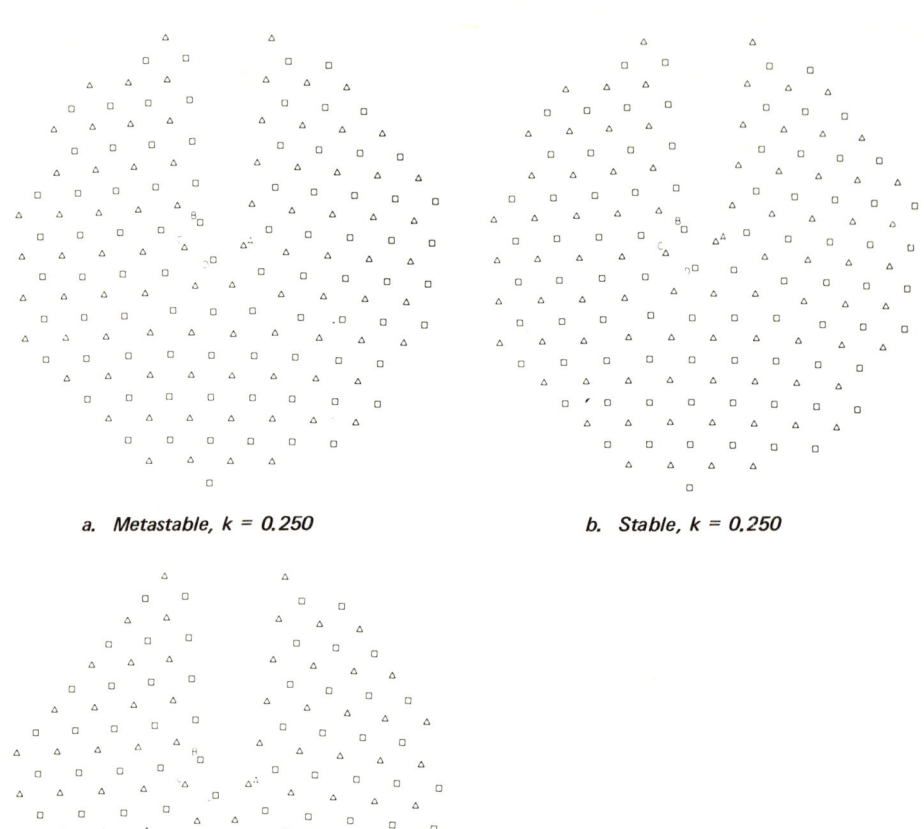

Fig. 1. Atomic configurations for (100) crack. Different symbols represent consecutive 100 layers.

a. Metastable, k = 0.250

b. Stable, k = 0.250

c. Stable, k = 0.300

When the two blocks of atoms are placed together, the crack front, which is straight in each block, becomes jogged. With the exception of the special case of $k = k_c$, the relaxation proceeds by moving the jog out of the crystallite so that the crack front is once again straight. Specifically, when $k < k_c$ this occurs by atomic pairs *making* new bonds. For $k > k_c$, it occurs by *breaking* bonds. The final equilibrium configuration in either case is one in which the crack front is straight. In particular, the result for $k = 0.250$, a value less than k_c, is that shown in Fig. 1a, substantiating that crack closure has occurred. For $k = 0.300$, a value greater than k_c, crack advance occurs with the final configuration shown in Fig. 1c. Here AB is broken and AD is the crack tip pair. Notice that bond AC is broken in Figs. 1-a,b,c.

To describe this process in more detail it is necessary to define a crack front in a discrete model. A convenient definition is as follows. Let the "crack plane" be the plane which in the undeformed lattice lies midway between the two atomic layers which separate to form the crack. The positions at which the forces exerted between atoms intersect the crack plane form a set of points on the crack plane. Because the potential has a finite cutoff distance, no points will appear in the cracked region. Then, the line joining the peripheral points is called the crack line. The atoms associated with these points are referred to as "crack tip pairs".

The processes of crack closure and crack advance by jog movement are then illustrated in Fig. 2. These processes take place sequentially: the crack tip pairs associated with the edge of the jog move in or out of range to each other, the remaining atoms move a comparatively small amount to accommodate the shift, whereupon the jog can be considered to have jumped to a new position. The self-sustaining process will repeat itself ad infinitum in an infinite crystal. Clearly, a jogged-crack-front configuration is unstable under any k level with the possible exception of k_c itself.

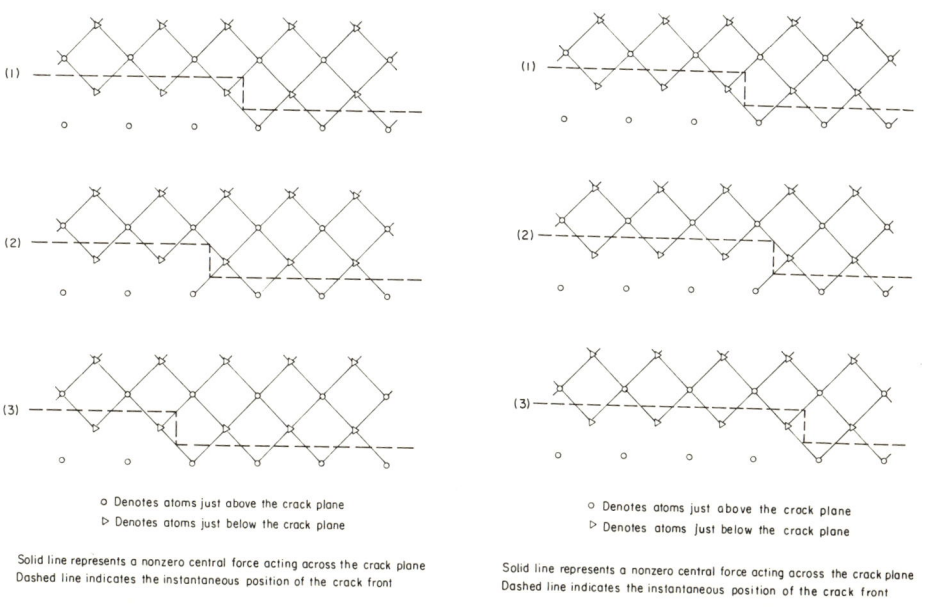

a. Crack Closure

b. Crack Growth

Fig. 2. Three successive stages in jog movement on (100) plane in bcc material.

A verification of the latter part of this statement can be obtained by comparing the separation distances of the crack tip pairs as a function of relaxation time for stress-intensity-factor values that are above, below, and just at $k = k_c$. Typical data are given in Table I. It can be seen that for an amount of relaxation quite sufficient for the $k \neq k_c$ cases to reach stable equilibrium configurations with a straight crack line, the intermediate level has retained a jogged-crack-front configuration. Note that, because these computations were carried out by the quasidynamic method, the time scale has no direct physical significance here. Furthermore, because energy is taken out of the system, there is probably a range of k values for which no discernable propensity toward either closure or growth could be obtained within reasonable computation times. Hence, while this technique is appropriate for verifying an estimate of k_c, it does not provide a realistic technique for calculating k_c.

4 DISCUSSION

The process of crack closure and the reverse process of crack extension (Fig. 2) indicate that the effect of jog movement can be likened to

Table I. Crack Tip Bond Lengths in a Jogged Crack Front for Three Different k Values

Relaxation Time [a]	Bond Lengths of Selected Crack Tip Pairs [c]			
	$k = 0.250$			
0	2.911	2.914	3.691 [b]	3.697 [b]
100	2.861	2.933	3.458 [b]	3.602 [b]
200	2.796	2.806	3.360	3.564 [b]
300	2.786	2.786	2.789	2.789
400	2.789	2.789	2.789	2.789
	$k = 0.274$			
0	2.868	2.871	3.825 [b]	3.831 [b]
100	2.889	3.011	3.641 [b]	3.797 [b]
200	2.903	3.039	3.646 [b]	3.805 [b]
300	2.912	3.074	3.673 [b]	3.817 [b]
400	2.910	3.075	3.675 [b]	3.817 [b]
	$k = 0.300$			
0	2.940	2.944	4.006 [b]	4.015 [b]
100	3.063	3.466 [b]	3.934 [b]	3.990 [b]
200	4.046 [b]	4.044 [b]	4.042 [b]	4.043 [b]
300	4.062 [b]	4.062 [b]	4.063 [b]	4.062 [b]
400	4.062 [b]	4.062 [b]	4.062 [b]	4.062 [b]

(a) Central processor time (sec) for 252 movable atoms, 420 atoms total.
(b) Distance greater than out-of-range separation distance (3.44 A).
(c) Corresponding to AB type pairs in Fig. 1.

the action of a zipper; the crack being zipped closed when $k < k_c$ and zipped open when $k > k_c$. Barenblatt[10] has also used the zipper analogy, but referred to one which "unzips" in the direction of crack advance. In this feature, the present model is inadequate because it permits crack movement by one-half lattice parameter only. This is a consequence of the constraints imposed by the rigid positioning of the boundary atoms. In reality, once crack movement has begun, by whatever mechanism, it would behave like a zipper in Barenblatt's sense as well.

While it is important to incorporate "flexible" boundary conditions into the crack simulation, programming complexities have dictated that the development be carried out independently. A method which is in the spirit of the crystallite-continuum interaction used by Gibson, et al.[11] is being developed and its application to the simulation of an edge-dislocation will be reported soon. Subsequently, the technique will be introduced into crack simulation as the next stage in the development of the work reported here. Once this has been accomplished, the effects of thermal vibrations will be investigated — one purpose being to determine whether such effects could provide an alternative crack-advance mechanism. Ultimately, a model will be developed to determine from first principles how the fracture process occurs in bcc metals as a function of temperature.

When Fig. 1 is viewed at an angle of 45 degrees the outline of an edge dislocation may be seen. The presence of these dislocations may be the beginning of plastic deformation at the crack tip. Conceivably, when flexible boundaries are introduced, these dislocations will move under the stress and, in light of the paper by Basinski, et al. (this volume), more experiments with longer models and/or flexible boundaries will then be required.

Finally, while the existence of an out-of-range separation distance is exploited in this work, this feature of the potential, while convenient, is not at all essential to the success of the computations. The studies of Goodier and Kanninen[12] and of Chang[8] have indicated that the configurations computed using a long-range potential are not much different from those obtained with a short-range potential, for example. The entire question of how the potential affects crack *movement*, where small differences may be crucial, has not been fully explored, however. In addition, the effect of noncentrality in the interatomic potential must eventually be considered.

ACKNOWLEDGMENT

This work was supported by the Office of Naval Research and the Battelle-Columbus Computation Center. Dr. G. T. Hahn contributed many useful comments and suggestions.

REFERENCES

1. P. C. Gehlen and M. F. Kanninen, *Inelastic Behavior of Solids,* M. F. Kanninen, W. A. Adler, A. R. Rosenfield, and R. I. Jaffee, eds., McGraw-Hill, New York, p. 587 (1970).
2. P. C. Gehlen, this volume.
3. J. P. Hirth, *Inelastic Behavior of Solids,* M. F. Kanninen, W. A. Adler, A. R. Rosenfield, and R. I. Jaffee (editors), McGraw-Hill, New York, p. 605 (1970).
4. A. A. Griffith, Phil. Trans. Roy. Soc. (London), **A221**, 163 (1920).
5. G. R. Irwin, *Handbuch der Physik,* **79**, 551 (1958).
6. G. C. Sih and H. Liebowitz, *Fracture,* Vol. II, H. Liebowitz, editor, Academic Press Inc., New York (1968).
7. R. A. Johnson, Physical Review, **145**, 423 (1966).
8. R. Chang, Int. J. Fracture Mech., **6**, 111 (1970).
9. W. R. Tyson and L.C.R. Alfred, presentation at the Corrosion Fatigue Conference, the University of Connecticut, June 14-18, 1971.
10. G. I. Barenblatt, Advances in Applied Mechanics, **7**, 55 (1962).
11. J. B. Gibson, A. N. Goland, M. Milgram and G. H. Vineyard, Phys. Rev., **120**, 1229 (1960).
12. J. N. Goodier and M. F. Kanninen, Technical Report No. 165, Division of Engineering Mechanics, Stanford University (1966).

DISCUSSION on Paper by M. F. Kanninen and P. C. Gehlen

KAUSCH: The treatment you have discussed suggests a treatment that could be applied to propagation of cracks in polymeric solids. The reason is that while the fracture of polymeric chains is definitely a local event, the loading of chains is not — because of the comparatively small forces transmitted in shear. Therefore, within a diameter of several hundred angstroms the continuum stress field loses its significance. What do you consider the smallest model size for which the continuum stresses are valid?

KANNINEN: While our model is presently applicable only to crystalline materials, the possibility of extending it to polymers is quite intriguing. With regard to the fidelity of the continuum stress field near the crack tip, the computations performed so far (employing rigid-boundary conditions and only moderately large array sizes) have not permitted such a determination. We expect to examine this and related questions later.

TYSON: You did not observe bond breaking in tension for the straight crack even under very large applied stresses. This must mean that the tip is blunting by plastic flow; in other words by dislocation nucleation. As you have said, the boundary conditions used are very important in this problem and one must see a sufficiently large crystallite to determine unambiguously the failure mode.

KANNINEN: As we have emphasized, the simulation of failure by dislocation nucleation and movement is as important as simulating crack extension by bond rupture. In order that these processes not be unnaturally constrained, the crystallite must be extensive in all three dimensions. We have demonstrated the results of removing the periodicity conditions and ultimately expect to remove the constraints arising from rigidly positioning the boundary atoms as well.

VITEK: I would think that the reason that the crack has not propagated, even at very high stresses, is that this process should be associated with one or more dislocations moving in front of the crack. However, the applied boundary conditions will probably prevent nucleation of such dislocations. Wouldn't it be, therefore, physically reasonable to introduce such dislocations into the calculations from the beginning?

KANNINEN: While it is true that in the simulation of a straight crack-front bond rupture does not occur, we do not believe that the reason is a *lack* of dislocation nucleation. Nevertheless, artificially introducing various different dislocations into the crystallite to study their interaction with the crack tip may well be instructive. We will consider this possibility when the improvements in the model have been implemented.

SEEGER: The results obtained from linear elastic fracture mechanics and those of the jog-picture are surprisingly close, although the physical pictures used are quite different. Is this true because the Griffith-Irwin theory is essentially based on energy balance considerations, which may be insensitive to details of the mechanism?

KANNINEN: Because the boundary atoms are fixed into the positions given by the linear elastic displacement field in the atomic simulation of the crack tip, it will a priori yield a result of the Griffith type. The model has the freedom only to determine a multiplicative constant which turns out to be about 12 percent greater than that appearing in the Griffith formula. This value is likely to depend on the potential and the array size used, although we have not performed enough computations to determine these variations explicitly.

SEEGER: A general theoretical framework enables us to treat crack propagation by the jog mechanism and the propagation of dislocations in crystals from a common viewpoint. The crack may be represented by distribution of edge dislocations (over the crack edge) or by a wedge disclination. Crack propagation may be described by adding edge dislocations to the disclination. The dislocation itself then is created by jog motion (for cracks) or waves by kink motion (for plastic deformation).

In the hierarchy: disclination – dislocation – jog or kink – one simply has to start one grade higher for crack propagation than for plastic deformation.

HIRTH: With regard to Seeger's comment about a transition from crack growth by kink motion to rapid straight crack growth, the situation is exactly analogous to the kink-type glide motion of a dislocation changing to straight dislocation motion. In fact, the same nucleation and growth models, well known for the glide case, can be used directly in the crack case with the appropriate replacement of shear stress by normal stress, etc. In brief, at low stresses, slow crack growth would occur by kink nucleation and growth; and, at high stresses at the crack tip, the crack could move as rapidly as a straight crack. The transition will occur when the stress is about equal to the analog of the Peierls stress in the crack case.

BULLOUGH: Do you think that your inability to propagate the straight crack is model or potential sensitive? You also had difficulty in moving the straight screw dislocation whereas Duesbery et al., were able to move their dislocation. They suggested that this movement was achieved because they used a larger model. Could the situation be analogous for the unjogged crack?

KANNINEN: Preliminary computational results of our own and those of Chang indicate that the relaxed configurations for a given value of k are not very dependent on the array size or the potential used. We recognize, of course, that small differences could well be significant in determining whether or not the crack will move and intend to study this point very closely. In particular, we will attempt to reproduce the results of Duesbery et al., for the dislocation in order to thoroughly understand the differences between their work and our own.

AGENDA DISCUSSION: SURFACES AND INTERFACES

P. G. Shewmon[*]

Materials Science Division
Argonne National Laboratory
9700 South Cass Avenue
Argonne, Illinois 60439

W. R. Tyson[**]

Department of Physics
Trent University
Peterborough, Ontario Canada

Interatomic potentials available for use in the bulk have a reasonable theoretical basis and can be calibrated by fitting elastic constant, phonon spectra, and defect energy data. Interatomic potentials suitable for surface problems appear to have a weaker theoretical basis than those for the bulk and, also, are more difficult to calibrate. The chairman opened the discussion by focusing on the need to develop a stronger theoretical basis for surface potentials and for methods to calibrate them.

[*]Chairman.
[**]Secretary.

Such considerations divide into two categories: internal surfaces and external surfaces. The potentials being used to describe defect properties and behavior in the bulk seem to be quite satisfactory for calculations of internal surface properties and behavior. Several types of comparisons with experimental data are possible, such as internal surface energy as a function of misorientation and impurity content. Also, simulations of atomic diffusion at grain boundaries and of boundary motion studies are possible, although the simulation of thermally activated processes in grain-boundary regions requires more computer time and computer experiment technique development than does simulation of defects in the bulk.

No satisfactory potential seems to exist for simulation of external surface properties and behavior. Discussion of interaction potential calibration was made somewhat difficult by the fact that pertinent experimental data are obtained for several different surface configurations: for atoms in smooth, low-index surfaces; for atoms at bumps or steps on surfaces; for adatoms on an otherwise smooth surface. The discussion started and remained centered primarily on the question of how one might develop a potential for the atoms within a free surface.

POTENTIALS

Ashcroft began by outlining the essential features of the problem by considering the simple case of a cubic lattice which had been cut apart to form two flat surfaces. Along a traverse from the crystal interior out through a crystal surface the electron density must drop from a characteristic interior average value within the lattice to zero outside the surface. The resulting charge redistribution is shown schematically in Fig. 1. It is known that the interatomic potential ϕ is sensitive to the electron density ρ. Therefore, the charge redistribution that must occur at a surface may cause ϕ to be different for atoms at the surface than for those in the bulk. The magnitude of this difference could be estimated by calculating the dependence of ϕ on ρ in the crystal interior. If one could also obtain an estimate of the variation of ρ at the surface, then an estimate of ϕ for surface atoms would be immediately at hand. This can now be done fairly readily for simple metals on a first-principles basis. More complicated metals, however, require the use of empirical potentials, even for the simulation of behavior in the bulk. No fundamental basis for adjusting these empirical potentials so as to describe a surface has yet been developed. Ashcroft felt that for the transition metals there is some justification for a separation of the interaction into a d-state part, possibly describable by a simple Lennard-Jones or (m,n) interaction, and a free electron part. It is not clear how the screened dispersion-type forces might be altered at a surface.

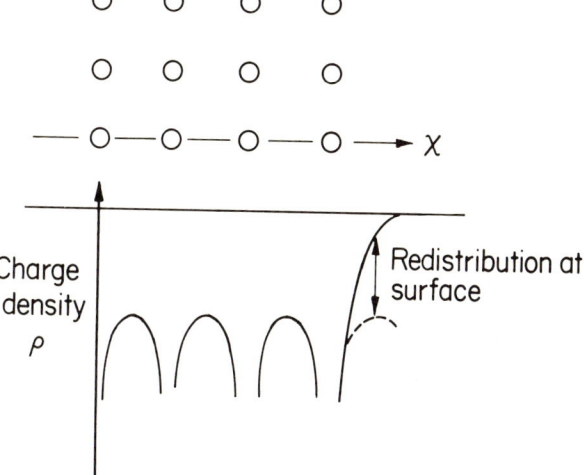

Fig. 1. Schematic Charge distribution at crystalline surface.

Enderby suggested that there were three cases to be distinguished:
- The case exemplified by argon wherein ϕ does not change significantly at the surface.
- The case exemplified by sodium wherein the dominant change in the potential energy interaction is in the repulsive part of the interaction since this part is due primarily to screening effects.
- The case exemplified by nickel wherein the free electron aspects of the potential energy interaction are sensitive to surface effects, but not the d-electron aspects.

Taylor stated that he has looked at the problem of the variation of the sodium potential with electron density and finds that it is remarkably insensitive to changes in density. Despite the fact that the repulsive core potential is largely derived from screening, it does not seem to change in the manner that Enderby suggested. At that point, Harrison commented optimistically that there is a good chance that the interactions remain unchanged at the surface due to relaxation effects such as those around a vacancy, and that we should not assume there is a change unless we can put forward a good theoretical or experimental reason for one. March, on the other hand, supported Enderby's point that whereas the Ar-Ar potential varies little from gas to liquid, the Li-Li interaction in the liquid and in the Li_2 molecule can differ by a factor of 50 due to the Mott transition; however, it is not clear how far one must be from the metal substrate to eliminate the screening effects.

Seeger suggested that the bonding of an adatom to a metal surface should be treated like a case of chemical bonding as the quantum chemists do for molecules. It was agreed during the subsequent discussion that the interaction must vary considerably with the number of nearest neighbors. For example, the interaction between atoms in a flat surface differs from the interaction for an adatom or with an atom in a kink site; moreover, interactions with atoms outside of a surface which are relevant for forces across a crack plane must differ from interactions within it. Huntington suggested that the free electron gas might serve to lubricate the motion of atoms along channels but pin adatoms to the surface.

Regarding empirical potentials, Ehrlich felt that guidance should be sought from theory to find correlations that enable one to build relevant experimental information into the potentials rather than simply making guesses as to their form.

EXPERIMENTS

Ashcroft next raised the possibility of using information about surface restructuring as found in LEED experiments to test the predictions of pair potentials. Ehrlich pointed out the disagreements in surface-restructuring results from different laboratories, concluding that most "restructured" metal surfaces are heavily contaminated. Ashcraft mused that "errors, like straws upon the surface flow; he who would search for pearls must dive below" (Lord Byron) and retired for coffee. Available evidence indicates that the lattice structure within surface planes in metals is the same as that within the bulk, although this may not be true for other types of solids. Wilson suggested ionic crystal surfaces as a good subject for study, but according to Tosi the theoretically predicted surface effects are very model dependent. Unfortunately, present LEED theory is insufficiently precise to allow determination of the spacing normal to the surface. Hirth pointed out, however, that electron diffraction data from small metal droplets yields such information; there seems to be an increase in lattice spacing for very small particles, which is also found in X-ray work on very fine powders.

Later in the discussion, Ehrlich reported that analysis of intensities in LEED work may soon progress to the point where one could deduce the location of the atoms in an adsorbed layer over substrates, as well as their structure. In connection with surface restructuring, Cowley pointed out the possibility of the formation of an ordered 2-dimensional superlattice under the action of long-range oscillating potentials that may be different within the surface than those within the bulk.

March reported some unpublished work with R. C. Brown on the surface tension of liquid metals. It appears that ordering influences are sometimes greater at the surface than in the bulk, and these bear a remarkable correlation to the temperature dependence of the surface tension. Hopefully, the ordering may be explained by noncentral forces that can be related to the dielectric response function. And, as pointed out by Frenkel years ago, there is a relation, $K_T \sigma$ = constant, between the isothermal compressibility, K_T, and the surface tension, σ, of liquids. March then proceeded to develop a theoretical explanation of a relationship between these two properties. Beginning with the expression for the energy expansion of the liquid which contains terms $E_O(\rho) + \lambda \, (\mathrm{grad}\, \rho)^2/\rho$, [where $E_O(\rho)$ is the energy density of a free electron gas, λ is a constant, and ρ is the electron density], an expression may be derived for the surface tension $\sigma = \int_0^{\rho_O} f(E_O(\rho), E_W) \, d\rho$ [where ρ_O is the mean electron density in the bulk, and f is a function of E_O and the electronic work function E_W of the Jellium model] which does not depend on the form of variation of ρ at the surface of the liquid. Thus, σ may be related directly to properties of the bulk liquid rather than the properties of the density profile with the single assumption that the energy of the electron gas due to its inhomogeneity at the surface may be treated as the leading term in a gradient expansion.

COMPUTER SIMULATION

Discussion moved to the calculation of grain-boundary configuration. Weins commented on the need to incorporate entropy effects at high temperatures, and suggested that a study of impurity effects in boundary diffusion could give interesting results and allow comparison with experiment. On sensitivity of results to the potential used, Doran and Weins commented that truncation distance can be important, especially for point defects, but Weins felt that boundary structures are relatively insensitive to this parameter.

Answering a question of Shewmon's about electron-microscope observations, Dahl stated that there was agreement of his computer results with the ledge structures observed and showed some slides demonstrating boundary stability against thermal excursions. Bullough suggested that work be done on systems that expose geometrical principles, such as low-angle and martensitic boundaries, particularly a study of motion of the latter.

OTHER PROBLEMS

The chairman invited discussion of other phenomena involving surfaces and interfaces that might be fruitfully tackled using computer techniques. Hirth suggested a problem first raised by J. R. Rice that may be amenable to such an approach, which involves, in essence, crack motion by successive nucleation of crack nuclei. In the stress field of a crack such as that shown in Fig. 2, dislocations A and B will tend to run toward the advancing tip and could coalesce to form a dislocation C which, by virtue of its strain field, will increase the tensile stresses, σ, ahead of the advancing crack tip. This could be an important propagation mechanism.

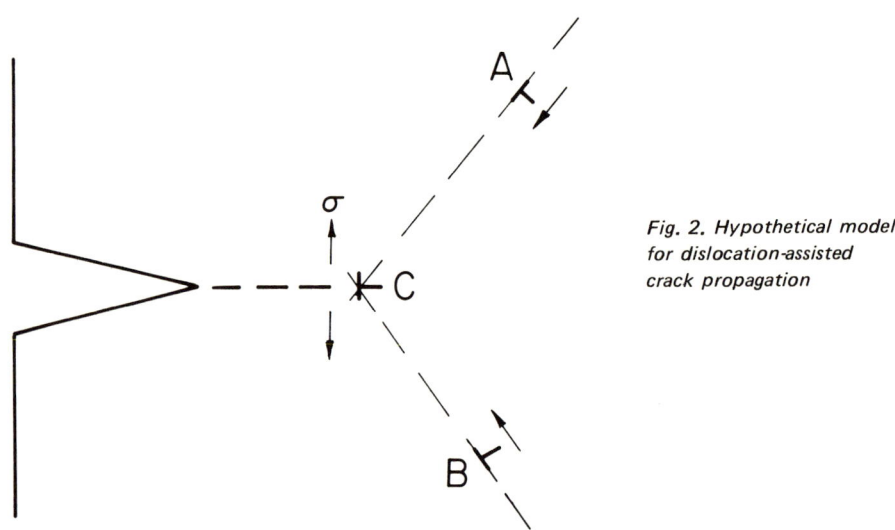

Fig. 2. Hypothetical model for dislocation-assisted crack propagation

Hirth made a second suggestion relating to Ehrlich's work on adatom diffusion and reflection from a ledge. This effect might be material sensitive and could be interesting experimentally. Other evidence from growth of platelets or whiskers indicates that diffusion *around* kink corners may be easy for some materials, i.e. potassium, cadmium, silver, and mercury.

deWette reported that he had investigated crystallization of a two-dimensional array of particles interacting via a (6,12) potential and showed a movie of this phenomenon. Starting with a random array (and hence a high temperature) of 400 particles with free surfaces and allowing relaxation with incorporation of a damping term which served as an energy drain that could be turned on or off, the formation of voids and their annealing out with formation of crystallites could be clearly seen as the array "cooled".

SUMMARY

Interatomic potentials devised for bulk problems such as crystal structure determination and point defect configuration have been used fruitfully in the simulation of internal surfaces such as grain boundaries.

However, the local disruption of electron density is much greater at a free surface. Considerable work remains to be done in developing a theory of interatomic forces for this case, and until this is done the results of computer simulation of surface problems will suffer from uncertainty in the empirical potentials used. Theoretical approaches that predict correlations between measurable surface properties, and which therefore have an experimental touchstone, such as the correlation between compressibility and surface tension outlined by March, could be a most useful starting point in approaching this problem.

Part Six

COMPUTER TECHNIQUES

AGENDA DISCUSSION: COMPUTER TECHNIQUES

Joe R. Beeler, Jr. *
North Carolina State University
Raleigh, North Carolina

G. L. Kulcinski **
Battelle, Pacific Northwest Laboratories
Richland, Washington

The previous Agenda Discussions dealt largely with problems and ideas concerned with the construction of atomic potential energy interaction functions (potential functions) and the results of computer experiments on crystal lattice defect properties. In these discussions the primary interest was centered on the intrinsic merit of the potential functions considered and the realism of the results obtained for defect properties based on these potentials. In contrast, the Computer Experiment Techniques session was concerned with the practical aspects of using computer experiments to compute crystal lattice defect properties and to simulate the lattice dynamics behavior of a crystal. The primary interest was focussed on how to perform a computer experiment correctly, given a set of potential functions, rather than on the merit of the potential functions. Said in another way, the primary interest was focussed on how to be

---
*Chairman.
**Secretary.

certain that one was actually computing what one intended to compute. It is usually easy to compute the properties of a defect that disturbs crystal perfection in a simple and local way because, in this instance, there are only a few possible configuration states and the equilibrium state configuration develops quickly in a simulation of the defect. However, when the defect disturbs the crystal perfection either in a complex way or extensively, then there are usually many possible configuration states with comparable state energies. In these instances it is difficult to be certain that one has indeed simulated the *intended* equilibrium state, in either a static or a dynamic simulation.

It is important to know and use the correct techniques for computing defect properties in at least two areas of great current interest: (1) application of computer experiment methods in solving engineering problems, test design problems and experiment design problems; and (2) construction of potential functions for use in defect property simulations. Computer experiments on defect properties are useful in engineering applications and in experiment design, for example, because they enjoy a range of applicability beyond that available to continuum models. The matching of a potential function to experimental data on defect properties always entails a series of defect property calculations in order to discover the appropriate value assignments for adjustable parameters in the potential function. In each instance it is vital that the computer experiment technique used be correct.

The computer techniques agenda discussion was devoted to six topics. Each topical discussion was introduced and led by a person or persons currently active in the area concerned. The topics discussed were the following:

1. Accelerated convergence of defect equilibrium configuration calculations. (Quasidynamical and dynamical methods)
2. Dynamical method integration schemes.
3. Simulation of temperature effects. (Dynamical method)
4. Migration energy computation techniques. (Variational, dynamical, and quasidynamical methods)
5. Impurity atom calculations. (Variational, dynamical, quasidynamical methods)
6. Phase transformation simulation. (Dynamical and Monte Carlo methods)

In order to define terminology, it is necessary to describe briefly each type of computer experiment method that was concerned in the Symposium. As the computer experiment techniques available are not often described comparatively in detail, it is usually difficult to discuss computer experiment results (obtained using different computer methods for the same general problem) in the proper perspective. Curiously enough, even though most of the members of this Symposium were professionally involved in computer experiments, each individual was usually immediately concerned only with one or two computer experiment methods. More and more workers will realize, through experience, the advantages of being able to employ all of the available techniques.

1 COMPUTER EXPERIMENT METHODS

Five different types of computer experiment methods were used to obtain the defect property results reported at this Symposium. They were the dynamical, quasidynamical, variational, Monte Carlo, and lattice statics methods. An additive, pairwise atomic interaction model is assumed for computing structure-dependent forces and energy contributions. Structure-independent forces and their associated energy contributions usually are accounted for by specification of appropriate boundary forces and the atomic volume. Detailed descriptions of the first four methods and their applications are given in Ref. 1. The lattice statics method is explained in Refs. 2-6.

1.1 Dynamical Method

Dynamical method computer experiments are used to simulate the movement of individual atoms in an assembly of atoms on the basis of Newtonian mechanics. Given the initial position and velocity vectors for each atom in an assembly of atoms, the dynamical history of the assembly is generated. This is done by numerically integrating the $3N$ simultaneous equations of motion for the coordinate components of the N atoms in the assembly. In this numerical integration process, the $3N$ velocity components are also determined. Hence the dynamical method generates a phase space point trajectory as a function of time for the atom assembly concerned.

1.2 Quasidynamic Method

In the dynamical simulation of a crystal, each atom vibrates continually about its static equilibrium position. The array of static equilibrium atom positions can be obtained by computing the time-average position of each atom. In instances where only the time-average positions are desired, the dynamical method can be amended so that the atom motion is progressively damped as the atoms approach their static equilibrium positions. This type of computational approach is called the quasi-dynamical method.

1.3 Variational Method

Static equilibrium configurations can be obtained also by using the variational, lattice statics, or Monte Carlo methods. In the variational method, each atom is treated in succession, one at a time, and moved to a position of zero 'extrapolated' force. In this process, a force F_1 is computed at the current position of the atom, r, and a force F_2 is computed, also, at a nearby point, $r + \Delta r F_1/|F_1|$, on a ray emanating from r in the direction $F_1/|F_1|$. The constant Δr is usually assigned to be about 0.6 percent of the interatomic distance. A zero force position can then be estimated via linear extrapolation or interpolation, on the basis of r, Δr, F_1 and F_2. This process is repeated until each atom experiences a negligible net force.

1.4 Monte Carlo Method

At thermal equilibrium, the probability that an assembly of atoms will achieve the energy ΔE required to transform from one particular equilibrium configuration to another is given by the Boltzmann factor $\exp(-\Delta E/kT)$. This circumstance is utilized as the basis for the Monte Carlo method for computing equilibrium defect configuration states as a function of temperature. Each atom in the crystal is considered in turn, one at a time. A new 'trial' position is sampled at random for the atom, within a small spherical region about its current position, and the change in crystal energy associated with movement to this trial position is computed. A random number is then selected from the interval (0,1) and compared with the Boltzmann factor for the trial move. If the random number is smaller than the Boltzmann factor the move is made; otherwise the atom is retained at its original position. This process is repeated until a sufficiently large ensemble of configuration states has been obtained.

1.5 Lattice-Statics Method

This method was originally introduced by Kanzaki[2], modified by Hardy[3], and reformulated by Flinn and Maradudin[4]. Its advantage is that all atom displacements in a defect static equilibrium configuration are solved simultaneously, in the process of minimizing the defect crystal energy, without using an iterative process. Other methods mentioned above for obtaining static equilibrium configurations require an iterative process. Hence, the lattice-statics method is a relatively fast and inexpensive method. However, as its solution is exact only in the harmonic approximation, the validity is questionable for large relaxation displacements.

In this approach, the computation of the atom displacements in a single defect configuration is converted into the calculation of atom displacements in an infinite superlattice of noninteracting defects identical to the single defect of interest. Each defect is located at the center of a 'supercell' containing N primitive unit cells of the crystal structure concerned. Since the supercells are identical, this is tantamount to considering one defect at the center of an isolated supercell and imposing periodic boundary conditions.

The direct space atom displacements in the defect atom configuration are expressed formally as Fourier series in terms of the static phonon amplitudes for the supercell. These static phonons are the N allowed Bloch states in the first Brillouin zone. The phonon amplitudes are solved for as follows: First, the energy of the supercell is expressed formally in terms of the phonon amplitudes to harmonic order. Second, the energy is minimized with respect to each of the N amplitudes as required by the equilibrium condition of the static lattice. In the harmonic approximation, each phonon amplitude becomes decoupled and can be solved for independently. In this way, the original problem of solving a $3N \times 3N$ array of linear equations for the direct space atom displacement components is reduced to the solution of N 3×3 arrays, one for each phonon amplitude, each of which is explicitly soluble. This is the crux of the method. Having solved for the phonon amplitudes, the N-term Fourier series for each direct space displacement can be evaluated to provide the atom configuration for the defect. The defect relaxation energy is an explicit function of these displacements.

2 ACCELERATED CONVERGENCE (ARTIFICIAL DAMPING OF ATOM MOTION)

Two general circumstances arise in equilibrium-state simulation in which it is convenient to artificially damp atom motion as the system comes progressively closer to an equilibrium state. One instance is simulation of a static equilibrium defect configuration state. The second instance is simulation of condensation from a fluid to a solid state.

Gibson et al[7] introduced the quasidynamical method for static equilibrium configuration calculations wherein the velocity of each atom is set to zero each time the kinetic energy of the entire crystallite reaches a maximum. This procedure is suggested by the circumstance that the velocity magnitude of a mass point in simple harmonic motion is maximum at the time the mass point is passing through the static equilibrium position (zero net force position). Several variations of this basic idea have evolved. As a mass point, in simple harmonic motion, approaches the static equilibrium position, the dot product of its velocity and the restoring force is a monotonically decreasing positive quantity which vanishes at the equilibrium position and then becomes negative. On this basis, Evans and Beeler introduced the idea of setting the velocity of any atom to zero whenever the velocity-force dot product becomes negative, in a static equilibrium calculation. This individual atom criterion for damping leads to a significantly faster convergence than the maximum total kinetic energy criterion. An individual atom criterion for damping is especially necessary when two or more markedly different atom masses are involved.

De Wette showed that damping is usually required even in a purely dynamic simulation, when one wishes to simulate condensation from a random to a regular array. An unphysical, local potential energy excess can be introduced when a random initial array is constructed via straightforward statistical sampling. In such instances, damping is introduced directly into the equations of motion to selectively depress large-atom velocities. As the system approaches a thermal equilibrium dynamical state, damping is removed from the equations of motion. De Wette emphasized that condensation simulations seem to require a long time. This circumstance motivates the use of artificial damping.

3 INTEGRATION SCHEMES FOR THE DYNAMICAL METHOD

The four different integration schemes for use in dynamical calculations that were discussed include the central difference, Euler-Cauchy, simple predictor-corrector, and Nordsieck methods. The essence of each method can be described adequately in one dimension with the notation, x as the position coordinate, v as the velocity, $a = F/m$ as the acceleration where F is the net force and m is the atom mass. The net force on a given atom is a function of the coordinates of all atoms in the assembly. The one-dimensional integration scheme can be applied immediately to each of the $3N$ coordinate components for the N atoms in a three-dimensional crystal.

3.1 Central Difference Method

The central difference method used by Gibson et al, computes the atom coordinate components at integer multiples of the integration time step Δt and atom velocity components at half-integer multiples of Δt. The fundamental known quantities on which each integration step is based are $x(t)$ and $v(t - \Delta t/2)$. The net force $F(t)$ is determined by the positions of all the atoms at time t, and $a(t) = F(t)/m$. Given $a(t)$, the advanced time quantities $v(t + \Delta t/2)$ and $x(t + \Delta t)$ are given by

$$v(t + \Delta t/2) = v(t - \Delta t/2) + a(t)\,\Delta t$$

$$x(t + \Delta t) = x(t) + v(t + \Delta t/2)\,\Delta t$$

3.2 Euler-Cauchy Method

In the Euler-Cauchy method, an iterative cycle is repeated until no change occurs in the coordinates and velocities of an atom. The fundamental known quantities are $x(t)$ and $v(t)$, both evaluated at integer multiples of the integration time step Δt. For the first iteration in the calculation of $x(t + \Delta t)$ and $v(t + \Delta t)$ one has,

$$x^{(1)}(t + \Delta t) = x(t) + v(t)\Delta t + (1/2)a(t)(\Delta t)^2$$

then for the n-th iteration ($n \geqslant 2$)

$$a^{(n)}(t + \Delta t) = F(x^{(n-1)}(t + \Delta t))$$
$$v^{(n)}(t + \Delta t) = v(t) + (\Delta t/2)(a(t) + a^{(n)}(t + \Delta t))$$
$$x^{(n)}(t + \Delta t) = x(t) + (\Delta t/2)(v(t) + v^{(n)}(t + \Delta t)) \quad .$$

3.3 Simple-Predictor Corrector Method

The simple-predictor corrector method used in Rahman's initial paper on liquids is a good example of this method.[8] The notation used is that $x_p(t)$ and $v_p(t)$ are predicted values and $x_c(t)$ and $v_c(t)$ are corrected values. The fundamental information is $x(t - \Delta t)$ and $v(t)$. On this basis, for the first iteration,

$$x_p(t + \Delta t) = x(t - \Delta t) + 2\Delta t\, v(t)$$
$$a^{(1)}_p(t + \Delta t) = F(x_p(t + \Delta t))$$
$$v_c^{(1)}(t + \Delta t) = v(t) + (\Delta t/2)[a_p(t + \Delta t) + a(t)]$$
$$x_c^{(1)}(t + \Delta t) = x(t) + (\Delta t/2)[v_c(t + \Delta t) + v(t)]$$

For the n-th iteration, $n \geqslant 2$,

$$a^{(n)}_p(t + \Delta t) = F[x_c^{(n-1)}(t + \Delta t)]$$
$$v_c^{(n)}(t + \Delta t) = v(t) + (\Delta t/2)[a_p^{(n)}(t + \Delta t) + a(t)]$$
$$x_c^{(n)}(t + \Delta t) = x(t) + (\Delta t/2)[v_c^{(n)}(t + \Delta t) + v(t)]$$

3.4 Nordsieck Method for Newton's Equations

Nordsieck[9] developed a method for integrating ordinary differential equations which is highly suitable for automatic computation. It proceeds as follows: Let $q_o(t)$ be one of the 3N coordinate components $x_1, \ldots, x_{3N}$ for an assembly of N atoms. The first five scaled derivatives of $q_o(t)$ are,

$$q_1(t) = (dq_o/dt)\Delta t$$

$$q_2(t) = (1/2)(d^2q_o/dt^2)(\Delta t)^2$$

$$q_3(t) = (1/6)(d^3q_o/dt^3)(\Delta t)^3$$

$$q_4(t) = (1/24)(d^4q_o/dt^4)(\Delta t)^4$$

$$q_5(t) = (1/120)(d^5q_o/dt^5)(\Delta t)^5 \quad .$$

In the Nordsieck method, as applied to Newton's equations of motion, the predicted values for the q_n at $(t+\Delta t)$ are obtained through a Taylor series expansion,

$$q_o(t + \Delta t) = q_o(t) + q_1(t) + q_2(t) + q_3(t) + q_4(t) + q_5(t)$$

$$q_1(t + \Delta t) = \phantom{q_o(t) + {}} q_1(t) + 2q_2(t) + 3q_3(t) + 4q_4(t) + 5q_5(t)$$

$$q_2(t + \Delta t) = \phantom{q_o(t) + q_1(t) + {}} q_2(t) + 3q_3(t) + 6q_4(t) + 10q_5(t)$$

$$q_3(t + \Delta t) = \phantom{q_o(t) + q_1(t) + q_2(t) + {}} q_3(t) + 4q_4(t) + 10q_5(t)$$

$$q_4(t + \Delta t) = \phantom{q_o(t) + q_1(t) + q_2(t) + q_3(t) + {}} q_4(t) + 5q_5(t)$$

$$q_5(t + \Delta t) = \phantom{q_o(t) + q_1(t) + q_2(t) + q_3(t) + q_4(t) + {}} q_5(t) \quad .$$

The predicted value of each q_o is then used to compute the associated predicted force component F for the atom concerned, in the usual way, and this force is used to evaluate a displacement function

$$\phi = [(1/2)(F/m)(\Delta t)^2 - q_2(t+\Delta t)] \quad ,$$

where m is the atom mass and $q_2(t+\Delta t)$ the predicted value of q_2. The corrected values for the q_n are then evaluated using the 'dynamical' information contained in ϕ. They are as follows:

$$q_0^c(t + \Delta t) = q_0(t + \Delta t) + c_0\phi$$

$$q_1^c(t + \Delta t) = q_1(t + \Delta t) + c_1\phi$$

$$q_2^c(t + \Delta t) = q_2(t + \Delta t) + c_2\phi$$

$$q_3^c(t + \Delta t) = q_3(t + \Delta t) + c_3\phi$$

$$q_4^c(t + \Delta t) = q_4(t + \Delta t) + c_4\phi$$

$$q_5^c(t + \Delta t) = q_5(t + \Delta t) + c_5\phi$$

The constants, $c_0, c_1, \ldots, c_5$ are as follows: $c_0 = 3/16$, $c_1 = 251/360$, $c_2 = 1$, $c_3 = 11/18$, $c_4 = 1/6$ and $c_5 = 1/60$. Two points need to be emphasized. (1) Calling the scheme above a fifth order method, the values cited for the constants $c_0, \ldots, c_5$ are pertinent only to that order. If for any reason a lower or higher order method were used, the associated constants $\{c_1\}$ would be entirely different. A table of $\{c_1\}$ for fourth, fifth, sixth and seventh order methods is given in Appendix C of Gear's report[10]. (2) The memory requirements of this fifth-order method are 6 x 3N for the q_n and 3N for ϕ. A lower-order method is indicated when the available computer memory is small.

The constants $c_0, c_1, \ldots, c_5$ are determined so as to provide stability with a minimum degradation of accuracy. They are modifications of the constants which result directly when the standard Taylor series remainder functions are used in the expansions for the q_n.

Accurate integration of Newton's equations is required in the dynamical simulation of lattice vibrations, in computing entropy contributions to the crystal free energy, in the dynamical simulation of nucleation events, and in the dynamical simulation of phase transformation.

Beeler has tested the central-difference, Euler-Cauchy, Simple predictor-corrector, and Nordsieck methods against the exact solution for a simple harmonic oscillator. Integration time steps ranging from 0.25×10^{-14} sec to 1.0×10^{-14} sec were used for dynamical histories of 142 cycles. In all instances (all Δt), the central-difference method gave vibrational frequencies greater than the true frequency. In contrast, the Euler-Cauchy and Simple predictor-corrector methods gave vibrational frequencies less than the true frequency. Energy was not conserved (constant) in the central-difference method. The kinetic energy change per time step was not equal to the potential energy change per time step. Note that an interpolated value of the kinetic energy at integer multiples of the time step was used in the comparison, not the kinetic energy at half-integer multiples of the time step. The magnitudes of the kinetic and potential energy changes per time step were equal when the Euler-Cauchy and Simple predictor-corrector schemes were used, but these methods gave the wrong magnitude for the change. The Nordsieck method was tested using $\Delta t = 0.25 \times 10^{-14}$ sec and the results were the same as the exact solution results.

Gear has recently reviewed numerical methods for integrating ordinary differential equations and concludes that Nordsieck-type methods (N-methods) are best for high-accuracy problems. Rahman has adopted these integration methods in his simulations of liquid states[11]. On the basis of Gear's analysis of high-accuracy methods and the inadequacy of the other methods in describing a harmonic oscillator over a long time span, it would seem that new dynamical programs should be written to include a Nordsieck-type integration scheme option. The Nordsieck methods have the added advantages of being self-starting and capable of utilizing a variable integration time step without loss of stability or accuracy.

4 INITIAL CONDITIONS FOR TEMPERATURE SIMULATION

Temperature-simulation discussions were introduced by Robinson, Perrin, and Gehlen. Robinson described the method used by Torrens and Robinson (this conference page 423) in estimating the effect of temperature on the development of a collision cascade. In this instance, a Gaussian distribution for the atom displacements from their time-average equilibrium positions was assumed. Each Cartesian coordinate of the displacement for an atom was selected independently using a straightforward Monte Carlo procedure. The amplitude of the displacement was usually selected according to Debye theory. When actual empirical data, such as X-ray Debye-Waller factors, were available, they could be used to specify the atom displacement amplitudes.

Perrin described the method used by Tsai, Bullough, and Perrin[12] in their simulation of point defect migration. They initially assigned

Maxwellian velocity components to each atom, choosing the atoms in random order, with a mean kinetic energy of kT in each degree of freedom. The initial atom coordinates assignments were the equilibrium position coordinates. The total energy of $3kT$ became quickly distributed between the kinetic and the potential energies of the lattice as the simulation proceeded in time from these initial conditions. After equilibration, the total mean kinetic energy was about $(3/2)kT$ and the velocity distribution remained Maxwellian, i.e., the mean kinetic energy in each degree of freedom was reduced to about $(1/2)kT$. At the same time, the atom displacement distribution quickly assumed a Gaussian form.

Gehlen described the method used, by himself and Beeler, to introduce temperature effects in the study of dislocation kink motion (this conference page 475). This method immediately achieves the correct partitioning between the average kinetic and potential energies. Each atom was assumed to be an Einstein oscillator. An effective "harmonic" force constant was obtained for the assumed anharmonic potential interaction between the atom and its first and second neighbors as a function of displacement amplitude. This was done using a straightforward Monte Carlo method. For the polynomial potential function form used, the partitioning fractions for the average potential and kinetic energies can be obtained using the Virial Theorem. Given the temperature, the force constant, and partitioning function information, one can then compute maximum oscillator displacement r_o and velocity v_o for the Einstein oscillator. The initial displacement and velocity were selected for each atom via a Monte Carlo method, from $r = r_o \sin \theta$ and $v = \omega v_o \cos \theta$. The angle $\theta = \omega t$ was selected at random between 0 and 2π. The direction for the colinear initial displacement and velocity assignments was selected at random. This method allows one to specify the temperature and always have the correct partitioning between average kinetic and potential energies. The initial velocity distribution is initially an arc cos distribution but rapidly assumes the form of a Maxwellian distribution. Gehlen and Beeler worked with a bare 855-atom bcc crystallite and used Johnson's α-iron potential.

5 MIGRATION ENERGY CALCULATIONS

The discussion on migration energy calculations was introduced by Johnson who also led the discussion on migration energy calculation techniques.

Johnson pointed out that most workers in point defect physics had assimilated the results of the early computer experiments on atom migration paths in an almost subliminal way. He then remarked that, as a consequence of this, many of the original results are now regarded as being intuitively obvious. However, the persons who first did computer experiments on defect motion did not have the advantage of this immediate "intuitive" group of atom migration details. As a consequence, much effort was devoted in the early computer experiment work to computing energy contour maps associated with the movement of an itinerant atom in regions between normal lattice sites. These mappings are the foundations for much of the currently "intuitively obvious" knowledge on atom migration paths.

Johnson restricted his discussion to migration energy calculations for vacancies and interstitials, predominantly in the fcc crystal structure; he included ample reference to the bcc structure to give a complete exposition. Simulation of the positional interchange between a vacancy and one of its nearest-neighbor atoms provides an example in which the computational technique is straightforward. In all monatomic cubic systems, the itinerant atom concerned moves along the close-packed line direction joining its initial center position and that of the vacancy. Johnson tacitly assumed the use of a nonoscillatory potential function. Later in the session, he amplified this aspect of migration energy calculations to state that for complicated potentials, the appropriate crystal configuration state could not always be specified by defining the position of a single atom.

Johnson described the essential procedure required when one enters any previously unexplored region of defect migration study by using interstitial atom configuration and migration calculations as an example. In short, this procedure consists of two parts: careful energy contour mapping, and utilization of several computer experiment method approaches.

Given a potential function, there is a minimum distance of approach for two atoms in a static equilibrium configuration which, essentially, is a stability limit. Johnson noted that in an fcc crystal, no two atoms ever seemed to approach each other closer than 1.2 hlc, with $\sqrt{2}$ hlc being the normal interatomic distance. Hence, in the case of interstitial atom migration, he assigned a sphere of radius 0.6 hlc about each normal lattice site. He then defined any atom whose center was positioned within such a sphere as being a normal atom and any atom whose center laid outside any of these spheres as being an interstitial atom. This defines a simple interstitial atom position. The so-called split-interstitial atom configurations are those in which two atoms have their centers at the end of a diametrical chord through a normal lattice site sphere and in which no atom center is contained within the normal atom sphere. In any interstitial migration event, then, no individual atom ever moves more than 0.6 hlc during the movement of the center of the interstitial configuration from one lattice site to another.

In his own work, Johnson approached the problem of interstitial motion by mapping the energy contours for itinerant atom positions within the interstitial region between normal-site spheres.[13] This was done on a three-dimensional basis and the results were plotted on a series of transparent plastic sheets. A complete three-dimensional view of the energy contours was obtained by stacking the transparent sheets with the appropriate intervening spatial elevation, between sheets. In this way, it was possible to define the interstitial atom path associated with the transformation from one interstitial form to another, as well as the probable central structure of the associated saddle-point configuration.

Johnson's mapping data were obtained by placing the interstitial atom at an intermediate point in the interstitial space and then following the "motion" of the atom through the subsequent iterations of the variational calculation as the crystal relaxed to a lower energy configuration. Although this method is an estimate, its use is justified by the close similarity of the results thus obtained with those run in a fully dynamic simulation of the same event with the same initial starting atom positions. This is true for nonoscillatory potentials.

In simple interstitials of the tetrahedral and octahedral types, an octahedral interstitial migrates from one octahedral site to another by passing through a tetrahedral interstitial saddle point. The normal thing is to associate a [111] split-interstitial in fcc with the tetrahedral simple interstitial and a [100] split-interstitial with the octahedral simple interstitial. This reasoning would lead to a [100]-[111]-[100] migration sequence for the [100] split, in analogy with the simple interstitial migration sequence.

The migration mechanism which Johnson discovered from the energy contours and interstitial motion simulation, however, was a cornering mechanism in which the movement of the interstitial atom is largely confined to a (100) plane and the center of the interstitial configuration moves a first-neighbor distance.

Vineyard noted that this cornering mechanism for [100] split-interstitial migration was also observed in fully dynamic simulations of Frenkel pair production and annihilation. This note and a subsequent question by Dahl underscore the general utility of approaching a migration event simulation by using two or more different simulation techniques. Dahl emphasized the difficulty of knowing how to proceed in simulating interstitial motion in a disordered region such as that near a grain boundary or a dislocation. This point was seconded by Doran.

Johnson responded that for disordered regions and complicated potentials, one could not rely on trying, in effect, to define a configuration state by specifying the position of a single atom. This circumstance makes it very difficult to map energy contours using only the variational method. This is true because it is very difficult to know how to arrange intermediate positions for more than one atom, in a self-consistent way, using the variational method.

As shown by Beeler (this conference page 339), a multiple-valued configuration energy for the vacancy-carbon complex occurred when the configuration was specified by stating the carbon atom position relative to the vacancy. In this instance, mapping failed to disclose the lowest energy configuration. The lowest energy configuration was, in fact, found by employing randomized initial atom positions (a pseudo-temperature effect) and a dynamical method simulation. This type of mixed procedure probably would be generally advantageous in disordered region problems.

Nearly always, several migration channels (paths) are available to an atom when it starts out from its initial position on a migration journey. When migration energies are computed on a semiautomatic basis, without mapping, by holding the migrating atom in each of a succession of parallel planes between the initial and final position, the results may be, therefore, a mixture of data for two or more channels. For example, depending on the scheme used, automated variational calculations for vacancy motion in a bcc structure give data for migration in a [001] channel during, roughly, the first third of the migration simulation. Then, just below the saddle point, there is an abrupt shift to the correct [111] channel. This correct channel is followed during the middle third of the migration but there is an abrupt shift back to the [001] channel during the final third of the migration simulation. Dynamical simulations show that the [111] channel actually is followed during the complete migration process, and so does a mapping procedure. Hence, automated procedures for migration energy calculations must be carefully designed and used.

6 IMPURITY ATOM CALCULATIONS

The discussion on impurity atom calculations was introduced and led by Wilson. The methods used in impurity atom calculations turn out to be a subset of the general procedure outlined by Johnson for the movement of atoms in the interstitial region. Again care is required to avoid specifying a configuration in terms of the position of a single atom, and exploration for configurations and migration modes should be performed using at least two different computer experiment methods. In the case of strong chemical binding between the impurity and the host atoms, randomized initial atom positions are recommended to fully sample the interaction space.

As impurity-atom-host atom potentials had not been discussed in detail, Wilson elaborated on the methods used to obtain and test impurity-atom-host atom potentials. In response to questions from the audience, he stated that the state of the art is such that the arithmetic mean or geometric mean models for constructing an interaction potential for two different atom types are the approximations used most frequently. In addition he outlined his procedure for constructing an interaction potential for two different atom types using Wedepohl potentials and also using molecular orbital calculations.

7 PHASE TRANSFORMATIONS

A discussion of phase-transformation simulation techniques was initiated by a question from Roger Chang on whether or not the Johnson Potential could be modified such that α-iron (bcc) would convert to γ-iron (fcc) at 910 C in a dynamical, temperature effect simulation calculation. This naturally opened the way to questions on what are appropriate boundary conditions for such a calculation and what are appropriate mechanisms for initiating such a transformation.

Both a bare bcc crystal and a bare fcc crystal can be sustained by Johnson's potential in either static or dynamical equilibrium states. The crystal energy in the fcc case is larger than that in the bcc case for the same number of atoms at 0 degrees K. The interatomic distance for the fcc structure is about 5.5 percent larger than that in the bcc structure. Experimentally it is observed that at the bcc-fcc transition temperature the fcc interatomic distance of iron is about 2.4 percent larger than the bcc interatomic distance. On this basis, it would seem that the Johnson potential, as it stands, is close to describing the two crystal structures appropriately in a qualitative sense, and could allow conversion to a fcc structure simply on the basis of large thermally induced displacements.

In this regard, Bullough mentioned that additional information on the interatomic interaction pertinent to a bcc-fcc transformation in iron might be obtained from high-temperature phonon measurements in manganese-doped iron. In this material, the γ-phase can be retained down to a low temperature.

Beeler suggested that a bare crystal should be used in simulating such a transformation because the imposition of other than free boundary conditions would influence the transformation in unknown ways. Maradudin proposed that the adoption of periodic boundary conditions would be permissible and pointed out that this gave good results for order-disorder transformations. De Wette reminded the audience of the motion picture he had shown earlier on the transformation from a liquid to a solid state in which free boundary conditions had been used for a slab geometry. His argument was that periodic boundary conditions were too restrictive to allow the extensive nonperiodic rearrangements which were observed in his dynamical simulations of a structural transformation. De Wette commented that his group had attempted a fully three-dimensional liquid-crystal simulation starting with a liquid configuration due to Rahman, in the shape of a cube. The density was almost that needed for the solid crystalline form. They then removed the periodic boundary constraints and replaced them with completely free boundary conditions. The effect was that the atom assembly first formed itself into a ball in the initial process of transformation, while slightly raising the density. Upon lowering the temperature, the system appeared to be frozen into an amorphous configuration. Transformation to further order seemed to proceed exceedingly slowly and was hard to diagnose. Fluctuation phenomena are probably required to reach the configuration of lowest energy. On the basis of this experience, de Wette felt that an actual three-dimensional crystallization process takes a real time period, unattainable by present-day computer dynamic techniques. Coming back to the question of boundary conditions in these problems, he felt that if one impresses a certain symmetry and orientation on the system, as is done with periodic boundary conditions, such conditions are most likely to introduce unnatural features and constraints into the transformations.

Vineyard pointed out that the selection of particular boundary conditions depended upon what one wished to calculate. If the goal were to watch the process of a phase transformation, which involves watching nucleation and growth, then one must do a dynamical calculation and one is faced with lots of problems on what particular boundary conditions to adopt. He felt that the problem could be done, in principle, with either free boundary conditions or periodic boundary conditions. The use of periodic boundary conditions forces one to work at a fixed volume determined by the periodic distance. Such a constant-volume system should eventually transform provided a sufficient amount of heat is added to it. In such an instance, he went on, the transformation would be observed at constant volume and that might be quite unphysical. The only way out of that difficulty would be to study the transformation at several different fixed volumes.

Vineyard then commented on the case where one wants to know what the equilibrium crystal structure is as a function of temperature but is not interested in the dynamics of the process. In this instance he recommended that a Monte Carlo calculation on the partition function for free energy be performed to determine the minimum free energy and the associated crystal structure. He suggested that this procedure would probably be much more economical in computer time and that, given appropriate potential functions, one could determine equilibrium phases as a function of temperature.

Hirth pointed out that a number of phase transformations can be represented by partial dislocation motion giving rise to shears, usually accompanied by atomic relaxations normal to the glide plane. By the use of boundary conditions that anticipate this partial dislocation motion by forcing the relaxation to take place normal to the prospective glide plane, one might induce a transformation.

In some special transformations, such as the cobalt fcc to hcp transformation, one can send in partial dislocations of alternating sign such that there is no net macroscopic shear of the lattice. This could be accomplished by putting in some kind of periodically oscillating shear waves at the surface. This would perhaps induce a transformation without getting into the severe boundary condition problems associated with simulating a large macroscopic shear.

REFERENCES

1. J. R. Beeler, Jr., in *Advances in Materials Research Vol. 4,* edited by H. Herman, John Wiley & Sons 1970, pp. 295-476.
2. H. Kanzaki, J. Phys. Chem. Solids, **2**: 24 (1957).
3. J. R. Hardy, J. Phys. Chem. Solids, **15**: 39 (1960).
4. P. A. Flinn and A. A. Maradudin, Ann. Phys. (N.Y.), **18**: 81 (1962).
5. P. S. Ho, Phys. Rev., **B3**: 4035 (1971).
6. J. W. Flocken and J. R. Hardy, Phys. Rev., **177**: 1054 (1969); Phys. Rev., **175**: 919 (1968).
7. J. B. Gibson, A. N. Goland, M. Milgram and G. H. Vineyard, Phys. Rev., **120**: 1229 (1960).
8. A. Rahman, Phys. Rev., **136**: A405 (1964).
9. A. Nordsieck, Math. of Comp., **16**: 22 (1962).
10. C. W. Gear, "The Numerical Integration of Ordinary Differential Equations of Various Orders", ANL 7126 (1966), Argonne National Laboratory Report.
11. A. Rahman and F. H. Stillinger, J. Chem. Phys., **55**: 3336 (1971).
12. D. H. Tsai, R. Bullough and R. C. Perrin, J. Phys. C: Solid St. Phys., **3**: 2022 (1970).
13. R. A. Johnson and E. Brown, Phys. Rev., **127**: 446 (1962).

Part Seven

CRITICAL ISSUES

CONCLUDING AGENDA DISCUSSION: CRITICAL ISSUES

A. Seeger [*]
Max-Planck-Institut fur Metallforschung
Institut fur Physik
7000 Stuttgart 1
Azenbergstrasse 12, Germany

R. G. Hoagland and E. W. Collings [**]
Metal Science Group
Battelle, Columbus Laboratories
Columbus, Ohio USA

The Concluding Agenda Discussion afforded an opportunity for the Colloquium to collect its thoughts and to attempt to gain an overview of various important issues and subjects which had been, or perhaps should have been discussed. The chairman suggested that during the course of the discussion we could perhaps see better where the work on interatomic potentials and computer simulations is going, and what role this information can play in helping us understand the properties of solids.

[*]Chairman.
[**]Secretaries.

Accordingly the material, which in fact was discussed in two sessions, has been collected together roughly into two groupings entitled respectively: "Philosophy of Potentials and Computer Simulation", and "Determination and Applications of Potentials".

In introducing the first subject the chairman speculated on the extent to which experiments can be replaced by computer simulation, and how one might maintain a proper balance between experiment and analysis. He pointed out that one outstanding issue underlying most of the presentations and subsequent discussions throughout the Colloquium was the question of just how faithful the potential must be for any given application. If every effort has been applied to the construction of a potential that is accurate in full detail in the range of interest, then one possibly has the right to expect that some reliable predictions can be derived from its use in a computer experiment. The chairman stressed that a critical test of such a potential would be to determine whether its use could be extended beyond the scope for which it was originally designed. The justification for taking these steps is philosophical and raises the question of model-dependence. To be specific, if a potential obtained from, say, some experiment on a liquid is successful in one situation, what is the guarantee that it will be applicable in another situation? Clearly, the new predictions could be verified only by experiment, in which case the computational step would seem to be an unnecessary duplication of a full experimental investigation. What then is the usefulness of a computer experiment? The answer is that, first, valuable insights into the validity of the potential and the dominant physical factors can be achieved and second, that such step-by-step verification of a potential lends credence to the results of its continued use, perhaps leading toward the solution of *experimentally inaccessible* problems.

In opening the session, the chairman suggested that there were several points which, although within the scope of the conference, had not yet been specifically discussed. One of these concerned potentials in polar crystals. For example, although we classify both the silver halides together with the alkali-metal halides as polar materials, there are marked and important differences between the two groups, e.g., in their mechanical and point-defect properties. He offered to the conference the challenging problem of elucidating these differences in terms of the atomic potential approach. Recently, R. Bauer[1] has looked into the problem from a quantum-mechanical point of view, and has come up with the suggestion that the above-mentioned differences are related to a strong contribution with cubic symmetry to the interionic potential in the silver halides, as distinct from the contribution with spherical symmetry, which is by far the dominant symmetry in the alkali halides. The contribution with cubic symmetry reflects the important role of the d-electrons in the bonding of the silver halides.

The chairman pointed out that, in the long run, it would be necessary to obtain a deeper understanding of atomic potentials from a quantum-mechanical standpoint. He felt that curve-fitting (or experimental adjustment) of atomic potentials will remain an important technique during the foreseeable future; but even so it would be helpful if this could be done within the framework of a full quantum-mechanical theory (as distinct from, say, second-order perturbation theory). In this regard it is possible that a great deal can be learned from an already established field of study, quantum chemistry. In order to link quantum chemistry with the metallic systems in which we are interested one might look to metallic molecules such as Cu_2, Al_2, Na_2. These are amenable to experimental study (c.f., Drowart[2]), and they are simple enough for quantitative calculations to be made of their properties. This represents a new approach to the problem of bonding between metal ions and may in particular provide information on the relative importance of s- and d-electrons for the bonding properties. At present this approach is being pursued by K. Differt[3]. An interesting and important characteristic possessed by these metallic molecules is their interatomic separation, which is some 15 percent smaller than that in the metal crystal. This property might be of help in probing the r-dependence of the ion-core interaction potential at very small r. The obvious flaws in the approach were the subject of later discussions.

The chairman recalled that since a potential $V(r)$ is a function of r_{nm}, the vector separations of all the atoms, it could, following Vineyard, be thought of as a surface in 3n-dimensional configuration space. Some of the important unresolved problems facing us today in the field of computer simulation could be stated as involving the variation of the potential from one local region in this multidimensional r-space to a distant one. The values of potentials that control collision processes, or control the forces between atoms in strongly distorted crystals (e.g., in dislocation cores), are generally quite different from those that determine the positions and vibrational frequencies of atoms in an undistorted lattice. Consequently any information, whether experimental or theoretical, in addition to that provided by experiments whose results are responsive to the properties of atoms in their normal positions, will be of considerable value. In this spirit the chairman offered the suggestion that as the Colloquium went on to a detailed discussion of atomic potentials and critical problems associated with them, an attempt might be made to bridge the gap between the information obtainable from what are essentially collision processes (i.e., small interatomic separations) and that obtained in the more usual experiments on crystals in equilibrium. Indeed the significance of atom-pair interactions, and of the role played by atomic scattering experiments, were subsequently discussed in detail.

Philosophy of Potentials and Computer Simulation

Discussion from the floor opened with a response by Harrison to Seeger's remarks on the relative merits of *"simple"* versus *"nonsimple"* potentials. With due regard to the obvious inaccuracies inherent in simple potentials, Harrison pointed out that, in one respect, their use could be advantageous since their shortcomings could be rapidly seen, and their effects on the final result could be relatively easily assessed. The converse would be true for a more-sophisticated potential that had been subjected to detailed adjustment for exchange, boundary conditions, etc. In this regard the Lennard-Jones potential should have continuing utility as a simple standard type of potential. Seeger did not disagree with this viewpoint, particularly since he felt in fact that we seemed in no imminent danger from being misled by the use of over-sophisticated potentials. He stressed, however, that computer simulation must be recognized as a calculation, the results of which can never be a more faithful representation of reality than the data initially supplied. Johnson pointed out that the potentials his group were using had been constructed to be as simple as possible, consistent with their being effective in the kinds of problems they were interested in, and that there seems to be a threshold of simplicity below which a potential is of no real value. Harrison's point was essentially that if it is inevitable for our input potential to be deficient in one aspect or another, it is better to be able to assess the inadequacies as easily as possible, rather than fall victim to a hidden artifact in the potential. Later Bullough restated this view, in slightly different terms. He pointed out the hazards associated with the existence of a very complex, but yet still incomplete potential. Such a potential, because of its complexity, will tend rather naturally to "fossilize" and to be used noncritically. Accordingly, Bullough recommended that as a next step from the rather primitive potentials that we now have, we should proceed to *complete* potentials. To further stress this point, Vineyard suggested that it would be worthwhile for people working in this field to distinguish between a *model calculation* and a *simulation* and to adopt this terminology accordingly.

Torrens then recommended that any calculation or computer simulation should be carried out using at least two different potentials. This should be a feasible procedure to adopt, particularly if multiple runs on a computer are to be made in any case. According to Torrens, the use of such a procedure provides some confidence that a calculated result is more than just a computational artifact. Additionally, as suggested later by Wilson, it would be helpful if, for one of these potentials, a universally recognized model potential were chosen. Such a procedure would result in at least one *standard element* being present in all calculations; an obvious advantage in a field beset with a large number of variables. Following Torrens' suggestion, the question was raised by Yoshikawa as to what guidelines should be followed in choosing the two potentials for a calculation; or how should an initially chosen potential be meaningfully adjusted before performing the second calculation. The following discussion indicated that it was impossible to give a useful answer to this question in general terms, since the direction in which one should proceed in modifying a potential would be different for different kinds of problems or classes of materials. A specific answer to the question was given by Rahman within the context of the problem of self-diffusion in liquid metals, and how it responds to certain types of modifications of the potential. Rahman referred to the work of Schiff[4] at Orsay who changed the hard core, or repulsive part of the potential, and studied the effect that this has on the velocity autocorrelation and the process of self-diffusion. Rahman said that there was very little to be learned by repeating a calculation with a new potential if it has little relationship to the first. Modifications of a potential must be made systematically and in a manner appropriate to the particular problem under study. As a second point Rahman stressed that one of the goals of computer simulation need not necessarily be the defining of a unique potential for a given material, but rather a potential, some of the characteristics of which would lead to experimentally recognizable behavior. A very good example of this, as March pointed out, would be a study of the response of the calculated liquid metal structure factor to variation of the steepness of the hard-core repulsive part of the potential, which is responsible for the long-range oscillations in q-space of the structure-factor.

Shewmon, mildly critical of the study of potentials as a pure art form, contended that their existences could best be justified if they were to be used eventually in the treatment of otherwise intractable problems. He mentioned, as an example of a particularly appropriate application of computer methods using potentials, the study of postirradiation annealing processes in which many mechanisms operate simultaneously.

Seeger referred to the inherent lack of generality associated with existing potentials and suggested that experiments and calculations should always be closely coupled. Basinski emphasized that in calculations of point defects, dislocations, etc., since the energies involved are fairly small, great care must be taken in comparing experiment with theory. Unfortunately, calculations are limited to systems for which experimental data are relatively scarce. Accordingly, a plea was made to the experimentalist to help alleviate the situation by working on systems accessible to theory.

Determination and Application of Potentials

In his opening remarks, the chairman had introduced the subject of metallic-molecules and atomic-scattering experiments, particularly of the kind in which energetic metallic atoms are scattered from dense metallic vapors. March had discussed the influence of the form of the potential at short distances and the liquid structure-factor at large q. As was pointed out by Robinson the atomic-scattering technique is beset with both interpretational and experimental difficulties. An example of the former is the difficulty of deriving accurate information from the scattering at low angles; and of the latter, the elaborate instrumentation required to deal with the effects of inelastic collision processes. However, it was Vineyard's opinion that although some anomalous scattering does take place, the dominant strong interactions are governed by potentials which can, in part, be determined to considerable accuracy. The results of atomic-scattering experiments, and the related problem – binding in metallic diatomic molecules – became the subject of considerable discussion. A controversy arose with regard to the kind of information obtainable from such studies, and its relevance to the solid-state situation. There were clearly two schools of thought, the pragmatic and the conceptual: i.e., those who felt that no model of binding in diatomic metallic molecules would help in the solution of an atomic-potential problem in bulk metal, and those who were willing to search for and discuss what useful information could be gleaned from a study of metallic molecules or atomic scattering.

We must, of course, recognize at the start that metallic vapor does not represent the metallic state; and the question, as posed by Ashcroft, is how information from vapor-scattering experiments can be employed in the construction of a potential for a metal. In solid metal, the ion cores are screened by the conduction electrons, in the absence of which the potential well is extremely deep. As March commented, the well-depth in Li_2 of about 2eV is entirely different from the corresponding depth of the minimum in bulk metal which is of the order 0.05 eV. An atomic potential well is, of course, necessary for the cohesion of Li_2; but not for metallic Li, in which cohesion is supplied by the conduction electrons. Wilson believed that it was not possible to derive information for a crystal from the results of molecular studies. Ashcroft and March both said that, while information from atomic scattering experiments might be applied with reasonable validity to nonmetallic solids, the absence of screening electrons in isolated molecules prevented such applicability of the scattering data to the bulk metal. Ashcroft pointed out that the hard-core-like potential associated with atomic scattering originates in the exchange interactions between the core electrons, whereas the hard-core potential in metals derives partially from screening. At about this time it became apparent to the meeting that we were to some extent describing the properties of d-electron metals. March had already suggested that the results of scattering experiments might have significance and some limited usefulness for dealing with metals such as copper which possess hard cores. As emphasized by Enderby, the essential point to be realized, in attempting to couple the results of scattering experiments to the properties of bulk metal, is as follows: scattering experiments provide information on the detailed functional form of the atomic potential for close core-core interactions; accordingly the results are applicable to those bulk metals in which this part of the potential plays the dominant role. Enderby pointed to the results of diffraction experiments on liquid metals. For example, the damping of oscillations of the liquid structure factor in metals such as copper is very much less than that for pseudopotential metals such as sodium. In fact, it can be stated generally that there is a qualitative difference between the liquid structure factor for transition metals (where the results of core-core interactions can be seen) and those for "screening-type" metals. It follows that, although, as Ashcroft pointed out, contributions by s-electrons in the solid cannot be ignored, the results of atomic-scattering experiments on d-electron metals can to some extent carry over to the solid, and provide information on at least a part of the atomic potential very close to the ion core.

Seeger pointed out the technique of inelastic neutron scattering, in which the line width contains information on the anharmonicity of the potential, as a little-used technique for providing information on the potential for moderate displacements from equilibrium. This technique seems capable of filling a gap, since for strong, close-range interactions as in atomic collisions, we seem to be on reasonably good theoretical and experimental ground with pairwise interactions as a valid approximation. On the other hand, for large and intermediate displacements from equilibrium in the solid, there are problems which have not yet been solved. Here, as March indicated, we should think in terms of a nonrigid-ion model; or, as Seeger pointed out, we must learn from electron theorists how to take many-body effects into account.

A number of other techniques and measurements that offer some information on potentials were discussed. Torrens brought up the subject of shock-wave techniques, which provide access to the potential for r values that are 80-90 percent of 1-atmosphere value, provided, as Doran pointed out, a suitable equation of state is available to correct the data for the high temperatures that accompany the high-pressure pulses. An example of the use of high-pressure data to help define the short-range part of the potential is given in Reference 5.

March emphasized the importance of X-ray measurements on pure metals, and pointed out that intensity measurements of the Bragg reflections can provide information on the charge distribution among the ions. Thus, the degree of applicability of a rigid-ion or pseudo-atom picture in specific cases can be indicated. It follows from general electron theory, transcending pseudopotential theory, that slight displacements of atoms from their equilibrium lattice positions can be described within the framework of a pseudoatom or rigid-ion model. However this is no longer possible when atoms are moving large distances from their points of origin, as they are during melting or when dislocations are introduced. To give a microscopic description of such process we must be prepared to use a much more complex "deformable-ion" model. While speaking on the subject of x-ray techniques March also underscored some earlier remarks made by Cowley on the importance of X-ray diffuse scattering data, which can also be interpreted to give information on the electronic relaxation around defects. Seeger agreed with this, and mentioned that certainly for alkali halides, the potentials that have been worked out are in good accordance with X-ray charge density data for the crystals.[1] As a further example of the use of X-ray techniques, Huntington referred to the work of Edelheit et al[6] who had made a study of strain in the vicinity of point defects. Further indirect contributions to this discussion were made by Robinson who referred to the work of Young and his colleagues[7] at Oak Ridge; and by Seeger who referred to the X-ray work by Schilling and co-workers[8]. Parsons commented on the applicability of the electron-microscope technique not only for the "direct visual observation" of strain fields in the vicinity of a dislocation but as a means of estimating the potentials for conditions of large displacements. However, Seeger felt that, because a complicated imaging problem was involved, there was still room for interpretational progress.

Ho mentioned that field ion microscopy, in conjunction with field pulse evaporation, made it possible to map out features such as voids and radiation-damage cascades. The great advantage of the technique is of course its extremely high resolution. Difficulties are encountered in the interpretation of the results; and a disadvantage, from some standpoints, is that only surface atoms can be studied since, as Torrens pointed out, we are not able to take possible special surface effects into account. As a more unusual application of field ion microscopy Seeger suggested that, since the results depend to some extent on the potential energy along the path taken by, say, a helium atom as it leaves the surface, the field effect might be utilized as a means of exploring the potential between the projectile atoms and the metal surface.

On the subject of vacancy-formation-energy (VFE) measurements Seeger mentioned that data of unprecedented accuracy were now being obtained through the use of positron annihilation. Excellent results have been obtained for Zn, Cd, Al, Pb, and data are forthcoming on the noble metals. Unfortunately it turns out that those metals for which calculations would be easiest (the alkali metals) do not show an effect of vacancies on positron annihilation; the excessive near-neighbor relaxation that occurs in these metals prevents the positron from becoming trapped. Interpretational difficulties associated with the positron-annihilation experiment were discussed by Seeger and Taylor. Strong model dependence is associated with the interpretation of the capture cross-section data; and some assumptions must be made about possible volume-dependence of the VFE. A distinct advantage of the positron-annihilation method of measuring VFE is that the data are insensitive to the behavior of the interstitial. March stated that it would be very helpful if values of the interstitial-formation energy (IFE) were available; and he inquired about the status of such measurements. Seeger replied that possibly the only way to get at this quantity is to measure the heat released (less the VFE) when Frenkel pairs, introduced during irradiation, recombine. In such measurements on a number of metals (but not bcc, in reply to a question by Bullough) the IFE was found to be about three times greater than that required for the formation of a vacancy. For this reason the interstitial does not occur as frequently (in thermal equilibrium) as has sometimes been supposed. While experiments on sodium have demonstrated that the dominant equilibrium defect is the monovacancy, an analysis of self-diffusion and high-temperature equilibrium data indicates that the next most important defect is the divacancy rather than the interstitial.[9]

The problem of vacancy relaxation is especially interesting. March pointed out that in bcc metals, particularly those with small cores, the relaxations have been shown to be quite potential sensitive, as evidenced by the considerable variability in the calculated near-neighbor relaxation behavior, and asked how experiment could help to illuminate this problem. Seeger's reply indicated that it would be difficult to design an experiment that would be sufficiently refined to yield more information than just the change of volume associated with a relaxed vacancy; although a temperature-dependent contribution to the vacancy-formation energy coming from the temperature dependence of the near-neighbor configurations also enter into the jump probability. In reply to a question by Torrens, Seeger indicated that vacancy volumes can be obtained from the pressure dependences of the self-diffusion coefficients, either directly or through measurement of the atomic-jump frequency by NMR techniques under pressure. In sodium, experiment has shown the relaxation around both vacancies and divacancies to be rather complete, as one would have expected from pseudopotential considerations, leaving only about 0.3 atomic volume for a monovacancy, and about twice that volume for a divacancy.

The session had begun in a philosophical vein with a discussion on the relative merits of simple and complicated potentials and related topics. It concluded in a somewhat similar mood with some comments, within the framework of dislocation calculations, on the spacial regions of applicability and validity of computer calculations.

Seeger observed that the role of computation in dislocation theory is somewhat different from that in point-defect work. In the former we have a general analytical technique valid for large distances and applicable to whole classes of materials. With computational methods we can approach the dislocation core, but that technique requires computation for specific materials and is limited by the fact that reliable potentials are available for only a few materials. In dislocation work we must choose some dividing line between problems which are best solved analytically, and those for which we might deem it preferable to use computational methods.

As indicated in general terms at the beginning, one of the roles of computer simulation is to lead us to new ideas about the structure of dislocations and the mechanisms of dislocation motion. It then becomes the task of the experimentalist to derive experimental tests of individual components of the model; for example, the existence of, and activation energies associated with, jogs, constriction processes, etc. We then go back to the computer, try different potentials, and see whether it is possible to account for the experimental values in detail. If this is not possible it may mean, not necessarily that the interpretation of the experiment is wrong, but that the theory is missing some important ingredient. This approach is illustrated nicely by Gehlen's experiences with a computer study of dislocation movement in α-iron.

The session concluded with a short discussion on the capacities of various computational procedures to maintain their descriptive validity as the dislocation core is approached.

Bullough asserted that continuum methods will not take one closer toward the dislocation core than will the linear model; the nonlinear solution is perturbed off the linear solution and it goes wrong just at the point where you must not go any further. Seeger replied that the attitude toward these questions depends on what you mean by "going wrong". He was of the opinion that a 10 percent improvement in the values of stresses and strains, which may easily be obtained by perturbation theory in the range where continuum theory is still applicable, may be worthwhile. Furthermore, he emphasized that there some properties simply cannot be derived from any linear solution. One example is the volume expansion associated with a dislocation. Here you may go to any trouble you want in deriving a potential for use in a computer solution for the dislocation core, but as long as you put it into the straight-jacket of the linear solution, you will not get out any volume expansion. In other words, in this case, one is forced to invoke the nonlinear elastic constants. One has to match a computer solution in the core to an anisotropic second-order elasticity calculation outside the core. This is at present being undertaken for edge-dislocations in NaCl by a joint effort of F. Granzer and his group in Frankfurt am Main and C. Teodosiu in Stuttgart.

Bullough restated his assertion by saying that one cannot penetrate any nearer to the singularity of the dislocation core by including higher-order elasticity terms. Seeger replied that this was indeed possible by the technique of universal solutions of the equations of elasticity, which is a technique not based on perturbation theory and therefore free from some of the objections to series solutions. This technique is admittedly of limited applicability, but its application to screw-dislocation in isotropic media[10] shows that the method is available. The point raised by Bullough is therefore, in Seeger's opinion, a matter of technique and not of principle. Of course, the continuum approach can never succeed when discrete effects come in, but apart from that, it is mainly limited by our lack of knowledge of the elastic energy function at large strains. We could calculate the elastic energy function if we had good potentials near the dislocation core. But just because we do not have them, we are looking, according to Seeger, to analytical solutions applicable to wide classes of materials — so this is a kind of vicious circle of which a way out will be found only if and when better potentials have been obtained.

REFERENCES

1. R. Bauer, Doctoral thesis, University of Stuttgart, 1971.
2. J. Drowart, "Phase Stability in Metals and Alloys", Eds., P. S. Rudman, J. Stringer, and R. I. Jaffee, McGraw-Hill, N. Y., 1967, p. 305.
3. K. Differt, work in progress for Doctoral thesis at University of Stuttgart.
4. D. Schiff, Phys. Rev. 186, **151** (1969).
5. J. R. Beeler, Jr., "The Role of Computer Experiments in Materials Research", in Advances in Materials Research, Vol. 4, H. Herman, ed., J. Wiley and Sons, New York, 1970, p. 295-476; also pp. 458-460 and 469-470.
6. L. S. Edelheit, J. C. North, J. G. Ring, J. S. Koehler and F. W. Young, Jr., Phys. Rev. **B2**: 2913 (1970).
7. F. W. Young, Jr., T. O. Baldwin, and P. H. Dederichs, "Vacancies and Interstitials in Metals", eds., A. Seeger, D. Schumacher, W. Schilling, and J. Diehl (North Holland Publishing Company, Amsterdam, 1970), pp. 619-636.
8. W. Schilling and P. Erhard Doctoral Thesis, Technische Hochschule, Aachen, 1971 (Huang Tails of Bragg Reflections from Self-Interstitials in Al), W. Schilling and P. Müller (Diffuse Scattering from Self-Interstitials), unpublished work at Kernforschungsanlage, Julich, Germany.
9. P. Kunz, A. Seeger, and H. Mehrer, unpublished results.
10. A. Seeger and Z. Wesotowski, "Physics of Strength and Plasticity" (65th Anniversary Volume for Egon Orowan), edited by Ali S. Argon, The M.I.T. Press, Cambridge, Mass., 1969, p. 15-17; Z. Wesotowski and A. Seeger, "Finite Elasticity, in Mechanics of Generalized Continua" (Proceedings of the IUTAM-Symposium Freudenstadt/Stuttgart, 1967), edited by E. Kröner, Springer-Verlag, Berlin-Heidelberg-New York, 1968, p. 294-297.

AUTHOR INDEX

Ashcroft, N. W., author of paper, 91-110
   discussion of Bullough-Tewary paper, 176
      Cowley-Wilkins paper, 278
      de Wette paper, 669
      Ehrlich paper, 617
      Enderby-Howells paper, 231-32
      Girifalco-Di Vincenzo paper, 215
      Ho paper, 338
      Johnson-Wilson paper, 317
      March-Rousseau paper, 138
      Rahman paper, 248
      Tosi paper, 153
      Vineyard paper, 24
Askar, A., co-author of paper, 177-88
Basinski, Z. S., co-author of papers, 525-36; 537-52
   discussion of Johnson-Wilson paper, 318
      Parsons paper, 472
      Gehlen paper, 490
Beeler, J. R., author of paper, 339-74
   co-author of paper, 673-94
   discussion of Doran-Burnett paper, 420
      Ho paper, 336
   discussion leader, 735-51
Bourquin, R. D., co-author of paper, 673-94
Bowen, D. K., co-author of paper, 493-508
Bullough, R., co-author of papers, 155-76; 509-24
   discussion of Basinski et al. paper, 535-6
      Beeler paper, 372
      Chang paper, 400
      Enderby-Howells paper, 232
      Gehlen paper, 489
      Johnson-Wilson paper, 316
      Kanninen-Gehlen paper, 724
      Rahman paper, 248
      Torrens-Robinson paper, 437
      Vitek et al. paper, 508
      Wynblatt paper, 651
Burnett, R. A., co-author of paper, 403-22

Chang, R., author of paper, 391-401
   discussion of Gehlen paper, 489-90
      Johnson-Wilson paper, 317
      Tosi paper, 152
Collings, E. W., discussion secretary, 755-65
Cotterill, R.M.J., co-author of paper, 439-49
Cowley, J. M., co-author of paper, 265-80
   discussion of Parsons paper, 473
Dahl, R. E., co-author of paper, 673-94
   discussion of Perrin et al. paper, 523
      Weins paper, 712
      Wilson-Johnson paper, 387
de Wette, F. W., author of paper, 653-71
   discussion of Cotterill-Pedersen paper, 448
      Rahman paper, 248
      Wilson-Johnson paper, 389
Di Vincenzo, T. M., co-author of paper 189-215
Doran, D. G., co-author of paper, 403-22
   discussion of Johnson-Wilson paper, 318
      Perrin et al. paper, 523
      Wilson-Johnson paper, 387
      Wynblatt paper, 650
Duesbery, M. S., co-author of papers, 525-36; 537-52
   discussion of Chang paper, 400
      Gehlen paper, 490
      Girifalco-Di Vincenzo paper, 214
      Ho paper, 337
      Vitek et al. paper, 507
Ehrlich, G., author of paper, 573-618
   discussion of Ho paper, 338
      Wynblatt paper, 649
Englert, A., co-author of paper, 509-24
   discussion of Beeler paper, 374
      Johnson-Wilson paper, 317-18
Gehlen, P. C., author of paper, 475-91
   co-author of paper, 713-24
   discussion of Basinski et al. paper, 552
      Cowley-Wilkins paper, 279
Girifalco, L. A., co-author of paper, 189-215

discussion of Weins paper, 712
    Wynblatt paper, 649
Hahn, G. T., discussion secretary, 561-70
Harrison, W. A., author of paper, 69-90
  discussion of Basinski et al. paper, 536
    Enderby-Howells paper, 231
    Girifalco-Di Vincenzo paper, 213
    Ho paper, 337
    Johnson-Wilson paper, 317
    Weins paper, 711
    Wynblatt paper, 649
Hirth, J. P., discussion leader, 561-70
  discussion of Kanninen-Gehlen paper, 724
    Vitek et al. paper, 508
    Wynblatt paper, 650
Ho, P. S., author of paper, 321-38
  discussion of Beeler paper, 372-73
    Enderby-Howells paper, 230-31
    Girifalco-Di Vincenzo paper, 214
    Harrison paper, 88
    Torrens-Robinson paper, 438
    de Wette paper, 570
    Wilson-Johnson paper, 388
Hoagland, R. G., discussion secretary, 755-765
Huntington, H. B., discussion of Dahl et al. paper, 694
    Johnson-Wilson paper, 315-16
    Wynblatt paper, 649
Jackson, D. P., author of paper, 621-32
  discussion of Wilson-Johnson paper, 388
    Torrens-Robinson paper, 437
    Weins paper, 712
Johnson, R. A., co-author of papers, 301-19; 375-90
  discussion of Chang paper, 400
    Harrison paper, 89
    Ho paper, 338
    Perrin et al. paper, 522
Kanninen, M. F., co-author of paper, 713-24
Kausch, H. H., discussion of Kanninen-Gehlen paper, 722
Kulcinski, G. L., discussion secretary, 735-51
  discussion of Beeler paper, 374
    Torrens-Robinson paper, 437
    Wilson-Johnson paper, 389
Lejček, L., co-author of paper, 493-508
Maradudin, A. A., author of paper, 27-68
  discussion of Bullough-Tewary paper, 175
March, N. H., co-author of paper, 111-38
  discussion of Girifalco-Di Vincenzo paper, 214
    Enderby-Howells paper, 229
    Perrin et al. paper, 522
    Tosi paper, 152

Parsons, J. R., author of paper, 463-73
  discussion of Perrin et al. paper, 524
    Torrens-Robinson paper, 437
Pedersen, L. B., co-author of paper, 439-49
Perrin, R. C., co-author of paper, 509-24
Rahman, A., author of paper, 233-48
Robinson, M. T., author of paper, 281-97
  co-author of paper, 423-38
  discussion of Jackson paper, 632
    Parsons paper, 472-3
    Wilson-Johnson paper, 386-7
Rousseau, J. S., co-author of paper, 111-38
Seeger, A., discussion leader, 755-65
  discussion of Basinski et al. paper, 552
    Cotterill-Pedersen paper, 448
    Cowley-Wilkins paper, 277
    Enderby-Howells paper, 229
    Girifalco-Di Vincenzo paper, 215
    Ho paper, 336
    Kanninen-Gehlen paper, 723
    Maradudin paper, 67
    Perrin et al. paper, 522
    Rahman paper, 246-47
    Vineyard paper, 25
Shewmon, P. G., discussion leader, 725-31
  discussion of Beeler paper, 373
    Dahl et al. paper, 692
    Gehlen paper, 490
    Johnson-Wilson paper, 318
    Perrin et al. paper, 522
    Rahman paper, 247
    de Wette paper, 670
    Wynblatt paper, 650
Taylor, R., co-author of papers, 525-36; 537-52
  discussion of Ho paper, 337
    Johnson-Wilson paper, 316
    Tosi paper, 151
Tewary, V. K., co-author of paper, 155-76
Torrens, I. M., co-author of paper, 423-38
  discussion of Ashcroft paper, 110
    Beeler, paper, 373
    Cotterill-Pedersen paper, 449
    Doran-Burnett paper, 421-22
    Ehrlich paper, 617
    Enderby and Howells paper, 229-30
    Harrison paper, 90
    Ho paper, 335
    Johnson-Wilson paper, 319
    Rahman paper, 247
    Robinson paper, 297
    Wilson-Johnson paper, 389
Tosi, M. P., author of paper, 141-54

Tyson, W. R., author of paper, 553-60
   discussion secretary, 725-31
      Cotterill-Pedersen paper, 449
      Kanninen-Gehlen paper, 722
      Perrin et al. paper, 524
      Vitek et al. paper, 507
Vineyard, G. H., author of paper, 3-25
   discussion of Basinski et al. paper, 551
      Doran-Burnett paper, 421
      Enderby and Howells paper 230
      Parsons paper, 473
      Rahman paper, 247
Vitek, V., co-author of paper, 493-508
   discussion of Basinski et al. paper, 551
      Chang paper, 400
      Ho paper, 336
      Kanninen-Gehlen paper, 723
Weiner, J. H., co-author of paper, 177-88
   discussion of Gehlen paper, 490

Weins, M. J., author of paper 695-712
   discussion of Dahl et al. paper, 692-3
      Wynblatt paper, 650
Wilkins, S., co-author of paper, 265-80
Wilson, W. D., co-author of papers 301-19; 375-90
   discussion of Beeler paper, 371-2
      Chang paper, 401
      Ehrlich paper, 616
      Gehlen paper, 488-9
      Girifalco-Di Vincenzo paper, 215
      Torrens-Robinson paper, 438
      Weins paper, 711
Wynblatt, P., author of paper, 633-51
   discussion of Dahl et al. paper, 693
      Ehrlich paper, 617
      Perrin et al. paper, 523
      Wilson-Johnson paper, 389

SUBJECT INDEX

A

Adatom, binding energy, 592-602
   migration, 602-609
Adiabatic approximation, 51
   for Green's function construction, 157
   for metals, 58
Alkali-halide,
   focused collision sequences, 6
   substitutional H^- ion, 57
Alkaline-earth fluoride, interstitial H, 57
Alloy, phase stability, 105
   potential functions, 225, 217-32
Al-Ni, normalized intensity patterns, 222
   partial interference functions, 223
   partial radial distribution functions, 223
   potential functions, 225
Aluminum, coupling coefficients, 260
   dislocation core, 470
   elastic constants, 332
   force constants, 332
   interstitial formation energy, 396
   interstitial strain field, 396, 455
   lattice images, 463
   phonon dispersion curves, 227
   plane spacing oscillations, 464
   potential function, 218-19
   pseudopotential, 77, 332
   tilt boundary, 470-71; 701-712
   vacancy formation energy, 396
   vacancy relaxation energy, 334
   vacancy strain field, 334, 396
Amorphous-layer model, for grain boundary, 696
Anharmonicity, 12-13; 440-41; 654-55
   dynamical method computer experiment, 15; 233, 439, 653
   effect on atom vibrations, 13, 653
   effect on crystal energy, 13, 440
   effect on defect formation energy, 440, 448

   effect on pressure, 13
   ionic crystal energy, 37
   ionic crystal infrared absorption, 36
   lead, 259-60
   phonon local modes, 17
   from radial distribution function, 256-57
Anharmonic oscillator, energy levels, 37
Annealing, short-term, 404
Annihilation region, 404
   Fe(fcc), 409-10; 414
Argon, dynamical liquid method computer experiment, 238-40
   interference function, 228
   mean square displacement, 239
   neutron diffraction, 239
   radial distribution function, 239
   velocity autocorrelation function, 239
Artificial damping, dynamical method, 655, 739
Atomic collision cascade, 4-9; 423-38
Atomic collisions, focused sequence, 4
   inelastic, 430
   with a surface, 623-31
Atomic force constants, see Force constants
Atomic fracture mechanics, 717-21
Atomic interactions, see Force constants; Potential function
Atomic interactions, insulators, 86-88
   ionic crystals, 82-86
   simple metals, 73-80
   transition metals, 80-82; 301-19; 375-90
   valence crystals, 82-86
   Wannier representation, 199-201
Atomic vibrations, at surface, 581
Atomic x-ray scattering factor, 249-51
Atom migration, at surface, 573
Atom-pair correlation parameters, defined, 266
Auger spectroscopy, 581

SUBJECT INDEX

B

BaF_2, 57
Band structure energy, 75, 98, 218
Binary collision model, collision cascades, 7, 423
B_2O_3, coupling coefficients, 262
Boltzmann impurity atmosphere, 510
Born-Green theory, 221
Born-Mayer potential, 6, 93, 157, 427
    Cu, 171
    KCl, 54
Born-von Karman method, 226
Born-von Karman model, 157, 162
Boundary conditions,
    for dislocation computer experiment, 478; 565-67
    flexible, 484-85
    for phase transformation computer experiment, 749-51
    for point defect computer experiment, 380

C

CaF_2, 57
    U-center in, 45
Carbon, clusters, at grain boundary, 689-90
    effect on grain-boundary structure, 687-89
    effect on grain-boundary interaction, 687-89
Carbon-free surface complex, Fe(bcc), 356-366
    strain field, 358, 361
Carbon interstitial, iron(bcc), 356-66
    (100) free surface, 363-66
    (110) free surface, 356-63
    strain field, 357, 358, 361
CASCADE Program, 406
Cauchy relations, volume forces, 71
Central difference method, 740
Central force model, limitations, 71
Cesium
    vacancy formation energy, 333
    vacancy formation volume, 333
Channeling, defined, 9, 281-2
    potential function construction, 281-97
CLUSTER Program, 406
Cobalt, coupling coefficients, 260
    (hcp), interstitial carbon, 366-68
    mean force potential, 259
Cohesive energy, alkali metals,
    Wannier theory, 201-9
    band structure energy, 75
    electron-electron interaction, 79
    electron-ion potential, 148-50
    perfect crystal, 28
    simple metals, 95-101
    structural contributions, 392
    Umklapp effects, 149
    volume dependent terms, 78
    volume force contribution, 75
    Wannier function method, 189-215
Coincidence site, grain boundary, 673, 695-712
    model, 695-712
Collision cascade, binary collision model, 7, 424-31
    Cu, 404, 423
    Au, 423
    Fe(bcc), 7-8; 423
Combination approximation, 7
Compound defects, 339-74
Compressibility, dynamical method computer experiment, 13
Compressibility sum rule, 146
Computer experiment,
    anharmonicity effects, 15
    crack propagation, 11, 713-24
    crystal lattices, 3-25
    defined, 3
    dislocations, 18, 475-560
        core structure, 493-508
        kink motion, 475-91
        partial, 525-36
        point defect interactions, 509-24
        screw motion, 537-52
    dynamical method, 3-25, 233-48; 339-74; 653-71
    fluids, 12, 233-48
    future prospects, 22
    grain boundaries, 18, 673-94; 695-712
    heat conduction, 16
    lattice statics method, 155-76; 321-38; 738
    methods, 735-51
    Monte Carlo method, 4, 339-74; 403-22; 738, 750
    quasidynamic method, 339-74; 475-91; 673-94; 737
    stacking faults, 18, 493-508
    surfaces, 18, 621-32; 633-51
    thermal conductivity, 17
    variational method, 339-74; 633-51; 737
    defined, 737
Computer Simulation, see Computer Experiment
Conductivity, thermal, see Thermal conductivity
Copper, channeling, 9
    collision cascades, 4, 404, 423
    dislocation-point defect interaction, 509, 513-20

SUBJECT INDEX

displaced charge about vacancy, 126
displacement cascades, 4, 404, 423
edge dislocation, 509
electronic properties, pseudopotential theory, 81
focused collision sequences, 4
Frank dislocation, 509
helium, interstitial,
    formation energy, 385
    migration energy, 380
helium, substitutional, detrapping energy, 385
    formation energy, 385
interstitial relaxation energy, 172
interstitial strain field, 168, 172
interstitial volume change, 172
ion bombardment, 9-10
lattice Green's function, 165
potential function, noncentral, 307
radiation damage, 4, 404, 423
static Green's function values, 164-65
surface (001), 633-51
vacancy strain field, 166
Core-core interaction, effective valence, 304
Correlation, time-displaced, 20-21; 233-48
Coulomb potential, 6
Coupling coefficients, 260
Covalent crystal, long-range interactions, 71
Crack front, discrete model, 719
    jogged, 714
Crack propagation, 11, 713-24
    jog mechanism for, 717-21
    mechanisms, 723-4
Crack propagation computer experiment, 11, 713-24
    boundary conditions, 721
    quasidynamic method, 713-24
Cu_3Au, focused collision sequences, 6
    radiation damage, 6
Cu_6Sn_5, partial interference functions, 221

D

Damage efficiency, 432
Darwin function, 330
Debye temperature, surface, 582-3
Debye-Waller factor, 163
    LEED data on, 660
Defect formation, energy, pseudopotential formulation of lattice statics method, 323-29
    volume, pseudopotential of lattice statics method, 323-333
Defect Green's function, 166
DEFECT Program, 343
Defect relaxation energy, minimization, 171
    pseudopotential formulation of lattice statics method, 327-28
Defects, effect on radial distribution function, 263
Defect space, semicontinuous Green's function method, 167
Defect strain field, Green's function method, 156, 164-66; 168-73
    Kanzaki method, 156
Deformation dipole model, phonon-dispersion, 28
Deformation mechanisms, 568-69
Degrees of freedom, internal, 19
Desorption energy, 592
Dielectric constant, electron liquid, 143
Dielectric function, 142-48
    calculation of, 143
    phase stability rules, 104
Dielectric function method, electron screening, 218
Dielectric response function, valence and ionic crystals, 84
Diffuse scattering, ordering energy, 265-68
Diffusion, liquids, dynamical method complete experiment, 236-46
Diffusion mechanisms, liquids, 240, 246
Direct correlation function, defined, 222, 252
Direct Green's function method, defect strain field, 156
Dislocation core structure, 463-79; 475-91; 493-508
    bcc crystals, 475-91; 493-508
    diamond structure crystals, 463
    fcc crystals, 463
    hcp crystals, 494
    Na, 539-46
    Peierls model, 497-500
Dislocation dynamics, 475-91; 537-52
    bcc crystal, 475, 537
    kink model, 475
Dislocation kink, 475, 480
Dislocation motion
    kink model, 475
    screw dislocation, 475, 537
Dislocation-point defect interaction, 509-24
    edge dislocation, 513-20
Dislocation simulation, boundary conditions, 478, 489, 512, 565-67; 764-5
    Cu, 509
    strain field, 463, 469, 470
Disordered lattices, dynamical method, 16-17
Displacement cascade, annealing, 403-22
    defined, 403-4; 423
    production, 423-38
    temperature effects, 431-2
Displacement efficiency, 432
Divacancy binding energy, bcc metals, 315

(Fe, Mo, Ta, W, V)
Dynamical correlations, liquid, 234
Dynamical matrix, 157
    perturbed crystal, 32
Dynamical method computer experiment,
    anharmonicity, 15
    artificial damping, 17, 339-74; 653-71;
        739-40
    collision cascades, 4-9; 423
    compound defects, 339-74
    compressibility, 13
    defect migration, 18; 744-5
    defined, 737
    diffusion in liquid, 236-46
    dislocation motion, 475
    disordered lattices, 16-17
    extended defects, 18, 475
    focused collision sequences, 4-6
    grain boundary, 673-94
    Grüneisen constant, 13
    heat conduction, 16
    integration schemes compared, 740-44
    ion bombardment, 9-10; 423
    ionic crystals, 6
    lattice defects, 17, 339-74; 423-38;
        475-91; 439-49; 673-94; 713-24
    lattice dynamics, 654-55
    liquid, 12; 233-48; 653-71
    magnetism, 19
    metals, 3-23; 339-74; 423-38; 439-49;
        475-91; 653-71; 673-94; 713-24
    normal mode lifetime, 14
    phase transformation, 12, 749-51
    phonon frequency distribution, 13-15;
        653-71
    pressure, 12
    radiation damage, 4-10, 350-2; 423-38
    review, 3-25
    shock waves, 10
    surface dynamics, 653-71
    thermal conductivity, 17
    thermal expansion, 13
    time-displaced correlations, 20-21; 233-48
    transport phenomena, 16; 233-48; 673-94
    vacancy formation energy, 439-49
    velocity autocorrelation function, 13;
        233-48
Dynamical properties, electron-ion potential,
    148-50
Dynamic lattice model, 15
DYNAMIC Program, 343

E

Elastic constants, from mean-force potential,
    259

Electron, conduction, 217
    core, 217
    coulomb correlations, 143
    exchange diagram, 143
    gas, energy of, 96
    -ion, interaction, pseudopotential
        representation, 74
        potential, Ashcroft, 142, 145, 148
    liquid, density-density response
        function, 142
        dielectric function, 142
    in liquid metals, effect on pressure, 241
    microscope, atomic level images, 463
        lattice image formation, 463
    screening, dielectric function method, 218
    self-energy corrections, 143
Electrons, short-range exchange, 143
Energy contour mapping, migration energy
    computer experiment, 746
Equipartition theorem, 12
Euler-Cauchy method, 741
Ewald electrostatic energy, 218
Exchange interaction, Hartree-Fock-Slater
    approximation, 377
Extended defects, *see* dislocations, stacking
    faults, grain boundaries, tilt boundaries, twin
    boundaries
Extended defects, dynamical method, 18,
    475-91; 493-508; 509-24; 525-36; 537-52;
    553-60; 561-70; 673-94; 695-712; 713-24

F

Fermi energy, 59
Feynman's theorem, 199
Field desorption, 588
Field emission microscope, 576
Field ion microscope, 586-92
Finite crystal, 494, 745, 749
Firsov potential, 426
Flexible boundary conditions, 478, 721
Fluctuation-dissipation theorem, 143
Focused collision sequences, alkali halides, 6
    Cu, 4
    Cu_3Au, 6
    Fe(bcc), 4
Force constants, atomic, 27, 32
    crystal distortion contribution, 32
    empirical, 29
    independent, 28, 29
    invariance properties, 27
    normal mode frequencies, 28
    perturbed crystal, 32, 33
    potential function, 28
    rotational invariance, 34
    symmetry operation condition, 35

transformation properties, 27
U-center, 53
Force laws, see Potential function, Force constants
Form factor, 74
    pseudopotential selection criteria, 392, 395
Fracture mechanics, atomic, 717-21
    continuum, 714-17
Fracture, metals, 475
Frank dislocation, structure, 520-1
Frank loop, Cu, 513
Free surface, 356-66; 573-617; 621-32; 633-51; 653-71
    atom collisions with, 621
    carbon at, 356
    dynamics, 653
    impurity atom at, 356
    review, 573
    structure, 633
Frenkel pair production, effect of carbon on, 350-52
Friedel oscillations, 69, 225

G

Generalized direct correlation function, defined, 222
Germanium, dislocation core, 469
    inclined dislocations in, 464
    lattice images, 463, 469
    plane spacing oscillations, 464
Gold, channeling in, 281
    displacement cascades, 423
    potential function, noncentral, 307
    tilt boundary, 701-712
Grain boundary, coincidence site model, 695-712
    configuration, 673-94; 695-712
    diffusion, 685, 711
    effects on vacancies, 681-85
    energy, 678-81
    high-angle, Al, 695; Fe(fcc), 673; Au, 695
    Fe(fcc), 673-94
    low-angle, Fe(fcc), 673-94
    structure, 675-78; 697, 709
GRAPE Program, xiv-xv, 477
Green's function construction, for point defect, 157-75
Green's function, lattice, 156-76
    uniqueness, 174
Green's function method, defect strain field, 166-73
Green's function, spin, 21

Green's function, vibrational, perturbed crystal, 32
Griffith crack model, 714-17
Grüneisen constant, dynamical method, 13

H

HAPFCC Program (annealing), 405-7
Hard core repulsion, 81
Hard sphere model, 12, 228; alloys, 221; regime of validity, 103-4
Harmonic approximation 15, 27, 155, 321, 738; Green's function construction, 157
Harmonic potential, 15
Hartree approximation, 58
Hartree-Fock-Roothaan method, 56
Hartree-Fock-Slater exchange, 377
Hartree potential, 142
Heat capacity, see Specific heat
Heat conduction, see Thermal conductivity
Heine-Abarenkov bare ion potential, 330
Helium in metals, 375-90
    interstitial impurity, 380-83
        formation energy, 385
        migration energy, 380
    potential functions, 376, 378
    substitutional impurity, 383-5
        detrapping energy, 385
        formation energy, 385
    void nucleation, 376
High-angle grain boundary, 673, 695
Hubbard screening, 144
Hubbard-Sham screening, 330
Hume-Rothery valence, 304
Hybrid computer, defect annealing computer experiment, 404-5
Hybride ion, see U-center
Hybridization, 81, 109
Hyper-netted chain approximation, 251

I

Impurity atom, 27-68; 217-32; 265-80; 339-74; 375-90; 575-85; 686-90
    alloying effects, 217-32; 265-80
    calculations, 748-9
    effect on force constants, 29-50
    -free surface complexes, 356-66
    at grain boundaries, 686-90
    interstitial, 340-68; 380-2
    -interstitial complexes, 349-50, 354-6
    substitutional, 369-71; 383-5
    at surfaces, 356-66; 575-6; 583-5
    -vacancy complexes, 343-49; 353-4; 369-71; 383-5

776 SUBJECT INDEX

Impurity-host interactions, 27-68
   first-principles calculation, 50-7
   ionic crystals, 50-7
   metals, 57-66
Impurity ion, ionic crystal, 27-68
   -host ion interaction, from localized
      mode data, 50; from resonance mode
      data, 50
Indirect ion-ion potential, 144
Inert-gas solid, potential function, 57
   compared with gas phase potential, 57
Infrared absorption, impurity ions, 36
Interatomic forces, *see* Atomic interactions;
   Force constants; Potential function
Interatomic interactions, *see* Atomic
   interactions
Interatomic potential, *see* Potential function
Interference function, damping, 228
   defined, 219
   inherent uncertainties, 219
Interionic potential, 224
   construction, 225
   screening, 141-54
Interstitial-carbon complexes,
   binding energy, 341
   configuration, Fe(bcc), 355; Fe(fcc), 352;
      Ni, 350
   Fe, 352, 355-6
   Ni, 349-50
Interstitial configuration,
   Ag, 313-14
   Cu, 313-14; 520
   Ni, 313-14
   Pd, 313-14
Interstitial, equilibrium separation
   distance, copper, 172
Interstitial formation energy, fcc metals,
   313 (Ag, Cu, Ni, Pd)
Interstitial formation volume, Cu, 172
Interstitial, introduction via the structure
   factor, 394
Interstitial loops, 521
Interstitial migration, correlated, 407-8
Interstitial relaxation energy, Cu, 172
Interstitial strain field, Cu, 157, 168-73
Ion bombardment, 424
   Cu, 9-10
Ion, deformation dipole model, 28
Ionic crystal, 29-68; 70
   collision cascades, 6
   focused collision sequences, 6
   impurity-host interaction, 29
   pseudopotential, 82-6
   radiation damage, 6
   effective ion charge, 40, 41
Ionicity concept, 70, 85

Ion-ion interaction, ionic crystals, 35-57
   metals, 91-110
   temperature dependence, 106-7
Iron(bcc), collision cascades, 4, 7-8; 423
   carbon at (001) surface, 363-66
   carbon at (110) surface, 356-63
   crack propagation, 713-24
   displacement cascade, 7-8; 404, 423
   focused collision sequences, 4
   interstitial-carbon complex, 354-56
   interstitial helium, formation energy, 385
   kink configuration, 479
   potential function, 310
   radiation damage, 4, 7-8; 404, 423
      annealing, 8-9
   screw dislocation motion, 479
   substitutional He, detrapping energy,
      385
      formation energy, 385
   temperature simulation, 486
   vacancy-carbon complex, 352-54
Iron(fcc), displacement cascade annealing,
   403-22
   grain boundary, 673-94

J

Jog mechanism, crack propagation, 719,
   723-24
Jones zone, semiconductors, 83

K

Kanzaki, force, 166; defined, 113
   lattice, defined, 113
   method, defect strain field, 156, 166
KBr, resonance modes, 48-50
KCl, impurity ion vibrational amplitudes, 36
   radiation damage, 6
   U-center, 54-55
Kink,
   configuration, Fe(bcc), 479-83
   energy, Fe(bcc), 483-4
   motion, critical stress, 486
      temperature dependence, 486
      velocity, 487
   nucleation, double kink, 490-1
Kramers-Kronig relations, 159, 162
Krypton, solid, phonon frequency distribution,
   14
   vacancy formation, temperature
      dependence, 439-49

L

Lattice defect, computer experiment, 17
   displaced charge, 112-29

electron distribution, 113
energy, quantum effects, 453-4
-impurity complexes, 339-74
migration, 17-8; 301-19; 375-90; 403-22; 685
phonon local modes, 18
Lattice dynamics, imperfect crystal, 28-68
perfect crystal, 28
at surfaces, 581; 653-71
Lattice Green's function, 156-76
Cu, 165
Lattice image contrast calculations, 466
Lattice image electron microscopy, 463-73
Lattice statics method, 112-3; 155-76; 321-38; 510, 738
defect strain field, 32; 164-73; 327-330
defined, 327-30; 738
pseudopotential formulation, 322-31
vacancy formation energy, 324; 327-29
vacancy formation volume, 329
vacancy relaxation energy, 327-8
Lattice vibrations, see Lattice dynamics
Lead, anharmonicity, 257
coupling coefficients, 260
displaced charge about vacancy, 126
liquid, interference function, 220
radial distribution function, 220
mean force potential, 259
potential function, 224
radial distribution function, 257
total radial density function, 258
Ledge, configuration, 637
energy, 637-42
interactions, 640
LEED, atom arrangement results, 579
Lennard-Jones crystal, 12
surface dynamics, 656-63
surface phonon dispersion, 656-59
surface phonon frequency distribution, 658-60
thermal motions, 11-17
violation of Cauchy relations, 71
Lennard-Jones liquid, 238-40
Lennard-Jones potential, 12, 654; 698-99
argon, 238
water molecule, 244
Lindemann's rule, 262
Lindhard function, 99
Liquid, Lennard-Jones, 238-40
mean square displacement, 239
non—Gaussian behavior, 239
quasicrystalline model, 240
radial distribution function, 239
velocity autocorrelation function, 239
Lithium, vacancy formation energy, 333

vacancy formation volume, 333
Local modes, KCl, impurity ion, 36
NaCl, impurity ion, 36
phonon, 17
Low-angle grain boundary, 673

M

Madelung energy, 96
Magnesium, displaced charge about vacancy, 126
Magnetism, internal degrees of freedom, 19-22
Man-computer interaction, importance, 405
Matrix-partitioning technique, Green's function construction, 166-68
Mean force potential, defined, 254
from radial distribution function, 254
Metallic crystals, 69-90; 91-110; 141-154; 301-19; 321-38
Method of long waves, 161
Microdeformation, 491
Microkink configuration, Fe(bcc), 475
Mie Potential, 576
Migration energy calculations,
computer experiment techniques for, 745-48
contour mapping, 383
dependence on potential function, 454
Molecular-beam-scattering data, for potential function construction, 377
Molecular crystals, 92
Molecular dynamics, 233, 653
Molecular dynamics method, see Dynamical method
Moliere potential, 427
Molybdenum, interstitial helium formation energy, 385
potential function, 310
substitutional helium, detrapping energy, 385
formation energy, 385
Moment trace method, 161
Monte Carlo method, 4, 12, 347, 403-22; 738, 744-5; 750
compound defects, 347
defect annealing, 403-22
defined, 738
equation of state, 12, 738
phase transformation, 12; 749-51
radiation damage annealing, 8-9; 403-22
thermal equilibrium state, 4; 744-5
Morse potential, 625, 635; 698-9
shock-wave propagation study, 10

N

NaCl, impurity ion vibrational amplitudes, 36

radiation damage, 6
surface dynamics, 664-68
surface phonon dispersion, 664-66
surface thermodynamical functions, 666-68
NaI:NaCl, 47
Neon, solid, 12
Neutron diffraction,
 time-of-flight method, 226
 triple-axis spectrometer, 226
Nickel,
 He, interstitial
  formation energy, 385
  migration energy, 380
 He, substitutional
  detrapping energy, 385
  formation energy, 385
 interstitial-carbon complex, 349-50
 interstitial formation energy, 313
 potential function, noncentral, 307
 vacancy-carbon complex, 341-49
Noble metals, pseudopotentials, 69
Noncentral forces, 228
 x-ray scattering criterion for, 454
Noncentral interaction, radial distribution function for, 228
Noncentral potential functions, 301-19
Nordsieck method, 742
Normal modes, force constants, 28
Normal mode lifetime, dynamical method, 14

O

Ordering energy, from short-range order diffuse scattering, 265-68
Oscillator, anharmonic, energy levels, 37
Oxygen ion channeling, Au, 288

P

Pair correlation function, see radial distribution function
Pair correlation parameters, defined, 266
Pairwise interaction concept, 80, 92; 189-215; 249-64; 321-38
 defined, 92-4
 failure at surfaces, 634
 pseudopotential theory basis, 69, 80, 91
 quantum theory basis, 190
 Wannier function theory basis, 212
Palladium,
 He, interstitial
  formation energy, 385
  migration energy, 380
 He, substitutional
  detrapping energy, 385
  formation energy, 385
 interstitial formation energy, 313
 potential function, noncentral, 307

Partial dislocation interactions, fcc crystals, 525
Partial interference function, defined, 220
Partial radial distribution function, defined, 220
PbI_2, radiation damage, 6
Peierls-Nabarro stress computer experiment, 553-60
Percus-Yevick approximation, 222, 252
Periodic boundary conditions,
 Green's function construction, 157
 phase transformation, 749-50
 thermal motions, 12
Phase stability, dielectric function criterion for, 104
Phase transformation computer experiment, 749-50
 dynamical method, 12; 749-50
 Monte carlo method, 12, 750
Phonon dispersion,
 ab initio calculations, 29
 defect Green's function from, 155-76
 from neutron spectroscopy, 28
 relations, 226
 (100) surface in fcc, 656
Phonon frequency distribution,
 influence of extended defects, 33
 normalized, 13
 at surfaces, 658
 temperature dependence, 15
 volume dependence, 14
Phonon frequency spectrum, see Phonon frequency distribution
Phonon Green's function,
 construction, 157
 relation to static Green's function, 158
Phonon local modes, 17
Planar channeling, 281-96
Plasmon resonance, 143
Point defect, dynamical properties, 27
 dynamic displacements, 31
 static displacements, 31
Potassium,
 vacancy formation energy, 333
 vacancy formation volume, 333
Potential function construction,
 bcc metals, 308-11
 from channeling data, 281-96
 from diffraction data, 225-6; 245, 254; 269-71
  experimental accuracy required, 245, 252
 from elastic constant data, bcc metals, 308-11; 331; fcc metals, 303-7; 331
 from field ion microscopy observations; 609-14
 inner cutoff, 304

noncentral potential, 303-7
　outer cutoff, 304
　from phonon dispersion data, 304
　from radial distribution function,
　　225-6; 239, 255
　from vacancy formation energy, 304
　from vacancy migration energy, 304
　Wedepohl method, 377
Potential function, empirical, 22-3; 35, 69,
　217, 233, 249, 265, 281, 301, 321, 375,
　609; 760-64
　lack of uniqueness, 302
Potential function, host-impurity, 27
　long-range, usefulness, 69
　noncentral, radial distribution for, 228
　power law, 6
　range, effect on migration energy, 382
　shape, quantum mechanical basis, 94
　short-range, 302-11, usefulness, 72
Potential functions, alloy, 27, 104-5; 225-6;
　269-71
　elemental,
　　Ag, 305
　　Ag-He, 378
　　Al, 77, 225, 332, 699
　　Al-Ni, 225
　　Ar, 238
　　Au, 305, 427, 699
　　Au-He, 378
　　Cs, 210-11
　　Cu, 305, 427, 512, 635
　　Cu-He, 378
　　Fe, 309, 342, 427
　　Fe-C, 342
　　Fe-He, 378
　　K, 210-11
　　Kr, 442
　　Li, 210-11
　　Mo, 309
　　Mo-He, 378
　　Na, 107, 145; 210-11; 241
　　Ni, 225, 305
　　Ni-He, 378
　　Pd, 305
　　Pd-He, 378
　　Rb, 210-11
　　Ta, 309
　　Ta-He, 378
　　V, 309
　　V-He, 378
　　W, 309-613
　　W-He, 378
　Power law potential, 6
　Predictor-corrector method, 741
　Pressure, dynamical method, 12
　Pressure, liquids, electronic effects, 241

Primary knock-on atom, defined, 404, 424
Proton channeling, gold, 289
Pseudopotential, 69, 91, 141, 217, 321, 391
　electron-electron interaction
　　approximations, 79
　form factor, selection criteria, 392, 395
　inherent arbitrariness, second order
　　pseudopotential, 78
　limitations, 452-3
　method, 217-18

Q

Quasicrystalline model, liquid, 240
Quasidynamic method, 655
　crack propagation computer experiment,
　　713-24
　defined, 737
　grain boundary computer experiment,
　　673-94
Quasiharmonic approximation, 655

R

Radial density function, 257-8
Radial distribution function, as basis for
　potential function, 249-64
　defined, 219, 236
　direction-dependent information in, 257
　effect of atom vibrations on, 255-6
　effect of defects on, 255; 263-64
　for hard spheres, 228
　for Lennard-Jones Liquid, 239
　for real liquid, 228
　for sodium, liquid, 241
Radiation damage annealing, 8-9; 403-22
Radiation damage cascade, 4-9; 423-38
　Cu, 4, 423
　Fe(bcc) 4; 7-8; 423
Radiation damage production, effect of
　carbon on, 350-52
Random walk, correlated,
　defect annealing, 405
　surface diffusion, 603
Rare gases in metals, 375-90
Rayleigh modes, 659
Read-Shockley model, grain boundary
　energy, 678-80
Resonance modes, U-center, 45
Response function, screened, 143
　density-density, 142
Root sampling technique, Green's function
　construction, 164
Rubidium, liquid, interference function, 228
　vacancy formation energy, 333
　vacancy formation volume, 333

S

Screened response function, 143
Screening charge, metals, isotropy, 86
Screening function,
    from planar channeling data, 293-4
    simple metals, 141-54
Screening models,
    Geldart-Taylor, 145-6
    Hubbard-Sham, 143-5
    Singwi et al., 145-6
    Toigo-Woodruff, 145, 147
Screening, Umklapp effects, 147-49
Selenium, coupling coefficients, 261
    liquid, interference function, 228
Self-diffusion coefficient,
    computer experiment on, 233
    in liquid, 236-46
    from velocity autocorrelation function, 237
Self-diffusion, liquid, mechanisms, 233, 237
Semicontinuum Green's function method,
    defect strain field, 156-76
Semidynamical approximation, 7
    channeling, 9-10
    collision cascades, 7-8
    defined, 7
    ion bombardment, 9-10; 423
Shell model, phonon dispersion, 28; 41-5
Shell parameters, effect of defects on radial distribution function, 263
Shockley partial dislocations, 511
Shock front, velocity oscillations, 10
Shock wave propagation, dynamical method, 10-11
Short-term annealing, 403
Silver,
    He, interstitial,
        formation energy, 385
        migration energy, 380
    He, substitutional,
        detrapping energy, 385
        formation energy, 385
    interstitial formation energy, 313
    potential function, noncentral, 307
Simple metal, defined, 95
    pseudopotential, 69
    screening function, 144-54
Sintering, 573
Sodium, cohesive energy, Umklapp effects, 149
    ion-ion interaction, 103
    interionic potential, 145, 147
    liquid, dynamical method computer experiment, 240-43
    mean square displacement, 242
    potential function, 241
    radial distribution function, 241
    self-diffusion, 240
    velocity autocorrelation function, 242
Peierls stress, 549-50
screw dislocation, core, 539-40
    motion, 550
slip mechanisms, 550
vacancy formation energy, 333, 397
vacancy formation volume, 333
vacancy strain field, 397
Solid-Vacuum Interface, see Free surface
Specific heat, surface, 668
    temperature dependence, 163
    temperature fluctuation, relation to, 236
Spin correlations, time displaced, 20
Spin Green's function, 21
SrF_2, 57
Short-range order diffuse scattering, connection with ordering energy, 265-68
Stacking faults, 18, 493
Static Green's function, parameter values, 164-5
    relation to phonon Green's function, 158
Strain aging, 510
Strain field, effect on diffraction data, 255-64
    effect on radial distribution function, 256-7
Stress-intensity-factor, 715-6
Structure factor, 74
Surface, 18, 339, 573, 621, 633, 653
    adsorbed atom, 575
    atom arrangement, 579
        migration, 602-9
    atom binding energy, 592-602
    -atom collisions, 621
    diffusion, 573, 603
    dynamical properties, 28, 653
    dynamics, 581; 653-71
    energy, anisotropy, 634; 642-8
    free, 573, 339
    ledge, Cu, 633
        configuration, 633
    self-adsorbed atoms, 583
    thermal expansion, 663
    vibration amplitudes, 660
    vibrations, 653-71
Surface phonon dispersion, Lennard-Jones crystal(fcc), 656-9
    NaCl, 664-6
Surface phonon frequency distribution, 658
Surface simulation computer experiment, 633-51

Surface specific heat, 668
Surface structure, 575; 578-81
    Terrace-Ledge-Kink model, 634
Surface tension, 575
    data, 577
    orientation dependence, 577
    potential function, 575
Surface terrace, Cu, 633
Szigeti effective charge, for ions in an ionic crystal, 40-1

T

Tantalum,
    He, interstitial, formation energy, 385
    He, substitutional,
        detrapping energy, 385
        formation energy, 385
    potential function, 310
Tellurium, 228
Temperature, effect on phonon frequency distribution, 15
Temperature effects, thermal motions, 13, 20, 423, 478, 567; 744-5
Terrace-Ledge-Kink model, 633-51
    defined, 634
Thermal conductivity, 16-7
Thermal diffuse scattering, x-ray, 256
Thermal equilibrium, test for, 237
    Monte Carlo computer experiment, 4, 750
Thermal expansion, dynamical method, 13, 655
    effect on vacancy formation energy, 445
    at surface, 663
Thermal motion, in crystals, 11-17; 20, 486; 655-64; 744-45
    dynamical method, 11-17; 20; 233-48; 439-49; 653-71; 744-45; 749
    mean square amplitude, 13, 660
Thermodynamic properties computer experiment, 12, 666
Thin film growth, 573
Third estate of physics, 3
Thomas-Fermi screening parameter, 144
Tilt boundary, asymmetric, Fe(fcc), 673-94
Tilt boundary, symmetric, 695-712
Time-displaced correlations,
    liquids, 237
    magnetism, 20-21
Total correlation function, defined, 219
Total radial density function, 257-58
Transition metal, interactions, 80
    pseudopotentials, 69; 80-82
Transport coefficients, 237
Transport phenomena, dynamical method, 16, 233, 439, 653

Tungsten,
    He, interstitial, formation energy, 385
    He, substitutional,
        detrapping energy, 385
        formation energy, 385
    potential function, 310
Twin-boundary grooving, relation to surface tension, 575-76

U

U-center, anharmonicity induced by, 36
    deformability, 45
    energy levels, 35-50
    force constant, 53
    local modes induced by, 35
        anharmonicity, 37
        energy levels, 37-40
    polarizability, 45
    resonance modes, 45
    rigid-ion model, 41
    transverse effective charge, 45

V

Vacancy-carbon complex,
    binding energy, 341
    configuration, Fe(bcc), 353-54
        Ni, 344, 349
    configuration energy, 341
    Fe(bcc), 353-54
    Fe(fcc), 352
    Ni, 343-49
Vacancy, displaced charge in metals, 124-26
    introduction by structure factor, lattice statics method, 326, 394
Vacancy formation,
    grain boundary effects, 683
    temperature effects, 444
    volume effects, 452; 456-58; 458-59
Vacancy-grain boundary interaction, 681-86
Vacancy loops, 521
Vacancy migration, correlated jumps, 409
    energy, bcc metals, 314 (Fe, Mo, Ta, V W)
    grain boundary effects, 685-86
Vacancy strain field, Cu, 157; 166-67
Vacancy-substitutional impurity complex, 369-71; 383-85
Valence crystal, 69
    pseudopotential, 82-86
Vanadium,
    He, interstitial, formation energy, 385

He, substitutional,
    detrapping energy, 385
    formation energy, 385
  potential function, 310
Van der Waals forces, 93
Van Hove function, 237, 242
Vapor pressure, size dependence, 576
Variational method,
  compound defects, 339-74
  defined, 737
  surface, 633-51
  surface ledge, 637
  symmetric tilt-boundary, 695-712
Velocity autocorrelation function,
  defined, 236
  relation to phonon frequency
    distribution, 13
  relation to self-diffusion, 236
Virial theorem, 12; 744-45
Void, formation, 376
  growth, biased diffusion model, 521
  nucleation, 521
    effect of He, 376

W

Wannier functions, defined, 191
Water, liquid,
  diffusion coefficient, 245
  dynamical method computer experiment,
    243-45
  mean square displacement, 244-45
  oxygen-oxygen radial distribution
    function, 244
Water, molecule, four-point-charge model,
  243-45
Wedepohl method, potential function
  construction, 377
    inner cutoff, 308

Z

Zinc, Fermi surface, change with distortion,
  73

QD
931
B38
1971

MAR 4 1974